Life

The Science of Biology

FIFTH EDITION

Life

The Science of Biology

William K. Purves
Harvey Mudd College
Claremont, California

Gordon H. Orians
The University of Washington
Seattle, Washington

H. Craig Heller
Stanford University
Stanford, California

David Sadava
The Claremont Colleges
Claremont, California

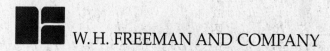
SINAUER ASSOCIATES, INC.

W.H. FREEMAN AND COMPANY

The Cover

Grizzly bears (*Ursus arctos*) take many years to mature reproductively, and cubs remain with their mothers for several years. If their populations are to persist, adult bears must have access to rich food resources. Grizzly bears that live in coastal Alaska depend on salmon that swim up the rivers to spawn. Given that abundant, high-quality food, they grow to become the world's largest carnivorous mammal. Photograph by Michio Hoshino/Minden Pictures.

The Frontispiece

A sunset scene with nesting painted storks (*Mycteria leucocephalus*) taken in Bhakatpur, India. Photograph by Mike Powles/Woodfall Wild Images.

LIFE: The Science of Biology, Fifth Edition

Volume I: The Cell and Heredity ISBN: 0-7167-3326-9

Copyright © 1998 by Sinauer Associates, Inc. All rights reserved.

Address editorial correspondence to Sinauer Associates, Inc., 23 Plumtree Road, Sunderland, Massachusetts 01375 U.S.A.

www.sinauer.com

Address orders to W. H. Freeman and Co. Distribution Center, 4419 West 1980 South, Salt Lake City, Utah 84104 U.S.A.

Examination copy information: 1-800-446-8923
Orders: 1-800-877-5351

www.whfreeman.com

Library of Congress Cataloging-in-Publication Data

Life, the science of biology / William K. Purves ... [et al.] —
  5th ed.
             p.   cm.
  Rev. ed. of: Life, the science of biology / William K. Purves,
Gordon H. Orians, H. Craig Heller. 4th ed. c1995.
  Includes bibliographical references and index.
  ISBN 0-7167-2869-9 (hardcover)
  1. Biology   I. Purves, William K. (William Kirkwood), 1934–   .
II. Purves, William K. (William Kirkwood), 1934– Life, the science
of biology

QH308.2.L565   1997   79-34772
579—dc21   CIP

Printed in U.S.A.
Second printing 1998 Courier Companies, Inc.

To Jean, Betty, Renu, and Angeline

About the Authors

Bill Purves is Professor Emeritus of Biology as well as founder and former chair of the Department of Biology at Harvey Mudd College in Claremont, California. He received his Ph.D. from Yale University in 1959 under Arthur Galston. A fellow of the American Association for the Advancement of Science, Professor Purves has served as head of the Life Sciences Group at the University of Connecticut, Storrs, and as chair of the Department of Biological Sciences, University of California, Santa Barbara, where he won the Harold J. Plous Award for teaching excellence. His research interests focus on the chemical and physical regulation of plant growth and flowering. Professor Purves elected early retirement in 1995, after teaching introductory biology for 34 consecutive years, in order to turn his skills to writing and producing multimedia for introductory biology students.

Gordon Orians is Professor Emeritus of Zoology at the University of Washington. He received his Ph.D. from the University of California, Berkeley, in 1960 under Frank Pitelka. Professor Orians has been elected to the National Academy of Sciences, the American Academy of Arts and Sciences, and is a Foreign Fellow of the Royal Netherlands Academy of Arts and Sciences. He was President of the Organization for Tropical Studies, 1988–1994, and President of the Ecological Society of America, 1995–1996. He is a recipient of the Distinguished Service Award of the American Institute of Biological Sciences. Professor Orians is a leading authority in ecology and evolution, with research experience in behavioral ecology, plant–herbivore interactions, community structure, the biology of rare species, and environmental policy. He elected early retirement to be able to devote more time to writing and environmental policy activities.

Craig Heller is the Lorry Lokey/Business Wire Professor of Biological Sciences and Human Biology at Stanford University. He has served as Director of the popular interdisciplinary undergraduate program in Human Biology and is now Chairman of Biological Sciences. He is a fellow of the American Association for the Advancement of Science and received the Walter J. Gores Award for Excellence in Teaching. Dr. Heller received his Ph.D. from Yale University in 1970 and did postdoctoral work at Scripps Institution of Oceanography on how the brain regulates body temperature of mammals. His current research is on the neurobiology of sleep and circadian rhythms. Over the years Dr. Heller has done research on systems ranging from sleeping college students to diving seals to hibernating bears to meditating yogis. He teaches courses on animal and human physiology and neurobiology in Stanford's introductory core curriculum.

David Sadava is the Pritzker Family Foundation Professor of Biology at Claremont McKenna, Pitzer, and Scripps, three of the Claremont Colleges. He received his Ph.D. from the University of California, San Diego in 1972 and has been at Claremont ever since. The author of textbooks on cell biology and on plants, genes and agriculture, Professor Sadava has done research in many areas of cell biology and biochemistry, ranging from developmental biology, to human diseases, to pharmacology. His current research concerns human lung cancer and its resistance to chemotherapy. Virtually all of the research articles he has published have undergraduates as coauthors. Professor Sadava teaches introductory biology and has recently developed a new course on the biology of cancer. For the last 15 years, Dr. Sadava has been a visiting professor in the Department of Molecular, Cellular, and Developmental Biology at the University of Colorado, Boulder.

Preface

This is an exciting time to be a biologist: our knowledge of living systems is expanding rapidly and our technologies for research improve daily. This fifth edition of *Life: The Science of Biology* has been an opportunity for us to communicate to students the excitement of modern biology by expanding and refining our coverage, by finding new ways to make important concepts more understandable and memorable, and by conveying the sense of adventure in biological research.

Our overriding goal continues to be to stimulate students' interests in biology. We have tried to do this by making underlying concepts clear and easy to grasp and showing their relevance to medical, agricultural, and environmental issues. Also, we want students to appreciate *how* we know rather than just *what* we know. To that end, we discuss scientific methods and show how experiments, field observations, and comparative methods help biologists formulate and test hypotheses. In the preparation of this edition, we have tried to introduce opportunities for students to think about concepts rather than just learning facts.

Themes and approaches that characterize the new edition

Throughout the book, we use several themes to link chapters and provide continuity. These themes, which are introduced in Chapter 1, include evolution, the experimental foundations of our knowledge, the flow of energy in the living world, the application and influence of molecular techniques, and human health considerations

One of our approaches is to show how basic principles presented in earlier chapters apply in later chapters. For example, programmed cell death, also called apoptosis, has been a major focus of biological research in the past few years. This process is first presented in the context of cell reproduction (Chapter 9). Then we show its applications in development (Chapter 15), cancer (Chapter 17), and the immune system (Chapter 18). Another example is cladistics, introduced first in Chapter 22, and applied in subsequent chapters to show how evolutionary relationships help us understand a wide variety of biological problems.

A new organization enhances accessibility

In chapter after chapter, we have concentrated on making the descriptions, explanations, and applications more accessible to student readers. We have rewritten obscure or difficult passages, deleted some details, simplified the writing and illustrations, and shortened both paragraphs and sections. We have tried to tighten connections, improve transitions, and sharpen the focus. Many changes have been made in how information is distributed among the text, captions, figure labels, and a new feature of the illustrations—"balloon captions."

We have also taken a new approach to headings. We have tried to offer the reader more guidance in identifying, understanding, and interrelating key topics. We use two levels of heads (although occasionally a third level is introduced). Major heads divide the chapters into discrete topics, and second-level heads, now full sentences, identify the explicit focus of each subsection. In addition to providing a clear outline and introduction to covered topics, these "sentence heads" are useful to students for study and review.

To further guide the reader, we have provided explicit forecasts of concepts about to be discussed, both as part of the introduction to each chapter and as part of the introductions to most of the major sections within each chapter. This forecasting allows students to read with expectation and direction, better equipped to appreciate the implications of early topics and to see relationships among topics across the entire chapter.

Different students have different learning styles: some are more image-focused, others more text-focused. Line drawings and photographs have the advantages of directness, emphasis, and drama; on the other hand, text explanations provide explicit information and better describe events that occur through time. We have combined the strengths of both text and graphics through the abundant use of what we're calling "balloon captions." These brief statements are incorporated directly into the graphics and go beyond mere labeling to describe, define, or explain graphic elements. Thus, text becomes more intimately related to graphic representations and the graphics take on more significance. Balloon captions, sometimes numbered to clarify a sequence, guide the reader through the inevitable complexities of some figures; in other figures, balloons emphasize the most important features. This new feature has drawn extensive praise during the development of this edition, and we believe that students will find them highly effective aids to their learning.

A new format for the chapter summaries emphasizes the chapter outline, using major heads to distinguish and identify summary statements. The summary emphasizes major points but also includes specific references to key figures and tables where supporting details are found.

The seven parts: Content, changes, and themes

Each section of the book has undergone important changes. In Part One, The Cell, we eliminated some details and advanced topics, notably in Chapter 6 (Energy, Enzymes, and Metabolism), allowing us to develop certain key concepts such as allostery and cooperativity more clearly. New developments in such areas as protein folding are now introduced in a broad context so the student can relate them to other topics. When appropriate, we have tried to link biochemical and cellular phenomena to specific conditions and diseases that affect human health and well-being.

In Part Two, Information and Heredity, the first six chapters (Chapters 9–14) describe what we know and how we have gained some of this knowledge, and the final four (Chapters 15–18) describe its biological applications. The expression of DNA is dealt with separately in prokaryotes (Chapter 13) and eukaryotes (Chapter 14), and these principles are then used to describe the molecular analysis of development (Chapter 15), the manufacture of useful products via biotechnology (Chapter 16), the diagnosis and treatment of human genetic diseases (Chapter 17), and the production of antibodies (18). Because of its centrality to genetics and molecular biology, we now devote separate chapters to the structure and the role of DNA (Chapters 11 and 12, respectively).

The chapter on development (Chapter 15) in Part Two now concentrates entirely on molecular and genetic aspects of development; the cellular and tissue aspects of embryology are presented in Chapter 40. In addition to applying the principles of molecular biology to recombinant DNA technologies, Chapter 16 emphasizes how these technologies are being used in agriculture and medicine. The "molecular revolution" that is just beginning in medicine, including the Human Genome Project, is the subject of an extensively updated chapter (Chapter 17).

In Part Three, Evolutionary Processes, we have expanded the treatment of cladistic methods to assess evolutionary relationships and show how cladograms are constructed and why knowing evolutionary relationships helps us better understand a wide array of biological problems, including human health problems. With this background, we are able to use phylogenetic trees in subsequent chapters to illustrate evolutionary patterns that range from individual molecules to phyla.

Part Three also includes an entirely new chapter (Chapter 23) on molecular evolution, one of the most exciting and vigorous fields in contemporary biology. Contributed by Peg Riley of Yale University, this chapter emphasizes both detailed molecular comparisons among species and their implications as to why and how molecules change over evolutionary time as organisms encounter and survive environmental challenges.

The results of molecular evolutionary studies have led us to a new emphasis on lineages in Part Four, The Evolution of Diversity, especially in our treatment of bacteria, archaea, and protists. Systematics is in ferment, and we try to impart some sense of current controversies in the field in Chapters 25 and 26. We explicitly treat today's diversity of organisms as the product of evolution.

In Part Five of the fourth edition, we introduced a new chapter, The Biology of Flowering Plants, on plant responses to environmental challenges. It was so well received that we have enriched it with an up-to-date treatment of plant–pathogen interactions. This topic and others continue to emphasize the theme of evolution. Part Five also includes new findings on multiple phytochromes and on developmental mutants in *Arabidopsis*.

In response to requests from instructors, Part Six, The Biology of Animals, now features a chapter (Chapter 40) on animal embryology, which follows the chapter on animal reproduction. The coverage of neurobiology (Chapters 41–44) has been redesigned and expanded to include a new chapter (Chapter 43) on the organization and higher functions of the mammalian brain.

Our theme of human health concerns is manifest throughout Part Six. Chapter 47, on animal nutrition, includes new material on environmental toxicology, an emerging discipline we feel will be of increasing importance to the well-being of our planet.

In Part Seven, Ecology and Biogeography, we have further expanded our coverage of the role of experiments in helping biologists understand the complex interactions among organisms that structure ecological systems. New materials illustrate the role of phylogenetic analyses in behavioral ecology and biogeography. In Chapter 54 we have designed an original graphic method of displaying material on Earth's biomes. This new and striking presentation enables students to visualize and quantify the differences and similarities in the dominant features of Earth's major biomes.

We wish to thank a lot of people

We were all students and teachers long before we were textbook authors, and we want to help students in every way possible. In the next section, "To the Student," we offer some advice that many of our own students have told us they found helpful.

Again, we have been fortunate to receive cogent and significant advice from the more than 60 colleagues who reviewed chapters or whole sections of the book. Their names are listed after this Preface. Their reviews helped to shape many of the changes described above, ranging from the addition of new chapters to the many ways in which we worked to sharpen our story. We thank them all, and hope this new edition measures up to their expectations.

We were already indebted to J/B Woolsey Associates for the elegance and effectiveness of the art programs they developed for the third and fourth editions of this textbook. They have, of course, produced many new illustrations for this edition. However, rather than limiting ourselves to incremental changes in the existing art program, we have taken a major step forward this time with the introduction of the balloon captions. The success of this approach is the result of many factors. James Funston worked with authors and illustrators, offering input to virtually every pixel in the entire art program. John Woolsey and a dedicated team of artists led by Michael Demaray turned our ideas and suggestions into exciting new art.

James Funston, the developmental editor we chose to work with us on this edition, paid close attention to clarity and pedagogical focus. Stephanie Hiebert provided rigorous copy editing from beginning to end. Her sharp eye extended to the illustrations, and her polishing of and additions to the balloon copy often enhanced the clarity of the presentation. Carol Wigg once again coordinated and checked every change made by editors, artists and authors—indeed, she coordinated the entire preproduction process, and she applied her knowledge and talent to writing captions that tightly link the illustations to key points in the text. We owe her more than we can say for her patience, persistence, and skill. Jane Potter, as photo researcher, found many new and exciting photographs to enhance the learning experience and enliven the appearance of the book as a whole.

We wish to thank the dedicated professionals in W. H. Freeman's marketing and sales group. Their efficiency and enthusiasm has helped bring *Life* to a wider audience. We appreciate their constant support and valuable marketing feedback. A large share of *Life*'s success is due to their efforts in this publishing partnership.

Sinauer Associates provided the best publishing environment we can imagine. Their years of success in publishing biology books at the introductory, intermediate, and advanced levels result from their ability to envision a product and to guide, assist, and motivate authors through the long, demanding process. Remarkably, Andy Sinauer never ceases to extend helpful, and, above all, warm support to his authors.

Bill Purves   *Gordon Orians*   *Craig Heller*   *David Sadava*

November, 1997

Reviewers for the Fifth Edition

Kraig Adler, Cornell University

Henry W. Art, Williams College

Carla Barnwell, University of Illinois

Judith L. Bronstein, University of Arizona

Robert J. Brooker, University of Minnesota

Steven B. Carroll, Northeast Missouri State University

James J. Champoux, University of Washington

William A. Clemens, University of California/Berkeley

Frederick M. Cohan, Wesleyan University

Newton Copp, The Claremont Colleges

D. Andrew Crain, University of Florida

Joe W. Crim, University of Georgia

Rowland H. Davis, University of California/Irvine

Patrick E. Elvander, University of California/Santa Cruz

Wayne R. Fagerberg, University of New Hampshire

Michael Feldgarden, Yale University

Rachel D. Fink, Mt. Holyoke College

Barbara Fishel, University of Arizona

William Fixsen, Harvard University

Cecil H. Fox, Molecular Histology, Inc.

Stephen A. George, Amherst College

Wayne Goodey, University of British Columbia

Deborah Gordon, Stanford University

David M. Green, McGill University

Adrian Hayday, Yale University

Joseph Heilig, University of Colorado

Walter S. Judd, University of Florida

Mark V. Lomolino, University of Oklahoma

Michael A. Lydan, University of Toronto/Erindale

Denis H. Lynn, University of Guelph

Laura MacIntosh, Stanford University

James Manser, Harvey Mudd College

John M. Matter, Juniata College

Larry R. McEdward, University of Florida

Michael Meighan, University of California/Berkeley

Melissa Michael, University of Illinois

Charles W. Mims, University of Georgia

Anthony G. Moss, Auburn University

Shahid Naeem, University of Minnesota

Peter Nonacs, University of California/Los Angeles

Barry M. O'Connor, University of Michigan

Ron O'Dor, Dalhousie University

Richard Olmstead, University of Washington

Laura J. Olsen, University of Michigan

Judith A. Owen, Haverford College

Randall W. Phillis, University of Massachusetts

Lorraine Pillus, University of Colorado

Ellen Porzig, Stanford University

Thomas L. Poulson, University of Illinois at Chicago

Loren Reiseberg, Indiana University

Wayne C. Rosing, Middle Tennessee State University

Albert Ruesink, Indiana University

C. Thomas Settlemire, Bowdoin College

Joan Sharp, Simon Fraser University

Esther Siegfried, Pennsylvania State University

Anne Simon, University of Massachusetts

Mitchell L. Sogin, Marine Biological Laboratory, Woods Hole

Collette St. Mary, University of Florida

Millard Susman, University of Wisconsin

Elizabeth Vallen, Swarthmore College

Elizabeth Van Volkenburgh, University of Washington

Gary Wagenbach, Carleton College, Minnesota

Bruce Walsh, University of Arizona

Mark Wheelis, University of California/Davis

Brian White, Massachusetts Institute of Technology

Fred Wilt, University of California/Berkeley

Gregory A. Wray, State University of New York at Stony Brook

To the Student

Welcome to the study of life! In our student days—and ever since—we have enjoyed studying the fascinating and fast-changing field of biology, and we hope that you will, too.

There are a few things you can do to help you get the most from this book and from your course. For openers, read the book actively—don't just read passively, but do things that force you to think as you read. If we pose questions, stop and think about them. If a passage reminds you of something that has gone before, think about that, or even check back to refresh your memory. Ask questions of the text as you go. Do you understand what is being said? Does it relate to something you already know? Is it supported by experimental or other evidence? Does that evidence convince you? How does this passage fit into the chapter as a whole? Annotate the book—write down comments in the margins about things you don't understand, or about how one part relates to another, or even when you find an idea particularly interesting. The point of doing these things is that they will help you learn. People remember things they think about much better than they remember things they have read passively. Highlighting is passive; copying is drudge work; questioning and commenting are active and well worthwhile.

"Read" the illustrations actively too. You will find the balloon captions in the illustrations especially useful—they are there to guide you through the complexities of some topics and to highlight the major points.

The chapter summaries will help you quickly review the high points of what you have read. A summary identifies particular illustrations that you should study to help organize the material in your mind. It is essential that you study the cited illustrations and their captions as you review because important information that is covered in illustrations has been left out of the summary statements. Add concepts and details to the framework by reviewing the text. A way to review the material in slightly more detail after reading the chapter is to go back and look at the boldfaced terms. You can use the boldfaced terms to pose questions—and see if you can answer those questions. The boldfacing will probably be more useful on a second reading than on the first.

Use the self-quizzes and "Applying Concepts" questions at the end of each chapter. The self-quizzes are meant to help you understand some of the more detailed material and to help you sort out the information we have laid before you. Answers to all self-quizzes are in the back of the book. The concept questions, on the other hand, are often fairly open-ended and are intended to cause you to reflect on the material.

Two parts of a textbook that are, unfortunately, often underused or even ignored are the glossary and the index. Both can help you a great deal. When you are uncertain of the meaning of a term, check the glossary first—there are more than 1,500 definitions in it. If you don't find a term in the glossary, or if you want a more thorough discussion of the term, use the index to find where it's discussed.

What if you'd like to pursue some of the topics in greater detail? At the end of each chapter there is a short, annotated list of supplemental readings. We have tried to choose readings from books and magazines, especially *Scientific American*, that should be available in your college library.

To provide another kind of help for students, we commissioned a CD-ROM (*Life 5.0*) covering the subject matter of Parts One and Two of this textbook. *Life 5.0* introduces and illustrates (often with unique animations) over 1700 key terms and concepts. You can access this information in several ways: via *Life* chapter reviews; via minicourses such as "Molecular Structure," "The Cell Cycle," and "DNA Replication"; or via a hyperlinked index. There are also several hundred self-quiz items and dozens of thought problems. You may have a copy of the disk inside the front cover of this book; if not, and if you would like to purchase one, contact **www.mona-group.com**. If you use the disk, explore its contents to see which of its tools best correspond to your needs.

Most students occasionally have difficulty in courses, including biology courses. If you find that you are slipping behind in the course, or if a particular topic is giving you an unreasonable amount of trouble, here are some useful steps you might take. First, the basics: attend class, take careful lecture notes, and read the textbook assignments. Second, note that one of the most important roles of studying is to discover what you don't know, so that you can do something about it. Use the index, the glossary, the chapter summaries, and the text itself to try to answer any questions you have and to help you organize the material. Make a habit of looking over your lecture notes within 24 hours of when you take them—find out right away what points are unclear, and get them straightened out in your mind. The CD-ROM can help by providing a different perspective.

If none of these self-help remedies does the trick, get help! Other students are often a good source of help, because they are dealing with the material at the same level as you are. Study groups can be very useful, as long as the participants are all committed to learning the material. Tutors are almost always helpful and useful, as are faculty members. The main thing is to get help when you need it. It is not a good idea to be strong and silent and drift into a low grade.

But don't make the grade the point of this or any other course. You are in college to learn, to pursue interesting subjects, and to enjoy the subjects you are pursuing. We hope you'll enjoy the pursuit of biology.

Bill Purves   Gordon Orians   Craig Heller   David Sadava

Contents in Brief

Contents

Part One

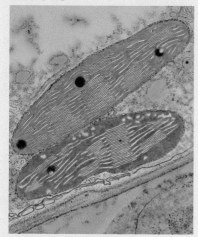

The Cell

Part Two

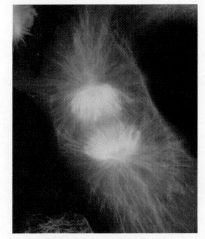

Information and Heredity

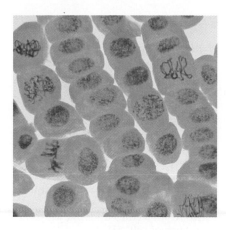

Part Three

Evolutionary Processes

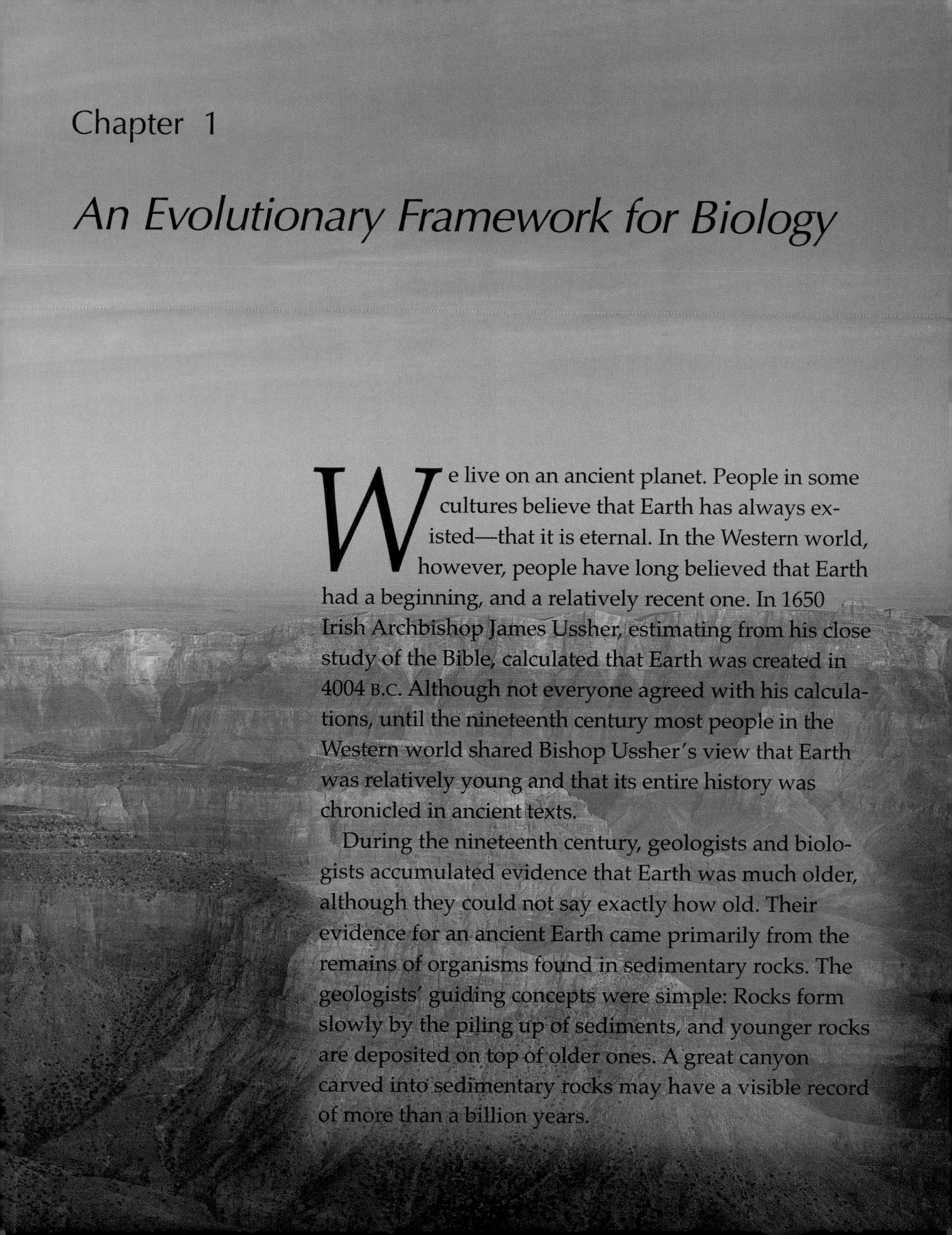

Chapter 1

An Evolutionary Framework for Biology

We live on an ancient planet. People in some cultures believe that Earth has always existed—that it is eternal. In the Western world, however, people have long believed that Earth had a beginning, and a relatively recent one. In 1650 Irish Archbishop James Ussher, estimating from his close study of the Bible, calculated that Earth was created in 4004 B.C. Although not everyone agreed with his calculations, until the nineteenth century most people in the Western world shared Bishop Ussher's view that Earth was relatively young and that its entire history was chronicled in ancient texts.

During the nineteenth century, geologists and biologists accumulated evidence that Earth was much older, although they could not say exactly how old. Their evidence for an ancient Earth came primarily from the remains of organisms found in sedimentary rocks. The geologists' guiding concepts were simple: Rocks form slowly by the piling up of sediments, and younger rocks are deposited on top of older ones. A great canyon carved into sedimentary rocks may have a visible record of more than a billion years.

Preserved within some rocks were **fossils**—the remains of organisms that lived while the sediments were accumulating. When they compared older rocks with younger ones, geologists could detect slight but significant differences among similar fossil organisms. Furthermore, they found fossils of similar organisms at widely separated locations. By assuming that rocks at different locations containing the same type of fossil were of approximately the same age, early geologists determined the general order of events in the history of life on Earth. Although they could establish a sequence, these geologists had no method for determining the absolute ages of fossils.

One of the triumphs of twentieth-century science has been the development of methods to date materials formed in the past. The discovery that unstable forms (radioactive isotopes) of familiar atoms such as carbon and phosphorus decay at constant rates made it possible to date materials. Radioactive isotopes are incorporated into rocks and fossils in proportion to their presence in the environment when the rock solidified. Each type of radioactive isotope then begins to decay at its own constant rate, eventually becoming stable. Scientists can calculate the absolute ages of rocks from the proportions of radioactive and stable isotopes they contain.

Meanwhile, scientists in the fields of astronomy and physics, using data from the powerful telescopes and space probes that became available in the latter half of the twentieth century, have come to believe that our planet formed approximately 4 billion years ago. The earliest known fossils have been dated, using radioisotopes, as being 3.8 billion years old, so we know that life arose early in the history of Earth.

Evolutionary Milestones

The fullness of time is difficult for people to grasp. We all understand time spans measured in seconds, minutes, hours, days, years, and decades, but we find it difficult to comprehend millions, much less billions, of years. The following overview of the major evolutionary milestones is intended to provide a framework that presents life's characteristics as they will be covered in this book, and an overview of how these characteristics evolved during the history of life on Earth.

Life arose from nonlife

The first life must have come from nonlife. All matter, living and nonliving, is made up of chemicals. The smallest chemical units are atoms, which bond together into molecules (the properties of these molecules are the subject of Chapter 2). We think that the processes leading to life began nearly 4 billion years ago with interactions among small molecules that stored information in easy-to-copy sequences.

Chemical information became more complex when the information stored in these simple sequences resulted in the synthesis of larger molecules with complex but relatively stable shapes. Because they were both complex and stable, these molecules could participate in increasing numbers and kinds of chemical reactions. Certain types of large molecules—carbohydrates, lipids, proteins, and nucleic acids—are formed only by living systems, and they are found in all living systems. The properties and functioning of these complex *organic molecules* are the subject of Chapter 3.

About 3.8 billion years ago interacting systems of molecules came to be enclosed in compartments surrounded by *membranes*. Within these membrane-enclosed units, or *cells*, control was exerted over the entrance and retention of molecules, the chemical reactions taking place within the cell, and the exit of molecules. Cells and membranes are the subjects of Chapters 4 and 5.

Cells are so effective at capturing energy and replicating themselves—two fundamental characteristics of life—that since they evolved, cells have apparently outcompeted any noncellular life. The cell is the unit on which all life has been built.

An Englishman, Robert Hooke, built a simple microscope in 1665 and was the first person to observe cells. The cells he saw were those of cork, wood, and other dead plant materials, and they were empty. Living organisms that were fully contained in a single cell were first observed a few years later by the Dutch naturalist Antoni van Leeuwenhoek.

By 1839, microscopes had improved and enough living material had been observed that the German physiologist Theodor Schwann could assert that *all organisms consist of cells*. In 1858, the German physician Rudolf Virchow suggested that *all cells come from preexisting cells*. Experiments by the French chemist and microbiologist Louis Pasteur between 1859 and 1861 convinced most scientists that cells do not arise from noncellular material, but must come from other cells. In the modern world, life no longer arises from nonlife.

The first organisms were single cells

For 2 billion years all cells were small. They lived mostly autonomous lives, each separate from the other. Their lives were confined to the oceans, where they were shielded from lethal ultraviolet light. The relatively small amounts of genetic information that allowed these **prokaryotic cells** to replicate themselves and the biochemical machinery by which they obtained energy floated loose within an outer membrane. Some prokaryotes living today are similar to those that existed early in the evolution of living cells, several billion years ago (Figure 1.1).

To maintain themselves, to grow, and to reproduce, all organisms—whether they consist of one cell or tril-

1.1 Early Life May Have Resembled These Cells These "rock-eating" bacteria, appearing red in the artificially colored micrograph, were discovered in pools of water trapped between layers of rock more than 1,000 meters below Earth's surface. Deriving chemical nutrients from the rocks and living in an environment devoid of oxygen, they may resemble some of the earliest prokaryotic cells.

lions of cells—must obtain raw materials and energy from the environment. These raw materials are chemicals that are digested; the products are used to synthesize large carbon-based molecules. The energy obtained from chemical digestion is used to power the synthetic reactions. These conversions of matter and energy are called **metabolism**.

All organisms can be viewed as devices to capture, process, and convert matter and energy from one form to another; these conversions are the subjects of Chapters 6 and 7. *A major theme in the evolution of life is the development of increasingly diverse ways of capturing external energy and using it to drive biologically useful reactions.*

The earliest cells derived their energy from simple chemical compounds because complex molecules were scarce in their environment. On early Earth, volcanoes poured large quantities of methane and hydrogen sulfide into the atmosphere. Early prokaryotes evolved the ability to ingest these molecules and use them as sources of energy.

Photosynthesis and sex changed the course of evolution

Two powerful evolutionary events took place in the first billion years. One, the evolution of photosynthesis, created new metabolic pathways. The other, the evolution of sex, stimulated the evolution of the almost unimaginable diversity of organisms on Earth.

PHOTOSYNTHESIS CHANGED EARTH'S ENVIRONMENT. About 2.5 billion years ago, some prokaryotes evolved the ability to use the energy of sunlight to power their metabolism. Although raw chemicals were still taken up from the environment, the energy used to metabolize these chemicals came directly from the sun.

The early photosynthetic prokaryotes were probably similar to present-day cyanobacteria (Figure 1.2). The energy-capturing process they used—**photosynthesis**—is the basis of nearly all life on Earth today. As you will learn in Chapter 8, photosynthesis is a complex process made up of many chemical reactions. The ability to perform the photosynthetic reactions probably accumulated gradually during the first billion years or so of evolution, but once the ability had evolved, the effects of photosynthesis were dramatic.

Photosynthetic prokaryotes were so successful that they released vast quantities of oxygen gas (O_2) into the atmosphere. The presence of oxygen opened up new avenues of evolution. Metabolic pathways based on O_2—*aerobic metabolism*—came to be used by most organisms on Earth. The air we breathe today would not exist without photosynthesis.

Over a much longer time frame, the vast quantities of oxygen liberated by photosynthesis had another effect. A form of oxygen we call ozone (O_3) began to accumulate along with the O_2 in the atmosphere. The ozone slowly formed a dense layer that acted as a shield, intercepting much of the sun's deadly ultraviolet radiation. Eventually (although only within the last 800 million years of evolution) the presence of this shield allowed organisms to leave the protection of the ocean and find new lifestyles on Earth's land surfaces.

SEX CHANGED EVOLUTIONARY RATES. The earliest unicellular organisms reproduced by dividing. Progeny cells were identical to parent cells. But **sexual recombination**—the combining of genes from two cells in one cell—appeared early during the evolution of life. Sex is advantageous because an organism that receives genetic information from another individual produces

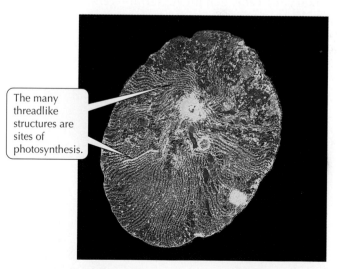

The many threadlike structures are sites of photosynthesis.

1.2 Oxygen Produced by Prokaryotes Changed Earth's Atmosphere This modern cyanobacterium is probably very similar to early photosynthetic prokaryotes.

offspring that are more variable. *Reproduction with variation is a major characteristic of life.* Because environments continuously vary, individuals that produce variable offspring rather than genetically identical "clones" are more likely to produce at least some offspring that *adapted* to changes in the environment.

Adaptation to environmental change is one of life's most distinctive features. An organism is adapted to its environment when it possesses features that enhance its survival and ability to reproduce in a given environment. Sex is so adaptive that today nearly all organisms on Earth engage in sex at least occasionally. By creating increased variation, sexual recombination increased the rate of evolutionary change.

Early prokaryotes engaged in sex (exchanging genetic material) and reproduction (cell division) at different times. Even today in many unicellular organisms, sex and reproduction occur at different times (Figure 1.3). But a different kind of organism evolved that would require a more complicated sex life.

Eukaryotes are "cells within cells"

As the ages passed, some prokaryotic cells became large enough to attack and consume smaller cells, becoming the first *predators*. Usually the smaller cells were destroyed within the predators' cells, but some of these smaller cells survived and became permanently integrated into the operation of their hosts' cells. In this manner, cells with complex internal compartments arose. We call these cells **eukaryotic cells**. Their appearance slightly more than 1.5 billion years ago opened more new evolutionary pathways.

Prokaryotic cells—including all the early bacteria and archaea—have only one membrane, the one that surrounds them. Eukaryotic cells, on the other hand, are filled with membrane-enclosed compartments. In eukaryotic cells, genetic material—*genes* and *chromo-*

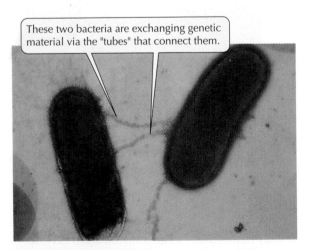

These two bacteria are exchanging genetic material via the "tubes" that connect them.

1.3 Sex Between Prokaryotes   Genetic exchange produces variation that leads to adaptive evolution.

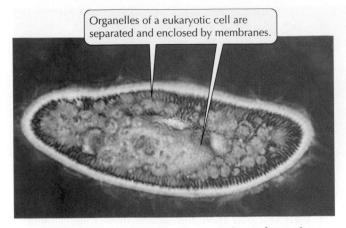

Organelles of a eukaryotic cell are separated and enclosed by membranes.

1.4 Multiple Compartments Characterize Eukaryotic Cells   The nucleus and other specialized compartments (known as organelles) probably evolved from small prokaryotes that were ingested by a larger prokaryotic cell.

somes—became contained within a discrete **nucleus** and became increasingly complex. Other compartments became specialized for other purposes, such as photosynthesis (Figure 1.4). We refer to these specialized compartments as **organelles**.

Cells evolved the ability to change their structures and specialize

Until slightly more than 1 billion years ago, only unicellular organisms existed. Two key developments made the evolution of **multicellular organisms**—organisms consisting of more than one cell—possible. One was the ability of a cell to change its structure and functioning to meet the challenges of a changing environment. Prokaryotes accomplished this when they evolved the ability to change from rapidly growing cells into resting *spores* that could survive harsh environmental conditions. The second development allowed cells to stick together after they divided, forming a multicellular organism.

Once organisms began to be composed of many cells, it became possible for the cells to specialize. Certain cells, for example, could be specialized to perform photosynthesis. Other cells might become specialized to transport chemical raw materials, such as oxygen, from one part of an organism to another. Very early in the evolution of multicellular life, certain cells began to be specialized for sex. As multicellular life evolved, sex and reproduction became linked. In almost all present-day multicellular organisms, sex and reproduction occur together.

With more complicated and specialized sex cells, sex itself became more complicated. Simple cell division, which we know as **mitosis**, was and is sufficient for the needs of most cells. But a whole new method of cell division—**meiosis**—evolved that opened up new

1.5 Organisms May Change Dramatically during Their Lives The caterpillar, pupa, and adult are all stages in the life cycle of a monarch butterfly. The transition from one stage to another is triggered by internal signals.

realms of recombination possibilities for the specialized sex cells, or *gametes*. Mitosis and meiosis are explained and compared in Chapter 9.

The cells of an organism are constantly adjusting

Both the emergence of multicellular life and the changes in Earth's atmosphere that allowed life to move out of the oceans and exploit the environments of the land masses quickened the pace of evolution. Photosynthetic green plants colonized the land and provided a rich source of energy for a vast array of organisms that consumed them. But whether it is made up one cell or many, an organism must respond appropriately to many signals emanating from its external and internal environments.

The external environment can change rapidly and unpredictably in ways that are outside of the organism's control. An organism can remain healthy only if its internal environment remains within a given range of physical and chemical conditions. Organisms maintain a relatively constant internal environment by adjusting their metabolism in response to external and internal signals indicating such things as a change in temperature, the presence or absence of sunlight, the presence or absence of specific chemicals, the need for nutrients (food) and water, or the presence of a foreign agent inside the organism's body. Maintenance of a relatively stable internal condition is called **homeostasis**.

The adjustments that organisms make to maintain constant internal conditions are usually minor; they are not obvious, because nothing appears to change. However, at some time during their lives many organisms respond to signals not by maintaining their status, but by undergoing major physical reorganization. We mentioned in the previous section the ability of prokaryotes to change from rapidly growing cells into dormant spores in response to environmental stresses. A striking example that evolved much later is *metamorphosis*, seen in many modern insects, such as butterflies. In response to internal chemical signals, a caterpillar changes into a pupa and then into an adult butterfly (Figure 1.5).

A major theme in the evolution of life is the development of increasingly complicated systems for responding to signals and maintaining homeostasis. Indeed, some animals exhibit a widespread and important biological process called *learning*, in which important changes in the internal environment result from responses to external signals (such as this textbook).

Multicellular organisms develop and grow

Multicellular organisms cannot achieve their adult shapes or function effectively unless their growth is carefully regulated. Uncontrolled growth, one example of which is cancer, ultimately destroys life. The functioning of a multicellular organism requires a sequence of events leading from a single cell to a multicellular adult. *A vital characteristic of living organisms is regulated growth.*

The activation of information within cells, and the exchange of information among many cells, produce the well-timed events that are required by the transition from single cell to adult form. Genes control the metabolic processes necessary for life. The astounding nature of the genetic material that controls these lifelong events has been understood only within the twentieth century; it is the story to which much of Part Two of this book is devoted.

Altering the timing of developmental processes can produce striking changes. Just a few genes can control processes that result in dramatically different adult organisms. Chimpanzees and humans share more than 97 percent of their genes, but the differences between

1.6 Genetically Similar but Quite Distinct   By looking at the two, you would never guess that chimpanzees and humans share more than 97 percent of their genes.

the two in form and in behavioral abilities, most notably speech, are dramatic (Figure 1.6). When we realize how little information it sometimes takes to create major transformations, the still-mysterious process of *speciation* becomes a little less of a mystery.

Speciation has resulted in the diversity of life

All organisms on Earth today are the descendants of a unicellular organism that lived almost 4 billion years ago. The preceding sections of this chapter described the changes that led to the diversity present in life today. The course of this evolution has been accompanied by the storage of larger and larger quantities of information and increasingly complex mechanisms for using it. But if that were the entire story, only one kind of organism might exist on Earth today. Instead, Earth is populated by many millions of kinds of organisms that are genetically different from one another. We call these genetically independent groups **species**.

As long as individuals within a population mate at random and reproduce, structural and functional changes may occur, but only one species will exist. However, if a population becomes divided into two or more groups, individuals can mate only with individuals in their own group. When this happens, differences may accumulate with time, and the groups may evolve into different species.

The splitting of groups of organisms into separate species has resulted in the great richness and variety of life found on Earth today, as described in Chapter 19. How species form is explained in Chapters 20 and 21. From a single ancestor, many species may arise as a result of the repeated splitting of populations. How bi-

ologists determine which species have descended from a particular ancestor is discussed in Chapter 22.

Sometimes humans refer to species as "primitive" or "advanced." These and similar terms, such as "lower" and "higher," are best avoided because they imply that some organisms function better than others. The abundance of prokaryotes—all of which are relatively simple—readily demonstrates that they are highly functional, despite their relative simplicity. Therefore, in this book, we usually use the terms "ancestral" and "derived" to describe characteristics that appeared earlier and later in the evolution of lineages of life, respectively, recognizing that all organisms that have survived are successfully adapted to their environments. The wings that allow a bird to fly or the structures that allow green plants to survive in environments where water is either scarce or overabundant are examples of the rich array of adaptations found among organisms (Figure 1.7).

Biological Diversity: Domains and Kingdoms

As many as 30 million species of organisms inhabit Earth today. Many times that number lived in the past but are now extinct. To help us understand the past and current diversity of organisms, biologists use classification systems that group organisms according to their evolutionary relationships.

In the classification system used by most modern biologists, organisms are grouped into three **domains** and six **kingdoms** (Figure 1.8). Organisms belonging to a particular domain have been evolving separately from organisms in other domains for more than a billion years. Organisms in the domains **Archaea** and **Bacteria** are prokaryotes, single cells that lack a nucleus and other internal compartments found in the cells from other kingdoms.

Archaea and Bacteria differ so fundamentally from one another in the chemical pathways by which they function and in the products they produce that they are believed to have separated into distinct evolutionary lineages very early during the evolution of life. Each of these domains consists of a single kingdom, Archaebacteria and Eubacteria, respectively. These kingdoms are covered in Chapter 25.

Members of the other domain—**Eukarya**—have eukaryotic cells with nuclei and complex cellular compartments called *organelles*. The Eukarya are divided into four kingdoms—Protista, Plantae, Fungi, and Animalia. The kingdom **Protista** (protists), the subject of Chapter 26, contains mostly single-celled organisms. The remaining three kingdoms, nearly all of whose members are multicellular, are believed to have arisen from ancestral protists.

Most members of the kingdom **Plantae** (plants) convert light energy to chemical energy by photosynthesis.

1.7 Adaptations to the Environment (a) The long, pointed wings of the peregrine falcon allow it to accelerate rapidly as it dives on its prey. (b) The wings of an Arctic tern allow it to hover above the water while searching for fish. (c) In a water-limited environment, this saguaro cactus stores water in its fleshy trunk. Its roots spread broadly to quickly extract water immediately after it rains. (d) The above-ground root system of red mangroves is a modification that allows them to thrive while inundated by water—an environment that would kill most terrestrial plants.

The biological molecules that they and some protists synthesize are the primary food for nearly all other living organisms. This kingdom is covered in Chapter 27.

The kingdom **Fungi**, the subject of Chapter 28, includes molds, mushrooms, yeasts, and other similar organisms, all of which are *heterotrophs*—that is, they require a food source of energy-rich molecules synthesized by other organisms. Fungi absorb food substances from their surroundings and digest them within their cells. Many are important as decomposers of the dead bodies of other organisms.

Members of the kingdom **Animalia** (animals) are heterotrophs that ingest their food source, break down (digest) the food outside their cells, and then absorb the products. Animals eat other forms of life for their raw materials and energy. Perhaps because we are animals ourselves, we are often drawn to study members of this kingdom, which is covered in this book in Chapters 29 and 30.

Biologists recognized that organisms were adapted for life in differing environments long before they understood how adaptation came about. Nearly 150 years ago, Charles Darwin and Alfred Russel Wallace proposed the first scientifically testable theory about adaptation. Their suggestion—that *adaptation is the result of evolution by natural selection*—has guided biological investigations ever since.

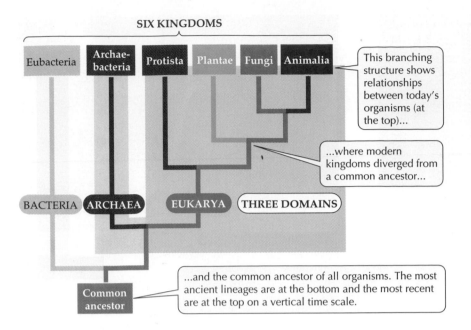

SIX KINGDOMS

Eubacteria | Archae-bacteria | Protista | Plantae | Fungi | Animalia

This branching structure shows relationships between today's organisms (at the top)...

...where modern kingdoms diverged from a common ancestor...

BACTERIA   ARCHAEA   EUKARYA   THREE DOMAINS

...and the common ancestor of all organisms. The most ancient lineages are at the bottom and the most recent are at the top on a vertical time scale.

Common ancestor

1.8 Domains and Kingdoms In the classification system used in this book, Earth's organisms are divided into three domains and six kingdoms.

The World Into Which Darwin Led Us

Long before scientists understood how biological evolution happened, they suspected that living organisms had evolved from organisms no longer alive on Earth. In the 1760s, the French naturalist Count George-Louis Leclerc de Buffon (1707–1788) wrote his *Natural History of Animals*, which contained a clear statement of the possibility of evolution.

Buffon originally believed that each species had been divinely created for different ways of life, but as he studied animal anatomy, doubts arose. He observed that the limb bones of all mammals, no matter what their way of life, were remarkably similar in many details (Figure 1.9). Buffon also noticed that the legs of certain animals, such as pigs, have toes that never touch the ground and appear to be of no use. He found it difficult to explain the presence of these seemingly useless small toes by special creation.

However, both these troubling facts could be explained if mammals had not been specially created in their present forms but had been modified from a common ancestor. Buffon suggested that the limb bones of mammals might all have been inherited, and that pigs might have functionless toes because they inherited them from ancestors with fully formed and functional toes. This was an early statement of evolution (descent with modification), although Buffon did not attempt to explain how such changes took place.

Buffon's student Jean Baptiste de Lamarck (1744–1829) wrote extensively about evolution and was the first person to propose a mechanism of evolutionary change. Lamarck suggested that lineages of organisms may change gradually over many generations as offspring inherit structures that have become larger and more highly developed as a result of continued use or, conversely, have become smaller and less developed as a result of disuse.

For example, Lamarck suggested that aquatic birds extend their toes while swimming, stretching the skin between them. This stretched condition, he thought, could be inherited by the offspring, who would further stretch their skin during their lifetimes and would also pass this condition along to their offspring. According to Lamarck, birds with webbed feet would thereby evolve over a number of generations. He explained many other examples of adaptations in a similar way.

Today scientists do not believe that evolutionary changes are produced by inheritance resulting from use and disuse, as Lamarck suggested. But Lamarck did realize that species evolve with time, and his ideas

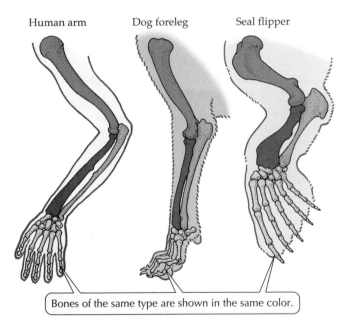

Human arm   Dog foreleg   Seal flipper

Bones of the same type are shown in the same color.

1.9 All Mammals Have Similar Limb Bones Mammalian forelimbs have different purposes: Humans use theirs for manipulating objects, dogs use theirs for walking, and seals use theirs for swimming. But the number and type of their bones are similar, indicating that they have been modified over time from a common ancestor.

deserved more attention than they received from his contemporaries, most of whom believed in a relatively young and unchanging universe. After Lamarck, other naturalists, scientists, and thinkers speculated that with time species change.

By 1858, the climate of opinion (among biologists, at least) was receptive to a theory of evolutionary processes proposed independently by Charles Darwin and Alfred Russel Wallace. By then geologists had shown that Earth had changed over millions of years, not merely a few thousand years. Thus, the presentation in the latter half of the nineteenth century of a well-documented and thoroughly scientific argument for evolution triggered a transformation of biology.

Darwin initiated the scientific study of evolution

Charles Darwin (1809–1882) based his approach to evolution on the following hypotheses:

1. Earth is very old, and organisms have been changing steadily throughout the history of life.
2. All organisms are descendants of a single common ancestor.
3. Species multiply by splitting into daughter species; such speciation has resulted in the great diversity of life found on Earth.
4. Evolution proceeds via gradual changes in populations, not by the sudden production of individuals of dramatically different types.
5. The major agent of evolutionary change is natural selection.

These five hypotheses have all been supported by the mass of research that has been conducted since Darwin published his book *The Origin of Species* in 1859.

Darwin's major insight was to perceive the significance of facts that were familiar to most of his fellow biologists. He understood that populations of all species have the potential for exponential increases in numbers. To illustrate this point, Darwin used the following example:

> Suppose … there are eight pairs of birds, and that only four pairs of them annually … rear only four young, and that these go on rearing their young at the same rate, then at the end of seven years (a short life, excluding violent deaths for any bird) there will be 2048 birds instead of the original sixteen.

Charles Darwin

1.10 Many Organisms Have High Birth and Death Rates
This pair of sergeant majors is guarding the thousands of reddish eggs the female laid in a compact mat. If all the offspring of sergeant majors grew to adulthood and reproduced, the world's population would be overwhelming within a few years. However, most of these eggs and the fish that hatch from them will not survive.

Yet such rates of increase are rarely achieved in nature; the numbers of individuals of most species are relatively stable through time. Therefore, death rates in nature must be high (Figure 1.10). Without high death rates, even the most slowly reproducing species would quickly reach enormous population sizes.

Darwin also observed that, although offspring tend to resemble their parents, the offspring of most organisms are not identical to each other or to their parents (Figure 1.11). He suggested that slight variations among individuals significantly affect the chance that a given individual will survive and reproduce. He called this differential reproductive success of individuals **natural selection**.

Darwin probably used the words "natural selection" because he was familiar with the artificial selection practices of animal and plant breeders. Many of Darwin's observations on the nature of variation came from domesticated plants and animals. Darwin himself was a pigeon breeder, and he knew firsthand the astonishing diversity in color, size, form, and behavior that could be achieved by artifical human selection of which pigeons to mate (Figure 1.12). He recognized close parallels between artificial selection by breeders and selection in nature.

1.11 Offspring Differ from Their Parents These ducklings are members of a brood that hatched from a single clutch of eggs laid by this patterned female and fathered by a single male. Genetic variability among offspring of two parents is the norm.

Darwin states his case for natural selection

In *The Origin of Species* Darwin argued his case for natural selection:

> How can it be doubted, from the struggle each individual has to obtain subsistence, that any minute variation in structure, habits or instincts, adapting that individual better to the new conditions, would tell upon its vigour and health? In the struggle it would have a better chance of surviving; and those of its offspring which inherited the variation, be it ever so slight, would have a better chance.

That statement, written more than 100 years ago, still stands as a good expression of the idea of evolution by natural selection. Since Darwin wrote these words, biologists have developed a much deeper understanding of the genetic basis of evolutionary change and

have assembled a rich array of examples of how natural selection acts.

Biology began a major conceptual shift a little more than a century ago with the general acceptance of long-term evolutionary change and the recognition that natural selection is the primary agent that adapts organisms to their environments. The shift took a long time because it required abandoning many components of an earlier worldview. The pre-Darwinian view held that the world was young, and that organisms had been created in their current forms. In the Darwinian view, the world is ancient, and both Earth and its inhabitants have been changing from forms very different from the ones they now have.

Accepting this new world view means accepting not only the processes of evolution, but also the view that the living world is constantly evolving, but without any "goals." The idea that evolutionary change is not directed toward a final goal or state has been more difficult for some people to accept than the process of evolution itself.

Asking the Questions "How?" and "Why?"

Biological processes and products can be viewed from two different but complementary perspectives. We ask functional questions: How does it work? We also ask adaptive questions: Why has it evolved to work in that way?

Suppose, for example, some marine biologists out on their research vessel are suddenly surrounded by dolphins leaping completely out of the ocean (Figure 1.13). Two obvious questions to ask are, *How* do these marine mammals achieve such a jump?, and *Why* do they do it? An answer to the how question would deal with the molecular mechanisms underlying muscular contraction, nerve and muscle interactions, and the receipt of stimuli by the dolphins' brains.

An investigation to answer the why question would attempt to determine why leaping out of the water is

1.12 Many Types of Pigeons Have Been Produced by Artificial Selection
Charles Darwin, who raised pigeons as a hobby, saw similar forces at work in artificial and natural selection.

1.13 How and Why do Dolphins Leap? Scientists from different disciplines focus on only one of these questions, and their answers are certain to be very different.

adaptive—that is, why it improves the survival and reproductive success of dolphins. For this particular instance, the why question is more obscure, and scientists still have no clear-cut answer. Some animals, such as frogs, evolved jumping behavior because it increased their chances of escape from predators; for others, such as mountain goats, jumping allowed them to cross obstacles like ravines and ridges. Neither of these explanations seems to apply to dolphins, who do not encounter barriers in the ocean and who jump when no predators are chasing them. Jumping may allow them to see better, or it may help dislodge parasites from their bodies; but neither of these possible benefits is established. Perhaps *you* can come up with an explanation of the behavior; in the meantime, scientists (along with most other humans) will enjoy watching it.

Is either of the two types of question more basic or important than the other? Is any one of the answers more fundamental or more important than the others? Not really. The richness of possible answers to apparently simple questions makes biology a complex science, but also an exciting field. Whether we're talking about molecules bonding, cells dividing, blood flowing, dolphins leaping, or forests growing, we are constantly posing both how and why questions. To answer these questions, scientists generate hypotheses that can be tested.

Hypothesis testing guides scientific research

Underlying all scientific research is the **hypothetico-deductive approach** with which scientists ask ques-

tions and test answers. This method allows scientists to modify and correct their beliefs as new observations and information become available. The method has five stages:

1. Making observations.
2. Asking questions.
3. Forming **hypotheses**, which are tentative answers to the questions.
4. Making predictions based on the hypotheses.
5. Testing the predictions by making additional observations or conducting experiments. The additional data gained may support or contradict the predictions being tested.

If the data support the hypothesis, it is subjected to additional predictions and tests. If they continue to support it, confidence in its correctness increases and the hypothesis comes to be considered a theory. If the data do not support the hypothesis, it is abandoned or modified in accordance with the new information. Then new predictions are made, and more tests are conducted. An example will illustrate this process.

APPLYING THE HYPOTHETICO-DEDUCTIVE METHOD. Biologists have long known that some caterpillars are conspicuously colored but that other caterpillars blend in with their backgrounds (Figure 1.14). Conspicuously colored caterpillars often live in groups but are seldom attacked by birds. These initial observations suggested questions that were used to develop hypotheses, make predictions, and devise an experiment to test the predictions. Let's examine how this was done.

GENERATING A HYPOTHESIS. The bright colors of some caterpillars, together with the observation that potential predators usually avoid brightly colored caterpillars, suggested a question that became a hypothesis: The bright color patterns of these caterpillars signal to potential predators that the caterpillars are distasteful or toxic. A companion hypothesis is that inconspicuous caterpillars are good to eat (palatable), and their coloration thus reduces the chance that predators will discover and eat them.

For each hypothesis of an effect there is a corresponding **null hypothesis** asserting that the proposed effect is absent. The null hypothesis for the hypotheses we have just stated is that there is no difference in palatability between colorful and camouflaged caterpillars.

Notice that these hypotheses depend on certain assumptions or on previous knowledge. For example, we assume that birds have color vision and can learn about the qualities of their prey by encountering and tasting them. If such assumptions are uncertain, they should be tested before other experiments are performed.

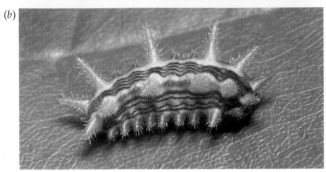

1.14 Caterpillars Can Be Easy or Hard to See (a) Many caterpillars blend into their surroundings, like these catocala moth larvae camouflaged by the bark of an oak tree. (b) This colorful butterfly larva, a stinging rose, contrasts with its leafy environment. Its name implies that its effect on a predator is unpleasant.

MAKING AND TESTING PREDICTIONS. The hypotheses about colorful and inconspicuous caterpillars led to predictions that were tested by an experiment. Captive birds, blue jays, were presented with both brightly colored monarch butterfly caterpillars and caterpillars that blended into their backgrounds. The blue jays were first deprived of food long enough to make them hungry, so they readily attacked the caterpillars. Ingesting even part of one monarch caterpillar caused a blue jay to vomit.

Because the birds were housed individually, the experimenters knew which ones had previously tasted monarchs and which ones had not. They found that a single experience with a monarch caterpillar was enough to cause a blue jay to reject all other monarch caterpillars presented to it. The camouflaged caterpillars, on the other hand, were readily attacked and eaten, and the jays continued to eat these caterpillars without showing signs of sickness or discontent.

These results supported the palatability hypothesis. The null hypothesis, that there is no difference in palatability between colorful and camouflaged caterpillars, was thus rejected.

Experiments are powerful tools

Scientists use a variety of methods to test predictions from their hypotheses. Among these are laboratory and field experiments and carefully focused observations. Each method has its strengths and weaknesses. The key feature of **experimentation** is the control of most factors so that the influence of a single factor can be seen clearly. In the experiments with blue jays, all birds were equally hungry and they were presented with caterpillars in the same way. By controlling these conditions, the experimenters could reject alternative explanations. Their results, for example, could not have been due to lack of hunger on the part of the birds or to the birds' failure to see the caterpillars.

The advantage of working in a laboratory is that control of the environment is easier. Field experiments are more difficult because it is usually impossible to control more than a small part of the total environment. But field experiments have one important advantage: Their results are more readily applicable to what happens where the organisms actually live and evolve. Just because an organism does something in the laboratory does not mean that it behaves the same way in nature. Because biologists usually wish to explain nature, not processes in the laboratory, combinations of laboratory and field experiments are needed to test most hypotheses about what organisms do (Figure 1.15).

A single piece of evidence supporting a hypothesis rarely leads to widespread acceptance of the hypothesis. Similarly, a single contrary result rarely leads to abandonment of a hypothesis. Negative results can be obtained for many reasons, only one of which is that the hypothesis is wrong. For example, the error may be that incorrect predictions were made from a correct hypothesis. A negative result can result from poor experimental design or because an inappropriate organism was chosen for the test. For example, a predator lacking color vision, or one that uses primarily its sense of smell, would not be appropriate for testing hypotheses about the colors of caterpillars.

A general textbook like this one presents hypotheses and theories that have been extensively tested and that are generally accepted. When possible in this text, we illustrate hypotheses and theories with observations and experiments that support them, but we cannot, because of space constraints, detail all the evidence. Remember as you read that statements of biological "fact" are mixtures of observations, predictions, and interpretations.

SOMETIMES ORGANISMS MUST BE SACRIFICED. Obtaining answers to many of the questions posed by biologists requires manipulating and sometimes sacrificing living organisms. To study the antipredator adaptations of caterpillars, the investigators had to keep blue jays in cages, make them hungry by depriving them of food, and then feed them caterpillars. This procedure resulted in the deaths of some caterpillars and temporary stress for some of the birds. To study their

(a)

(b)

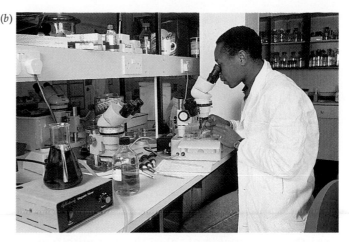

1.15 Experimentation Is Essential in Biology Research and experimentation in biology are carried out in the field and in laboratories. (a) Biologists who study the canopies of rainforest trees use special climbing equipment that allows them to collect data and carry out vital studies in the field. (b) Many scientific experiments take place within the laboratory. Work with cells and many tiny organisms requires the use of microscopes.

detailed structure and functioning, scientists must often kill organisms.

No amount of observation without intervention could possibly substitute for experimental manipulation. However, this does not mean that scientists are insensitive to the welfare of the organisms with which they work. Most scientists who work with animals are continually alert to finding ways of getting answers that use the smallest number of experimental subjects and that cause the subjects the least pain and suffering.

Not all forms of inquiry are scientific

If you understand the methods of science, you can distinguish science from nonscience. Recently some people have claimed that "creation science," sometimes called "scientific creationism," is a legitimate science that deserves to be taught in schools together with the evolutionary view of the world presented in this book. In spite of these claims, creation science is not science.

Science begins with observations and the formulation of testable hypotheses that can be rejected by contrary evidence. Creation "science" begins with the unsubstantiated assertion that Earth is only about 4,000 years old and that all species of organisms were created in approximately their present forms. This assertion is not presented as a hypothesis from which testable predictions are derived. Advocates of creation science do not believe that tests are needed, because they assume the assertion to be true, nor do they suggest what evidence would refute it.

In this chapter we have outlined the hypothesis that Earth is about 4 billion years old, that today's living organisms evolved from single-celled ancestors, and that many organisms dramatically different from those we see today lived on Earth in the remote past. The rest of this book will provide evidence supporting this scenario. To reject this view of Earth's history, a person must reject not only evolutionary biology, but also modern geology, astronomy, chemistry, and physics. All of this extensive scientific evidence is rejected or misinterpreted by proponents of creation "science" in favor of a religious belief held by a very small proportion of the world's people.

Evidence gathered by scientific procedures does not diminish the value of religious accounts of creation. Religious beliefs are based on faith—not on falsifiable hypotheses, as science is. They serve different purposes, giving meaning and spiritual guidance to human lives. They form the basis for establishing values—something science cannot do. The legitimacy and value of both religion and science is undermined when a religious belief is called science.

Life's Emergent Properties

Biologists study structures and processes ranging from the simple to the complex and from the small to the large. They study the structure and functioning of small parts of cells, as well as the interactions among the hundreds or thousands of different types of organisms that live together in a particular region. Biology can be visualized as ordered into a hierarchy in which the units, from the smallest to the largest, are molecules, cells, tissues, organs, organisms, populations, communities, and biomes (Figure 1.16).

We have already identified the cell as the fundamental unit of life. A **tissue** is a group of cells with similar and coordinated functions. Several different tissues are usually joined to form larger structures called **organs**, such as hearts, brains, kidneys, and lungs. Organs often are joined to form **organ systems**, such as the nervous system of animals and the vascular system of plants. Organs and organ systems perform distinct functions for the **organism** in which they are found.

The organization of living systems extends beyond the individual organism. Organisms living in the same area that are capable of interbreeding with one another form a **population**. All of the populations of a particular kind of organism, whether or not they live in the same area, constitute a species. Individuals of many different species typically live together and interact to form an ecological **community**. Ecological communities are grouped by their distinctive vegetation into **biomes**. All the biomes on Earth constitute the **biosphere**.

The organism is the central unit of study in biology, and Parts Five and Six of this book discuss organismal biology in detail. But to understand organisms, biologists must study life at all its levels of organization. Biologists must study molecules, chemical reactions, and cells to understand the operations of tissues and organs. They study organs and organ systems to determine how whole organisms function and maintain internal homeostasis. At higher levels in the hierarchy, biologists study how organisms interact with one another to form social systems, populations, ecological communities, and biomes, which are the subjects of Part Seven of this text.

Each level of biological organization has properties, called **emergent properties**, that are not found at lower levels. For example, cells and multicellular organisms have processes and characteristics that are not shown by the molecules of which they are composed. Emergent properties arise in two ways.

First, many emergent properties of systems result from interactions among their parts. For example, at the organismal level, developmental interactions of cells result in a multicellular organism whose adult features are vastly richer than those of the single cell from which it developed. Other examples of properties that emerge through complex interactions are memory, learning, consciousness, and emotions such as hate, fear, envy, anger, and love. These properties result from interactions in the human brain among the 10^{12} (trillion) cells with 10^{15} (a quadrillion) connections. No single cell, or even small group of cells, possesses them.

Second, emergent properties arise because aggregations have collective properties that their individual units lack. For example, individuals are born and they die—they have a life span. An individual does not have a birth rate or a death rate, but a population does. Death rate is an emergent property of a population. Other emergent properties of populations include age distribution and density. Evolution is an emergent property of populations that depends on differences in birth and death rates of individuals. Ecological communities possess emergent properties such as species richness.

Emergent properties do not violate principles that operate at lower levels of organization. However, emergent properties usually cannot be detected or even suspected by studying lower levels. Biologists could never discover the existence of human emotions by studying single nerve cells, even though they may eventually be able to explain those emotions in terms of interactions among many nerve cells.

Curiosity Motivates Scientists

The most important motivator of most biologists is curiosity. People are fascinated by the richness and diversity of life and want to learn more about organisms and how they function. Curiosity is probably an adaptive trait. Humans who were motivated to learn about their surroundings are likely to have survived and reproduced better, on average, than their less curious relatives. We hope this book will help you share the excitement biologists feel as they develop and test hypotheses. There are vast numbers of how and why questions for which we do not have answers, and new discoveries usually engender questions no one thought to ask before. Perhaps *your* curiosity will lead to an important new idea.

Summary of "An Evolutionary Framework for Biology"

- Earth is an ancient planet.
- Geologists in the nineteenth century provided the first evidence of Earth's antiquity and determined the sequence of events in life's evolution.
- The discovery of radioisotopes in the twentieth century enabled evolutionary events to be dated accurately.

Evolutionary Milestones

- Life arose from nonlife about 3.8 billion years ago when interacting systems of molecules became enclosed in membranes to form cells.
- All living organisms contain the same types of large molecules—carbohydrates, lipids, proteins, and nucleic acids. These organic molecules are formed only by living systems.
- All organisms consist of cells, and all cells come from pre-existing cells. Life no longer arises from nonlife.

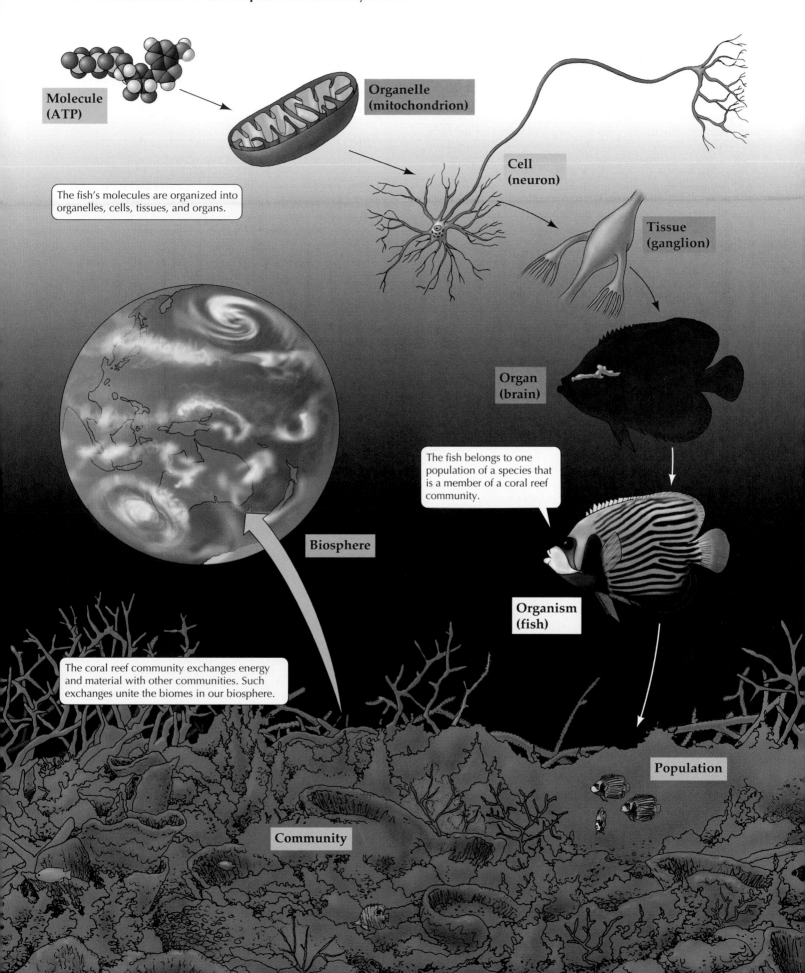

Molecule (ATP)

Organelle (mitochondrion)

Cell (neuron)

Tissue (ganglion)

The fish's molecules are organized into organelles, cells, tissues, and organs.

Organ (brain)

The fish belongs to one population of a species that is a member of a coral reef community.

Biosphere

Organism (fish)

The coral reef community exchanges energy and material with other communities. Such exchanges unite the biomes in our biosphere.

Population

Community

• A major theme in the evolution of life is the development of increasingly diverse ways of capturing external energy and using it to drive biologically useful reactions.

• Photosynthetic single-celled organisms released large amounts of O_2 into Earth's atmosphere, making possible the oxygen-based metabolism of large cells and, eventually, multicellular organisms.

• Reproduction with variation is a major characteristic of life. The evolution of sex greatly increased the rate of life's evolution.

• Complex eukaryotic cells evolved and were able to "stick together" after they divided, forming multicellular organisms. The individual cells of multicellular organisms became modified for specific functions within the organism.

• A major theme in the evolution of life is the development of increasingly complicated systems for responding to signals from the internal and external environments and for maintaining homeostasis.

• Regulated growth is a vital characteristic of life.

• Speciation resulted in the millions of species living on Earth today.

• Adaptation to environmental change is one of life's most distinctive features and is the result of evolution by natural selection. This principle guides virtually all biological investigation today.

Biological Diversity: Domains and Kingdoms

• Species are classified into three domains: Archaea, Bacteria, and Eukarya. The domains Archaea and Bacteria consist of prokaryotic cells and each contains one kingdom, Archaebacteria and Eubacteria, respectively. The domain Eukarya contains the kingdoms Protista, Plantae, Fungi, and Animalia, all of which have eukaryotic cells. **Review Figure 1.8**

The World Into Which Darwin Led Us

• Evolution is the theme that unites all of biology. The idea and evidence for evolution existed before Darwin.

• Darwin based his approach to evolution on five major hypotheses: (1) Earth is very old, (2) all organisms are descendants of a single common ancestor, (3) species multiply by splitting into daughter species, (4) evolution proceeds via gradual changes in populations, and (5) the major agent of evolutionary change is natural selection.

Asking the Questions "How?" and "Why?"

• Biologists ask two kinds of questions. Functional questions concern *how* organisms work. Adaptive questions concern *why* they evolved to work in that way.

• Both how and why questions are usually answered using a hypothetico-deductive approach. Hypotheses are tentative answers to the questions. Predictions are made on the basis of a hypothesis; then the predictions are tested by observations and experiments, which may support or refute the hypothesis.

• Science is based on the formulation of testable hypotheses that can be rejected by contrary evidence. The acceptance on faith of already refuted, untested, or untestable assumptions is not science.

Life's Emergent Properties

• Biology is organized into a hierarchy from molecules to the biosphere. Each level has emergent properties that are not found at the lower levels. **Review Figure 1.16**

Curiosity Motivates Scientists

• The most important motivator of most biologists is curiosity. There are vast numbers of unanswered—and in many cases, unasked—questions still to be explored.

Applying Concepts

1. According to the theory of evolution by natural selection, an organism evolves certain features because they improve the chances that it will survive and reproduce. There is no evidence, however, that evolutionary mechanisms have foresight or that organisms can anticipate future conditions. What do biologists mean when they say, for example, that wings are "for flying"?

2. Why is it so important in science that we design and perform tests capable of rejecting a hypothesis?

3. One hypothesis about the conspicuous coloration of caterpillars was described in this chapter, and some tests were mentioned. Suggest some other plausible hypotheses for conspicuous coloration in these animals. Develop some critical tests for one of these alternatives. What are the appropriate associated null hypotheses?

4. Some philosophers and scientists believe that it is impossible to prove any scientific hypothesis—that we can only fail to find a cause to reject it. Evaluate this view. Can you think of reasons why we can be more certain about rejecting a hypothesis than about accepting it?

Readings

Cziko, G. 1995. *Without Miracles: Universal Selection Theory and the Second Darwinian Revolution*. MIT Press, Cambridge, MA. An in-depth analysis of how an evolutionary view of the world can account for all of the features of life, including human self-awareness.

Darwin, C. 1859. *The Origin of Species by Means of Natural Selection*. John Murray, London. The book that set the world to thinking about evolution; still well worth reading. Many reprinted versions are available.

Futuyma, D. J. 1995. *Science on Trial: The Case for Evolution*, Revised Edition. Sinauer Associates, Sunderland, MA. A thorough presentation, for a general audience, of the scientific arguments that support evolution and its status as the single most fundamental principle in biology.

Margulis, L. and K. V. Schwartz. 1998. *Five Kingdoms: An Illustrated Guide to the Phyla of Life on Earth*, 3rd Edition. W. H. Freeman, New York. A good introduction to the kingdoms of organisms, in which the two kingdoms of prokaryotes are united into one. Excellent examples and illustrations.

Mayr, E. 1991. *One Long Argument: Charles Darwin and the Genesis of Modern Evolutionary Thought*. Harvard University Press, Cambridge, MA. An excellent account of the history of evolutionary thinking during the past century, written by a prominent exponent of the modern, or neo-Darwinian, synthesis of evolution.

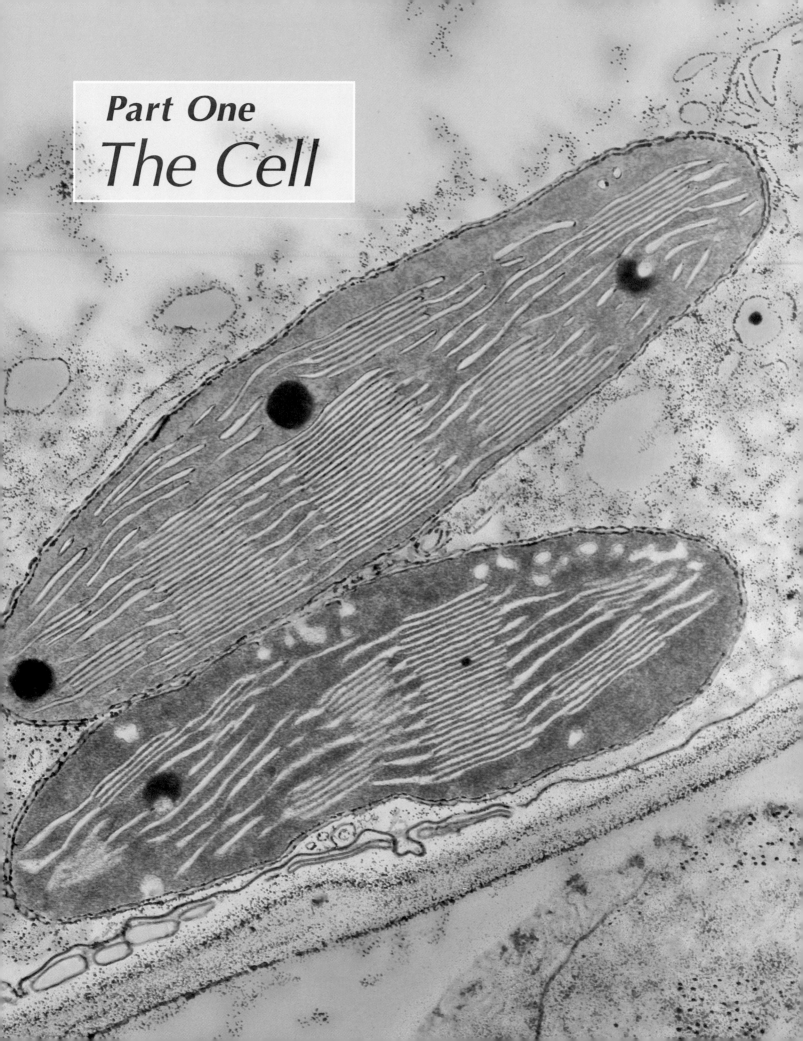

Part One
The Cell

Chapter 2

Small Molecules: Structure and Behavior

The Source of Life
Water, most of it in the oceans, covers three-fourths of Earth's surface. The oceans teem with life.

Water spews from the Earth as geysers. It circles the globe as clouds and falls from the sky as the gentle rain or in thundering torrents. Frozen, it covers parts of Antarctica to a depth of 3,000 meters (10,000 feet) or more. Vast oceans of it submerge much of the planet. Water is one of the key ingredients that makes life on Earth possible. It makes up as much as 95 percent of the weight of some living things.

Water is a simple substance containing only three atoms—two of hydrogen and one of oxygen. What does the composition of water tell us about its characteristics, and why is water so important to living systems? We can't answer these questions without more information about atoms in general, and about hydrogen and oxygen atoms in particular. But it is not only the constituents of matter that are important. To understand the behavior of something as apparently simple as water, we need to know how its constituent atoms are linked together. The same is true of the other small molecules that are essential to living systems.

The first part of this chapter will address the constituents of matter: atoms—their variety, properties, and capacity to combine with other atoms. Then we'll consider how matter changes. In addition to changes in state (solid to liquid to gas), substances undergo changes that transform both their composition and their characteristic properties. When cells use oxygen to "burn" glucose, the products are water, carbon dioxide, and energy to power life activities. This transformation is similar to what happens when the fuel propane is burned in a stove, a combustion reaction.

By studying simple systems such as the combustion of propane, we can understand better what happens in systems as complicated as living cells. The discussion of general chemical principles and their application to small molecules aids our understanding of the large molecules that form the basis for the life of an organism.

Later in this chapter, we return to a consideration of the structure and properties of water and its relationship to acids and bases. We close with a bridge to the next chapter—a consideration of characteristic groups of atoms that contribute specific properties to larger molecules of which they are part.

Atoms: The Constituents of Matter

More than a million million (1×10^{12}) atoms could fit in a single layer over the period at the end of this sentence. Each atom consists of a dense, positively charged nucleus, around which one or more negatively charged electrons move. The nucleus contains one or more protons and may contain one or more neutrons. Atoms and their component particles have mass. Mass is a property of all matter. Measuring mass measures the quantity of matter present.* The greater the mass, the greater the quantity of matter.

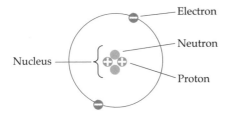

The mass of a proton serves as a standard unit: the atomic mass unit (amu), or dalton (named after the English chemist John Dalton). A single proton or neutron has the mass of 1 dalton, which is 1.7×10^{-24} grams (0.0000000000000000000000017 g). The mass of an electron is 9×10^{-28} g (0.0005 dalton). Because the mass of an electron is so much less than the mass of a proton or a neutron, the contribution of electrons to the mass of an atom can usually be ignored.

The positive electric charge on a proton is defined as a unit of charge. An electron has a charge equal and opposite to that of a proton. Thus the charge of a proton is +1 unit, that of an electron is –1 unit. Unlike charges attract each other; like charges repel. The neu-

tron, as its name suggests, is electrically neutral, so its charge is 0 unit. Because the number of protons in an atom equals the number of electrons, the atom itself is electrically neutral.

An element is made up of only one kind of atom

The element hydrogen consists only of hydrogen atoms; the element iron consists only of iron atoms. An element is a pure substance that contains only one type of atom. The atoms of each element have certain characteristics or properties that distinguish them from the atoms of other elements. The more than 100 elements found in the universe are arranged in the periodic table (Figure 2.1). The periodic table arranges elements with similar properties in vertical columns in order of their increasing size. Although there are more than 100 elements in the world, about 98 percent of the mass of a living organism (bacterium, turnip, or human) is composed of just six elements—carbon, hydrogen, nitrogen, oxygen, phosphorus, and sulfur. The chemistry of these elements will be our primary concern.

A substance (such as oxygen gas) that contains only one kind of atom is an **elemental substance**. A substance that contains more than one kind of atom is a **compound**. Most substances of biological interest are compounds.

The number of protons identifies the element

An atom is distinguished from other atoms by the number of its protons, which does not change. This number is called the **atomic number**. An atom of hydrogen contains 1 proton, a helium atom has 2 protons, carbon has 6 protons, and plutonium has 94. The atomic numbers of these elements are thus 1, 2, 6, and 94, respectively.

Every atom except hydrogen has one or more neutrons in its nucleus. The **mass number** of an atom equals the total number of protons and neutrons in its nucleus. Because the mass of an electron is infinitesimal compared with that of a neutron or proton, electrons are ignored in calculating the mass number. The nucleus of a helium atom contains 2 protons and 2 neutrons; oxygen has 8 protons and 8 neutrons. Helium, therefore, has a mass number of 4 and oxygen a mass number of 16. The mass number may be thought of as the weight of the atom, in daltons.

Each element has its own one- or two-letter symbol. For example, H stands for hydrogen, He for helium, and O for oxygen. Some symbols come from other languages: Fe (from Latin *ferrum*) stands for iron, Na (Latin *natrium*) for sodium, and W (German *Wolfram*) for tungsten. The periodic table (Figure 2.1) gives the symbols for all of the 92 natural elements, as well as those for 14 elements that do not occur naturally. In text, the atomic number and mass number of

---

* The term "weight" is sometimes substituted for the term "mass," but the two concepts are not identical. Weight is the measure of the Earth's gravitational attraction for mass. On another planet, the same quantity of mass will have a different weight. On Earth, however, the term "weight" is often used as a measure of mass.

The six elements highlighted in yellow make up 98% of the mass of any living organism.

Chemical symbol
Atomic number
Atomic weight

Elements shown in orange are present in tiny amounts in many organisms.

Vertical columns have elements with similar properties.

1 H 1.0079																	2 He 4.003
3 Li 6.941	4 Be 9.012											5 B 10.81	6 C 12.011	7 N 14.007	8 O 15.999	9 F 18.998	10 Ne 20.179
11 Na 22.990	12 Mg 24.305											13 Al 26.982	14 Si 28.086	15 P 30.974	16 S 32.06	17 Cl 35.453	18 Ar 39.948
19 K 39.098	20 Ca 40.08	21 Sc 44.956	22 Ti 47.88	23 V 50.942	24 Cr 51.996	25 Mn 54.938	26 Fe 55.847	27 Co 58.933	28 Ni 58.69	29 Cu 63.546	30 Zn 65.38	31 Ga 69.72	32 Ge 72.59	33 As 74.922	34 Se 78.96	35 Br 79.909	36 Kr 83.80
37 Rb 85.4778	38 Sr 87.62	39 Y 88.906	40 Zr 91.22	41 Nb 92.906	42 Mo 95.94	43 Tc (99)	44 Ru 101.07	45 Rh 102.906	46 Pd 106.4	47 Ag 107.870	48 Cd 112.41	49 In 114.82	50 Sn 118.69	51 Sb 121.75	52 Te 127.60	53 I 126.904	54 Xe 131.30
55 Cs 132.905	56 Ba 137.34	57–71 La–Lu	72 Hf 178.49	73 Ta 180.948	74 W 183.85	75 Re 186.207	76 Os 190.2	77 Ir 192.2	78 Pt 195.08	79 Au 196.967	80 Hg 200.59	81 Tl 204.37	82 Pb 207.19	83 Bi 208.980	84 Po (209)	85 At (210)	86 Rn (222)
87 Fr (223)	88 Ra 226.025	89–103 Ac–Lr	104	105	106	107	108	109									

Lanthanide series	57 La 138.906	58 Ce 140.12	59 Pr 140.9077	60 Nd 144.24	61 Pm (145)	62 Sm 150.36	63 Eu 151.96	64 Gd 157.25	65 Tb 158.924	66 Dy 162.50	67 Ho 164.930	68 Er 167.26	69 Tm 168.934	70 Yb 173.04	71 Lu 174.97
Actinide series	89 Ac 227.028	90 Th 232.038	91 Pa 231.0359	92 U 238.02	93 Np 237.0482	94 Pu (244)	95 Am (243)	96 Cm (247)	97 Bk (247)	98 Cf (251)	99 Es (252)	100 Fm (257)	101 Md (258)	102 No (259)	103 Lr (260)

2.1 The Periodic Table The periodic table of the elements reflects a periodicity of physical and chemical properties among the elements. Elements with similar properties are found in vertical columns.

an element are written to the left of the element's symbol:

Mass number **12**
6 Atomic number
C Symbol of element

Thus hydrogen, carbon, and oxygen are written as ^1_1H, $^{12}_6C$, and $^{16}_8O$.

Isotopes differ in number of neutrons

We have been speaking of hydrogen and oxygen as if each had only one atomic form. But this is not true. Not all atoms of an element have the same mass number. The different atomic forms of a single element are called **isotopes** of the element (Figure 2.2). Isotopes of an element differ in the number of neutrons in the atomic nucleus. The common form of hydrogen is ^1H, but about one out of every 6,500 hydrogen atoms on Earth has a neutron as well as a proton in its nucleus and is thus ^2H, called deuterium. Furthermore, it is possible to create ^3H, tritium, which has *two* neutrons and a proton in its nucleus. Because all three types of

hydrogen atoms have only one proton, they all have the atomic number 1. Deuterium, tritium, and common hydrogen have virtually identical chemical properties, although ^2H is twice and ^3H three times as heavy as ^1H.

In nature, many elements exist as several isotopes. For example, the natural isotopes of carbon are ^{12}C, ^{13}C, and ^{14}C. Unlike the hydrogen isotopes, the isotopes of most other elements do not have distinct names. Rather they are written in the form shown here and are referred to as carbon-12, carbon-13, and carbon-14, respectively. Most carbon atoms are ^{12}C, about 1.1 percent are ^{13}C, and a tiny fraction are ^{14}C. An element's **atomic mass** (**atomic weight**) is the average of the mass numbers of a representative sample of atoms of the element, with all isotopes in their normal proportions. For example, the atomic mass of carbon is 12.011. In biology, one encounters the terms "weight" and "atomic weight" more frequently than "mass" and "atomic mass"; therefore, we will use "weight" for the remainder of this book. Thus we say that the atomic weight of carbon is 12.011.

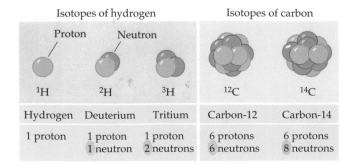

Isotopes of hydrogen | Isotopes of carbon

Hydrogen	Deuterium	Tritium	Carbon-12	Carbon-14
^{1}H	^{2}H	^{3}H	^{12}C	^{14}C
1 proton	1 proton 1 neutron	1 proton 2 neutrons	6 protons 6 neutrons	6 protons 8 neutrons

2.2 Isotopes Have Different Numbers of Neutrons

Some isotopes, called **radioisotopes**, are unstable and spontaneously give off energy as α (alpha), β (beta), or γ (gamma) radiation from the atomic nucleus. Such radioactive decay transforms the original atom into another type, usually of another element. For example, uranium-238 loses an alpha particle to form thorium-234, and carbon-14 loses a beta particle to form nitrogen-14. Biologists can incorporate radioisotopes into molecules and use the emitted radiation as a tag to identify changes that the molecules undergo in the body or to identify the locations of molecules within the cell (Figure 2.3). Some radioisotopes commonly used in biological experiments are ^{3}H (tritium),^{14}C (carbon-14), and ^{32}P (phosphorus-32).

Black dots identify the location of radioisotope.

2.3 Probing an Embryo with a Radioisotope Developmental biologists treated a fruit fly embryo with a substance labeled with a radioisotope. Later, they pressed the embryo against X-ray film and left the embryo and film in the dark for several days. Wherever a radioactive atom decayed, the emitted particle exposed the film. When the film was developed, silver grains appeared over the parts of the embryo that contained the radioactive substance.

Although radioisotopes are useful for experiments and medicine, even low doses of radiation from radioisotopes have the potential to damage molecules and cells. Gamma radiation from cobalt-60 (^{60}Co) is used medically to damage or kill rapidly dividing cancer cells. In addition to these applications, radioisotopes can be used to date fossils (see Chapter 19).

Electron behavior determines chemical bonding

In atoms, biologists are concerned primarily with electrons. To understand living organisms, biologists must study chemical changes that occur in living cells. These changes, called **chemical reactions** or just **reactions**, are changes in the atomic composition of substances. They occur because of the way in which electrons behave. The characteristic number of electrons in each atom of an element determines how the atom reacts with other atoms. All chemical reactions involve changes in the relationships of electrons with each other.

The location of a given electron in an atom at any given time is impossible to determine. We can only describe a volume of space within the atom where the electron is likely to be. The region of space within which the electron is found at least 90 percent of the time is the electron's **orbital** (Figure 2.4). In an atom, a given orbital can be occupied by at most two electrons. Thus any atom larger than helium (atomic number 2) must have electrons in two or more orbitals. As Figure 2.4 shows, the different orbitals have characteristic forms and orientations in space.

The orbitals constitute a series of **electron shells**, or energy levels, around the nucleus (Figure 2.5). The innermost electron shell, called the K shell, consists of only one orbital, called an *s* orbital. The *s* orbital fills first, and its electrons have the lowest energy. Hydrogen ($_1$H) has one K shell electron; helium ($_2$He) has two. All other atoms have two K shell electrons, as well as electrons in other shells. The L shell is made up of four orbitals (an *s* orbital and three *p* orbitals) and hence can hold up to eight electrons. The M, N, O, P, and Q shells have different numbers of orbitals, but the outermost orbitals can usually hold only eight electrons.

In any atom, the outermost shell determines how the atom combines with other atoms; that is, these outermost electrons determine how an atom behaves chemically. When an outermost shell consisting of four orbitals contains eight electrons, the atom is stable and will not react with other atoms. Examples of some chemically inert elements are helium, neon, and argon, in each of which the outermost shell contains eight electrons (see Figure 2.5). The atoms of other elements seek to attain the stable condition of having eight electrons in their outer orbitals. They attain this stability by sharing electrons with other atoms or by gaining or

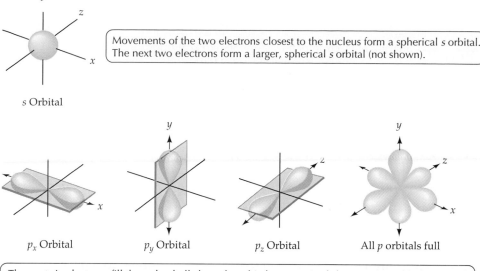

s Orbital

Movements of the two electrons closest to the nucleus form a spherical *s* orbital. The next two electrons form a larger, spherical *s* orbital (not shown).

p_x Orbital   *p_y* Orbital   *p_z* Orbital   All *p* orbitals full

The next six electrons fill three dumbell-shaped *p* orbitals, one pair of electrons per orbital, oriented on the *x*, *y*, and *z* axes through a point in the center of the atom.

2.4 Electron Orbitals

Orbitals are the regions around an atom's nucleus where electrons are most likely to be found.

Many atoms in biologically important molecules follow the octet rule—for example, carbon (C) and nitrogen (N). However, some biologically important atoms such as hydrogen and phosphorus are exceptions to the rule. Hydrogen (H) attains stability when two electrons occupy its outermost orbital; phosphorus (P) is stable when its outermost orbitals contain ten electrons.

losing one or more electrons from their outermost orbitals.

When they share electrons, atoms are bonded together. Such bonds create stable associations of atoms called molecules. A **molecule** can be defined as two or more atoms linked by chemical bonds. The tendency of atoms in stable molecules to have eight electrons in their outermost orbitals is known as the **octet rule**.

Chemical Bonds: Linking Atoms Together

A **chemical bond** is an attractive force that links two atoms to form a molecule. There are different kinds of chemical bonds (Table 2.1), but all strong chemical

2.5 Electron Shells Determine the Reactivity of Atoms

Each shell can hold a specific maximum number of electrons. The K shell holds two; the L and M shells each hold eight. An atom with room for more electrons in its outermost shell may react with other atoms.

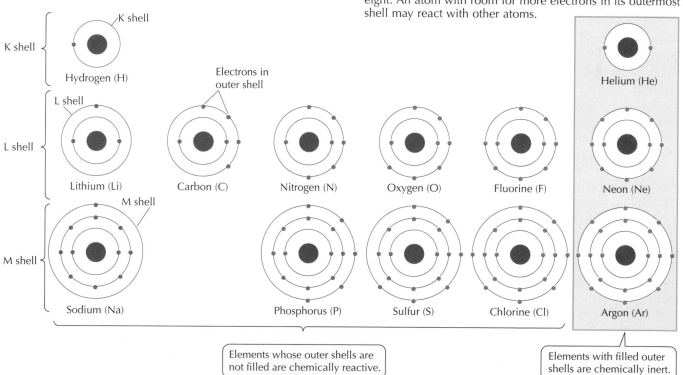

K shell

L shell

M shell

K shell — Hydrogen (H)

L shell — Lithium (Li)   Carbon (C)   Nitrogen (N)   Oxygen (O)   Fluorine (F)   Neon (Ne)

Electrons in outer shell

M shell — Sodium (Na)   Phosphorus (P)   Sulfur (S)   Chlorine (Cl)   Argon (Ar)

Helium (He)

Elements whose outer shells are not filled are chemically reactive.

Elements with filled outer shells are chemically inert.

TABLE 2.1 Chemical Bonds

TYPE OF BOND	BASIS OF BONDING	ENERGY	BOND LENGTH
Covalent bond	Sharing of electron pairs	50–110 kcal/mol[a]	≈0.1 nm[b]
Ionic bond	Attraction of opposite charges	3–7 kcal/mol	0.28 nm (optimal)
Hydrogen bond	Sharing of H atom	3–7 kcal/mol	0.26–0.31 nm (between atoms that share H)
van der Waals interaction	Interaction of electron clouds	≈1 kcal/mol	0.24–0.4 nm

[a]kcal/mol = kilocalories per mole; for other abbreviations of units of measurement see the inside front cover.
[b]For comparison, the radii of the H and C atoms are 0.2 and 0.08 nm, respectively.

bonds result from an atom's tendency to attain stability by filling its outermost electron orbitals. Atoms can gain stability in the outermost orbitals by sharing electrons or by losing or gaining one or more electrons. In this section, we will first discuss covalent bonds, the strong bonds that result from sharing of electrons. Then we'll examine hydrogen bonds, which are weaker than covalent bonds but enormously important to biology. Finally, we'll consider ionic bonding, which results when ions form as a consequence of the complete loss or gain of electrons by atoms.

Covalent bonds consist of shared pairs of electrons

When two atoms attain stable electron numbers in their outer shells by sharing one or more pairs of electrons, a **covalent bond** forms (Figure 2.6). A hydrogen atom has one electron in its only shell, but two electrons would be a more stable condition. Imagine two hydrogen atoms, initially far apart but coming closer and closer, until they begin to interact. The negatively charged electron in each hydrogen atom is attracted by the positively charged proton in the nucleus of the other hydrogen atom. When the two atoms are close enough, the two electrons spend time between both nuclei, and the two atoms are covalently bonded together, forming a molecule of hydrogen gas (H_2). The two hydrogen nuclei share the two electrons equally and completely.

The two atoms do not come *too* close together, because their positively charged nuclei strongly repel each other. A certain distance between the coupled atoms gives the most stable arrangement. Pulling the atoms slightly farther apart would require an input of energy because of the "gluing" effect of the shared electrons. Pushing the atoms closer together would require energy because of the mutual repulsion of the protons. So the most stable arrangement of the covalently bonded hydrogens can also be described as an arrangement that has a minimum amount of energy and that is less reactive than are the individual atoms alone, each of which has an incompletely filled orbital in the K shell.

A carbon atom has a total of six electrons; two electrons fill its inner shell and four are in its outer L shell. Because the L shell can hold up to eight electrons, this atom can share electrons with up to four other atoms. Thus it can form four covalent bonds. When an atom of carbon reacts with four hydrogen atoms, a substance called methane (CH_4) forms, resulting from the overlapping of electron orbitals (Figure 2.7a and b). Thanks to electron sharing, the outer shell of methane's carbon atom is filled with eight electrons, and the outer shell of each hydrogen atom is also filled. Thus four covalent bonds—each bond consisting of a shared pair of electrons—hold methane together. Table 2.2 shows the covalent bonding capacities for some biologically significant elements.

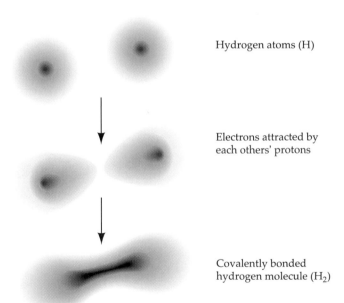

Hydrogen atoms (H)

Electrons attracted by each others' protons

Covalently bonded hydrogen molecule (H_2)

2.6 Electrons Are Shared in Covalent Bonds Two hydrogen atoms combine to form a hydrogen molecule. Each electron is attracted to both protons, but the two protons cannot come too close together, because they strongly repel each other. A covalent bond forms when the electron orbitals in the K shells of the two atoms overlap.

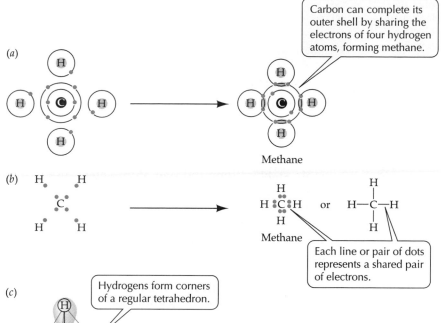

Carbon can complete its outer shell by sharing the electrons of four hydrogen atoms, forming methane.

(a)

Methane

(b)

Methane

Each line or pair of dots represents a shared pair of electrons.

(c)

Hydrogens form corners of a regular tetrahedron.

Methane

2.7 Covalent Bonding with Carbon Different representations of covalent bond formation in methane (CH_4). (a) A representation illustrating filling of a carbon molecule's outer electron shells. (b) Two common ways of representing bonds. (The two electrons in the K shell of carbon are not shown; these renderings indicate only the initially unfilled shells.) (c) Representation of the spatial orientation of methane's bonds.

ORIENTATION OF BONDS IN SPACE. Not only the number, but also the spatial orientation, of bonds is important. The four filled orbitals around the carbon nucleus of methane distribute themselves in space so that the bonded hydrogens are directed to the corners of a regular tetrahedron with carbon in the center (Figure 2.7c). Although the orientation of orbitals and shapes of molecules differ depending on the kinds of atoms and how they are linked together, it is essential to remember that all molecules occupy space and have three-dimensional shapes. The shapes of molecules contribute to their biological functions, as we will see in Chapter 3.

MULTIPLE COVALENT BONDS. A covalent bond is represented by a line between the chemical symbols for the atoms. Bonds in which a single pair of electrons is shared are called **single bonds** (for example, H—H, C—H). When four electrons (two pairs) are shared, the link is a **double bond** (C=C). In the gas ethylene ($H_2C=CH_2$), two carbon atoms share two pairs of electrons. **Triple bonds** (six shared electrons) are rare,

but there is one in nitrogen gas (N≡N), the chief component of the air we breathe. In the covalent bonds in these five examples, the electrons are shared more or less equally between the nuclei; consequently all regions of the bonds are identical. However, when electrons are shared unequally in a covalent bond, regions of partial electric charge exist.

UNEQUAL SHARING OF ELECTRONS. So far we have discussed the covalent bonds that result from the equal sharing of electrons between two nuclei. Now we want to consider the kind of covalent bond that results from unequal sharing of the electrons.

Some atoms hold electrons to themselves more firmly than other atoms do. This characteristic is called **electronegativity**. Highly electronegative atoms that form covalent bonds include oxygen and nitrogen. When these atoms are covalently bonded to atoms with weaker electronegativity, such as carbon and hydrogen, the bonding pair of electrons is unequally shared between the two atoms, and the result is a **polar covalent bond** (Figure 2.8). For example, when oxygen is bonded to hydrogen, the bonding electrons spend much more time near the oxygen nucleus than near the hydrogen nucleus. Consequently, the oxygen end of the bond is slightly negative (symbolized δ– and spoken as "delta negative," meaning a partial unit charge), and the hydrogen end is slightly positive (δ+). The bond is polar because these opposite charges are separated at the two ends of the bond. The partial charges that result from polar covalent bonds produce **polar molecules** or polar regions of large molecules. Polar bonds greatly influence the interactions between molecules that contain them, as we see in the interaction of water molecules in the liquid state.

TABLE 2.2 Typical Bonding Capabilities of Some Biologically Important Elements

ELEMENT	NUMBER OF COVALENT BONDS
Hydrogen (H)	1
Oxygen (O)	2
Sulfur (S)	2
Nitrogen (N)	3
Carbon (C)	4
Phosphorus (P)	5

2.8 The Polar Covalent Bond in the Water Molecule

A covalent bond between atoms with different electronegativities is a polar covalent bond with partial (δ) electric charges at the ends. Two ways of representing the water molecule are shown. (*a*) The lines between the symbols represent covalent bonds. (*b*) Four pairs of electrons are identified by dots, and the dots for the covalent bonds are displaced toward oxygen and away from the hydrogens.

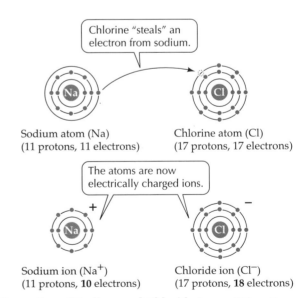

(*a*)   (*b*)

Water has polar covalent bonds.

Water's bonding electrons are shared unequally; electron density is greatest around the oxygen atom.

Hydrogen bonds may form between molecules

In liquid water, the negatively charged oxygen (δ–) atom of one water molecule is attracted to the positively charged hydrogen (δ+) of another water molecule. (Remember, negative charges attract positive charges.) The bond resulting from this attraction is called a **hydrogen bond** and is usually symbolized by a series of dots (Figure 2.9). Hydrogen bonds are not restricted to water molecules. They may form between any covalently bonded hydrogen and an electronegative atom, usually oxygen or nitrogen: —H···O— or —H···N—. Hydrogen bonds form between small molecules or between different parts of large molecules. Covalent bonds and polar covalent bonds, on the other hand, are always found *within* molecules.

A hydrogen bond is a weak bond; it has about one-twentieth (5 percent) of the strength of a covalent bond between a hydrogen atom and an oxygen atom. However, where many hydrogen bonds form, they have considerable strength and greatly influence the structure and properties of substances. Later in this chapter we'll discuss further how hydrogen bonding in water contributes to many of the properties of water that are significant for living systems. Hydrogen bonds also play important roles in determining and maintaining the three-dimensional shapes of giant molecules such as DNA and protein (see Chapter 3).

Ions form bonds by electrical attraction

When one interacting atom is much more electronegative than the other, a complete transfer of one or more electrons may take place. For example, a sodium atom has only one electron in its outermost shell; this condition is unstable. A chlorine atom has seven electrons in its outer shell, another unstable condition. The reaction between sodium and chlorine makes both atoms more stable. When the two atoms meet, the highly electronegative chlorine atom takes the single unstable electron from the sodium atom (Figure 2.10). The result is two electrically charged particles, called ions. **Ions** are electrically charged particles that form when atoms gain or lose one or more electrons.

The sodium ion (Na⁺) has a +1 unit charge because it has one less electron than it has protons. The outermost electron shell of the sodium ion is full, with eight electrons, so the ion is stable. The chloride ion (Cl⁻) has a –1 unit charge because it has one more electron than it has protons. This additional electron gives Cl⁻ an outer shell with a stable load of eight electrons. Negatively charged ions are called **anions**; positively charged ions are called **cations**.

Some elements form ions with multiple charges by losing or gaining more than one electron to achieve a stable electron configuration in their outer shell. Examples are Ca²⁺ (the calcium ion, created from a cal-

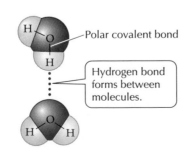

2.9 Hydrogen Bonding

Hydrogen bonds form between a slightly positive hydrogen and a slightly negative oxygen in separate molecules. The same type of bond could form between hydrogen and nitrogen.

Polar covalent bond

Hydrogen bond forms between molecules.

Chlorine "steals" an electron from sodium.

Sodium atom (Na)
(11 protons, 11 electrons)

Chlorine atom (Cl)
(17 protons, 17 electrons)

The atoms are now electrically charged ions.

Sodium ion (Na⁺)
(11 protons, **10** electrons)

Chloride ion (Cl⁻)
(17 protons, **18** electrons)

2.10 Formation of Sodium and Chloride Ions
When it reacts with a sodium atom, a chlorine atom, because it is much more electronegative, "steals" an electron and becomes a chloride ion (Cl⁻). This ion is negatively charged because it contains one more electron than it does protons. The sodium atom, upon losing the electron, becomes a sodium ion (Na⁺).

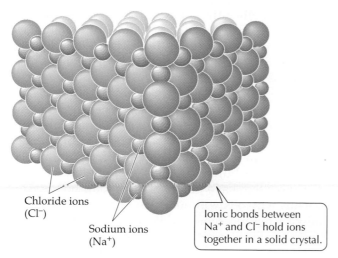

Chloride ions
(Cl⁻)

Sodium ions
(Na⁺)

Ionic bonds between
Na⁺ and Cl⁻ hold ions
together in a solid crystal.

2.11 Ionic Bonding in a Solid   Ionic bonds are electrical attractions between ions with opposite electric charges.

IONIC BONDS: ELECTRICAL ATTRACTIONS.   **Ionic bonds** are the bonds formed by electrical attractions between ions bearing opposite charges. In solids such as table salt (NaCl), the cations and anions are held together by ionic bonds (Figure 2.11). In solids, the ionic bonds are strong because the ions are close together. However, when ions are dispersed in water, the distance between them can be large; the strength of their attraction is thus greatly reduced. Under the conditions that exist in the cell, an ionic bond is less than one-tenth as strong as a covalent bond that shares electrons equally, so an ionic bond can be broken much more readily than a covalent bond.

Not surprisingly, ions with one or more unit charges can interact with polar substances as well as with other ions. Such interaction results when table salt or any other ionic solid dissolves in water (Figure 2.12). The hydrogen bond that we described earlier is a weak type of ionic bond, because it is formed by electrical attractions. However, it is weaker than most ionic bonds because the hydrogen bond is formed by partial charges ($\delta-$ and $\delta+$) rather than by whole unit charges (+1 unit, −1 unit).

Nonpolar substances have no attraction for polar substances

We have been discussing the bonds that result from electrical attractions between positive and negative charges (ionic bonds and hydrogen bonds). Now let's return to a brief consideration of substances that have "pure" covalent bonds (Figure 2.13). These bonds

cium atom that has lost two electrons), Mg^{2+} (magnesium ion), and Al^{3+} (aluminum ion). Two biologically important elements each yield more than one stable ion: Iron yields Fe^{2+} (ferrous ion) and Fe^{3+} (ferric ion), and copper yields Cu^+ (cuprous ion) and Cu^{2+} (cupric ion). Groups of covalently bonded atoms that carry an electric charge are called **complex ions**; examples include NH_4^+ (ammonium ion), SO_4^{2-} (sulfate ion), and PO_4^{2-} (phosphate ion).

Once they form, ions are usually stable, and no more electrons are lost or gained. As stable entities, ions can enter into stable associations through ionic bonding. Thus stable solids such as sodium chloride (NaCl) and potassium phosphate (K_3PO_4) are formed. Although a very complex solid, bone has as one of its major components the simple ionic compound $Ca_3(PO_4)_2$.

2.12 Water Molecules Surround Ions
Polar water molecules cluster around cations or anions in solutions, blocking their reassociation into a solid.

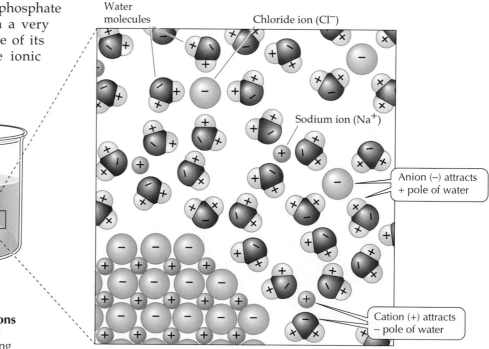

Water molecules

Chloride ion (Cl⁻)

Sodium ion (Na⁺)

Anion (−) attracts + pole of water

Cation (+) attracts − pole of water

Undissolved sodium chloride

Water, a polar molecule   Ethane, a nonpolar molecule

2.13 Nonpolar Molecules: Hydrocarbons   The electrons are uniformly distributed around the entire hydrocarbon molecule (ethane, CH_3—CH_3). The molecule is nonpolar, with no electrical attraction to substances with electric charges or polar bonds.

form between atoms that have equal or nearly equal electronegativities—such as carbon and hydrogen—which share the bonding electrons equally. Such bonds are abundant in the compounds of hydrogen and carbon—the **hydrocarbons**. Molecules such as ethane (CH_3—CH_3) and butane (CH_3—CH_2—CH_2—CH_3) are small hydrocarbons, but in living systems, molecules exist with hydrocarbon chains consisting of 16 or more carbon atoms.

ATTRACTIONS BETWEEN NONPOLAR MOLECULES.   Nonpolar substances such as oils and fats show **van der Waals attractions** between molecules. These attractive forces operate only when nonpolar substances come very close to each other. The random variations in the electron distribution in one molecule create an opposite charge distribution in the adjacent molecule, and the result is a brief, weak attraction. Although each such interaction is brief and weak at any one site, the summation of many such interactions over the entire span of a nonpolar molecule can produce substantial attraction. Thus van der Waals interactions are important in holding together the long hydrocarbon chains that make up the inner portion of biological membranes. They also stabilize portions of the DNA double helix and the intricate folded structure of proteins (see Chapters 3 and 11).

POLAR AND NONPOLAR INTERACTIONS.   When electrons are shared equally, the resultant covalent bonds are nonpolar and they do not interact with the charges of polar covalent bonds. Substances with only nonpolar bonds, such as butane and oils, will not interact with substances that have polar bonds or ionic bonds. This explains why oils will not dissolve in water. Oils are nonpolar hydrocarbons, while water is a highly polar substance. Substances with nonpolar covalent bonds are said to be **hydrophobic** (literally "water-fearing"), which refers to the fact that there is no attraction between nonpolar substances and water or other electrically charged substances. Hydrophobic substances are also called **nonpolar substances**.

When hydrocarbons are dispersed in water, they slowly come together—dispersed molecules form droplets that form larger droplets. The forces that bring about this combining of molecules are sometimes called *hydrophobic interactions*, but this term is somewhat misleading. The interactions that combine nonpolar substances have less to do with forces between the nonpolar molecules than with the hydrogen bonding of the water that surrounds the molecules (Figure 2.14).

When nonpolar substances such as hydrocarbons are introduced into water, they cause a disruption in the usual hydrogen bonding between water molecules. In the vicinity of the hydrocarbon, the water molecules form a hydrogen-bonded "cage" that surrounds the nonpolar hydrocarbons and pushes them together.

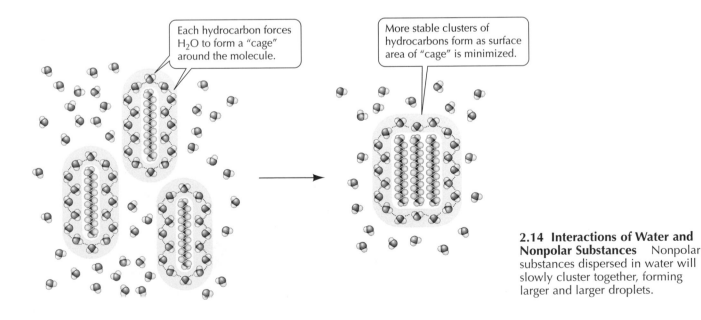

Each hydrocarbon forces H_2O to form a "cage" around the molecule.

More stable clusters of hydrocarbons form as surface area of "cage" is minimized.

2.14 Interactions of Water and Nonpolar Substances   Nonpolar substances dispersed in water will slowly cluster together, forming larger and larger droplets.

These water cages can bring together dispersed nonpolar molecules into larger groups (see Figure 2.14).

Eggs by the Dozen, Molecules by the Mole

In modern biology, as in chemistry, the question "How much?" is as important as "What kind?" In this section, we will examine briefly how chemists and biologists deal quantitatively with atoms and molecules.

The **molecular formula** uses chemical symbols to identify different atoms, and subscript numbers to show how many atoms are present. For example, the molecular formula for methane is CH_4 (each molecule contains one carbon atom and four hydrogen atoms), that for oxygen gas is O_2, and that for sucrose (table sugar) is $C_{12}H_{22}O_{11}$. The hormone insulin is represented by the molecular formula $C_{254}H_{377}N_{65}O_{76}S_6$! Although molecular formulas tell us what kinds of atoms and how many of each kind are present in the molecule, they tell us nothing about which atoms are linked to which. **Structural formulas** give us this information (see Figure 2.7).

Each compound has a **molecular weight** (molecular mass): the sum of the atomic weights of the atoms in the molecule. The atomic weights of hydrogen, carbon, and oxygen are 1.008, 12.011, and 16.000, respectively. Thus the molecular weight of water (H_2O) is $(2 \times 1.008) + 16.000 = 18.016$, or about 18. What is the molecular weight of sucrose ($C_{12}H_{22}O_{11}$)? Your calculations should tell you that the answer is approximately 342. If you remember the molecular weights of a few representative biological compounds, you will be able to picture the relative sizes of molecules that interact with one another (Figure 2.15). Experiments require quantitative information. Suppose we want to compare how sodium chloride (NaCl), potassium chloride (KCl), and lithium chloride (LiCl) affect a biological process. At first you might think we could simply give, say, 2 grams (g) of NaCl to one set of subjects, 2 g of KCl to another, and 2 g of LiCl to the third. But because the molecular weights of NaCl, KCl, and LiCl are different, 2-gram samples of each of these substances contain different numbers of molecules. The comparison would thus not be legitimate. Instead, we want to give *equal numbers of molecules* of each substance so that we can compare the activity of one molecule of one substance with that of one molecule of another. How can we measure out equal numbers of molecules?

We can calculate numbers of molecules by weighing

Measuring the number of molecules is essential to studying how many molecules take part in chemical reactions. Consider two large barrels, one filled with bolts and the other with pennies. How can we determine the number of units in each barrel? We could count them, but that would be tedious (and it is impossible to count individual molecules directly). However, if we know the weight of one bolt and one penny, we can calculate the number of units by dividing the total weight by the weight of each unit. The same principles apply to determining the number of molecules in a weighed sample.

To determine the number of molecules in a sample of a pure substance, we first determine the weight of the substance in grams, then we divide the grams by the relative weight of one molecule (the molecular weight defined earlier). Therefore, to measure out quantities of substances containing equal numbers of molecules, we weigh out in grams a quantity of each substance equivalent to its molecular weight. The

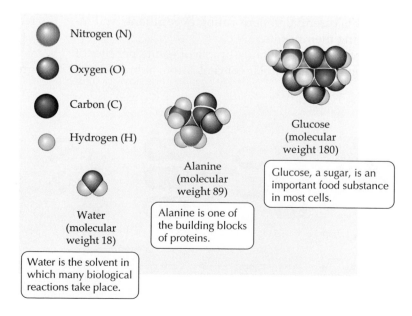

2.15 Molecular Weight and Size The relative sizes of three common molecules and their molecular weights. These space-filling models are the most realistic representations of molecules we can create; you will continue to see them in this and subsequent chapters. The color conventions are standard for the atoms (yellow is used for sulfur and phosphorus atoms, which are not shown here).

Nitrogen (N)

Oxygen (O)

Carbon (C)

Hydrogen (H)

Glucose (molecular weight 180)

Glucose, a sugar, is an important food substance in most cells.

Alanine (molecular weight 89)

Alanine is one of the building blocks of proteins.

Water (molecular weight 18)

Water is the solvent in which many biological reactions take place.

molecular weight of NaCl is 58.45. Therefore 58.45 g of NaCl will have the same number of molecules as 74.55 g of KCl, whose molecular weight is 74.55.

If we divide the 74.55 g by the molecular weight of 74.55, the answer is 1. But one what? Surely not one molecule! The calculated numbers here are not the exact number, because we did not divide by the actual weight in grams of a molecule of potassium chloride (which would be about 1×10^{-22} g). Instead we divided by the relative weight. Therefore, instead of knowing the exact number of molecules, we know the relative number, which is called a **mole**. *One mole of a substance is an amount whose weight in grams is numerically equal to the molecular weight of the substance.* Potassium chloride (KCl) has a molecular weight of 74.55, so 1 mole of KCl weighs 74.55 g. Likewise, 1 mole of NaCl weighs 58.45 g, and 1 mole of LiCl, 42.40 g.

A mole of one substance contains the same number of molecules as does a mole of any other substance. This number, known as **Avogadro's number**, is 6.023×10^{23} molecules per mole. The concept of the mole is important for biology because it enables us to work easily with known numbers of molecules. Since we can neither weigh nor count individual molecules, we work with moles. The mole concept is analogous to the concept of a dozen or a gross. We buy some things, such as eggs, by the dozen (or another similar unit, such as the gross) because the individual items are inconvenient to count. Just as a dozen of anything contains twelve of that thing, a mole of any substance contains Avogadro's number of molecules of that substance.

Reactions take place in solutions

In cells and in the laboratory, reactions take place in solutions formed when substances (solutes) are dissolved in water (the solvent). The amount of substance dissolved in a given amount of water is the concentration of the solution. Knowing the concentrations of solutions is essential in performing experiments and interpreting their results.

A solution that contains 1 mole of solute per liter of solution has a one-molar concentration, abbreviated 1 *M*. A solution containing a half mole per liter is referred to as 0.5 *M*, or half-molar. How would you make 100 milliliters (ml) of a 0.5 *M* sucrose solution? The molecular weight of sucrose is 342, so 1 liter (1,000 ml) of a 1 *M* sucrose solution contains 1 mole, or 342 g, of sucrose. You were asked to make just 100 ml of 0.5 *M* solution. Since 34.2 g of sucrose would make 100 ml of 1 *M* sucrose, to make 100 ml of a 0.5 *M* sucrose solution, you would use 0.5×34.2 g = 17.1 g of sucrose.

Chemical Reactions: Atoms Change Partners

When atoms combine or change bonding partners, a chemical reaction is occurring. Consider the combustion reaction that takes place in the flame of a propane stove. When propane (C_3H_8) reacts with oxygen gas (O_2), the carbon atoms become bonded to oxygen atoms instead of to hydrogen atoms, and the hydrogen atoms become bonded to oxygen instead of to carbon (Figure 2.16). As the covalently bonded atoms change bonding partners, the composition of the matter changes, and propane and oxygen gas become carbon dioxide and water. This chemical reaction can be represented by the balanced equation

$$C_3H_8 + 5\,O_2 \rightarrow 3\,CO_2 + 4\,H_2O$$

In this equation, the propane and oxygen are the **reactants**, and the carbon dioxide and water are the **products**. The arrow symbolizes the chemical reaction. The numbers preceding the molecular formulas balance the equation and indicate how many molecules react or are produced. In this and all other chemical reactions, matter is neither created or destroyed. The total number of carbons on the left equals the total number on the right. However, there is another product of this reaction: energy.

The heat of the stove's flame and its blue light reveal that the reaction of propane and oxygen releases a great deal of energy. **Energy** is defined as the capacity

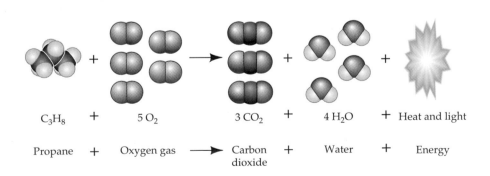

C_3H_8 + $5\,H_2O$ → $3\,CO_2$ + $4\,H_2O$ + Heat and light

Propane + Oxygen gas → Carbon dioxide + Water + Energy

2.16 Bonding Partners and Energy May Change in a Chemical Reaction
One molecule of propane reacts with five molecules of oxygen gas to give three molecules of carbon dioxide and four molecules of water. This reaction releases energy, in the form of heat and light.

to do work, but on a more intuitive level, it can be thought of as the capacity to change. Chemical reactions do not create or destroy energy, but *changes* in energy usually accompany chemical reactions. The energy released as heat and light was present in the reactants in another form, called *potential chemical energy*. In some chemical reactions, energy must be supplied from the environment (for example, some substances will react only after being heated), and some of this supplied energy becomes stored as potential chemical energy in the bonds formed in the reactants.

We can measure the energy associated with chemical reactions using the unit called a **calorie (cal)**. A calorie is the amount of heat energy needed to raise the temperature of 1 g of pure water from 14.5°C to 15.5°C. (The nutritionist's Calorie, with a capital C, is what biologists call a kilocalorie (kcal) and is equal to 1,000 heat-energy calories.) Another unit of energy that is increasingly used is the **joule (J)**. When you compare data on energy, always compare joules to joules and calories to calories. The two units can be interconverted: 1 J = 0.239 cal, and 1 cal = 4.184 J. Thus, for example, 486 cal = 2033 J, or 2.033 kJ. Although defined in terms of heat, the calorie and the joule are measures of any form of energy—mechanical, electric, or chemical. We'll discuss energy changes further in Chapter 6.

Within living cells, chemical reactions called oxidation–reduction reactions take place that have much in common with this combustion of propane. The fuel for these biological reactions is different (the sugar glucose, rather than propane), and the reactions proceed by many intermediate steps that permit the energy released from the glucose to be harvested and put to use by the cell. But the products are the same: carbon dioxide and water. We will present and discuss oxidation–reduction reactions and several other types of chemical reactions that are prevalent in living systems in the chapters that follow.

Water: Structure and Properties

Water, like all other matter, can exist in three states: solid (ice), liquid, and gas (vapor) (Figure 2.17). Liquid water is the medium in which life originated on Earth more than 3.5 billion years ago, and it is in water that life evolved for about a billion years. Today water covers three-fourths of Earth's surface, and all active organisms contain between 45 and 95 percent water. No organism can remain biologically active without water, which is important in both the outer and inner environments. Within cells, water participates directly in many chemical reactions, and it is the medium (or solvent) in which most reactions take place. In this section we will consider the structure and interactions of water molecules, exploring how these generate properties essential to life.

Water has a unique structure and special properties

The shape of a water molecule is determined by the distribution in space of the four pairs of electrons in the outer shell of the oxygen atom. Each pair of electrons is confined to an orbital. Two of these pairs are bonded to hydrogens, but the other two pairs (nonbonding pairs) also influence shape. Because of their negatively charged electrons, the four orbitals, each containing a pair of electrons, repel one another. In seeking to be as far apart as possible, they give the water molecule a nearly tetrahedral shape (Figure 2.18).

The shape of the water molecule, its polar nature, and the formation of hydrogen bonds between water molecules or with other substances give water its unusual properties. For example, ice floats, and compared to other liquids, water is an excellent solvent, making it an ideal medium for biochemical reactions. Water is both cohesive (sticking to itself) and adhesive (sticking to other things). And the energy changes that

2.17 Water: Solid and Liquid Solid water from a glacier floats in its liquid form. The clouds are also water, but not in its gaseous phase: They are composed of fine drops of liquid water.

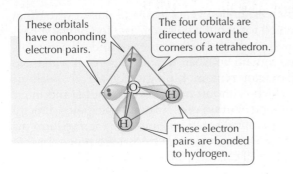

These orbitals have nonbonding electron pairs.

The four orbitals are directed toward the corners of a tetrahedron.

These electron pairs are bonded to hydrogen.

2.18 The Shape of the Water Molecule The four pairs of electrons in the outer shell of oxygen repel each other, producing a tetrahedral shape.

accompany its transitions from solid to liquid to gas are significant for living systems.

ICE FLOATS. In its solid state (ice), water is held by its hydrogen bonds in a rigid, crystalline structure in which each water molecule is hydrogen-bonded to four other molecules (Figure 2.19a). Although these molecules are held firmly in place, they are not as tightly packed as they are in liquid water (Figure 2.19b). In other words, *solid water is less dense than liquid water*, which is why ice floats in water. If ice sank in water, as almost all other solids do in their corresponding liquids, ponds and lakes would freeze from the bottom up, becoming solid blocks of ice in winter and killing most of the organisms living in them. Once the whole pond had frozen, its temperature could drop well below the freezing point of water. However, because ice floats, it forms a protective insulating layer on the top of the pond, reducing heat flow to the cold air above. Thus fish, plants, and other organisms in

the pond can survive the winter at temperatures no lower than 0°C, the freezing point of pure water.

MELTING AND FREEZING. Compared to other nonmetallic substances of the same size, ice requires a great deal of heat energy to melt. Melting 1 mole of water molecules requires the addition of 5.9 kJ of energy. This value is high because more than a mole of hydrogen bonds must be broken for 1 mole of water to change from solid to liquid. In the opposite process, freezing, a great deal of energy must be lost for water to transform from liquid to solid. These properties help make water a moderator of temperature changes.

Another property of water that moderates temperature is the high heat capacity of liquid water. **Heat capacity** is the amount of heat energy that is required to raise the temperature of a substance 1°C. Raising the temperature of liquid water takes a relatively large amount of heat. The temperature of a given quantity of water is raised only 1°C by an amount of heat that would increase the temperature of the same quantity of ethyl alcohol by 2°C, or of chloroform by 4°C. This phenomenon contributes to the surprising constancy of the temperature of the oceans and other large bodies of water through the seasons of the year. The temperature changes in coastal land masses are also moderated by large bodies of water. Indeed, water helps minimize variations in atmospheric temperature throughout the planet.

COHESION AND SURFACE TENSION. In liquid water, the molecules are free to move about. The hydrogen bonds between the water molecules continually form and break. In other words, liquid water has a dynamic structure. On the average, every water molecule forms 3.4 hydrogen bonds with other water molecules. This number represents fewer bonds than exist in ice, but it is still a high number.

(a) Ice (solid water)

(b) Liquid water

Hydrogen bonds continually break and form as water molecules move.

In ice, water molecules are held in a rigid state by hydrogen bonds.

2.19 Hydrogen Bonding in Liquid Water and Ice
Hydrogen bonding exists between the molecules of water in both its liquid and solid states. (a) Solid water. (b) Liquid water. Although more structured, ice is less dense than liquid water, so it floats.

These hydrogen bonds explain the cohesive strength of water. The **cohesive strength** of water is what permits narrow columns of water to stretch from the roots to the leaves of trees more than 100 meters high. When water evaporates from leaves, the entire column moves upward in response to the pull of the molecules at the top.

Water also has a high **surface tension**, which means that the surface of water exposed to the air is difficult to puncture. The water molecules in this surface layer are hydrogen-bonded to other water molecules below. The surface tension of water permits a container to be filled slightly above its rim without overflowing, and it permits small animals to walk on the surface of water (Figure 2.20).

EVAPORATION AND COOLING. Water has a high **heat of vaporization**, which means a lot of heat is required to change water from its liquid state to its gaseous state (the process of evaporation). This heat is absorbed from the environment in contact with the water. Evaporation thus has a cooling effect on the environment—whether a leaf, a forest, or an entire land mass. This effect explains why sweating cools the human body: As the sweat evaporates off the skin, it takes with it some of the adjacent body heat.

Water molecules sometimes form ions

The water molecule has a slight but significant tendency to come apart into a hydroxide ion (OH^-) and a hydrogen ion (a proton, H^+). Actually, *two* water molecules participate in ionization. One of the two molecules "captures" a hydrogen ion from the other, forming a hydroxide ion and a hydronium ion:

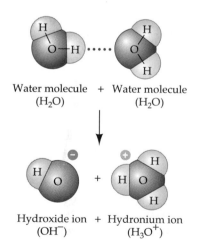

Water molecule + Water molecule
(H_2O)   (H_2O)

↓

Hydroxide ion + Hydronium ion
(OH^-)   (H_3O^+)

The hydronium ion is in effect a hydrogen ion bound to a water molecule. For simplicity, biochemists tend to use a modified representation of the ionization of water:

$$H_2O \rightarrow H^+ + OH^-$$

2.20 Surface Tension Water striders "skate" along, supported by the surface tension of the water that is their home.

Even though only about one water molecule in 500 million is ionized at any given time, the transformation is significant because H^+ and OH^- ions participate in many important biochemical reactions.

Acids, Bases, and the pH Scale

The ionization of water is very important for all living creatures. This fact may seem surprising, since the ionization is so slight—only one ionization out of 5×10^8 water molecules. But we are less surprised if we focus on the abundance of water in living systems and the reactive nature of H^+ produced by that ionization. Remember that H^+ is a proton, a tiny bit of charged matter—smaller than any atom or molecule in the cell. Usually it attaches to another water molecule, but in the cell it can attach to other molecules and substantially change their properties. Because acids and bases donate and accept H^+, they can profoundly alter the water environment in which the chemical reactions and processes of life take place.

In this section we'll examine acids, bases, and their measurement using the pH scale. We'll close by looking at how buffers limit the changes in pH.

Acids donate H⁺, bases accept H⁺

In pure water, the concentration of hydrogen ions exactly equals that of hydroxide ions (OH^-), and this "solution" is said to be **neutral**. Now suppose we add some HCl (hydrochloric acid). As it dissolves, the HCl ionizes, releasing H^+ and Cl^- ions:

$$HCl \rightarrow H^+ + Cl^-$$

Now there are more H^+ than OH^- ions. Such a solution is *acidic*. A *basic*, or alkaline, solution is one in which there are more OH^- than H^+ ions. A basic solution can be made from water by adding, for example, sodium hydroxide (NaOH), which ionizes to yield OH^- and Na^+ ions, thus making the concentration of OH^- ions greater than that of H^+ ions.

$$NaOH \rightarrow Na^+ + OH^-$$

An **acid** is any compound that can *release* H^+ ions in solution. HCl is an acid, as is H_2SO_4 (sulfuric acid). One molecule of sulfuric acid may ionize to yield two H^+ ions and one SO_4^{2-} ion. Biological compounds such as acetic acid and pyruvic acid, which contain —COOH (the carboxyl group; see Figure 2.23) are also acids, because —COOH $\rightarrow$ —COO$^-$ + H^+.

Bases are compounds that can *accept* H^+ ions. These include the bicarbonate ion (HCO_3^-), which can accept a H+ ion and become carbonic acid (H_2CO_3); ammonia (NH_3), which can accept a H+ ion and become an ammonium ion (NH_4^+); and many others.

Note that although —COOH is an acid, —COO$^-$ is a base, because —COO$^-$ + H^+ $\rightarrow$ —COOH. Acids and bases exist as pairs, such as —COOH and —COO$^-$, because any acid becomes a base when it releases a proton, and any base becomes an acid when it gains a proton.

You may have noticed that the two reactions just discussed are the opposites of each other. The reaction that yields —COO$^-$ and H^+ is reversible and may be expressed as

$$—COOH \rightleftharpoons —COO^- + H^+$$

A **reversible reaction** is one that can proceed in either direction—left to right or right to left—depending on the relative starting concentrations of reacting substances and products. In principle, *all* chemical reactions are reversible. We will look at some consequences of this reversibility in Chapter 6.

pH is the measure of hydrogen ion concentration

The terms "acid*ic*" and "bas*ic*" refer only to *solutions*. How acidic or basic a solution is depends on the relative concentrations of H^+ and OH^- ions in it. "Acid" and "base" refer to *compounds* and *ions*. A compound or ion that is an acid can donate H^+ ions; one that is a base can accept H^+ ions.

How do we specify how acidic or basic a solution is? First, let's look at the H^+ ion concentrations of a few contrasting solutions. In pure water, the H^+ concentration is $10^{-7} M$. In 1 *M* hydrochloric acid, the H^+ concentration is 1 *M*; and in 1 *M* sodium hydroxide, the H^+ concentration is $10^{-14} M$. Because its values range so widely—from more than 1.0 *M* to less than $10^{-14} M$—

the H^+ concentration itself is an inconvenient quantity. It is easier to work with the logarithm of the concentration, because logarithms compress this range. We indicate how acidic or basic a solution is by its **pH** (a term derived from "*potential of Hydrogen*"). The pH value is defined as the negative logarithm of the hydrogen ion concentration in moles per liter (molar concentration). In chemical notation, molar concentration is often indicated by putting brackets around the symbol for a substance; thus [H^+] stands for the molar concentration of H^+. The equation for pH is

$$pH = -\log_{10}[H^+]$$

Since the H^+ concentration of pure water is $10^{-7} M$, its pH is $-\log(10^{-7}) = -(-7)$, or 7. A smaller negative logarithm means a larger number. In practical terms, a lower pH means a higher H^+ concentration, or greater acidity. In 1 *M* HCl, the H^+ concentration is 1 *M*, so the pH is the negative logarithm of 1 ($-\log 10^0$), or 0. The pH of 1 *M* NaOH is the negative logarithm of 10^{-14}, or 14. A solution with a pH of less than 7 is acidic: It contains more H^+ ions than OH^- ions. A solution with a pH of 7 is neutral. And a solution with a pH value greater than 7 is basic. Because the pH scale is logarithmic, the values are exponential: A solution with a pH of 5 is 10 times more acidic than one with a pH of 6 (it has ten times as great a concentration of H^+); a solution with a pH of 4 is 100 times more acidic than one with a pH of 6. Figure 2.21 shows the pH values of some common substances.

Buffers minimize pH changes

An organism must control the chemistry of its cells—in particular, the pH of the separate compartments within cells. Animals must also control the pH of their blood. The normal pH of human blood is 7.4, and deviations of even a few tenths of a pH unit can be fatal. The control of pH is made possible in part by **buffers**, systems that maintain a relatively constant pH even when substantial amounts of acid or base are added. A buffer is a mixture of an acid that does not ionize completely in water and its corresponding base—for example, carbonic acid (H_2CO_3) and bicarbonate ions (HCO_3^-). If acid is added to this buffer, not all the H^+ ions from the acid stay in solution. Instead, many of the added H^+ ions combine with bicarbonate ions to produce more carbonic acid, thus using up some of the H^+ ions in the solution and decreasing the acidifying effect of the added acid:

$$HCO_3^- + H^+ \rightleftharpoons H_2CO_3$$

If base is added, the reaction reverses. Some of the carbonic acid ionizes to produce bicarbonate ions and more H^+, which counteracts some of the added base.

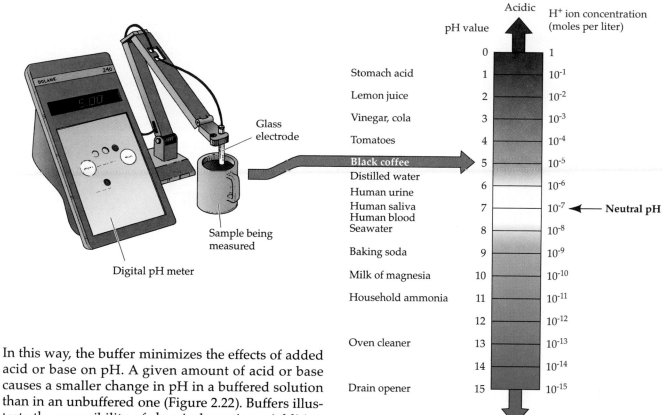

2.21 pH Values of Some Familiar Substances A pH meter such as the one on the left tells us the pH of a solution. This scale reads from low (acidic) pH values at the top to high (basic) pH values at the bottom.

In this way, the buffer minimizes the effects of added acid or base on pH. A given amount of acid or base causes a smaller change in pH in a buffered solution than in an unbuffered one (Figure 2.22). Buffers illustrate the reversibility of chemical reactions: Addition of acid drives the reaction in one direction; addition of base drives it in the other.

The Properties of Molecules

Molecules vary in size. Some are small, such as H_2 and CH_4. Others are larger, such as a molecule of table sugar (sucrose), which has 45 atoms. Still other molecules, such as proteins, are gigantic, sometimes containing tens of thousands of atoms bonded together in specific ways. Whether large, medium, or small, most of the molecules in living systems contain carbon atoms and are thus referred to as **organic molecules**. Most organic molecules include hydrogen and oxygen atoms, and many also include nitrogen and phosphorus.

All molecules have a specific three-dimensional shape. For example, the orientation of the bonding orbitals around the carbon atom gives the molecule of methane (CH_4) the shape of a regular tetrahedron (see Figure 2.7c), while in carbon dioxide (CO_2) the three atoms are in line with one another. Larger molecules have specific complex shapes that result from the number and kinds of atoms present and the ways in which these atoms are linked together. Some large molecules have compact ball-like shapes. Others are long, thin, ropelike structures. The shapes relate to the roles these molecules play in living cells.

In addition to size and shape, molecules have certain properties that characterize them and determine

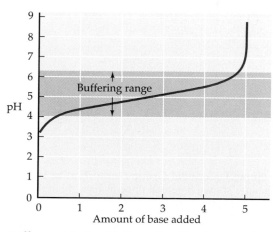

2.22 Buffers Minimize Changes in pH Adding a base increases the pH of a solution. With increasing amounts of added base, the overall slope of a graph of pH is upward. In the buffering range, however, the slope is shallow. At high and low values of pH, where the buffer is ineffective, the slopes are much steeper.

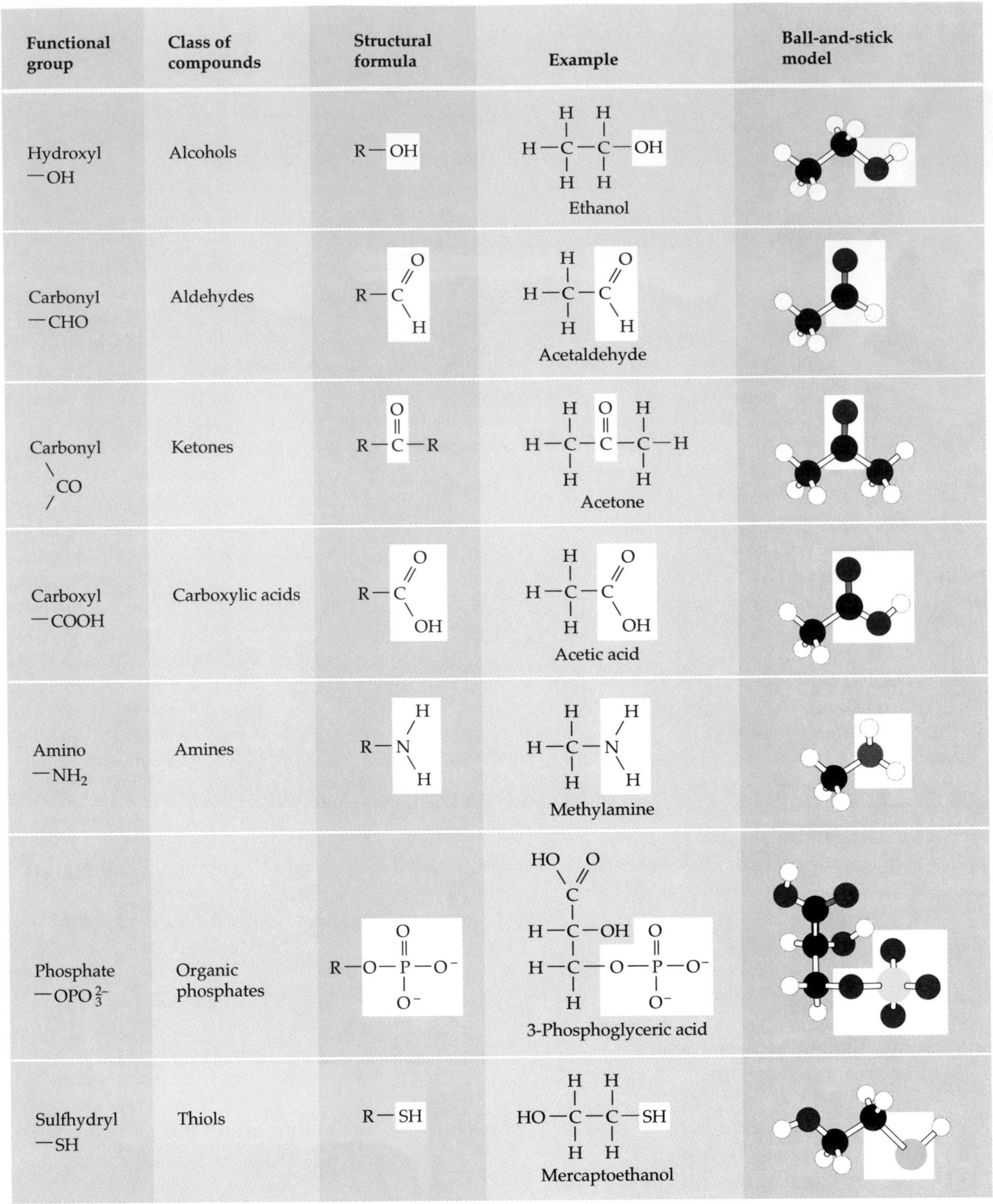

Functional group	Class of compounds	Structural formula	Example	Ball-and-stick model
Hydroxyl —OH	Alcohols	R—OH	Ethanol	
Carbonyl —CHO	Aldehydes	R—C(=O)—H	Acetaldehyde	
Carbonyl \CO/	Ketones	R—C(=O)—R	Acetone	
Carboxyl —COOH	Carboxylic acids	R—C(=O)—OH	Acetic acid	
Amino —NH$_2$	Amines	R—N(H)—H	Methylamine	
Phosphate —OPO$_3^{2-}$	Organic phosphates	R—O—P(=O)(O$^-$)—O$^-$	3-Phosphoglyceric acid	
Sulfhydryl —SH	Thiols	R—SH	Mercaptoethanol	

2.23 Some Functional Groups Important to Living Systems Compounds of the types shown here will appear throughout this book. The functional groups (highlighted) are the most common ones found in biologically important molecules. The letter *R* represents the remainder of the molecule, which may consist of any of a large number of carbon skeletons or other chemical groupings.

their biological roles. Chemists can use the characteristics of composition, structure (three-dimensional shape), reactivity, and solubility to distinguish a sample of one pure compound from another. For example, a compound such as sugar is soluble in water, which means that the solid disperses to form a uniform homogeneous mixture, but the same compound is insoluble in oil (the solid sugar remains a solid), and the resulting mixture is heterogeneous. Another substance, such as butane, is insoluble in water but highly soluble in oil. These solubility differences are due to the different atoms present and to their arrangement in these two kinds of molecules.

The presence of polar or charged sites on a molecule plays important roles in determining the molecule's solubility in water. Such sites can also determine the kinds of chemical reactions in which the molecule participates. The sizes, shapes, solubilities, and reactivities of molecules are significant to understanding the structures and operations of cells and organisms. Certain groups of atoms found together in a variety of molecules simplify our understanding of the reactions that molecules undergo.

Functional groups give specific properties to molecules

On the basis of atomic composition, structure, and reactivity, we can distinguish families of molecules from other families. Each member within a family of compounds has some characteristics in common with the other members of that family but differs from them in other characteristics. For example, organic acids are a family of carbon compounds that are all acidic but differ in other ways. They all contain a characteristic group of atoms called the **carboxyl group**,

$$
\begin{array}{c}
\text{O} \\
\parallel \\
-\text{C}-\text{OH, or COOH}
\end{array}
$$

that is the source of the H^+ that defines an acid:

$$
\begin{array}{cc}
\text{O} & \text{O} \\
\parallel & \parallel \\
-\text{C}-\text{OH} & \rightarrow -\text{C}-\text{O}^- + \text{H}^+
\end{array}
$$

The carboxyl group is a functional group. **Functional groups** are groups of atoms that are part of a larger molecule and have particular reactive characteristics. The same functional group may be part of very different molecules. In addition to the carboxyl group, you will encounter several other functional groups in your study of biology (Figure 2.23).

Several classes of biologically important compounds are defined by the functional groups they contain. When the functional group is a **hydroxyl group** (—OH), the product is an **alcohol**. Perhaps the most familiar alcohol is **ethanol** (also called ethyl alcohol, CH_3CH_2OH). Small alcohols like ethanol are soluble in water because of the hydrogen bonding possible between the polar hydroxyl group and water molecules, but larger alcohols are not soluble in water because of their long hydrocarbon chains.

Sugars contain both hydroxyl and carbonyl groups. The **carbonyl group** has a central carbon atom with a double bond to an oxygen atom (C=O). If one of the other two bonds of the carbon atom in a carbonyl group is attached to a hydrogen atom, the compound is an **aldehyde**.

Carboxyl groups are found in organic acids. Some organic bases, called amines, possess an **amino group** (—NH_2), which has a tendency to react with H^+ to produce the positively charged ammonium group (—NH_3^+). This H^+-accepting characteristic accounts for the classification of amines as bases.

Amino acids are important compounds that possess both a carboxyl group *and* an amino group attached to the same carbon atom, the α (alpha) carbon. Also attached to the α carbon atom are a hydrogen atom and a side chain designated by the letter R (Figure 2.24). Different side chains have different chemical compositions, structures, and properties. Each of the 20 amino acids found in proteins has a different side chain that gives it its distinctive chemical properties, as we'll see in Chapter 3. Because they possess both carboxyl and amino groups, amino acids are simultaneously acids and bases. At the pH values commonly found in cells, both the carboxyl and the amino groups are ionized: The carboxyl group has lost a proton, and the amino group has gained one (see Figure 2.24).

2.24 Amino Acid Structure Two depictions of the general structure of an amino acid. The side chain attached to the α carbon differs from one amino acid to another. At pH values found in living cells, both the carboxyl group and the amino group of an amino acid are ionized, as shown in the right-hand model.

Conventional depiction

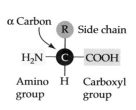

Three-dimensional depiction

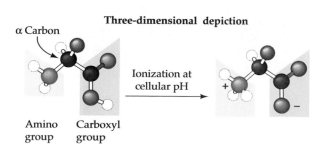

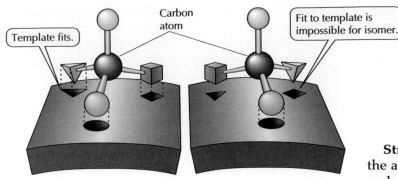

Template fits.

Carbon atom

Fit to template is impossible for isomer.

2.25 Optical Isomers Optical isomers are mirror images of each other. They result when four different groups are attached to a single carbon atom. If a template is laid out to match the groups on one carbon atom, the groups on the mirror-image isomer cannot be rotated to fit the same template.

Two other functional groups deserve our attention: the sulfhydryl group and the phosphate group (see Figure 2.23). The **sulfhydryl group** (—SH) is an important constituent of the side chains of two amino acids. When two of these sulfhydryl groups react, the two hydrogen atoms are lost and a covalent bond, called a disulfide bond, is formed between the sulfur atoms (—S—S—). As we'll see in the next chapter, disulfide bonds help maintain the three-dimensional structure of proteins that is required for normal protein functioning.

The **phosphate group** ($—OPO_3^{2-}$) is found on many different kinds of molecules. It plays important structural roles in DNA and RNA and participates in reactions that transfer energy. The removal (hydrolysis) of phosphate groups from some molecules releases energy that can be used to fuel other energy-requiring reactions.

Isomers have different arrangements of the same atoms

Isomers are compounds that have the same chemical formula but different arrangements of the atoms. (The prefix "iso-" means "same" and is encountered in many technical terms.) Of the different kinds of isomers, we will consider two: structural isomers and optical isomers.

Structural isomers are isomers that differ in how the atoms are joined together. Consider two simple molecules, each composed of four carbon and ten hydrogen atoms bonded covalently, with the formula C_4H_{10}. These atoms can be linked in two alternative ways in which carbon can form four bonds and hydrogen one bond. These two forms are called butane and isobutane. Their different bonding relationships are distinguished in structural formulas, and they have different chemical properties.

$$
\begin{array}{cc}
\text{H} \quad \text{H} & \text{CH}_3 \\
| \quad | & | \\
\text{H}_3\text{C} — \text{C} — \text{C} — \text{CH}_3 & \text{H}_3\text{C} — \text{C} — \text{CH}_3 \\
| \quad | & | \\
\text{H} \quad \text{H} & \text{H} \\
\text{Butane} & \text{Isobutane}
\end{array}
$$

Many molecules of biological importance, particularly the sugars and amino acids, have **optical isomers**. Optical isomers (also called enantiomers) are related to each other in the way an object is related to its mirror image (Figure 2.25). Optical isomers occur whenever a carbon atom has four *different* atoms or groups attached to it. These instances allow two different ways of making the attachments, each the mirror image of the other. Such a carbon atom is an asymmetric carbon, and the pair of compounds are optical isomers of each other. Your right and left hands are optical isomers. Just as a glove is specific for a particular hand, so some biochemical molecules can interact with a specific optical isomer of a compound but are unable to "fit" the other.

The α carbon in an amino acid is an asymmetric carbon because it is bonded to four different groups. Therefore, amino acids exist in two isomeric forms, called D-amino acids and L-amino acids (Figure 2.26). "D" and "L" are abbreviations for right and left, respectively. Only L-amino acids are commonly found in most proteins of living things.

The compounds discussed in this chapter include some of the more common ones found in organisms.

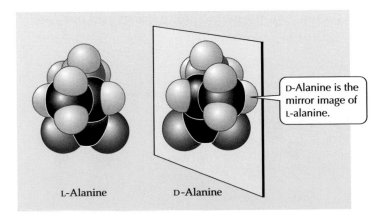

D-Alanine is the mirror image of L-alanine.

L-Alanine   D-Alanine

2.26 Isomers of an Amino Acid Space-filling models of the D and L forms of the amino acid alanine. Only L-alanine is commonly found in living things.

Between these small molecules and the world of the living stands another level, that of the giant macromolecules. These huge molecules—the proteins, lipids, carbohydrates, and nucleic acids—are the subject of the next chapter.

Summary of "Small Molecules: Structure and Behavior"

Atoms: The Constituents of Matter

• Matter is composed of atoms. Each atom is a positively charged nucleus of protons and neutrons surrounded by electrons bearing negative charges. Electrons are distributed in shells consisting of orbitals. Each orbital contains a maximum of two electrons. **Review Figures 2.4, 2.5**
• Isotopes of an element differ in their numbers of neutrons. Some isotopes are radioactive, emitting radiation as they decay. **Review Figure 2.2**
• In losing, gaining, or sharing electrons to become more stable, an atom can combine with other atoms to form molecules. **Review Figures 2.6, 2.7, 2.8**

Chemical Bonds: Linking Atoms Together

• Covalent bonds are strong bonds formed when two atomic nuclei share one or more pairs of electrons. Covalent bonds have spatial orientations that give molecules three-dimensional shapes. **Review Figure 2.7**
• Nonpolar covalent bonds occur when the electronegativity of both atoms is equal. When atoms with high electronegativity (e.g., oxygen) bond to atoms with weaker electronegativity (e.g., hydrogen), a polar covalent bond is formed in which one end is $\delta+$ and the other is $\delta-$. **Review Figure 2.8**
• Hydrogen bonds are weak electrical attractions that form between a $\delta+$ hydrogen and a $\delta-$ nitrogen or oxygen atom in another molecule or in another part of a large molecule. Hydrogen bonds are abundant in water. **Review Figure 2.9**
• Ions are electrically charged bodies that form when an atom gains or loses one or more electrons. Ionic bonds are electrical attractions between oppositely charged ions. Ionic bonds are strong in solids, but weaker when the ions are separated from each other in a water solution. **Review Figures 2.10, 2.11, 2.12**
• Molecules with no electric charge are nonpolar and do not interact directly with polar or charged substances, including water. Nonpolar molecules are attracted to each other by very weak bonds called van der Waals attractions. In water, nonpolar molecules are pushed together because of cages of water molecules that form around them. **Review Figures 2.13, 2.14**

Eggs by the Dozen, Molecules by the Mole

• Molecular formulas identify the kind and number of atoms in a molecule; structural formulas tell which atoms are linked together.

• A mole of any substance equals its molecular weight in grams and contains the same number of molecules (6.023×10^{23}) as a mole of any other substance.
• Solutions are produced when substances dissolve in water. The concentration of a solution is the amount of substance in a given amount of solution.

Chemical Reactions: Atoms Change Partners

• In chemical reactions, substances change their atomic compositions and properties. Energy is released in some reactions, whereas in others energy must be provided. Neither matter nor energy is created or destroyed in a chemical reaction, but both change form.
• Combustion reactions are oxidation–reduction reactions in which a fuel is converted to carbon dioxide and water, and energy is released as heat and light. In living cells, the same kind of reaction takes place in multiple steps so that the energy released can be harvested for cellular activities. **Review Figure 2.16**

Water: Structure and Properties

• Water's molecular structure and its capacity to form hydrogen bonds give it unusual properties that are significant for life. Water is an excellent solvent; solid water floats in liquid water; and water loses a great deal of heat when it condenses from vapor or freezes, a property that moderates environmental temperature changes.
• The cohesion of water molecules permits liquid water to rise to great heights in narrow columns and produces a high surface tension. Water's high heat of vaporization assures effective cooling when water evaporates.
• The ionization of water molecules produces negatively charged hydroxide ions and positive hydrogen ions. These ions participate in many important chemical reactions. **Review Figures 2.18, 2.19**

Acids, Bases, and the pH Scale

• Acids are substances that donate hydrogen ions. Bases are substances that accept hydrogen ions.
• Hydrogen ion concentration is measured by the pH scale, which has values from less than 0 to more than 14.
• The pH of a solution is the negative logarithm of the hydrogen ion concentration. Values lower than pH 7 indicate an acidic solution; values above pH 7 indicate a basic solution. **Review Figure 2.21**
• Buffers are systems of weak acids and bases that limit the change in pH when H^+ ions are added or removed. **Review Figure 2.22**

The Properties of Molecules

• Molecules vary in size, shape, complexity, reactivity, solubility, and other chemical properties.
•Functional groups are part of a larger molecule and have particular chemical properties. The consistent chemical behavior of functional groups helps us understand the properties of molecules that contain them. **Review Figure 2.23**
• Structural and optical isomers have the same kinds and numbers of atoms, but differ in their structures and properties. **Review Figures 2.25, 2.26**

Self-Quiz

1. The atomic number of an element
   a. equals the number of neutrons in an atom.
   b. equals the number of protons in an atom.
   c. equals the number of protons minus the number of neutrons.
   d. equals the number of neutrons plus the number of protons.
   e. depends on the isotope.

2. The atomic weight (atomic mass) of an element
   a. equals the number of neutrons in an atom.
   b. equals the number of protons in an atom.
   c. equals the number of electrons in an atom.
   d. equals the number of neutrons plus the number of protons.
   e. depends on the relative abundances of its isotopes.

3. Which of the following statements about all the isotopes of an element is *not* true?
   a. They have the same atomic number.
   b. They have the same number of protons.
   c. They have the same number of neutrons.
   d. They have the same number of electrons.
   e. They have identical chemical properties.

4. Which of the following statements about a covalent bond is *not* true?
   a. It is stronger than a hydrogen bond.
   b. One can form between atoms of the same element.
   c. Only a single covalent bond can form between two atoms.
   d. It results from the sharing of electrons by two atoms.
   e. One can form between atoms of different elements.

5. Hydrophobic interactions
   a. are stronger than hydrogen bonds.
   b. are stronger than covalent bonds.
   c. can hold two ions together.
   d. can hold two nonpolar molecules together.
   e. are responsible for the surface tension of water.

6. Which of the following statements about water is *not* true?
   a. It releases a large amount of heat when changing from liquid into vapor.
   b. Its solid form is less dense than its liquid form.
   c. It is the most effective solvent known.
   d. It is typically the most abundant substance in an active organism.
   e. It takes part in some important chemical reactions.

7. A solution with a pH of 9
   a. is acidic.
   b. is more basic than a solution with a pH of 10.
   c. has 10 times the hydrogen ion concentration of a solution with a pH of 10.
   d. has a hydrogen ion concentration of 9 M.
   e. has a hydroxide ion concentration of 9 M.

8. Which of the following compounds is an alcohol?
   a. O_2
   b. $CH_3CH_2CH_2OH$
   c. CH_3COOH
   d. C_3H_8
   e. CH_3COCH_3

9. Which of the following statements about the carboxyl group is *not* true?
   a. It has the chemical formula —COOH.
   b. It is an acidic group.
   c. It can ionize.
   d. It is found in amino acids.
   e. It has an atomic weight of 45.

10. Which of the following statements about amino acids is *not* true?
    a. They are the building blocks of proteins.
    b. They contain carboxyl groups.
    c. They contain amino groups.
    d. They do not ionize.
    e. They have both L and D isomers.

Applying Concepts

1. Would you expect the elemental composition of Earth's crust to be the same as that of the human body? How could you find out? Try to find and discuss the answer.

2. Lithium (Li) is the element with atomic number 3. Draw the structures of the Li atom and of the Li^+ ion.

3. Draw the structure of a pair of water molecules held together by a hydrogen bond. Your drawing should indicate the covalent bonds.

4. The molecular weight of sodium chloride (NaCl) is 58.45. How many grams of NaCl are there in 1 liter of a 0.1 M NaCl solution? How many in 0.5 l of a 0.25 M NaCl solution?

5. The two optical isomers of alanine are shown in Figure 2.27. The side chain of the amino acid glycine is simply a hydrogen atom (—H). Are there two optical isomers of glycine? Explain.

Readings

Atkins, P. W. and L. L. Jones. 1997. *Chemistry: Molecules, Matter, and Change*, 3rd Edition. W. H. Freeman, Inc., New York. This and the next two listings are popular and accessible textbooks.

Brown, T. L., H. E. LeMay, Jr. and B. E. Bursten. 1997. *Chemistry: The Central Science*, 7th Edition. Prentice Hall, Englewood Cliffs, NJ.

Chang, R. 1994. *Chemistry*, 5th Edition. McGraw-Hill, New York.

Henderson, L. J. 1958. *The Fitness of the Environment*. Beacon Press, Boston. An essay written in 1912 about physical properties of water and carbon dioxide in relation to life. Includes a thought-provoking introduction.

Lancaster, J. R. 1992. "Nitric Oxide in Cells." *American Scientist*, vol. 80, pages 248–259. A readable account of the multiple biological effects of this very small molecule.

Mertz, W. 1981. "The Essential Trace Elements." *Science*, vol. 213, pages 1332–1338. A review of the roles of more than a dozen elements that are necessary in small amounts for animals to function normally.

Zumdahl, S. 1995. *Chemical Principles*, 2nd Edition. D. C. Heath, Lexington, MA. A higher-level text for students with strong math backgrounds.

Chapter 3

Macromolecules: Their Chemistry and Biology

A Gigantic Molecule
The protein actin, shown here in a computer-rendered graphic, is an example of a macromolecule. The thousands of atoms that make up the protein are shown in different colors.

The lives of cells are dances and dramas with tens of thousands of different kinds of molecules. Their dramatic choreography is what scientists reveal as they investigate the molecules of living systems, their physical properties, and their chemical reactions. What molecules are present? What are their chemical structures and properties? And what biological functions do these molecules perform? In different ways, we will be concerned with these questions throughout much of this book.

In this chapter, we'll look at gigantic molecules called macromolecules (*macro-*, "large") and the subunits from which they are constructed. Macromolecules may contain hundreds or thousands of atoms and have very large molecular weights. For example, human hemoglobin has a molecular weight of 64,500 and contains more than 6,000 atoms. The molecular complexity of such a structure can be understood in terms of the structures and properties of its subunits, which are fewer in number and easier to understand. Macromolecules are chains of small, individual units called monomers (*mono-*, "one"; *-mer*, "unit") covalently bonded together to form a polymer (*poly-*, "many").

In living systems there are three types of macromolecules: polysaccharides, proteins, and nucleic acids. *Polysaccharides* are constructed from sugar monomers such as glucose; *proteins* are constructed from amino acids; and *nucleic acids* are composed of nucleotides. Another important group of molecules that can form large structures is the *lipids*. But because the individual components

3.1 Building Blocks Diverse structures can be assembled from a few simple units. In a child's play, these units may be Lego® pieces. In the cell, these building blocks are monomers such as glucose, amino acids, and nucleotides.

(the lipids) do not covalently bond together, the resulting large structures are not true macromolecules.

The monomers that form the basis of polysaccharides, proteins, and nucleic acids are identical in different species. For example, a molecule of glucose in human blood is identical to a molecule of glucose from a reptile or a cabbage plant. The same group of 20 different amino acids is used to construct proteins for all living things—from bacteria to birds, bats, and humans. However, this cannot always be said of the larger molecular forms built from these monomers.

Proteins and nucleic acids in particular are informational macromolecules that are different in different species. For example, human hemoglobin is different from the hemoglobin found in fish or birds. No animal acquires its macromolecules directly from its food. Instead, it uses the subcomponents of its food to construct new macromolecules suited to its unique needs. For example, we eat proteins constructed by other animals and plants, but we break these proteins down into their amino acids and then reassemble the amino acids into the chemically different proteins of our own bodies. This process is like picking up Lego toys that somebody else has made, taking them apart, and putting the parts together again to make the toys *we* want (Figure 3.1).

Before we turn to the main focus of this chapter—the macromolecules: carbohydrates, proteins, and nu-cleic acids—let's first take a look at lipids and the large macromolecule-like structures that they form.

Lipids: Water-Insoluble Molecules

Lipids are a chemically diverse group of hydrocarbons. The property they all share is an insolubility in water that is due to the presence of many nonpolar covalent bonds. Nonpolar molecules can associate together and form massive structures, but these structures are not considered macromolecules, because the bonds between the separate molecules are not covalent bonds. As we saw in Chapter 2, these nonpolar hydrocarbon molecules are literally pushed together by surrounding water molecules, which are not attracted to the non-polar substance. When the nonpolar molecules are sufficiently close together, weak but additive van der Waals forces hold them together. Although insoluble in water, lipids are soluble in other lipids or in nonpolar solvents such as ether or benzene.

In addition to their diverse chemical structures, lipids have many different biological roles. Some of them store energy (the fats and oils). Others play important structural roles in cell membranes (phospholipids) (Figure 3.2). The carotenoids help plants capture light energy, and the steroids and some lipids play regulatory roles. We'll begin our discussion by considering the structures and properties of the fats and oils.

Fats and oils store energy

Chemically, fats and oils are **triglycerides**, also known as *simple lipids*. Triglycerides that are solid at room temperature (20°C) are called **fats**; those that are liquid at room temperature are called **oils**. Triglycerides

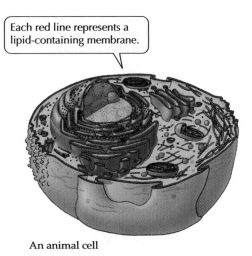

Each red line represents a lipid-containing membrane.

An animal cell

3.2 Lipids Form Cellular Membranes Membranes made of lipids separate the cell from its environment; they also separate the contents of internal compartments from the rest of the cell. Materials that do not dissolve in lipids usually cannot pass through membranes, but lipid-soluble materials move through with relative ease.

(a) Palmitic acid

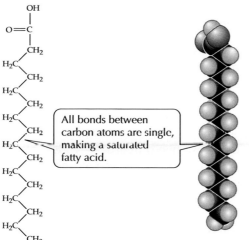

All bonds between carbon atoms are single, making a saturated fatty acid.

(b) Oleic acid

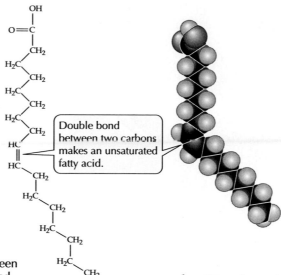

Double bond between two carbons makes an unsaturated fatty acid.

3.3 Fatty Acids (a) The straight-chain configuration seen in the model of palmitic acid is characteristic of saturated fatty acids. (b) The double bond between two carbons in the chain of an unsaturated fatty acid such as oleic acid causes the molecule to bend, obstructing the packing of the hydrocarbon chains.

are composed of two types of building blocks: fatty acids and glycerol. **Glycerol** is a small molecule with three hydroxyl (—OH) groups. **Fatty acids** have long nonpolar hydrocarbon tails and a polar carboxyl functional group (—COOH). Fatty acids can be of different lengths (usually 16 to 20 carbon atoms) and may be either saturated or unsaturated.

In **saturated fatty acids**, all the bonds between the carbon atoms in the hydrocarbon chain are single bonds—there are no double bonds. That is, all the bonds are *saturated* with hydrogen atoms (Figure 3.3a). Saturated fatty acids include palmitic acid, which has 16 carbon atoms, and stearic acid, which has 18. These molecules are relatively rigid and straight, and they pack together tightly, like pencils in a box.

In **unsaturated fatty acids**, the hydrocarbon chain contains one or more double bonds. Oleic acid is a monounsaturated fatty acid that

has 18 carbon atoms, and one double bond near the middle of the hydrocarbon chain causes a kink in the molecule (Figure 3.3b). Fatty acids, such as linoleic acid, that have more than one double bond are **polyunsaturated** and have multiple kinks. These kinks prevent the molecules from packing together tightly. The two fatty acids that humans cannot synthesize and must obtain from their diet are both unsaturated fatty acids. Diets favoring unsaturated fatty acids over saturated ones tend to reduce the incidence of arteriosclerosis, which is a narrowing of the arterial wall that restricts blood flow to the heart and may lead to heart attacks (see Chapter 46). Diets rich in saturated fatty acids have the opposite effect.

Three fatty acid molecules combine with a molecule of glycerol to form a molecule of a triglyceride (Figure 3.4). The carboxyl group of each fatty acid reacts with a hydroxyl group of the glycerol to form an

3.4 A Triglyceride and Its Components Three fatty acid molecules and a molecule of glycerol react to form a triglyceride molecule and three molecules of water (H_2O). The long hydrocarbon chains from the fatty acid make triglyceride a very hydrophobic substance. In living things the reaction is more complex, but the end result is as shown here.

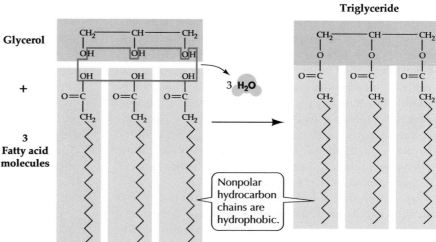

Glycerol

3 Fatty acid molecules

3 H_2O

Triglyceride

Nonpolar hydrocarbon chains are hydrophobic.

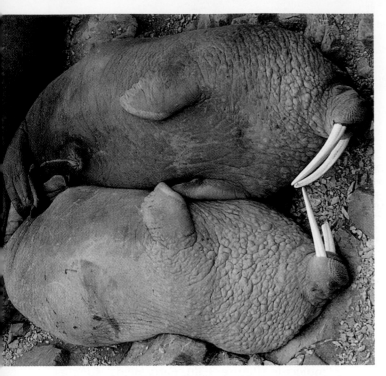

3.5 Energy to Fight Cold Temperatures These Alaskan walrus spend much of their time in frigid water. Layers of body fat insulate their bodies against the cold and store energy efficiently.

ester (the reaction product of an acid and an alcohol) and water. The ester linkage looks like this:

$$\begin{matrix} & O & \\ & \parallel & \\ -C & -O- & C- \end{matrix}$$

The three fatty acids in one triglyceride molecule need not all have the same length, nor do they all have to be either saturated or unsaturated. The kinks associated with double bonds are important in determining the fluidity and melting point of a lipid. Animal fats such as lard and tallow are usually solids, and their triglycerides tend to have many long-chain saturated fatty acids. These fats are solids at room temperature because their fatty acids pack well together. The triglycerides of plants tend to have short or unsaturated fatty acids. Because of their kinks, these fatty acids pack poorly together and these triglycerides are usually liquid at room temperature. Natural peanut butter, for example, contains a great deal of oil. However, to market a more convenient and solid product, manufacturers often subject the oil to hydrogenation, which adds hydrogens to double bonds and converts unsaturated fatty acids to saturated fatty acids.

Lipids are marvelous storehouses for energy. By taking in excess food, many animal species deposit fat (lipid) droplets in their cells as a means for storing energy (Figure 3.5). Some plant species, such as olives, avocados, sesame, castor beans, and all nuts, have substantial amounts of lipids in their seeds or fruits that serve as energy reserves for the next generation.

Phospholipids form the core of biological membranes

Because lipids and water do not interact, a mixture of water and lipids forms two distinct phases. Many biologically important substances—such as ions, sugars, and free amino acids—that are soluble in water are insoluble in lipids. These two properties—water solubility and lipid insolubility—have proved essential to distinguishing the interior of cells from their external environment.

Suppose that you must design water-filled compartments, separated from each other and from their environment by barriers that limit the passage of materials. Given the properties of lipids, a seemingly effective way to accomplish this task is to use membranes that contain a special class of lipid called the phospholipid (Figure 3.6). This is the system that has evolved in nature. Molecular traffic within an organism or into and out of its compartments is constrained by the properties of the lipid portion of the surrounding membrane (see Figure 3.2). Compounds that dissolve readily in lipids can move rapidly through biological membranes, but compounds that are insoluble in lipids are prevented from passing through the membrane or must be transported across the membrane by specific proteins.

Like triglycerides, **phospholipids** have fatty acids bound to glycerol by ester linkages. In phospholipids, however, any one of several phosphate-containing compounds may replace one of the fatty acids (see Figure 3.6). Many phospholipids are important constituents of biological membranes. If you think carefully about the structure and properties of phospholipids, you will find it easy to understand how they are oriented in membranes. The phosphate functional group has negative electric charges, so this portion is **hydrophilic,** attracting polar water molecules. But the two fatty acids are **hydrophobic**, so they are pushed together by water (see Chapter 2).

In a biological membrane, phospholipids line up in such a way that the nonpolar, hydrophobic "tails" pack tightly together to form the interior of the membrane, and the phosphate-containing "heads" face outward (some to one side of the membrane and some to the other), where they interact with water, which is excluded from the interior of the membrane. The phospholipids thus form a bilayer, a sheet two molecules thick (Figure 3.7). Because membranes are so important, we will devote all of Chapter 5 to them.

Because the word "lipid" defines compounds in terms of their solubility rather than their structural

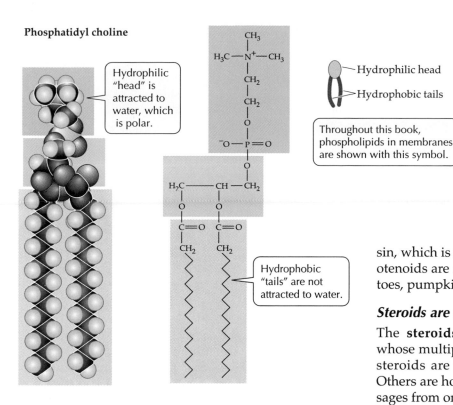

Phosphatidyl choline

Hydrophilic "head" is attracted to water, which is polar.

Hydrophilic head

Hydrophobic tails

Throughout this book, phospholipids in membranes are shown with this symbol.

Hydrophobic "tails" are not attracted to water.

3.6 Phospholipid Structure Every phospholipid molecule has two hydrophobic "tails" and a hydrophilic "head." The two fatty acids bonded to a glycerol provide the hydrophobic tails, while the phosphorus-containing compound linked to the third —OH of the glycerol provides the hydrophilic head. In phosphatidyl choline, the hydrophilic head consists of phosphate and choline. In other phospholipids, the amino acid serine, the sugar alcohol inositol, or another compound can replace choline.

similarity, a great variety of different chemical structures are included as lipids. The next two lipid classes we'll discuss—the carotenoids and the steroids—have chemical structures very different from the structures of triglycerides and phospholipids and from the structures of each other.

Carotenoids trap light energy

The **carotenoids** are a family of light-absorbing pigments found in plants and animals (Figure 3.8). Beta-carotene (β-carotene) is one of the pigments that traps light energy in leaves during photosynthesis (see Chapter 8). It is the β-carotene in plants that senses light and causes their parts to grow toward or away from the light (a behavior called phototropism, discussed in Chapter 35). In humans, a molecule of β-carotene can be broken down into two vitamin A molecules, from which we make the pigment rhodop-

sin, which is required for vision (see Chapter 42). Carotenoids are responsible for the color of carrots, tomatoes, pumpkins, egg yolks, and butter.

Steroids are signal molecules

The **steroids** are a family of organic compounds whose multiple rings share carbons (Figure 3.9). Some steroids are important constituents of membranes. Others are hormones, chemical signals that carry messages from one part of the body to another. *Testosterone* and the *estrogens* are steroid hormones that regulate sexual development in vertebrates. *Cortisol* and related hormones play many regulatory roles in the digestion of carbohydrates and proteins, in the maintenance of salt balance and water balance, and in sexual development.

Cholesterol is synthesized in the liver and contributes to the structure of some cellular membranes. It is the starting material for making testosterone and other steroid hormones, as well as the bile salts that help break down dietary fats so that they can be digested. We absorb cholesterol from foods such as milk, butter, and animal fats. When we have too much cholesterol in our blood, it is deposited in our arteries (along with other substances), a condition that may lead to arteriosclerosis and heart attack.

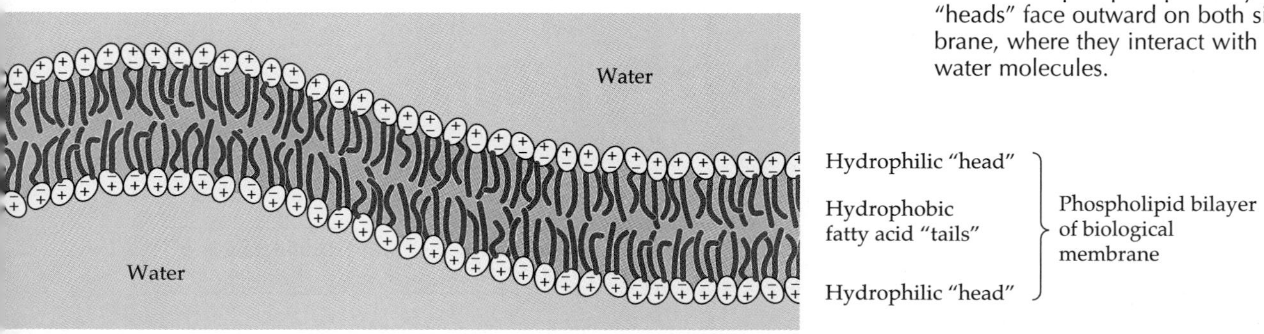

3.7 Phospholipids in Biological Membranes
In a water environment, hydrophobic interactions bring the "tails" of phospholipids together in the interior of a phospholipid bilayer. The hydrophilic "heads" face outward on both sides of the membrane, where they interact with the surrounding water molecules.

Water

Water

Hydrophilic "head"

Hydrophobic fatty acid "tails"

Hydrophilic "head"

Phospholipid bilayer of biological membrane

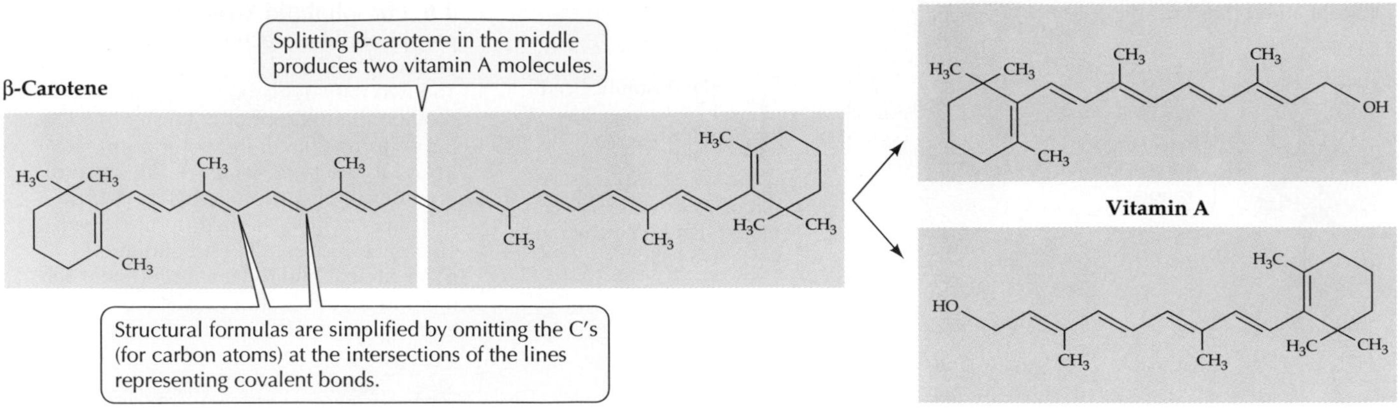

Splitting β-carotene in the middle produces two vitamin A molecules.

Structural formulas are simplified by omitting the C's (for carbon atoms) at the intersections of the lines representing covalent bonds.

β-Carotene

Vitamin A

3.8 β-Carotene Is the Source of Vitamin A The carotenoid β-carotene is symmetrical around the central double bond; when split, β-carotene becomes two vitamin A molecules. The simplified structural formula is standard chemical shorthand for large structures with many carbon atoms.

Some lipids are vitamins

A large group of fat-soluble substances, including the carotenoids and steroids, are synthesized by covalent linking and chemical modification of isoprene to form a series of isoprene units:

$$CH_2 = \overset{\overset{\displaystyle CH_3}{|}}{C} - CH = CH_2$$

The fat-soluble vitamins (A, D, E, and K) are formed in this manner by plants and bacteria. These substances are not synthesized by humans and must be acquired from dietary sources. **Vitamin A** is formed from β-carotene found in green and yellow vegetables (see Figure 3.8). Among other roles, vitamin A is directly involved in the reception of light by our eyes. Deficiency in vitamin A leads to dry skin, eyes, and internal body surfaces; retarded growth and development; and night blindness, which is a diagnostic symptom for the deficiency. **Vitamin D** regulates the absorption of calcium from the intestines. It is necessary for the proper deposition of calcium in bones; a deficiency of vitamin D can lead to rickets, a bone-softening disease.

Vitamin E is not a single vitamin, but a group of related lipids that seem to protect cells from damaging effects of oxidation–reduction reactions. These lipids appear to have an important role in preventing unhealthy changes in the double bonds in the unsaturated fatty acids of membrane phospholipids. Commercially, vitamin E is added to some foods to slow spoilage. **Vitamin K** is found in green leafy plants and is also synthesized by bacteria normally present in the human intestine. This vitamin is essential to the formation of blood clots. Predictably, a deficiency of vitamin K leads to slower clot formation and potentially fatal bleeding from a wound.

Because of their insolubility in water, in the body some lipids may require water-soluble carrier proteins in order to be transported in the blood, which is mostly water. Among the lipids, only triglycerides and phospholipids assemble into large structures in cells:

3.9 All Steroids Have the Same Ring Structure These steroids, all important in vertebrates, are composed of carbon and hydrogen and are highly hydrophobic. However, small chemical variations, such as the presence or absence of a methyl or hydroxyl group, can produce enormous functional differences.

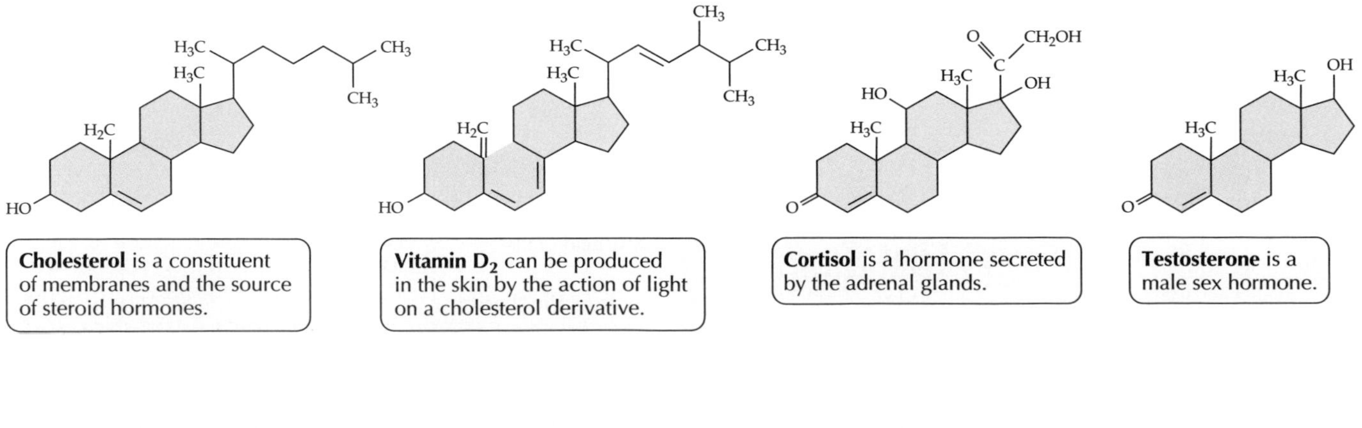

Cholesterol is a constituent of membranes and the source of steroid hormones.

Vitamin D₂ can be produced in the skin by the action of light on a cholesterol derivative.

Cortisol is a hormone secreted by the adrenal glands.

Testosterone is a male sex hormone.

triglycerides form droplets of stored fat, and phospholipids form the bilayers of membranes. But in spite of their size, these droplets and bilayers are not usually considered macromolecules.

Now that we have seen the macromolecule-like structures that lipids form, let's turn to the large structures that constitute true macromolecules.

Macromolecules: Giant Polymers

As noted earlier, **macromolecules** are giant *polymers* constructed by the covalent linking of smaller molecules called *monomers*. These monomers may or may not be identical, but they always have similar chemical structures. Molecules with molecular weights exceeding 1,000 are usually considered macromolecules, and the polysaccharides, proteins, and nucleic acids of living systems certainly fall into this category.

In the cell, each type of macromolecule performs some combination of a diversity of functions: energy storage, structural support, protection, catalysis, transport, defense, regulation, movement, and heredity. These roles are not necessarily exclusive. For example, both carbohydrates and proteins can play structural roles—supporting and protecting cells and organisms. However, only nucleic acids specialize in information and function as hereditary material, carrying both species and individual traits from generation to generation.

The functions of macromolecules are directly related to their shapes and the chemical properties of their monomers. Some macromolecules, such as catalytic and defensive proteins, fold into compact spherical forms with surface features that make them water-soluble and capable of intimate interaction with other molecules. Other proteins and carbohydrates form long, fibrous systems that provide strength and rigidity to cells and organisms. Still other long, thin assemblies of proteins can contract and cause movement.

Because macromolecules are so large, they contain many different functional groups. For example, a large protein may contain hydrophobic, polar, and charged groups. These groups give specific properties to local sites on a macromolecule. As we will see, this diversity of properties determines the shapes of macromolecules and their interactions with both other macromolecules and smaller molecules.

Macromolecules form by condensation reactions

The polymers of living things are constructed by a series of reactions called **condensation reactions,** or dehydration reactions (both words refer to the loss of water). Condensation reactions covalently bond monomers. Water is lost in the reaction:

$$A\text{---}H + B\text{---}OH \rightarrow A\text{---}B + H_2O$$

A—H is a molecule with a reactive hydrogen; B—OH is a molecule with a hydroxyl group. The products of the reaction are A—B and H_2O. The atoms that make up the water molecule are derived from the reactants: one hydrogen atom from one reactant, and an oxygen atom and the other hydrogen atom from the other reactant. Condensation reactions are not limited to polymer formation; for example, we encountered them in the linkage (esterification) of fatty acids to glycerol (see Figure 3.4).

The condensation reactions that produce the different kinds of macromolecules differ in detail, but in all cases polymers will form only if energy is added to the system. In living systems, specific energy-rich molecules supply the energy, and there are additional steps to the reaction.

The reverse of a condensation reaction is a **hydrolysis reaction** (*hydro-*, "water"; *-lysis*, "breakage"). These reactions digest polymers and produce monomers. Water reacts with the bonds that link the monomers together, and the products are free monomers. The elements of the reactant H_2O become part of the products. Hydrolysis reactions of all sorts are very important in cellular functioning and are not limited to the digestion of polymers.

Carbohydrates: Sugars and Sugar Polymers

Carbohydrates are a diverse group of compounds based on the general formula CH_2O. Some are relatively small, with molecular weights less than 100. Others are true macromolecules, with molecular weights of hundreds of thousands. There are four categories of biologically important carbohydrates. *Monosaccharides* (*mono-*, "one"; *saccharide*, "sugar")—such as glucose or fructose—are the monomers out of which the larger forms are constructed. *Disaccharides* (*di-*, "two") consist of two monosaccharides. *Oligosaccharides* (*oligo-*, "several") have several monosaccharides (3 to 20). *Polysaccharides* (*poly-*, "many")—such as starch, glycogen, and cellulose—are composed of hundreds of thousands of glucose units.

The relative proportions of carbon, hydrogen, and oxygen indicated by the general formula for carbohydrates, CH_2O, is true for monosaccharides. However, for disaccharides, oligosaccharides, and polysaccharides, these proportions differ slightly from this general formula because two hydrogens and an oxygen are lost during the condensation reactions.

Monosaccharides are simple sugars

All living cells contain the monosaccharide **glucose**, whose formula is $C_6H_{12}O_6$. Green plants produce glucose by photosynthesis, and other organisms acquire it directly or indirectly from plants. Cells use glucose as an energy source, changing it through a series of reac-

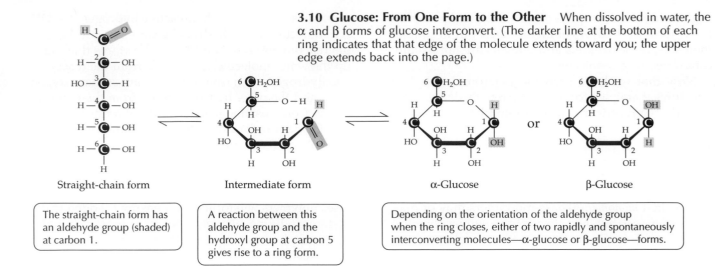

3.10 Glucose: From One Form to the Other When dissolved in water, the α and β forms of glucose interconvert. (The darker line at the bottom of each ring indicates that that edge of the molecule extends toward you; the upper edge extends back into the page.)

Straight-chain form

The straight-chain form has an aldehyde group (shaded) at carbon 1.

Intermediate form

A reaction between this aldehyde group and the hydroxyl group at carbon 5 gives rise to a ring form.

α-Glucose

β-Glucose

Depending on the orientation of the aldehyde group when the ring closes, either of two rapidly and spontaneously interconverting molecules—α-glucose or β-glucose—forms.

tions that release stored energy and produce water and carbon dioxide.

Two forms of glucose, the straight chain and the ring, exist in equilibrium with each other when dissolved in water, but the ring form predominates (>99%) (Figure 3.10). The two distinct ring forms (α- and β-glucose) differ only in the placement of the —H and —OH attached to carbon 1. (The convention for numbering carbons shown in Figure 3.10 is used throughout this book.)

Most of the monosaccharides found in living systems belong to the D series of optical isomers (see Chapter 2). But there are structural isomers—composed of the same kinds and numbers of atoms, but with the atoms combined differently in each. All **hex-**

oses (*hex-*, "six"), a group of structural isomers, have the formula $C_6H_{12}O_6$. Included among the hexoses are fructose (so named because it was first found in fruits), mannose, and galactose (Figure 3.11).

Pentoses (*pent-*, "five") are five-carbon sugars. Some pentoses are found primarily in the cell walls of plants, as are several of the hexoses. Two pentoses are of particular importance: **Ribose** and **deoxyribose** form part of the backbones of RNA and of DNA, respectively. These two pentoses are not isomers; rather, one oxygen atom is missing from carbon 2 in deoxyribose (*de-*, "absent") (see Figure 3.11). As we will see in Chapter 12, the absence of this oxygen atom has enormous consequences for the functional distinction of RNA and DNA.

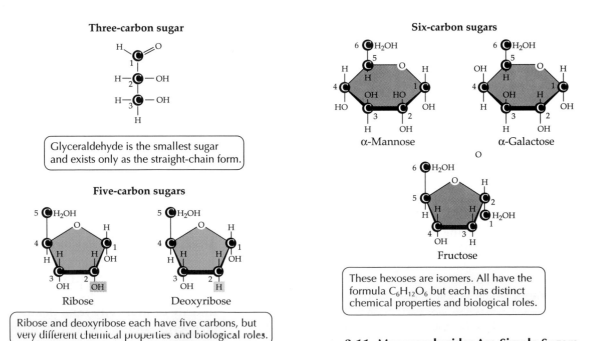

Three-carbon sugar

Glyceraldehyde is the smallest sugar and exists only as the straight-chain form.

Five-carbon sugars

Ribose

Deoxyribose

Ribose and deoxyribose each have five carbons, but very different chemical properties and biological roles.

Six-carbon sugars

α-Mannose

α-Galactose

Fructose

These hexoses are isomers. All have the formula $C_6H_{12}O_6$ but each has distinct chemical properties and biological roles.

3.11 Monosaccharides Are Simple Sugars

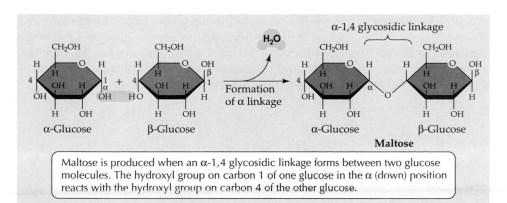

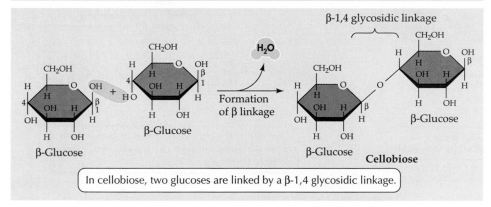

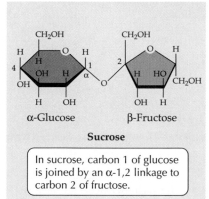

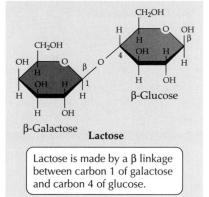

In sucrose, carbon 1 of glucose is joined by an α-1,2 linkage to carbon 2 of fructose.

Lactose is made by a β linkage between carbon 1 of galactose and carbon 4 of glucose.

3.12 Disaccharides Are Formed by Glycosidic Linkages

Glycosidic linkages between two monosaccharides create many different disaccharides. Which disaccharide is formed depends on the nature of the linked monosaccharides (glucose, galactose, fructose) and the site (carbon atoms) and form (α or β) of the linkage. These are condensation reactions, creating a water molecule as each linkage forms.

linkage with carbon 4 of a second glucose molecule gives *maltose*, whereas a β linkage gives *cellobiose*. Maltose and cellobiose are disaccharide isomers, and both have the formula $C_{12}H_{22}O_{11}$. However, they are different compounds with different properties. They undergo different chemical reactions and are recognized by different catalytic proteins (enzymes).

Oligosaccharides contain several monosaccharides linked by glycosidic linkages at various sites. Many oligosaccharides gain additional functional groups, which give them special properties. Oligosaccharides are often covalently bonded to proteins and lipids on the outer cell surface, where they serve as cell recognition signals.

Polysaccharides are energy stores or structural materials

Polysaccharides are giant chains of monosaccharides connected by glycosidic linkages. **Starch** is a polysaccharide of glucose with linkages in the α orientation. **Cellulose** is a similar giant polysaccharide made up solely of glucose, but its individual units are connected by β linkages instead of α linkages (Figure 3.13a). Cellulose is the predominant component of plant cell walls and is by far the most abundant organic compound on this planet. Both starch and cellulose are composed of nothing but glucose, but their

Glycosidic linkages bond monosaccharides together

Monosaccharides are covalently bonded together by condensation reactions that form **glycosidic linkages.** Such a linkage between two simple sugars forms a **disaccharide** (Figure 3.12). For example, a molecule of *sucrose* (table sugar)—the sugar that is transported to different parts of plants—is formed from a glucose and a fructose molecule, while *lactose* (milk sugar) contains glucose and galactose. The disaccharide *maltose* contains two glucose molecules.

But maltose is not the only disaccharide that can be made from two glucoses. When glucose molecules form glycosidic linkages, as shown in Figure 3.12, the disaccharide product must be one of two types: α-linked or β-linked, depending on whether the molecule that bonds by its 1 carbon is α-glucose or β-glucose. An α biological functions and chemical and physical properties are entirely different.

Starch can be more or less easily degraded by the actions of chemicals or catalytic proteins (enzymes). Cellulose, however, is very stable because of its β glycosidic linkages. Thus starch is a good storage form that can be easily degraded to supply glucose for energy-producing reactions, while cellulose is an excellent structural component that can withstand harsh environmental conditions without changing.

(*a*) **Molecular structure**

Cellulose

Starch and glycogen

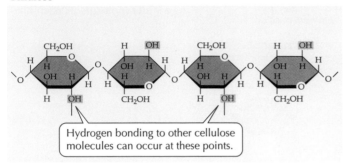

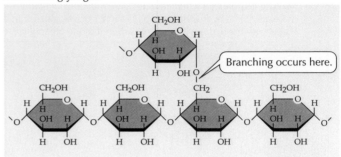

Branching occurs here.

Hydrogen bonding to other cellulose molecules can occur at these points.

Cellulose is an unbranched polymer of glucose with β-1,4 glycosidic linkages that are chemically very stable.

Glycogen and starch are polymers of glucose with α-1,4 glycosidic linkages. α-1,6 glycosidic linkages produce branching at carbon 6.

(*b*) **Macromolecular structure**

Linear strands of cellulose molecules

Branched starch molecule

Highly branched glycogen molecule

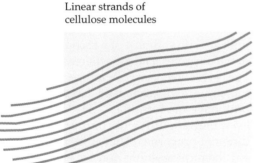

Parallel cellulose molecules hydrogen-bond together to form long thin fibrils.

Branching limits the number of hydrogen bonds that can form in starch molecules, making starch less compact than cellulose.

The high amount of branching in glycogen makes its solid deposits less compact than starch.

(*c*) **Polysaccharides in cells**

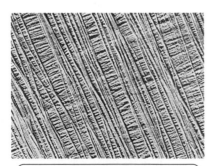

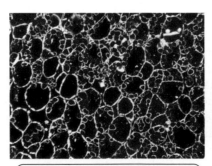

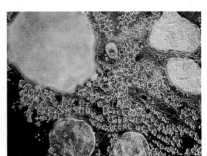

Layers of cellulose fibrils, as seen in this scanning electron micrograph, give plant cell walls great strength.

Dyed red in this micrograph, starch deposits have a large granular shape within cells.

Colored pink in this electron micrograph of human liver cells, glycogen deposits have a small granular shape.

3.13 Representative Polysaccharides Cellulose, starch, and glycogen demonstrate different levels of branching and compaction in polysaccharides.

Starch is not just one chemical substance, but a large family of giant molecules of broadly similar structure. All starches are large polymers of glucose with α linkages, but different starches can be distinguished by the amount of branching between carbons 1 and 6 (Figure 3.13*b*). Some starches are highly branched; others are not. Plant starches, called **amylose**, are not highly branched. The polysaccharide **glycogen**, which stores glucose in animal livers and muscles, is highly branched.

What do we mean when we say that starch and glycogen are storage compounds for energy? Very simply, these compounds can readily be hydrolyzed to yield glucose monomers. Glucose, in turn, can enter into a series of reactions that liberate its stored energy in forms that can be used for cellular activities. However, glucose can also serve the cell's need for raw materials—that is, carbon atoms. Glucose can undergo other chemical reactions that disassemble parts of the molecule and rearrange its carbon atoms to form the skeletons of other compounds needed by cells. Glycogen and starch are thus storage depots for carbon atoms as well as for energy.

Derivative carbohydrates contain other elements

Sometimes carbohydrates are modified by chemical changes in their structure or by the addition of functional groups such as phosphate and amino groups, thus becoming **derivative carbohydrates** (Figure 3.14). For example, carbon 6 in glucose may be oxidized from —CH_2OH to a carboxyl group (—COOH), producing glucuronic acid. Or a phosphate group may be added to one or more of the —OH sites. Some of these *sugar phosphates*, such as fructose 1,6-bisphosphate, are important intermediates in cellular energy reactions (see Chapters 7 and 8). When an amino group is substituted for an —OH group, *amino sugars* such as glucosamine and galactosamine are produced. Galactosamine is a major component of cartilage, the material that forms caps on the ends of bones and stiffens the protruding parts of the ears and nose. A derivative of glucosamine produces the polymer **chitin**, which is the principal structural polysaccharide in the skeletons of insects, crabs, and lobsters, as well as in the cell walls of fungi. Fungi and insects (and their relatives) constitute more than 80 percent of the species ever described, and chitin is one of the most abundant substances on Earth.

Proteins: Amazing Polymers of Amino Acids

Proteins are an extraordinary group of macromolecules. Constructed of amino acids, proteins are fascinating because they have such intricate and diverse structures and because they perform so many different functions for cells. Among the cellular functions of macromolecules listed earlier, only energy storage and

(a) **Sugar phosphate**

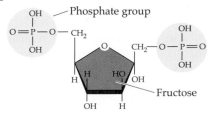

Fructose 1,6-bisphosphate is involved in the reactions that liberate energy from glucose. (The numbers in its name refer to the carbon sites of phosphate bonding; *bis-* indicates that two phosphates are present.)

(b) **Amino sugars**

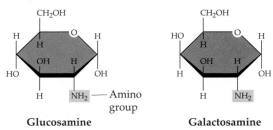

The monosaccharides glucosamine and galactosamine are amino sugars with an amino group in place of a hydroxyl group.

(c) **Chitin**

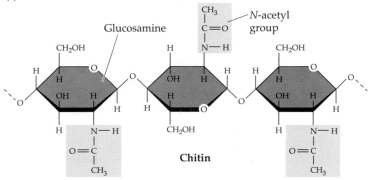

Chitin is a polymer of *N*-acetylglucosamine; *N*-acetyl groups provide additional sites for hydrogen bonding between the polymers.

3.14 Derivative Carbohydrates Added functional groups modify the form and properties of a carbohydrate.

heredity are not functions of proteins. Proteins are involved in structural support, protection, catalysis, transport, defense, regulation, and movement. Of particular importance are the catalytic proteins, called **enzymes**, that increase the rates of chemical reactions in cells. In general, each chemical reaction requires a different enzyme, because proteins show great specificity for the smaller molecules with which they will interact.

Proteins range in size from the small RNA-digesting enzyme *ribonuclease A*, which has a molecular weight of 5,733 and 51 amino acid residues, to gigantic molecules such as the cholesterol transport protein *apolipoprotein B,* which has a molecular weight of 513,000 and 4,636 amino acid residues. (The word "residue" is used for a monomer when it is part of a polymer.) Each of these proteins consists of one chain of amino acids folded into a specific three-dimensional shape that is required for protein function. Some proteins have more than one polymer chain. For example, the oxygen-carrying protein *hemoglobin* has four chains that are folded separately and associate together to make the functional protein. The largest protein complex known has more than 40 separate chains of amino acids.

All of these different proteins have a characteristic amino acid composition, but every protein contains neither all 20 amino acids nor an equal number of different amino acids. Nor is there a simple regular sequence in which the amino acids are linked. The diversity in amino acid content and sequence is the source of the diversity in protein structures and functions.

In some proteins, different kinds of chemical substances, called **prosthetic groups**, may be attached to the protein. These prosthetic groups include carbohydrates, lipids, phosphate groups, the iron-containing heme group, and metal ions such as copper and zinc. Whether they are small or large, and whether or not they have prosthetic groups, there is nothing "casual" about the structures of proteins. Proteins have specific three-dimensional shapes that are necessary for their specific functions.

To understand this stunning variety of functions, we must explore protein structure. First, we will examine the properties of the 20 amino acids and the characteristics of how they link to form proteins. Then we will systematically examine the four levels of protein structure and look at how a linear chain of amino acids is consistently folded into a compact three-dimensional shape.

Proteins are composed of amino acids

The 20 different amino acids commonly found in proteins show a wide variety of properties. In Chapter 2, we considered the structure of amino acids and identified four different groups attached to a central carbon atom: a hydrogen atom, an amino group, a carboxyl group, and a side chain, or R group (see Figure 2.24). Since amino acids are linked together by reactions between their amino and carboxyl groups, these groups are not exposed and do not give the protein its distinguishing properties and functional specificity—the side chains fill this role.

The **side chains** of amino acids are a protein's reactive groups and show a wide variety of chemical properties. Side chains control the function of a protein and contribute to its structure. The sequence of side chains determines how a protein folds into a three-dimensional shape (which we will discuss shortly). Although side chains are very important, they are often omitted from diagrams of proteins when the focus is on other aspects of structure. In these cases, the side chains are identified by the letter *R* (for "residue") and are sometimes called R groups. Side chains are highlighted in white in Table 3.1.

As Table 3.1 shows, one useful classification of amino acids is based on whether their side chains are electrically charged (+1, −1), polar (δ+, δ−), or nonpolar and hydrophobic. The five amino acids that have electrically charged side chains attract water and oppositely charged ions of all sorts. The four amino acids that have polar side chains tend to form weak hydrogen bonds with water and with other polar or charged substances. Eight amino acids have side chains that are nonpolar hydrocarbons or very slightly modified hydrocarbons. In the watery environment of the cell, the hydrophobic side chains may cluster together.

Three amino acids—cysteine, glycine, and proline—are special cases, although their side chains are generally hydrophobic. Two *cysteine* side chains, which have terminal —SH groups, can react to form a covalent bond in a **disulfide bridge** (—S—S—) (Figure 3.15). Hydrogen bonds and disulfide bridges help determine how a protein chain folds. When cysteine is not part of a disulfide bridge, its side chain is very hydrophobic. The *glycine* side chain consists of a single hydrogen atom; thus glycines may fit into tight corners in the interior of a protein molecule, where a larger side chain could not fit. *Proline* differs from other amino acids because it possesses a modified amino group (see Table 3.1).

Peptide linkages covalently bond amino acids together

When amino acids polymerize, the carboxyl group of one amino acid reacts with the amino group of another, undergoing a condensation reaction that forms a **peptide linkage**. Figure 3.16 gives a simplified description of the reaction. (In living cells, other molecules must activate the reactants in order for this reaction to proceed, and there are intermediate steps.) A linear polymer of amino acids connected by peptide linkages is a **polypeptide**. A **protein** is made up of one or more polypeptides.

At one end of the polypeptide is a free amino group. This end is the N terminus, named for the nitrogen atom in the amino group. At the other end of the polypeptide—the C terminus—is a free carboxyl group. The other amino and carboxyl groups are bound in peptide linkages. Thus a protein has direction. For example, the dipeptide glycine–alanine, in which glycine has the free amino group, differs from alanine–glycine, in which alanine has the free amino

TABLE 3.1 Twenty amino acids found in proteins

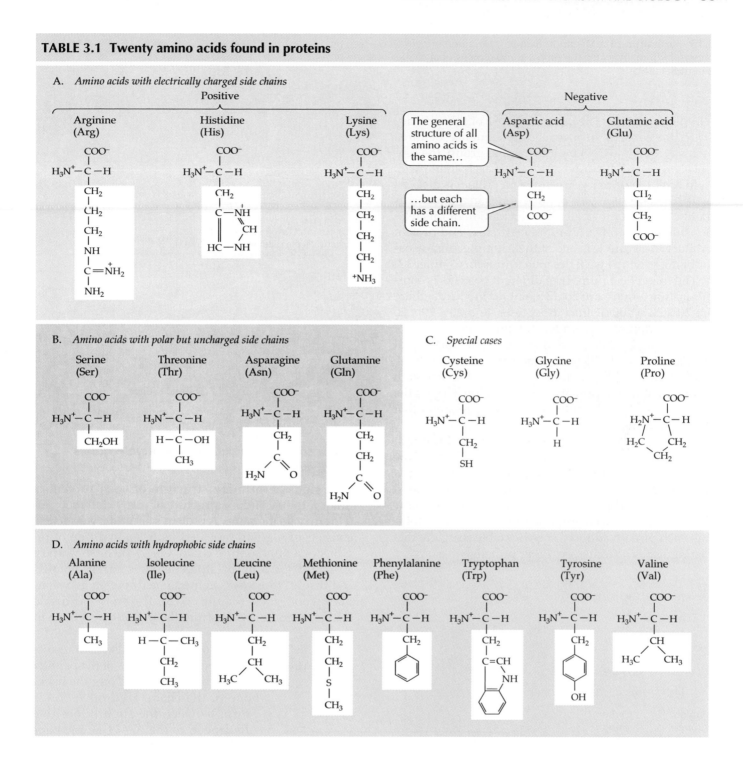

A. *Amino acids with electrically charged side chains*

B. *Amino acids with polar but uncharged side chains*

C. *Special cases*

D. *Amino acids with hydrophobic side chains*

group. In cells, the synthesis of polypeptide chains begins with the N terminus.

In the peptide linkage, the C=O oxygen carries a slight negative charge ($\delta-$), whereas the N—H hydrogen is slightly positive ($\delta+$). This asymmetry of charge favors hydrogen bonding (see Chapter 2) within the protein molecule itself and with other molecules, contributing to both the structure and the function of many proteins.

The primary structure of a protein is its amino acid sequence

Protein structure is elegant and complex—so complex that it is described as consisting of four different levels: *primary*, *secondary*, *tertiary*, and *quaternary* (Figure 3.17). However, proteins are always linear chains; there are no branches. The precise sequence of amino acids in a polypeptide constitutes the **primary structure** of a pro-

3.15 Formation of a Disulfide Bridge Disulfide bridges (—S—S—) are important in maintaining the proper three-dimensional shapes of some protein molecules.

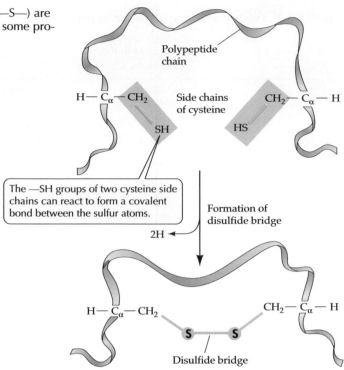

The —SH groups of two cysteine side chains can react to form a covalent bond between the sulfur atoms.

Formation of disulfide bridge

Disulfide bridge

tein (Figure 3.17a). The *peptide backbone* of this primary structure consists of a repeating sequence of three atoms (—N—C—C—) from the amino group, the central carbon, and the carboxyl group of each amino acid.

In cells, the primary structure of a protein is dictated by the precise sequence of nucleotides in a linear segment of a DNA molecule. The elucidation of this relationship between DNA primary structure and protein primary structure was one of the triumphs of molecular biology, which we'll describe further in Chapter 12.

The theoretical number of different proteins is enormous. Since there are 20 different amino acids, there are $20 \times 20 = 400$ distinct dipeptides, and $20 \times 20 \times 20 = 8,000$ different tripeptides. Imagine this process of multiplying by 20 extended to a protein made up of 100 amino acids (which is considered a small protein): There could be 20^{100} of these small proteins, each with its own distinctive primary structure. How large is the number 20^{100}? In the entire universe there aren't that many electrons!

At the higher levels of protein structure, local coiling and folding give the final functional shape of the molecule, but all of these levels derive from the primary structure. The different properties associated with a precise sequence of amino acids determine how the protein can twist and fold. By twisting and folding, each protein adopts a specific stable structure that distinguishes it from every other protein.

The secondary structure of a protein requires hydrogen bonding

Although the primary structure of each protein is unique, the secondary structure of many different proteins may be the same. A protein's **secondary structure** consists of regular, repeated patterns in different regions of a polypeptide chain. One type of secondary structure, the **α helix** (alpha helix), is a right-handed coil that is "threaded" in the same direction as a standard wood screw (Figure 3.17b). The amino acid side chains extend outward from the peptide backbone of the helix.

This helical structure of a polypeptide chain results from hydrogen bonds between elements of the peptide bonds that are distributed along the backbone of the chain. Hydrogen bonds form between the slightly positive hydrogen of the N—H of one peptide bond and the slightly negative oxygen of the C=O of another peptide bond (Figure 3.17b). When this pattern of hydrogen bonding is established repeatedly over a segment of the protein, it stabilizes the twisted form, resulting in an α helix. However, the ability of a protein to

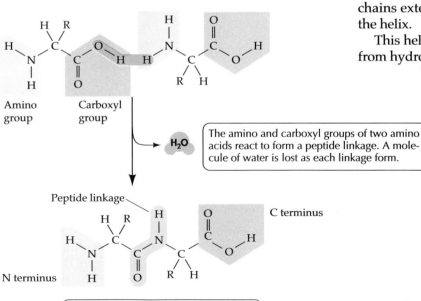

Amino group   Carboxyl group

The amino and carboxyl groups of two amino acids react to form a peptide linkage. A molecule of water is lost as each linkage form.

Peptide linkage   C terminus

N terminus

Repetition of this reaction links many amino acids together into a polypeptide.

3.16 Formation of Peptide Linkages In living things the reaction has many intermediate steps, but the reactants and products are the same as shown here.

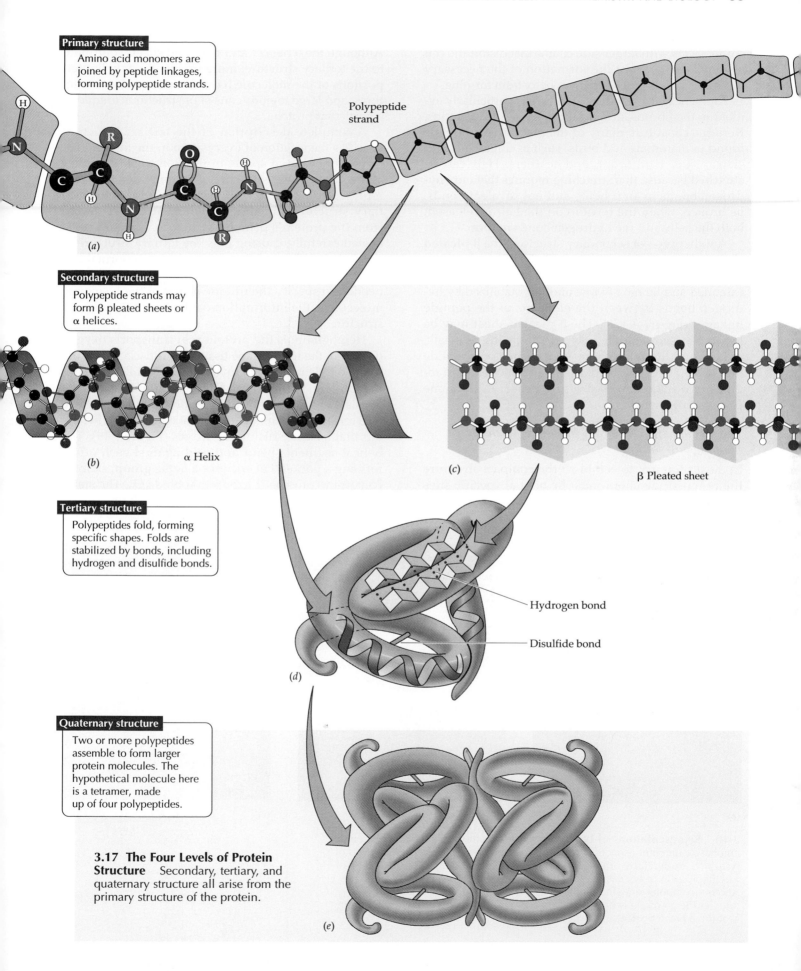

Primary structure
Amino acid monomers are joined by peptide linkages, forming polypeptide strands.

Polypeptide strand

(a)

Secondary structure
Polypeptide strands may form β pleated sheets or α helices.

(b)   α Helix

(c)   β Pleated sheet

Tertiary structure
Polypeptides fold, forming specific shapes. Folds are stabilized by bonds, including hydrogen and disulfide bonds.

Hydrogen bond

Disulfide bond

(d)

Quaternary structure
Two or more polypeptides assemble to form larger protein molecules. The hypothetical molecule here is a tetramer, made up of four polypeptides.

3.17 The Four Levels of Protein Structure   Secondary, tertiary, and quaternary structure all arise from the primary structure of the protein.

(e)

form an α helix depends on its primary structure. Amino acids with larger side chains that distort the coil or otherwise prevent the formation of the necessary hydrogen bonds will keep the α helix from forming.

Alpha helical secondary structure is particularly evident in the fibrous structural proteins called keratins. Keratins constitute many of the protective materials found in mammals and birds, such as fingernails and claws, skin, hair, wool, and feathers. Hair can be stretched because this stretching requires that only hydrogen bonds in an α helix, and not covalent bonds, be broken; when the tension on the hair is released, both the helix and the hydrogen bonds re-form.

Another type of secondary structure, the **β pleated sheet**, is found in the protein silk. In silk, rather than being coiled, the protein chains are almost completely extended and lie next to one another, stabilized by hydrogen bonds between the elements of the peptide linkages (Figure 3.17c). The β pleated sheet may be found between separate polypeptide chains, as in silk, or between different regions of the same polypeptide that is bent back on itself. Many enzymes contain regions of α helix and of β pleated sheet in the same polypeptide chain.

The tertiary structure of a protein is formed by bending and folding

In most proteins, to establish the compact structure the polypeptide chain must be bent at specific sites and folded back and forth. The resulting overall shape of a protein is its **tertiary structure** (Figure 3.17d). Although the α helices and β pleated sheets contribute to the tertiary structure, more frequently only limited portions of the molecule have these secondary structures, and large regions consist of structures unique to a particular protein.

A complete description of the tertiary structure specifies the location of every atom in the molecule in three-dimensional space, in relation to all the other atoms. The tertiary structure of the protein *lysozyme* is represented in Figure 3.18. Bear in mind that this tertiary structure and the secondary structure derive from the protein's primary structure. If lysozyme is heated carefully, causing only the tertiary structure to break down, the protein will return to its normal tertiary structure when it cools. The only information needed to specify the unique shape of the lysozyme molecule is the information contained in its primary structure.

Hemoglobin is the protein that transports oxygen (O_2) from the lungs to the tissues. In muscle tissue, hemoglobin delivers O_2 to myoglobin, a smaller but similar protein, for storage. The hemoglobin molecule consists of four similar polypeptide chains that have tertiary structures made up almost entirely of α helices but that lack any disulfide bridges (Figure 3.19). The helical segments bend and fold against each other, forming a pocket that encloses a **heme group**, an iron-containing prosthetic group that binds O_2. The stabilizing interaction of hydrophobic side chains on the

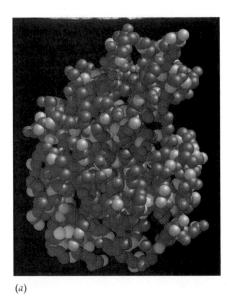

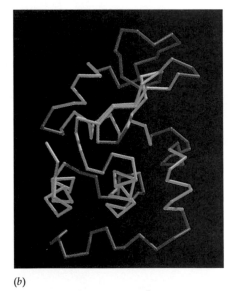

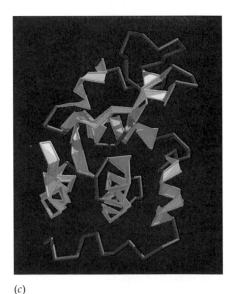

(a) (b) (c)

3.18 Representations of Lysozyme Different molecular representations emphasize different aspects of tertiary structure. These three representations of lysozyme are similarly oriented. (a) This computer drawing gives the most realistic impression of lysozyme's tertiary structure, which is densely packed. (b) Another computer drawing emphasizes the backbone of the folded polypeptide. (c) The green coils here represent the α helices, and orange arrows represent the β pleated sheet.

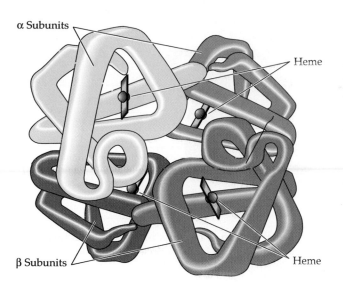

3.19 Quaternary Structure of a Protein Hemoglobin consists of four folded polypeptide subunits that assemble themselves into the quaternary structure shown here. In these two graphic representations, each subunit is a different color. The heme groups contain iron and are the oxygen-carrying sites.

inner sides of the α helices helps ensure that the helices fold against one another correctly as each polypeptide assumes its tertiary structure.

The quaternary structure of a protein consists of subunits

As we mentioned earlier, some proteins have two or more polypeptide chains folded into their own unique tertiary structures. Each of these folded polypeptides is considered a protein subunit. **Quaternary structure** results from the ways in which these protein subunits fit together and interact; it can be illustrated by hemoglobin (see Figure 3.19). Hydrophobic interactions, hydrogen bonds, and ionic bonds all help hold the four subunits together to form the functional hemoglobin molecule. As the hemoglobin molecule takes up one O_2 molecule, the four subunits shift their relative positions slightly, changing the quaternary structure. Ionic bonds are broken, exposing buried side chains that enhance the binding of additional O_2 molecules.

Each subunit of hemoglobin is folded like a myoglobin molecule, suggesting that both hemoglobin and myoglobin are evolutionary descendants of the same oxygen-binding ancestral protein. But on the surfaces where its subunits come in contact with each other—regions that on myoglobin are exposed to aqueous surroundings and are hydrophilic—hemoglobin has hydrophobic side chains. Again, the chemical nature of side chains on individual amino acids determines how the molecule folds and packs in three dimensions.

Molecular chaperones help shape proteins

The primary structure of a protein constrains the secondary, tertiary, and quaternary structures (if subunits exist). By determining the primary structure, DNA also determines the higher levels of structure. However, other factors also affect the tertiary structure that is required for proper protein function.

Elevated temperatures, pH changes, or altered salt concentrations can cause a protein to adopt a different, biologically inactive tertiary structure. Biological function depends on a specific three-dimensional structure. Increased temperature causes more rapid molecular movement and thus can break weak hydrogen bonds and hydrophobic interactions. Altered pH can change the pattern of ionization of carboxyl and amino groups in the side chains of amino acids, thus disrupting the pattern of ionic attractions and repulsions that contribute to normal tertiary structure.

The loss of appropriate tertiary structure is called **denaturation**, and it is always accompanied by a loss of the normal biological function of the protein (Figure 3.20). Denaturation can be caused by heat or high concentrations of polar substances such as urea that disrupt the hydrogen bonding that is crucial to protein structure. Nonpolar solvents may also disrupt normal structure. Usually denaturation is irreversible, particularly if many denatured proteins interact in random nonspecific ways. However, in some cases denaturation is reversible, and upon return to normal environmental conditions, the protein may return to its active form, as does lysozyme (see Figure 3.18). This return is called **renaturation**.

How can such a complicated molecule spontaneously fold into its normal tertiary structure? Biologists are just beginning to be able to predict how a protein will fold, given its primary structure. The study of protein folding has a long way to go. The sit-

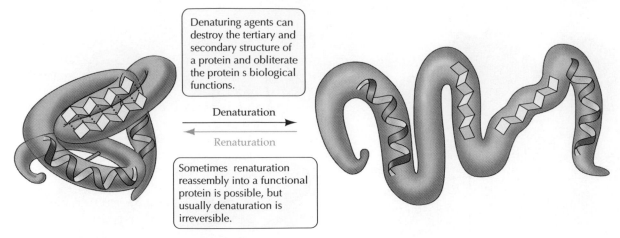

Denaturing agents can destroy the tertiary and secondary structure of a protein and obliterate the protein s biological functions.

Denaturation →

← Renaturation

Sometimes renaturation reassembly into a functional protein is possible, but usually denaturation is irreversible.

3.20 Denaturation Is the Loss of Protein Structure and Function Agents that can cause denaturation include high temperatures and certain chemicals.

uation in the living cell is complex—even more complex than we first imagined. The protein doesn't form all at once. In a human cell it may take minutes from the beginning of synthesis until the entire protein squeezes into the compartment where it folds—and things are very crowded in that compartment. There, other proteins present inappropriate potential partners for interaction (in the form of amino acid side chains).

A special group of proteins, called the **chaperone proteins**, at least in part help prevent newly formed proteins from reacting inappropriately. The chaperones attach to a new protein while it is forming and protect it from interactions with other proteins until the new molecule has its correct shape. Most chaperone proteins can act on many different forming proteins. Some act on only a few, and others may act on only a single, specific protein.

What if the chaperone proteins fail to do their job, or if for other reasons a protein folds abnormally? Evidence suggests that abnormal protein folding underlies certain infectious diseases, including mad cow disease and, in humans, Creutzfeldt–Jakob disease. Alzheimer's disease may also result from protein misfolding.

Nucleic Acids: Informational Macromolecules

The nucleic acids are linear polymers specialized for the storage, transmission, and use of information. There are two types of nucleic acids: DNA (deoxyribonucleic acid) and RNA (ribonucleic acid). **DNA** molecules are giant polymers that encode hereditary information and pass it from generation to generation (through reproduction). The information encoded in

DNA is also used to make specific proteins through the intermediate RNA. **RNA** molecules of various types copy the information in segments of DNA to specify the sequence of amino acids in proteins. Information flows from DNA to DNA in reproduction. But for nonreproductive activities of the cell, information flows from DNA to RNA to proteins, which ultimately carry out these functions. What compositions, structures, and properties of nucleic acids permit them to play these fundamental roles in living systems?

The nucleic acids have characteristic structures and properties

Nucleic acids are composed of monomers called **nucleotides**, each of which consists of a pentose sugar, a phosphate group, and a nitrogen-containing base—either a pyrimidine or a purine (Figure 3.21). Molecules consisting of a pentose sugar and a nitrogenous base, but no phosphate group, are called nucleo*sides*. In DNA, the pentose sugar is deoxyribose, which differs from the ribose found in RNA by only one oxygen atom (see Figure 3.11).

In both RNA and DNA, the backbone of the molecule consists of alternating sugars and phosphates (—sugar—phosphate—sugar—phosphate—). Bases are attached to the sugars and project from the chain (Figure 3.22). The nucleotides are joined by covalent bonds in what are called **phosphodiester linkages** between the sugar of one nucleotide and the phosphate of the next ("-diester" refers to the two bonds formed by reacting —OH groups with acidic phosphate groups). The phosphate groups link carbon 3' in one ribose to carbon 5' in the adjacent ribose.

Most RNA molecules are *single-stranded*, consisting of only one polynucleotide chain. DNA, however, is usually *double-stranded*; it has two polynucleotide chains held together by hydrogen bonding between their complementary nitrogenous bases. The two strands of DNA run in opposite directions. You can see what this means by drawing an arrow through the

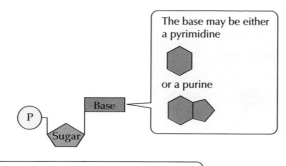

The base may be either a pyrimidine

or a purine

Base

P

Sugar

A nucleotide consists of a phosphate, a pentose sugar, and a nitrogen-containing base.

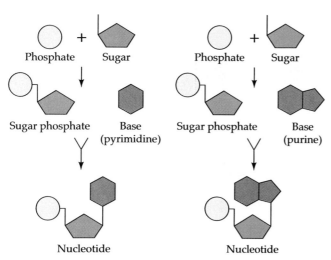

Phosphate + Sugar

Phosphate + Sugar

Sugar phosphate   Base (pyrimidine)

Sugar phosphate   Base (purine)

Nucleotide

Nucleotide

3.21 Nucleotides Have Three Components A nucleotide consists of a phosphate group, a pentose sugar, and a nitrogen-containing base—all linked together by covalent bonds. The nitrogenous bases fall into two categories: Purines have two fused rings, and the smaller pyrimidines have a single ring.

phosphate group from carbon 3′ to carbon 5′ in the next ribose. If you do this for both strands, the arrows point in opposite directions. This antiparallel orientation is necessary for the strands to fit together.

The uniqueness of a nucleic acid resides in its base sequence

Only four nitrogenous bases—and thus only four nucleotides—are found in DNA. The DNA bases and their abbreviations are: adenine (A), cytosine (C), guanine (G), and thymine (T). A key to understanding the structures and functions of nucleic acids is the principle of **complementary base pairing** through hydrogen bond formation. In double-stranded DNA, adenine and thymine always pair (AT), and cytosine and guanine always pair (CG).

Base pairing is complementary because of two factors: the corresponding sites for hydrogen bonding and the molecular sizes of the paired bases. Adenine

and guanine are both purines consisting of two fused rings. Thymine and cytosine are both pyrimidines consisting of only one ring. The pairing of a large purine with a small pyrimidine ensures a stable and consistent dimension to the double-stranded molecule of DNA. In Chapter 11, we'll discuss in more detail how complementary strands of DNA separate and are faithfully copied.

Ribonucleic acids also have four different monomers, but the nucleotides differ from those of DNA. In RNA the nucleotides are termed **ribonucleotides.** They contain ribose rather than deoxyribose, and instead of the base thymine, RNA uses the base uracil (Table 3.2). The other three bases are the same as in DNA.

Although RNA is generally single-stranded, complementary hydrogen bonding between ribonucleotides can take place. These bonds play important roles in determining the shapes of some RNA molecules and in the associations between RNA molecules during protein synthesis. During the DNA-directed synthesis of RNA, complementary base pairing also takes place between ribonucleotides and the bases of DNA. In RNA, guanine and cytosine pair as in DNA, but adenine pairs with uracil. Adenine in an RNA strand can pair either with uracil (in another RNA strand) or with thymine (in a DNA strand).

The three-dimensional appearance of DNA is strikingly regular. The segment shown in Figure 3.23 could be from any DNA molecule. Through hydrogen bonding, the two complementary polynucleotide strands pair and twist to form a **double helix.** Compared to the complex and varied tertiary structures of different proteins, this formation seems amazingly regular. But this structural contrast makes sense in terms of the functions of these two classes of macromolecules.

DNA is a purely informational molecule. *The information in DNA is encoded in the sequence of bases carried in its chains.* This sequence is, in a sense, like the tape of a tape recorder. The message must be read easily and reliably, in a specific order. A uniform molecule like DNA can be interpreted by standard molecular

TABLE 3.2	Distinguishing RNA from DNA	
NUCLEIC ACID	**SUGAR**	**BASES**
RNA	Ribose	Adenine
		Cytosine
		Guanine
		Uracil
DNA	Deoxyribose	Adenine
		Cytosine
		Guanine
		Thymine

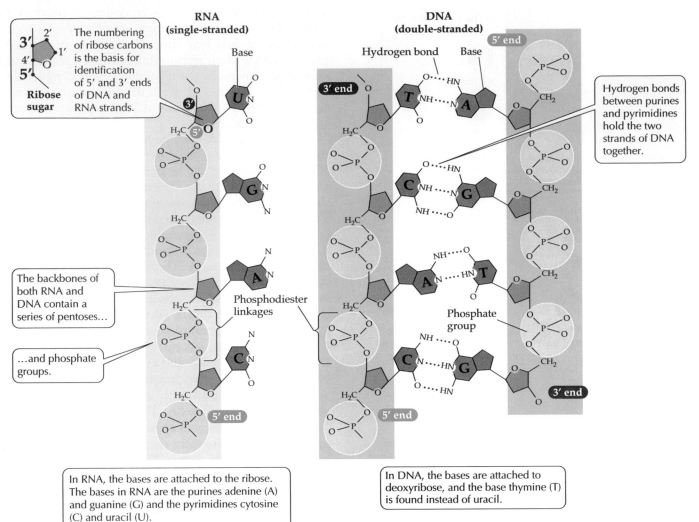

RNA (single-stranded)

The numbering of ribose carbons is the basis for identification of 5' and 3' ends of DNA and RNA strands.

Ribose sugar

Base

The backbones of both RNA and DNA contain a series of pentoses...

Phosphodiester linkages

...and phosphate groups.

In RNA, the bases are attached to the ribose. The bases in RNA are the purines adenine (A) and guanine (G) and the pyrimidines cytosine (C) and uracil (U).

DNA (double-stranded)

Hydrogen bond

Base

Hydrogen bonds between purines and pyrimidines hold the two strands of DNA together.

Phosphate group

In DNA, the bases are attached to deoxyribose, and the base thymine (T) is found instead of uracil.

3.22 Common and Distinguishing Characteristics of DNA and RNA RNA is usually a single strand. DNA usually consists of two strands running in opposite directions.

machinery, and any cell's machinery can read any molecule of DNA—just as a tape player can play any tape of the right size.

Proteins, on the other hand, have good reason to be so varied. In particular, different enzymes must recognize their own specific "target" molecules. They do this by having a unique three-dimensional form that can match at least a portion of the surface of their target molecules. In other words, structural diversity in the molecules with which enzymes react requires corresponding diversity in the structure of the enzymes themselves. *In DNA the information is in the sequence of bases; in proteins the information is in the shape.*

DNA is a guide to evolutionary relationships

Because DNA carries hereditary information between generations, a series of DNA molecules with changes in base sequences stretches back through time. Of course, we cannot study all of these DNA molecules, because many of their organisms have become extinct. However, we can study the DNA of living organisms, some of which are judged to belong to lineages that are more ancient than those of other living forms. Comparisons and contrasts of these DNAs add to the evolutionary record.

Closely related living species should have more similar base sequences than species judged by other criteria to be more distantly related. Indeed this is the case. The examination of base sequences confirms many of the evolutionary relationships that have been inferred from the study of microscopic or macroscopic structures or studies of biochemistry and physiology. For example, the closest living relative of humans (*Homo sapiens*) is the chimpanzee (genus *Pan*), which shares more than 98 percent of its DNA base sequence with human DNA.

Confirmation of well-established evolutionary relationships gives credibility to the use of DNA to elucidate relationships when studies of structure are not

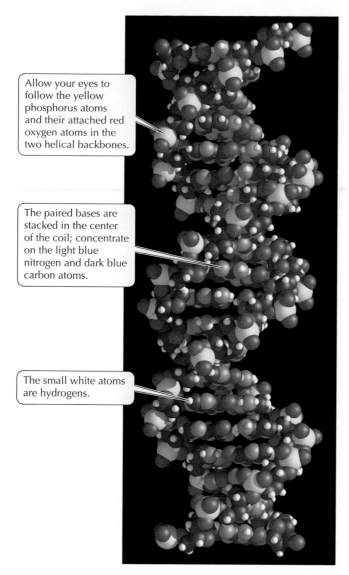

> Allow your eyes to follow the yellow phosphorus atoms and their attached red oxygen atoms in the two helical backbones.

> The paired bases are stacked in the center of the coil; concentrate on the light blue nitrogen and dark blue carbon atoms.

> The small white atoms are hydrogens.

3.23 The Double Helix of DNA The backbones of the two strands in a DNA molecule are coiled in a double helix.

possible or are not conclusive. For example, DNA studies revealed a close evolutionary relationship between starlings and mockingbirds that was not expected on the basis of studies of anatomy or behavior. DNA studies support the division of the prokaryotes into two kingdoms (Eubacteria and Archaebacteria). Each of these two groups of prokaryotes is as distinct from the other as either is from the other four kingdoms into which living things are classified. In addition, DNA comparisons support the hypothesis that certain subcellular compartments of eukaryotes (the organelles called mitochondria and chloroplasts) evolved from early bacteria that established a stable and mutually beneficial way of life inside larger cells.

The Interactions of Macromolecules

We have been treating the classes of macromolecules as if each were completely separate from the others. In cells, however, certain macromolecules of different classes may be covalently bonded to one another. Proteins with attached oligosaccharides are called *glycoproteins* (*glyco-*, "sugar"). The specific oligosaccharide chain determines the placement of a glycoprotein within the cell.

Other carbohydrate chains bind to lipids, resulting in *glycolipids*, which reside in the cell surface membrane, with the carbohydrate chain extending out into the cell's environment. In humans, the substances that determine the ABO blood types are carbohydrates attached to either proteins or lipids. When a human body cell becomes cancerous, the glycolipids and glycoproteins on the cell surface are modified, and these changes serve as a recognition signal for the body defenses to destroy the abnormal cell.

Not all the associations among macromolecules or between a macromolecule and a smaller molecule are covalent. Weaker linkages such as ionic bonds, hydrogen bonds, and even van der Waals forces may also be involved. Enzymes bind to their reactants using a variety of weak bonds. Ionic bonds link proteins to DNA, forming *nucleoproteins* that regulate the activities of DNA. Still other proteins, in combination with cholesterol and other lipids, form *lipoproteins*. The lipoproteins make it possible to move very hydrophobic lipids through water-rich environments such as the blood and tissue fluid and deliver cholesterol to all the body's cells.

Summary of "Macromolecules: Their Chemistry and Biology"

Lipids: Water-Insoluble Molecules
• Although lipids can form gigantic structures such as lipid droplets and membranes, these aggregations are not considered macromolecules because they are not linked by covalent bonds.
• Saturated fatty acids have a reactive carboxyl group and a nonpolar hydrocarbon chain with no double bonds. The hydrocarbon chains of unsaturated fatty acids have one or more double bonds that bend the chain, making close packing less possible. **Review Figure 3.3**
• In certain cells, deposits of fat or oil store energy. Fats and oils are composed of three fatty acids covalently bonded to a glycerol molecule by ester linkages. **Review Figure 3.4**
• Phospholipids form the core of biological membranes. Phospholipids have a hydrophobic "tail" and a hydrophilic "head." **Review Figures 3.2, 3.6**

• In water, the interactions of the hydrophobic tails and hydrophilic heads in membranes generate a phospholipid bilayer that is two molecules thick. The head groups are directed outward, where they interact with the surrounding water. The tails are packed together in the interior of the bilayer. **Review Figure 3.7**

• Carotenoids trap light energy in green plants. β-Carotene can be split to form vitamin A, a lipid vitamin. The other lipid vitamins are D, E, and K. **Review Figure 3.8**

• Vitamins are substances that are required for normal functioning but that must be acquired from the diet.

• Some steroids, such as testosterone, function as hormones. Cholesterol is synthesized by the liver and has a role in some cell membranes, as well as in the digestion of other fats. Too much cholesterol in the diet can lead to arteriosclerosis. **Review Figure 3.9**

Macromolecules: Giant Polymers

• Macromolecules are constructed by the formation of covalent bonds between smaller molecules called monomers. Macromolecules include polysaccharides, proteins, and nucleic acids.

• Macromolecules have specific, characteristic three-dimensional shapes that depend on the structures, properties, and sequence of their monomers. Different functional groups give local sites on macromolecules specific properties that are important for their biological functioning and interactions with other macromolecules.

• Monomers join by condensation reactions, which lose a molecule of water for each bond formed. Hydrolysis reactions use water to digest polymers into monomers.

Carbohydrates: Sugars and Sugar Polymers

• All carbohydrates approximate multiples of the general formula CH_2O.

• Hexoses contain six carbon atoms. Examples of hexoses include glucose, galactose, and fructose, which can exist as chains or rings. **Review Figures 3.10, 3.11**

• The pentoses are five-carbon monosaccharides; two pentoses, ribose and deoxyribose, are components of the nucleic acids RNA and DNA, respectively. **Review Figure 3.11**

• Glycosidic linkages may have either α or β orientation in space. They covalently link monosaccharides into larger units such as disaccharides (for example, sucrose, lactose, maltose, and cellobiose), oligosaccharides, and polysaccharides. **Review Figures 3.12, 3.13**

• Cellulose, a very stable glucose polymer, is the principal component of the cell walls of plants. It is formed by glucose units linked together by β-glycosidic linkages between carbons 1 and 4. **Review Figure 3.13**

• Starches, less dense and less stable than cellulose, store energy in plants. Starches are formed by α-glycosidic linkages between carbons 1 and 4 and are distinguished by the amount of branching that occurs through glycosidic bond formation at carbon 6. **Review Figure 3.13**

• Glycogen contains α-1,4 glycosidic linkages and is highly branched. Glycogen stores energy in animal livers and muscles. **Review Figure 3.13**

• Derivative monosaccharides include the sugar phosphates and amino sugars. One such derivative, N-acetylglucosamine, polymerizes to form the polysaccharide chitin, which is found in the cell walls of fungi and the exoskeletons of insects. **Review Figure 3.14**

Proteins: Amazing Polymers of Amino Acids

• The functions of proteins include support, protection, catalysis, transport, defense, regulation, and movement. Protein function sometimes requires an attached prosthetic group, such as the heme group.

• There are 20 amino acids found in proteins. Each amino acid consists of an amino group, a carboxyl group, a hydrogen, and a side chain bonded to a central carbon atom. **Review Table 3.1 and Figure 2.24**

• The side chains of amino acids may be charged, polar, or hydrophobic; there are also "special cases," such as the —SH groups that can form disulfide bridges. The side chains give different properties to each of the amino acids. **Review Table 3.1 and Figure 3.15**

• Amino acids are covalently bonded together by peptide linkages that form by condensation reactions between the carboxyl and amino groups. **Review Figure 3.16**

• The polypeptide chains of proteins are folded into specific three-dimensional shapes. Four levels of structure are possible: primary, secondary, tertiary, and quaternary.

• The primary structure of a protein is the sequence of amino acids bonded by peptide linkages. This primary structure determines both the higher levels of structure and protein function. **Review Figure 3.17a**

• Secondary structures of proteins, such as α helices and β pleated sheets, are maintained by hydrogen bonding between atoms of the peptide linkages. **Review Figure 3.17b**

• The tertiary structure of a protein is generated by bending and folding of the polypeptide. **Review Figures 3.17d, 3.18**

• The quaternary structure of a protein is the arrangement of polypeptides in a single functional unit consisting of more than one polypeptide subunit. **Review Figures 3.17e, 3.19**

• Molecular chaperones assist protein folding by preventing random, nonspecific interactions that have no biological function.

• Proteins denatured by heat, acid, or reagents lose tertiary and secondary structure as well as biological function. Renaturation is not always possible. **Review Figure 3.20**

Nucleic Acids: Informational Macromolecules

• In cells, DNA is the hereditary material. Both DNA and RNA play roles in the formation of proteins. Information flows from DNA to RNA to protein.

• Nucleic acids are polymers of nucleotides that consist of a phosphate group, a sugar (ribose in RNA and deoxyribose in DNA), and a nitrogen-containing base. In DNA the bases are adenine, guanine, cytosine, and thymine, but in RNA uracil substitutes for thymine. **Review Figure 3.21 and Table 3.2**

• In the nucleic acids, the bases extend from a sugar–phosphate backbone. The information content of DNA and RNA resides in their base sequences.

• RNA is single-stranded. DNA is a double-stranded helix in which there is complementary, hydrogen-bonded base pairing between adenine and thymine (AT) and guanine and cytosine (GC). The two strands of the DNA double helix run in opposite directions. **Review Figures 3.22, 3.23**

• Comparing of DNA base sequences of different living species provides information on their evolutionary relatedness.

The Interactions of Macromolecules

• Both covalent and noncovalent linkages are found between the various classes of macromolecules.

• Glycoproteins contain an oligosaccharide "label" that directs the protein to the proper cell destination. The carbohydrate groups of glycolipids are displayed on the cell's outer surface, where they serve as recognition signals.

• Weaker forces bind a catalytic protein (enzyme) to a reactant. Hydrophobic interactions bind cholesterol to the protein that transports it in the blood.

Self-Quiz

1. All lipids
   a. are triglycerides.
   b. are polar.
   c. are hydrophilic.
   d. are polymers.
   e. are more soluble in nonpolar solvents than in water.

2. Which of the following is *not* a lipid?
   a. A steroid
   b. A fat
   c. A triglyceride
   d. A biological membrane
   e. A carotenoid

3. All carbohydrates
   a. are polymers.
   b. are simple sugars.
   c. consist of one or more simple sugars.
   d. are found in biological membranes.
   e. are more soluble in nonpolar solvents than in water.

4. Which of the following is *not* a carbohydrate?
   a. Glucose
   b. Starch
   c. Cellulose
   d. Hemoglobin
   e. Deoxyribose

5. All proteins
   a. are enzymes.
   b. consist of one or more polypeptides.
   c. are amino acids.
   d. have quaternary structures.
   e. are more soluble in nonpolar solvents than in water.

6. Which of the following statements about the primary structure of a protein is *not* true?
   a. It may be branched.
   b. It is determined by the structure of the corresponding DNA.
   c. It is unique to that protein.
   d. It determines the tertiary structure of the protein.
   e. It is the sequence of amino acids in the protein.

7. The amino acid leucine (see Table 3.1)
   a. is found in all proteins.
   b. cannot form peptide linkages.
   c. is likely to appear in the part of a membrane protein that lies within the phospholipid bilayer.
   d. is likely to appear in the part of a membrane protein that lies outside the phospholipid bilayer.
   e. is identical to the amino acid lysine.

8. The quaternary structure of a protein
   a. consists of four subunits—hence the name *quaternary*.
   b. is unrelated to the function of the protein.
   c. may be either α or β.
   d. depends on covalent bonding among the subunits.
   e. depends on the primary structures of the subunits.

9. All nucleic acids
   a. are polymers of nucleotides.
   b. are polymers of amino acids.
   c. are double-stranded.
   d. are double-helical.
   e. contain deoxyribose.

10. Which of the following statements about condensation reactions is *not* true?
    a. Protein synthesis results from them.
    b. Polysaccharide synthesis results from them.
    c. Nucleic acid synthesis results from them.
    d. They consume water as a reactant.
    e. Different condensation reactions produce different kinds of macromolecules.

Applying Concepts

1. Phospholipids make up a major part of every biological membrane; cellulose is the major constituent of the cell walls of plants. How do the chemical structures and physical properties of phospholipids and cellulose relate to their functions in cells?

2. Suppose that, in a given protein, one lysine is replaced by aspartic acid (see Table 3.1). Does this change occur in the primary structure or in the secondary structure? How might it result in a change in tertiary structure? In quaternary structure?

3. If there are 20 different amino acids commonly found in proteins, how many different dipeptides are there? How many different tripeptides? How many different trinucleotides? How many different single-stranded RNAs composed of 200 nucleotides?

4. Contrast the following three structures: hemoglobin, a DNA molecule, and a protein that spans a biological membrane.

Readings

Doolittle, R. F. 1985. "Proteins." *Scientific American*, October. A strikingly illustrated treatment of protein structure and evolution.

Horgan, J. 1993. "Stubbornly Ahead of His Time." *Scientific American*, March. About Linus Pauling who developed the theory of covalent bonding, discovered α-helical protein structure, and won Nobel prizes for chemistry and peace.

Lehninger, A. L., D. L. Nelson and M. M. Cox. 1993. *Principles of Biochemistry*, 2nd Edition. Worth, New York. A balanced textbook that uses energetic, evolutionary, regulatory, and structure-function themes to place biochemistry in physical, chemical, and biological contexts.

Stryer, L. 1995. *Biochemistry*, 4th Edition. W. H. Freeman, New York. A relatively advanced but beautiful reference on the subjects of this chapter; outstanding illustrations, concise descriptions, clear prose.

Taubes, G. 1996. "Misfolding the Way to Disease." *Science*, vol. 271, pages 1493–1495. A brief overview of protein misfolding, with good references.

Chapter 4

The Organization of Cells

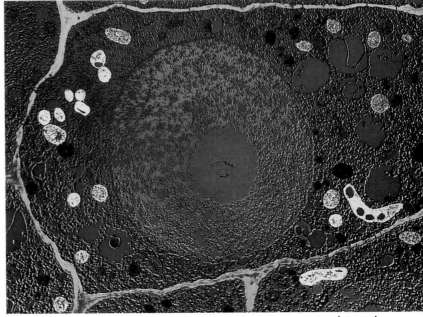

Efficient Organization at the Cellular Level
This cell in the tip of a corn root is surrounded by a firm wall. The circular nucleus is separated from the rest of the cell by membranes, and other membranes enclose other functional parts of the cell.

2 μm

Y ou enter a room on campus and are greeted by a tuba player practicing full blast, workers from a dining hall washing dishes, a professor lecturing earnestly, a karate class in full kick, and six students holding their ears and trying to study while another group plans a ski trip. Is this any way to run a campus?

Campuses are organized so that their many functions can proceed efficiently. They have special rooms for special purposes and different employees to provide different services. Without these divisions, there would be chaos. Like the campus, cells are also divided into special structures that carry out special functions. A cell is not simply a bag of enzymes and other molecules; it is a highly ordered, efficient workplace.

In this chapter, we will examine the general structure of cells. We'll start with the distinctions between *prokaryotic cells* and *eukaryotic cells* and describe the microscopic methods that reveal their structure. Then we will turn to a detailed examination of the components of eukaryotic cells that are involved in information storage and processing, energy transformations, and support and protection. After contrasting cells to *viruses*, we'll close the chapter with a description of the experimental techniques by which cells are taken apart and their components isolated for study. Such chemical studies complement the structural knowledge from microscopy and have revealed much about how living cells function.

64

The Cell: The Basic Unit of Life

All organisms are composed of cells, and all cells come from preexisting cells. These two statements constitute the **cell theory**. Cells show the characteristics by which living systems are recognized: They use DNA as hereditary material and proteins as catalysts. In addition, most cells reproduce, transform matter and energy, and respond to their environment.

Viruses, on the other hand, show only a few of these characteristics and thus are not usually considered alive. Viruses depend entirely on living cells to reproduce, and they neither transform matter and energy nor respond to the environment. Some cells can develop into entire multicellular organisms, but no *part* of a cell can do that—nor can viruses.

Most cells are tiny. The volume of cells ranges from 1 to 1,000 μm³ (Figure 4.1). The eggs of some birds are enormous exceptions, to be sure, and individual cells of several types of algae are large enough to be viewed with the unaided eye. And although neurons (nerve cells) have a volume that is within the "normal cell" range, they often have fine projections that may extend for meters, carrying signals from one part of a large animal to another. But by and large, cells are minuscule. Why?

Why are cells so small?

Whether they are individual unicellular organisms or parts of multicellular organisms consisting of millions and millions of cells, cells are small. Why can't a single cell be the size of a human body? The answer relates to the change in the surface area-to-volume ratio (*SA/V*) of any object as it increases in size.

As a cell increases in volume, its surface area increases also, but not to the same extent. Consider the example illustrated in Figure 4.2. The surface area of a cube with sides measuring 4 mm is 96 mm² ($SA = 6 \times 4^2$ mm²), and its volume is 64 mm³ ($V = 4^3$ mm³). However, if the same volume—64 mm³—is contained in 64 smaller cubes, each of which has sides measuring 1 mm, the surface area is 384 mm² ($SA = 64 \times 6 \times 1^2$ mm²). In other words, although the total volume remains unchanged, the surface area has increased fourfold (384 = 4 × 96). And the surface area-to-volume ratio has increased from 1.5/1 to 6/1. This phenomenon has great biological significance.

The volume of a cell is related to the amount of chemical activity the cell carries out per unit time. However, the surface area of the cell limits the exchange of nutrients and waste products with its environment. As the living cell grows larger, its rate of pro-

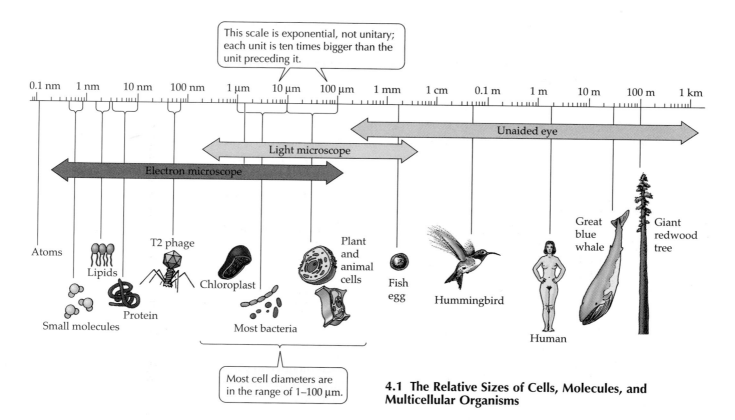

4.1 The Relative Sizes of Cells, Molecules, and Multicellular Organisms

4.2 The Ratio of Surface Area to Volume Explains Why Cells Are So Small Although they contain equal volumes, a cube measuring 4 mm on an edge has a smaller surface area than the total surface area of 64 cubes each measuring 1 mm on an edge. In cells, a large surface-area-to-volume ratio is needed for adequate exchange of nutrients and wastes with the environment.

	One 4-mm cube	Eight 2-mm cubes	Sixty-four 1-mm cubes
Surface area (mm^2)	96	192	384
Volume (mm^3)	64	64	64
Surface area-to-volume ratio	1.5/1	3/1	6/1

duction of wastes and its need for resources increase faster than the surface area through which the cell must obtain resources and excrete wastes. The more limited increase in surface area restricts the increase in volume as the cell grows—this explains why large organisms must consist of many small cells. *Cells are small in volume and thus maintain a large surface-area-to-volume ratio.*

In a multicellular organism, the large surface area represented by the multitude of small cells that make up the whole organism enables the myriad functions required for survival of the organism. Special structures transport food, oxygen, and waste materials to and from the small cells that are distant from the external surface of the organism.

Cells show two organizational patterns: Prokaryotic and eukaryotic

Cells must carry out many functions in order to survive. They must obtain and process energy; they must convert the genetic information of DNA into protein; and they must keep certain biochemical reactions separate from other reactions—incompatible reactions that must occur simultaneously. Specific structures within the cell—organized into two general patterns: *prokaryotic* and *eukaryotic*—perform these functions.

Prokaryotic cell organization is characteristic of the kingdoms Eubacteria (Domain bacteria) and Archaebacteria (Domain Archaea). Organisms in these kingdoms are called *prokaryotes.* Their cells do not have membrane-enclosed internal compartments.

Eukaryotic cell organization is found in the other four kingdoms of living things (Protista, Plantae, Fungi, and Animalia, which comprise the domain Eukarya). The DNA of eukaryotic cells is contained in a special membrane-enclosed compartment called the nucleus. Eukaryotic cells also contain other membrane-enclosed compartments in which specific chem-

ical reactions take place. These compartments isolate certain molecules and chemical reactions from other molecules and reactions. Organisms with this type of cell are known as *eukaryotes*.

Both prokaryotes and eukaryotes have prospered through many hundreds of millions of years of evolution, and both are great success stories. Let's look first at prokaryotic cells.

Prokaryotic Cells

Prokaryotes are very diverse in their metabolic capabilities. They can live off a greater diversity of energy sources than any other living creatures, and they inhabit greater environmental extremes. The vast diversity within the prokaryotic kingdoms is the subject of Chapter 25. Some archaea are found in sulfurous hot springs at temperatures that would denature the proteins of most other life-forms. Some bacteria are photosynthetic and use light energy to synthesize needed materials from CO_2. Other prokaryotes are able to oxidize inorganic ions to obtain energy for the synthesis of cell-specific materials from raw materials.

Prokaryotic cells are generally smaller than eukaryotic cells, ranging in dimensions from 0.25×1.2 μm to 1.5×4 μm (see Figure 4.1). Although each prokaryote is a single cell, many types of prokaryotes are usually seen in chains, small clusters, or even colonies containing hundreds of individuals.

In this section, we will first consider the features that cells in the kingdoms Eubacteria and Archaebacteria have in common. Then we will examine structural features that are found in some, but not all, prokaryotes.

All prokaryotic cells share certain features

All prokaryotic cells have a plasma membrane, a nucleoid, and cytoplasm filled with ribosomes. The

plasma membrane encloses the cell, separating it from its environment and regulating the traffic of materials into and out of the cell. The **nucleoid** is a relatively clear area (as seen under the electron microscope) that contains the hereditary material (DNA) of the cell. The rest of the material within the plasma membrane—called the **cytoplasm**—is made up of two parts: the cytosol and insoluble suspended particles.

The **cytosol** consists mostly of water that contains dissolved ions, small molecules, and soluble macromolecules such as enzymes. The insoluble suspended particles are revealed at high magnification. For example, high magnification shows that the cytoplasm contains many minute, roughly spherical structures called **ribosomes**. Although the ribosomes seem small in comparison to the cell in which they are contained, on a macromolecular scale they are gigantic structures assembled from both RNA and proteins. The ribosomes coordinate the synthesis of proteins.

Although structurally less complicated than eukaryotic cells, prokaryotic cells are functionally complex. The enzymes in prokaryotic cells catalyze thousands of chemical reactions. In addition to making these thousands of enzymes, prokaryotic cells are capable of shutting off the synthesis of enzymes that are not needed in a particularly rich nutrient environment.

Some prokaryotic cells have specialized features

Many prokaryotic cells have at least a few more structural complexities. For example, many prokaryotes have a **cell wall** located outside the plasma membrane (Figure 4.3). The rigidity of the cell wall supports the cell and determines its shape. The cell walls of most bacteria, but not archaea, contain **peptidoglycan**, a polymer of amino sugars, cross-linked to form a single molecule around the entire cell. In some bacteria, another layer—the **outer membrane** (a polysaccharide-rich phospholipid membrane)—encloses the cell wall. Unlike the plasma membrane, this outer membrane is not a major permeability barrier, and some of its polysaccharides are disease-causing toxins.

Enclosing the cell wall and outer membrane in many bacteria is a layer of slime composed mostly of polysaccharide and referred to as a **capsule**. The capsules of some bacteria may protect them from attack by white blood cells in the animals they infect. The capsule helps keep the cell from drying out, and sometimes it traps other cells for the bacterium to attack. Many prokaryotes produce no capsule at all, and those that do have capsules can survive even if they lose them, so the capsule is not a structure essential to cell life.

Some groups of bacteria—the cyanobacteria and some others—carry on photosynthesis. In *photosynthesis*, the energy of sunlight is converted to chemical energy that can be used for a variety of energy-requiring reactions, such as the synthesis of cellular proteins and DNA. In these photosynthetic bacteria, the plasma membrane folds into the cytoplasm—often

4.3 A Prokaryotic Cell The bacterium *Pseudomonas aeruginosa* illustrates typical prokaryotic cell structures. The electron micrograph on the left is magnified about 80,000 times. Note the existence of several protective structures external to the plasma membrane.

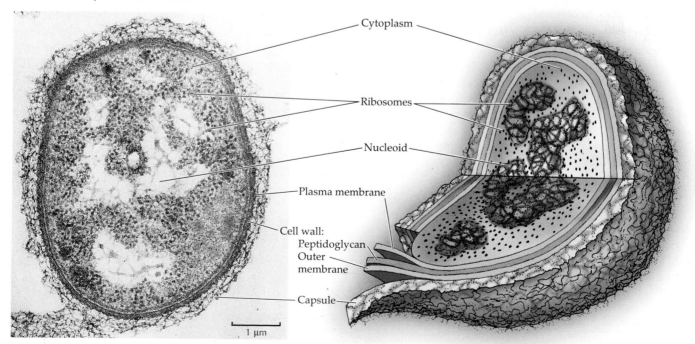

Cytoplasm

Ribosomes

Nucleoid

Plasma membrane

Cell wall:
Peptidoglycan
Outer membrane

Capsule

1 µm

(a)

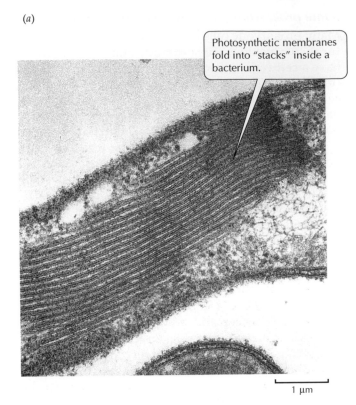

> Photosynthetic membranes fold into "stacks" inside a bacterium.

(b)

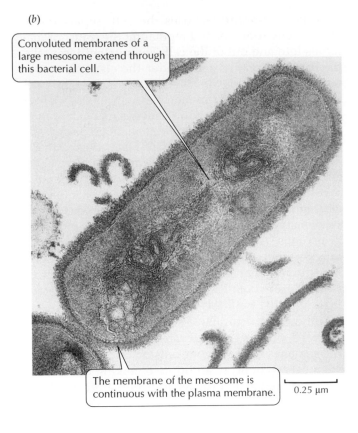

> Convoluted membranes of a large mesosome extend through this bacterial cell.

> The membrane of the mesosome is continuous with the plasma membrane.

1 μm

0.25 μm

4.4 Some Prokaryotes Have Internal Systems of Membranes The presence of internal membranes in prokaryotic cells contradicts the mistaken notion that prokaryotes are nothing more than tiny bags of molecules. (a) Photosynthetic membranes contain compounds needed for photosynthesis. (b) The convoluted membranes of a mesosome contain enzymes involved in cellular respiration.

very extensively—to form an internal membrane system that contains bacterial chlorophyll and other compounds needed for photosynthesis (Figure 4.4a).

Other groups of prokaryotes possess different sorts of membranous structures called **mesosomes**, which may function in cell division or in various energy-releasing reactions (Figure 4.4b). Like the photosyn-

4.5 Prokaryotic Projections Surface projections such as these bacterial flagella (a,b) and pili (c) contribute to movement, to adhesion, and to the complexity of prokaryotic cells.

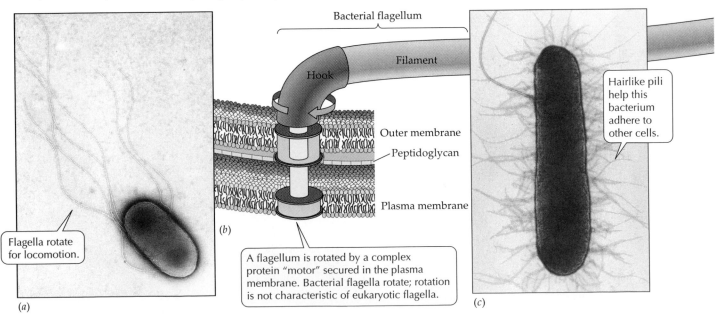

Bacterial flagellum

Filament

Hook

Outer membrane

Peptidoglycan

Plasma membrane

> Flagella rotate for locomotion.

(a)

(b)

> A flagellum is rotated by a complex protein "motor" secured in the plasma membrane. Bacterial flagella rotate; rotation is not characteristic of eukaryotic flagella.

> Hairlike pili help this bacterium adhere to other cells.

(c)

thetic membrane systems, mesosomes are formed by infolding of the plasma membrane. They remain attached to the plasma membrane and never form the free-floating, isolated, membranous organelles that are characteristic of eukaryotic cells.

Some prokaryotes swim by using appendages called **flagella** (Figure 4.5*a*). A single flagellum, made of a protein called flagellin, looks at times like a tiny corkscrew. It spins on its axis like a propeller, driving the cell along. Ring structures anchor the flagellum to the plasma membrane and, in some bacteria, to the outer membrane of the cell wall (Figure 4.5*b*). Flagella are known to cause the motion of the cell because if they are removed, the cell cannot move.

Pili project from the surface of some groups of bacteria (Figure 4.5*c*). Shorter than flagella, these thread-like structures seem to help bacteria adhere to one another during mating, as well as to animal cells for protection and food.

Microscopes: Revealing the Subcellular World

The preceding discussion included images of prokaryotes produced by microscopes. How are these images produced? In this section, we examine the principles and technologies of microscopy by means of visible light and by beams of electrons: light and electron microscopy, respectively (Figure 4.6).

Many significant advances in our knowledge of cells have depended on the **resolution**, or resolving power, of the instruments we use to magnify tiny objects. The resolving power of a lens or microscope is the smallest distance separating two objects that allows them to be seen as two distinct things rather than as a single entity. For example, most humans see two fine parallel lines as distinct markings if they are separated by at least 0.1 mm; if they are any closer together, we see them as a single line. Thus, the resolving power of the human eye is about 0.1 mm, which is the approximate diameter of the human egg. To see anything smaller, we must use some form of microscope.

Development of the light microscope made the study of cells possible

The **light microscope** uses glass lenses and visible light to form a magnified image of an object (see Figure 4.6*a*). In its contemporary form, the light microscope has a resolving power of about 200 nm (that is, 0.2 µm, or 0.0002 mm), so it gives a useful view of cells and can reveal features of some of the subcellular organelles of eukaryotes. Today, half a century after the invention of the electron microscope, which has more resolving power than the light microscope, the light microscope still remains an important tool for the biologist. Many of the illustrations in this book are pho-

4.6 Light and Electron Microscopes Both light and electron microscopes magnify a specimen, permitting more detail to be observed. Light microscopes use glass lenses to focus visible light, creating an enlarged image. Electron microscopes use magnets to focus a beam of electrons to create an image. Because of their greater resolution, electron microscopes reveal more internal detail than light microscopes can.

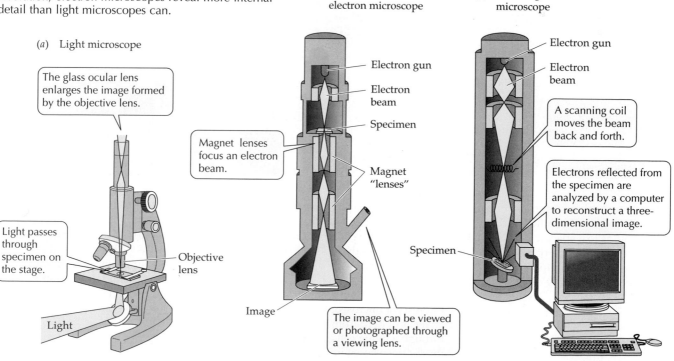

(*a*) Light microscope

The glass ocular lens enlarges the image formed by the objective lens.

Light passes through specimen on the stage.

Objective lens

Light

(*b*) Transmission electron microscope

Electron gun

Electron beam

Specimen

Magnet "lenses"

Magnet lenses focus an electron beam.

Image

The image can be viewed or photographed through a viewing lens.

(*c*) Scanning electron microscope

Electron gun

Electron beam

A scanning coil moves the beam back and forth.

Electrons reflected from the specimen are analyzed by a computer to reconstruct a three-dimensional image.

Specimen

tographs taken through the light microscope; they are called photomicrographs or just *micrographs*.

The light microscope has its limitations—principally its 200-nm resolving power. This resolving power cannot be improved by the addition of lenses or by the enlargement of micrographs. Although micrographs can be enlarged, such enlargements do not increase the resolution; as the images become larger, they simply become fuzzier.

One way to make images clearer is to kill the cells and stain them before examining them in the light microscope. But killing is not always necessary. One of

---

TABLE 4.1   Systems of Light Microscopy

THE FOLLOWING FORMS OF LIGHT MICROSCOPY CAN BE USED TO VIEW LIVING CELLS.[a]

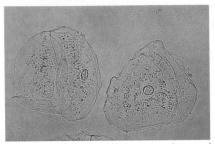

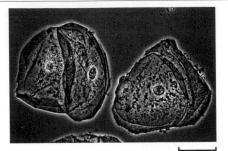

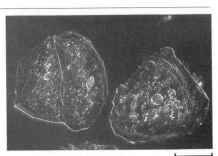

25 μm | 25 μm | 25 μm

In **bright-field microscopy**, light passes directly through the cells. Unless natural pigments are present, there is little contrast and details are not distinguished.

In **phase-contrast microscopy**, contrast in the image is increased by emphasizing differences in refractive index (the capacity to bend light), thereby enhancing light and dark regions in the cell.

Differential interference-contrast microscopy (Nomarski optics) uses two beams of plane-polarized light. Changes in the phase of these two beams as they pass through adjacent parts of a cell result in a bright image (if the beams are in phase) or a dark image (if the beams are out of phase). The observed image looks as if the cell were casting a shadow on one side.

[a]Certain dyes, called vital dyes, can sometimes be used to stain cells to enhance contrast without killing the cells.

THE FOLLOWING STAINING PROCEDURES USUALLY REQUIRE NONLIVING, PRESERVED CELLS.

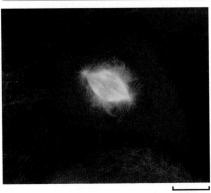

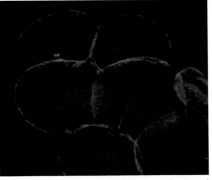

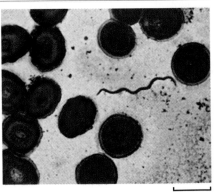

40 μm | 40 μm | 75 μm

In **fluorescent microscopy**, a natural substance in the cell or a fluorescent dye that binds to a specific cell material is stimulated by a beam of light, and the longer wavelength fluorescent light is observed coming directly from the dye.

Confocal microscopy uses fluorescent materials but adds a system of focusing both the stimulating and emitted light so that a single plane through the cell is viewed. The result is a sharper two-dimensional image than is obtained with standard fluorescent microscopy. Electronic analysis of light from multiple planes through a living cell can provide a three-dimensional image.

In **stained bright-field microscopy**, a stain added to preserved cells enhances contrasts and reveals detail not otherwise visible. Stains differ greatly in their chemistry and capacity to bind to cell materials. so many choices are available.

the great advantages of light microscopy over electron microscopy is that *living* cells can be examined. Many different kinds of light microscopy have been invented and used in the past 50 years in order to examine living cells more effectively (Table 4.1).

Electron microscopy has expanded our knowledge of cellular structures

Some cellular structures are far too small to be resolved with the light microscope. Ribosomes, for example, are 20 nm or less in diameter and cannot be resolved as individual objects under the light microscope. The electron microscope, however, can readily resolve ribosomes.

An **electron microscope** uses powerful magnets as lenses to focus an electron beam (see Figure 4.6*b*), much as the light microscope employs glass lenses to focus a beam of light. Since we cannot see electrons, the electron microscope directs them at a fluorescent screen or a photographic film to create an image we can see. These images are called *electron micrographs.* Figures 4.3, 4.4, and 4.5 are examples of electron micrographs.

The resolving power of modern electron microscopes is about 0.2 nm, but no biological specimen has yet been seen in such detail. One reason is that the energy of the electron beam at that power is so great that it destroys biological molecules before they can be seen. Because of this and other technical limitations, most electron micrographs resolve detail no finer than 2 nm, and even the best micrographs rarely resolve detail as fine as 1 nm. The corresponding resolving power is about 100,000 times finer than that of the human eye.

There are two types of electron microscopy: transmission and scanning. In **transmission electron microscopy**, which produced the electron micrographs in Figures 4.4 and 4.5, electrons pass *through* a sample. Transmission electron microscopy is used to examine thin slices of objects. Extremely thin sections are shaved off the material to be examined and placed on a grid, which supports the material in the beam of electrons. The grid is comparable to the glass slide that supports the material observed in the light microscope.

In **scanning electron microscopy**, electrons are directed at the surface of the sample, where they cause other electrons to be emitted; the scanning electron microscope focuses these secondary electrons on a viewing screen. Scanning electron microscopy reveals the *surface* structures of three-dimensional objects, such as the egg and the amoeba shown in Figure 4.22. A scanning electron microscope has a resolving power no better than 10 nm, so scanning electron micrographs are usually at a somewhat lower magnification than transmission electron micrographs.

You might think that with such resolving power the electron microscope would be used for all microscopic studies, but this is not so. For some applications, electron microscopy would be sheer overkill—like using a magnifying glass to look at an elephant. A significant limitation of electron microscopy is that biological samples must be killed and dehydrated before they can be examined. In addition, for transmission electron microscopy, cells must be stained or shadowed with heavy metals that will deflect electrons. And the cells must be sliced in thin sections before being placed in the electron beam.

To obtain a reasonable three-dimensional view of large cells or tissues with a microscope, one must look at many successive slices, rather like examining successive slices of Swiss cheese in order to "see" one of the holes. When you look at a transmission electron micrograph, keep in mind that the image is a two-dimensional slice through a three-dimensional reality. The imaginative reconstruction of the third dimension is particularly important with eukaryotic cells because, as we will see in the next section, they have many internal compartments.

Eukaryotic Cells

Animals, plants, fungi, and protists have cells that are larger and structurally more complex than those of the prokaryotes. To get a sense of the most prominent differences, compare the plant and animal cells in Figure 4.7 (eukaryotic cells) with Figure 4.3 (a prokaryotic cell).

Eukaryotic cells generally have dimensions ten times greater than those of prokaryotes; for example, the spherical yeast cell has a diameter of 8 μm. Like prokaryotic cells, eukaryotic cells have a plasma membrane, cytosol, and ribosomes. Unlike prokaryotes, however, eukaryotic cells have an internal cytoskeleton that maintains cell shape and moves materials. And eukaryotic cells contain many different kinds of membranous compartments that carry on particular biochemical functions.

The compartmentalization of eukaryotic cells is the key to their function

The compartments of eukaryotic cells are defined by one or two membranes that regulate what enters and leaves the interior. The membranes ensure that conditions inside the compartment are different from those in the surrounding cytoplasm. Some of the compartments are like little factories that make specific products. Others are like power plants that take energy in one form and convert it to a more useful form. These membranous compartments, as well as other structures (such as ribosomes) that lack membranes but possess distinctive shapes and functions, are called **organelles** (see Figure 4.7).

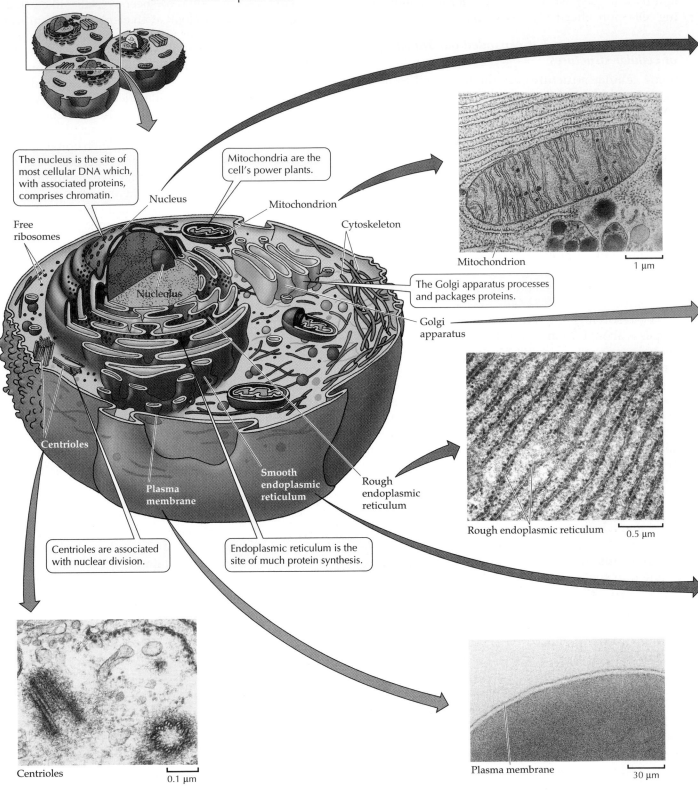

AN ANIMAL CELL

4.7 Eukaryotic Cells In electron micrographs, many plant cell organelles are nearly identical in form to those observed in animal cells. Cellular structures unique to plant cells include the cell wall and the chloroplasts. Animal cells contain centrioles, which are not found in plant cells.

The nucleus is the site of most cellular DNA which, with associated proteins, comprises chromatin.

Free ribosomes

Nucleus

Nucleolus

Mitochondria are the cell's power plants.

Mitochondrion

Cytoskeleton

Mitochondrion

1 μm

The Golgi apparatus processes and packages proteins.

Golgi apparatus

Centrioles

Plasma membrane

Smooth endoplasmic reticulum

Rough endoplasmic reticulum

Rough endoplasmic reticulum

0.5 μm

Centrioles are associated with nuclear division.

Endoplasmic reticulum is the site of much protein synthesis.

Centrioles

0.1 μm

Plasma membrane

30 μm

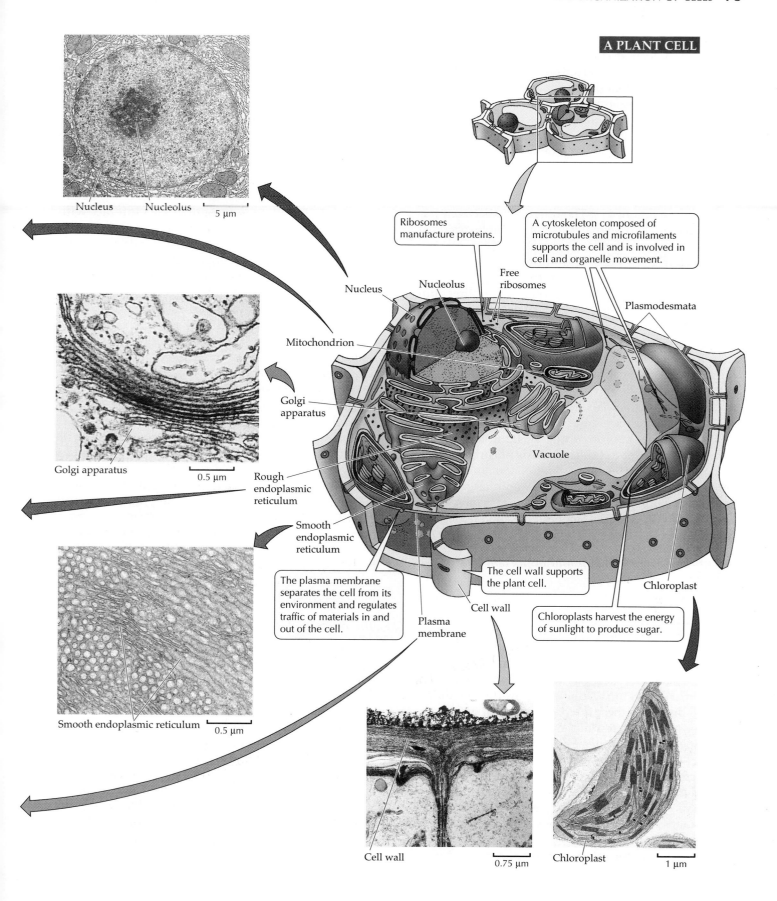

A PLANT CELL

Nucleus   Nucleolus   5 µm

Golgi apparatus   0.5 µm

Smooth endoplasmic reticulum   0.5 µm

Ribosomes manufacture proteins.

A cytoskeleton composed of microtubules and microfilaments supports the cell and is involved in cell and organelle movement.

Nucleus   Nucleolus   Free ribosomes

Plasmodesmata

Mitochondrion

Golgi apparatus

Rough endoplasmic reticulum

Smooth endoplasmic reticulum

Vacuole

The plasma membrane separates the cell from its environment and regulates traffic of materials in and out of the cell.

Plasma membrane

Cell wall

The cell wall supports the plant cell.

Chloroplast

Chloroplasts harvest the energy of sunlight to produce sugar.

Cell wall   0.75 µm

Chloroplast   1 µm

The organelles of cells from different eukaryotes are similar in molecular composition and structure, and they look similar in the electron microscope. Thus, if you know the characteristic appearance of the *nucleus, mitochondrion, endoplasmic reticulum,* and *Golgi apparatus,* you can recognize them in electron micrographs of different eukaryotic cells.

The membranes of all these organelles have similar structure (a phospholipid bilayer), dimensions, and appearance. But there are important differences between the organelles of a plant cell and those of an animal cell (see Figure 4.7). The differences between nuclei from plant and animal cells, for example, reside in the informational content of their DNA. The Golgi apparatuses in both plant and animal cells look very similar and always function in the storage, internal transport, and processing of proteins and other substances. However, the nature of these proteins and other substances cannot be determined by electron microscopy; instead, chemical methods of extraction, purification, and identification are necessary to answer questions about function.

Not all organelles are present or equally abundant in all cells. For example, some plant cells have a specialized organelle, called the *chloroplast* (a type of plastid), that is the site for capturing light energy and using it to bring about the synthesis of carbohydrate from carbon dioxide and water. Cells in animals lack chloroplasts. In another example, animal cells that synthesize and secrete digestive enzymes have more ribosomes and associated membranes than do less active cell types.

Membranes are abundant and important in eukaryotic cells

In 1952, the first people to look at clear electron micrographs of eukaryotic cells were stunned by the complexity of what they saw. No one expected that cells would contain so many internal membranes or that these membranes would be arranged in such diverse forms. What do all these membranes do? How do they function? What is their structure? How do they differ? We will deal with these questions in detail in Chapter 5; here we discuss a few of the most basic ideas.

Biological membranes form compartments whose internal contents and conditions differ from those of the surrounding environment. Although they are a barrier, membranes also permit the passage of certain materials. Thus transport is a major function of all membranes: They regulate molecular traffic between an inside and an outside.

The hydrophobic interior of the membrane serves as a barrier to the passage of charged or polar materials that are readily soluble in water (see Figure 3.2).

Many polar molecules and ions are transported through the membrane with the help of highly specific protein molecules inserted in the phospholipid bilayer. In addition, larger quantities of materials enter the cell by a process in which part of the plasma membrane folds inward and detaches from the rest of the membrane to form a small spherical compartment, a *vesicle,* in the cytoplasm.

Membranes participate in many activities besides transport. They are staging areas for interactions between cells. For example, in the human body, defensive white blood cells recognize and interact with their targets by means of recognition signals built into their plasma membranes (see Chapter 18). And the proper development and organization of multicellular animals depends on recognition of such surface recognition signals (see Chapter 15).

The mitochondrial membranes carry enzymes responsible for energy transformations in cells. In the chloroplasts of green plant cells, tightly stacked membranes contain chlorophyll and carotenoid molecules that capture light energy for photosynthesis. In many respects, a discussion of eukaryotic cells is a discussion of membranes and compartments that are specialized for various cellular functions.

Organelles That Process Information

Living things depend on accurate, appropriate information: internal signals, environmental cues, and stored instructions. Information is *stored* as the sequence of bases in DNA molecules. Most DNA in eukaryotic cells resides in the nucleus. Information is *translated* from the language of DNA into the language of proteins on the surface of the ribosomes. (This process is described in detail in Chapter 12.)

The nucleus stores most of the cell's information

The **nucleus** is usually the largest organelle in a cell (see Figure 4.7). The nucleus of most animal cells is approximately 5 μm in diameter—substantially larger than most entire prokaryotic cells.

The possession of a membrane-enclosed nucleus is the defining property of the eukaryotic cell—prokaryotes have no membranes separating their DNA from the surrounding cytoplasm. As viewed under the electron microscope, a nucleus is surrounded by *two* membranes, which together are called the **nuclear envelope** (Figure 4.8). During most of the life cycle of the cell, the nuclear envelope is a stable structure. However, before duplicated DNA is distributed to separate nuclei during division, the nuclear envelope fragments into vesicles. It re-forms when nuclear division is completed.

4.8 The Nucleus Is Enclosed by a Double Membrane The electron micrograph shows the nucleus of a nondividing animal cell. The double-membraned nuclear envelope, the nucleolus, nuclear lamina, and nuclear pores are common features of all cell nuclei.

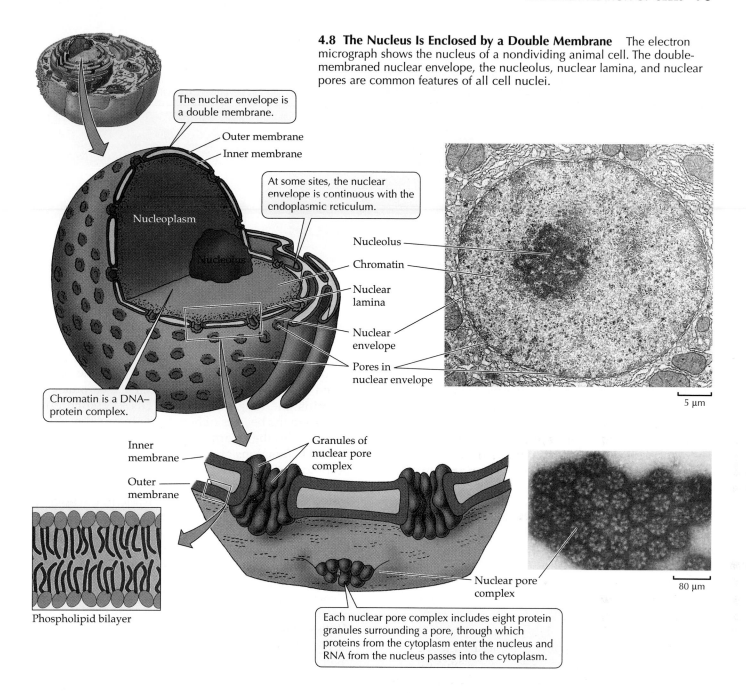

The nuclear envelope is a double membrane.

Outer membrane
Inner membrane

At some sites, the nuclear envelope is continuous with the endoplasmic reticulum.

Nucleoplasm

Nucleolus

Chromatin is a DNA–protein complex.

Nucleolus
Chromatin
Nuclear lamina
Nuclear envelope
Pores in nuclear envelope

5 µm

Inner membrane
Outer membrane

Granules of nuclear pore complex

Phospholipid bilayer

Nuclear pore complex

80 µm

Each nuclear pore complex includes eight protein granules surrounding a pore, through which proteins from the cytoplasm enter the nucleus and RNA from the nucleus passes into the cytoplasm.

The membranes of the nuclear envelope are separated by only a few tens of nanometers and are perforated by **nuclear pores** approximately 9 nm in diameter that connect the interior of the nucleus with the cytoplasm (see Figure 4.8). At these pores the outer membrane of the nuclear envelope is continuous with the inner membrane. Each pore is surrounded by eight large protein granules arranged in an octagon where the inner and outer membranes merge. RNA and water-soluble molecules pass through these pores to enter or leave the nucleus.

At certain sites, the outer membrane of the nuclear envelope folds outward into the cytoplasm and is continuous with the membrane of another organelle, the endoplasmic reticulum (discussed later in the chapter). The endoplasmic reticulum and, to a lesser extent, the cytoplasmic surface of the nuclear envelope often carry great numbers of ribosomes. But there are no ribosomes on the other membrane surfaces of the nuclear envelope.

Inside the nucleus, DNA combines with proteins to form a fibrous complex called **chromatin**. Surround-

4.9 The Nuclear Lamina The shape of the nucleus is maintained by a meshwork of proteins, the nuclear lamina. Both the lamina and the nuclear envelope break down before the nucleus divides, then re-form after nuclear division.

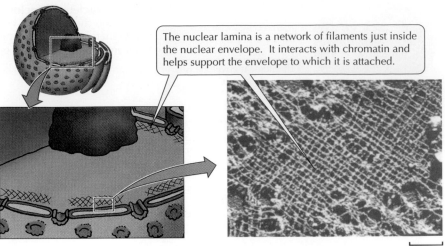

The nuclear lamina is a network of filaments just inside the nuclear envelope. It interacts with chromatin and helps support the envelope to which it is attached.

0.5 µm

ing the chromatin are water and dissolved substances collectively referred to as the **nucleoplasm**. At the periphery of the nucleus, chromatin attaches to a protein meshwork, called the **nuclear lamina**, which is formed by the polymerization of proteins called *lamins* into filaments (Figure 4.9). The nuclear lamina maintains the shape of the nucleus by its attachment to both chromatin and the nuclear envelope. When the nuclear envelope breaks down in preparation for nuclear division, the nuclear lamina also depolymerizes, but after nuclear division it re-forms just as the nuclear envelope re-forms.

Throughout most of the life cycle of the cell, chromatin exists as exceedingly long, thin, entangled threads that cannot be clearly distinguished by any

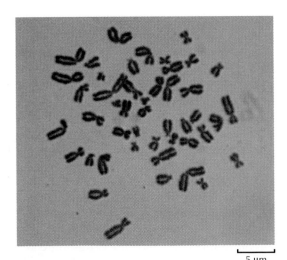

4.10 Humans Have 46 Chromosomes The chromosome complement of a eukaryotic cell can be observed early in the process of nuclear division. If a nucleus about to divide is ruptured and treated with certain stains, its chromosomes are readily visible under the light microscope, as this human cell shows.

5 µm

microscope. However, when the nucleus is about to divide (that is, to undergo mitosis or meiosis; see Chapter 9), the chromatin condenses and coils tightly to form a precise number of readily visible objects called **chromosomes** (Figure 4.10). Each chromosome contains one long molecule of DNA. The chromosomes are the bearers of hereditary instructions; their DNA carries the information required to perform the synthetic functions of the cell and to endow the cell's descendants with the same instructions.

Usually, dense, roughly spherical bodies called **nucleoli** (singular nucleolus) are visible in the nucleus (see Figure 4.8). Taken together, the nucleoli contain from 10 to 20 percent of a cell's RNA. Ribosomal subunits are assembled in the nucleolus from ribosomal RNA and specific proteins. Then they move out of the nucleus into the cytoplasm, where ribosome assembly is completed. Each nucleus must have at least one nucleolus, and the nuclei of some species have several. The exact number of nucleoli in its cells is characteristic of a species. As nuclear and cell division approach, the nucleoli become smaller and disappear, but after division they reappear.

Ribosomes are the sites of protein synthesis

In both eukaryotic and prokaryotic cells, proteins are synthesized on ribosomes. Ribosomes reside in three places in almost all eukaryotic cells: free in the cytoplasm, attached to the surface of endoplasmic reticulum (as will be described later in this chapter), and contained in the mitochondria, where energy is processed. Ribosomes are also found in chloroplasts, the photosynthetic organelles of plant cells. In each of these locations, the ribosomes provide the site where proteins are synthesized under the direction of nucleic acids (see Chapter 12).

The ribosomes of prokaryotes and of eukaryotes are similar in that both consist of two different-sized subunits. Eukaryotic ribosomes are somewhat larger, but the structure of prokaryotic ribosomes is better understood. Chemically, ribosomes consist of a special type of RNA, called ribosomal RNA, to which more than 50 different protein molecules are bound. The ribosome temporarily binds two other types of RNA molecules (messenger RNA and transfer RNA) as hereditary information from the DNA is translated into the primary structure of protein.

Organelles That Process Energy

In addition to information, cells require energy and raw materials. A cell uses energy to transform raw materials into cell-specific materials that it can use for ac-

tivities such as growth, reproduction, and movement. Energy is transformed from one form to another in the mitochondria found in all eukaryotic cells and in the chloroplasts of cells that harvest energy from sunlight (see Figure 4.7). In contrast, energy transformations in prokaryotic cells are associated with enzymes attached to the inner surface of the plasma membrane or extensions of the plasma membrane that protrude into the cytoplasm.

Mitochondria are energy transformers

In eukaryotic cells, utilization of food molecules such as glucose begins in the cytosol (the liquid part of the cytoplasm). The fuel molecules that result from partial degradation of this food enter **mitochondria** (singular mitochondrion), whose primary function is to convert the potential chemical energy of fuel molecules into a form that the cell can use: the energy-rich molecule called **ATP**, or *adenosine triphosphate* (see Chapter 6).

ATP is not a long-term energy storage form but a kind of energy currency. Its role in the cell is analogous to the role of paper money and coins in an economy. Chemically, ATP can participate in a great number of different cellular reactions and processes that require energy. In the mitochondria, the production of ATP using fuel molecules and O_2 is called *cellular respiration*.

Typical mitochondria are small—somewhat less than 1.5 μm in diameter and 2 to 8 μm in length—about the size of many bacteria. Mitochondria are visible with a light microscope, but almost nothing was known of their precise structure until they were examined with the electron microscope. Electron micrographs show that mitochondria have two membranes: an outer membrane and an inner membrane. The **outer membrane** is smooth and protective, and it offers little resistance to the movement of substances into and out of the mitochondrion.

Immediately inside the outer mitochondrial membrane is an **inner membrane** that folds inward in many places, giving it a much greater surface area than that of the outer membrane (Figure 4.11). In animal cells, these folds tend to be quite regular, giving rise to shelflike structures called **cristae**. The mitochondria of plants also have cristae, but plant cristae tend to be much less regular in size and structure. Special techniques and electron microscopy show that the inner

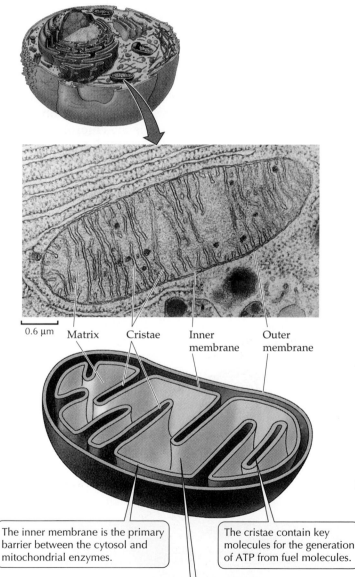

0.6 μm   Matrix   Cristae   Inner membrane   Outer membrane

The inner membrane is the primary barrier between the cytosol and mitochondrial enzymes.

The cristae contain key molecules for the generation of ATP from fuel molecules.

The matrix, the space enclosed by the inner membrane, contains several of the enzymes for cellular respiration. It also contains ribosomes and DNA.

4.11 A Mitochondrion Converts Energy from Fuel Molecules into ATP The electron micrograph is a two-dimensional slice through a three-dimensional reality. As this drawing emphasizes, the cristae are extensions of the inner mitochondrial membrane.

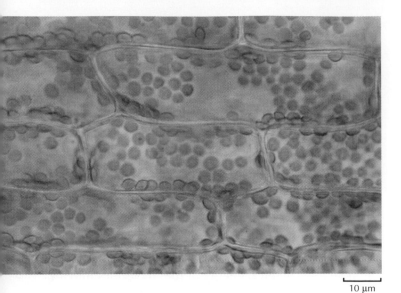

some enzymes, the matrix contains some ribosomes and DNA that make some of the proteins needed for cellular respiration. In some ways, the mitochondria function as semiautonomous "organisms" within the cytoplasm of eukaryotic cells. With their own DNA and the capacity to use its coded information for the synthesis of some of their own proteins, these organelles show striking similarities to bacteria. We will explore the evolutionary origin of mitochondria and chloroplasts a little later in this chapter.

Almost all eukaryotes have mitochondria. The number of mitochondria per cell ranges from one contorted giant in some unicellular protists to a few hundred thousand in large egg cells. An average human liver cell contains more than a thousand mitochondria. Cells that require the most chemical energy tend to have the most mitochondria per unit volume. In Chapter 7 we will see how different parts of the mitochondrion work together in cellular respiration.

10 μm

4.12 Chloroplasts in Cells   These plant cells contain many chloroplasts, the green organelles that carry on photosynthesis.

mitochondrial membrane contains many large protein molecules that participate in cellular respiration and the production of ATP. The inner membrane exerts much more control over what enters and leaves the mitochondrion than does the outer membrane.

The region enclosed by the inner membrane is referred to as the **mitochondrial matrix**. In addition to

Plastids photosynthesize or store materials

CHLOROPLASTS.   One class of organelles—the **plastids**—is produced only in plants and certain protists. The most familiar of the plastids is the **chloroplast**, which contains the green pigment *chlorophyll* and is the site of photosynthesis (Figure 4.12). In photosynthesis, light energy is converted into the energy of chemical bonds. The molecules formed in photosynthesis provide food

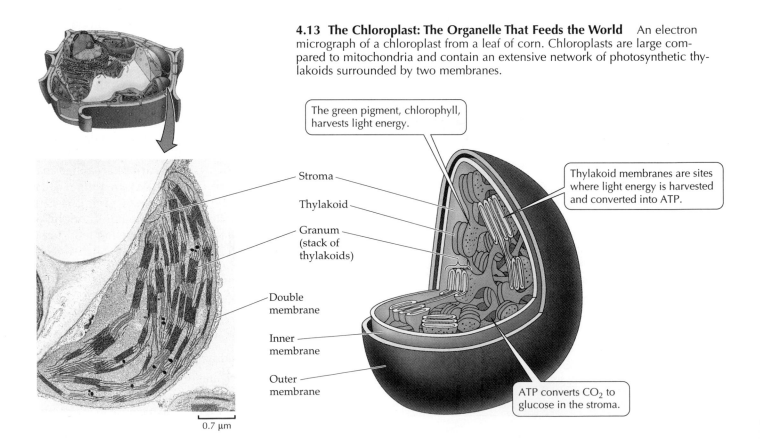

4.13 The Chloroplast: The Organelle That Feeds the World   An electron micrograph of a chloroplast from a leaf of corn. Chloroplasts are large compared to mitochondria and contain an extensive network of photosynthetic thylakoids surrounded by two membranes.

The green pigment, chlorophyll, harvests light energy.

Thylakoid membranes are sites where light energy is harvested and converted into ATP.

Stroma

Thylakoid

Granum (stack of thylakoids)

Double membrane

Inner membrane

Outer membrane

ATP converts CO_2 to glucose in the stroma.

0.7 μm

for the plant itself and for other organisms that eat plants. Directly or indirectly, photosynthesis is the energy source for most of the living world.

Like the mitochondrion, the chloroplast is surrounded by two membranes. Arising from the inner membrane is a series of discrete internal membranes whose structure and arrangement vary from one group of photosynthetic organisms to another. As an introduction, we concentrate on the chloroplasts of the flowering plants. Even these show some variation, but the pattern shown in Figure 4.13 is typical.

As seen in electron micrographs, chloroplasts contain membrane structures that look like stacks of pancakes. These stacks, called **grana** (singular granum), consist of a series of flat, closely packed, circular sacs called **thylakoids**. Thylakoids are surrounded by a single membrane composed of the usual membrane components (phospholipids and proteins), to which have been added chlorophyll and carotenoids, which are needed to harvest light energy and use it to produce glucose from CO_2 and water. All the cell's chlorophyll is contained in the thylakoid membranes. Thylakoids of one granum may be connected to those of other grana (see Figure 4.13), making the interior of the chloroplast a highly developed network of membranes.

The fluid in which the grana are suspended is referred to as **stroma**. Like the mitochondrial matrix, the chloroplast stroma contains ribosomes and DNA, and these are used to synthesize some, but not all, of the proteins that make up the chloroplast.

Not all plant cells contain chloroplasts. Most root cells, for example, lack chloroplasts, although the DNA in their nuclei has the information for chloroplast construction. By not forming chloroplasts in cells that do not receive light, the plant conserves energy.

Animal cells do not *produce* chloroplasts, but some do *contain* functional chloroplasts. These are taken up either as free chloroplasts derived from the partial digestion of green plants, or as bound chloroplasts contained within unicellular algae that live within the animal's tissues. The green color of some corals and sea anemones results from chloroplasts in algae that live within the animals (Figure 4.14). The animals derive some of their nutrition from the photosynthesis that these chloroplast-containing "guests" carry out.

OTHER TYPES OF PLASTIDS. Chloroplasts are not the only plastids found in plants. The red color of a ripe tomato results from the presence of legions of plastids called **chromoplasts**. Just as chloroplasts derive their color from chlorophyll, chromoplasts are red, orange, or yellow depending on the kinds of carotenoid pigments present (see Chapter 3). The chromoplasts have no known chemical function in the cell, but the colors they give to some petals and fruits probably help attract ani-

4.14 Living Together: Anemone–Alga Symbiosis
Symbiosis is the coexistence of organisms, sometimes for mutual benefit, as in this case. This giant sea anemone, an animal, owes its green color to the chloroplasts in a unicellular alga that lives and carries on photosynthesis within the anemone's tissues.

mals that assist in pollination or seed dispersal. (On the other hand, carrot roots gain no apparent advantage from being orange.) Other plastids, called **leucoplasts**, are storage depots for starch and fats.

All plastids develop from small proplastids, which are very simple in structure. And all plastid types are related to one another. For example, chromoplasts are formed from chloroplasts that lose their chlorophyll and undergo changes in internal structure.

Some organelles have an endosymbiotic origin

Chloroplasts and mitochondria are about the size of whole prokaryotes; they contain DNA and have ribosomes that are similar to prokaryotic ribosomes. And these organelles divide within the cell to produce additional mitochondria and chloroplasts. Given these facts, might they not be treated like little cells in their own right?

At one time, biologists believed that it might be possible to grow chloroplasts or mitochondria in culture, outside the cells they normally inhabit. These efforts failed because organelles depend on the cell's nucleus and cytoplasm for some essential components. But the experiments did help nurture thoughts about another important question: How did the eukaryotic cell with its organelles arise in the first place?

As we have seen, prokaryotic cells are generally simpler in structure than eukaryotic cells, precisely because prokaryotes *lack* membrane-enclosed organelles. Prokaryotic fossils have been found in sediments well over 3 billion years old, whereas the earliest known eukaryotic fossils date back to only 1.4 billion years ago. Biologists thus generally agree that eukaryotes evolved from prokaryotes. But how?

An explanation that has gained acceptance in recent decades is the **endosymbiosis theory** of the origin of mitochondria and chloroplasts. An important current champion of and contributor to this theory is Lynn Margulis of the University of Massachusetts, Amherst, who proposed the following idea.

Picture a time, around 2 billion years ago, when only prokaryotes inhabited Earth. Some of them absorbed their food directly from the environment. Others were photosynthetic. Still others fed on smaller prokaryotes by engulfing them (Figure 4.15). Under these conditions, suppose that a small, photosynthetic prokaryote was *ing*ested by a larger one but was not *dig*ested. Instead it survived trapped within a vesicle in the cytoplasm of the larger cell.

Suppose further that the smaller prokaryote divided at about the same rate as the larger one, so successive generations of the larger cell also contained the offspring of the smaller one. We would call this phenomenon *endosymbiosis* (*endo-*, "within"; *symbiosis*, "living together"), comparable to the algae that live within sea anemones seen in Figure 4.14. The endosymbiosis described here provided benefits for both organisms: The larger cell obtained the photosynthetic products from the smaller cell, and the smaller cell was protected by the larger one.

Could the little green prokaryote that took up residence in the larger prokaryote have been the first chloroplast? Present-day chloroplasts are surrounded by a double membrane, a structure that might have arisen when, in the process of engulfing the photosynthetic cell, the plasma membrane of the larger cell extended around the plasma membrane of the smaller cell (see Figure 4.15). But there is even stronger evidence for the endosymbiosis theory.

The fact that chloroplasts contain ribosomes and DNA is consistent with the endosymbiosis theory. Additional evidence comes from studies of ribosome structures. In composition, structure, and size, the ribosomes of chloroplasts resemble the smaller ribosomes of prokaryotes more than they resemble the ribosomes in the eukaryotic cytoplasm.

Similar evidence and arguments also support the proposition that mitochondria are the descendants of respiring prokaryotes engulfed by, and ultimately endosymbiotic with, larger prokaryotes. In addition to having prokaryote-like ribosomes, the mitochondrial inner membrane shows striking similarities to some of the energy-transforming bacterial membranes, and certain mitochondrial enzymes have primary structures similar to those of prokaryotic enzymes. In the case of mitochondria, the benefits of the initial endosymbiotic relationship might have been due to the

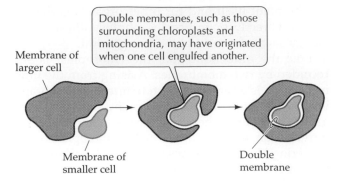

Double membranes, such as those surrounding chloroplasts and mitochondria, may have originated when one cell engulfed another.

Membrane of larger cell

Membrane of smaller cell

Double membrane

4.15 Origin of an Organelle's Double Membrane
The double membrane that encloses mitochondria and chloroplasts may have arisen from two different sources: the outer membrane from the engulfing cell's plasma membrane and the inner membrane from the engulfed cell's plasma membrane.

capacity of the engulfed prokaryote to detoxify molecular oxygen (O_2), which was increasing in the atmosphere because of photosynthesis.

The fact that a few modern cells contain other, smaller cells as endosymbionts supports the endosymbiosis theory of the origin of the eukaryotic cell by showing that engulfment can lead to a stable condition. But mitochondria and chloroplasts are not enough to make a prokaryote into a eukaryote. The theory is still incomplete. For example, the origins of the nuclear envelope and other important structures—including those responsible for nuclear division—still need to be understood better.

We discuss further aspects of the origin of the eukaryotic cell in Chapter 26. Is the endosymbiosis theory true? Almost certainly. The endosymbiosis theory is a good example of creative biological thinking that makes logical sense out of a variety of facts about organelles and bacteria.

The Endomembrane System

Much of the volume of a eukaryotic cell is taken up by its extensive membrane systems. All of these membranes look similar in electron micrographs. In addition, when organelles are extracted from whole cells and separated from other cells, their lipid compositions are found to be similar, if not identical. These observations suggest that some of the organelles are parts of a single system, called the **endomembrane system**. Evidence for the interrelationship of the cellular membranes exists in electron micrographs that show connections between different parts of the endomembrane system.

In this section, we'll examine the functional significance of these interrelationships and discover that materials synthesized in the endoplasmic reticulum can be transferred to another organelle, the Golgi apparatus, for further processing, storage, or transport. We will also describe the role of the lysosome in cell digestion.

The endoplasmic reticulum is a complex factory

Electron micrographs reveal a network of membranes branching throughout the cytoplasm. These membranes form tubes and flattened sacs called the **endoplasmic reticulum**, or **ER**, in which the interior compartment, referred to as the *lumen*, is separate and distinct from the surrounding cytoplasm (Figure 4.16). At certain sites, the ER is continuous with the outer membrane of the nuclear envelope.

Parts of the ER are liberally sprinkled with ribosomes, which are attached to the outer faces of the flattened sacs. Because of their appearance in the electron microscope, these regions are called **rough ER** (see the top halves of Figure 4.16). The attached ribosomes are sites for the synthesis of proteins that function outside the cytosol—that is, proteins that are to be exported from the cell, incorporated into membranes, or moved into organelles of the endomembrane system. These proteins enter the lumen of the ER as they are synthesized, directed there by a special sequence of amino acids known as the *signal sequence*.

Once in the lumen of the ER, these proteins undergo several changes, including the formation of disulfide bridges and folding into their tertiary structures (see Figure 3.18). Some proteins gain carbohydrate groups in the rough ER, thus becoming glycoproteins. The carbohydrate groups are part of an "addressing" system that ensures that the right proteins are directed to the right parts of the cell.

Proteins that remain within the cytosol or move into mitochondria and chloroplasts are synthesized on "free" ribosomes in the cytosol—ribosomes that are *not* attached to the ER. These proteins lack the signal sequence that would otherwise direct them into the ER (see Chapter 12).

Some parts of the endoplasmic reticulum, called the **smooth ER**, are more tubular (less like flattened sacs) and lack ribosomes (see the bottom halves of Figure 4.16). Within the lumen of the smooth ER, proteins that have been synthesized on the rough ER are chemically modified. The smooth ER is also the site at which phospholipids, steroids, and fatty acids are synthesized, some carbohydrates are metabolized, and toxic substances such as drugs are rendered inert. Lots of things happen in the smooth ER.

Cells that synthesize a lot of protein for export are usually packed with ER. Examples include glandular cells that secrete digestive enzymes (see Chapter 47) and plasma cells that secrete antibodies (see Chapter

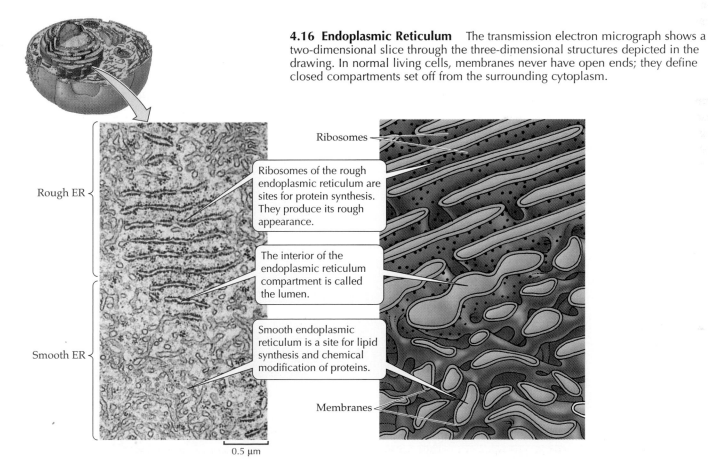

4.16 Endoplasmic Reticulum The transmission electron micrograph shows a two-dimensional slice through the three-dimensional structures depicted in the drawing. In normal living cells, membranes never have open ends; they define closed compartments set off from the surrounding cytoplasm.

Rough ER

Smooth ER

0.5 μm

Ribosomes

Ribosomes of the rough endoplasmic reticulum are sites for protein synthesis. They produce its rough appearance.

The interior of the endoplasmic reticulum compartment is called the lumen.

Smooth endoplasmic reticulum is a site for lipid synthesis and chemical modification of proteins.

Membranes

18). In contrast, cells with less work to do (such as storage cells) contain very little ER.

The Golgi apparatus stores, modifies, and packages proteins

In 1898 the Italian microscopist Camillo Golgi discovered a delicate structure in nerve cells, which came to be known as the **Golgi apparatus**. Because of the resolution limits of light microscopy, and because the staining technique often failed to reveal the structure, many biologists regarded the structure as a figment of Golgi's imagination. In the late 1950s, however, the electron microscope showed clearly that the Golgi apparatus does exist—and not just in nerve cells, but in most eukaryotic cells.

The exact appearance of the Golgi apparatus varies from species to species, but it always consists of flattened membranous sacs called *cisternae* and small membrane-enclosed *vesicles*. The flattened sacs are always seen lying together like a stack of saucers (Figure

4.17*a*). In the cells of plants, protists, fungi, and many invertebrate animals, these stacks are individual units scattered throughout the cytoplasm. In vertebrate cells, a few such stacks usually form a larger, more complex Golgi apparatus. The bottom saucers, constituting the *cis* region of the Golgi apparatus, lie nearest the nucleus or a patch of rough ER (Figure 4.17*b*). The top saucers, constituting the *trans* region, lie closest to the surface of the cell. The saucers in the middle make up the *medial* region of the complex. These three parts of the Golgi apparatus contain different enzymes and perform different functions.

What are the functions of the organelle that Golgi discovered? The first clue comes from observing the relationships between the Golgi apparatus and other parts of the cell. Vesicles form from the rough ER, move through the cytoplasm, and fuse with the *cis* region of the Golgi apparatus, where their contents are released into the lumen of the Golgi. Other small vesicles move between the flattened sacs, always in the direction *cis* to *trans*, transporting proteins. Associated with the sacs, particularly those toward the *trans* region, are tiny vesicles that pinch off from the sacs and move to other sacs or away from the Golgi (see Figure 4.17*b*).

The membranes of two vesicles can sometimes make contact with each other and fuse, resulting in a larger vesicle and a mixing of the contents. Vesicles may also fuse with other organelles or with the plasma membrane, where they release their contents

(*a*)

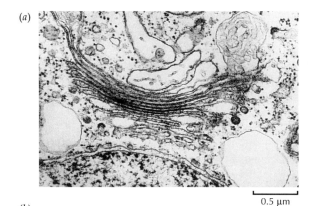

0.5 µm

(*b*)

4.17 The Golgi Apparatus The Golgi apparatus modifies incoming proteins and "targets" them to the correct addresses.

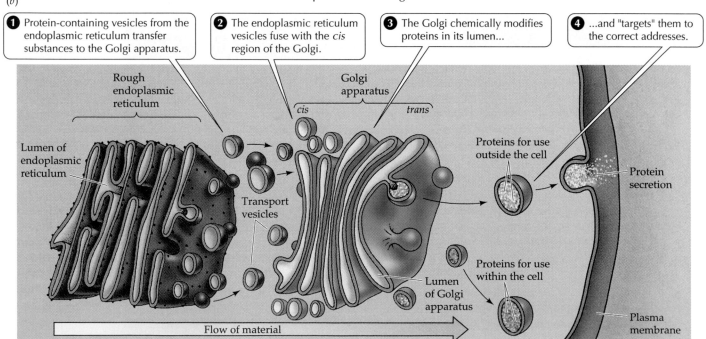

❶ Protein-containing vesicles from the endoplasmic reticulum transfer substances to the Golgi apparatus.

❷ The endoplasmic reticulum vesicles fuse with the *cis* region of the Golgi.

❸ The Golgi chemically modifies proteins in its lumen...

❹ ...and "targets" them to the correct addresses.

Rough endoplasmic reticulum

Golgi apparatus

cis   *trans*

Lumen of endoplasmic reticulum

Transport vesicles

Proteins for use outside the cell

Protein secretion

Proteins for use within the cell

Lumen of Golgi apparatus

Plasma membrane

Flow of material

to the outside of the cell. The formation, transport, and fusing behavior of vesicles is essential to the function of the Golgi apparatus: The Golgi apparatus serves as a sort of postal depot in which some of the proteins synthesized by ribosomes on the rough ER are stored, chemically modified, and packaged for delivery to the outside of the cell or to other organelles within the cell.

How does the Golgi apparatus send the right proteins to the right destinations? The delivery system consists of a series of chemical reactions in which proteins gain specific chemical "address tags." For example, a protein destined for use in an organelle called a lysosome (discussed in the next section) is given a certain signal sequence that directs it into the lumen of the rough ER as the protein is synthesized. In the lumen of the ER, the signal sequence is removed, and an oligosaccharide "address label" is added in the form of a glycoprotein. This tag identifies the proteins that are to be secreted from the cell. But before the Golgi secretes these proteins, its enzymes modify the oligosaccharide by adding a phosphate group (that is, by phosphorylating it).

When a phosphorylated glycoprotein reaches the *trans* region of the Golgi apparatus, it binds to a specific receptor protein in the Golgi membrane. A vesicle containing the bound and phosphorylated glycoprotein separates from the Golgi apparatus and delivers its contents to the developing lysosome. Comparable mechanisms deliver other proteins to other parts of the cell. The Golgi apparatus exports some proteins constantly but retains others, releasing them only at the appropriate time. In these ways the Golgi apparatus directs the molecular "mail" of the cell.

Lysosomes contain digestive enzymes

Originating in part from the Golgi apparatus, organelles called **lysosomes** contain and transport digestive enzymes that accelerate the breakdown of proteins, polysaccharides, nucleic acids, and lipids. Lysosomes are surrounded by a single membrane and have a densely staining, featureless interior (Figure 4.18*a*).

Lysosomes are sites for the breakdown of food and foreign objects taken up by phagocytosis. In **phagocytosis** (*phago-*, "eating"; *cytosis*, "cellular"), a pocket

(a)

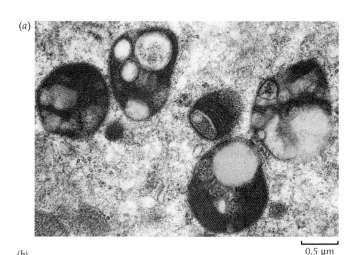

0.5 μm

(b)

4.18 Lysosomes Isolate Digestive Enzymes from the Cytoplasm (a) In this electron micrograph of a rat cell, the darkly stained organelles are secondary lysosomes in which digestion is taking place. (b) The origin and action of lysosomes and lysosomal digestion.

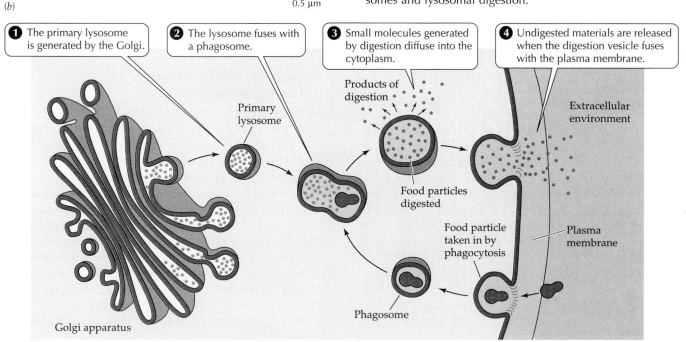

❶ The primary lysosome is generated by the Golgi.

❷ The lysosome fuses with a phagosome.

❸ Small molecules generated by digestion diffuse into the cytoplasm.

❹ Undigested materials are released when the digestion vesicle fuses with the plasma membrane.

Products of digestion

Extracellular environment

Primary lysosome

Food particles digested

Food particle taken in by phagocytosis

Plasma membrane

Phagosome

Golgi apparatus

forms in the plasma membrane and eventually deepens and encloses material from outside the cell. The pocket becomes a small vesicle and breaks free of the plasma membrane to move into the cytoplasm as a **phagosome** that contains food or other material for chemical digestion (Figure 4.18*b*). The phagosome fuses with a primary lysosome pinched off from the Golgi apparatus, to form a **secondary lysosome**, in which digestion takes place.

The effect of this fusion is rather like releasing hungry foxes into a chicken coop. The enzymes in the secondary lysosome quickly hydrolyze the food particles. The activity of the enzymes is enhanced by the mild acidity of the lysosome's interior, where the pH is lower than in the surrounding cytoplasm. The products of digestion exit through the membrane of the lysosome, providing fuel molecules and raw materials for other cell processes. The "used" secondary lysosome containing undigested particles then moves to the plasma membrane, fuses with it, and releases the undigested contents to the environment.

Phagocytosis and lysosomal digestion are important activities of some of the white blood cells in humans and other vertebrates. These cells identify and attack foreign cells, abnormal cells, and cell debris resulting from trauma, disease, or normal wear and tear.

Lysosomal compartmentalization is an effective arrangement to prevent the digestive enzymes from attacking the contents of the cytosol and the other organelles. These enzymes are isolated in the lysosome and cannot escape. The consequences of digestive enzymes escaping from the lysosomes can be severe. But such digestive activity is sometimes appropriate, as during the development of a frog from a tadpole. The fleshy tail of the tadpole, which no longer exists on the mature frog, disappears in part because lysosomes within the tail cells of the tadpole break down, releasing enzymes into the cytoplasm that digest the cells themselves.

Other Organelles

In addition to the information-processing organelles (nucleus and ribosomes), the energy-processing organelles (mitochondria and chloroplasts), and the organelles that form the endomembrane system of the cell (endoplasmic reticulum, Golgi apparatus, and lysosomes), there are two other kinds of membrane-enclosed organelles: peroxisomes and vacuoles.

Peroxisomes house specialized chemical reactions

As seen with the electron microscope, **peroxisomes** are small organelles—0.2 to 1.7 μm in diameter—and they have a single membrane and a granular interior (Figure 4.19). Peroxisomes form on the rough ER and

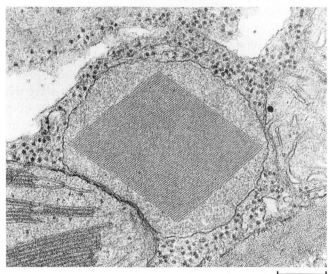

4.19 A Peroxisome A diamond-shaped crystal, composed of an enzyme, almost entirely fills this rounded peroxisome in a leaf cell. The enzyme catalyzes one of the reactions fulfilling the special function of the peroxisome.

are found at one time or another in at least some of the cells of almost every eukaryotic species.

Peroxisomes are organelles within which toxic peroxides (such as hydrogen peroxide, H_2O_2) are formed as unavoidable side products of chemical reactions. Subsequently, the peroxides are safely broken down within the peroxisomes without mixing with other parts of the cell.

A structurally similar organelle, the **glyoxysome**, is found only in plants. Glyoxysomes, which are most prominent in young plants, are the sites where stored lipids are converted into carbohydrates for transport to growing cells.

Vacuoles are filled with water and soluble substances

Many eukaryotic cells, but particularly those of plants and protists, contain membrane-enclosed organelles that look empty under the electron microscope. These organelles are called **vacuoles** (Figure 4.20). They are not actually empty; rather they are filled with aqueous solutions that contain many dissolved substances.

Despite their structural simplicity, vacuoles have a variety of functions in the lives of cells. For example, like animals and other organisms, plant cells produce a number of toxic by-products and waste materials. Animals have specialized excretory mechanisms for getting rid of such wastes, but plants are not equipped in the same way. Although plants can secrete some wastes to their environment, many toxic and waste

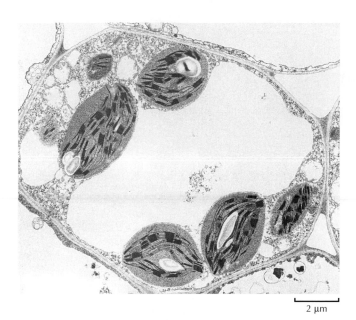

2 µm

4.20 Vacuoles in Plant Cells Are Usually Large The large central vacuole in this cell is typical of mature plant cells. Smaller vacuoles are visible toward each end of the cell.

materials they simply store within vacuoles. And since they are poisonous or distasteful, these stored materials deter some animals from eating the plants. Thus stored wastes may contribute to plant survival for reproduction.

In many plant cells, enormous vacuoles take up more than 90 percent of the cellular volume and grow as the cell grows. But vacuoles are by no means a waste of space, for the dissolved substances in the vacuole, working together with the vacuolar membrane, provide the turgor, or stiffness, of the cell, which in turn provides support for the structure of nonwoody plants. Vacuoles even play a role in the sex life of plants. Some pigments (especially blue and pink ones) in petals and fruits are contained in vacuoles. These pigments—the anthocyanins—are visual cues that encourage animals to visit flowers and thus aid in pollination, or to eat fruits and thus aid in seed dispersal.

Some unicellular protists, simple multicellular organisms such as sponges, and some of the other ancient invertebrate animals obtain nutrients directly by phagocytosis. Particles from the environment are trapped and engulfed by phagosomes, which in these cells are called *food vacuoles.*

Many freshwater protists also have a highly specialized **contractile vacuole** (see Chapter 26). Its function is to rid the cell of excess water that rushes in because of the imbalance in salt concentration between the relatively salty interior of the cell and its freshwater environment. The contractile vacuole visibly enlarges as water enters, then abruptly contracts, forcing the water out of the cell through a special pore structure.

The Cytoskeleton

As you have discovered, membranes divide the cytoplasm of eukaryotic cells into numerous compartments. But membrane-enclosed organelles and ribosomes are not the only large constituents of the cytoplasm. Also present are a dynamic set of long, thin fibers called the **cytoskeleton**, which fills at least two important roles: maintaining cell shape and support, and providing for various types of cell movement (Figure 4.21). Some fibers act as tracks or supports for "motor proteins" that help a cell move or that move things within the cell. In the discussion that follows, we'll look at three components of the cytoskeleton that are visible in electron micrographs: microfilaments, intermediate filaments, and microtubules. These structures are illustrated in detail in Figure 4.21.

Microfilaments function in support and movement

Microfilaments (actin filaments) exist as single filaments, in bundles, or in networks. They stabilize cell shape and help the entire cell or parts of the cell contract. In muscle cells, actin fibers are associated with myosin fibers, and their interactions account for the contraction of muscles. In other cells, actin fibers are associated with localized changes of shape in cells. For example, microfilaments are involved in a flowing movement of the cytoplasm called *cytoplasmic streaming*, in movements of specific organelles and particles within cells, and in "pinching" contractions that divide an animal cell into two daughter cells (Figure 4.22*a*). Microfilaments are also involved in the formation of cellular extensions, called *pseudopodia* (*pseudo-*, "false;" *podia*, "feet"), that enable amoebas to move (Figure 4.22*b*).

Microfilaments are assembled from a protein called *G actin*. G actin is a globular protein that has distinct "head" and "tail" sites for interacting with other molecules of G actin to assemble into a long chain (see Figure 4.21). Two of these chains interact to form the double helical structure that is a microfilament, which is about 7 nm in diameter and several micrometers long. The polymerization of G actin into microfilaments is reversible, and microfilaments can disappear from cells by breaking down into free units of G actin that are too small to see with the electron microscope.

In some cells (animal muscle cells, for example) microfilaments are very stable. For instance, microfilaments provide an internal structure to maintain the form of tiny extensions called *microvilli* (singular microvillus) that increase the surface area of cells special-

The cytoskeleton

Components of the cytoskeleton maintain cell shape, reinforce the cell, and contribute to cell movements.

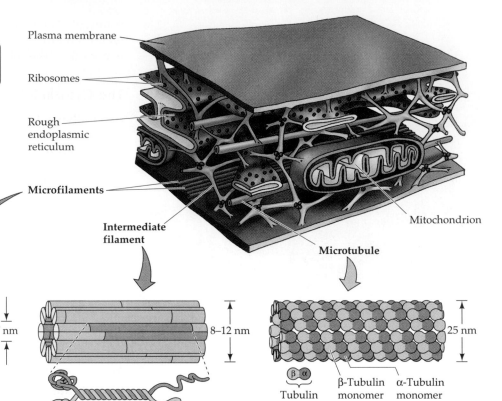

Plasma membrane

Ribosomes

Rough endoplasmic reticulum

Microfilaments

Intermediate filament

Mitochondrion

Microtubule

7 nm

Actin monomer

8–12 nm

Fibrous subunit

25 nm

β α
Tubulin dimer

β-Tubulin monomer

α-Tubulin monomer

Microfilaments are made up of strands of the protein actin and often interact with strands of other proteins. Microfilaments may occur singly, or in bundles or networks. They change cell shape and drive cellular motion, including contraction, cytoplasmic streaming, and the "pinched" shape changes that occur during cell division. Microfilaments and myosin strands together drive muscle action.

Intermediate filaments are made up of fibrous proteins organized into tough, ropelike assemblages that stabilize a cell's structure and help maintain its shape. Some intermediate filaments hold neighboring cells together. Others make up the nuclear lamina.

Microtubules are long, hollow cylinders made up of many molecules of the protein tubulin. Tubulin consists of two subunits, α-tubulin and β-tubulin. Microtubules lengthen or shorten by adding or subtracting tubulin dimers. Microtubule shortening moves chromosomes. Interactions between microtubules drive the movement of cells. Microtubules serve as "tracks" for the movement of vesicles.

4.21 The Cytoskeleton Three highly visible and important structural components of the cytoskeleton are shown in detail.

ized for absorption (Figure 4.23). Microfilaments can form networks when other specific proteins, called *actin-binding proteins*, interact with G actin units in specific ways. Such a network forms directly beneath the plasma membrane and helps determine cell shape, as described in more detail in the next section.

MEMBRANE INTEGRITY UNDER STRESS. Red blood cells appear fragile, yet they survive repeated compression and deformation as they squeeze through the finest of capillaries. The red blood cell gets this surprising resilience from certain proteins associated with its plasma membrane (Figure 4.24). The protein spectrin

4.22 Microfilaments for Motion
The contraction of microfilaments contributes (*a*) to the division of animal cells and (*b*) to amoeboid motion.

(*b*)

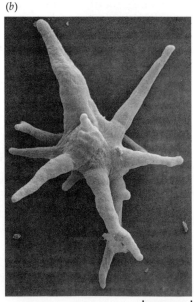

25 μm

(*a*)

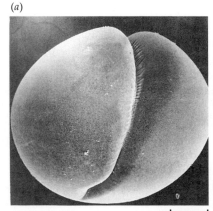

200 μm

Structure of a microvillus

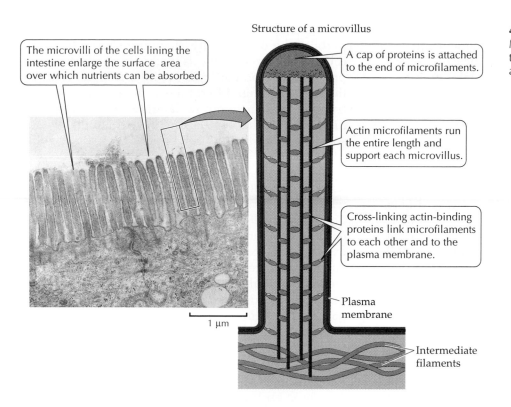

The microvilli of the cells lining the intestine enlarge the surface area over which nutrients can be absorbed.

A cap of proteins is attached to the end of microfilaments.

Actin microfilaments run the entire length and support each microvillus.

Cross-linking actin-binding proteins link microfilaments to each other and to the plasma membrane.

Plasma membrane

Intermediate filaments

1 μm

4.23 Microfilaments for Support
Microfilaments form the backbone of the microvilli that increase the surface area of some cells.

forms a meshwork of microfibrils on the cytoplasmic surface of the plasma membrane. This spectrin meshwork provides structural support. It is anchored to the actin filaments of the cytoskeleton. Another peripheral (surface) protein, ankyrin, anchors the spectrin to the membrane at many points by binding both the spectrin and an anion channel that is a protein embedded in the membrane.

Genetic defects in spectrin and others of these proteins result in abnormal red blood cells and thus in various diseases. Mice with hemolytic anemia have spherical, fragile red blood cells. Their red blood cells have very little spectrin, but the cells take on a normal shape if provided with spectrin.

Other linkages between the cytoskeleton and the plasma membrane can be very important. For exam-

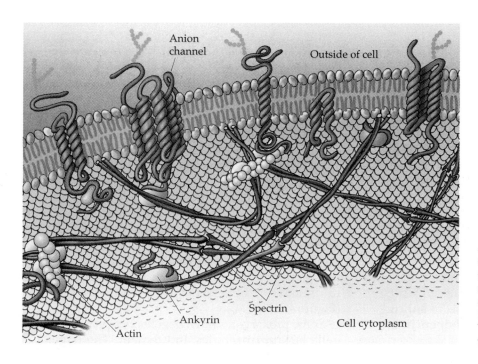

Anion channel

Outside of cell

Ankyrin

Spectrin

Cell cytoplasm

Actin

4.24 Some Proteins of the Red Blood Cell Membrane Many proteins contribute to this resilient structure. Spectrin does not bind the membrane directly but connects by way of linker proteins; it also binds actin filaments of the cytoskeleton. The structures of this cell membrane depiction will be discussed in detail in Chapter 5.

ple, a genetic deficiency in one protein, called *dystrophin*, that normally links actin filaments to the plasma membrane in muscle cells, is the cause of the most common form of muscular dystrophy, an inherited human disease that involves progressive loss of muscle cells and affects one out of every 3,500 male children.

Intermediate filaments are tough supporting elements

Filaments of another type, the **intermediate filaments**, are found only in multicellular organisms and play more static roles: They stabilize cell structure and resist tension (see Figure 4.21). Although there are at least five distinct types of intermediate filaments, all share the same general structure and are composed of fibrous proteins of the keratin family similar to the protein that makes up hair and fingernails. In cells, these proteins are organized into tough, ropelike assemblages 8 to 12 nm in diameter.

In some cells, intermediate filaments end at the nuclear envelope and may maintain the positions of the nucleus and other organelles in the cell. The lamins of the nuclear lamina are intermediate filaments. Other kinds of intermediate filaments help hold a complex apparatus of microfilaments in place in muscle cells (see Chapter 44). Still other kinds stabilize and help maintain rigidity in surface tissues by connecting "spot welds" called *desmosomes* between adjacent cells (see Figure 5.6*b*). Rapidly growing or newly formed cells do not contain intermediate filaments.

Microtubules are long and hollow

Microtubules are long, hollow, unbranched cylinders about 25 nm in diameter and up to several micrometers long. Many of the microtubules in a cell radiate from a region called the *microtubule organizing center*. They are assembled from molecules of the protein **tubulin**.

Tubulin itself is a dimer made up of two polypeptide subunits, called α-tubulin and β-tubulin. Thirteen rows, or protofilaments, of tubulin dimers surround the central cavity of the microtubule (see Figure 4.21). Microtubules have polarity: One end is called the + end, the other the – end. Tubulin dimers can be added or subtracted from the + end, lengthening or shortening the microtubule to affect the cell in various ways. The capacity to change length rapidly makes microtubules dynamic structures.

In plants, microtubules help control the arrangement of the fibrous components of the cell wall. Electron micrographs of plants frequently show microtubules lying just inside the plasma membrane of cells that are forming or extending their cell walls. Disruption of the cell's microtubules leads to a disordered arrangement of newly synthesized fibers in the cell wall.

In animal cells, microtubules are often found in the parts of the cell that are changing shape. In some cells, microtubules serve as tracks along which motor proteins carry protein-laden vesicles from one part of the cell to another. Microtubules are essential in distributing chromosomes to daughter cells during cell division (see Chapter 9). And they are intimately associated with movable cell appendages: the flagella and cilia.

FLAGELLA AND CILIA. Many eukaryotic cells possess whiplike appendages, the flagella and cilia. These organelles push or pull the cell through its aqueous environment, or they may move surrounding liquid over the surface of the cell (Figure 4.25*a*). Cilia and eukaryotic flagella are both assembled from specialized microtubules and have identical internal structures, but they differ in their relative lengths and their patterns of beating.

Flagella are longer than cilia and are usually found singly or in pairs; waves of bending propagate from one end of a flagellum to the other in snakelike undulation. **Cilia** are shorter appendages and are usually present in great numbers. They beat stiffly in one direction and recover flexibly in the other direction (like a swimmer's arm), so the recovery stroke does not undo the work of the power stroke (see Chapter 44).

In cross section, a typical cilium or eukaryotic flagellum is seen to be covered by the plasma membrane and to contain what is usually called a "9 + 2" array of microtubules. As Figure 4.25*b* shows, there are actually nine fused pairs of microtubules—called **doublets**—forming an outer cylinder, and one pair of unfused microtubules running up the center. The motion of cilia and flagella results from the sliding of the microtubules past one another (as described in Chapter 44).

But what is the "motor" that drives this sliding? It is a protein called **dynein**, which can undergo changes in tertiary structure driven by energy from ATP. Dynein molecules on one microtubule bind a neighboring microtubule. Then, as the dynein molecules change shape, they "row" one microtubule past its neighbor (Figure 4.26*a*).

Dynein and another motor protein, **kinesin**, are responsible for moving cell organelles in opposite directions along microtubules. Recall that microtubules have a + end and a – end. The dynein moves attached vesicles and other organelles toward the – end of the tubule, while the kinesin moves them toward the + end. These motor proteins attach both to the microtubules, which guide their motion, and to the vesicles that transport materials from the ER to the Golgi apparatus and to the plasma membrane for secretion (Figure 4.26*b*).

Some prokaryotes have flagella, but prokaryotic flagella lack microtubules and dynein. The flagella of

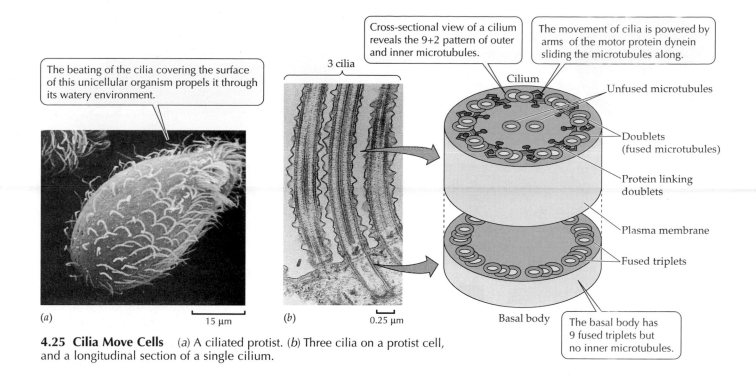

The beating of the cilia covering the surface of this unicellular organism propels it through its watery environment.

Cross-sectional view of a cilium reveals the 9+2 pattern of outer and inner microtubules.

The movement of cilia is powered by arms of the motor protein dynein sliding the microtubules along.

3 cilia

Cilium

Unfused microtubules

Doublets (fused microtubules)

Protein linking doublets

Plasma membrane

Fused triplets

Basal body

The basal body has 9 fused triplets but no inner microtubules.

(a)   15 μm

(b)   0.25 μm

4.25 Cilia Move Cells   (a) A ciliated protist. (b) Three cilia on a protist cell, and a longitudinal section of a single cilium.

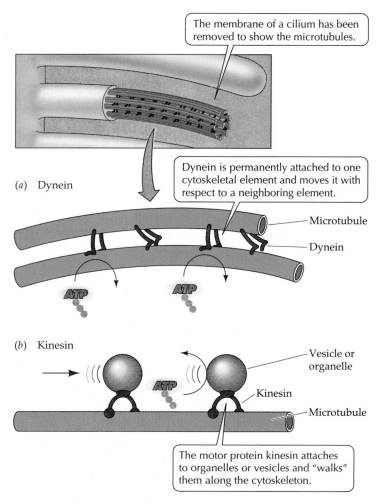

The membrane of a cilium has been removed to show the microtubules.

(a)   Dynein

Dynein is permanently attached to one cytoskeletal element and moves it with respect to a neighboring element.

Microtubule

Dynein

ATP

ATP

(b)   Kinesin

Vesicle or organelle

Kinesin

Microtubule

The motor protein kinesin attaches to organelles or vesicles and "walks" them along the cytoskeleton.

prokaryotes are neither structurally nor evolutionarily related to those of eukaryotes. The prokaryotic flagellum is assembled from a protein called flagellin, and it has a much simpler structure and a smaller diameter than those of a single microtubule. And whereas eukaryotic flagella beat in a wavelike motion, prokaryotic flagella rotate (see Figure 4.5).

At the base of every eukaryotic flagellum or cilium is an organelle called a **basal body** (see Figure 4.25*b*). The nine microtubule doublets extend into the basal body. In the basal body, each doublet is accompanied by another microtubule, making nine sets of *three* microtubules. The central, unfused microtubules of the cilium or flagellum do not extend into the basal body.

Centrioles are organelles that are almost identical to basal bodies. Centrioles are found in all eukaryotes except for cells of the flowering plants, the pines and their relatives, and some protists. Under the light microscope, a centriole looks like a small, featureless particle, but the electron microscope reveals that it is made up of a precise bundle of microtubules, arranged as nine sets of three

4.26 Motor Proteins Use Energy from ATP to Move Things   (a) Dynein operates in muscle contraction and flagellar movement. (b) Kinesin delivers vesicles to various parts of the cell. All motor proteins work by undergoing reversible shape changes powered by energy from ATP.

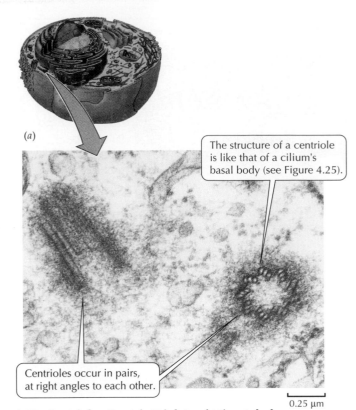

(a)

> The structure of a centriole is like that of a cilium's basal body (see Figure 4.25).

> Centrioles occur in pairs, at right angles to each other.

0.25 μm

4.27 Centrioles Contain Triplets of Microtubules
Centrioles are found in the microtubule organizing center, a region near the nucleus. The electron micrograph (a) shows a pair of centrioles at right angles to each other. Nine sets of three fused microtubules are evident in the centriole on the right, which is seen in cross section. The diagram (b) emphasizes the three-dimensional structure of a centriole.

fused microtubules each (Figure 4.27). Centrioles lie in the microtubule organizing center in cells that are about to undergo division.

Extracellular Structures

Although the plasma membrane is the functional barrier between the inside and outside of a cell, many structures outside the plasma membrane are produced by the cell and play essential roles in protecting, supporting, or attaching cells. These structures are said to be **extracellular**. In bacteria, the peptidoglycan cell wall is the extracellular structure that plays these roles. In eukaryotes, a variety of extracellular structures play these roles: in plants, the cellulose cell wall; in animals, the extracellular matrix between cells in multicellular organisms and tissues.

The plant cell wall consists largely of cellulose

The plant **cell wall** is a semirigid structure outside the plasma membrane (Figure 4.28). It consists primarily of polysaccharides, the most prominent of which is cellulose (see Chapter 3). The cell wall provides sup-

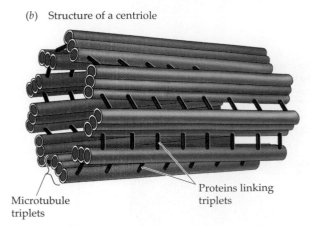

(b) Structure of a centriole

Microtubule triplets

Proteins linking triplets

port for the cell and limits its volume by remaining rigid. In some instances in which hydrophobic substances are added to the cellulose, the cell wall can prevent water from reaching the plasma membrane. Modifications of the cell wall such as the addition of substances are important in determining the functional roles of some plant cells (see Chapter 31).

Because of their thick cell walls, under the light microscope plant cells appear entirely isolated from each other, but electron microscopy reveals that this is not the case. The cytoplasm of adjacent plant cells is connected by numerous plasma membrane-lined channels, called **plasmodesmata**, that are about 20 to 40 nm in diameter and extend through the walls of adjoining cells (see Figure 4.28). These connections permit the diffusion of water, ions, small molecules, and many proteins between connected cells. Such diffusion ensures that cells have uniform concentrations of these substances.

Multicellular animals have an elaborate extracellular matrix

In multicellular animals, cells lack the semirigid cell wall that is characteristic of plant cells, but many animal cells are surrounded by or are in contact with an **extracellular matrix**. This matrix is composed of proteins such as collagen (the most abundant protein in mammals) and glycoproteins. These proteins and other substances are secreted by cells that are present in or near the matrix. In the human body, some tissues, such as those in the brain, have very little extracellular matrix. Other tissues, such as bone and cartilage, have large amounts of extracellular matrix.

The cells embedded in bone and cartilage secrete and maintain the characteristic composition and structure of the extracellular material. Bone cells are embedded in an extracellular matrix that consists primarily of collagen and substantial amounts of the ionic solid calcium phosphate. This matrix gives bone its familiar rigidity. Epithelial cells, which line body cavi-

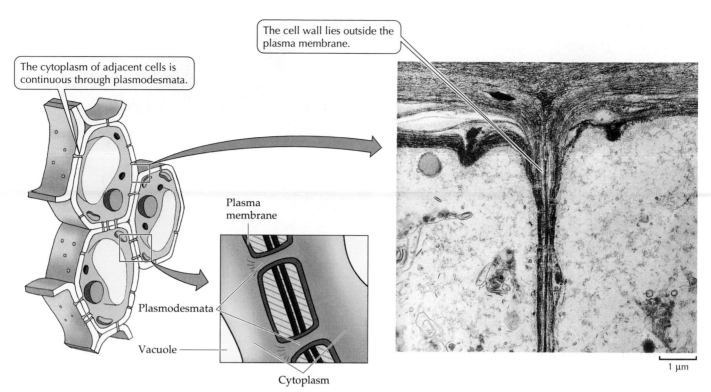

The cell wall lies outside the plasma membrane.

The cytoplasm of adjacent cells is continuous through plasmodesmata.

Plasma membrane

Plasmodesmata

Vacuole

Cytoplasm

1 µm

4.28 The Plant Cell Wall This semirigid structure provides support for plant cells. Plasmodesmata allow water and other molecules to cross the wall and move from cell to cell.

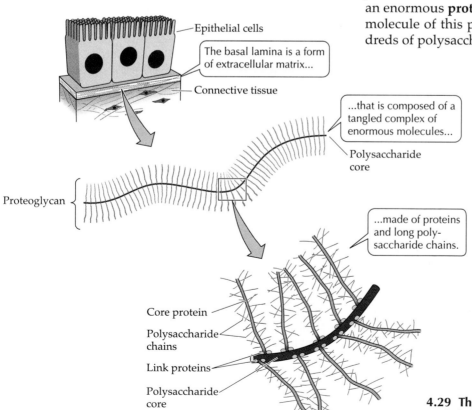

Epithelial cells

The basal lamina is a form of extracellular matrix...

Connective tissue

...that is composed of a tangled complex of enormous molecules...

Polysaccharide core

Proteoglycan

...made of proteins and long poly- saccharide chains.

Core protein

Polysaccharide chains

Link proteins

Polysaccharide core

ties, lie together as a sheet spread over the **basal lamina**, or basement membrane, a form of extracellular matrix (Figure 4.29).

In some cases the extracellular matrix is made up, in part, of one of the most spectacular molecules ever seen, an enormous **proteoglycan** (see Figure 4.29). A single molecule of this proteoglycan consists of many hundreds of polysaccharides attached to about a hundred proteins, all of which are attached to one enormous polysaccharide. The molecular weight of the proteoglycan can exceed 100 million; the molecule takes up as much space as an entire prokaryotic cell. What do such fantastic molecules do for the organism?

The proteoglycans and other components of the extracellular matrix contribute to the physical properties of cartilage, skin, and other tissues. They help filter materials passing between the blood and urine. They help orient cell migration during embryonic development and during tissue regeneration following injury. They play roles in chemi-

4.29 The Extracellular Matrix of Animal Cells Epithelial cells secrete a basal lamina, which anchors them to the connective tissue.

Nonenveloped virus

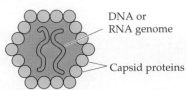

DNA or
RNA genome

Capsid proteins

4.30 Viruses Are Incapable of Reproducing on Their Own Viruses lack the ribosomes, enzymes, and other cellular constituents necessary for independent reproduction and processing of raw materials and energy. Some viruses are enveloped by a membrane, whereas others lack such membranes.

Enveloped virus

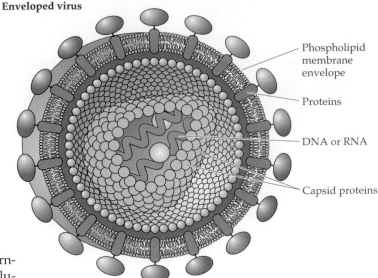

Phospholipid
membrane
envelope

Proteins

DNA or RNA

Capsid proteins

cal signaling from one cell to another. We are still learning about the structure and functions of the extracellular matrix.

Cells versus Viruses

As you have discovered, in contrast to prokaryotes, eukaryotes have a true nucleus and other membrane-enclosed organelles that allow various cellular activities to be concentrated in specialized compartments. In addition, eukaryotes have many specialized molecules, such as tubulin and actin, that are not found in prokaryotes. The cell walls of prokaryotes differ structurally and chemically from those of eukaryotes. And, as we'll see in Chapter 14, prokaryotic DNA and eukaryotic DNA have very different organization.

Do such differences mean that eukaryotes are more advanced, or "higher," or more successful than prokaryotes? Not at all. Every surviving species is the product of eons of natural selection and is superbly adapted to its environment. Each species has characteristics that enable it to live where and how it does, and to compete successfully against other species. But where do viruses fit into this picture?

Viruses are "gigantic" macromolecular assemblies of proteins and nucleic acid, some of which are surrounded by a phospholipid membrane (Figure 4.30). But they are not cells. Viruses lack many of the important characteristics of living systems. They cannot reproduce outside of the living cells that supply the raw materials and energy from which new viruses are constructed. Specific kinds of viruses infect animal, plant, fungal, protist, and prokaryotic cells. When a virus infects a host cell, the viral nucleic acid enters the cell and subverts the host's metabolic machinery to make new viruses (see Chapter 13).

Unlike cells, viruses do not transform energy, nor are they responsive to environmental stimuli as are organisms. But viruses do reproduce, and they evolve. Many different viruses infect humans, causing both mild diseases such as the common cold and life-threat-

ening conditions such as AIDS. Some viruses have been implicated in certain forms of cancer in humans.

Isolating Organelles for Study

During the early days of cell biology in the nineteenth century, all that anyone could do with organelles— and only the largest ones at that—was to observe them with a light microscope. The refinement of the electron microscope in the middle of the twentieth century made it possible to view cells and their organelles at higher magnification and greater resolution.

However, to answer questions about the composition and function of such organelles, it was necessary to perform chemical analyses, and such analyses required that relatively large quantities of the different organelles be obtained in pure form. To accomplish this task, the processes of cell fractionation were developed. In **cell fractionation**, cells are first ruptured; then a system of centrifugation separates the organelles from one another.

Depending on the type of cell or tissue, various methods are available to break cells open without excessively damaging organelles or proteins. The simplest methods use an old-fashioned mortar and pestle or a hand-operated glass homogenizer (which squeezes and shears cells between two tightly fitting, counter-rotating ground-glass surfaces). Motor-driven homogenizers or blenders are also commonly used.

All these methods break open plasma membranes and (if present) cell walls, liberating the cytoplasm into a solution that contains a substantial concentration of solutes. This solution prevents the organelles from bursting as they would in a more dilute solution or in pure water (see Figure 5.12). Cooling with ice further minimizes damage to the organelles and proteins.

The techniques described here reduce biological tissue to a crude suspension of mixed organelles, unbroken cells, and debris. To separate the components of such a suspension requires one of two types of *centrifugation*. The **centrifuge** is a laboratory instrument that can spin materials extremely rapidly about a fixed axis. This rapid rotation generates large centrifugal forces that cause components of the suspension to sediment according to mass, size, or density. The two common types of centrifugation are differential centrifugation and equilibrium centrifugation.

Differential centrifugation is a process of repeated centrifugations at increasing speeds, at higher forces (designated as multiples of the force of gravity, *g*), and for longer periods of time (Figure 4.31). After cells are ruptured in an appropriate solution, the mixture is centrifuged briefly at low speed (and hence low relative centrifugal force). The largest and densest parti-

cles sediment out, forming a *pellet* in the bottom of the tube and a liquid *supernatant fluid* above. This pellet is left behind for further study when the supernatant fluid and its suspended contents are poured off. The supernatant fluid is collected and spun at a higher speed and for a longer time, causing other organelles to sediment to the bottom. By repeating this procedure with ever-increasing centrifugal force, one can separate out many organelles.

After cell fractionation is completed, the nuclei, mitochondria, components of the ER, and free ribosomes are isolated from each other in separate tubes for further study. The contents of each tube can be purified further by repeating the centrifugation routine. The contents and purity of each pellet are confirmed by the electron microscope or by enzyme assays. Variations on this procedure have revealed extensive information about the activities and interrelationships of the different organelles and about the life of the cell.

Equilibrium centrifugation differs from differential centrifugation not in its goals but in its use of a solution that varies in density from the top to the bottom

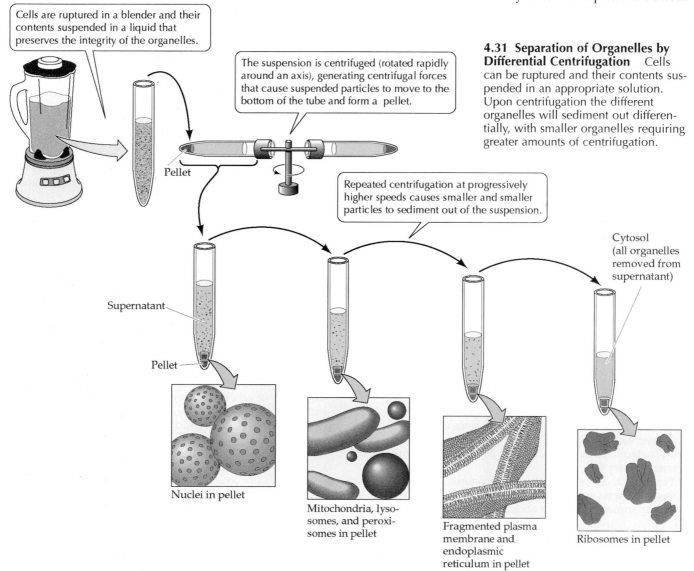

Cells are ruptured in a blender and their contents suspended in a liquid that preserves the integrity of the organelles.

The suspension is centrifuged (rotated rapidly around an axis), generating centrifugal forces that cause suspended particles to move to the bottom of the tube and form a pellet.

4.31 Separation of Organelles by Differential Centrifugation Cells can be ruptured and their contents suspended in an appropriate solution. Upon centrifugation the different organelles will sediment out differentially, with smaller organelles requiring greater amounts of centrifugation.

Repeated centrifugation at progressively higher speeds causes smaller and smaller particles to sediment out of the suspension.

Pellet

Supernatant

Pellet

Cytosol (all organelles removed from supernatant)

Nuclei in pellet

Mitochondria, lysosomes, and peroxisomes in pellet

Fragmented plasma membrane and endoplasmic reticulum in pellet

Ribosomes in pellet

4.32 Equilibrium Centrifugation A density gradient of sucrose solution is prepared as described in the text and a sample of the suspension to be fractionated is added. During centrifugation, particles of different density move to discrete bands in the gradient and can be recovered in separate tubes.

Organelles in suspension

A density gradient of sucrose solution is prepared. A sample of the suspension to be fractionated is gently added to the top of the tube.

During centrifugation, organelles of different density will move to the location in the tube where the sucrose density is equal to that of the organelle, forming discrete bands.

Sucrose gradient

Centrifuge

Lower density

Higher density

A small hole is punched in the bottom of the tube and successive layers or fractions are collected in separate tubes.

of a centrifuge tube (Figure 4.32). This system permits the separation of substances that have different densities. A density gradient is prepared in a plastic centrifuge tube by first adding a small amount of a highly concentrated (and hence very dense) sugar solution—say, 60 percent sucrose. Next, a small amount of a 50 percent sucrose solution is layered on top, then 40 percent, and so on. In practice, such a **density gradient** is usually constructed by an automatic device, so changes in density are smooth rather than abrupt.

After the gradient is established, the mixture of organelles to be separated is carefully layered on the top of the sugar solution, and the tube is centrifuged. Under the forces generated by centrifugation, organelles sediment into the gradient as long as they are more dense than the surrounding liquid. Once an organelle reaches that part of the gradient where its density matches the density of the solution, it stops sedimenting—it has reached buoyant equilibrium. If two kinds of organelles have different average densities, they form two separate bands, which can be collected by using a pipette or by punching a hole in the bottom of the tube and collecting drops from it in separate tubes.

Summary of "The Organization of Cells"

The Cell: The Basic Unit of Life
• All cells come from preexisting cells and have certain processes, types of molecules, and structures in common.
• To maintain adequate exchanges with its environment, a cell's surface area must be large compared to its volume. **Review Figure 4.2**
• Prokaryotic cell organization is characteristic of the kingdoms Eubacteria and Archaebacteria. Prokaryotes lack internal compartments such as a nucleus. **Review Figure 4.3**

• Eukaryotic cell organization is characteristic of cells in the other four kingdoms. Eukaryotic cells have many membrane-enclosed compartments, including a nucleus that contains DNA. **Review Figure 4.7**

Prokaryotic Cells
• All prokaryotic cells have a plasma membrane, a nucleoid region with DNA, and a cytoplasm that contains ribosomes, dissolved enzymes, water, and small molecules. Some prokaryotes have additional protective structures: cell wall, outer membrane, and capsule. Some prokaryotes contain photosynthetic membranes, and some show mesosomes. **Review Figures 4.3, 4.4**
• Projecting from the surface of some prokaryotes, rotating flagella move prokaryotic cells from place to place. Pili are sites at which prokaryotic cells attach to one another or to environmental surfaces. **Review Figure 4.5**

Microscopes: Revealing the Subcellular World
• Because of their greater resolving power, electron microscopes enable observation of greater detail than can be seen with light microscopes. Whereas light microscopy can be used for viewing either dead or living cells, electron microscopy can be used only with preserved dead material. **Review Table 4.1**
• Scanning electron microscopy gives a three-dimensional view of surfaces. Transmission electron microscopy gives a two-dimensional view of a three-dimensional reality. **Review Figures 4.3, 4.4**

Eukaryotic Cells

• Like prokaryotic cells, eukaryotic cells have a plasma membrane, cytoplasm, and ribosomes. However, eukaryotic cells are larger and contain many membrane-enclosed organelles, such as the nucleus (containing DNA), mitochondria, endoplasmic reticulum, and Golgi apparatus. **Review Figure 4.7**

• Plant cells have chloroplasts and cellulose cell walls not found in animal cells.

• The membranes that envelop organelles in the eukaryotic cell are partial barriers, ensuring that the chemical composition of the interior of the organelle differs from the chemical composition of the surrounding cytoplasm.

Organelles That Process Information

• The nucleus is usually the largest organelle in a cell and is bounded by two membranes—together called the nuclear envelope, which disassembles during nuclear division. Within the nucleus, the nucleoli are the source of the ribosomes found in the cytoplasm. Ribosomes participate in protein synthesis and are sometimes attached to the endoplasmic reticulum. **Review Figure 4.8**

• The nucleus contains most of the cell's DNA, which is associated with protein to form chromatin. Chromatin has a diffuse appearance, but before nuclear division it condenses to form chromosomes. **Review Figure 4.10**

• Nuclear pores have complex structures that govern what enters and leaves the nucleus. **Review Figure 4.8**

Organelles That Process Energy

• Mitochondria are enclosed by an outer membrane and an inner membrane that folds inward to form cristae. Mitochondria contain the enzymes for cellular respiration and the generation of ATP. **Review Figure 4.11**

• Almost all eukaryotic cells contain mitochondria. However, green plant cells also contain chloroplasts, which are enclosed by two membranes and contain an internal system of thylakoids organized as grana. **Review Figures 4.7, 4.13**

• Thylakoids within chloroplasts contain the chlorophyll and proteins that harvest light energy for the synthesis of glucose from carbon dioxide. **Review Figure 4.13**

• Both mitochondria and chloroplasts contain their own DNA and ribosomes and are capable of making some of their own proteins.

• The endosymbiosis theory of the evolutionary origin of mitochondria and chloroplasts states that mitochondria and chloroplasts originated when larger prokaryotes engulfed but did not digest smaller prokaryotes. Mutual benefits permitted this symbiotic relationship to be maintained and to evolve into the eukaryotic organelles observed today. **Review Figure 4.15**

The Endomembrane System

• The endomembrane system is a series of interrelated membranes and compartments.

• The rough ER has attached ribosomes that synthesize proteins. These proteins enter the lumen of the ER, where they are chemically processed and sorted into vesicles. The smooth ER lacks ribosomes and is associated with the synthesis of lipids. **Review Figures 4.7, 4.17**

• The Golgi apparatus receives materials from the rough ER vesicles that fuse with the *cis* region of the Golgi. Within the lumen of the Golgi, signal molecules are added to proteins, directing them to their proper destinations. Vesicles originating from the *trans* region of the Golgi contain proteins for different cellular functions. **Review Figures 4.7, 4.17**

• Some vesicles fuse with the plasma membrane and release their contents outside the cell. Other vesicles, such as lysosomes, are retained within the cell. **Review Figure 4.17**

• Lysosomes contain many digestive enzymes. Lysosomes fuse with the phagosomes produced by phagocytosis to form secondary lysosomes in which engulfed materials are digested. Undigested materials are excreted from the cell when the secondary lysosome fuses with the plasma membrane. **Review Figure 4.18**

Other Organelles

• Membrane-enclosed organelles include peroxisomes and glyoxysomes. These organelles contain special enzymes and carry out specialized chemical reactions for the cell.

• Vacuoles are prominent in many plant cells and consist of a membrane-enclosed compartment that contains water and dissolved substances. By receiving water, vacuoles enlarge and provide the pressure needed to stretch the cell wall during plant cell growth.

The Cytoskeleton

• The cytoskeleton within the cytoplasm of eukaryotic cells provides shape, strength, and movement. It consists of three interacting types of protein fibers with different diameters. **Review Figure 4.21**

• Microfilaments consist of two helical chains of G actin units. Microfilaments strengthen cellular structures such as microvilli and provide movement involved in animal cell division, cytoplasmic streaming, and pseudopod extension. Microfilaments are found as independent fibers, bundles of fibers, or networks of fibers joined by linking proteins. **Review Figures 4.21, 4.23**

• Intermediate filaments are formed of proteins such as keratin and are organized into tough ropelike structures that add strength to cell attachments in multicellular organisms. **Review Figure 4.21**

• Microtubules are composed of dimers of the protein tubulin. They can lengthen and shorten by adding and losing tubulin dimer units and are involved in the distribution of chromosomes during nuclear division. They are involved in the structure and function of cilia and flagella, both of which have a characteristic 9 + 2 pattern of microtubules. The movements of cilia and flagella are due to the motor protein dynein. Basal bodies and centrioles are also constructed of microtubules. Microtubules also move cellular organelles through the action of motor proteins, including kinesin and dynein, that use energy from ATP. **Review Figures 4. 25, 4.26, 4.27**

Extracellular Structures

• Materials external to the plasma membrane provide protection, support, and attachment for cells in multicellular systems.

• The cell wall of plants consists principally of cellulose and is pierced by plasmodesmata that join the cytoplasm of adjacent cells. **Review Figure 4.28**

- In multicellular animals, the extracellular matrix consists of different kinds of proteins, including proteoglycan. In bone and cartilage the protein collagen predominates. **Review Figure 4.29**

Cells versus Viruses

- Although prokaryotes and eukaryotes differ in specific ways, they share characteristics that identify them as living cells. Viruses are not living. They are smaller than the smallest cells and consist of nucleic acid, protein, and sometimes a membrane. Like cells, viruses use nucleic acid as hereditary material, but they require a living cell in order to reproduce. And viruses do not transform matter and energy or respond to the environment. **Review Figure 4.30**

Isolating Organelles for Study

- To study an organelle chemically, one must isolate it from other cell contents by rupturing the cells and suspending their contents in a liquid medium, then subjecting this suspension to centrifugation.
- In differential centrifugation, a series of rotations at increasing speeds followed by separation of supernatant fluid and pellet results in the isolation of organelles in separate tubes. **Review Figure 4.31**
- In equilibrium centrifugation, cellular materials are added to a sucrose solution that has a density gradient and centrifuged so that cellular materials with different densities come to reside at different sites in the gradient. **Review Figure 4.32**

Self-Quiz

1. Which statement is true of both prokaryotic and eukaryotic cells?
   a. They contain ribosomes.
   b. They have peptidoglycan cell walls.
   c. They contain membrane-enclosed organelles.
   d. They contain true nuclei.
   e. Their flagella have the 9 + 2 structure.

2. Which statement about the nuclear envelope is *not* true?
   a. It is continuous with the endoplasmic reticulum.
   b. It has pores.
   c. It consists of two membranes.
   d. RNA and some proteins pass through it to move in and out of the nucleus.
   e. Its inner membrane bears ribosomes.

3. Which statement about mitochondria is *not* true?
   a. Their inner membrane folds to form cristae.
   b. They are usually 1 μm or less in diameter.
   c. They are green because of the chlorophyll they contain.
   d. Energy-rich substances from the cytosol are oxidized in them.
   e. Much ATP is synthesized in them.

4. Which statement about plastids is true?
   a. They are found in prokaryotes.
   b. They are surrounded by a single membrane.
   c. They are the sites of cellular respiration.
   d. They are found in fungi.
   e. They are of several types, with different functions.

5. Which statement about the endoplasmic reticulum is *not* true?
   a. It is of two types: rough and smooth.
   b. It is a network of tubes and flattened sacs.
   c. It is found in all living cells.
   d. Some of it is sprinkled with ribosomes.
   e. Parts of it modify proteins.

6. The Golgi apparatus
   a. is found only in animals.
   b. is found in prokaryotes.
   c. is the appendage that moves a cell around in its environment.
   d. is a site of rapid ATP production.
   e. packages and modifies proteins.

7. Which organelle is *not* surrounded by one or more membranes?
   a. Ribosome
   b. Chloroplast
   c. Mitochondrion
   d. Peroxisome
   e. Vacuole

8. Eukaryotic flagella
   a. are composed of a protein called flagellin.
   b. rotate like propellers.
   c. cause the cell to contract.
   d. have the same internal structure as cilia.
   e. cause the movement of chromosomes.

9. Microfilaments
   a. are composed of polysaccharides.
   b. are composed of actin.
   c. provide the motive force for cilia and flagella.
   d. make up the spindle that aids the movement of chromosomes.
   e. maintain the position of the nucleus in the cell.

10. Which statement about the plant cell wall is *not* true?
    a. Its principal chemical components are polysaccharides.
    b. It lies outside the plasma membrane.
    c. It provides support for the cell.
    d. It completely isolates adjacent cells from one another.
    e. It is semirigid.

Applying Concepts

1. Which organelles and other structures are found in both plant and animal cells? Which are found in plant but not animal cells? Which in animal but not plant cells? Discuss, in relation to the activities of plants and animals.

2. Through how many membranes would a molecule have to pass in going from the interior of a chloroplast to the interior of a mitochondrion? From the interior of a lysosome to the outside of a cell? From one ribosome to another?

3. How does the possession of double membranes by chloroplasts and mitochondria relate to the endosymbiosis theory of the origins of these organelles? What other evidence supports the theory?

4. What sorts of cells and subcellular structures would you choose to examine by transmission electron microscopy? By scanning electron microscopy? By light microscopy? What are the advantages and disadvantages of each of these modes of microscopy?

5. Some organelles that cannot be separated from one another by equilibrium centrifugation can be separated by differential centrifugation. Other organelles cannot be separated from one another by differential centrifugation but can be separated by equilibrium centrifugation. Explain these observations.

Readings

Alberts, B., D. Bray, J. Lewis, M. Raff, K. Roberts and J. D. Watson. 1994. *Molecular Biology of the Cell*, 3rd Edition. Garland, New York. An outstanding book in which to pursue the topics of this chapter in greater detail; authoritative treatment of modern cell biology and its experimental basis.

Allen, R. D. 1987. "The Microtubule as an Intracellular Engine." *Scientific American*, February. A description of how microtubules enable two-way transport of materials in cells.

Brandt, W. H. 1975. *The Student's Guide to Optical Microscopes*. William Kaufmann, Los Altos, CA. A short, programmed guide for those interested in learning how to use a light microscope.

Cooper, G. M. 1997. *The Cell: A Molecular Approach*. ASM Press/Sinauer Associates, Sunderland, MA. An excellent textbook—up-to-date, well illustrated, and concise.

De Duve, C. 1975. "Exploring Cells with a Centrifuge." *Science*, vol. 189, pages 186–194. A discussion by a Nobel laureate of the uses of centrifugation in studies of cells.

Fawcett, D. W. 1981. *The Cell*, 2nd Edition. Saunders, Philadelphia. Beautiful electron micrographs of subcellular structures in animal cells.

Glover, D. M., C. Gonzalez and J. W. Raff. 1993. "The Centrosome." *Scientific American*, June. A report of new findings about the structure and function of the organelle that directs the assembly of the cytoskeleton and controls cell division—when it is present.

Howells, M. R., J. Kirz and D. Sayre. 1991. "X-Ray Microscopes." *Scientific American*, February. A description of novel methods of microscopy that afford striking improvements in resolution.

Lodish, H., D. Baltimore, A. Berk, S. L. Zipursky, P. Matsudaira and J. Darnell. 1995. *Molecular Cell Biology*, 3rd Edition. Scientific American Books, New York. Another excellent middle-level book; fine illustrations.

Margulis, L. 1993. *Symbiosis in Cell Evolution*, 2nd Edition. W. H. Freeman, New York. An authoritative and thought-provoking reference on the origin and evolution of eukaryotic cells by a leading student of the problem.

Rothman, J. E. 1985. "The Compartmental Organization of the Golgi Apparatus." *Scientific American*, September. A discussion of the structure and function of the Golgi apparatus.

Weber, K. and M. Osborn. 1985. "The Molecules of the Cell Matrix." *Scientific American*, October. A clear treatment of microfilaments, intermediate filaments, tubulin, and the ways in which they are studied.

Chapter 5

Cellular Membranes

Complex Membranes That Process Solar Energy
The distinct proteins that appear as dots embedded in these thylakoids from a spinach chloroplast, magnified about 80,000 times, are necessary for photosynthesis.

*B*iological membranes are both abundant and essential in cells. Membranes form the compartments that isolate the interior of cells from their outside environments, and they isolate internal cell processes in different organelles. Biological membranes function in this capacity because their lipid composition and hydrophobic bilayer structure make them effective barriers to the passage of many hydrophilic substances. However, some substances move across membranes in spite of these restrictions, and this chapter will describe both passive and active transport processes.

Biological membranes are more than just barriers that regulate the passage of substances into and out of the cell. In addition to defining compartments and restricting movement between them, membranes process materials, energy, and information. Many plasma membranes are veritable antennae for information in their environment. They respond to some signals by puckering up and nibbling at bits of the environment. Other signals—chemical messengers from other parts of the body—cause dramatic changes in the "receiving" cell. A flash of light falling on a rod cell in our eye causes the cell's plasma membrane to become less permeable to sodium ions; this is one of the first steps leading to our seeing the flash. Poke an electrode through the plasma membrane of any living cell and you discover that the membrane is electrically charged, with the interior of the cell electrically more negative than the exterior. This electric charge permits nerve cell membranes to carry messages.

In this chapter, we'll examine the molecular composition and structure of membranes and consider specialized structures in plasma membranes. Then we'll explore the passive and active processes by which substances move across membranes to enter and leave cells and their compartments. Finally we'll con-

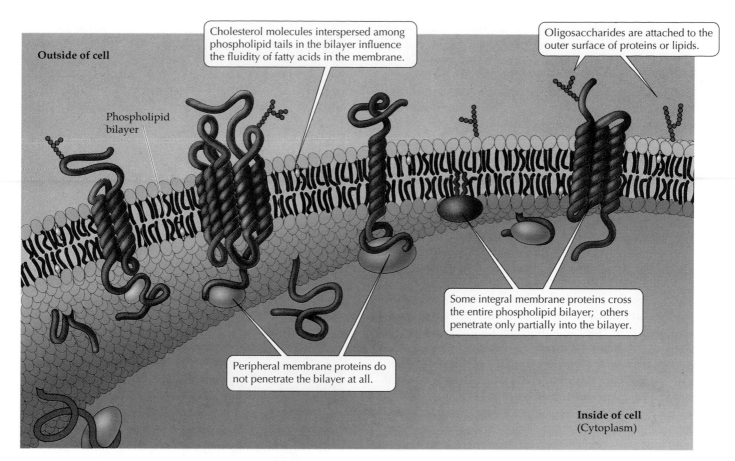

Outside of cell

Cholesterol molecules interspersed among phospholipid tails in the bilayer influence the fluidity of fatty acids in the membrane.

Oligosaccharides are attached to the outer surface of proteins or lipids.

Phospholipid bilayer

Some integral membrane proteins cross the entire phospholipid bilayer; others penetrate only partially into the bilayer.

Peripheral membrane proteins do not penetrate the bilayer at all.

Inside of cell
(Cytoplasm)

5.1 The Fluid Mosaic Model The general molecular structure of biological membranes is a continuous phospholipid bilayer in which proteins are embedded. On the outer surface, carbohydrates may be attached to proteins or phospholipids.

sider how membranes contribute to processing information and transforming energy, as well as the dynamic interrelationships between membranes.

Membrane Composition and Structure

The plasma membrane and other cellular **membranes** are thin, pliable bilayers of phospholipids with embedded proteins (Figure 5.1). (The phospholipid composition and bilayer architecture of biological membranes were introduced in Chapter 3, and membrane-enclosed organelles were described in Chapter 4.) As we consider membranes, the relationship between what a membrane does (its functions) and its chemical composition and physical structure is particularly obvious. Because of their composition and structure, biological membranes are **selectively permeable**; that is, they permit some substances to pass through but block others.

The chemical makeup, physical organization, and functioning of a biological membrane depend on three classes of biochemical compounds: lipids, proteins, and carbohydrates (see Figure 5.1). The phospholipids establish the physical integrity of the membrane and are an effective barrier to the passage of many hydrophilic materials. In addition to serving as a barrier, the phospholipid bilayer is a lipid "lake" in which a variety of proteins "float." This general design is known as the **fluid mosaic model** of the membrane. It applies to all biological membranes, although the membranes of archaea differ from the model in certain details.

Membrane proteins stretch across the phospholipid bilayer and protrude on both sides. They are responsible for many of the specific tasks membranes perform. Certain membrane proteins allow materials to pass through the membrane that cannot pass through the pure lipid bilayer. Other proteins receive chemical signals from the cell's external environment and respond by regulating certain processes inside the cell. Still other proteins function as enzymes and accelerate chemical reactions on the membrane surface.

Like some proteins, carbohydrates—the third class of compounds important in membranes—are crucial in recognizing specific molecules. The carbohydrates attach either to lipid or to protein molecules on the outside of the plasma membrane, where they protrude into the environment, away from the cell.

Let's look at each of the three major components of membranes—lipids, proteins, and carbohydrates—in more detail.

Lipids constitute the bulk of a membrane

Nearly all lipids in biological membranes are **phospholipids**. Recall from our discussion in Chapter 3 that some compounds are hydrophilic ("water loving") and others are hydrophobic ("water fearing"). Phospholipids are both: They have both hydrophilic regions and hydrophobic regions.

The long, nonpolar fatty acid parts of phospholipids are hydrophobic and associate easily with other nonpolar materials, but they do not dissolve in water or associate with hydrophilic substances. The phosphorus-containing region of the phospholipid is electrically charged and hence very hydrophilic. As a consequence, one way for phospholipids and water to coexist is for the phospholipids to form a double layer, with the fatty acids of the two layers interacting with each other and the polar regions facing the outside water environment (Figure 5.2). It is easy to make artificial membranes with the same two-layered arrangement in the laboratory. Both artificial and natural membranes form continuous sheets. Because of the tendency of the fatty acids to associate with one another and exclude water, small holes or rips in a membrane seal themselves spontaneously. This property helps membranes fuse during vesicle fusion, phagocytosis, and related processes.

The phospholipid bilayer stabilizes the entire membrane structure. At the same time, the fatty acids of the phospholipids make the membrane somewhat fluid—about as fluid as lightweight machine oil—so some material can move laterally within the plane of the membrane. As we will see, some membrane proteins migrate about relatively freely, and individual phospholipid molecules may also "travel."

A given phospholipid molecule in the plasma membrane of a bacterium may travel from one end of the bacterium to the other in a little more than a second. On the other hand, seldom does a phospholipid molecule in one half of the bilayer flop over to the other side and trade places with another phospholipid molecule. For such a swap to happen, the polar part of each molecule would have to move through the hydrophobic interior of the membrane. Since phospholipid flip-flops are rare, the two halves of the bilayer may be quite different in the kinds of phospholipids present.

All biological membranes are similar, but membranes from different cells or organelles may differ greatly in their lipid composition. For example, 25 percent of the lipid in many membranes is *cholesterol* (see Chapter 3), but some membranes have no cholesterol at all. When present, cholesterol is commonly situated

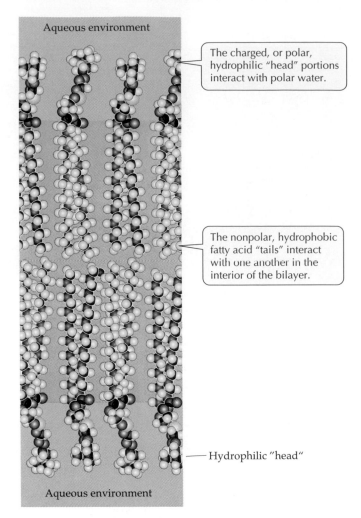

The charged, or polar, hydrophilic "head" portions interact with polar water.

The nonpolar, hydrophobic fatty acid "tails" interact with one another in the interior of the bilayer.

Hydrophilic "head"

5.2 A Phospholipid Bilayer Separates Two Aqueous Regions The eight phospholipid molecules shown here represent a small section of a membrane bilayer.

next to an unsaturated fatty acid, and its polar region extends into the surrounding aqueous layer (see Figure 5.1). Cholesterol plays more than one role in determining the fluidity of the membrane: It may either increase or decrease membrane fluidity, depending on the circumstances. Shorter fatty acid chains make for a more fluid membrane, as do unsaturated fatty acids.

Organisms can modify their membrane lipid composition, thus changing membrane fluidity, to compensate for changes in temperature. For example, some houseplants can survive both indoor and outdoor temperatures. But when accustomed to indoor temperatures, they can die if suddenly placed outdoors at cooler, but nonfreezing, temperatures. The sudden change in temperature does not give the plants sufficient time to adjust their membrane lipid composition. Lipids constitute a major fraction of all membranes, and they always form the continuous matrix into which the other chemical components become inserted.

Membrane components are revealed by freeze-fracturing

To determine how proteins and other components are inserted in the membrane, we use techniques that begin by freezing the tissue sample. The frozen tissue is then fractured (broken), splitting the lipid bilayer of the membrane and exposing the integral membrane proteins as bumps. The figure at the beginning of this chapter was produced by this *freeze-fracturing* technique. As an analogy, consider that slicing a chocolate-almond candy bar with a sharp razor gives one view of the interior; but if you break the bar instead, you reveal the almonds as protruding lumps.

We can reveal further detail in tissue by putting a freeze-fractured sample under a high vacuum and allowing water to evaporate. This *freeze-etching* technique reveals more texture. Freeze-etched samples can also be sprayed with metals such as platinum (shadowcasting) to reveal shadow patterns that further enhance contrast (Figure 5.3).

Membrane proteins are asymmetrically distributed

Biological membranes possess two types of proteins: integral membrane proteins and peripheral membrane proteins (see Figure 5.1). **Integral membrane proteins** penetrate the phospholipid bilayer, and many are *transmembrane* proteins, extending from one side of the membrane to the other. **Peripheral membrane**

5.3 Membrane Proteins Revealed by Freeze-Etching
The outer membrane of this mitochondrion has been fractured away, exposing the inner membrane. The image has been magnified about 65,000 times. The particles giving the inner membrane a grainy appearance are proteins necessary for cellular respiration.

proteins are not embedded in the bilayer. Instead, they are attached to exposed parts of the integral membrane proteins or phospholipid molecules by weak (noncovalent) bonds. These two types of proteins play a variety of different roles in membrane function and cellular processes.

The membranes of the various organelles differ sharply in protein composition, and different organelles house quite different chemical reactions (many of them requiring membrane-bound enzymes). In cellular respiration and photosynthesis, membrane-bound enzymes carry electrons from a donor to an acceptor molecule. Accordingly, both mitochondria and chloroplasts have highly specialized internal membranes, and these differ markedly (see Figure 5.3 and the figure at the beginning of the chapter).

Many membrane proteins move relatively freely within the phospholipid bilayer. Experiments using the technique of cell fusion illustrate this migration dramatically. In the laboratory, specially treated cells from humans and mice can be fused so that one continuous plasma membrane surrounds the combined cytoplasm and both nuclei. Initially, the membrane proteins from the two different cells reside in distinguishable halves of the joint plasma membrane. However, the membrane proteins of the two cells migrate, and after about 40 minutes they are uniformly dispersed (Figure 5.4). Although many membrane proteins are mobile in the membrane, there is also good evidence that other membrane proteins are not free to migrate. These proteins are "anchored" by components of the cytoskeleton as described in Chapter 4 (see Figure 4.25).

Proteins are asymmetrically distributed in membranes. Many transmembrane proteins show different "faces" on the two membrane surfaces. Such proteins have certain specific domains (or regions) of their primary structure on one side of the membrane, other domains within the membrane, and still other domains on the other side of the membrane. Peripheral membrane proteins are localized on one side of the membrane or the other, but not both. This arrangement gives the two sides or surfaces of the membrane different properties.

In addition to surface differences, there are regional differences. Some membrane proteins are confined to one part of the cell surface rather than being scattered evenly. This segregation of different proteins contributes to the functional specialization of various regions on the cell surface. For example, in certain muscle cells, the plasma membrane protein receptor for the chemical signal from nerve cells is normally found only at the site where a nerve cell meets the muscle cell. None of this protein is found elsewhere on the surface of the muscle cell. However, if the nerve is experimentally separated from the muscle, the protein receptor molecules become evenly distributed through-

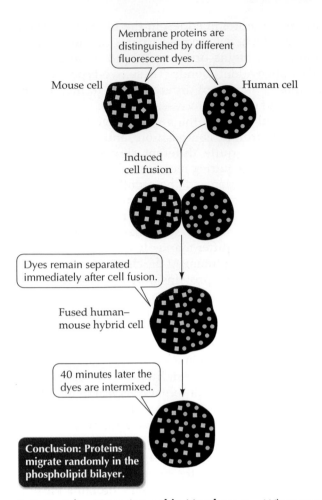

Membrane proteins are distinguished by different fluorescent dyes.

Mouse cell   Human cell

Induced cell fusion

Dyes remain separated immediately after cell fusion.

Fused human–mouse hybrid cell

40 minutes later the dyes are intermixed.

Conclusion: Proteins migrate randomly in the phospholipid bilayer.

5.4 Proteins Move Around in Membranes When treated mouse and human cells join, one continuous plasma membrane surrounds the fused hybrid cell and membrane proteins become randomly distributed.

out the plasma membrane of the muscle cell. If the nerve regenerates its attachment to the muscle, then the protein is once again limited to the junction area. This experiment not only attests to the mobility of membrane proteins and their regional grouping; it also shows how cell interactions can determine membrane characteristics of a cell.

What determines whether a particular membrane protein is integral or peripheral? If it is integral, what controls whether it reaches all the way through the membrane or is limited to one side? What keeps it in the bilayer, and what determines how far in it reaches? All these questions are answered in terms of the location of particular amino acids in the tertiary structure of the protein. Recall that the side chains of the various amino acids in a protein differ chemically—some of the side chains are hydrophilic and others are hydrophobic.

After a polypeptide chain folds into its tertiary structure, the protein may have both hydrophilic an-

hydrophobic surfaces. If one end of a folded protein is hydrophilic and the other hydrophobic, it will be an integral membrane protein, sticking out of one side of the membrane. Many integral membrane proteins that stretch across the entire membrane have long hydrophobic α-helical regions that penetrate the entire depth of the phospholipid bilayer. They also have hydrophilic ends that protrude into the aqueous environments on either side of the membrane (Figure 5.5). Integral membrane proteins like the one illustrated in Figure 5.5 resist being removed from the membrane. If the hydrophobic surface of such a protein is withdrawn partway out of the phospholipid bilayer, it is repelled by the aqueous environment. If an integral membrane protein is pushed farther into the membrane, its hydrophilic end is pushed back by the hydrophobic fatty acid region of the lipids. Thus such a protein may migrate laterally in the membrane sheet, but it may not push through the membrane or pop out of it.

How does an integral membrane protein get into the membrane in the first place? Clearly, it cannot fold into its final shape and then be pushed into the membrane—for the same reasons that it cannot push through or pop out of the membrane. Thus an integral protein is usually inserted into the membrane while the protein is being synthesized.

Specific sequences of hydrophobic amino acids in the growing protein interact with receptors on the endoplasmic reticulum membrane, causing the protein to penetrate the hydrophobic bilayer. However, the presence of additional sequences of hydrophilic amino acids may prevent the growing protein from extending farther into the bilayer and entering the lumen of the ER. Instead, it remains lodged in the membrane. Final folding is achieved after the protein is completely synthesized and all its regions have taken up residence either in the hydrophobic bilayer or in the hydrophilic regions on either surface.

Membrane carbohydrates are recognition sites

All plasma membranes and some other membranes contain significant amounts of carbohydrate along with the lipids and proteins. For example, the plasma membrane of the human red blood cell consists by weight of approximately 40 percent lipid, 52 percent protein, and 8 percent carbohydrate. The carbohydrates are located on the outer surface and serve as recognition sites (see Figure 5.1).

Some of the membrane carbohydrate is bound to lipids, forming *glycolipids*. These carbohydrate units often serve as recognition signals for interactions between cells. For example, the carbohydrate component of some glycolipids changes when a cell becomes cancerous. This change may identify the cancer cell for destruction by white blood cells.

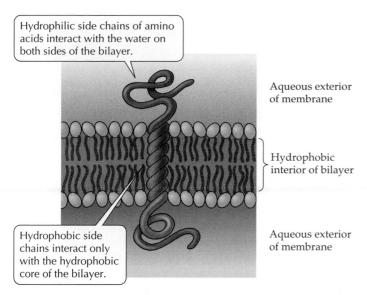

Hydrophilic side chains of amino acids interact with the water on both sides of the bilayer.

Aqueous exterior of membrane

Hydrophobic interior of bilayer

Hydrophobic side chains interact only with the hydrophobic core of the bilayer.

Aqueous exterior of membrane

5.5 Interactions of Integral Membrane Proteins This integral membrane protein is held in the membrane by the distribution of the hydrophilic and hydrophobic side chains of its amino acids.

Most of the carbohydrate in membranes is bound to proteins, forming *glycoproteins*. These bound carbohydrates are *oligosaccharide chains*, usually not exceeding 15 monosaccharide units in length. Glycoproteins enable a cell to recognize foreign substances. The oligosaccharide chains are added to the membrane proteins inside the endoplasmic reticulum and are modified in the Golgi apparatus.

A few different monosaccharides can provide an alphabet to generate a diversity of messages. Different messages are formed when different kinds of monosaccharides bond together at different sites and in different numbers. Recall that monosaccharides can link together at any of several different carbons to form branched oligomers. The possibility of different branching patterns in itself greatly increases the specificity and diversity of signals that oligosaccharides can provide.

Carbohydrates associated with the plasma membrane are always on the outer surface. In this location, their structural diversity is important for binding reactions at the cell surface whereby cells recognize and react with specific substances (see Figure 5.1).

Animal Cell Junctions

Animal cells form special junctions to help them adhere to other cells. In a multicellular system such as a tissue, direct links between the cells and sometimes between the cells and components of the extracellular matrix establish stable cohesion. The integral membrane proteins that accomplish this cohesion are called *cell adhesion proteins*. In a kind of "hook and eye" inter-

action, the external domain of a cell adhesion protein from one cell forms a strong link with that of an adjacent cell.

Generally, such adhesions are too small to be directly observed in electron micrographs, but the *intercellular space* between adjacent cells can be observed as an open region between the cells. Materials moving out of one cell and into the adjacent cell must cross this space. This intercellular space is also the region where all sorts of molecules can move around cells. In some cases, adjacent cells show specialized structures that link them or that restrict movement through the intercellular space.

Specialized structures that join cells are observed in electron micrographs of epithelial tissues, which are layers of cells that line body cavities or cover body surfaces. We will examine three general types of **cell surface junctions** that enable cells to make direct physical contact and link with one another: tight junctions, desmosomes, and gap junctions (see Figure 5.6).

Tight junctions seal tissues and prevent leaks

Tight junctions are specialized structures that link adjacent epithelial cells lining a hollow structure such as the intestine. Tight junctions result from the mutual binding of strands of specific membrane proteins that form belts encircling the epithelial cells in the region surrounding the lumen or cavity (Figure 5.6*a*). Tight junctions have two functions: to limit the passage of materials from the lumen through the intercellular space, and to restrict the migration of proteins in the cells' plasma membranes.

Because they are restricted from moving through the intercellular space, substances from the lumen must pass through the epithelial cell. This restriction gives epithelial cells control over what enters the body. The tight junctions also restrict the migration of membrane proteins and phospholipids from one region of the cell to other regions.

Thus, the membrane proteins and phospholipids in the *apical region* facing the lumen are different from the proteins and phospholipids in the *basolateral regions* facing the sides and bottom of the cell. By forcing materials to enter some cells, and by allowing different ends of cells to have different membrane proteins, tight junctions help ensure the directional movement of materials into the body.

Desmosomes rivet cells together

Desmosomes are specialized structures associated with the plasma membrane at certain sites in epithelial tissues. They hold adjacent cells firmly together, acting like spot welds or rivets at individual points (Figure 5.6*b*). Each desmosome consists of a dense *plaque* on the cytoplasmic surface of the plasma membrane that is attached to *keratin fibers* in the cytoplasm and *cell ad-*

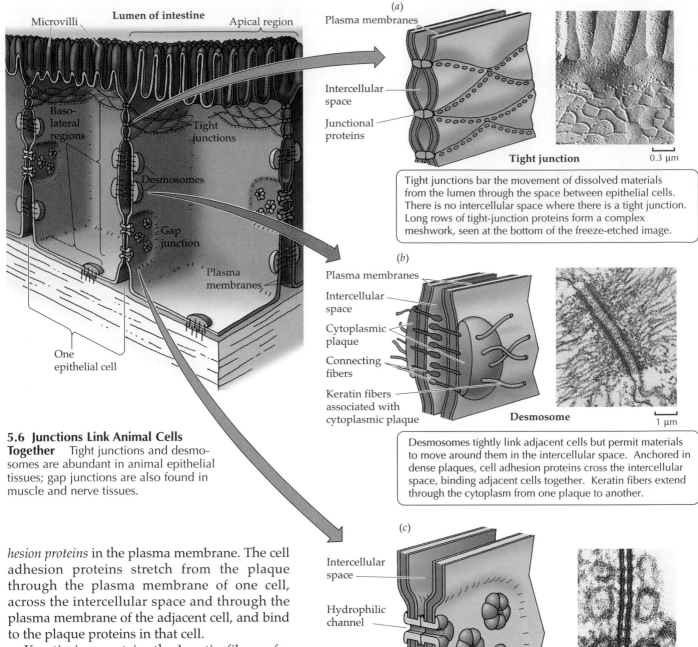

Microvilli — Lumen of intestine — Apical region

Baso-lateral regions

Tight junctions

Desmosomes

Gap junction

Plasma membranes

One epithelial cell

(a)

Plasma membranes

Intercellular space

Junctional proteins

Tight junction

0.3 μm

Tight junctions bar the movement of dissolved materials from the lumen through the space between epithelial cells. There is no intercellular space where there is a tight junction. Long rows of tight-junction proteins form a complex meshwork, seen at the bottom of the freeze-etched image.

(b)

Plasma membranes

Intercellular space

Cytoplasmic plaque

Connecting fibers

Keratin fibers associated with cytoplasmic plaque

Desmosome

1 μm

Desmosomes tightly link adjacent cells but permit materials to move around them in the intercellular space. Anchored in dense plaques, cell adhesion proteins cross the intercellular space, binding adjacent cells together. Keratin fibers extend through the cytoplasm from one plaque to another.

5.6 Junctions Link Animal Cells Together Tight junctions and desmosomes are abundant in animal epithelial tissues; gap junctions are also found in muscle and nerve tissues.

(c)

Intercellular space

Hydrophilic channel

2.7 nm space

Connexons

Plasma membranes

Gap junction

0.1 μm

Gap junctions let adjacent cells communicate. Dissolved molecules and electric signals may pass from one cell to the other through the channels formed by two connexons extending from adjacent cells.

hesion proteins in the plasma membrane. The cell adhesion proteins stretch from the plaque through the plasma membrane of one cell, across the intercellular space and through the plasma membrane of the adjacent cell, and bind to the plaque proteins in that cell.

Keratin is a protein; the keratin fibers of a desmosome are intermediate filaments of the cytoskeleton (see Figure 4.22).These fibers stretch from one cytoplasmic plaque through the cytoplasm to connect with another cytoplasmic plaque on the other side of the cell. Anchored thus on both sides of the cell, these extremely strong protein fibers provide great mechanical stability to epithelial tissues, which often receive rough wear in protecting the organism's body integrity.

Gap junctions are a means of communication

Whereas tight junctions and desmosomes have mechanical roles, **gap junctions** facilitate communication between cells. Gap junctions are made up of specialized protein channels, called *connexons*, that span the plasma membranes of adjacent cells and the intercellular space between them (Figure 5.6c). Through these channels pass a variety of small molecules and ions but not proteins, nucleic acids, or cellular organelles.

In some nerve cells and in vertebrate heart and smooth muscle, gap junctions allow the direct passage of an electric signal from one cell to the next. Gap junctions are also important in normal embryonic development, for they appear at a specific developmental stage.

If animal tissues are experimentally disrupted to dissociate the cells, the cells quickly form new gap junctions as they reassociate. In fact, cells isolated from one species of vertebrate readily form gap junctions with cells from other vertebrate species. Cancer cells, however, never develop gap junctions, and this failure of cancer cells to communicate with other cells contributes to their abnormal uncontrolled growth.

We have examined the general structure of biological membranes and the specialized junctions found between some animal cells: tight junctions, desmosomes, and gap junctions. Now we want to consider the more general functions of biological membranes. One of the most important of these functions is *selective permeability*—the ability to allow some substances, but not others, to pass through the membrane to enter or leave the cell.

Passive Processes of Membrane Transport

There are two fundamentally different classes of processes by which substances cross biological membranes to enter and leave cells or organelles: passive processes and active processes. We'll discuss active processes later in the chapter; here we focus on the passive processes, all of which are different types of diffusion: simple diffusion through the phospholipid bilayer, diffusion through channel proteins, and facilitated diffusion by means of carriers. However, before considering diffusion as it works across a membrane, we must understand the basic principles of diffusion.

The physical nature of diffusion

Nothing in this world is ever absolutely at rest. Everything is in motion, though the motions may be very small. The constant jiggling of molecules and ions in solution increases as the temperature rises. An important consequence of the random jiggling is that all the components of a solution tend eventually to become evenly distributed throughout the system. For example, if a drop of ink is allowed to fall into a container of water, the pigment molecules of the ink are initially very concentrated. Without human intervention such as stirring, the pigment molecules of the ink move about at random, spreading slowly through the water until eventually the concentration of pigment—and thus the intensity of color—is exactly the same in every drop of liquid in the container. A solution in which the particles are uniformly distributed is said to be at equilibrium because there will be no future net change in concentration. **Diffusion** is the

process of random movement toward a state of equilibrium.

Although the motion of each individual particle is absolutely random, in diffusion the *net movement* of particles is directional until equilibrium is reached. Diffusion is this net movement from regions of *greater concentration* to regions of *lesser concentration* (Figure 5.7). In a complex solution (one with many different solutes), the diffusion of each substance is independent of that of the other substances. How fast substances diffuse depends on four factors: (1) the diameter of the molecules or ions; (2) the temperature of the solution; (3) the electric charge, if any, of the diffusing material; and (4) the **concentration gradient** in the system. The concentration gradient is the change in concentration with distance in a given direction. The greater the concentration gradient, the more rapidly a substance diffuses.

DIFFUSION WITHIN CELLS. Within cells, where distances are very short, solutes distribute rapidly by diffusion. Small molecules and ions may move from one end of an organelle to another in a millisecond (10^{-3} s). The diffusion of a chemical signal from one nerve cell to another takes less than one millionth of a second ($<10^{-6}$ s). On the other hand, the usefulness of diffusion as a transport mechanism declines drastically as distances become greater. In the absence of mechanical stirring, diffusion across more than a centimeter may take an hour or more, and diffusion across meters may take years! Diffusion would not be adequate to distribute materials over the length of the human body, but within our cells or across layers of one or two cells, diffusion is rapid enough to distribute small molecules and ions almost instantaneously.

DIFFUSION ACROSS MEMBRANES. In a solution without barriers, all the solutes diffuse at rates determined by their physical properties and the concentration gradient of each solute. If a biological membrane is introduced as a barrier, the movement of the different solutes can be affected. The membrane is said to be *permeable* to solutes that can cross it more or less easily, but *impermeable* to substances that cannot move across it. Molecules to which the membrane is permeable diffuse from one compartment to the other until their concentrations are equal on both sides of the membrane. Molecules to which the membrane is impermeable remain in distinct compartments, so their concentrations remain different on the two sides of the membrane.

Equilibrium is reached when the concentrations of the diffusing substance are identical on both sides of the membrane. Individual molecules are still passing through the membrane when equilibrium is established, but equal numbers of molecules are moving in each direction, so there is no net change in concentra-

(a)

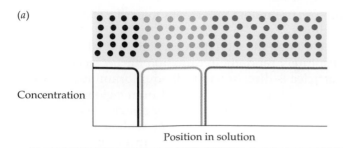

Concentration

Position in solution

Initially, three different solutes are concentrated in different parts of the solution.

(b)

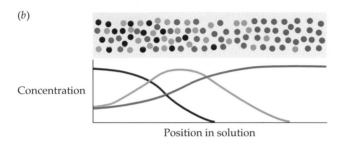

Concentration

Position in solution

As the solutes diffuse, each is at a higher concentration—its peak on the graph—in one part of the solution, but it is also present elsewhere.

(c)

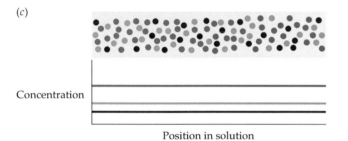

Concentration

Position in solution

At equilibrium, all three solutes are uniformly distributed throughout the solution, as shown by the lack of peaks on the graph.

5.7 Diffusion Leads to Uniform Distribution of Solutes
Diffusion is the net movement of a solute from regions of greater concentration to regions of lesser concentration. The speed of diffusion varies with the substances involved, but it continues until the solution reaches an equilibrium.

tion. The two primary ways in which substances diffuse through biological membranes are simple diffusion and facilitated diffusion.

Simple diffusion is passive and unaided

In **simple diffusion**, small, nonpolar molecules pass through the lipid bilayer of the membrane. The more lipid-soluble the molecule, the more rapidly it diffuses through a biological membrane (Figure 5.8). This statement holds true over a wide range of molecular weights. Only certain ions and the smallest of molecules seem to deviate from this rule; materials such as

water, potassium ions (K^+), and chloride ions (Cl^-) pass through membranes much more rapidly than their solubilities in lipid would predict. How does a membrane's composition and molecular structure affect diffusion?

The key feature of membrane architecture, as we have seen, is the phospholipid bilayer that forms the framework of the membrane. The inner portion of the bilayer consists of the fatty acid chains of the phospholipids, along with cholesterol and other highly hydrophobic, nonpolar materials. When a hydrophilic molecule or ion moves into such a hydrophobic region, it is "rejected" by the lipid layer and forced back out again. Such a molecule seldom enters the hydrophobic region; it penetrates the membrane only when energy is available to push it in. On the other hand, a molecule that is itself hydrophobic, and hence soluble in lipids, enters the membrane readily and is thus able to pass through it.

This explanation accounts for most of the information in Figure 5.8, but the problem of how polar water molecules and charged ions can move across biological membranes remains. The passive diffusion of water into and out of cells occurs through the process of *osmosis* and will be discussed later in the chapter. The rapidity of osmosis is still something of a mystery, and several alternative mechanisms have been suggested to explain it. However, we *do* understand the rapid movement of certain ions through biological membranes. These ions pass through specific, water-filled protein pores in the membrane. Such **channel proteins** are known to allow the diffusion of ions such as potassium, sodium, calcium, and chloride.

Membrane transport proteins are of several types

The channel proteins through which certain ions diffuse and the carriers for facilitated diffusion and active transport are both **membrane transport proteins**—integral membrane proteins that stretch from one side of the membrane to the other. Different membrane transport proteins allow specific substances to pass through in various ways (Figure 5.9). The pore size and other properties of different channel proteins are different, permitting certain ions to pass through and excluding others. Some channel proteins have an aqueous region through which ions can diffuse.

There are two classes of membrane transport facilitated by **carrier proteins**: uniport and coupled transport. In **uniport**, a single type of solute is transported in one direction. In **coupled transport**, two or more different solutes are transported, but neither solute can move through the membrane unless the other solute is also present. Coupled transport is further divided into two types: symport and antiport.

In **symport**, the two coupled solutes are transported in the same direction. For example, one system for the

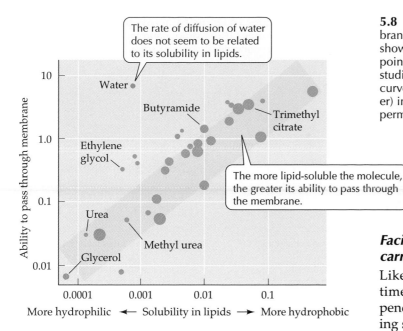

The rate of diffusion of water does not seem to be related to its solubility in lipids.

The more lipid-soluble the molecule, the greater its ability to pass through the membrane.

5.8 Membrane Permeability Most substances cross membranes at rates determined by their solubility in lipids, as shown by the data points on this graph. The sizes of the points correspond roughly to the sizes of the molecules studied. The assortment of molecules of all sizes along the curve (that is, the fact that like sizes are not clumped together) indicates that size alone is not an important factor for permeability.

Facilitated diffusion is passive but uses carrier proteins

Like simple diffusion, **facilitated diffusion** (sometimes called *carrier-mediated diffusion* because it depends on the carrier proteins described in the preceding section) involves movement down a concentration gradient, from the side of higher concentration to the side of lower concentration. Eventually both processes produce equal concentrations of solute on the two sides of a membrane. The difference is that in facilitated diffusion the solute molecules do not diffuse through the phospholipid bilayer or through pores formed by channel proteins. Rather, they combine with carrier proteins in the membrane.

Carrier proteins have an abundance of hydrophobic amino acid side chains on the surface of the molecule in contact with the hydrophobic core of the membrane bilayer. However, within each carrier protein is a highly specific hydrophilic region to which a solute molecule binds. This binding appears to cause the protein to undergo a slight but significant change in shape (tertiary or quaternary structure) that moves the solute

uptake of glucose by the cell is a symport system in which the entrance of glucose is coupled to the entrance of a sodium ion. In **antiport**, the two substances are transported in opposite directions. For example, the plasma membrane of red blood cells contains about a million antiport transport proteins for the exchange of bicarbonate and chloride ions. This particular exchange is important in transporting carbon dioxide, which is present in the bloodstream as bicarbonate ions, from working tissues where carbon dioxide is produced to the lungs where it is released.

The specific tertiary structures of membrane transport proteins contribute to their transport specificities. A given membrane transport protein will carry only one particular solute or only closely related solutes. The great diversity of protein structures allows this high specificity for different transported solutes, just as it allows enzymes to recognize only specific reactants in the chemical reactions they accelerate.

5.9 Membrane Transport Proteins Substances cross biological membranes by many mechanisms, most of which involve membrane transport proteins.

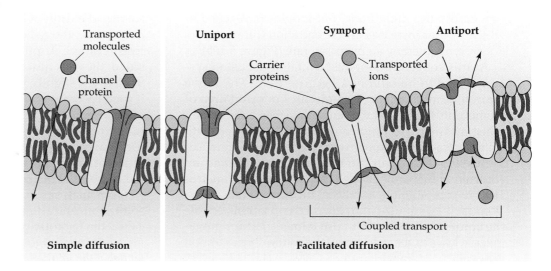

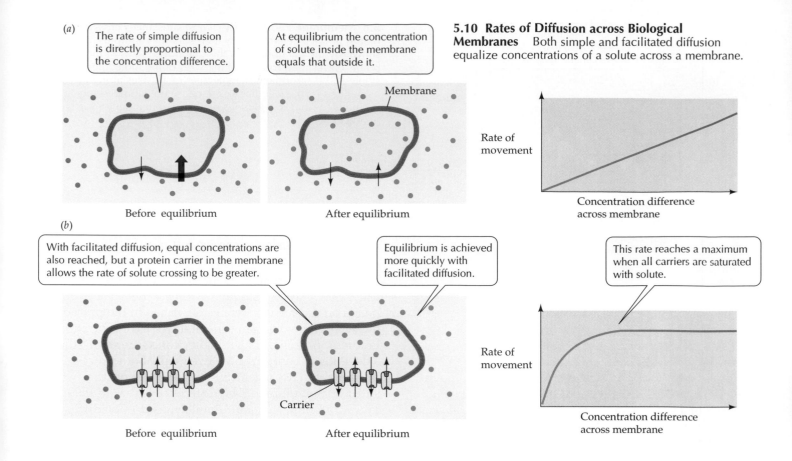

5.10 Rates of Diffusion across Biological Membranes Both simple and facilitated diffusion equalize concentrations of a solute across a membrane.

(a) The rate of simple diffusion is directly proportional to the concentration difference.

At equilibrium the concentration of solute inside the membrane equals that outside it.

Membrane

Before equilibrium

After equilibrium

Rate of movement

Concentration difference across membrane

(b) With facilitated diffusion, equal concentrations are also reached, but a protein carrier in the membrane allows the rate of solute crossing to be greater.

Equilibrium is achieved more quickly with facilitated diffusion.

This rate reaches a maximum when all carriers are saturated with solute.

Carrier

Before equilibrium

After equilibrium

Rate of movement

Concentration difference across membrane

across the bilayer and releases it on the other side. By this mechanism, still not completely understood, these carrier molecules speed (or facilitate) the passage of the solute molecules across the membrane.

In both simple diffusion and facilitated diffusion, the *rate* of movement depends on the concentration difference across the membrane. In simple diffusion, the net rate of movement is directly proportional to the concentration difference across the membrane (Figure 5.10*a*). In facilitated diffusion, the rate of movement also increases with the difference in solute concentration across the membrane, but a point is reached at which further increases in concentration difference are not accompanied by an increased rate (Figure 5.10*b*).

At this concentration, the facilitated diffusion system is said to be *saturated*. Because there are only a limited number of carrier molecules per unit area of membrane, the rate of movement reaches a maximum when all the carrier molecules are fully loaded with solute molecules. In other words, when the differences in the solute concentration across the membrane are sufficiently high, not enough carrier molecules are free at a given moment to handle all the solute molecules.

In facilitated diffusion, the carrier proteins enable the solutes to pass in both directions. The net movement is toward the side where the solute concentration is lower.

Osmosis is passive water movement through a membrane

Water moves through membranes by a diffusion process called **osmosis**. This completely passive process uses no metabolic energy and can be understood in terms of the solute concentrations of solutions. Osmosis depends on the *number* of solute particles present—not the kind of particles. We will describe osmosis using red blood cells and plant cells.

Red blood cells are normally suspended in a fluid called *plasma*, which contains salts, proteins, and other solutes. If a drop of blood is examined under the light microscope, the red cells are seen to have their characteristic shape. If pure water is added to the drop of blood, the cells quickly swell and burst (Figure 5.11*a*). Similarly, if slightly wilted lettuce is put in pure water, it soon becomes crisp; by weighing it before and after, we can show that it has taken up water (Figure 5.11*b*). If, on the other hand, the red blood cells or crisp lettuce leaves are placed in a relatively concentrated solution of salt or sugar, the leaves become limp and the red blood cells pucker and shrink.

From analyses of such observations, we learn that solute concentration is the principal factor in what is called the **osmotic potential** of a solution. *The greater the solute concentration, the more negative the osmotic potential of the solution.* Pure water has nothing dissolved

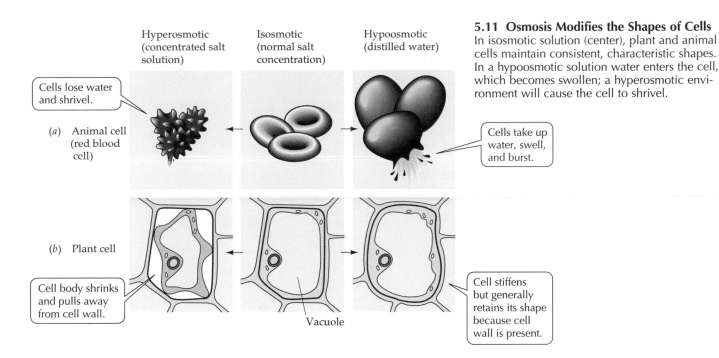

Hyperosmotic (concentrated salt solution)

Isosmotic (normal salt concentration)

Hypoosmotic (distilled water)

5.11 Osmosis Modifies the Shapes of Cells
In isosmotic solution (center), plant and animal cells maintain consistent, characteristic shapes. In a hypoosmotic solution water enters the cell, which becomes swollen; a hyperosmotic environment will cause the cell to shrivel.

Cells lose water and shrivel.

(a) Animal cell (red blood cell)

Cells take up water, swell, and burst.

(b) Plant cell

Cell body shrinks and pulls away from cell wall.

Cell stiffens but generally retains its shape because cell wall is present.

Vacuole

in it, so its osmotic potential equals zero. Other things being equal, if two unlike solutions are separated by a membrane that allows water to pass through but not solutes, water molecules will move across the membrane toward the solution with the more negative osmotic potential. This movement is called osmosis. In other words, water will move from a region of its higher concentration (lower concentration of solutes) to a region of its lower concentration (higher concentration of solutes).

If two solutions have identical osmotic potentials, they are **isosmotic** to one another (*iso-*, "same"). (This is true even if their chemical compositions are very different.) If the two solutions are not isosmotic, then solution A with a more negative osmotic potential (with a higher concentration of solutes) is said to be **hyperosmotic** to solution B, and solution B is **hypoosmotic** to solution A (*hyper-*, "more, high"; *hypo-*, "less, low").

All three of these terms are strictly relative. They can be used only in comparing the osmotic potentials of two specific solutions. For instance, no solution can be called hyperosmotic except in comparison with another solution that has a less negative osmotic potential. A related set of terms—iso-, hypo-, and hypertonic—refers to *solute concentrations* of solutions. Two solutions with equal total solute concentrations are *isotonic*. If solution A has a higher solute concentration than solution B, solution A is *hypertonic* to B, which is *hypotonic* to A.

Osmotic potentials determine the direction of osmosis in all animal cells. A red blood cell takes up water from a solution that is hypoosmotic to the cell's contents. The cell bursts because its plasma membrane cannot withstand the swelling of the cell (see Figure 5.11a). The integrity of red blood cells and other blood cells is absolutely dependent on the maintenance of a constant osmotic potential in the plasma in which they are suspended: The plasma must be isosmotic with the cells if the cells are not to burst or shrink.

In contrast to animal cells, the cells of plants, archaea, bacteria, fungi, and some protists have cell walls that limit the volume of the cells and keep them from bursting. Unlike the red blood cell, the wilted lettuce leaf becomes firm and crisp when placed in pure water (see Figure 5.11b). Cells with sturdy cell walls take up a limited amount of water and, in so doing, build up an internal pressure against the cell wall that prevents further water from entering. This pressure within the cell, called the **turgor pressure**, is the driving force for growth of plant cells—it is a normal and essential component of plant development.

In other words, in cells with walls, osmosis is regulated not only by osmotic potentials but also by the opposed turgor pressure. A salt solution sitting in a glass has a turgor pressure of zero. Enclose some of it with a rigid membrane that is impermeable to salt but permeable to water, drop this package into pure water, and the enclosed salt solution will have a turgor pressure that increases as water moves by osmosis through the membrane. Osmotic phenomena in plants are discussed in Chapter 32.

Active Processes of Membrane Transport

Earlier in this chapter we distinguished passive transport processes from active transport processes. As we have seen, the passive processes are driven by a concentration difference for the material being transported. Passive transport processes always operate from regions of greater to regions of lesser concentration, and no external source of energy is required.

Active transport systems, on the other hand, require a source of energy to move materials. In this section we will examine two different classes of transport phenomena that require the input of energy. First, we'll consider highly specific systems located in membranes that operate on one or a few molecules (or ions) at a time. Second, we'll consider systems of membrane fusion that enable a cell to engulf or secrete bulk quantities of material.

Active transport requires energy and carriers

Like facilitated diffusion, **active transport** relies on carrier molecules that are proteins. Unlike facilitated diffusion, this process requires energy to move ions or molecules across a biological membrane from regions of lower concentration to regions of greater concentration—that is, to move solutes *up* a concentration gradient. Sometimes, but not always, the immediate source of energy for this process is energy provided by the hydrolysis of ATP. There are two basic types of active transport: primary and secondary.

Primary active transport requires the direct participation of adenosine triphosphate (ATP). In primary active transport, energy released from ATP drives the movement of specific ions against a concentration dif-

TABLE 5.1   Concentration of Major Ions Inside and Outside the Nerve Cell of a Squid

ION	CONCENTRATION (MOLAR)	
	INSIDE	OUTSIDE
K^+	0.400	0.020
Na^+	0.050	0.440
Cl^-	0.120	0.560

ference. For example, we can compare the concentrations of potassium ions (K^+) and sodium ions (Na^+) inside a nerve cell and in the fluid bathing the nerve (Table 5.1). The K^+ concentration is much higher inside the cell, whereas the Na^+ concentration is much higher outside. Nevertheless, the nerve cells continue to pump Na^+ out and K^+ in, against these concentration differences, ensuring that the differences are maintained.

This **sodium–potassium pump** is found in all animal cells and is an integral membrane *glycoprotein*. Breaking down a molecule of ATP to ADP and phosphate (P_i), the sodium–potassium pump brings two K^+ ions into the cell and exports three Na^+ ions (Figure 5.12). The sodium–potassium pump is thus an antiport transport system. Different pumps are responsible for the transport of several other ions, but only cations are

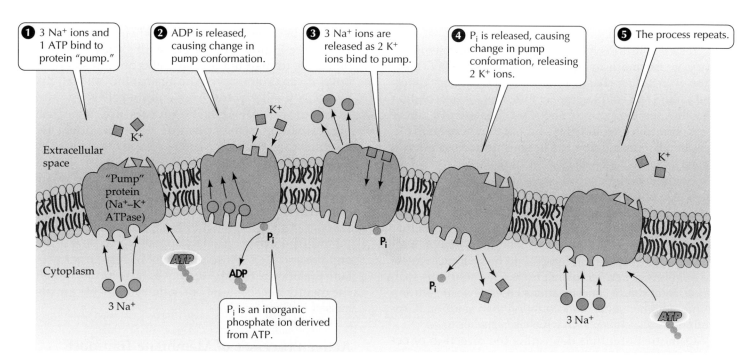

1 3 Na^+ ions and 1 ATP bind to protein "pump."

2 ADP is released, causing change in pump conformation.

3 3 Na^+ ions are released as 2 K^+ ions bind to pump.

4 P_i is released, causing change in pump conformation, releasing 2 K^+ ions.

5 The process repeats.

Extracellular space

"Pump" protein (Na^+–K^+ ATPase)

K^+

Cytoplasm

ATP

ADP

P_i

P_i is an inorganic phosphate ion derived from ATP.

3 Na^+

P_i

K^+

ATP

5.12 Primary Active Transport: The Sodium–Potassium Pump   In active transport, energy is used to move a solute against its concentration gradient. Even though the Na^+ concentration is higher outside the cell and the K^+ concentration is higher inside the cell, for each molecule of ATP used, two K^+ ions are pumped into the cell and three Na^+ ions are pumped out of the cell.

transported directly by pumps in primary active transport. Other solutes are transported by secondary active transport.

Unlike primary active transport, **secondary active transport** does not use ATP directly; rather, the transport of the solute is tightly coupled to the difference in ion concentration established by primary active transport. The movement of particular solutes, such as sugars and amino acids, is regulated by coupled transport systems that move these specific solutes against their concentration difference, using energy "regained" by letting Na^+ or other ions diffuse across the membrane from regions of higher to regions of lower concentration—that is, to move *down* their concentration gradient (Figure 5.13).

Both types of coupled transport—symport and antiport—are used for secondary active transport. Putting primary and secondary active transport together, we see that energy from ATP is used in one example of primary active transport to establish concentration differences of potassium and sodium ions; then the passive transport of some sodium ions in the opposite direction provides energy for the secondary active transport of the sugar glucose (see Figure 5.13). Other secondary active transporters aid in the uptake of amino acids and sugars, which are essential raw materials for cell maintenance and growth.

Macromolecules and particles enter the cell by endocytosis

A process called **endocytosis** brings macromolecules, large particles, and small cells into the eukaryotic cell (Figure 5.14*a*). There are three types of endocytosis: phagocytosis, pinocytosis, and receptor-mediated endocytosis. In all three, the plasma membrane folds inward, making a small pocket. The pocket deepens, forming a vesicle whose contents are materials from the environment. This vesicle separates from the surface of the cell and migrates to the cell's interior.

We encountered phagocytosis briefly in Chapter 4. In this process, part of the plasma membrane engulfs fairly large particles or even entire cells. Phagocytosis is a cellular feeding process found in unicellular protists and in some white blood cells that defend the body against foreign cells and substances (see Chapter 18). The food vacuole or phagosome formed usually fuses with a lysosome, and its contents are digested (see Figure 4.19). Vesicles also form in pinocytosis ("cellular drinking"). However, these vesicles are smaller, and the process operates to bring in small dis-

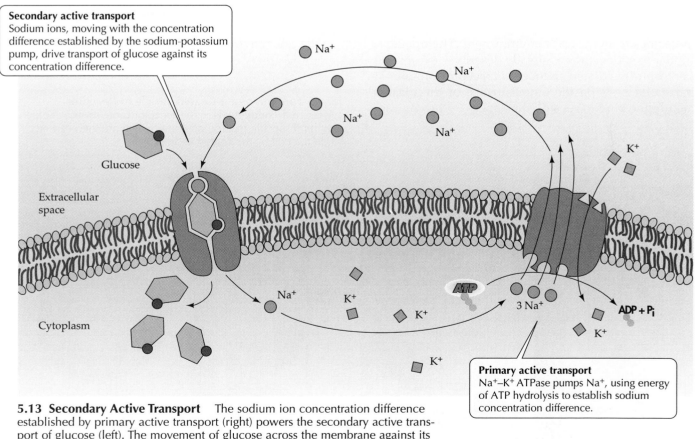

Secondary active transport
Sodium ions, moving with the concentration difference established by the sodium-potassium pump, drive transport of glucose against its concentration difference.

Primary active transport
Na^+–K^+ ATPase pumps Na^+, using energy of ATP hydrolysis to establish sodium concentration difference.

5.13 Secondary Active Transport The sodium ion concentration difference established by primary active transport (right) powers the secondary active transport of glucose (left). The movement of glucose across the membrane against its concentration difference is coupled by a symport protein to the diffusion of Na^+ into the cell.

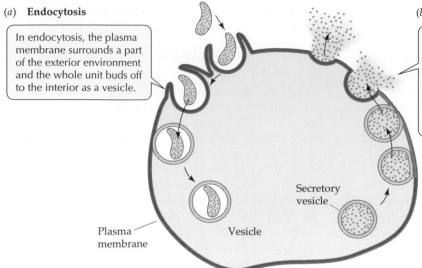

(a) **Endocytosis**

In endocytosis, the plasma membrane surrounds a part of the exterior environment and the whole unit buds off to the interior as a vesicle.

(b) **Exocytosis**

In exocytosis, a membrane-enclosed vesicle containing substances for export fuses with the plasma membrane. The contents of the vesicle scatter, and the vesicle membrane becomes part of the plasma membrane.

Secretory vesicle

Plasma membrane

Vesicle

5.14 Endocytosis and Exocytosis Endo- and exocytosis are used by all eukaryotic cells to take up and eliminate substances from and to the outside environment.

solved substances or fluids. The third type of endocytosis, receptor-mediated endocytosis, involves specific reactions at the cell surface that trigger uptake. Let's take a closer look at this process.

Receptor-mediated endocytosis is highly specific

Most animal cells have a mechanism called **receptor-mediated endocytosis** that captures specific macromolecules from the cell's environment. The uptake is similar to endocytosis as already described, except that in receptor-mediated endocytosis receptor proteins at particular sites on the outer surface of the plasma membrane bind to specific substances in the environ-

ment outside the cell. These sites are called **coated pits** because they have a slight depression of the plasma membrane whose inner surface is coated by fibrous proteins, such as *clathrin*.

When a receptor protein binds the appropriate macromolecule outside the cell, the associated coated pit invaginates (folds inward) and forms a **coated vesicle** around the bound macromolecules. Strengthened and stabilized by clathrin molecules, this vesicle carries the macromolecules into the cell (Figure 5.15). Once inside, the vesicle loses its coat and may fuse with another vesicle for the processing and release into the cytoplasm of the engulfed material. Because of its specificity for particular macromolecules, receptor-mediated endocytosis is a more rapid and efficient method of taking up what may be minor constituents of the cell's environment.

5.15 Formation of a Coated Vesicle in Receptor-Mediated Endocytosis The receptor proteins in a coated pit bind specific macromolecules, which are then carried into the cell by the coated vesicle.

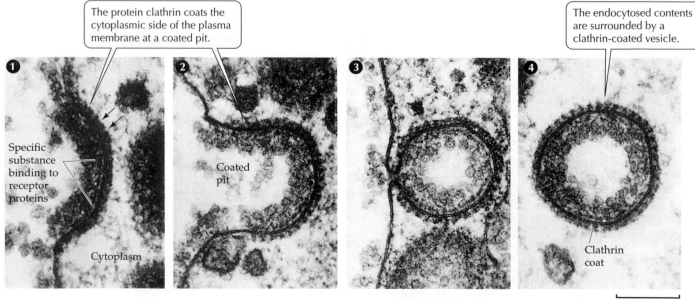

The protein clathrin coats the cytoplasmic side of the plasma membrane at a coated pit.

The endocytosed contents are surrounded by a clathrin-coated vesicle.

Specific substance binding to receptor proteins

Cytoplasm

Coated pit

Clathrin coat

0.1 μm

Receptor-mediated endocytosis is the method by which cholesterol is taken up by most mammalian cells. Water-insoluble cholesterol is synthesized in the liver and transported in the blood attached to a protein, forming a lipoprotein such as **low-density lipoprotein**, or **LDL** (Figure 5.16). The uptake mechanism for cholesterol includes the binding of LDL to specific receptor proteins in coated pits. After being engulfed by endocytosis, LDL particles are freed from the receptors, which then segregate to a region that buds off to form a new vesicle that is recycled to the plasma membrane. The freed LDL particles remain in a vesicle that fuses with a lysosome in which the LDL particles are digested and the cholesterol made available for cell use. In the inherited disease *hypercholesterolemia* (*-emia*, "blood"), patients have dangerously high levels of cholesterol in their blood because of a deficient receptor for LDL.

Exocytosis moves materials out of the cell

In eukaryotic cells (but not in prokaryotic cells) the plasma membrane regulates ongoing traffic: Membrane-enclosed "packages" enter the cell by endocytosis and leave by exocytosis. **Exocytosis** is the process by which materials packaged in vesicles are secreted from the cell when the vesicle membrane fuses with the plasma membrane (see Figure 5.14*b*). The phospholipid regions of the two membranes merge, and an opening to the outside of the cell develops. The contents of the vesicle are released to the environment, and the vesicle membrane is smoothly incorporated into the plasma membrane.

In Chapter 4 we encountered exocytosis as the last step in processing the material engulfed by phagocytosis: the excretion of indigestible materials to the environment. Exocytosis is also important in the secretion of many different substances, including digestive enzymes from the pancreas, neurotransmitters from nerve cells, and materials for plant cell growth.

Membranes Are Not Simply Barriers

We have discussed some functions of membranes—the compartmentalization of cells, the regulation of traffic between compartments, and the active pumping of solutes—but there are more. As discussed in Chapter 4, the membrane of rough endoplasmic reticulum serves as a site for ribosome attachment. Newly formed proteins are passed from the ribosomes through the membrane and into the interior of the ER for delivery to other parts of the cell. The membranes of nerve cells, muscle cells, some eggs, and other cells are also electrically excitable. In nerve cells, the plasma membrane is the conductor of the nerve impulse from one end of the cell to the other (see Chapter 41). Numerous other biological activities and properties

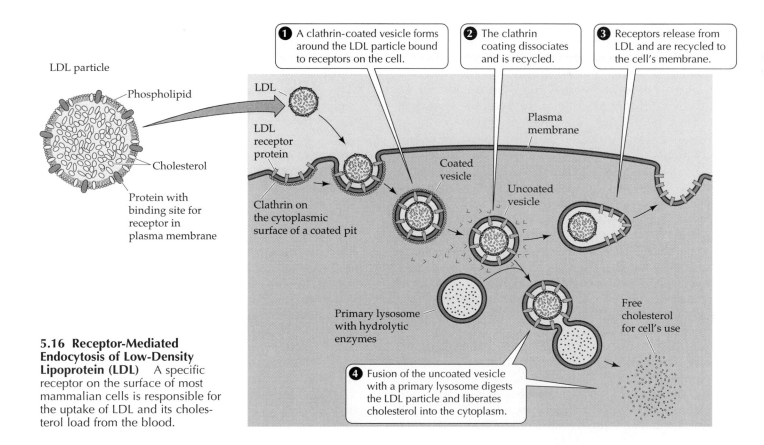

5.16 Receptor-Mediated Endocytosis of Low-Density Lipoprotein (LDL) A specific receptor on the surface of most mammalian cells is responsible for the uptake of LDL and its cholesterol load from the blood.

discussed in the chapters to follow are integrally associated with membranes.

To elaborate further on the versatility of biological membranes, we will describe three additional cellular functions of membranes: their roles in processing information from the environment, their roles in energy-trapping and energy-releasing reactions, and their roles in helping cells associate as tissues.

Some membranes process information

The plasma membranes at cell surfaces and the membranes within cells have protruding portions of integral membrane proteins or attached carbohydrates that recognize and bind to specific substances in their environment. The general term for such a substance is **ligand**.

The binding of a receptor to a ligand initiates specific change in the function of the cell or organelle to which the receptor is attached. In this type of information processing, specificity in binding is essential. We have already seen the role of a specific receptor in the endocytosis of LDL and its cargo of cholesterol (review Figure 5.16), but there are many other examples.

RECOGNITION AND BINDING BY PROTEINS AND CARBOHYDRATES. Antibodies are specific molecules synthesized by the body to defend against invading foreign substances and cells (see Chapter 18). Antibodies begin defensive operations by reacting with the specific proteins or carbohydrates protruding from the plasma membranes of many foreign or cancerous cells. Here, as in other membrane signaling, specificity is important; antibodies must not destroy normal body cells. Another example of the specificity of signals on cell surfaces is found in the initial steps of virus infection.

Viruses may begin attacking their intended host cells by attaching to carbohydrates on the surface of the host cell. Viruses show a high degree of specificity. For example, the viruses that infect bacteria do not infect plants or animals. A plant virus that infects tobacco plants will not infect begonias, and some animal viruses infect just one or a few species. The measles virus, for example, infects only humans, while the rabies virus infects humans and other mammals.

Many hormones, including insulin, are recognized by membrane proteins that serve as receptors on the hormone's target cells (see Chapter 38). And the passage of a nerve impulse from a nerve cell to a neighboring nerve or muscle cell by a neurotransmitter requires a receptor molecule on the plasma membrane of the target cell that is specific for the particular neurotransmitter that is released (see Chapter 41). Again specificity is essential; receptors for one transmitter will not accept another kind of transmitter. One of the most important classes of receptors consists of those that bind substances called growth factors, which regulate cell reproduction and differentiation (see Chapter 15).

Receptor proteins are integral membrane proteins that span the entire thickness of the plasma membrane. Receptor carbohydrates are bound to such proteins or to phospholipid molecules. The specificity of receptors (both proteins and carbohydrates) resides in their particular tertiary structure, that is, in their three-dimensional shape. Some portion of the receptor protein or carbohydrate fits the ligand like a hand fits a glove. We will deal with the molecular nature of such binding specificities in the next chapter, and this topic will arise again in Chapter 18 when we discuss immune responses.

When a ligand binds to its specific receptor on the cell surface, changes occur within the cell, on the cytoplasmic side of the membrane. How can such a visitor produce an effect inside the cell if it does not actually enter it? The answer is signal transduction.

SIGNAL TRANSDUCTION BY PROTEINS. *Signal transduction* is the conversion of one type of signal to another type of signal. When a ligand binds to a receptor protein on the outside surface of a cell, the receptor's tertiary structure changes—there is a slight *conformational* change. In an integral membrane protein that spans the entire bilayer, this change is transmitted to the part of the protein exposed to the cytoplasm (Figure 5.17). In many cases the change in protein activates some dormant function of the protein. For example, this change may allow the ligand–receptor complex to activate another membrane protein, called a G protein.

The **G protein**, once activated, diffuses in the membrane to activate or inactivate another membrane protein. This last protein may be an enzyme or a transport protein. Thus, the signal represented by the ligand is transduced by the membrane proteins into an altered flow of ions or a changed rate of chemical reaction. Both of these results have still broader consequences for cell functioning. Several different G proteins play roles in hormone action, vision, and other important processes to be discussed later. G proteins are so named because their activation requires GTP, a close chemical analog of ATP.

Some membranes transform energy

In a variety of cells, the membranes of organelles are specialized to process energy. For example, the thylakoid membranes of chloroplasts participate in the conversion of light energy to the energy of chemical bonds, and the inner mitochondrial membrane helps convert the energy of fuel molecules to the energy in ATP. The two characteristics of membranes that enable them to participate in these processesare their structural organization and the separation of their electric charges.

ORGANIZING CHEMICAL REACTIONS. Many processes in cells depend on a series of enzyme-catalyzed reac-

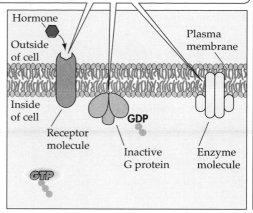

The actions of several membrane-associated proteins are required to convert the signal from a hormone to an amplified response in the cell.

Hormone

Outside of cell

Plasma membrane

Inside of cell

Receptor molecule

Inactive G protein

Enzyme molecule

GDP

GTP

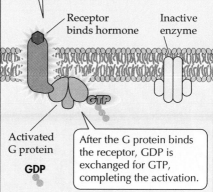

Hormone binding provides a signal that is converted to another type of signal: the activated G protein.

Receptor binds hormone

Inactive enzyme

Activated G protein

GTP

GDP

After the G protein binds the receptor, GDP is exchanged for GTP, completing the activation.

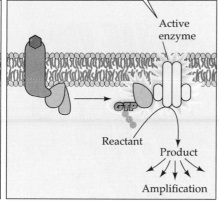

Part of the activated G protein activates an enzyme that converts thousands of reactants to products, thus amplifying the action of a single hormone molecule.

Active enzyme

Reactant

GTP

Product

Amplification

5.17 Signal Transduction by a G Protein In signal transduction, one type of signal (the binding of a hormone) is converted to another type of signal (the activation of a G protein). The G protein then activates an enzyme that repeatedly catalyzes a reaction.

tions in which the products of one reaction serve as the reactants for the next. For such a reaction to occur, all the necessary molecules must collide. In a solution, the reactants and enzymes are all randomly distributed and collisions are random. For this reason, a complete series of chemical reactions in solution may occur very slowly.

However, if the different enzymes are bound to a membrane in sequential order, the product of one reaction can be released close to the enzyme for the next reaction. With such an "assembly line," reactions proceed more rapidly and efficiently. In this sense, the membrane is a pegboard for the orderly attachment of specific proteins and other molecules.

SEPARATING ELECTRIC CHARGES. A biological membrane operates like an electric battery. Work can be obtained from a battery by letting electrons flow from one of its terminals to the other by way of a device, such as a motor or a light bulb, that uses the electric current. Something similar takes place in both photosynthesis and cellular respiration (see Chapters 7 and 8).

Because of the impermeability of the mitochondrial and chloroplast membranes to hydrogen ions (H^+) and because of the activities of certain electron carriers in those membranes, it is possible to separate charges and to concentrate H^+. In other words, a substantial gradient of both electric charge and pH is established across membranes. When electric charge flows back across the membrane, its diffusion is coupled to a protein that catalyzes the synthesis of ATP from ADP and phosphate, thus capturing the energy of a concentration gradient as chemical bond energy in ATP. Without

a membrane to allow the separation of charge, these reactions would not proceed.

Both the structure of membranes and their ability to separate electric charges relate to the properties of the two bulk components of membranes: lipids and proteins. The pegboard effect of the membrane's structure comes from the ability of the phospholipid bilayer to hold certain proteins in a defined plane so that they do not diffuse freely throughout the cell. Certain membrane proteins create the separation of charges, which is then maintained by the selective permeability of the phospholipid bilayer.

Cell adhesion molecules organize cells into tissues

During the growth of an animal embryo, the **cell adhesion proteins** mentioned earlier are crucial to the formation of the adult organism. As the embryo develops, its cells move about and associate with other specific cell types. This behavior is mediated by cell adhesion molecules in the plasma membranes, which play roles in organizing the cells into tissues. One type of cell adhesion molecule, for example, organizes individual nerve cells into nerve cell bundles. Groups of cells that are about to migrate within the embryo lose their specific cell adhesion molecules; when they reach their new location, they regain their cell adhesion molecules and reorganize into tissue.

Membranes Are Dynamic

As we have seen in this chapter, membranes participate in numerous physiological and biochemical processes. Membranes are dynamic in another sense as well: They are constantly forming, transforming from one type to another, fusing with one another, and breaking down.

In eukaryotes, phospholipids are synthesized on the surface of the endoplasmic reticulum and rapidly

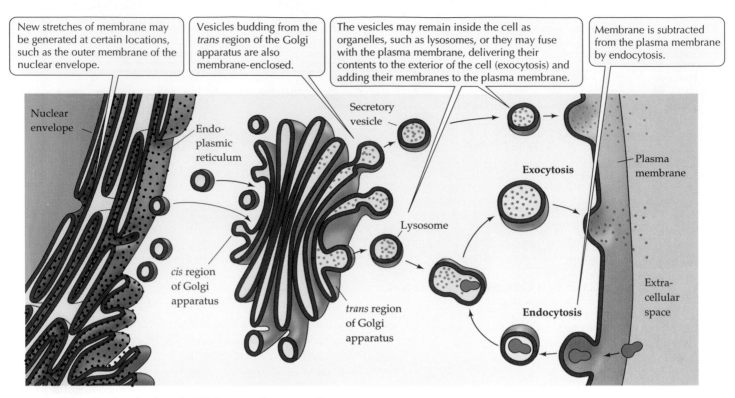

New stretches of membrane may be generated at certain locations, such as the outer membrane of the nuclear envelope.

Vesicles budding from the *trans* region of the Golgi apparatus are also membrane-enclosed.

The vesicles may remain inside the cell as organelles, such as lysosomes, or they may fuse with the plasma membrane, delivering their contents to the exterior of the cell (exocytosis) and adding their membranes to the plasma membrane.

Membrane is subtracted from the plasma membrane by endocytosis.

5.18 Dynamic Continuity of Cellular Membranes   The arrows trace how membranes form, move, and fuse in cells.

distributed to membranes throughout the cell as vesicles form from the ER, move away, and fuse with other organelles. Membrane proteins are inserted into the lumen of the rough endoplasmic reticulum as they form on ribosomes. Sugars may be added to the proteins while they are in the endoplasmic reticulum. Next the proteins are found in the Golgi apparatus, where other carbohydrates are added to some of them. The proteins then travel in Golgi-derived vesicles to the plasma membrane and are incorporated into it (see Figure 4.19).

Functioning membranes move about within eukaryotic cells. For example, portions of the rough ER bud away from the ER and join the *cis* faces of the Golgi apparatus (see Chapter 4). Rapidly—often in less than an hour—these segments of membrane find themselves in the *trans* regions of the Golgi, from which they bud away to join the plasma membrane (Figure 5.18). Bits of membrane are constantly merging with the plasma membrane by exocytosis, but this process is largely balanced by the removal of membrane in endocytosis. This removal by endocytosis affords a recovery path by which internal membranes are replenished. In sum, there is a steady flux of membranes as well as membrane components in cells.

Because we know about the constant interconversion of membranes, we might expect all subcellular membranes to be chemically identical. As you already

know, this is not the case, for there are major chemical differences among the membranes of even a single cell. Apparently membranes are changed chemically when they form parts of certain organelles. In the Golgi apparatus, for example, the membranes of the *cis* face closely resemble those of the endoplasmic reticulum, but the *trans*-face membranes are more similar in composition to the plasma membrane.

Ceaselessly moving, constantly carrying out functions vital to the life of the cell, biological membranes certainly are not the static, stodgy structures they once were thought to be.

Summary of "Cellular Membranes"

Membrane Composition and Structure
• Biological membranes consist of lipids, proteins, and carbohydrates. The fluid-mosaic model of membrane structure describes a phospholipid bilayer in which integral membrane proteins can move about in the plane of the membrane. **Review Figures 5.1, 5.4, 5.5**
• The two surfaces of a membrane have different properties because of their different phospholipid composition, exposed domains of integral membrane proteins, and peripheral membrane proteins. Defined regions of a plasma membrane may have different membrane proteins. **Review Figures 5.1, 5.2**

• Carbohydrates attached to proteins or phospholipids project from the external surface of the plasma membrane and function as recognition signals for interactions between cells. **Review Figure 5.1**

Animal Cell Junctions

• Tight junctions prevent the passage of molecules through the space around cells, and they define functional regions of the plasma membrane by restricting the migration of membrane proteins uniformly over the cell surface. Desmosomes allow cells to adhere strongly to one another. Gap junctions provide channels for chemical and electrical communication between adjacent cells. **Review Figure 5.6**

Passive Processes of Membrane Transport

• Solutes are passively transported across a membrane by diffusion from a region with a greater solute concentration to a region of lesser solute concentration. Equilibrium is reached when the concentrations of a given solute are identical on both sides of the membrane. **Review Figures 5.7, 5.8**

• Substances can diffuse across a membrane by three processes: unaided diffusion through the phospholipid bilayer, diffusion through protein channels that form specific pores, or diffusion by means of a carrier protein (facilitated diffusion). **Review Figure 5.9**

• Channel proteins and carrier proteins may function in uniport or in coupled transport systems (symport or antiport). **Review Figure 5.9**

• The rate of simple diffusion is directly proportional to the concentration difference. The rate of facilitated diffusion reaches a maximum when a solute concentration is reached that saturates the carrier proteins so that no further increase in rate is observed with increases in solute concentration. **Review Figure 5.10**

• In osmosis, water diffuses from regions of less negative osmotic potential to regions of more negative osmotic potential. Isosmotic environmental conditions must be maintained to prevent animal cells from destructive loss or gain of water. In hypoosmotic solutions cells tend to take up water, while in hyperosmotic solutions cells tend to lose water. **Review Figure 5.11a**

• The cell walls of plants and some other organisms prevent the cells from bursting under hypoosmotic conditions. The turgor pressure (internal pressure) that develops under these conditions keeps plants upright and stretches the cell wall during plant cell growth. **Review Figure 5.11b**

Active Processes of Membrane Transport

• Active transport always requires energy to move substances across a membrane. Active transport systems moving single solute molecules or ions include primary and secondary active transport.

• In primary active transport, energy from the hydrolysis of ATP is used to move Na^+ out of cells and K^+ into cells against their concentration gradients. **Review Figure 5.12**

• Secondary active transport couples the passive movement of Na^+ down its concentration gradient with the movement of a sugar molecule up its concentration gradient. Energy from ATP is used indirectly to establish the Na^+ concentration gradient. **Review Figure 5.13**

• Energy is required for the many cellular processes that contribute to endocytosis, which transports macromolecules, large particles, and small cells into eukaryotic cells by means of engulfment and vesicle formation from the plasma membrane. Phagocytosis and pinocytosis are both nonspecific types of endocytosis. In receptor-mediated endocytosis, a specific membrane receptor binds to a particular macromolecule. **Review Figures 5.14, 5.15, 5.16**

• In exocytosis, materials in vesicles are secreted from the cell when the vesicles fuse with the plasma membrane. **Review Figure 5.14**

Membranes Are Not Simply Barriers

• Membranes also function as sites for protein or carbohydrate recognition signals for hormones, neurotransmitters, and growth factors. Conformational changes in membrane proteins transmit signals from outside the cell to enzymes within the cell by means of G proteins. In mitochondria and chloroplasts, membranes binding enzymes ensure faster and more efficient operation of a series of chemical reactions. **Review Figure 5.17**

• Membranes impermeable to H^+ separate electric charges and provide the basis for synthesis of ATP from ADP and phosphate in mitochondria and chloroplasts.

• Cell adhesion proteins in plasma membranes play crucial roles in organizing cells into tissues during embryonic development.

Membranes Are Dynamic

• Although not all cellular membranes are identical, ordered modifications in membrane composition accompany the conversions of one type of membrane into another type. **Review Figure 5.18**

Self-Quiz

1. Which statement about membrane phospholipids is *not* true?
   a. They associate to form bilayers.
   b. They have hydrophobic "tails."
   c. They have hydrophilic "heads."
   d. They give the membrane fluidity.
   e. They flop readily from one side of the membrane to the other.

2. The phospholipid bilayer
   a. is readily permeable to large, polar molecules.
   b. is entirely hydrophobic.
   c. is entirely hydrophilic.
   d. has different lipids in the two layers.
   e. is made up of polymerized amino acids.

3. Which statement about membrane proteins is *not* true?
   a. They all extend from one side of the membrane to the other.
   b. Some serve as channels for ions to cross the membrane.
   c. Many are free to migrate laterally within the membrane.
   d. Their position in the membrane is determined by their tertiary structure.
   e. Some play roles in photosynthesis.

4. Which statement about membrane carbohydrates is *not* true?
   a. Most are bound to proteins.
   b. Some are bound to lipids.
   c. They are added to proteins in the Golgi apparatus.
   d. They show little diversity.
   e. They are important in recognition reactions at the cell surface.

5. Which statement about animal cell junctions is *not* true?
   a. Tight junctions are barriers to the passage of molecules between cells.
   b. Desmosomes allow cells to adhere strongly to one another.
   c. Gap junctions block communication between adjacent cells.
   d. Connexons are made of protein.
   e. The fibers associated with desmosomes are made of protein.

6. Which statement about diffusion is *not* true?
   a. It is the movement of molecules or ions to a state of even distribution.
   b. At the subcellular level it is a slow process.
   c. The motion of each molecule or ion is random.
   d. The diffusion of each substance is independent of that of other substances.
   e. Diffusion across meters takes years.

7. Which statement about membrane channels is *not* true?
   a. They are pores in the membrane.
   b. They are proteins.
   c. All ions pass through the same type.
   d. Movement through them is from high concentration to low.
   e. Movement through them is by simple diffusion.

8. Facilitated diffusion and active transport
   a. both require ATP.
   b. both require the use of proteins as carriers.
   c. both carry solutes in only one direction.
   d. both increase without limit as the solute concentration increases.
   e. both depend on the solubility of the solute in lipid.

9. Primary and secondary active transport
   a. both generate ATP.
   b. both are based on passive movement of sodium ions.
   c. both include the passive movement of glucose molecules.
   d. both use ATP directly.
   e. both can move solutes against their concentration gradients.

10. Which statement about osmosis is *not* true?
    a. It obeys the laws of diffusion.
    b. In animal tissues, water moves to the cell with the most negative osmotic potential.
    c. Red blood cells must be kept in a plasma that is hypoosmotic to the cells.
    d. Two cells with identical osmotic potentials are isosmotic to each other.
    e. Solute concentration is the principal factor in osmotic potential.

Applying Concepts

1. In Chapter 44 we will see that the functioning of muscles requires calcium ions to be pumped into a subcellular compartment against a calcium concentration gradient. What types of chemical substances are required for this to happen?

2. Some algae have complex glassy structures in their cell walls. The structures form within the Golgi apparatus. How do these structures reach the cell wall without having to pass through a membrane?

3. Organisms that live in fresh water are almost always hyperosmotic to their environment. In what way is this a serious problem? How do some organisms cope with this problem?

4. Contrast simple endocytosis and receptor-mediated endocytosis with respect to mechanism and to performance.

Readings

Alberts, B., D. Bray, J. Lewis, M. Raff, K. Roberts and J. D. Watson. 1994. *Molecular Biology of the Cell*, 3rd Edition. Garland Publishing, New York. An outstanding general text in modern cell and molecular biology. Chapters 10 and 11 are particularly suitable for further study of biological membranes.

Bretscher, M. S. 1985. "The Molecules of the Cell Membrane." *Scientific American*, October. A fine treatment of membrane chemistry, cell junctions, endocytosis, and other topics.

Cooper, G. M. 1997. *The Cell: A Molecular Approach*. ASM Press, Washington, DC, and Sinauer Associates, Sunderland, MA. An excellent short cell biology text. See Chapter 12 in particular.

Horwitz, A. E. 1997. "Integrins and Health." *Scientific American*, May. These cell adhesion molecules play many roles in human cells; malfunctioning integrins result in disease.

Lodish, H., D. Baltimore, A. Berk, S. L. Zipursky, P. Matsudaira and J. Darnell. 1995. *Molecular Cell Biology*, 3rd Edition. Scientific American Books, New York. A fine general text. See Chapters 14 and 15; Chapter 16 is also relevant.

Stryer, L. 1995. *Biochemistry*, 4th Edition. W. H. Freeman, New York. A reference for the major types of molecules found in membranes, with outstanding illustrations of phospholipids.

Unwin, N. and R. Henderson. 1984. "The Structure of Proteins in Biological Membranes." *Scientific American*, February. A clear presentation of how the proteins in membranes transport molecules.

Chapter 6

Energy, Enzymes, and Metabolism

States of Energy
This rock climber's precarious position was achieved only with a great deal of expended energy. As she climbs, she gains potential energy of position with each upward step.

Change is everywhere. Wherever we look, we observe changes occurring in both the physical and the biological worlds. Some changes are slow, others are fast. Some are gigantic, others are small. Some are complex, others are simple. The concept of energy is profoundly helpful in understanding how changes are different and how they are all similar. In the preceding chapters, we often referred to energy, and in Chapter 2 we defined it as the capacity to do work. Now we want to focus on the different forms of energy and their role in accomplishing the activities characteristic of living systems.

The rock climber shown opposite has expended a lot of energy—done a lot of work—to get where she is. Looking at her, you can practically *feel* that energy expenditure. A climber has more reason than most people to be concerned about her potential energy—energy of position or state. As she climbs higher, her potential energy becomes greater. If she slips and falls, her potential energy will convert to kinetic energy—the energy of motion. Her climb has increased her potential energy; a fall would decrease it. To accomplish her climb, the cells in her body have converted the potential energy of chemical bonds (chemical energy) into the kinetic energy of muscle contractions.

In this chapter, we will examine the nature of energy and its relationship to work. We will apply physical principles to the chemical substances and chemical changes that constitute living systems. Then we will examine the role of en-

119

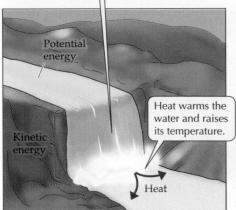

The water's potential energy is converted to kinetic energy, which is converted to heat at the base of the waterfall.

Potential energy

Kinetic energy

Heat warms the water and raises its temperature.

Heat

The dam converts the kinetic energy of a flowing river to potential energy by backing up the water.

Dam

Electric energy can be transmitted, stored, and used in a variety of ways to do work.

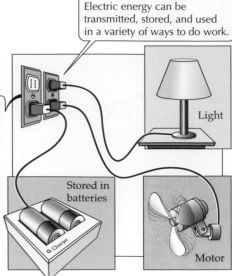

Light

Stored in batteries

Motor

A generator converts the movement of water released from the dam (kinetic energy) into electric energy.

6.1 Energy Conversions and Work
The kinetic energy of a flowing river can be converted to potential energy by a dam. Release of water from the dam converts the potential energy back into kinetic energy, which a generator can convert into electric energy.

zymes in speeding up chemical reactions in living systems. We'll close the chapter by focusing on metabolism, showing the roles of energy and enzymes in establishing and regulating the sequences of chemical reactions, called *pathways*, that transform matter and energy to produce the activities characteristic of life.

Energy Conversions: The Laws of Thermodynamics

All living things must obtain energy from the environment—no living cell manufactures energy. However, energy can be transformed from one kind into another. These energy transformations are linked to transformations of matter that occur as chemical reactions in cells, so changes in energy are associated with changes in matter. These changes in matter constitute metabolism. **Metabolism** is the total chemical activity of a living organism; at any instant, metabolism consists of thousands of individual chemical reactions.

Changes in energy are related to changes in matter

Energy comes in many forms (Figure 6.1). Heat, light, electric, and mechanical energy, among others, are forms of **kinetic energy**. Energy associated with position and gravity (think of the rock climber) is one type of **potential energy**. Potential energy is also associated with chemical bonds. If the rock climber jumps down to a ledge below, she loses some of her potential energy in a burst of movement—kinetic energy. Plants use light energy from the sun to drive the chemical re-

actions that convert the simple molecules CO_2 and H_2O into sugars, which are more complex. Some of the kinetic energy of light is stored as potential energy in the chemical bonds of the sugars. As you have seen in earlier chapters, these sugars can be stored as the polymer starch, or they can be degraded and the energy released in the process of cellular respiration.

In all cells (cells of plants, animals, fungi, protists, archaea, and bacteria), two types of metabolic reactions are found: anabolic and catabolic. **Anabolic reactions** (anabolism) link together simple molecules to form more complex molecules; the photosynthetic production of sugar is an anabolic reaction. Anabolic reactions store energy in the chemical bonds formed. **Catabolic reactions** (catabolism) break down complex molecules into simpler ones and release stored energy.

Catabolic and anabolic reactions are often linked: The energy released in the former is used to do biological work and drive the latter (Figure 6.2). In these energy conversions, there is a direction to the flow of energy. Energy flows through the biosphere in an irreversible conversion of light energy, by way of chemical energy, to heat. In general, the heat released by chemical reactions cannot be used for work in a biological system. Heat cannot do work unless it flows from one part of a system to another, which can happen only if one part of the system is hotter than another. Since all the parts of a cell are at the same temperature, heat cannot do work in cells.

The cellular activities of growth, motion, and active transport of ions across a membrane all require energy. None of these cellular activities or other forms of work would proceed without a source of energy. However,

6.2 Biological Energy Transformations   Cavorting lionesses convert chemical energy, obtained from the prey they have eaten, into a burst of kinetic energy of motion. Their prey obtained chemical energy by consuming plants. The plants trapped light energy and produced the prey's food by photosynthesis.

some kinds of energy cannot be used for these tasks. In the discussion that follows, you will discover the physical laws that govern all energy transformations, identify the energy available to do work, and consider the direction of energy flow.

The first law: Energy is neither created nor destroyed

When all forms of energy are accounted for, the total amount of energy in the universe is constant. In any conversion of energy from one form to another (light to chemical, mechanical to electric), *energy is neither created nor destroyed*. This is the **first law of thermodynamics** (Figure 6.3*a*). The first law applies to the universe as a whole or to any *closed system* within the universe. By "system" we mean any part of the universe with specified matter and energy. A **closed system** is one that is not exchanging energy with its surroundings. In considering the origin, evolution, and maintenance of living organisms on Earth, the appropriate closed system consists of the solar system, in particular the sun, Earth, and surrounding space (Figure 6.4).*

A closed system may contain parts that are not closed. **Open systems**, such as living cells, exchange

*Some energy does enter and leave the solar system—after all, we *can* see the light of stars—but the system is very nearly closed. In the same way, a thermos bottle is *almost* a closed system, although it does slowly gain or lose heat. The universe itself is a perfectly closed system.

matter and energy with their surroundings. Does this mean that cells disobey the first law or that the first law does not apply to living organisms? Not at all. It means that an open system is merely one part of a larger closed system and receives energy from other parts of that larger system.

The second law: Not all energy can be used, and disorder tends to increase

The second law concerns what happens when energy changes form. The law applies to all changes, but we will be concerned primarily with the changes that involve chemical reactions in living systems.

NOT ALL ENERGY CAN BE USED.   The **second law of thermodynamics** states that, although energy cannot be created or destroyed, *when energy is converted from one form to another, some of the energy becomes unavailable to do work* (see Figure 6.3*b*). In other words, no physical process or chemical reaction is 100 percent efficient, and not all the energy released can be converted to work. Some energy is lost as a form associated with disorder.

In any system, the total energy includes the *usable energy* that can be used to do work and the *unusable energy* that is lost to disorder.

total energy = usable energy + unusable energy

In biological systems, the total energy is called **enthalpy** (*H*). The usable energy that can do work is called the **free energy** (*G*). Free energy is what cells require for all the chemical reactions of cell growth, cell division, and the maintenance of cell health. The unusable energy is represented by **entropy** (*S*), which is the disorder of the system, multiplied by the absolute temperature (*T*). Thus we can rewrite the word equation above more precisely as

$$H = G + TS$$

Because we are interested in the usable energy, we rearrange this expression:

$$G = H - TS$$

Although we cannot measure *G*, *H*, or *S* absolutely, we can determine the *change* of each at a constant temperature. These energy changes are measured as calories (cal) or joules (J) (see Chapter 2). A change in a value is represented by the Greek letter delta (Δ), and it can be negative or positive. Therefore, the change in free energy (Δ*G*) of any reaction is defined in terms of the change in total energy (Δ*H*) and the change in entropy (Δ*S*):

$$\Delta G = \Delta H - T\Delta S$$

(a)

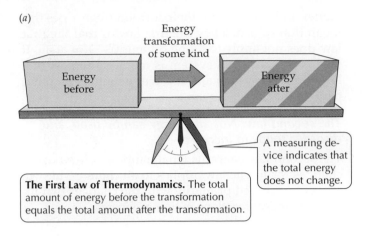

The First Law of Thermodynamics. The total amount of energy before the transformation equals the total amount after the transformation.

A measuring device indicates that the total energy does not change.

(b)

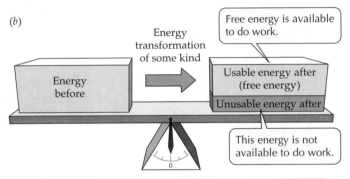

Free energy is available to do work.

This energy is not available to do work.

The Second Law of Thermodynamics. Although a transformation does not change the total amount of energy within a closed system, after any transformation the amount of free energy available to do work is always less than the original amount of energy.

(c)

Closed system

Unusable energy

Free energy

Another statement of the **Second Law** is that in any closed system, with repeated energy transformations free energy decreases and unusable energy–a function of entropy–increases.

This is an important relationship. It tells us whether free energy is *released* by a reaction, in which case ΔG is negative ($-\Delta G$), or free energy is *required* for a reaction, in which case ΔG is positive ($+\Delta G$). The sign and magnitude of ΔG depend on the two factors on the right of the equation: ΔH and $T\Delta S$.

For a chemical reaction, ΔH is the amount of energy added to the system ($+\Delta H$) or released ($-\Delta H$). In deter-

6.3 The Laws of Thermodynamics   Energy cannot be created or destroyed, but during energy transformations energy available to do work is lost.

mining the free energy released, a negative value for ΔH is obviously important. But the term $-T\Delta S$ must also be considered—it involves a change in entropy. If entropy increases for a chemical reaction, the products are more disordered or random. When are products more disordered?

If there are more products than reactants, as in the hydrolysis of a protein to its amino acids, the disorder and the entropy will be greater, and the change in entropy (ΔS) will be positive. If there are fewer products, and they are more constrained in their movements than the reactants, ΔS will be negative. Looking at the equation, we see that the negative sign in front of the entropy term means that a positive entropy (increasing disorder) will tend to make ΔG negative and large.

DISORDER TENDS TO INCREASE.   The second law of thermodynamics also states that *disorder tends to increase in the universe or a closed system*. Chemical changes, physical changes, biological processes—and anything else you can think of—all tend toward disorder, or randomness (Figure 6.3c). This tendency for disorder to increase gives a *directionality* to physical processes and chemical reactions. It explains why some reactions proceed in one direction rather than another. For example, consider the dissolving of table salt in a glass of water.

A solid crystal of sodium chloride is a highly ordered structure with each ion held rigidly in place; in other words, a crystal has low entropy. However, when a crystal dissolves in water, the ions have a more random relationship to one another (see Chapter 2): Disorder of the ions increases, randomness increases, and entropy increases ($+\Delta S$). A sodium chloride solution, on the other hand, will not spontaneously reorder itself into a crystal of salt and pure water. But if energy is added to the sodium chloride solution, the water can be evaporated and the salt crystals will reform. Without such intervention, crystals will not form spontaneously. This directionality is true of all reactions. For example, you know you must put energy into cleaning your room. Without that input of energy, disorder will increase.

Be sure you understand the second law correctly. It is *not* saying that ordered systems cannot be formed inside a large, complex closed system (Figure 6.4). In a closed system such as our solar system, free energy can be used to create order, just as an input of energy results in the formation of sodium chloride crystals from a solution, or an input of free energy maintains the order in your cells and in your room.

tants into unstable molecular forms called *transition-state species*. Transition-state species have higher free energies than either the reactants or the products. Although the activation energy needed for different reactions varies in size, it is often small compared to the change in free energy of the reaction. The activation energy that starts a reaction is recovered during the ensuing "downhill" phase of the reaction, so it is not a part of the free energy released, $-\Delta G$ (see Figure 6.11*a*).

With this energy barrier between reactants and products, how do reactions ever take place? In any collection of reactants, some molecules have more kinetic energy than others. Picture a mixture of reactant molecules with various kinetic energies. Some of the molecules in the mixture have enough energy to surmount the energy barrier and enter the transition state, and some of these molecules react, yielding products (Figure 6.12).

A reaction with a low activation energy proceeds more rapidly because more of the reactant molecules have enough energy to get over the initial hump. When activation energy is high, the reaction is slow unless more energy is provided, usually as heat. If the system is heated, all the reactant molecules move faster and have more kinetic energy. Since more of them have energy exceeding the required activation energy, the reaction speeds up.

Adding heat to increase the average kinetic energy of the molecules is not an effective option for living systems. Such a general, nonspecific approach would accelerate all the reactions, including destructive reactions, such as the denaturation of proteins that contribute to the structural integrity of the cell (see Chapter 3). Another, biologically more effective way to speed up a reaction is to lower the activation energy. In living cells, catalysts, most of which are enzymes, accomplish this task.

Enzymes bind specific reactant molecules

All types of catalysts speed chemical reactions. Most nonbiological catalysts are nonspecific; that is, they work on a diversity of reactants. For example, a powdered form of the metal platinum, called platinum black, catalyzes virtually any reaction in which molecular hydrogen (H_2) is a reactant. In contrast, most biological catalysts are members of the class of proteins called enzymes, and they are *highly specific*. An enzyme usually recognizes and binds to one or only a few closely related reactants, and it catalyzes only a single chemical reaction. The specificity of enzymes to the molecules they bind has been described as resembling the specificity of a lock and key.

The names of enzymes reflect the specificity of their functions. For example, the enzyme *RNA polymerase* will not work on DNA, and the enzyme *hexokinase* ac-

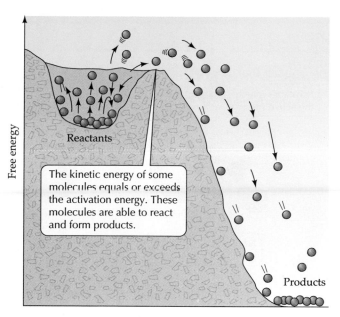

6.12 Over the Energy Barrier Some molecules surmount the energy barrier and react, forming products.

The kinetic energy of some molecules equals or exceeds the activation energy. These molecules are able to react and form products.

celerates the phosphorylation of hexose sugars but not pentose sugars. Most, but not all, names of enzymes end in the suffix "-ase."

In an enzyme-catalyzed reaction, the reactants are called **substrates**. Substrate molecules *bind* to a particular site on the protein surface, called the **active site**, where catalysis takes place (Figure 6.13). The binding of the substrate to the active site of the enzyme depends on the same kinds of forces that maintain the tertiary structure of the enzyme: hydrogen bonds, the attraction and repulsion of electrically charged groups, and hydrophobic interactions (see Chapter 2). The specificity of an enzyme results from the exact three-dimensional structure of its active site. Only one substrate fits precisely into the active site. Other molecules—with different shapes, different functional groups, and different properties—cannot properly fit and bind to the active site.

The binding of a substrate to the active site produces an **enzyme–substrate complex** held together by one or more means, such as hydrogen bonding, ionic attraction, or covalent bonding. The enzyme–substrate complex may form product and free enzyme:

$$E + S \rightarrow ES \rightarrow E + P$$

where E is the enzyme, S is the substrate, P is the product, and ES is the enzyme–substrate complex. Note that E, the free enzyme, is in the same chemical form at the end of the reaction as at the beginning. While bound to the substrate, it may change chemically, but by the end of the reaction it has been restored to its initial form.

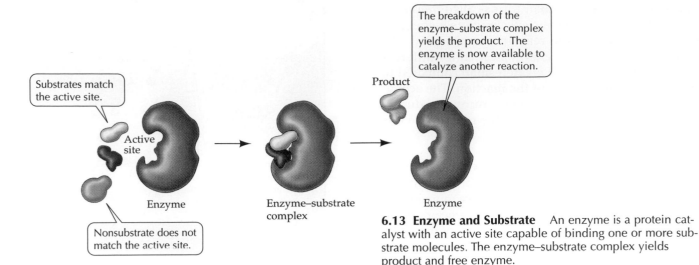

6.13 Enzyme and Substrate An enzyme is a protein catalyst with an active site capable of binding one or more substrate molecules. The enzyme–substrate complex yields product and free enzyme.

Enzymes lower the activation energy

The enzyme–substrate complex has a lower free energy than does the transition-state species of the corresponding uncatalyzed reaction (Figure 6.14). Thus the enzyme provides the reaction with a lower energy barrier—it offers an easier path. When an enzyme lowers the activation energy, both the forward and the reverse reactions speed up, so the enzyme-catalyzed overall reaction proceeds toward equilibrium more rapidly than the uncatalyzed reaction. Recall that the final equilibrium is the same with or without the enzyme. Adding an enzyme to a reaction does not change the difference in free energy (ΔG) between the reactants and the products; it changes only the activation energy and, consequently, the rate of reaction.

What are the chemical events at active sites of enzymes?

When the enzyme–substrate complex has formed, several types of chemical events in the active site contribute directly to the breaking of old bonds and the formation of new ones. We'll consider three types of catalysis: acid–base catalysis, covalent catalysis, and metal ion catalysis. In catalyzing a particular reaction, an enzyme may use more than one of these.

In *acid–base catalysis*, the acidic or basic side chains of amino acids forming the active site transfer H^+ to or from the substrate, destabilizing a covalent bond and

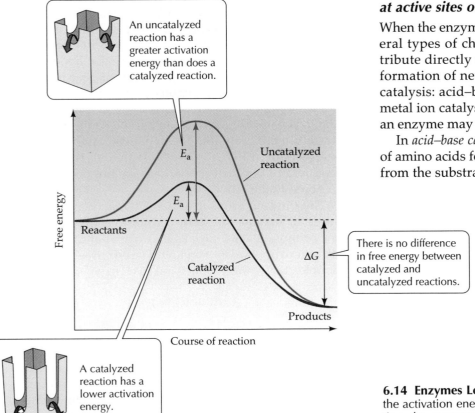

6.14 Enzymes Lower the Activation Energy Although the activation energy is lower in an enzyme-catalyzed reaction, the energy released is the same with or without catalysis. In other words, E_a is lower but ΔG is unchanged.

permitting it to break more readily. In *covalent catalysis,* a functional group in a side chain forms a temporary covalent bond with a portion of the substrate. For example, consider the general case of hydrolysis:

$$A—B + H_2O \rightarrow A—OH + BH$$

With an enzyme, this reaction may proceed in two steps:

$$A—B + enzyme—OH \rightarrow A—OH + enzyme—B$$
$$enzyme—B + H_2O \rightarrow enzyme—OH + BH$$

where the enzyme has a hydroxyl group that enters into the catalytic process. Such a path to reaction is effective only if the combined activation energies of its steps are lower than the total activation energy for the overall uncatalyzed reaction.

In *metal ion catalysis,* metal ions such as copper, zinc, iron, and manganese that are firmly bound to side chains of the protein contribute in several ways. All of these ions can lose or gain electrons without altering the bonds that hold them to the protein. This ability makes them important participants in oxidation–reduction reactions, which involve loss or gain of electrons. Metal ions are present in about a third of the enzymes that have been studied.

Substrate concentration affects reaction rate

For a reaction of the type A → B, the rate of the uncatalyzed reaction is directly proportional to the concentration of A (Figure 6.15). Addition of the appropriate enzyme speeds up the reaction, of course, but it also changes the shape of the plot of rate versus substrate concentration. At first, the rate of the enzyme-catalyzed reaction increases as the substrate concentration increases, but then the reaction rate levels off. Further increases in the substrate concentration do not increase the reaction rate. Since the concentration of the enzyme is usually much lower than that of the substrate, what we are seeing is a saturation phenomenon like the ones that occur in facilitated diffusion (see Figure 5.10*b*). When all the enzyme molecules are bound to substrate molecules, nothing is gained by adding more substrate, because no enzyme molecules are left to act as catalysts.

The study of the rates of enzyme-catalyzed reactions is called *enzyme kinetics.* As we will see later in this chapter, some graphs of rate versus substrate concentration are quite different from the one pictured in Figure 6.15. Such graphs tell us a great deal about the nature of the enzyme-catalyzed reaction.

Some enzymes couple reactions

Some of the most important reactions in living organisms are endergonic but proceed because specific enzymes couple them with other reactions that are exer-

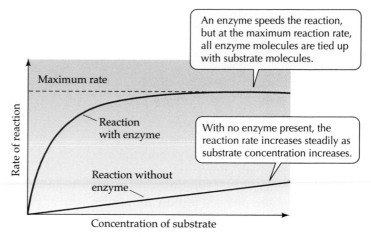

6.15 Enzymes Speed Up Reaction Rates Because there is usually less enzyme than substrate present, the reaction rate levels off when the enzyme becomes saturated.

gonic (recall Figure 6.10). Consider, for example, a pair of coupled reactions that occur in mitochondria. The first reaction, which converts succinate to fumarate, is highly exergonic. The second reaction, the hydrogenation of FAD (flavin adenine dinucleotide) to $FADH_2$, is endergonic, requiring a large input of free energy. The catalyst that couples these two reactions is the enzyme succinate dehydrogenase (Figure 6.16).

In a mitochondrion, the two hydrogen atoms that are removed from succinate are transferred to a molecule of a carrier substance, FAD. Succinate dehydrogenase couples the exergonic reaction to the endergonic one by ensuring that hydrogen atoms liberated by succinate are used to make $FADH_2$. One site on the enzyme surface binds succinate; a nearby second site binds FAD. Every time a succinate ion reacts with succinate dehydrogenase, much of the free energy that is released by this highly exergonic process is immediately trapped and used to synthesize $FADH_2$ from FAD. $FADH_2$ acts as a carrier of the hydrogen and the chemical free energy for use in an endergonic reaction (see Chapter 7).

In Chapter 5, you encountered other instances of coupled reactions. In animals the sodium–potassium pump (for primary active transport) is an enzyme that couples the exergonic hydrolysis of ATP to the endergonic pumping of Na^+ and K^+ against their concentration differences (see Figure 5.13):

$$ATP + H_2O \rightarrow ADP + P_i$$
$$3\ Na^+_{in} \rightarrow 3\ Na^+_{out}$$
$$2\ K^+_{out} \rightarrow 2\ K^+_{in}$$

The contractile proteins of muscle couple the exergonic breakdown of ATP to the performance of mechanical work against a load (see Chapter 44).

6.16 Succinate Dehydrogenase Couples Two Reactions

In mitochondria, the enzyme succinate dehydrogenase binds to both succinate and FAD, coupling an energy-producing reaction (succinate → fumarate) with an energy-requiring reaction (FAD → $FADH_2$).

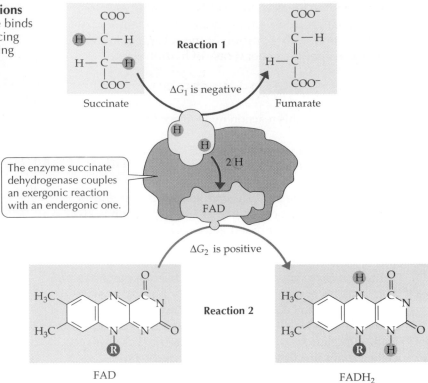

The enzyme succinate dehydrogenase couples an exergonic reaction with an endergonic one.

Overall reaction: $\Delta G = \Delta G_1 + \Delta G_2$
$\Delta G \approx 0$

ANOTHER WAY TO COUPLE REACTIONS. In metabolic pathways, there is another type of coupling, in which successive enzyme-catalyzed steps share compounds. A reaction A + B → C may be endergonic ($\Delta G = +10$ kJ/mol) but still proceed rapidly if the next step C + D → E is so exergonic ($\Delta G = -13$ kJ/mol) that the overall reaction (A + B + D → E) is exergonic (+10 kJ/mol – 13 kJ/mol = –3 kJ/mol).

Another way to look at this example is in terms of the effect of the second reaction on the equilibrium of the first reaction. As the highly exergonic reaction C + D → E proceeds toward its equilibrium (far to the right), it reduces the concentration of the substance C, so an equilibrium for the first reaction is never established. Thus more C is produced, since the reaction A + B → C progresses toward equilibrium.

These examples illustrate an important generalization: *Coupled reactions are the major means by which energy-requiring reactions are carried out in cells*. We will encounter many coupled reactions in the next two chapters.

Molecular Structure Determines Enzyme Function

Until the 1960s, biochemists knew little about the behavior of enzymes at the molecular level. It was generally agreed that the substrates of enzymes bind to an active site on the surface of the enzyme molecule, but the structure of an active site was not understood. The remarkable ability of an enzyme to select exactly the right substrate was explained by the assumption that the binding of the substrate to the site depends on a precise interlocking of molecular shapes. In 1894 the German chemist Emil Fischer compared the fit between an enzyme and substrate to that of a lock and key. Fischer's model persisted for more than half a century with only indirect evidence to support it.

The first direct evidence came in 1965, when David Phillips and his colleagues at the Royal Institution in London succeeded in crystallizing the enzyme lysozyme and determined its structure using the techniques of X ray crystallography (to be described in Chapter 11). The tertiary structure of lysozyme is shown in Figure 6.17. Lysozyme is an enzyme that protects the animals that produce it by destroying invading bacteria. To destroy the bacteria, it cleaves certain polysaccharide chains in their cell walls. Lysozyme is found in tears and other bodily secretions, and it is particularly abundant in the whites of bird eggs. In Figure 6.17, the active site of lysozyme appears as a large indentation filled with the substrate (shown in yellow).

In the discussion that follows, we'll consider how the binding of substrate to an enzyme may change the enzyme and contribute to catalysis. Then we will examine cofactors that are sometimes bound to proteins required for reactions.

Binding at the active site may cause enzymes to change shape

The entire structure of an enzyme molecule is the source of the impeccable specificity of enzymes, on which all metabolism depends. As Fischer suggested in his "lock and key" model, the structure of the active site fits the substrate molecule. However, we now know that after a substrate binds to an enzyme's active site, the entire enzyme may undergo a change in shape to accomplish the "fit" at the active site. This change in enzyme shape caused by substrate binding is called **induced fit**. Induced fit can be observed in the enzyme hexokinase when it is studied with and

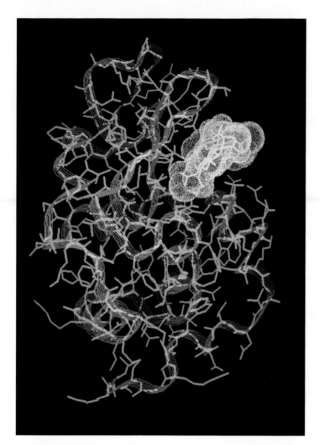

6.17 Tertiary Structure of Lysozyme The polysaccharide substrate, shown in yellow, is bound to a lysozyme molecule, shown in blue. The magenta ribbon highlights the backbone of the protein.

without its substrate glucose (Figure 6.18). Induced fit brings reactive side chains from the enzyme's active site into alignment with the substrate, facilitating the catalytic events described earlier (see Figure 6.13).

To operate, some enzymes require cofactors

Whether they consist of a single folded polypeptide chain or several chains, many enzymes require nonprotein molecules, called *cofactors*, in order to function. In addition to the bound metal ions mentioned earlier, some enzymes require other cofactors to catalyze the reaction; prosthetic groups and coenzymes are examples of such cofactors. *Prosthetic groups* are bound to the enzyme and include the heme groups that are attached to the protein in the oxygen-carrying protein hemoglobin (see Figure 3.19). *Coenzymes* are generally relatively small (compared to the enzyme) carbon-containing molecules that are required for the action of one or more enzymes (Figure 6.19).

Since coenzymes are not permanently bound to the enzyme, they must react with it as a substrate does. For the catalyzed reaction to proceed, coenzymes must collide with the enzyme and bind to its active site just as the substrate must. A coenzyme can be considered a "cosubstrate," because it changes chemically during the reaction and separates from the enzyme to participate in other reactions. Coenzymes move from enzyme molecule to enzyme molecule.

ATP and ADP can be considered coenzymes: They are necessary for reaction, are changed by the reaction, and bind and detach from the enzyme. In the next chapter, we will encounter coenzymes that function by accepting or donating electrons or hydrogen atoms. In animals, some coenzymes are produced from vitamins that must be obtained from food—they cannot be synthesized by the body.

6.18 Some Enzymes Change Shape When Substrate Binds to Them Shape changes result in an induced fit between enzyme and substrate, improving the catalytic ability of the enzyme–substrate complex.

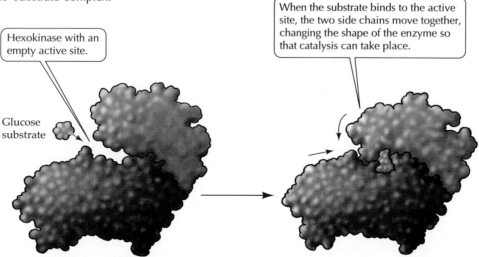

Hexokinase with an empty active site.

Glucose substrate

When the substrate binds to the active site, the two side chains move together, changing the shape of the enzyme so that catalysis can take place.

6.19 A Coenzyme Some enzymes require coenzymes in order to function. This illustration shows the relative sizes of the enzyme (red and blue) and coenzyme (white).

Metabolism and the Regulation of Enzymes

All organisms need to maintain stable internal conditions, or *homeostasis* (which will be covered in detail in Chapter 37). If homeostasis breaks down because the concentrations of some compounds rise or fall too much, illness results. Thus we and all other organisms must regulate our metabolism, and the regulation of the rates at which thousands of different enzymes operate contributes to metabolic homeostasis.

In the remainder of this chapter we'll investigate the role of enzymes in organizing and regulating metabolism. The activity of enzymes can be inhibited in various ways, so the presence of an enzyme does not necessarily ensure that it is functioning in a cell. We'll discover how the rate at which some enzymes catalyze reactions can be altered, making enzymes the target points at which entire pathways can be regulated. We'll close with an examination of how temperature and pH affect enzyme action.

Metabolism is organized into metabolic pathways

An organism's metabolism is the totality of the biochemical reactions that take place within it. Metabolism transforms raw materials and stored potential energy into forms that can be used by living cells. These reactions proceed down **metabolic pathways**, which are series of enzyme-catalyzed reactions. In these sequences, the product of one reaction is the substrate for the next:

$$A \rightarrow B \rightarrow C \rightarrow D$$

Some pathways synthesize the important chemical building blocks from which macromolecules are built. For example, one group of metabolic pathways forms the various amino acids. Another group produces the nucleotides for the nucleic acids. Some pathways harvest energy. The pathway called *glycolysis*, for example, partly degrades glucose and generates some ATP. The pathway called *cellular respiration* completes the catabolism of glucose and produces a great deal more ATP. Chapter 7 explores these pathways in more detail. To have adequate amounts of all the necessary building blocks without wasting raw materials or energy, a cell must regulate all its metabolic pathways constantly.

Enzyme activity is subject to regulation

Various substances, called **inhibitors**, bind to enzymes, decreasing the rates of enzyme-catalyzed reactions. Some inhibitors occur naturally in cells; others are artificial. Naturally occurring inhibitors regulate metabolism; the artificial ones are used either to treat disease or to study how enzymes work. Some inhibitors irreversibly inhibit the enzyme by permanently binding to it. Others have reversible effects; that is, these inhibitors can become unbound. The removal of a natural reversible inhibitor increases an enzyme's rate of catalysis.

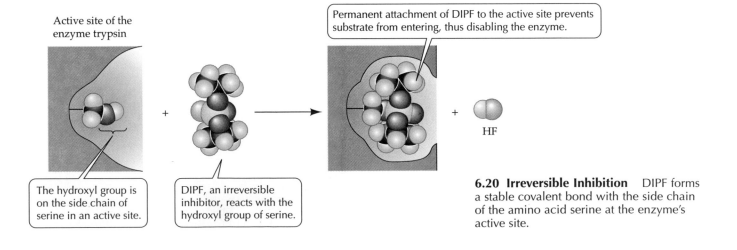

Active site of the enzyme trypsin

Permanent attachment of DIPF to the active site prevents substrate from entering, thus disabling the enzyme.

The hydroxyl group is on the side chain of serine in an active site.

DIPF, an irreversible inhibitor, reacts with the hydroxyl group of serine.

HF

6.20 Irreversible Inhibition DIPF forms a stable covalent bond with the side chain of the amino acid serine at the enzyme's active site.

6.21 Reversible Inhibition Enzyme inhibition is sometimes reversible. *(a,b)* In competitive inhibition, an inhibitor binds temporarily to the active site. For example, succinate dehydrogenase is subject to competitive inhibition by oxaloacetate. *(c)* A noncompetitive inhibitor binds away from the active site.

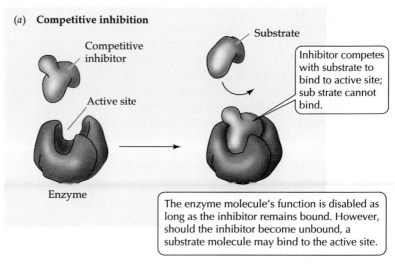

(a) **Competitive inhibition**

Competitive inhibitor

Active site

Substrate

Enzyme

Inhibitor competes with substrate to bind to active site; substrate cannot bind.

The enzyme molecule's function is disabled as long as the inhibitor remains bound. However, should the inhibitor become unbound, a substrate molecule may bind to the active site.

IRREVERSIBLE INHIBITION. Some inhibitors bond irreversibly to certain side chains at active sites of enzymes, thereby inactivating the enzymes by destroying their capacity to interact with the normal substrate. An example of such an **irreversible inhibitor** is DIPF (diisopropylphophorofluoridate). DIPF reacts with a hydroxyl group belonging to the amino acid serine (see Table 3.1) at an enzyme's active site, preventing the use of this side chain in the catalytic mechanism (Figure 6.20).

DIPF is an irreversible inhibitor for the protein-digesting enzyme trypsin and for many other enzymes whose active sites contain serine. Among these is acetylcholinesterase, an enzyme that is essential for the orderly propagation of impulses from one nerve cell to another (see Chapter 41). Because of their effect on acetylcholinesterase, DIPF and other similar compounds are classified as nerve gases.

REVERSIBLE INHIBITION. Not all inhibitory action is irreversible. Some inhibitor molecules are similar enough to a particular enzyme's natural substrate to bind to the active site, yet different enough that the enzyme catalyzes no chemical reaction. While such a molecule is bound to the enzyme, the natural substrate cannot enter the active site; thus the intruder effectively wastes the enzyme's time, inhibiting its catalytic action. These molecules are called **competitive inhibitors** because they compete with the natural substrate for the active site and block the action of the enzyme (Figure 6.21a); the blockage is reversible. When the concentration of the competitive inhibitor is reduced, it detaches from the active site and the enzyme is again active.

The enzyme *succinate dehydrogenase* is subject to competitive inhibition. Recall that this enzyme, found in all mitochondria, catalyzes the conversion of the compound succinate to another compound, fumarate (see Figure 6.16). The compound oxaloacetate is similar to succinate and can act as a competitive inhibitor of succinate dehydrogenase by binding to the active site (Figure 6.21b). However,

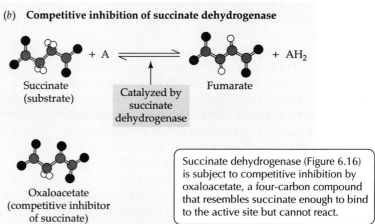

(b) **Competitive inhibition of succinate dehydrogenase**

Succinate (substrate) + A ⇌ Fumarate + AH$_2$

Catalyzed by succinate dehydrogenase

Oxaloacetate (competitive inhibitor of succinate)

Succinate dehydrogenase (Figure 6.16) is subject to competitive inhibition by oxaloacetate, a four-carbon compound that resembles succinate enough to bind to the active site but cannot react.

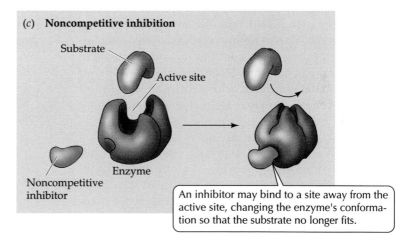

(c) **Noncompetitive inhibition**

Substrate

Active site

Enzyme

Noncompetitive inhibitor

An inhibitor may bind to a site away from the active site, changing the enzyme's conformation so that the substrate no longer fits.

having bound to oxaloacetate, the enzyme can do nothing more with it—no reaction occurs. An enzyme molecule cannot bind a succinate molecule until the inhibitor molecule has moved out of the active site. The inhibitor *can* move out of the site, because binding of a competitive inhibitor is reversible, *as is binding of the substrate.*

Some inhibitors that do not react with the active site are called **noncompetitive inhibitors**. Noncompetitive inhibitors bind to the enzyme at a site away from the active site. Their binding causes a conformational change in the protein that alters the active site (Figure 6.21c). The active site may still bind substrate molecules, but the rate of product formation is reduced. Noncompetitive inhibitors can become unbound, so their effects are reversible.

Allosteric enzymes have interacting subunits

Many important enzymes have a quaternary structure consisting of two or more polypeptide subunits, each with a molecular weight in the tens of thousands (see Chapter 3). These subunits are bound together by various weak bonds that permit changes in the shape of one subunit to influence the shape and properties of the other subunits. Multisubunit enzymes that undergo such changes in shape and function are called **allosteric enzymes** (*allo-*, "different"; *-steric*, "shape").

The activity of these complex enzymes is controlled by molecules called **effectors**, which may have no similarity either to the reactants or to the products of the reaction being catalyzed. Effectors bind to an **allosteric site** that is separate from the active site and enhance or diminish reactions at the active site. Thus, effectors are activators or inhibitors. Their binding changes the structure of the enzyme and thus its activity.

Allosteric enzymes and single-subunit enzymes differ greatly in their reaction rates when the substrate concentration is low. Graphs of rate plotted against substrate concentration show this relationship. For an enzyme with a single subunit, the plot looks like that in Figure 6.22a. The reaction rate first increases very sharply with increasing substrate concentration, then tapers off to a constant maximum rate as the supply of enzyme becomes saturated with substrate. The plot for many allosteric enzymes is radically different, with a sigmoidal (S-shaped) appearance (Figure 6.22b).

With sigmoid kinetics, the increase in rate with increasing substrate concentration is slight at low substrate concentrations, but within a certain range the reaction rate is extremely sensitive to relatively small changes in the substrate concentration. Because of this sensitivity, allosteric enzymes are important in fine-tuning the activities of a cell. We can understand this behavior in terms of interactions between the different kinds of subunits that make up an allosteric enzyme, which we'll examine next.

Catalytic and regulatory subunits interact and cooperate

An allosteric enzyme not only has more than one subunit, it has more than one *type* of subunit: A **catalytic subunit** has an active site that binds the enzyme's substrate; a **regulatory subunit** has one or more allosteric

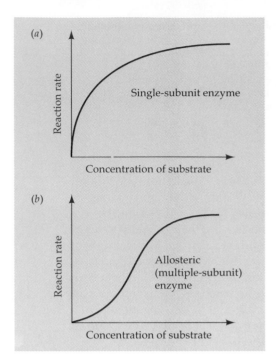

6.22 Allostery and Reaction Rate How the rate of an enzyme-catalyzed reaction changes with increasing substrate concentration depends on whether the enzyme consists of one or more than one polypeptide subunit.

sites that bind specific effector molecules (Figure 6.23). Binding of either a substrate or an effector affects the structure of the enzyme as a whole. An allosteric enzyme can exist in two forms. The *active form* has catalytic activity, whereas the *inactive form* lacks activity. An allosteric enzyme usually consists of two or more catalytic subunits and one or more regulatory subunits. The existence of two or more linked catalytic subunits allows for *cooperativity*, which works as follows.

When a molecule of substrate binds to the active site of one catalytic subunit of an allosteric enzyme, it causes a favorable, cooperative change in the other catalytic subunits, making it easier for substrate to bind them (Figure 6.23c). Conversely, when an allosteric inhibitor binds to the allosteric site of a regulatory subunit, it causes an unfavorable change in the catalytic subunits, making it harder for substrate to bind them.

Cooperativity makes an enzyme exquisitely sensitive to its environment. The binding of one substrate molecule makes it easier for further substrate molecules to react.

Allosteric effects control metabolism

Some metabolic pathways are branched. At a branch point an intermediate substance is acted on by more than one enzyme and thus is sent through more than one metabolic branch (Figure 6.24). Two different pathways emerge from a branch.

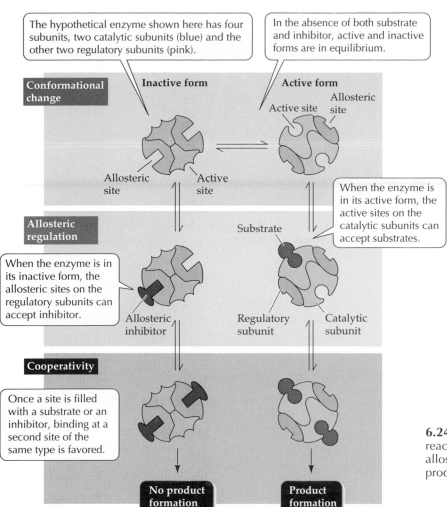

The hypothetical enzyme shown here has four subunits, two catalytic subunits (blue) and the other two regulatory subunits (pink).

In the absence of both substrate and inhibitor, active and inactive forms are in equilibrium.

Conformational change

Inactive form

Active form

Active site

Allosteric site

Allosteric site

Active site

Allosteric regulation

Substrate

When the enzyme is in its active form, the active sites on the catalytic subunits can accept substrates.

When the enzyme is in its inactive form, the allosteric sites on the regulatory subunits can accept inhibitor.

Allosteric inhibitor

Regulatory subunit

Catalytic subunit

Cooperativity

Once a site is filled with a substrate or an inhibitor, binding at a second site of the same type is favored.

No product formation

Product formation

6.23 Allosteric Regulation of Enzymes
The hypothetical enzyme shown here has four subunits: two catalytic (blue), the other two regulatory. When the enzyme is in its active form, the active sites on the catalytic subunits can accept substrate. When the enzyme is in its inactive form, the allosteric sites on the regulatory subunits can accept inhibitor.

6.24 Feedback in Metabolic Pathways   The first reaction following a branch point is catalyzed by an allosteric enzyme that can be inhibited by the end product of the pathway.

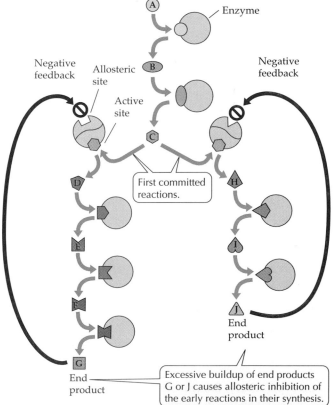

Enzyme

Negative feedback

Allosteric site

Active site

Negative feedback

First committed reactions.

End product

End product

Excessive buildup of end products G or J causes allosteric inhibition of the early reactions in their synthesis.

At the branching points where metabolic pathways diverge, **regulatory enzymes** catalyze reactions. These regulatory enzymes are like track switches in a railroad: Increasing or decreasing the rates of the reactions that they catalyze determines what fraction of the flow of material goes through which branch. What turns such a switch? The end product of a branch pathway may damp the initial step in that branch pathway, reducing the formation of the end product (see Figure 6.24). This is the principle of **feedback inhibition**, also called end product inhibition.

The end product of a particular pathway typically is an allosteric inhibitor of the regulatory enzyme that catalyzes the first **committed step** in its own synthesis—that is, the earliest step in the branched pathway that leads to the synthesis of only that end product and no other. The committed steps in metabolic pathways are particularly effective points for feedback control. For instance, inhibition of the C-to-D step in Figure 6.24 shunts all the reactants from the G pathway into the J pathway.

Allosteric regulation is very effective. It allows rapid adjustment to short-term changes in metabolism or in the environment. The activities of enzyme molecules are adjusted by their interactions with small molecules, the end products. However, if a particular enzyme were not needed, wouldn't it be a good idea simply to stop making it until it was needed? Wouldn't it be advantageous to regulate enzyme *production* as well as enzyme *activity*? The answer is yes, and the regulation of enzyme synthesis plays an important role in controlling metabolism and development (see Chapters 13, 14, and 15).

Enzymes are sensitive to their environment

Enzymes enable cells to perform chemical reactions and carry out complex processes without using the extremes of temperature and pH employed by chemists in the laboratory. Enzymes themselves are extremely sensitive to changes in the medium around them, particularly temperature and pH.

pH AFFECTS ENZYME ACTIVITY. The rates of most enzyme-catalyzed reactions depend on the pH of the medium in which they occur. Each enzyme is most active at a particular pH; its activity decreases as the solution is made more acidic or more basic than its ideal pH (Figure 6.25).

Several factors contribute to this effect. One is the ionization of carboxyl, amino, and other groups on either the substrate or the enzyme. Carboxyl groups (—COOH) ionize to become negatively charged carboxylate groups (—COO$^-$) in neutral or basic solutions. Similarly, amino groups (—NH$_2$) accept H$^+$ ions

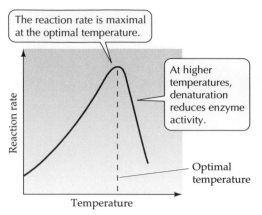

The reaction rate is maximal at the optimal temperature.

At higher temperatures, denaturation reduces enzyme activity.

Optimal temperature

Temperature

Reaction rate

6.26 Temperature Affects Enzyme Activity An enzyme is most active at a particular temperature.

in neutral or acidic solutions, becoming positively charged —NH$_3^+$ groups (see Chapter 2). Thus, in a neutral solution a molecule with an amino group is attracted electrically to another molecule that has a carboxyl group because both groups are ionized and they have opposite charges.

Evolution has matched enzymes to their environment. For example, the protein-digesting enzyme pepsin, found only in the stomach, works best at the very low pH values that prevail in the stomach after a meal. In contrast, salivary amylase works best at neutral pH, which is characteristic of the mouth.

TEMPERATURE AFFECTS ENZYME ACTIVITY. At low temperatures, warming increases the rate of an enzyme-catalyzed reaction because at higher temperatures a greater fraction of the reactant molecules have enough energy to provide the activation energy of the reaction (Figure 6.26). Temperatures that are too high, however, inactivate enzymes, because at high temperatures enzyme molecules vibrate and twist so rapidly that some of their noncovalent bonds break. When heat destroys their tertiary structure, enzyme molecules become inactivated, or **denatured** (see Chapter 3). Some enzymes denature at temperatures only slightly above that of the human body, but a few are stable even at the boiling point of water.

Individual organisms adapt to changes in the environment in many ways, one of which is based on groups of enzymes, called **isozymes**, that catalyze the same reaction but have different chemical compositions and physical properties. Within a given group, different isozymes may have different optimal temperatures. In rainbow trout, an example is the isozymes of the enzyme acetylcholinesterase, whose operation is essential to normal transmission of nerve impulses. If a rainbow trout is transferred from warm water to near-freezing water (2°C), the fish produces an isozyme of acetylcholinesterase that is different from

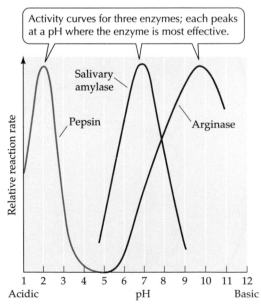

Activity curves for three enzymes; each peaks at a pH where the enzyme is most effective.

Salivary amylase

Pepsin

Arginase

Relative reaction rate

1 2 3 4 5 6 7 8 9 10 11 12
Acidic   pH   Basic

6.25 pH Affects Enzyme Activity Each enzyme catalyzes at a maximum rate at a particular pH.

the acetylcholinesterase produced at the higher temperature. The new isozyme has a lower optimal temperature, which helps the fish to perform normally in the colder water.

Summary of "Energy, Enzymes, and Metabolism"

Energy Conversions: The Laws of Thermodynamics

• Energy is the capacity to do work. Potential energy is the energy of state or position; it includes the energy stored in chemical bonds. Kinetic energy is energy of motion (and related forms such as electric energy, light, and heat).

• Potential energy can be converted to kinetic energy, which can do work. Biological work includes synthesis, growth, movement, transport, and maintenance. **Review Figure 6.1**

• The first law of thermodynamics tells us that energy cannot be created or destroyed. The second law of thermodynamics tells us that, in any closed system, the quantity of energy available to do work (free energy) decreases and unusable energy associated with entropy increases.

• Living things, like everything else, obey the laws of thermodynamics. Organisms are open systems that are part of a larger closed system. **Review Figures 6.3, 6.4**

• Changes in free energy, total energy, temperature, and entropy are related by the equation $\Delta G = \Delta H - T\Delta S$.

Chemical Reactions Release or Take Up Energy

• Spontaneous, exergonic reactions release free energy and have a negative ΔG. Nonspontaneous, endergonic reactions take up free energy and have a positive ΔG. Endergonic reactions proceed only if free energy is provided. **Review Figure 6.5**

• The change in free energy of a reaction determines its point of chemical equilibrium, at which the forward and reverse reactions proceed at the same rate. For spontaneous, exergonic reactions, equilibrium lies toward completion (products). **Review Figure 6.6**

ATP: Transferring Energy in Cells

• ATP (adenosine triphosphate) is an energy transfer molecule in cells. Hydrolysis of the terminal phosphate releases a relatively large amount of free energy ($\Delta G = -12$ kcal/mol). **Review Figure 6.7**

• The ATP cycle couples exergonic and endergonic reactions, transferring free energy from the exergonic to the endergonic reaction. **Review Figures 6.9, 6.10**

Enzymes: Biological Catalysts

• The rate of a chemical reaction is independent of ΔG but is determined by the size of the activation energy barrier. Catalysts speed reactions by lowering the activation energy. **Review Figures 6.11, 6.12**

• Enzymes are biological catalysts, proteins that are highly specific for their substrates. Substrates bind to the active site, where catalysis takes place, forming an enzyme–substrate complex. **Review Figure 6.13**

• At the active site, amino acid side chains and in some cases attached metal ions participate in chemical events such as acid–base catalysis, covalent catalysis, or metal ion catalysis that lower the activation energy for reactions. **Review Figure 6.14**

• Substrate concentration affects the rate of an enzyme-catalyzed reaction. **Review Figure 6.15**

• Some enzymes couple exergonic and endergonic reactions by catalyzing both reactions. Other reactions are coupled if a product of the endergonic reaction is a reactant for an exergonic reaction catalyzed by a second enzyme; this coupling works if the overall ΔG for the two reactions is negative. **Review Figures 6.10, 6.16**

Molecular Structure Determines Enzyme Function

• The active site where substrate binds determines the specificity of an enzyme. Upon binding to substrate, some enzymes change shape, facilitating catalysis. **Review Figures 6.13, 6.18**

• Some enzymes require cofactors to carry out catalysis. Prosthetic groups, such as heme, are permanently bound to the enzyme. Coenzymes are not usually bound to the enzyme, they enter into the reaction as a "cosubstrate," and they appear in modified form as a product. **Review Figure 6.19**

Metabolism and the Regulation of Enzymes

• Metabolism is organized into pathways, in which the product of one reaction is a reactant for the next reaction. Each reaction is catalyzed by an enzyme.

• Enzyme activity is subject to regulation. Some compounds react irreversibly with enzymes and reduce their catalytic activity. Others react reversibly, inhibiting enzyme action only temporarily. A compound closely similar in structure to an enzyme's normal substrate may competitively inhibit the action of the enzyme. **Review Figures 6.20, 6.21**

• For allosteric enzymes, the plots of reaction rate versus substrate concentration are sigmoidal, in contrast to plots of the same variables for single-subunit enzymes. **Review Figure 6.22**

• Allosteric inhibitors bind to a site different from the active site and stabilize the inactive form of the enzyme. The multiple catalytic subunits of many allosteric enzymes interact cooperatively. **Review Figure 6.23**

• Allosteric effects control metabolism. The end product of a metabolic pathway feeds back on the enzyme that catalyzes the first committed step in that branch, inhibiting that allosteric enzyme and preventing excessive buildup of the end product. **Review Figure 6.24**

• Enzymes are sensitive to their environment. Both pH and temperature affect enzyme activity. **Review Figures 6.25, 6.26**

Self-Quiz

1. Which statement about energy is *not* true?
   a. It can neither be created nor destroyed.
   b. It is the capacity to do work.
   c. All of its conversions are fully reversible.
   d. In the universe as a whole, the amount of free energy decreases.
   e. In the universe as a whole, the amount of entropy increases.

2. Which statement about thermodynamics is *not* true?
   a. Free energy is given off in an exergonic reaction.
   b. Free energy can be used to do work.
   c. A spontaneous reaction is exergonic.
   d. Free energy tends always to a minimum.
   e. Entropy tends always to a minimum.

3. In a chemical reaction,
   a. the rate depends on the value of ΔG.
   b. the rate depends on the activation energy.
   c. the entropy change depends on the activation energy.
   d. the activation energy depends on the value of ΔG.
   e. the change in free energy depends on the activation energy.

4. Which statement about enzymes is *not* true?
   a. They consist of proteins, with or without a nonprotein part.
   b. They change the rate of the catalyzed reaction.
   c. They change the value of ΔG of the reaction.
   d. They are sensitive to heat.
   e. They are sensitive to pH.

5. The active site of an enzyme
   a. never changes shape.
   b. forms no chemical bonds with substrates.
   c. determines, by its structure, the specificity of the enzyme.
   d. looks like a lump projecting from the surface of the enzyme.
   e. changes ΔG of the reaction.

6. A prosthetic group
   a. is a tightly bound, nonprotein part of an enzyme.
   b. is composed of protein.
   c. does not participate in chemical reactions.
   d. is present in all enzymes.
   e. is an artificial enzyme.

7. The rate of an enzyme-catalyzed reaction
   a. is constant under all conditions.
   b. decreases as substrate concentration increases.
   c. cannot be measured.
   d. depends on the value of ΔG.
   e. can be reduced by inhibitors.

8. Which statement about enzyme inhibitors is *not* true?
   a. A competitive inhibitor binds the active site of the enzyme.
   b. An allosteric inhibitor binds a site on the active form of the enzyme.
   c. A noncompetitive inhibitor binds a site other than the active site.
   d. Noncompetitive inhibition cannot be completely overcome by the addition of more substrate.
   e. Competitive inhibition can be completely overcome by the addition of more substrate.

9. Which statement about feedback inhibition of enzymes is *not* true?
   a. It is exerted through allosteric effects.
   b. It is directed at the enzyme that catalyzes the first committed step in a branch of a pathway.
   c. It affects the rate of reaction, not the concentration of enzyme.
   d. It acts very slowly.
   e. It is an example of negative feedback.

10. Which statement about temperature effects is *not* true?
   a. Raising the temperature may reduce the activity of an enzyme.
   b. Raising the temperature may increase the activity of an enzyme.
   c. Raising the temperature may denature an enzyme.
   d. Some enzymes are stable at the boiling point of water.
   e. The isozymes of an enzyme have the same optimal temperature.

Applying Concepts

1. How can endergonic reactions proceed in organisms?

2. Consider two proteins: One is an enzyme dissolved in the cytosol; the other is an ion channel in a membrane. Contrast the structures of the two proteins, indicating at least two important differences.

3. Plot free energy versus the course of an endergonic reaction and that of an exergonic reaction. Include the activation energy in both plots. Label E_a and ΔG on both graphs.

4. Consider an enzyme that is subject to allosteric regulation. If a competitive inhibitor (not an allosteric inhibitor) is added to a solution of such an enzyme, the ratio of enzyme molecules in the active form to those in the inactive form increases. Explain this observation.

Readings

Dickerson, R. E. and I. Geis. 1969. *The Structure and Action of Proteins.* W. A. Benjamin, Menlo Park, CA. This classic volume presents the structure of enzymes as high art.

Karplus, M. and J. A. MacCammon. 1986. "The Dynamics of Proteins." *Scientific American*, April. This article will correct any misconception of proteins as rigid molecules; it describes the constant, rapid changes in local shape that underlie the functioning of proteins.

Kauffman, S. A. 1993. *The Origins of Order.* Oxford University Press, New York. A discussion of thermodynamics, chaos, and life.

Koshland, D. E., Jr. 1973. "Protein Shape and Biological Control." *Scientific American*, October. This paper shows that the ability of proteins to change shape in specific circumstances underlies the control and coordination of biological processes.

Morowitz, H. J. 1978. *Foundations of Bioenergetics.* Academic Press, New York. An excellent advanced text on thermodynamics in biology.

Stryer, L. 1995. *Biochemistry*, 4th Edition. W. H. Freeman, New York. Good discussions of enzymes, the basic concepts of metabolism, and protein structure.

Chapter 7

Cellular Pathways That Harvest Chemical Energy

A Cold New World
Newborn humans experience a sudden drop in temperature when they emerge from the womb into the world. They adapt by metabolizing reserves of a special type of body fat.

A human fetus develops at the mother's body temperature. At birth, most newborns experience a drop in temperature, and their bodies must quickly do something about it. In fact what they do is the same thing a hibernating mammal does as it rouses itself from its winter "snooze." During hibernation, body temperature is low. In order to move about and take care of itself once awake again, an animal that has been hibernating must raise its body temperature.

Both the newly awakened mammal and the newborn baby respond to their temperature challenges—the low temperature of the hibernating body and the rapidly dropping temperature of the baby—by starting to metabolize stored "brown fat" reserves. Usually reserves are metabolized to generate ATP for work. In this case, however, a temperature-triggered signal tells the brown fat cells *not* to form ATP. Instead, these fat cells "waste" their stored energy as heat: They generate *lots* of heat and thus warm the body.

Cells convert energy from one form to another in order to carry out biological work (mechanical work, transport, and synthesis) and, sometimes, to generate heat. Most of the energy in the biosphere derives from the sun. Energy flows from the sun through photosynthetic *autotrophs* (plants and some protists and bacteria) to *heterotrophs* (organisms that must obtain their energy in the form of food—organic compounds) and back to the environment as heat. Photosynthetic autotrophs, as we will see in the next chapter, convert light energy into chemical energy—food for themselves and for the organisms that eat them. In this chapter we are concerned with how organisms process the chemical energy of food.

Cells of all living things can process this energy. Prokaryotes and eukaryotes share some energy-processing metabolic pathways. In this chapter we'll examine the pathways that operate to extract the energy from glucose by a series of oxidation–reduction reactions. The evolutionarily most ancient pathways probably operated without oxygen; the controlled chemical breakdown of glucose to pyruvate is part of this ancient process. However, as oxygen gas was produced by photosynthesis, chemical reactions using it proved to have a great advantage because they liberated more free energy, and thus aerobic metabolism evolved.

In this chapter, we'll emphasize energy metabolism that uses oxygen, starting with the chemical breakdown of glucose to produce pyruvate and the three sequential pathways by which pyruvate is completely oxidized, step-by-step, to water and carbon dioxide, and the released energy is captured as ATP. In the absence of oxygen, oxidation is less complete, and glucose molecules are converted to waste products, releasing only some of the potential energy. Toward the end of the chapter, we'll examine how the pathways of energy metabolism connect with other pathways for the catabolism and anabolism of other molecular raw materials of the cell. Then we'll see how these processes are regulated.

Obtaining Energy and Electrons from Glucose

What is the food that fuels all living cells? The most common food is the sugar glucose. Many other compounds serve as foods, but almost all of them yield their energy after being converted to glucose or to compounds intermediate in the metabolism of glucose. How do cells obtain energy from glucose?

Cells trap free energy while metabolizing glucose

Glucose burns readily, yielding carbon dioxide, water, and a lot of energy—but only if oxygen gas is present:

$$C_6H_{12}O_6 + 6\ O_2 \rightarrow 6\ CO_2 + 6\ H_2O + energy$$
$$(heat\ and\ light)$$

It wouldn't do cells a lot of good just to burn their glucose. They need to trap as much as possible of the chemical energy of glucose in usable form rather than as heat or light. Cells trap some of this energy in the energy storage compound ATP (adenosine triphosphate) (see Chapter 6).

Most kinds of organisms can metabolize their glucose completely in the processes of glycolysis and cellular respiration:

$$C_6H_{12}O_6 + 6\ O_2 \rightarrow 6\ CO_2 + 6\ H_2O + energy$$
$$(ATP\ and\ heat)$$

The change in free energy (ΔG) for the complete conversion of glucose and oxygen to carbon dioxide and water, whether by combustion or by complete metabolism, is –686 kcal/mol (–2,870 kJ/mol). Thus the overall reaction is highly exergonic and can drive the endergonic formation of a great deal of ATP. Some other kinds of organisms, unable to obtain or use oxygen gas, metabolize the glucose incompletely, thus obtaining less ATP per glucose molecule. Not all of the carbon atoms of glucose are converted to carbon dioxide in this incomplete metabolism, which is called *fermentation*.

Three metabolic processes play roles in the utilization of glucose for energy: glycolysis, cellular respiration, and fermentation (Figure 7.1). These processes consist of metabolic pathways made up of many distinct, but coupled, chemical reactions.

Glycolysis is a series of reactions that begins the metabolism of glucose in all cells and produces the product *pyruvate*. What happens to pyruvate depends on the type of organism that is extracting the energy and on whether the environment is **aerobic** (containing oxygen gas, O_2) or **anaerobic** (lacking oxygen gas). Glycolysis takes place under both conditions. However, fermentation occurs only under anaerobic conditions, and cellular respiration takes place only under aerobic conditions.

Less energy is captured as ATP during fermentation than during cellular respiration, and energy-rich carbon compounds such as lactic acid or ethanol (ethyl alcohol) are produced as waste products of fermentation. Cellular respiration releases much more energy from each glucose molecule than does fermentation. Importantly, glycolysis and cellular respiration operate together, but fermentation is an entirely separate process, operating only in the absence of O_2.

The combined operation of glycolysis and cellular respiration is the biological equivalent of burning glucose. When glucose is burned with a match, it releases energy rapidly as heat and light. If the burning is coupled to a mechanical device, some of the released energy can be used to do work. In cellular respiration, glucose is broken down to the same products (CO_2 and H_2O) as it is when it is simply burned, but much of the released energy is trapped in ATP. Both burning and cellular respiration are chemical reactions known as oxidation–reduction (redox) reactions.

Redox reactions transfer electrons and energy

Chemical reactions in which one substance transfers one or more electrons to another substance are called oxidation–reduction reactions, or **redox reactions**. The *gain* of one or more electrons by an atom, ion, or molecule is called **reduction**. The *loss* of one or more electrons is called **oxidation**. Although oxidation and reduction are always defined in terms of traffic in

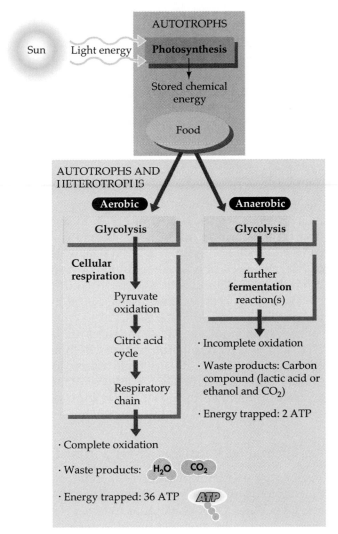

7.1 Energy for Life In photosynthesis, autotrophic organisms use light energy to synthesize food compounds. Heterotrophic and autotrophic organisms both process these food compounds by glycolysis, fermentation, and cellular respiration. Glycolysis precedes both fermentation and cellular respiration.

electrons, we must also think in these terms when hydrogen atoms (not hydrogen ions) are gained or lost, because transfers of hydrogen atoms involve transfers of electrons. Thus when a molecule loses hydrogen atoms, it becomes oxidized:

$$
\overset{\longleftarrow \text{oxidation} \longrightarrow}{AH_2 + B \rightarrow BH_2 + A}
$$
$$
\underset{\longleftarrow \text{reduction} \longrightarrow}{}
$$

Oxidation and reduction *always* occur together: As one material is oxidized, the electrons it loses are transferred to another material, reducing that material. In a redox reaction, we call the reactant that becomes re-

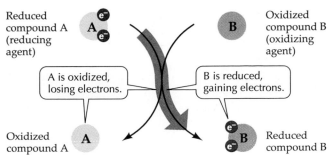

7.2 Oxidation and Reduction Are Coupled Compound A is oxidized and compound B reduced in a redox reaction. In the process, A loses electrons and B gains electrons.

duced an *oxidizing agent* and the one that becomes oxidized a *reducing agent* (Figure 7.2). An oxidizing agent accepts electrons; a reducing agent gives up electrons. In the process of oxidizing the reducing agent, the oxidizing agent itself becomes reduced. Conversely, the reducing agent becomes oxidized as it reduces the oxidizing agent. In the burning or metabolism of glucose, glucose is the reducing agent and oxygen gas the oxidizing agent.

Energy is transferred in a redox reaction: Some of the energy originally present in the reducing agent becomes associated with the reduced product. The overall ΔG is negative. As we will see, some of the key reactions of glycolysis and cellular respiration are highly exergonic redox reactions.

At some early stage in evolution, cells began to use reducing agents and oxidizing agents. Natural selection favored the use of certain of these agents (the ones whose redox reactions have suitable values of ΔG) as a system for the orderly exchange of electrons, analogous to the use of the ATP–ADP system for the orderly transfer of energy (see Figure 6.9). We have already encountered an example of such agents at work in cells: In Chapter 6 we saw that FAD accepts hydrogens during the respiratory conversion of succinate to fumarate (see Figure 6.16). In that reaction, FAD is an oxidizing agent and $FADH_2$ is a reducing agent. NAD is another important oxidizing agent that functions as a coenzyme in many reactions of glycolysis, cellular respiration, and fermentation.

NAD is a key electron carrier in redox reactions

The main pair of oxidizing and reducing agents in cells is based on the compound **NAD** (nicotinamide adenine dinucleotide). NAD exists in two chemically distinct forms, one oxidized (NAD^+) and the other reduced ($NADH + H^+$; Figure 7.3). NAD^+ and $NADH + H^+$ participate in biological redox reactions. The reduction

$$NAD^+ + 2\,H \rightarrow NADH + H^+$$

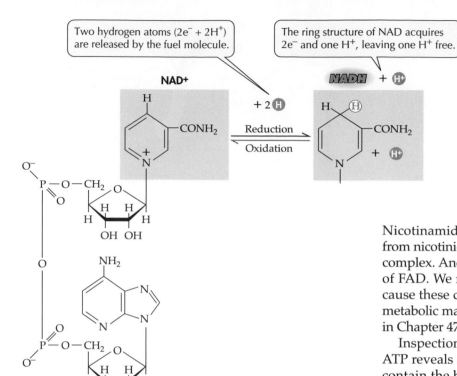

Two hydrogen atoms ($2e^- + 2H^+$) are released by the fuel molecule.

The ring structure of NAD acquires $2e^-$ and one H^+, leaving one H^+ free.

7.3 Oxidized and Reduced Forms of NAD NAD^+ is the oxidized form and NADH the reduced. As the shaded part of the NAD^+ molecule is reduced by acquiring two electrons (e^-) and a hydrogen ion (H^+), to yield a neutral NADH molecule, the second H^+ is released to the surroundings. The unshaded portion of the molecule remains unchanged by the reaction.

Nicotinamide, which is part of NAD, forms directly from nicotinic acid, or niacin, a member of the vitamin B complex. Another B vitamin is riboflavin, which is part of FAD. We need only small amounts of vitamins because these carrier molecules are recycled through the metabolic machinery. Vitamins are discussed more fully in Chapter 47.

Inspection of the chemical structures of NAD^+ and ATP reveals several common features. Both molecules contain the base adenine, the sugar ribose, and phosphate groups (see Figure 6.7). These components suggest a common evolutionary origin for both NAD^+ and ATP in an ancient and less efficient molecule.

is formally equivalent to the transfer of two hydrogen atoms ($2 H^+ + 2 e^-$). However, what is actually transferred is a hydride ion (H^-, a proton and two electrons), leaving a free proton (H^+). The oxidation of NADH + H^+ by oxygen gas is highly exergonic:

$$NADH + H^+ + \tfrac{1}{2} O_2 \rightarrow NAD^+ + H_2O$$

$$\Delta G = -52.4 \text{ kcal/mol} \ (-219 \text{ kJ/mol})$$

(Note that the oxidizing agent appears here as "$^{1}/_{2} O_2$" instead of "O." This notation emphasizes that it is oxygen gas, O_2, that acts as the oxidizing agent.) In the same way that ATP can be thought of as a means of packaging free energy in bundles of about 12 kcal/mol (50 kJ/mol), NAD can be thought of as a means of packaging free energy in bundles of approximately 50 kcal/mol (200 kJ/mol) (Figure 7.4).

Besides NAD and FAD, there are several other biologically important electron carriers. The molecular structures of NAD and some of the others include components that we humans need but cannot synthesize for ourselves; these are classified as vitamins.

An Overview: Releasing Energy from Glucose

The energy-extracting processes of cells may be divided into distinct pathways that we can consider one at a time. When O_2 is available as the final electron acceptor, four pathways operate: glycolysis, pyruvate oxidation, the citric acid cycle, and the respiratory chain. When O_2 is unavailable, pyruvate oxidation, the citric acid cycle, and the respiratory chain do not function, and additional *fermentation reactions* are added to the glycolytic pathway. Figure 7.5 summarizes the

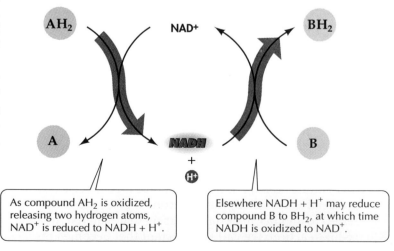

As compound AH_2 is oxidized, releasing two hydrogen atoms, NAD^+ is reduced to NADH + H^+.

Elsewhere NADH + H^+ may reduce compound B to BH_2, at which time NADH is oxidized to NAD^+.

7.4 NAD Is an Energy Carrier Thanks to its ability to carry free energy and electrons, NAD is a major and universal energy intermediary in cells.

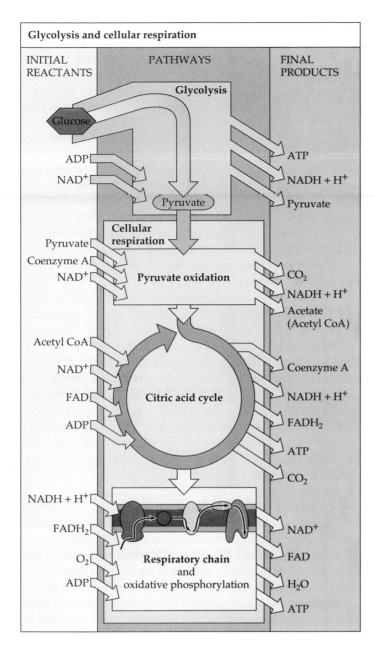

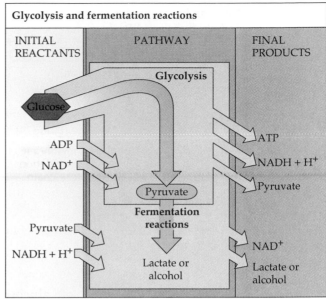

7.5 An Overview of the Cellular Energy Pathways The energy-producing reactions can be grouped into five pathways: glycolysis, pyruvate oxidation, the citric acid cycle, the respiratory chain, and fermentation. The three middle pathways occur only in the presence of oxygen and are collectively referred to as cellular respiration. Each pathway will be depicted in more detailed figures later in this chapter.

In prokaryotes, the enzymes used in glycolysis, fermentation, and the citric acid cycle are soluble in the cytosol. Enzymes involved in pyruvate oxidation and the respiratory chain are associated with the inner surface of the plasma membrane or inward elaborations of that membrane (see Chapter 2).

In the presence of O_2, glycolysis begins the breakdown of glucose

Glycolysis (the glycolytic pathway) is a sequence of 10 separate chemical reactions. This near-universal process was probably the first energy-releasing pathway to evolve; if any earlier pathway existed, it has disappeared from Earth. Today virtually all living cells, even the most evolutionarily ancient, use glycolysis.

Glycolysis is the pathway in which glucose is incompletely oxidized, to **pyruvate** (pyruvic acid).* Glycolysis may operate in the presence or the absence of O_2. In either case, it contains an oxidative step in which the electron carrier NAD^+ becomes reduced,

starting reactants and products of these five pathways.

These distinct chemical pathways are separated in the cell (Table 7.1). In eukaryotes, glycolysis and fermentation take place in the cytoplasm outside of mitochondria. The enzymes for these pathways were once believed to be soluble in the cytosol, but more recent discoveries suggest that at least some of them may be bound to components of the cytoskeleton. The other reactions are associated with the mitochondria. Pyruvate oxidation and the respiratory chain are both associated with the inner membrane of mitochondria, where their enzymes are bound. The enzymes and reactions of the citric acid cycle are found in the matrix of mitochondria. These different locations reflect separate evolutionary origins of these processes.

---

*We tend to use words like "pyruvate" and "pyruvic acid" interchangeably. However, at pH values commonly found in cells, the ionized form—pyruvate—is present rather than the acid—pyruvic acid. Similarly, all carboxylic acids are present as ions (the "-ate" forms) at these pH values.

TABLE 7.1 Cellular Locations for Energy Pathways in Eukaryotes and Prokaryotes

EUKARYOTES	PROKARYOTES
External to mitochondrion	**In cytoplasm**
Glycolysis	Glycolysis
Fermentation	Fermentation
	Citric acid cycle
	On inner face
Inside mitochondrion	**of plasma membrane**
Inner membrane	Pyruvate oxidation
Pyruvate oxidation	Respiratory chain
Respiratory chain	
Matrix	
Citric acid cycle	

acquiring electrons. Each molecule of glucose processed through glycolysis produces a net yield of two molecules of ATP. The major products of glycolysis are ATP (which the cell will use to drive endergonic reactions), pyruvate, and the electrons acquired by NAD. Both the pyruvate and the electrons must be processed further.

Cellular respiration operates when O_2 is available, yielding CO_2 and H_2O as products. It is made up of three pathways: pyruvate oxidation, the citric acid cycle, and the respiratory chain. In **pyruvate oxidation**, the end product of glycolysis (pyruvate) loses two hydrogen atoms and a carboxyl group ($—COO^-$) as CO_2, forming the two-carbon molecule acetate (acetic acid), which is activated by the addition of a coenzyme (coenzyme A).

The **citric acid cycle** (also called the Krebs cycle or the tricarboxylic acid cycle) is a cyclical series of reactions in which the product of pyruvate oxidation, acetate, becomes *completely* oxidized, forming CO_2 and transferring electrons (along with their hydrogen nuclei) to carrier molecules (FAD and NAD^+). The citric acid cycle produces many more electrons than are produced in glycolysis. And as we are about to see, harvesting more electrons means a greater ultimate harvest of ATP.

In glycolysis, pyruvate oxidation, and the citric acid cycle, the molecules that become reduced acquire hydrogen atoms. Through these pathways, energy originally present in the covalent bonds of glucose becomes associated with reduced forms of NAD and FAD ($NADH + H^+$ and $FADH_2$). Hydrogen is an outstanding fuel. When it reacts with O_2, a great deal of free energy is released; better still, the "waste" product of this reaction—water—is not toxic to the environment or to any organism that produces it.

The fourth energy-extracting pathway for aerobic cells is the **respiratory chain**, whose principal role is to release energy from reduced NAD in such a way that it may be used to form ATP. This pathway is a series of redox reactions in which electrons derived from hydrogen atoms are passed from one type of membrane carrier to another and finally are allowed to react with O_2 to produce water. In eukaryotes, the carriers (and associated enzymes) are bound to the folds of the inner mitochondrial membranes, the **cristae** (see Figure 4.13).

The transfer of electrons along the respiratory chain drives the active transport of hydrogen ions (protons) from the mitochondrial matrix into the space between the two mitochondrial membranes. This is active transport because energy is used to transport the protons *against* a concentration gradient—from a region of lower concentration to a region of higher concentration (see Chapter 5).

The subsequent diffusion of protons back into the matrix is coupled to the synthesis of ATP from ADP and P_i, as we will see later in this chapter. This is the way in which the vast majority of the ATP in our bodies is formed. The formation of ATP during operation of the respiratory chain is called **oxidative phosphorylation** because NADH is oxidized and ADP is phosphorylated.

Overall, the inputs to the respiratory chain are hydrogen atoms and O_2, and the outputs are water and energy captured as ATP.

In the absence of O_2, some cells carry on fermentation

If we are deprived of O_2 for too long, we die because the respiratory chain cannot function. Without oxygen molecules as receptors, the carriers in our mitochondrial cristae are unable to unload the electrons bound to them. Soon, no oxidized carriers are available to accept additional hydrogens. When that happens, glycolysis, pyruvate oxidation, and the citric acid cycle stop. Without these processes, and with no respiratory chain activity, we have insufficient ATP for our cells to maintain their structure and metabolism, and we die.

Not all cells require O_2. **Fermentation** utilizes glycolysis and an additional reaction. It produces only a fraction of the energy produced when cells utilize O_2. Our muscle cells have such an alternative way to rid themselves of the hydrogen atoms produced during glycolysis: The hydrogens produced by glycolysis are passed to the end product of glycolysis (pyruvate), and lactic acid (lactate) is formed.

$$\text{pyruvate} + \text{NADH} + \text{H}^+ \rightarrow \text{lactate} + \text{NAD}^+$$

This reaction recycles the NAD needed in glycolysis. Thus even in the absence of oxygen, glycolysis continues (often at an increased rate), without the activity of the citric acid cycle or the respiratory chain, and ATP continues to be produced. The cells that have the enzyme necessary for this reaction can function for a time in the absence of oxygen. However, eventually the concentration of lactic acid in muscles reaches a toxic level. Such anaerobic production of ATP is called fermentation. For some organisms that live entirely without oxygen, fermentation is the sole pathway to trap energy in ATP.

In the discussion that follows, we will examine in more detail the four pathways of aerobic energy metabolism (glycolysis, pyruvate oxidation, the citric acid cycle, and the respiratory chain) and fermentation.

Glycolysis: From Glucose to Pyruvate

In glycolysis, glucose is only partly oxidized. A molecule of glucose taken in by a cell enters the glycolytic pathway, which consists of 10 reactions that convert the six-carbon glucose molecule, step-by-step, into two molecules of the three-carbon compound pyruvic acid (Figure 7.6). These reactions are accompanied by the *net* formation of two molecules of ATP and by the reduction of two molecules of NAD^+ to two molecules of $NADH + H^+$. At the end of the pathway, then, ready energy is located in ATP, and four hydrogen atoms are located in $NADH + H^+$.

The fate of the pyruvic acid depends on the type of cell carrying out glycolysis and on whether the environment is aerobic or anaerobic. The fate of the NADH + H$^+$, too, varies. In most cases, $NADH + H^+$ is oxidized through the respiratory chain to yield H_2O and NAD^+—a chain of reactions that results in the formation of much more ATP (three molecules of ATP per molecule of $NADH + H^+$). In fermentation, however, $NADH + H^+$ is reoxidized to NAD^+ either by pyruvic acid itself or by one of its metabolites, with no further ATP production. In either case, glycolysis may be regarded as a series of *preparatory reactions*, to be followed either by the fermentation reactions or by cellular respiration (pyruvate oxidation, citric acid cycle, and respiratory chain).

Glycolysis can be divided into two groups of reactions: reactions that invest energy from ATP hydrolysis and energy-harvesting reactions that produce ATP.

The energy-investing reactions of glycolysis require ATP

Using Figure 7.6, we can trace our way through the glycolytic pathway. The first five reactions are endergonic, taking up free energy; that is, the cell is *investing* free energy rather than gaining it during the early reactions of glycolysis. In separate reactions, two molecules of ATP are invested in attaching two phosphate groups to the sugar (reactions 1 and 3), thereby raising its free energy by about 15 kcal/mol (62.7 kJ/mol)

(Figure 7.7). Later, these phosphate groups will be transferred to ADP to make new molecules of ATP.

Although both of these first steps of glycolysis use ATP as one of the substrates, each is catalyzed by a different, specific enzyme. The enzyme hexokinase catalyzes reaction 1, in which glucose receives a phosphate group from ATP. (A *kinase* is any enzyme that catalyzes the transfer of a phosphate group from ATP to another substrate.) In reaction 2, the six-membered glucose ring is rearranged to a five-membered fructose ring. Then, in reaction 3, the enzyme phosphofructokinase adds a second phosphate (taken from another ATP) to the sugar ring.

The fourth reaction opens the sugar ring with its two phosphates, and the six-carbon sugar bisphosphate* is cleaved to give two different three-carbon sugar phosphates. In reaction 5, one of these sugar phosphates (dihydroxyacetone phosphate) is converted into a second molecule of the other (glyceraldehyde 3-phosphate).

By this time, the halfway point in glycolysis, the following things have happened: Two molecules of ATP have been invested, and the six-carbon glucose molecule has been converted into two molecules of a three-carbon sugar phosphate. No ATP has been gained, and nothing has been oxidized.

The energy-harvesting reactions of glycolysis yield ATP and NADH + H$^+$

Now things begin to happen, including a key redox reaction. In what follows, remember that each step occurs twice for each glucose molecule going through glycolysis, because in the first five reactions of glycolysis each glucose molecule has been split into two molecules of three-carbon sugar phosphate, both of which go through the remaining steps of glycolysis.

Reaction 6 is a two-step reaction catalyzed by the enzyme triose phosphate dehydrogenase. The end product of reaction 6 is 1,3-bisphosphoglycerate (or 1,3-bisphosphoglyceric acid). A phosphate ion has been snatched from the surroundings (but not, this time, from ATP) and tacked onto the three-carbon compound. Figure 7.7 shows that this reaction is accompanied by an enormous drop in free energy—more than 100 kcal of energy per mole of glucose is released in this extremely exergonic reaction. What has happened here? Why the big energy change?

Reaction 6 is the conversion of a sugar to an acid:

$$R-\overset{\displaystyle O}{\overset{\displaystyle \|}{C}}-H + (O) \rightarrow R-\overset{\displaystyle O}{\overset{\displaystyle \|}{C}}-OH$$

*The root *bis-* means "two." A sugar bisphosphate has two phosphate groups.

The pathways of glycolysis and cellular respiration (pyruvate oxidation, citric acid cycle and respiratory chain) are represented by a "road map" of symbols that guide you to better understand the relationship of these pathways in the illustrations that follow.

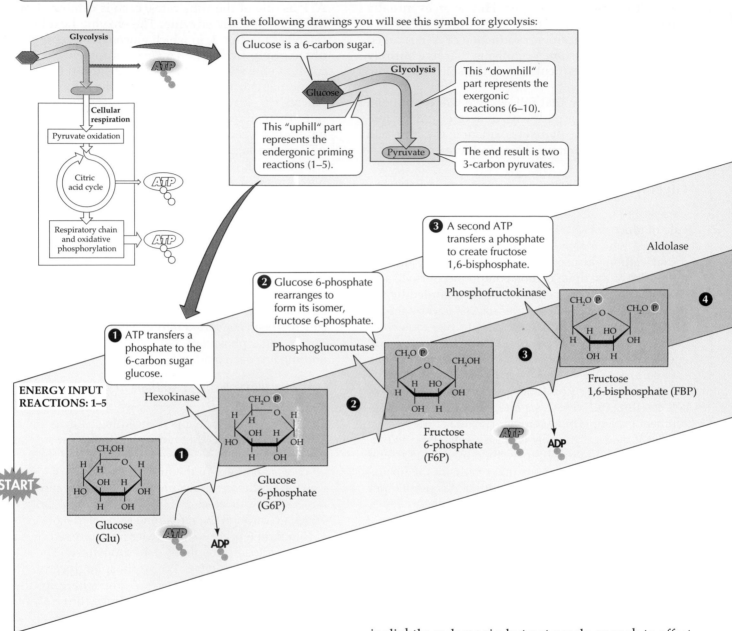

In the following drawings you will see this symbol for glycolysis:

Glucose is a 6-carbon sugar.

Glycolysis

This "downhill" part represents the exergonic reactions (6–10).

This "uphill" part represents the endergonic priming reactions (1–5).

The end result is two 3-carbon pyruvates.

Glycolysis

Cellular respiration

Pyruvate oxidation

Citric acid cycle

Respiratory chain and oxidative phosphorylation

3 A second ATP transfers a phosphate to create fructose 1,6-bisphosphate.

Aldolase

Phosphofructokinase

2 Glucose 6-phosphate rearranges to form its isomer, fructose 6-phosphate.

Phosphoglucomutase

1 ATP transfers a phosphate to the 6-carbon sugar glucose.

ENERGY INPUT REACTIONS: 1–5

Hexokinase

Fructose 1,6-bisphosphate (FBP)

Fructose 6-phosphate (F6P)

Glucose 6-phosphate (G6P)

Glucose (Glu)

START

Since this is an oxidation reaction, it is very exergonic. (Note that here we do *not* write $^{1}/_{2}$ O_2, because in this case oxygen gas does not participate in the reaction.) The formation of the phosphate ester from the acid:

$$R — \overset{\overset{O}{\|}}{C} — OH + HPO_4^{2-} \rightarrow R — \overset{\overset{O}{\|}}{C} — O — \overset{\overset{O}{\|}}{\underset{\underset{O^-}{|}}{P}} — O^- + H_2O$$

is slightly endergonic, but not nearly enough to offset the drop in free energy from the oxidation.

If this big energy drop were simply the loss of heat, glycolysis would not provide useful energy to the cell. However, rather than being lost, this energy is used to make two molecules of NADH + H$^+$ from two molecules of NAD$^+$. This stored energy is regained later—either in the respiratory chain, by the formation of ATP, or in the last step of fermentation, when pyruvate or its product is reduced and the two molecules of NADH + H$^+$ are restored once again to NAD$^+$. This cycling of NAD is necessary to keep glycolysis going; if all the NAD$^+$ is converted to NADH + H$^+$, glycolysis comes to a halt.

7.6 Glycolysis Converts Glucose to Pyruvate Starting with hexokinase, ten enzymes catalyze ten reactions in turn. Along the way, ATP is produced (reactions 7 and 10), and two NAD$^+$ are reduced to two NADH + 2 H$^+$ (reaction 6).

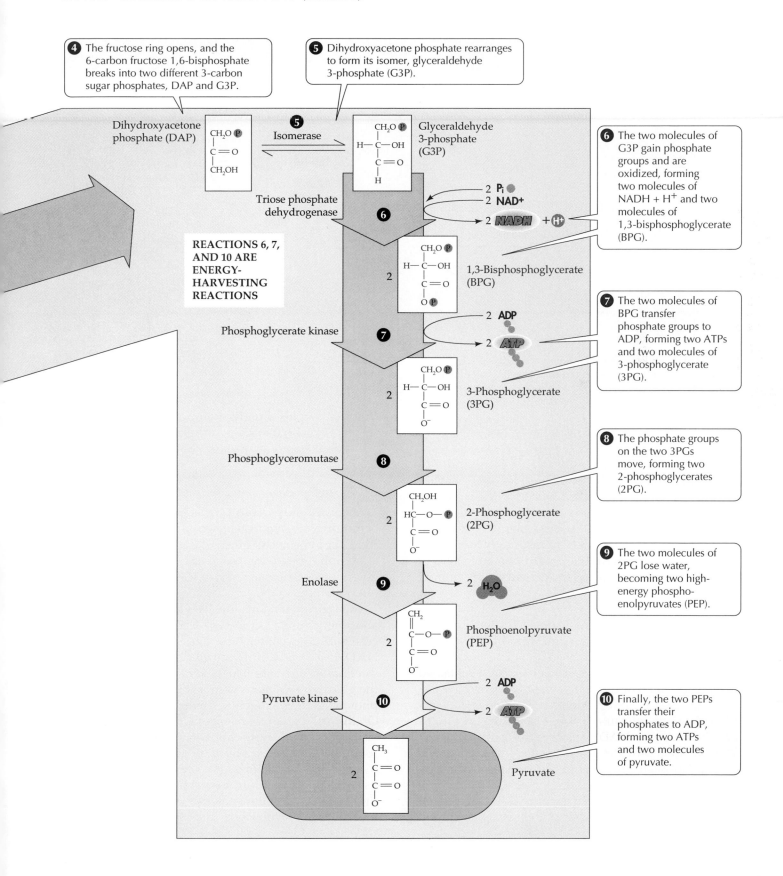

4 The fructose ring opens, and the 6-carbon fructose 1,6-bisphosphate breaks into two different 3-carbon sugar phosphates, DAP and G3P.

5 Dihydroxyacetone phosphate rearranges to form its isomer, glyceraldehyde 3-phosphate (G3P).

Dihydroxyacetone phosphate (DAP)

5 Isomerase

Glyceraldehyde 3-phosphate (G3P)

Triose phosphate dehydrogenase

6

2 P$_i$
2 NAD$^+$
2 NADH + H$^+$

6 The two molecules of G3P gain phosphate groups and are oxidized, forming two molecules of NADH + H$^+$ and two molecules of 1,3-bisphosphoglycerate (BPG).

REACTIONS 6, 7, AND 10 ARE ENERGY-HARVESTING REACTIONS

2 1,3-Bisphosphoglycerate (BPG)

Phosphoglycerate kinase

7

2 ADP
2 ATP

7 The two molecules of BPG transfer phosphate groups to ADP, forming two ATPs and two molecules of 3-phosphoglycerate (3PG).

2 3-Phosphoglycerate (3PG)

Phosphoglyceromutase

8

8 The phosphate groups on the two 3PGs move, forming two 2-phosphoglycerates (2PG).

2 2-Phosphoglycerate (2PG)

Enolase

9

2 H$_2$O

9 The two molecules of 2PG lose water, becoming two high-energy phospho-enolpyruvates (PEP).

2 Phosphoenolpyruvate (PEP)

Pyruvate kinase

10

2 ADP
2 ATP

10 Finally, the two PEPs transfer their phosphates to ADP, forming two ATPs and two molecules of pyruvate.

2 Pyruvate

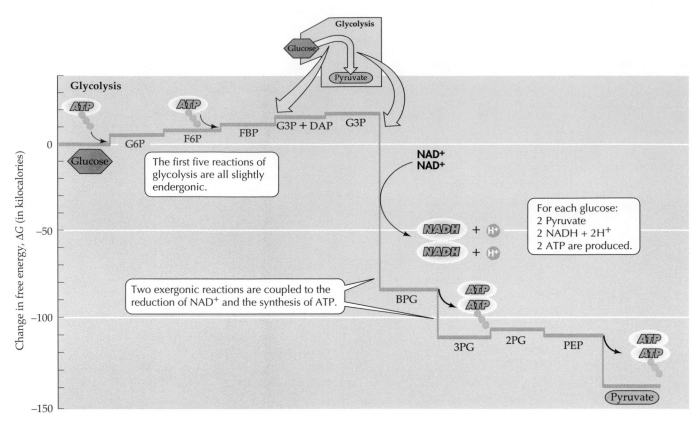

7.7 Changes in Free Energy During Glycolysis Each reaction of glycolysis changes the free energy available, as shown by the different energy levels of the series of reactants and products from glucose to pyruvate.

The remaining steps of glycolysis (see Figure 7.6) are simpler. The two phosphate groups of 1,3-bisphosphoglycerate are transferred, one at a time, to molecules of ADP, with a rearrangement in between. More than 20 kcal (83.6 kJ/mol) of free energy is stored in ATP for every mole of 1,3-bisphosphoglycerate broken down. Finally, we are left with two molecules of pyruvate for every molecule of glucose that entered glycolysis.

SUBSTRATE-LEVEL PHOSPHORYLATION. The enzyme-catalyzed transfer of phosphate groups from donor molecules to ADP molecules is called **substrate-level phosphorylation**. This process is driven by energy obtained from oxidation. For example, when glyceraldehyde 3-phosphate reacts with P_i and NAD^+, becoming 1,3-bisphosphoglycerate, an aldehyde is oxidized to a carboxylic acid, with NAD^+ acting as the oxidizing agent. The oxidation provides so much energy that the newly added phosphate group is linked to the rest of the molecule by a bond that has even higher energy than the high-energy bond of ATP (Figure 7.8*a*). A second enzyme catalyzes the transfer of this phosphate group from 1,3-bisphosphoglycerate to ADP, forming ATP (Figure 7.8*b*). Both reactions are exergonic, even though a substantial amount of energy is consumed in the formation of ATP.

Reviewing glycolysis and fermentation

A review of the glycolytic reactions shows that at the beginning of glycolysis, two molecules of ATP are used per molecule of glucose, but that ultimately four are produced (two for each of the two 1,3-bisphosphoglycerates)—a net gain of two ATP molecules and two $NADH + H^+$.

In fermentation (anaerobic conditions), the total usable energy yield is just two ATP molecules per glucose molecule. Under these anaerobic conditions, the $NADH + H^+$ is rapidly recycled to NAD^+ by the reduction of pyruvate. The NAD^+ is then available for the glycolytic reaction catalyzed by the enzyme triose phosphate dehydrogenase (reaction 6 in Figure 7.6). On the other hand, in the presence of oxygen (aerobic conditions), eukaryotes and some bacteria are able to reap far more energy by the complete oxidation of pyruvate and by oxidizing the $NADH + H^+$ of glycolysis through the respiratory chain, as we will see in the sections that follow. In eukaryotes, these reactions take place in the mitochondria.

(a) **Oxidation of substrate**

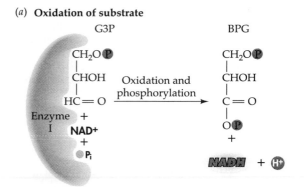

G3P   BPG

CH_2O Ⓟ

CHOH   Oxidation and   CHOH
          phosphorylation
HC$=$O   →   C$=$O

Enzyme   +   O Ⓟ
 I   **NAD⁺**
     +   +
     ● **P$_i$**   **NADH** + Ⓗ⁺

> Glyceraldehyde 3-phosphate is both oxidized and phosphorylated to become 1,3-bisphosphoglycerate.

(b) **Transfer of phosphate to ADP**

BPG   3PG

CH_2O Ⓟ CH_2O Ⓟ

CHOH   Hydrolysis and   CHOH
          phosphorylation
C$=$O   →   C$=$O

   O Ⓟ   O⁻

Enzyme   +
 II   **ADP**   +

                      ATP

> Hydrolysis of 1,3-bisphosphoglycerate releases enough energy to transfer a phosphate group to ADP, forming ATP.

7.8 Substrate-Level Phosphorylation   Two enzymes collaborate to catalyze substrate-level phosphorylation.

Pyruvate Oxidation

The oxidation of pyruvate to acetate is a complex multistep reaction catalyzed by an enormous enzyme complex that is attached to the mitochondrial inner membrane. The three-carbon compound, pyruvate, is oxidized to the two-carbon compound, acetate (CH_3COO^-), yielding free energy and CO_2. In this process, the acetate is linked to a coenzyme, called **coenzyme A (CoA)**, producing the energy-rich compound **acetyl coenzyme A**, or acetyl CoA (Figure 7.9). Acetyl CoA has 7.5 kcal/mol (31.4 kJ/mol) more energy than simple acetate. (Acetyl CoA can donate acetate to acceptors such as oxaloacetate, much as ATP can donate phosphate to various acceptors.)

There are three steps in this short pathway: (1) Pyruvate is oxidized to acetate, and CO_2 is released.

(2) Part of the energy from the oxidation in step 1 is saved by the reduction of NAD^+ to NADH + H^+. (3) Some of the remaining energy is stored temporarily by the combining of the acetate with CoA. An analogous three-step reaction occurs in glycolysis when glyceraldehyde 3-phosphate is converted to 1,3-bisphosphoglycerate (reaction 6 in Figure 7.6). In that reaction, an aldehyde group is oxidized to an acid, some of the energy released by oxidation is stored in NADH + H^+, and some of the remaining energy is preserved in a second phosphate bond in the molecule. As the similarity between these two three-step reactions shows, a good metabolic idea is likely to appear more than once; we will see this one yet again, in the citric acid cycle.

As you might suspect, a complex set of steps such as those in the reaction from pyruvate to acetyl CoA requires more than one type of catalytic protein. This reaction is catalyzed by the *pyruvate dehydrogenase com-*

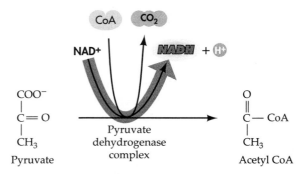

CoA CO_2

NAD⁺   **NADH** + Ⓗ⁺

COO⁻   O
 | ‖
C$=$O   C$-$CoA
 | Pyruvate |
CH_3 dehydrogenase CH_3
      complex
Pyruvate   Acetyl CoA

7.9 The Pyruvate Dehydrogenase Complex Catalyzes Pyruvate Oxidation   The massive complex transfers electrons, removes a carboxyl group, and adds a coenzyme.

plex, which consists of 72 subunits—24 each of three different types of protein, for a total molecular weight of 4.6 million (Figure 7.9*b*). The three component enzymes use a total of five different coenzymes—four of which are vitamins or contain vitamins (thiamin, riboflavin, pantothenic acid, and niacin). This enzyme complex is an impressive example of biological organization.

The Citric Acid Cycle

Acetyl CoA is the starting point for the citric acid cycle (Figure 7.10). In this section we examine the citric acid cycle, in which the two-carbon molecule of acetate is oxidized to two molecules of carbon dioxide.

Figure 7.7 shows that the metabolism of glucose to pyruvate is accompanied by a drop in free energy of about 140 kcal/mol (585 kJ/mol). About a third of this energy is captured in the formation of ATP and reduced NAD (NADH + H⁺). Oxidizing the pyruvate to acetate yields additional free energy for biological work. The citric acid cycle takes acetate and breaks it down to CO_2, using the hydrogen atoms to reduce carrier molecules and to pass chemical free energy to those carriers. The reduced carriers are later oxidized in the respiratory chain, which transfers an enormous amount of free energy to ATP.

The principal inputs to the citric acid cycle are acetate in the form of acetyl CoA, water, and oxidized electron carriers. The principal outputs are carbon dioxide and reduced electron carriers. Overall, for each molecule of acetate, during the citric acid cycle two carbons are removed as CO_2 and four pairs of hydrogen atoms are used to reduce carrier molecules that trap energy for use later in the synthesis of ATP. The energy-trapping reactions of the cycle are a major reason for its existence.

The citric acid cycle produces two CO₂ molecules and reduced carriers

At the beginning of the citric acid cycle, acetyl CoA, which has two carbon atoms in its acetate group, reacts with a four-carbon acid (oxaloacetate) to form the six-carbon compound citric acid (citrate). The remainder of the cycle consists of a series of enzyme-catalyzed reactions in which citric acid is degraded to a new four-carbon molecule of oxaloacetate. This new oxaloacetate can react with a second acetyl CoA, producing a second molecule of citrate and thus enabling the cycle to continue. Acetyl CoA enters the cycle from pyruvate, and CO_2 exits.

As we describe the citric acid cycle in detail, concentrate on how it is maintained in a steady state—that is, with material entering and leaving and with intermediate compounds like succinate and malate turning over constantly, but without changing concentration. Pay close attention to the numbered reactions

7.10 Pyruvate Oxidation and the Citric Acid Cycle ▶
Pyruvate diffuses into the mitochondrion and is oxidized to acetyl CoA, which enters the citric acid cycle. Notice that the two carbons from acetyl CoA are traced with color through reaction 4, after which they may be at either end of the molecule (note the symmetry of succinate and fumarate). For each glucose molecule, the cycle operates twice, producing 4 CO_2, 6 NADH + 6 H⁺, 2 FADH₂, and 2 ATP in all.

in Figure 7.10 as you read the next several paragraphs.

The energy temporarily stored in acetyl CoA helps drive the formation of *citrate* from *oxaloacetate* (reaction 1). During this reaction, the coenzyme A molecule falls away, to be recycled. In reaction 2, citrate is rearranged to *isocitrate*. In reaction 3, a CO_2 molecule and two hydrogen atoms are removed in the conversion of isocitrate to *α-ketoglutarate*. As Figure 7.11 indicates, this reaction produces a large drop in free energy. The released energy is stored in NADH + H⁺ and can be recovered later in the respiratory chain, when the NADH + H⁺ is reoxidized.

Like the oxidation of pyruvate to acetyl CoA, reaction 4 of the citric acid cycle is complex. The five-carbon α-ketoglutarate molecule is oxidized to the four-carbon molecule *succinate*, CO_2 is given off, some of the oxidation energy is stored in NADH + H⁺, and some of the energy is preserved temporarily by combining succinate with CoA to form *succinyl CoA*. In reaction 5, the energy in succinyl CoA is harvested to make GTP (guanosine triphosphate) from GDP and P_i, which is another example of substrate-level phosphorylation. Then GTP is used to make ATP from ADP.

Free energy is released in reaction 6, when two hydrogens are transferred to an enzyme that contains FAD. After a molecular rearrangement (reaction 7), one more NAD⁺ reduction occurs, producing oxaloacetate from *malate* (reaction 8). The oxaloacetate produced by all these reactions is ready to combine with another acetate from acetyl CoA and go around the cycle again. Bear in mind that the citric acid cycle operates twice for each glucose molecule that enters glycolysis.

Although most of the enzymes of the citric acid cycle are dissolved in the mitochondrial matrix, there are two exceptions: succinate dehydrogenase, which catalyzes reaction 6, and the enormous complex that catalyzes reaction 4. These enzymes are integral membrane proteins of the inner mitochondrial membrane.

The Respiratory Chain: Electrons, Proton Pumping, and ATP

Without the oxidizing agents NAD⁺ and FAD, the oxidative steps of glycolysis, pyruvate oxidation, and the citric acid cycle could not occur. Once reduced forms of these carriers have been produced, they must have

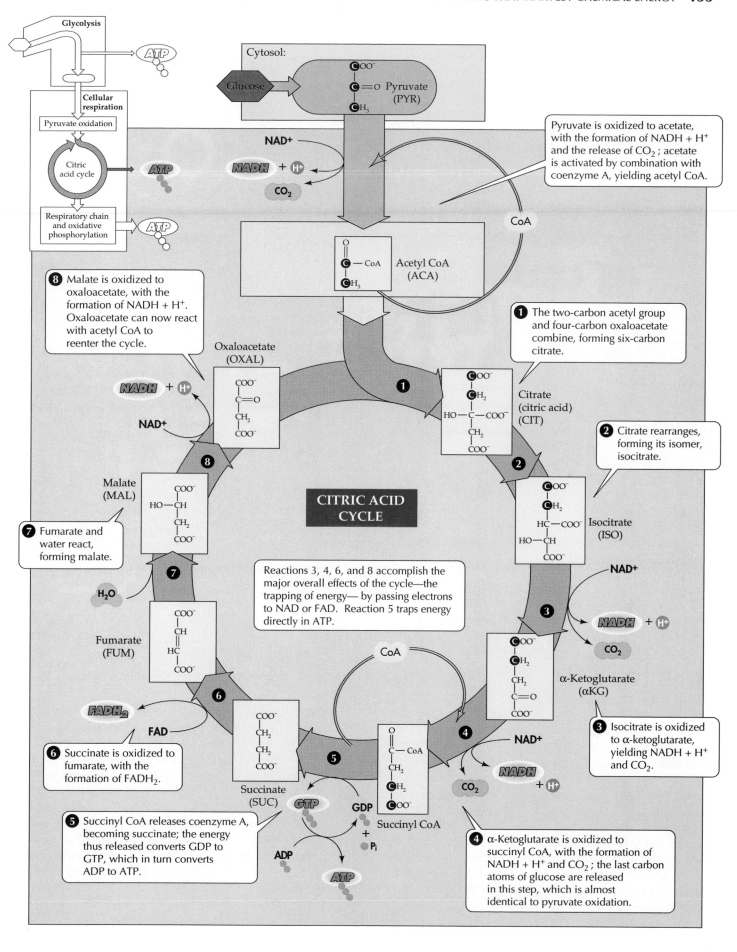

CITRIC ACID CYCLE

Glycolysis

Cellular respiration

Pyruvate oxidation

Citric acid cycle

Respiratory chain and oxidative phosphorylation

Cytosol:

Glucose → Pyruvate (PYR)

Pyruvate is oxidized to acetate, with the formation of NADH + H$^+$ and the release of CO_2; acetate is activated by combination with coenzyme A, yielding acetyl CoA.

Acetyl CoA (ACA)

❶ The two-carbon acetyl group and four-carbon oxaloacetate combine, forming six-carbon citrate.

❷ Citrate rearranges, forming its isomer, isocitrate.

Citrate (citric acid) (CIT)

Isocitrate (ISO)

❽ Malate is oxidized to oxaloacetate, with the formation of NADH + H$^+$. Oxaloacetate can now react with acetyl CoA to reenter the cycle.

Oxaloacetate (OXAL)

Malate (MAL)

❼ Fumarate and water react, forming malate.

Fumarate (FUM)

Reactions 3, 4, 6, and 8 accomplish the major overall effects of the cycle—the trapping of energy— by passing electrons to NAD or FAD. Reaction 5 traps energy directly in ATP.

α-Ketoglutarate (αKG)

❸ Isocitrate is oxidized to α-ketoglutarate, yielding NADH + H$^+$ and CO_2.

❻ Succinate is oxidized to fumarate, with the formation of FADH$_2$.

Succinate (SUC)

Succinyl CoA

❹ α-Ketoglutarate is oxidized to succinyl CoA, with the formation of NADH + H$^+$ and CO_2; the last carbon atoms of glucose are released in this step, which is almost identical to pyruvate oxidation.

❺ Succinyl CoA releases coenzyme A, becoming succinate; the energy thus released converts GDP to GTP, which in turn converts ADP to ATP.

7.11 The Citric Acid Cycle Releases Much More Free Energy Than Glycolysis Does Electron carriers (NAD in glycolysis; NAD and FAD in the citric acid cycle) are reduced and ATP is generated in reactions coupled to other reactions producing major drops in free energy.

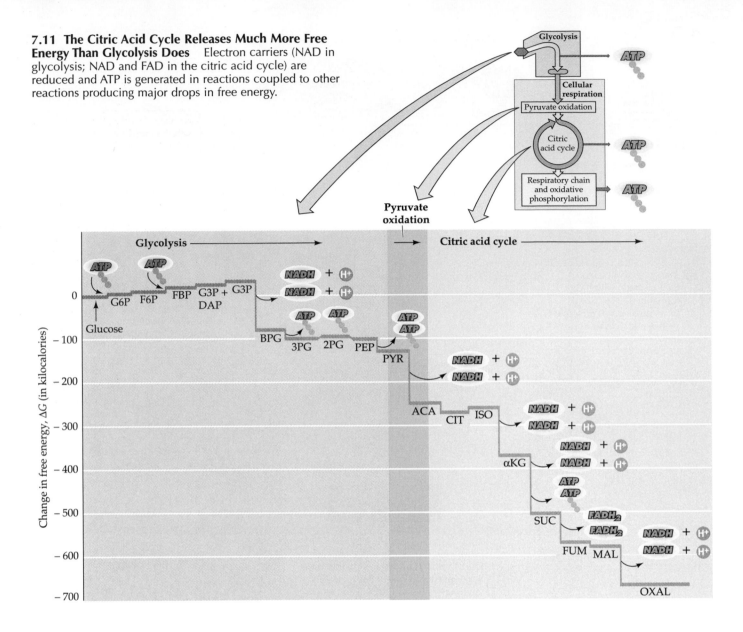

some place to donate their hydrogens ($H^+ + e^-$). The fate of these protons and electrons is the story of the remainder of cellular respiration.

The story has three parts: First, the electrons pass through a series of electron carriers called the respiratory chain. Second, the flow of electrons along the chain causes the active transport of protons across the inner mitochondrial membrane. Third, the protons diffuse back into the mitochondrial matrix, and this flow is coupled to the synthesis of ATP. The overall process of ATP synthesis resulting from electron transport through the respiratory chain is oxidative phosphorylation. The mechanism for oxidative phosphorylation involves proton transfer back and forth across the membrane and is called *chemiosmosis.*

The respiratory chain transports electrons and releases energy

The respiratory chain contains (1) three large protein complexes that span the inner mitochondrial membrane, (2) a small protein (cytochrome *c*), and (3) a nonprotein component called ubiquinone (abbreviated Q) (Figure 7.12). NADH + H^+ passes hydrogens to Q by way of **NADH-Q reductase**, a complex of 26 polypeptide subunits, with a total molecular weight of 850,000. **Cytochrome reductase**, with 10 subunits and a molecular weight of 280,000, lies between Q and cytochrome *c*. **Cytochrome oxidase**, with 8 subunits and a molecular weight of 160,000, lies between cytochrome *c* and oxygen. Different subunits within each of these protein complexes bear different electron carriers, so electrons are transported *within* each complex, as well as from complex to complex.

Cytochrome *c* is a peripheral membrane protein that lies in the space between the inner and outer mitochondrial membranes, loosely attached to the inner membrane. Ubiquinone is a small, nonpolar molecule that moves within the hydrophobic interior of the phospholipid bilayer of the inner membrane (see Figure 7.12).

Why should the respiratory chain have so many links? Why, for example, don't cells just use the following single step?

$$NADH + H^+ + \frac{1}{2} O_2 \rightarrow NAD^+ + H_2O$$

Wouldn't this accomplish the same thing, and more efficiently? To begin with, no enzyme will catalyze the direct oxidation of $NADH + H^+$ by oxygen. More fundamentally, this would be an untamable reaction. It would be terrifically exergonic—rather like setting off a stick of dynamite in the cell. There is no biochemical way to harvest that burst of energy efficiently and put it to physiological use (that is, no metabolic reaction is so endergonic as to consume a significant fraction of that energy in a single step).

To control the release of energy during oxidation of glucose in a cell, evolution has produced the lengthy respiratory chain we observe today: a *series* of reactions, each releasing a small, manageable amount of energy. Electron transport within each of the three protein complexes results, as we'll see, in the pumping of protons across the inner mitochondrial membrane, and the return of the protons across the membrane leads to the formation of ATP. Thus the energy originally contained in glucose and other foods is finally tucked into the cellular energy currency, ATP. For each pair of electrons passed along the respiratory chain from $NADH + H^+$ to oxygen, three molecules of ATP are formed.

The electron carriers of the respiratory chain (including those contained in the three protein complexes) differ as to how they change when they become reduced. NAD^+, for example, accepts H^- (a hydride ion—one proton and two electrons), leaving the proton from the other hydrogen atom to float free: $NADH + H^+$. Other carriers, including Q, bind both protons and both electrons, becoming, for example, QH_2. The remainder of the chain, however, is only an electron transport process. Electrons, but not protons, are passed

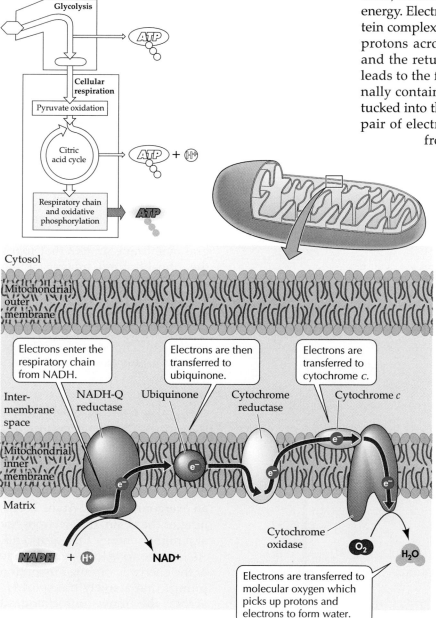

7.12 The Oxidation of NADH + H⁺
Electrons from $NADH + H^+$ are passed through the respiratory chain, a series of carrier molecules in the inner mitochondrial membrane (or the plasma membrane of an aerobic prokaryote). The carriers gain free energy when they become reduced and release free energy when they are oxidized.

from Q to cytochrome *c*. An electron from QH_2 reduces a cytochrome's Fe^{3+} to Fe^{2+}.

Electrons pour into the pool of Q molecules from the $NADH + H^+$ pathway, or they can come from another source: the succinate-to-fumarate reaction of the citric acid cycle (reaction 6 in Figure 7.10). Another protein complex, **succinate-Q reductase**, links the oxidation of succinate to the reduction of Q (Figure 7.13). The enzyme that constitutes the first part of succinate-Q reductase has attached to it an FAD carrier molecule, which is reduced by succinate to $FADH_2$. Later, hydrogen atoms are transferred to the Q molecules. No protons are pumped and hence no ATP is generated in the succinate-to-Q branch of the respiratory chain.

If only electrons are carried through the final reactions of the respiratory chain, what happens to H^+? And how is electron transport coupled to ATP production?

Active proton transport is followed by diffusion coupled to ATP synthesis

All the carriers and enzymes of the respiratory chain (except cytochrome *c*) are embedded in the inner mitochondrial membrane (see Figure 7.12). The operation of the respiratory chain results in the transport of protons (H^+), against their concentration gradient, across the inner membrane of the mitochondrion from inside to outside ("outside" being the space between the two mitochondrial membranes).

Three integral membrane protein complexes of the inner mitochondrial membrane (NADH-Q reductase, cytochrome reductase, and cytochrome oxidase) accomplish this active transport of protons across the membrane (Figure 7.14; see also Figure 7.12). Because of the charge on the proton (H^+), this transport also causes a difference in electric charge across the membrane. Together, the proton concentration gradient and the charge difference constitute a **proton-motive force** that tends to drive the protons back across the membrane, just as the charge on a battery drives the flow of electrons, discharging the battery.

The discharge of the proton-motive force is prevented by the proton impermeability of the phospholipid bilayer of the inner mitochondrial membrane. Protons can diffuse across the membrane only by passing through specific channel proteins, called **ATP synthases**, that couple proton movement to the synthesis of ATP from ADP and P_i. This coupling by an ATP synthase of the exergonic discharge of the proton-motive force to the endergonic formation of ATP is called the **chemiosmotic mechanism**. The ATP synthase complex is an integral membrane protein that spans the phospholipid bilayer of the inner mitochondrial membrane. The catalytic part of the complex is visible in electron micrographs as large knobs protruding from the inner mitochondrial membrane into the matrix (see Figure 7.14).

THE CHEMIOSMOTIC MECHANISM SUMMARIZED. The enzymes and other electron carriers of the respiratory chain are part of the inner mitochondrial membrane or are closely associated with it. The flow of electrons down the respiratory chain is an exergonic process. At the beginning of the chain, the electrons are at a higher energy than at its end. As the electrons lose energy, some of the energy is captured by proton pumps that actively transport H^+ across the inner mitochondrial membrane, whose phospholipid bilayer is impermeable to protons.

7.13 The Complete Respiratory Chain Electrons enter the chain from two sources, but they follow the same path from ubiquinone on.

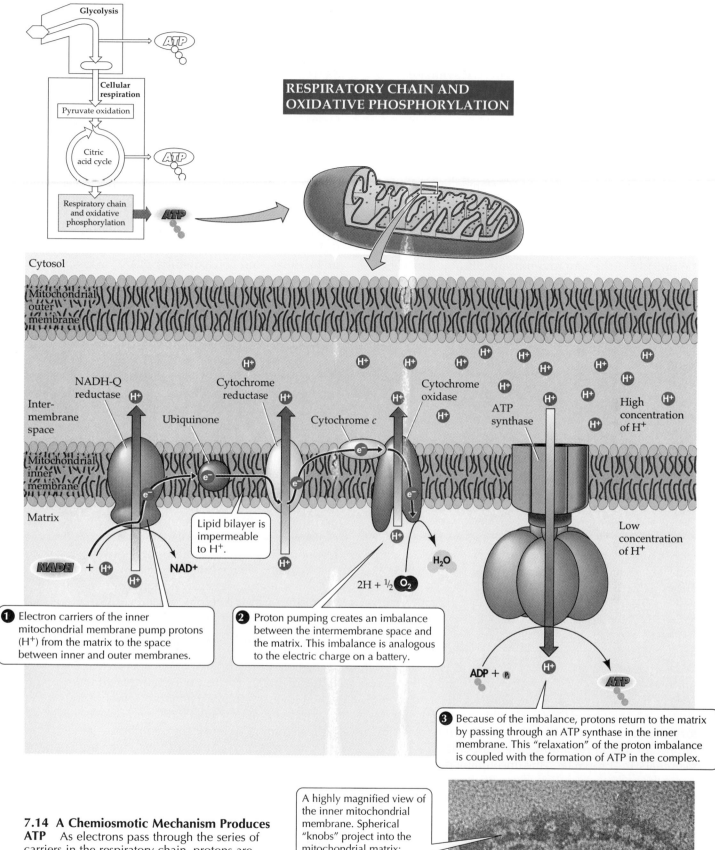

RESPIRATORY CHAIN AND OXIDATIVE PHOSPHORYLATION

Glycolysis

Cellular respiration

Pyruvate oxidation

Citric acid cycle

Respiratory chain and oxidative phosphorylation

Cytosol

Mitochondrial outer membrane

Intermembrane space

Mitochondrial inner membrane

Matrix

NADH-Q reductase

Cytochrome reductase

Cytochrome oxidase

ATP synthase

Ubiquinone

Cytochrome c

High concentration of H+

Low concentration of H+

Lipid bilayer is impermeable to H+.

NADH + H+ → NAD+

2H + ½ O₂

H₂O

ADP + Pᵢ

1 Electron carriers of the inner mitochondrial membrane pump protons (H+) from the matrix to the space between inner and outer membranes.

2 Proton pumping creates an imbalance between the intermembrane space and the matrix. This imbalance is analogous to the electric charge on a battery.

3 Because of the imbalance, protons return to the matrix by passing through an ATP synthase in the inner membrane. This "relaxation" of the proton imbalance is coupled with the formation of ATP in the complex.

7.14 A Chemiosmotic Mechanism Produces ATP As electrons pass through the series of carriers in the respiratory chain, protons are pumped from the mitochondrial matrix to the intermembrane space. As the protons return to the matrix through an ATP synthase, ATP forms.

A highly magnified view of the inner mitochondrial membrane. Spherical "knobs" project into the mitochondrial matrix; these knobs catalyze the synthesis of ATP .

This active transport establishes a proton concentration gradient. The potential energy of this proton gradient is used to synthesize ATP from ADP and P_i when protons diffuse back into the mitochondrial interior through the membrane protein ATP synthase.

An experiment demonstrates the chemiosmotic mechanism

According to the chemiosmotic model, one would expect that the mitochondrion could be "fooled" into making more ATP by the following clever trick. A sample of isolated mitochondria is maintained in a solution at pH 8 (slightly basic) until it is fully adapted; then suddenly the mitochondria are transferred into a second solution at pH 4 (fairly acidic) containing ADP and P_i. This transfer should lead to an excess of protons on the outside of the inner membrane, from where they should be able to proceed through the proposed ATP synthase channels, causing a burst of ATP production.

And that's exactly what happens. This acid-induced ATP production by isolated mitochondria stands as one of the stronger pieces of evidence favoring the chemiosmotic theory as the explanation for how oxidative phosphorylation proceeds in cells.

Proton diffusion can be uncoupled from ATP production

For the chemiosmotic mechanism to work, the phospholipid bilayer of the mitochondrial inner membrane must be quite impermeable to protons; otherwise, the protons would leak back across the bilayer as fast as they were pumped out by the respiratory chain. Cellular respiration would race along, and no ATP would form. The mechanism operates only because the diffusion of H^+ and the formation of ATP are tightly *coupled*: To move inward, the protons must pass through the channel protein ATP synthase.

Many years ago, biochemists discovered that certain hydrophobic compounds that are also very weak acids, such as *dinitrophenol*, uncouple respiratory metabolism from ATP formation. These **respiratory uncouplers** carry protons back across the bilayer, discharging the proton-motive force without capturing the energy in ATP. As a result, food is metabolized, but all the released energy is lost as heat. One naturally occurring respiratory uncoupler, a protein called *thermogenin*, plays an important role in regulating the temperature of some mammals: In "brown fat," the uncoupling of respiration raises the body temperature, as mentioned at the beginning of this chapter (see Chapter 37 for more detail).

Fermentation: ATP from Glucose, without O_2

Suppose that the supply of oxygen to a respiring cell is cut off, perhaps by drowning or by extreme exertion. As we can see in Figure 7.14, the first consequence of an insufficient supply of O_2 is that the cell cannot reoxidize cytochrome *c*, so all of that compound is soon in the reduced form. When this happens, there is no oxidizing agent to reoxidize QH_2, and soon all the Q is in the reduced form. So it goes, until the entire respiratory chain is reduced. Under these circumstances, no NAD^+ and no FAD are generated from their reduced forms; therefore, the oxidative steps in glycolysis, pyruvate oxidation, and the citric acid cycle also stop. If the cell has no other way to obtain energy from its food, it will die.

Some cells, such as muscle cells, have the enzymes necessary for fermentation and can thus switch from aerobic metabolism to fermentation. Fermentation has two defining characteristics. First, a fermentative reaction uses $NADH + H^+$ to reduce pyruvate or one of its metabolites, and consequently NAD^+ is regenerated. Once the cell has some NAD^+, it can carry more glucose through glycolysis (that is, through the early steps of fermentation).

The second characteristic of fermentation is that, by allowing glycolysis to continue, it enables a sustained production of ATP. Only as much ATP is produced as can be obtained from substrate-level phosphorylation in glycolysis—not the much greater yield of ATP obtained by the operation of pyruvate oxidation, the citric acid cycle, and the respiratory chain. However, this amount of ATP is enough to keep the cell alive, because the rate of glycolysis increases.

When cells capable of fermentation become anaerobic, the rate of glycolysis speeds up 10-fold or even more. Thus a substantial rate of ATP production is maintained, although the efficiency in terms of ATP molecules per glucose molecule is greatly reduced as compared to aerobic respiration. Some bacteria of the genus *Clostridium*, while growing anaerobically in the presence of glucose, grow and multiply as rapidly as the fastest-growing aerobic bacteria. This rapid growth is made possible by the fact that the *Clostridium* bacteria are running the glycolytic reactions much more rapidly than the aerobes do.

As noted earlier, some organisms are confined to totally anaerobic environments and use only fermentation. Other organisms carry on fermentation even in the presence of oxygen. And several bacteria carry on *cellular respiration*—not fermentation—*without using oxygen gas as an electron acceptor*. Instead, to oxidize their cytochromes these bacteria reduce nitrate ions (NO_3^-) to nitrite ions (NO_2^-). We'll put that observation into a broader context in Chapter 25.

Some fermenting cells produce lactic acid

Many different types of fermentation are found in different bacteria and some body cells. These different fermentations are identified by the final product produced. For example, in *lactic acid fermentation*, pyru-

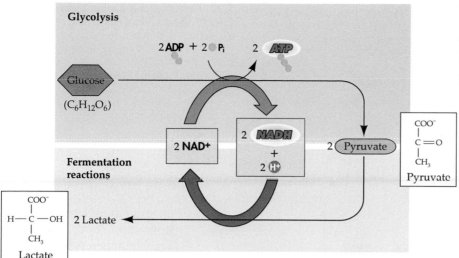

7.15 Lactic Acid Fermentation
Glycolysis produces pyruvate, as well as ATP and NADH + H$^+$, from glucose. Lactic acid fermentation, using NADH + H$^+$ as the reducing agent, then reduces pyruvate to lactic acid (lactate).

vate is reduced to lactic acid (Figure 7.15). Lactic acid fermentation takes place in many microorganisms and in our muscle cells. Unlike muscle cells, however, nerve cells (neurons) are incapable of fermentation because they lack the enzyme that reduces pyruvate to lactate. For this reason, without adequate oxygen our nervous system (including the brain) is rapidly destroyed; it is the first part of the body to die.

Other fermenting cells produce alcohol

Certain yeasts and some plant cells in anaerobic conditions carry on a process called **alcoholic fermentation** (Figure 7.16). This process requires two enzymes to metabolize pyruvate. First, carbon dioxide is removed from pyruvate, leaving the compound *acetaldehyde*. Second, the acetaldehyde is reduced by NADH + H$^+$, producing NAD$^+$ and *ethyl alcohol* (ethanol). Remember that recycling NAD in glycolysis (reaction 6 in Figure

7.6) allows the fermenting cell to produce ATP by substrate-level phosphorylation (see Figure 7.8). The brewing industry relies on alcoholic fermentation to produce wine and beer.

Contrasting Energy Yields

The total yield of stored energy from fermentation is two molecules of ATP per molecule of glucose oxidized. In contrast, the maximum yield that can be obtained from glycolysis followed by complete aerobic respiration of a molecule of glucose is much greater—about 36 molecules of ATP (Figure 7.17). (Study Figures 7.6, 7.10, and 7.14 to review where the ATP molecules come from.)

Why is so much more ATP produced by aerobic respiration? Because carriers (mostly NAD$^+$) are reduced in pyruvate oxidation and the citric acid cycle and then oxidized by the respiratory chain, with the accompanying production of ATP (three for each NADH + H$^+$ and two for each FADH$_2$) by the chemiosmotic mechanism. In an aerobic environment, a species capable of this type of metabolism will be at an advantage (in terms of energy availability per glucose molecule) over one limited to fermentation.

The total gross yield of ATP from one molecule of glucose taken through glycolysis and respiration is 38. However, we must subtract two from that gross—for a net yield of 36

7.16 Alcoholic Fermentation In alcoholic fermentation (the basis for the brewing industry), pyruvate from glycolysis is converted to acetaldehyde, and CO$_2$ is released. The NADH + H$^+$ from glycolysis acts as a reducing agent, reducing acetaldehyde to ethanol.

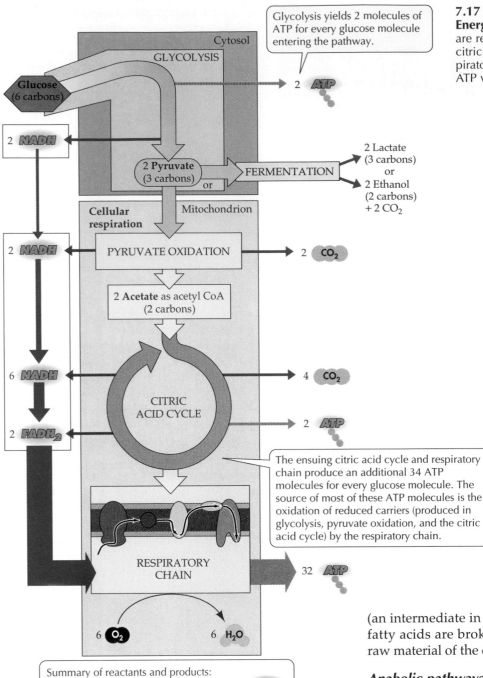

Glycolysis yields 2 molecules of ATP for every glucose molecule entering the pathway.

The ensuing citric acid cycle and respiratory chain produce an additional 34 ATP molecules for every glucose molecule. The source of most of these ATP molecules is the oxidation of reduced carriers (produced in glycolysis, pyruvate oxidation, and the citric acid cycle) by the respiratory chain.

Summary of reactants and products:
$$C_6H_{12}O_6 + 6\ O_2 \longrightarrow 6\ CO_2 + 6\ H_2O + 36\ \text{ATP}$$

7.17 Cellular Respiration Yields More Energy Than Gylcolysis Does Carriers are reduced in pyruvate oxidation and the citric acid cycle, then oxidized by the respiratory chain. These reactions produce ATP via the chemiosmotic mechanism.

an interchange, with traffic flowing in both directions. These energy pathways are connected to both catabolic (degradative) and anabolic (synthetic) pathways.

Catabolic pathways feed into respiratory metabolism

In addition to glucose, other sugars, fats, and even proteins can serve as the starting materials for respiratory ATP production. Other monosaccharides may be used, after being converted to glucose. Polymers such as starch and glycogen are **digested** (hydrolyzed) to glucose and subsequently metabolized to yield ATP. Fats are first digested to yield glycerol and fatty acids (see Chapter 3). The glycerol is then readily converted to glyceraldehyde 3-phosphate (an intermediate in glycolysis; see Figure 7.6), and the fatty acids are broken down to form acetyl CoA (the raw material of the citric acid cycle; see Figure 7.10).

Anabolic pathways use intermediates from energy pathways

Each of the reactions just described also operates in reverse. Thus, in the synthesis of fats, fatty acids form from acetyl CoA and glycerol forms from glyceraldehyde 3-phosphate. These reverse reactions occur only when the cell has an adequate energy supply; otherwise the acetyl CoA would be needed strictly for the citric acid cycle and ATP formation. However, with an abundant supply of glucose, the cell can divert some acetyl CoA to fatty acid production and some glyceraldehyde 3-phosphate to glycerol formation. The fat that forms on our bodies, adding baggage, is a result of this diversion.

ATP—because the inner mitochondrial membrane is impermeable to NADH, and a "toll" of one ATP must be paid for each NADH (produced in glycolysis) that is shuttled into the mitochondrial matrix.

Connections with Other Pathways

Glycolysis and the respiratory pathways do not operate in isolation from the rest of metabolism. Rather, there is

Some intermediates of the citric acid cycle are used in the synthesis of various important cellular constituents. Succinyl CoA (succinyl coenzyme A) is a starting point for chlorophyll synthesis, and α-ketoglutarate and oxaloacetate are starting materials for the synthesis of certain amino acids required for protein production. Still other amino acids derive from pyruvate, from glycolysis. Acetyl CoA has many different fates: In addition to its role in fatty acid production, it is a building block for various pigments, plant growth substances, rubber, and the steroid hormones of animals—among other functions.

Regulating the Energy Pathways

Glycolysis, cellular respiration, and fermentation are energy-harvesting pathways. Cells use the ATP produced by these pathways to do work: biosynthesis of macromolecules and other important compounds, active transport of materials across membranes, and movement of cellular components and entire cells. This work constitutes the life of the cell. To accomplish these activities, the cell must regulate its energy metabolism.

When a yeast cell switches from aerobic respiration to anaerobic fermentation at low concentrations of oxygen, it must metabolize glucose 18 times faster to obtain the same amount of energy. But as soon as aerobic respiration begins again, glycolysis in the yeast cell slows down. The amount of glucose used is only as much as is needed for energy production under the existing conditions, anaerobic or aerobic. The slowing down of glycolysis in response to the resumption of aerobic respiration is called the **Pasteur effect**, after its discoverer, Louis Pasteur. What is the mechanism that slows down glycolysis when the respiratory chain begins to operate?

Allostery regulates respiratory metabolism

Glycolysis, the citric acid cycle, and the respiratory chain are regulated by allosteric control of the enzymes involved (see Chapter 6). Some products of later reactions, if they are in oversupply, can suppress the action of enzymes that catalyze earlier reactions. On the other hand, an excess of the products of one branch of a synthetic chain can speed up reactions in another branch, diverting raw materials away from synthesis of the former (Figure 7.18).These negative and positive feedback control mechanisms are used at many points in the energy-extracting processes, which are summarized in Figure 7.19.

The main control point in glycolysis is the enzyme *phosphofructokinase* (reaction 3 in Figure 7.6). This enzyme is allosterically inhibited by ATP and activated by ADP or AMP. As long as fermentation proceeds, yielding a relatively small amount of ATP, phospho-

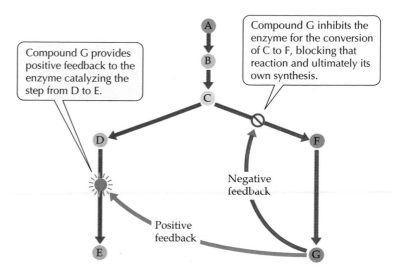

Compound G provides positive feedback to the enzyme catalyzing the step from D to E.

Compound G inhibits the enzyme for the conversion of C to F, blocking that reaction and ultimately its own synthesis.

Negative feedback

Positive feedback

7.18 Regulation by Negative and Positive Feedback
Allosteric regulation plays an important role in respiratory metabolism. Excess accumulation of some products can shut down their continued synthesis or stimulate the synthesis of other products.

fructokinase operates at full efficiency. But when aerobic respiration begins producing ATP 18 times faster than before, the excess ATP allosterically inhibits the conversion of fructose 6-phosphate to fructose 1,6-bisphosphate, and the rate of glucose utilization drops.

The main control point in the citric acid cycle is the enzyme *isocitrate dehydrogenase*, which converts isocitrate to α-ketoglutarate (reaction 3 in Figure 7.10). ATP and NADH + H$^+$ are feedback inhibitors of this reaction; ADP and NAD$^+$ are activators. If too much ATP is accumulating, or if NADH + H$^+$ is being produced faster than it can be used by the respiratory chain, the isocitrate reaction is almost completely blocked and the citric acid cycle is essentially shut down. A shutdown of the citric acid cycle would cause large amounts of isocitrate and citrate to accumulate, except that the conversion of acetyl CoA to citrate is also slowed by ATP and NADH + H$^+$.

The negative effects of halting the isocitrate reaction are thus spread backward through the chain of reactions. However, a certain excess of citrate does accumulate, and this excess acts as a negative-feedback inhibitor to slow the fructose 6-phosphate reaction early in glycolysis. Consequently, if the citric acid cycle has been slowed down because of an excess of ATP (and not because of a lack of oxygen), glycolysis is shut down as well. Both processes resume when the ATP level falls and they are needed. Allosteric control keeps the process in balance.

Another control point in Figure 7.19 involves a method for storing excess acetyl CoA. If too much ATP is being made and the citric acid cycle shuts down, the accumulation of citrate switches acetyl CoA to the syn-

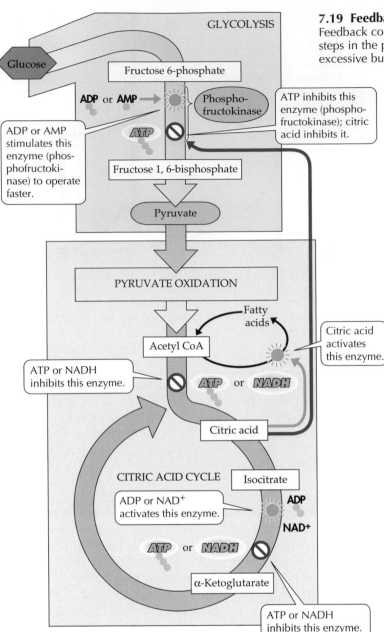

GLYCOLYSIS

Glucose

Fructose 6-phosphate

ADP or **AMP** →

Phospho-
fructokinase

ATP

ADP or AMP
stimulates this
enzyme (phos-
phofructoki-
nase) to operate
faster.

ATP inhibits this
enzyme (phospho-
fructokinase); citric
acid inhibits it.

Fructose 1, 6-bisphosphate

Pyruvate

PYRUVATE OXIDATION

Fatty
acids

Acetyl CoA

Citric acid
activates
this enzyme.

ATP or NADH
inhibits this enzyme.

ATP or *NADH*

Citric acid

CITRIC ACID CYCLE

Isocitrate

ADP or NAD+
activates this enzyme.

ADP

NAD+

ATP or *NADH*

α-Ketoglutarate

ATP or NADH
inhibits this enzyme.

7.19 Feedback Regulation of Glycolysis and the Citric Acid Cycle
Feedback controls glycolysis and the citric acid cycle at crucial early
steps in the pathways, increasing their efficiency and preventing the
excessive buildup of intermediates.

the energy-processing pathways present in cells
today, glycolysis and fermentation are the most ancient. These pathways appeared when the plane-
tary environment was strictly anaerobic and all life
was prokaryotic—before there were cells with
membrane-enclosed organelles. To this day, the enzymes of glycolysis and fermentation are located in
the cytoplasm.

Eventually some cells gained the capacity to
perform photosynthesis, a form of which added O_2
to the atmosphere, rendering most environments
aerobic. Evolution in the aerobic environment led
to the appearance of pyruvate oxidation and the
citric acid cycle. Elaboration of membranes, especially in eukaryotic cells, allowed the evolution of
chemiosmotic mechanisms for coupling electron
transport to ATP production, as in oxidative phosphorylation.

There have also been evolutionary refinements
within the pathways themselves. Eukaryotic cells,
but not prokaryotic ones, have a cytoskeleton
based on microtubules and actin microfilaments
(see Figure 4.23). With the appearance of a cytoskeleton, some glycolytic enzymes became attached to cytoskeletal components, which thus organized the enzymes into efficient associations that
allow molecules to move from one enzyme in the
pathway to the next. In eukaryotes hexokinase, the
first glycolytic enzyme, binds to the outer mitochondrial membrane, giving the enzyme immediate access to ATP produced within the mitochondria. In metabolism as in all the rest of biology,
evolution leads to adaptation.

thesis of fatty acids for storage. These fatty acids may
be metabolized later to produce more acetyl CoA.

Evolution has led to metabolic efficiency

Allosteric control of the sort illustrated in Figure 7.19
is one of the most impressive examples of the tight organization that can evolve through natural selection
when selection favors efficient operation in the competition among organisms for limited resources. Each of
the feedback controls regulates a part or various parts
of the energy-releasing pathways and keeps them operating in harmony and balance.

Not just the regulatory systems have evolved; the
pathways themselves are the products of evolution. Of

Summary of "Cellular Pathways That Harvest Chemical Energy"

Obtaining Energy and Electrons from Glucose

• When glucose burns, energy is released as heat and light:
$C_6H_{12}O_6 + 6 O_2 \rightarrow 6 CO_2 + 6 H_2O + energy$. The same equation applies to the complete aerobic metabolism of glucose
by cells, but the reaction is accomplished in many separate
steps so that the energy can be captured as ATP for cellular
work. **Review Figure 7.1**

• As a material is oxidized, the electrons it loses are transferred to another material, which is thereby reduced. Redox reactions transfer large amounts of energy. Much of the energy liberated in an oxidation is captured in the reduction of the oxidizing agent, such as NAD$^+$ or FAD. **Review Figures 7.2, 7.3, 7.4**

An Overview: Releasing Energy from Glucose

• Glycolysis is the most ancient energy pathway and operates in the presence or absence of O_2. Glycolysis breaks glucose down into two pyruvate molecules and produces two ATPs. Under aerobic conditions, cellular respiration continues the breakdown process. **Review Figure 7.5**

• Pyruvate oxidation and the citric acid cycle produce CO_2 and considerable numbers of hydrogen atoms carried by NADH and FADH$_2$. The respiratory chain combines these hydrogens with O_2, releasing enough energy for the synthesis of substantial amounts of ATP. **Review Figure 7.5**

• In some cells under anaerobic conditions, pyruvate can be reduced by NADH to form lactate and regenerate the NAD needed to sustain glycolysis. **Review Figure 7.5**

• In eukaryotes, glycolysis and fermentation take place in the cytoplasm outside of the mitochondria; pyruvate oxidation, the citric acid cycle, and the respiratory chain operate in association with mitochondria. In prokaryotes, glycolysis, fermentation, and the citric acid cycle take place in the cytoplasm; and pyruvate oxidation and the respiratory chain operate in association with the plasma membrane. **Review Table 7.1**

Glycolysis: From Glucose to Pyruvate

• Glycolysis is a pathway of 10 enzyme-catalyzed reactions located in the cytoplasm. Glycolysis provides starting materials for both cellular respiration and fermentation. **Review Figure 7.6**

• Energy-investing reactions use two ATPs per glucose molecule and eventually yield two glyceraldehyde 3-phosphate molecules. In the energy-harvesting reactions, two NADH molecules are produced, and four ATP molecules are generated by substrate-level phosphorylation. Two pyruvates are produced for each glucose molecule. **Review Figures 7.6, 7.7, 7.8**

Pyruvate Oxidation

• The pyruvate dehydrogenase complex, an assemblage of many polypeptide subunits, catalyzes three reactions: (1) Pyruvate is oxidized to acetate, releasing one CO_2 molecule and considerable energy. (2) Some of this energy is captured when NAD$^+$ is reduced to NADH + H$^+$. (3) The remaining energy is captured when acetate is combined with coenzyme A, yielding energy-rich acetyl CoA. **Review Figure 7.9**

The Citric Acid Cycle

• The energy in acetyl CoA drives the reaction of acetate with oxaloacetate to produce citric acid. The citric acid cycle is a series of reactions in which citrate is oxidized and oxaloacetate regenerated (hence a "cycle"). It produces two CO_2 , one FADH$_2$, three NADH, and one ATP for each acetyl CoA. **Review Figures 7.10, 7.11**

The Respiratory Chain: Electrons, Proton Pumping, and ATP

• NADH + H$^+$ and FADH$_2$ from glycolysis, pyruvate oxidation, and the citric acid cycle are oxidized by the respiratory chain, regenerating NAD$^+$ and FAD. With the exception of cytochrome c, the enzymes and other electron carriers of the respiratory chain are part of the inner mitochondrial membrane. Oxygen (O_2) is the final acceptor of electrons and protons, forming water (H$_2$O). **Review Figures 7.12, 7.13**

• The chemiosmotic mechanism couples proton transport to oxidative phosphorylation. As the electrons move along the respiratory chain, they lose energy, which is captured by proton pumps that actively transport H$^+$ out of the mitochondrion, establishing a gradient of both proton concentration and electric charge—the proton-motive force. This force is used to produce ATP when protons diffuse back into the mitochondrial interior through the membrane protein ATP synthase. The phospholipid bilayer of the inner membrane is impermeable to protons. **Review Figure 7.14**

Fermentation: ATP from Glucose, without O_2

• Many organisms and some cells live without O_2, deriving all their energy from glycolysis and fermentation reactions. Together, these pathways partly oxidize glucose and generate energy-containing waste molecules such as lactic acid or ethanol. Fermentation reactions anaerobically oxidize the NADH + H$^+$ produced in glycolysis. Some fermenting cells produce lactic acid; others produce ethanol and CO_2. **Review Figures 7.15, 7.16**

Contrasting Energy Yields

• For each molecule of glucose used, fermentation yields 2 molecules of ATP. In contrast, glycolysis operating with pyruvate oxidation, the citric acid cycle, and the respiratory chain yields up to 36 molecules of ATP per molecule of glucose. **Review Figure 7.17**

Connections with Other Pathways

• Catabolic pathways feed into respiratory metabolism. Starch and glycogen produce glucose. Other monosaccharides can be converted to glucose for respiratory metabolism. Glycerol from fats enters glycolysis; acetate from fatty acid degradation forms acetyl CoA to enter the citric acid cycle.

• Anabolic pathways use intermediate components of respiratory metabolism to synthesize fats, amino acids, and other essential building blocks for cellular structure and function.

Regulating the Energy Pathways

• When aerobic respiration is operating, glycolysis slows down (the Pasteur effect). The rates of glycolysis and the citric acid cycle are increased or decreased by the actions of ATP, ADP, NAD$^+$, or NADH + H$^+$ on allosteric enzymes.

• Inhibition of the glycolytic enzyme phosphofructokinase by abundant ATP from oxidative phosphorylation slows down glycolysis. ADP activates this enzyme, speeding up glycolysis. The citric acid cycle enzyme isocitrate dehydrogenase is inhibited by ATP and NADH and activated by ADP and NAD$^+$. **Review Figures 7.18, 7.19**

Self-Quiz

1. Which statement about ATP is *not* true?
   a. It is formed only under aerobic conditions.
   b. It is used as an energy currency by all cells.
   c. Its formation from ADP and P$_i$ is an endergonic reaction.
   d. It provides the energy for many different biochemical reactions.
   e. Some ATP is used to drive the synthesis of storage compounds.

2. Oxidation and reduction
   a. entail the gain or loss of proteins.
   b. are defined as the loss of electrons.
   c. are both endergonic reactions.
   d. always occur together.
   e. proceed only under aerobic conditions.

3. NAD$^+$
   a. is a type of organelle.
   b. is a protein.
   c. is an oxidizing agent.
   d. is a reducing agent.
   e. is formed only under aerobic conditions.

4. Glycolysis
   a. takes place in the mitochondrion.
   b. produces no ATP.
   c. has no connection with the respiratory chain.
   d. is the same thing as fermentation.
   e. reduces two molecules of NAD$^+$ for every glucose molecule processed.

5. Fermentation
   a. takes place in the mitochondrion.
   b. takes place in all animal cells.
   c. does not require O$_2$.
   d. requires lactic acid.
   e. prevents glycolysis.

6. Which statement about pyruvate is *not* true?
   a. It is the end product of glycolysis.
   b. It becomes reduced during fermentation.
   c. It is a precursor of acetyl CoA.
   d. It is a protein.
   e. It contains three carbon atoms.

7. The citric acid cycle
   a. takes place in the mitochondrion.
   b. produces no ATP.
   c. has no connection with the respiratory chain.
   d. is the same thing as fermentation.
   e. reduces two molecules of NAD$^+$ for every glucose molecule processed.

8. Which statement about the respiratory chain is *not* true?
   a. It operates in the mitochondrion.
   b. It uses O$_2$ as an oxidizing agent.
   c. It leads to the production of ATP.
   d. It regenerates oxidizing agents for glycolysis and the citric acid cycle.
   e. It operates simultaneously with fermentation.

9. Which statement about the chemiosmotic mechanism is *not* true?
   a. Protons are pumped across a membrane.
   b. Protons return through the membrane by way of a channel protein.
   c. ATP is required for the protons to return.
   d. Proton pumping is associated with the respiratory chain.
   e. The membrane in question is the inner mitochondrial membrane.

10. Which statement about oxidative phosphorylation is *not* true?
    a. It is the formation of ATP during operation of the respiratory chain.
    b. It is brought about by the chemiosmotic mechanism.
    c. It requires aerobic conditions.
    d. In eukaryotes, it takes place in mitochondria.
    e. Its functions can be served equally well by fermentation.

Applying Concepts

1. Trace the sequence of chemical changes that occurs in mammalian brain tissue when the oxygen supply is cut off. (The first change is that the cytochrome oxidase system becomes totally reduced, because electrons can still flow from cytochrome *c* but there is no oxygen to accept electrons from cytochrome oxidase. What are the remaining steps?)

2. Trace the sequence of chemical changes that occurs in mammalian *muscle* tissue when the oxygen supply is cut off. (The first change is exactly the same as that in Question 1.)

3. Some cells that use the citric acid cycle and the respiratory chain can also thrive by using fermentation under anaerobic conditions. Given the lower yield of ATP (per molecule of glucose) in fermentation, why can these cells function so efficiently under anaerobic conditions?

4. Describe the mechanisms by which the rates of glycolysis and of aerobic respiration are kept in balance with one another.

Readings

Alberts, B., D. Bray, J. Lewis, M. Raff, K. Roberts and J. D. Watson. 1994. *Molecular Biology of the Cell*, 3rd Edition. Garland Publishing, New York. Chapter 14 develops the themes introduced in this chapter; Chapter 2 is also useful as an introduction.

Hinkle, P. C. and R. E. McCarty. 1978. "How Cells Make ATP." *Scientific American*, March. Discussion of the chemiosmotic mechanism, in which ATP is formed by protons passing back through a membrane after being pumped out by the respiratory chain.

Lehninger, A. L., D. L. Nelson, and M. M. Cox. 1993. *Principles of Biochemistry*, 2nd Edition. Worth, New York. A balanced textbook stressing the themes of energy, evolution, regulation, and structure–function.

Lodish, H., D. Baltimore, A. Berk, S. L. Zipursky, P. Matsudaira and J. Darnell. 1995. *Molecular Cell Biology*, 3rd Edition. Scientific American Books, New York. Chapter 17 gives an excellent, more detailed treatment of the topics in this chapter.

Stryer, L. 1995. *Biochemistry*, 4th Edition. W. H. Freeman, New York. Although more advanced than this chapter, the section on glycolysis and respiration is straightforward and does not demand an advanced knowledge of chemistry.

Voet, D. and J. G. Voet. 1995. *Biochemistry*, 2nd Edition. John Wiley & Sons, New York. A general textbook with a full discussion of energy, enzymes, and catalysis. Outstanding illustrations.

Chapter 8

Photosynthesis: Energy from the Sun

Star of Life
Light energy from the sun sustains life on Earth.

We are creatures of the sun. Its light is the source—direct or indirect—of the free energy that powers life on Earth. This is the important message—perhaps already familiar to you—about photosynthesis.

As biologists, however, we might express things differently. We might say, "We are creatures of the chloroplasts." This abundant organelle, which gives green plants their color, captures the sun's energy for us. Inside plants, within stacks of connected, inner membranes that are wrapped in two outer membranes, chemical reactions convert and store the energy that we draw upon. **Photosynthesis** is the conversion of light energy to chemical energy. This process has profound significance for living systems on Earth.

Our own dependence on the sun, although absolute, is indirect. Like the other animals, the fungi, many protists, and most prokaryotes, humans depend on a ready supply of partly reduced, carbon-containing compounds as a food source. From such compounds we obtain all the free energy that keeps us alive and functioning. These compounds are also the source of the carbon atoms used in every organic molecule in our bodies. In a word, we are *heterotrophs*: We need to feed on something else. In a world populated exclusively by heterotrophs, all life would grind to an end as the food gradually disappeared. For all life to be self-sustained, some organisms must make food for others.

Our world owes the continued existence of life to the presence of *autotrophs*—organisms that do not need previously formed organic substances from their environment. For autotrophs, an energy source (such as light) and an inorganic carbon source (such as carbon dioxide gas) suffice as a diet. From these simple ingredients, autotrophs make the reduced carbon compounds from which their bodies are built and their food needs met. By feeding on autotrophs, the heterotrophs of the world meet their needs for energy and matter.

The principal autotrophs are photosynthetic organisms that use visible light as their energy source. From light, carbon dioxide, and water, they begin the chemistry that sustains almost the entire biosphere. The worldwide extent of photosynthetic activity is stunning: Each year, tens of billions of tons of carbon atoms are taken from carbon dioxide and incorporated into molecules of sugars, amino acids, and the other vital compounds of the biosphere.

In this chapter, we will examine the details of photosynthesis. What are the overall reactants and products of this process? What is the nature of light, how does it interact with chlorophyll, and what chemical changes does it cause? We'll examine the light-dependent production of ATP and reducing agents. Then we'll turn to the light-*independent* use of these chemical resources for the fixation of carbon dioxide and its reduction to sugars such as glucose. How are the pigments and enzymes for these processes arranged on the thylakoid membranes of the chloroplast? We close the chapter by looking at a key enzyme with a strange property that limits the effectiveness of photosynthesis, examining how different groups of plants have adapted to this limitation.

Identifying Photosynthetic Reactants and Products

By the beginning of the nineteenth century, scientists understood the broad outlines of photosynthesis. It was known to use three principal ingredients—water, carbon dioxide (CO_2), and light—and to produce not only food but also oxygen gas (O_2). Scientists had learned that the water for photosynthesis comes primarily from the soil (for plants living on land) and must travel from the roots to the leaves; that carbon dioxide is taken in and water and O_2 are released through tiny openings in leaves, called **stomata** (singular stoma), which can open and close (Figure 8.1); and that light is absolutely necessary for the production of oxygen and food.

The last of the important early discoveries, made during the first decade of the nineteenth century, was that carbon dioxide uptake and oxygen release are closely related and that both depend on light action.

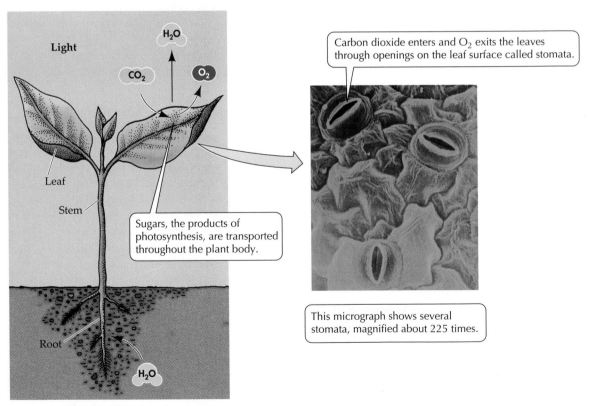

Carbon dioxide enters and O_2 exits the leaves through openings on the leaf surface called stomata.

Sugars, the products of photosynthesis, are transported throughout the plant body.

This micrograph shows several stomata, magnified about 225 times.

8.1 Ingredients for Photosynthesis A typical terrestrial plant uses light from the sun, water from the soil, and carbon dioxide from the atmosphere to form organic compounds by photosynthesis.

By 1804, scientists could summarize photosynthesis in plants as follows:

Carbon dioxide + water + light energy $\rightarrow$ sugar + oxygen

which turned into the balanced equation that is the reverse of the overall equation for cellular respiration given in Chapter 7:

$$6\ CO_2 + 6\ H_2O \rightarrow C_6H_{12}O_6 + 6\ O_2$$

Although true, these statements say nothing about the details of the process. What roles does light play? How do the carbons become linked? And where does the oxygen come from?

Almost a century and a half passed before the source of the O_2 released during photosynthesis—CO_2 or water—was determined. The direct demonstration depended on one of the first uses of an isotopic tracer in biological research. In the experiments, two groups of green plants were allowed to carry on photosynthesis. Plants in the first group were supplied with water containing the heavy-oxygen isotope ^{18}O and with CO_2 containing only the common oxygen isotope ^{16}O; plants in the second group were supplied with CO_2 labeled with ^{18}O and water containing only ^{16}O.

When oxygen gas was collected from each group of plants and analyzed, it was found that O_2 containing ^{18}O was produced in abundance by the plants that had been given ^{18}O-labeled water but not by the plants given labeled CO_2. From these results, scientists concluded that all the oxygen gas produced during photosynthesis comes from water (Figure 8.2). This discovery is reflected in a revised balanced equation:

$$6\ CO_2 + 12\ H_2O \rightarrow C_6H_{12}O_6 + 6\ O_2 + 6\ H_2O$$

Water appears on both sides of the equation because water is used as a reactant (the 12 molecules on the left) and released as a product (the 6 new ones on the right). In this equation there are now sufficient water molecules to account for all the oxygen gas produced.

The photosynthetic production of oxygen by green plants is an important source of atmospheric oxygen, which most organisms—including plants themselves—require in order to complete their respiratory chains and thus obtain the energy to live.

The Two Pathways of Photosynthesis: An Overview

The overall photosynthetic reaction just described cannot proceed in a single step. In fact, there is no precedent in all of chemistry for the completion of such a complex reaction in a single step. Rather, a whole series

Water and carbon dioxide provided	Photosynthesis	Oxygen released

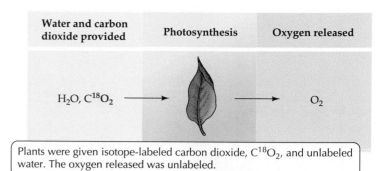

Plants were given isotope-labeled carbon dioxide, $C^{18}O_2$, and unlabeled water. The oxygen released was unlabeled.

Plants were given isotope-labeled water, $H_2^{18}O$, and unlabeled CO_2. The oxygen released was labeled.

8.2 Water Is the Source of the Oxygen Produced by Photosynthesis Because only plants given isotope-labeled water released labeled O_2, this experiment showed that water is the source of the oxygen released during photosynthesis.

of simpler steps is required. By the middle of the twentieth century, it was clear that photosynthesis includes two pathways: One pathway, driven by light, produces ATP and a reduced electron carrier (NADPH + H^+). The other pathway, which does not use light directly, uses the ATP, NADPH + H^+, and CO_2 to produce sugar (Figure 8.3).

Just as NAD (nicotinamide adenine dinucleotide) bridges the pathways of cellular respiration (see Chapter 7), a very similar compound, **NADP** (nicotinamide adenine dinucleotide phosphate) bridges the two pathways of photosynthesis. NADP is identical to NAD, except that it has another phosphate group attached to the ribose. Whereas NAD participates in metabolic breakdown reactions and energy transfers (catabolism), NADP participates in synthetic reactions (anabolism) that require energy and reducing power.

Like NAD, NADP exists in two forms. One (NADP$^+$) is an oxidizing agent; the other (NADPH + H^+) is a reducing agent (see Figure 7.4). NADPH + H^+ is an intermediary for energy and reducing power. ATP and NADPH + H^+ are carriers of reducing power because reduction is always an endergonic process that requires both energy and electrons.

Light energy is used to produce ATP from ADP and P_i. The reactions of this pathway, termed the *light reactions*, are mediated by molecular assemblies called **photosystems** (which we'll discuss in detail later), and

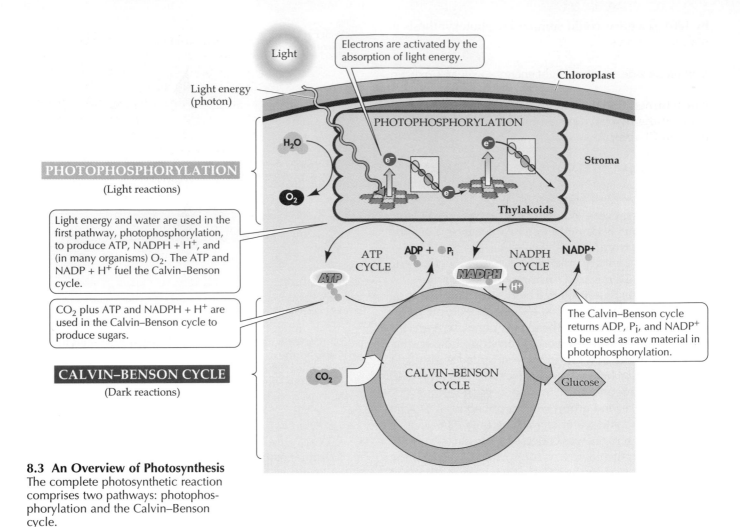

8.3 An Overview of Photosynthesis
The complete photosynthetic reaction comprises two pathways: photophosphorylation and the Calvin–Benson cycle.

the pathway itself is called **photophosphorylation** (see Figure 8.3). There are two different systems of photophosphorylation: *noncyclic photophosphorylation*, carried out by plants, which produces NADPH and ATP, and *cyclic photophosphorylation*, which produces only ATP.

The NADPH + H$^+$ and ATP produced in the first pathway of photosynthesis are used in the second pathway, where reactions trap CO_2 and reduce the resulting acid to sugar. These sugar-producing reactions constitute the Calvin–Benson cycle (see Figure 8.3), also known as the photosynthetic carbon reduction cycle, or simply the *dark reactions* (because none of them uses light directly). The reactions of both pathways proceed within the chloroplast, but, as we will see, they reside in different parts of the organelle. *Both pathways stop in the dark because ATP synthesis and NADP$^+$ reduction require light.* The rate of each set of reactions depends on the rate of the other. They are tied together by the exchange of ATP and ADP, and of NADP$^+$ and NADPH.

Properties of Light and Pigments

The living world makes marvelous use of light. Light is a source of both energy and information. In later chapters, we'll examine the many roles of light and pigments in the transmission of *information*. In this chapter, our focus is on light as a source of *energy*. However, before we can consider the reactions in the chloroplasts that use light energy, we need to examine the physical nature of light: its properties and energy content, and how it interacts with molecules.

Light comes in packets called photons

Light is a form of radiant energy. It comes in discrete packets called **photons**. Light also behaves as if it were propagated in waves. The **wavelength** of light is the distance from the peak of one wave to the peak of the next (Figure 8.4). Different colors result from different wavelengths. Light and other forms of radiant energy—cosmic rays, gamma rays, X rays, ultraviolet radiation, infrared radiation, microwaves, and radio

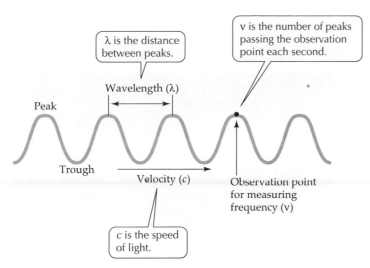

8.4 Light Has Wavelike Properties   Light can be envisioned as a series of waves whose peaks pass a fixed observation point with uniform frequency.

waves—are **electromagnetic radiation**. We have listed these forms of radiation in order of increasing wavelength and of decreasing energy per photon. Visible light fits into this **electromagnetic spectrum** of wavelengths between ultraviolet and infrared radiation (Figure 8.5).

The speed of light in a vacuum is one of the universal constants of nature. In a vacuum, light travels at 3×10^{10} cm/s (186,000 miles/s), a value symbolized as c. In air, glass, water, and other media, light travels slightly more slowly.

Let's consider light as a long train of waves moving in a straight line and see what the train would look like to a stationary observer. Successive peaks of the waves pass the observer with a uniform **frequency** determined by the wavelength and the speed of light. The exact relationship is $\nu = c / \lambda$, where ν (the Greek letter nu) is the frequency; c is the speed of light; and λ (Greek lambda) is the wavelength. Often ν is expressed in hertz (Hz), c in centimeters per second (cm/s), and λ in nanometers (nm). (1 nm $= 10^{-9}$ m or 10^{-7} cm; see the conversion table inside the back cover.)

Humans perceive light as having distinct colors (the reason for this will be explained in Chapter 42). The colors relate to the wavelengths of the light, as shown in Figure 8.5. Most of us can see electromagnetic radiation in the range of wavelengths from 400 to 700 nm. The wavelength at 400 nm marks the blue end of the visible spectrum, at 700 nm the red end. Wavelengths in the range from about 100 to 400 nm are ultraviolet radiation; those immediately above 700 nm are referred to as infrared.

The amount of energy, E, contained in a single photon is directly proportional to its frequency. The constant of proportionality that describes this relation-

ship, h, is named Planck's constant after Max Planck, who first introduced the concept of the photon. With this information we can write the equation $E = h\nu$, where ν is the frequency in Hz. Substituting c / λ for ν (from the earlier equation relating λ, ν, and c), we see that $E = hc / \lambda$. Thus shorter wavelengths mean greater energies; that is, energy is inversely proportional to wavelength. A photon of red light of wavelength 660 nm has less energy than a photon of blue light at 430 nm; an ultraviolet photon of wavelength 284 nm is much more energetic than either of these. For any light-driven biological process—such as photosynthe-

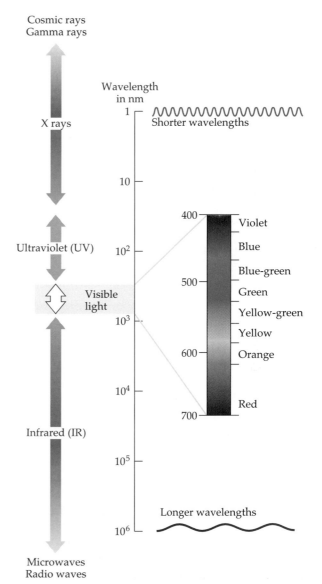

8.5 The Electromagnetic Spectrum   A portion of the electromagnetic spectrum in the vicinity of light that is visible to humans is represented here. Visible light comprises wavelengths between about 400 and 700 nm. Ultraviolet radiation extends from the short-wavelength end of the visible spectrum, infrared radiation from the long-wavelength end.

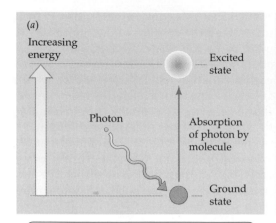

(a)

Increasing energy

Photon

Absorption of photon by molecule

Excited state

Ground state

When a molecule, initially in the ground state, absorbs a photon, the molecule is raised to an excited state and possesses more energy.

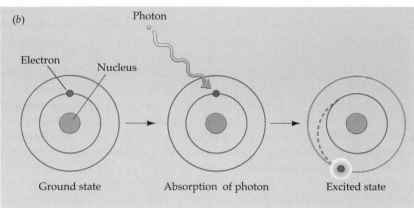

(b)

Photon

Electron

Nucleus

Ground state

Absorption of photon

Excited state

The absorption of the photon boosts one of the molecule's electrons to an orbital farther from the nucleus.

8.6 Exciting a Molecule After absorbing the energy of a photon (a), one of the molecule's electrons is boosted to a higher orbital (b) and, in this excited state, is held less firmly by the molecule.

sis—a photon can be active only if it consists of enough energy to perform the work required.

The brightness, or **intensity**, of light at a given point is the amount of energy falling on a defined area—such as 1 cm²—per second. Light intensity is usually expressed in energy units (such as calories) per square centimeter per second, but pure light of a single wavelength may also be expressed in terms of photons per square centimeter per second.

Absorption of a photon puts a pigment in an excited state

When a photon meets a molecule, one of three things happens. The photon may bounce off the molecule or it may pass through it; in other words, it may be reflected or transmitted. Neither of these causes any change in the molecule, and neither has any chemical consequences. The third possibility is that the photon may be *absorbed* by the molecule. In this case, the photon disappears. Its energy, however, cannot disappear, because energy is neither created nor destroyed.

The molecule acquires the energy of the absorbed photon and is thereby raised from a **ground state** (lower energy) to an **excited state** (higher energy) (Figure 8.6a). The difference in energy between this excited state and the ground state is precisely equal to the energy of the absorbed photon. The increase in energy boosts one of the electrons in the molecule into an orbital farther from the nucleus; this electron is now held less firmly by the molecule (Figure 8.6b). We will see the chemical consequence of this looser hold on the electron later in this chapter.

All molecules absorb electromagnetic radiation. The specific wavelengths absorbed by a particular molecule are characteristic of that type of molecule. Some molecules cannot absorb wavelengths in the visible region; those that can are called **pigments**.

When a beam of white light (light containing visible light of all wavelengths) falls on a pigment, certain wavelengths of the light are absorbed. The remaining wavelengths, which are reflected or transmitted, make the pigment appear to us to be colored. For example, if a pigment absorbs both blue and red light, as does the pigment chlorophyll, what we see is the remaining light—primarily green. The fact that chlorophyll absorbs light in both the blue and the red region of the spectrum indicates that it has two excited states of differing energy levels, both close enough to the ground state to be reached with the energy of photons of visible light.

Light absorption and biological activity vary with wavelength

A given type of molecule can absorb radiant energy of only certain wavelengths. If we plot a compound's absorption of light as a function of the wavelength of the light, the result is an **absorption spectrum** (Figure 8.7). Absorption spectra are good "fingerprints" of compounds; sometimes an absorption spectrum contains enough information to enable us to identify an unknown compound.

Light may be analyzed for its *biological effectiveness*, the magnitude of its effect on a particular activity such as photosynthesis. We may plot the effectiveness of light as a function of wavelength. The resulting graph is an **action spectrum**. Figure 8.8 shows the action spectrum for photosynthesis by *Anacharis*, a freshwater plant. All wavelengths of visible light are at least somewhat effective in causing photosynthesis, although some are more effective than others.

8.7 Photosynthetic Pigments Have Distinct Absorption Spectra

Photosynthesis uses most of the visible spectrum, because the participating pigments absorb photons most strongly at different wavelengths.

Because light must be absorbed in order to produce a chemical change or biological effect, action spectra are helpful in determining what pigment or pigments are used in a particular photobiological process, such as photosynthesis. We should be able to find which pigment or pigments have absorption spectra that match the action spectrum of the process.

Photosynthesis uses chlorophylls and accessory pigments

Certain pigments are important in biological reactions, and we will discuss them as they appear in the book. Here we discuss the pigments, found in leaves and in other parts of photosynthetic organisms, that play roles in photosynthesis. Of these, the most important ones are the **chlorophylls**. Chlorophylls occur universally in the plant kingdom, in photosynthetic protists, and in photosynthetic bacteria. A mutant individual that lacks chlorophyll is unable to perform photosynthesis and will starve to death.

In green plants, two chlorophylls predominate, **chlorophyll *a*** and **chlorophyll *b***; they differ only slightly in structure. Both have a complex ring struc-

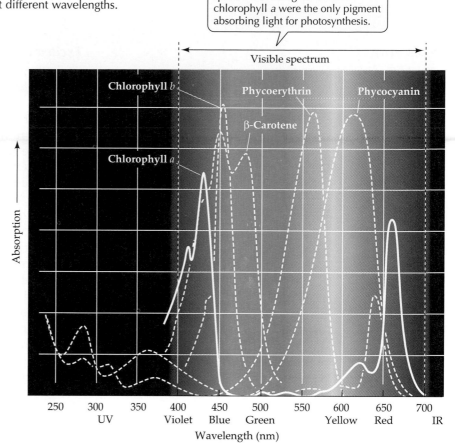

Notice how much of the visible spectrum would go to waste if chlorophyll *a* were the only pigment absorbing light for photosynthesis.

ture of a type referred to as a chlorin,* a lengthy hydrocarbon "tail," and a central magnesium atom in the chlorin ring (Figure 8.9).

We saw in Figure 8.7 that the chlorophylls absorb blue and red wavelengths, which are near the two ends of the visible spectrum. Thus if *only* chlorophyll pigments were active in photosynthesis, much of the visible spectrum would go unused. However, all photosynthetic organisms possess **accessory pigments** that absorb photons intermediate in energy between the red and the blue wavelengths and then transfer a portion of the energy to chlorophyll to use in photosynthesis.

*In Chapters 6 and 7 we learned about hemoglobin and cytochrome *c*. Both contain ring structures called porphyrins, which are very similar in structure to chlorins.

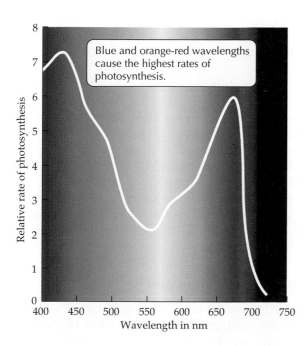

Blue and orange-red wavelengths cause the highest rates of photosynthesis.

8.8 Action Spectrum of Photosynthesis
An action spectrum plots the biological effectiveness of wavelengths of radiation against the wavelength. Here the rate of photosynthesis in the freshwater plant *Anacharis* is plotted against wavelengths of visible light. If we compare this action spectrum with the absorption spectra of specific pigments, such as those in Figure 8.7, we can identify which pigments are responsible for the process in *Anacharis*.

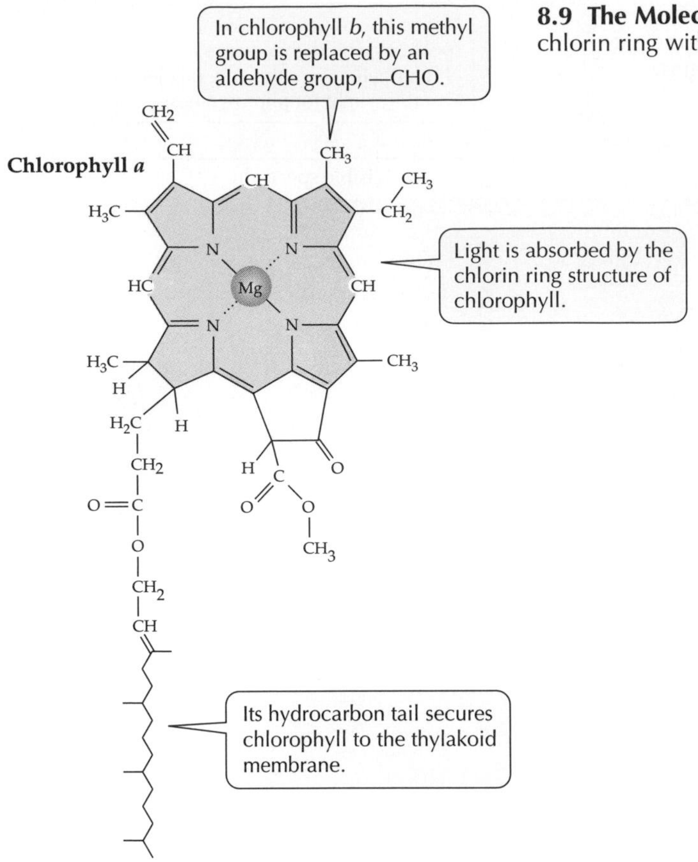

In chlorophyll *b*, this methyl group is replaced by an aldehyde group, —CHO.

8.9 The Molecular Structure of Chlorophyll Chlorophyll consists of a chlorin ring with magnesium (shaded area), plus a hydrocarbon "tail."

Chlorophyll *a*

Light is absorbed by the chlorin ring structure of chlorophyll.

Its hydrocarbon tail secures chlorophyll to the thylakoid membrane.

Among these accessory pigments are **carotenoids** such as β-carotene (see Figure 3.8); the carotenoids absorb photons in the blue and blue-green wavelengths and appear deep yellow. The **phycobilins** (phycocyanin and phycoerythrin), which are found in red algae and in cyanobacteria (contributing to their respective colors), absorb various yellow-green, yellow, and orange wavelengths (see Figure 8.7). Such accessory pigments, in collaboration with the chlorophylls, constitute an energy-absorbing *antenna system* covering much of the visible spectrum.

In the energy-absorbing antenna, any pigment molecule with a suitable absorption spectrum may absorb an incoming photon and become excited. The excitation passes from one pigment molecule to another in the antenna, moving to pigments that absorb longer wavelengths (lower energies) of light. Thus the excitation must end up in the one pigment molecule in the antenna that absorbs the longest wavelength—the molecule that occupies the **reaction center** of the antenna.

The reaction center is the part of the antenna that converts light absorption into chemical energy. In plants, the pigment molecule in the reaction center is always a molecule of chlorophyll *a*. There are many other chlorophyll *a* molecules in the antenna, but all of them absorb light at shorter wavelengths than does the molecule in the reaction center.

Excited chlorophyll acts as a reducing agent

A pigment molecule enters an excited state when it absorbs a photon (see Figure 8.6). The molecule usually does not stay in the excited state very long. It may return to the ground state, emitting light as **fluorescence**, or it may reduce another substance. In fluorescence, the boosted electron falls back from its higher orbital to the original, lower one. This process is accompanied by a loss of energy, which is given off as another photon (Figure 8.10*a*). The molecule absorbs one photon and within approximately 10^{-9} seconds emits another photon of longer wavelength than the one absorbed. The emitted light is fluorescence. If energy is simply absorbed and then rapidly returned as a photon of light, there can be no chemical changes or biological consequences—no chemical work is done.

For biological work to be done, something must happen to transfer energy in some other way during the billionth of a second before a photon is emitted as fluorescence. The excited pigment molecule may pass the energy to another pigment molecule (Figure 8.10*b*), as explained in the preceding section. Ultimately, however, photosynthesis conserves energy by using the excited chlorophyll molecule in the reaction center as a reducing agent (Figure 8.10*c*).

Ground state chlorophyll (symbolized as Chl) is not much of a reducing agent, but excited chlorophyll (Chl*) is a good one. To understand the reducing capability of Chl*, recall that in an excited molecule, one of the electrons is zipping about in an orbital farther from its nucleus than it was before. Less tightly held, this electron can be passed on in a redox reaction to an oxidizing agent. Thus Chl* (but not Chl) can react with an oxidizing agent A in a reaction like this:

$$Chl^* + A \rightarrow Chl^+ + A^-$$

This, then, is the first biochemical consequence of light absorption by chlorophyll in the chloroplast: The chlorophyll becomes a reducing agent and participates in a redox reaction that would not have occurred in the dark. As we are about to see, the further adventures of that electron (the one passed from chlorophyll to the oxidizing agent) produce ATP and a stable reducing agent (NADPH), both of which are required in the Calvin–Benson cycle.

Photophosphorylation and Reductions

In our earlier overview of photosynthesis, we distinguished the light-independent reactions (dark reac-

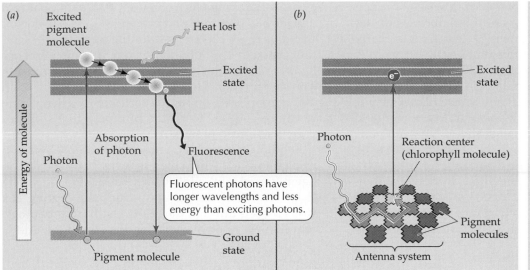

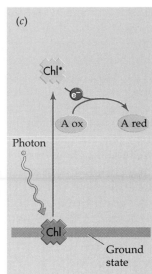

(a) When a pigment molecule absorbs a photon, boosting an electron to a higher orbital, the pigment moves to an excited state from its ground state. The molecule spends very little time in the excited state. Some of the absorbed energy is dissipated as heat, so that when the molecule returns to the ground state, a lesser amount of light energy is emitted as fluorescence.

(b) Rather than waste the absorbed energy in fluorescence, the excited pigment molecule may pass the excitation energy to another molecule, such as chlorophyll, in the antenna system of pigments. The excitation ends up in the reaction center—the molecule that absorbs the longest wavelength

(c) When a ground state chlorophyll molecule (Chl) becomes excited (Chl*), it becomes a reducing agent; the electron boosted to a higher orbital may pass to an oxidized electron carrier (A), reducing the carrier. Thus much of the energy of the excited state is chemically preserved rather than being lost, as it is in fluorescence.

8.10 Excitation, Energy Transfer, and Electron Transfer
A molecule spends very little time in the excited state. It may emit some of its acquired energy as fluorescence (a), but in order for work to be done, the energy must be transferred to the reaction center (b). In plants, the molecule in the reaction center is always chlorophyll *a*. This excited chlorophyll molecule is a reducing agent and participates in a redox reaction that is the initial step toward the production of the ATP and NADPH + H⁺ needed for the Calvin–Benson cycle.

tions) from the light-dependent reactions (light reactions). Now we want to return to focus on the light reactions and examine how they function to provide the ATP and NADPH + H⁺ needed to drive the dark reactions. We'll consider first the noncyclic reactions and cyclic reactions before considering the role of chemiosmosis in phosphorylation—a process that is very similar to that discovered for oxidative phosphorylation in mitochondria (see Chapter 7).

Noncyclic photophosphorylation produces ATP and NADPH

With the appearance of **noncyclic photophosphorylation,** the evolution of life on Earth made a crucial advance, because noncyclic photophosphorylation uses light energy not only to form ATP and NADPH + H⁺, but also to release O₂. The primitive atmosphere of Earth contained little O₂ before it began being released as a by-product of photosynthesis. The accumulation of O₂ in the atmosphere forever changed the Earth and

the course of evolution, making possible the use of aerobic metabolism and a remarkable diversification of life. The origin of O₂ from water requires us to look at the role of water in the production of NADPH and ATP.

In noncyclic photophosphorylation, water is oxidized, and the electrons from water replenish the electrons that chlorophyll molecules lose when they are excited by light. These electrons are transferred to oxidizing agents and ultimately to NADP⁺, reducing it to NADPH + H⁺. As the electrons are passed from water to chlorophyll, and ultimately to NADP, they go through a series of electron carriers. These spontaneous redox reactions are exergonic, and some of the free energy released is used ultimately to form ATP by a chemiosmotic mechanism.

Noncyclic photophosphorylation requires the participation of two distinct molecules of chlorophyll. These are associated with two different photosystems that consist of many chlorophyll molecules and accessory pigments in separate energy-absorbing antennas (Figure 8.11). **Photosystem I** makes a reducing agent strong enough to reduce NADP⁺ to NADPH + H⁺. The reaction center of the antenna for photosystem I contains a chlorophyll *a* molecule in a form called P₇₀₀ because it can absorb light of wavelength 700 nm.

Photosystem II uses light to oxidize water molecules, producing electrons, protons (H⁺), and O₂.

Electrons from water are passed to a series of redox carriers located in the thylakoid membranes of the chloroplast. Some of the energy lost in this redox process is used in the conversion of ADP + P_i to ATP. The reaction center of photosystem II contains a chlorophyll *a* molecule in a form called P_{680} because it absorbs light maximally at 680 nm. Thus photosystem II requires photons that are somewhat more energetic than those required by photosystem I. To keep noncyclic photophosphorylation going, both photosystems I and II must constantly be absorbing light, thereby boosting electrons to higher orbitals from which they may be captured by specific oxidizing agents.

We can follow the noncyclic pathway from water to NADP in Figure 8.11. Photosystem II (P_{680}) absorbs photons, sending electrons from P_{680} to an oxidizing agent (pheophytin-I) and causing P_{680} to become oxidized to P_{680}^+. Electrons from the oxidation of water are passed to P_{680}^+ of photosystem II, reducing it once again to P_{680}, which can absorb photons, and so on. The electron donated by photosystem II to its oxidizing agent passes through a series of exergonic redox reactions coupled to proton pumping that stores energy that is used to form ATP.

In photosystem I, P_{700} absorbs photons, becoming excited to P_{700}^*, which then reduces its own oxidizing agent (ferredoxin) while being oxidized to P_{700}^+. Then P_{700}^+ returns to the ground state by accepting electrons passed through the redox chain from photosystem II. Now photosystem II is accounted for, and we

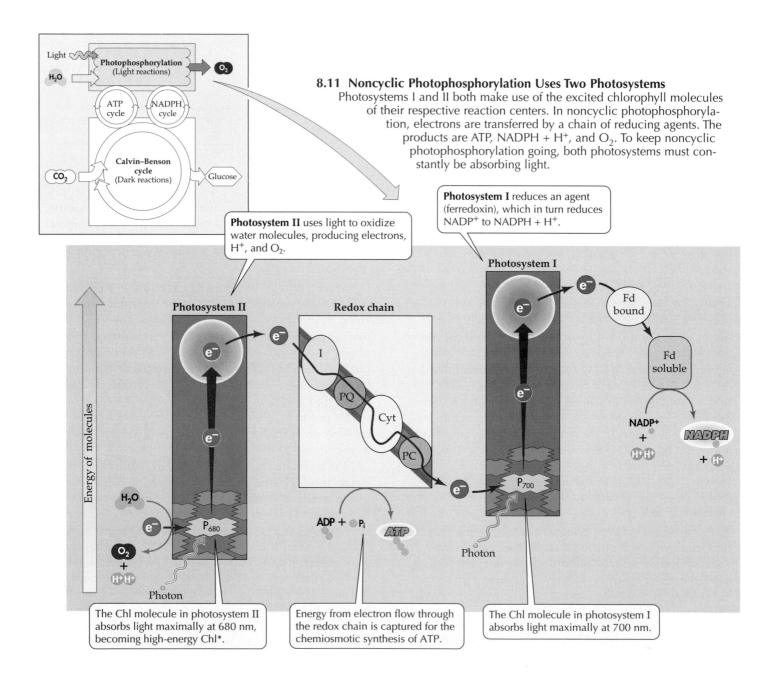

8.11 Noncyclic Photophosphorylation Uses Two Photosystems
Photosystems I and II both make use of the excited chlorophyll molecules of their respective reaction centers. In noncyclic photophosphorylation, electrons are transferred by a chain of reducing agents. The products are ATP, NADPH + H⁺, and O_2. To keep noncyclic photophosphorylation going, both photosystems must constantly be absorbing light.

Photosystem II uses light to oxidize water molecules, producing electrons, H⁺, and O_2.

Photosystem I reduces an agent (ferredoxin), which in turn reduces NADP⁺ to NADPH + H⁺.

The Chl molecule in photosystem II absorbs light maximally at 680 nm, becoming high-energy Chl*.

Energy from electron flow through the redox chain is captured for the chemiosmotic synthesis of ATP.

The Chl molecule in photosystem I absorbs light maximally at 700 nm.

must consider only the electrons from photosystem I. These are used in the last step of noncyclic photophosphorylation, in which two electrons and two protons (from two operations of the noncyclic scheme) are used to reduce a molecule of $NADP^+$ to $NADPH + H^+$.

In sum, noncyclic photophosphorylation uses a molecule of water, four photons (two each absorbed by photosystems I and II), one molecule each of $NADP^+$ and ADP, and one P_i. From these ingredients it produces one molecule each of $NADPH + H^+$ and ATP, and half a molecule of oxygen (review Figure 8.11). A substantial fraction of the light energy absorbed in noncyclic photophosphorylation is lost as heat, but another significant fraction is trapped in ATP and $NADPH + H^+$.

Cyclic photophosphorylation produces ATP but no NADPH

Noncyclic photophosphorylation produces equal quantities of ATP and $NADPH + H^+$. However, as we will see, the Calvin–Benson cycle uses more ATP than $NADPH + H^+$. In order to keep things in balance, plants sometimes make use of a supplementary form of photophosphorylation that does not generate $NADPH + H^+$.

Photophosphorylation that produces only ATP is called **cyclic photophosphorylation** because an elec-

tron passed from an excited chlorophyll molecule at the outset cycles back to the same chlorophyll molecule at the end of the chain of reactions (Figure 8.12). Water, which supplies electrons to restore chlorophyll molecules to the ground state in noncyclic photophosphorylation, does not enter these reactions; thus they produce no O_2.

Before cyclic photophosphorylation begins, P_{700}, the reaction center chlorophyll of photosystem I, is in the ground state. It absorbs a photon and becomes the reducing agent P_{700}^*. The P_{700}^* then reacts with oxidized ferredoxin (Fd_{ox}) to produce reduced ferredoxin (Fd_{red}). The reaction is spontaneous; that is, it is exergonic, releasing free energy.

In noncyclic photophosphorylation, Fd_{red} reduces $NADP^+$ to form $NADPH + H^+$. However, Fd_{red} can also pass its added electron to a *different* oxidizing agent, plastoquinone (PQ, a small organic molecule). This is what happens in cyclic photophosphorylation (see Figure 8.12), which occurs in some organisms when the ratio of $NADPH + H^+$ to $NADP^+$ in the chloroplast is high.

Fd_{red} reduces PQ (which is part of the redox chain that connects photosystems I and II; see Figure 8.11) in the reaction $Fd_{red} + PQ_{ox} \rightarrow Fd_{ox} + PQ_{red}$. Acting as if the electron passed to it came from pheophytin-I (as happens in noncyclic photophosphorylation), PQ_{red} passes the electron to a cytochrome complex (Cyt). The electron continues down the chain until it completes its cycle by returning to P_{700}. This cycle is a series of redox reactions, each exergonic, and the released energy is stored in a form that ultimately can be used to produce ATP.

Remember that when P_{700}^* passed its electron on to Fd, we were left with a molecule of positively charged P_{700}^+ (having

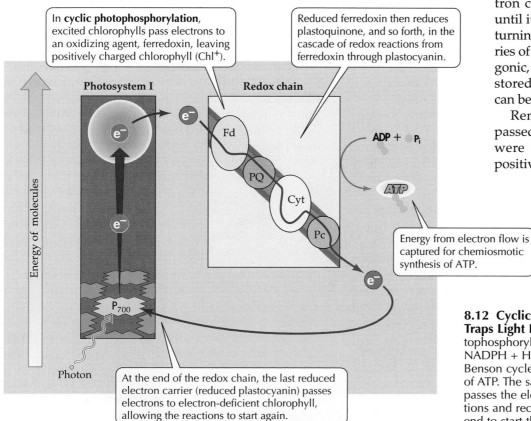

In **cyclic photophosphorylation**, excited chlorophylls pass electrons to an oxidizing agent, ferredoxin, leaving positively charged chlorophyll (Chl^+).

Reduced ferredoxin then reduces plastoquinone, and so forth, in the cascade of redox reactions from ferredoxin through plastocyanin.

Energy from electron flow is captured for chemiosmotic synthesis of ATP.

At the end of the redox chain, the last reduced electron carrier (reduced plastocyanin) passes electrons to electron-deficient chlorophyll, allowing the reactions to start again.

Photosystem I

Redox chain

Energy of molecules

Photon

P_{700}

Fd

PQ

Cyt

Pc

ADP + P_i

ATP

8.12 Cyclic Photophosphorylation Traps Light Energy as ATP Cyclic photophosphorylation produces ATP but no $NADPH + H^+$, thus balancing the Calvin–Benson cycle's need for greater amounts of ATP. The same chlorophyll molecule passes the electrons that start the reactions and receives the electrons at the end to start the process over again.

lost an electron, the chlorophyll has one unbalanced positive charge). In due course, P_{700}^+ interacts with a reducing agent that donates an electron, converting it back to uncharged P_{700}. This reducing agent, plastocyanin (PC), is the last member of the redox chain in Figure 8.12. By the time the electron (passed from P_{700}^* through the redox chain) comes back to P_{700}^+ and reduces it, all the energy from the original photon has been released. In each of the redox reactions, some free energy is lost, until all of the original energy has been converted to heat *except* for that used to form ATP.

Comparing the cartoon in Figure 8.13 with Figure 8.12 may help make the concept of cyclic photophosphorylation clearer to you.

Chemiosmosis is the source of ATP in photophosphorylation

How does electron transport in photosystems I and II form ATP? In Chapter 7 we considered the **chemiosmotic mechanism** for ATP formation in the mitochondrion. The chemiosmotic mechanism also operates in photophosphorylation. In chloroplasts, as in mitochondria, electrons move through a series of redox reactions releasing energy that is used to transport protons (H^+) across a membrane. This active proton transport results in a *proton-motive force*—a difference in pH and in electric charge across the membrane.

In the mitochondrion, protons are pumped from the matrix, across the internal membrane, and into the space between the inner and outer mitochondrial membranes (see Figure 7.14). In the chloroplast, the electron carriers are located in the thylakoid membranes (see Figure 4.13). The electron carriers are oriented so that protons move into the interior of the thylakoid, and the inside becomes acidic with respect to the outside. This difference in pH leads to the passive movement of protons back out of the thylakoid, through specific protein channels in the membrane. These proteins are the enzymes, ATP synthases, that couple the formation of ATP to the diffusion of protons back across the membrane, just as in mitochondria (Figure 8.14).

The hypothesis that this chemiosmotic mechanism is responsible for the formation of ATP in chloroplasts was tested by Andre Jagendorf (of Cornell University) and Ernest Uribe (now at Washington State University) in the following way. Chloroplast thylakoids were isolated from spinach leaves and then kept in the dark, so that there would be no light energy to drive the production of ATP (Figure 8.15). The thylakoids were moved from a neutral solution to one with a low pH so that by diffusion the interior of the thylakoids would become acidic. Then they were transferred to a solution that had a higher pH, so that the interiors of the thylakoids would be more acidic than the outsides—mimicking the situation created by light-driven pumping of protons into the interiors.

This final step immediately resulted in the formation of ATP, even though no light was available to serve as the energy source—precisely the result predicted by the chemiosmotic model. (Recall that a very similar experiment, using mitochondria, pH changes, and ATP formation, was described in Chapter 7.)

Photosynthetic pathways are the products of evolution

Photosystem I evolved before photosystem II; thus, cyclic photophosphorylation evolved before noncyclic photophosphorylation. Early in evolutionary history, photosynthetic bacteria used photosystem I and cyclic photophosphorylation to make ATP. This form of photosynthesis evolved long before Earth's atmosphere contained significant quantities of oxygen gas.

Nearly 3 billion years ago the cyanobacteria evolved photosystem II, thus gaining the ability to perform noncyclic photophosphorylation—and to extract electrons from water and use them to reduce $NADP^+$ while producing O_2. Over hundreds of millions of years, noncyclic photophosphorylation by cyanobacteria poured enough oxygen gas into the atmosphere to make possible the evolution of cellular respiration.

Today the evolutionarily ancient photosynthetic bacteria still have only cyclic photophosphorylation. Cyanobacteria, algae, and plants, which perform mostly noncyclic photophosphorylation, still perform cyclic photophosphorylation to produce ATP when their ratio of $NADPH + H^+$ to $NADP^+$ is high.

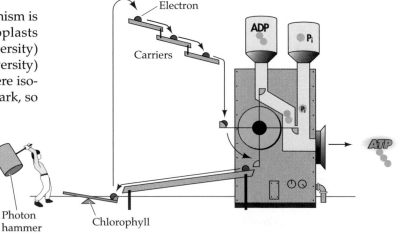

8.13 Cyclic Photophosphorylation Cycles Electrons A mythical "ATP machine" illustrates the concept of cyclic photophosphorylation.

Making Sugar from CO_2: The Calvin–Benson Cycle

The second main pathway of photosynthesis is the **Calvin–Benson cycle** of reactions, which incorporates CO_2 into sugars. In the chloroplast, most of the enzymes that catalyze the reactions of this pathway are dissolved in the stroma (see Figure 4.14), and this is where the reactions take place. These reactions are sometimes called the *dark reactions* because they do not directly require light energy. However, they require it indirectly, and *they take place only in the light*.

In this section, we will identify the methods and examine the experimental re-

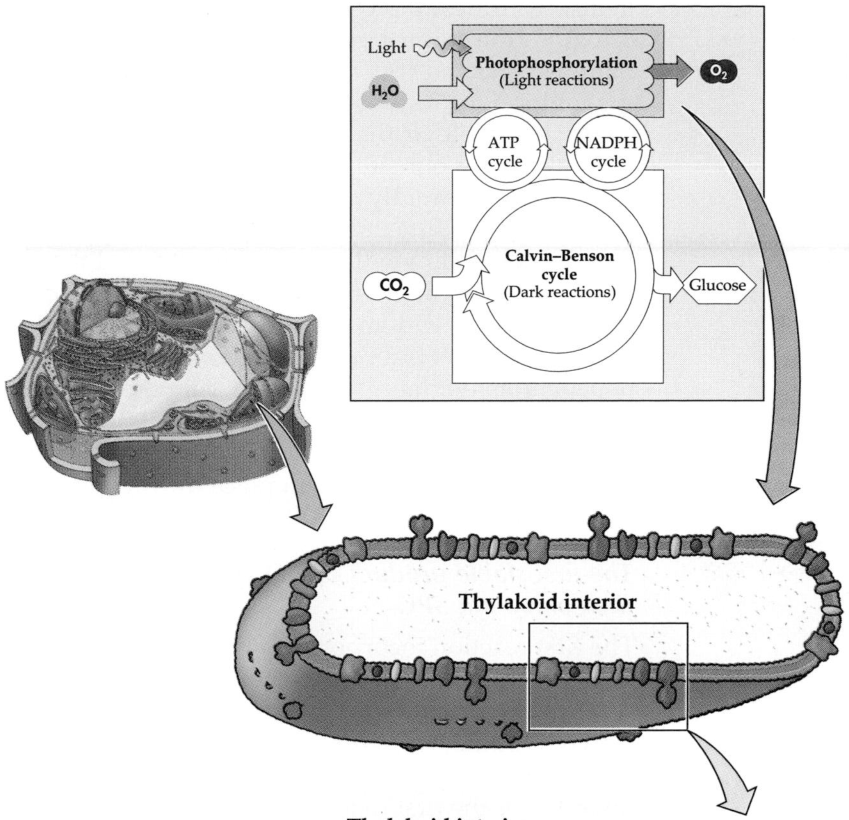

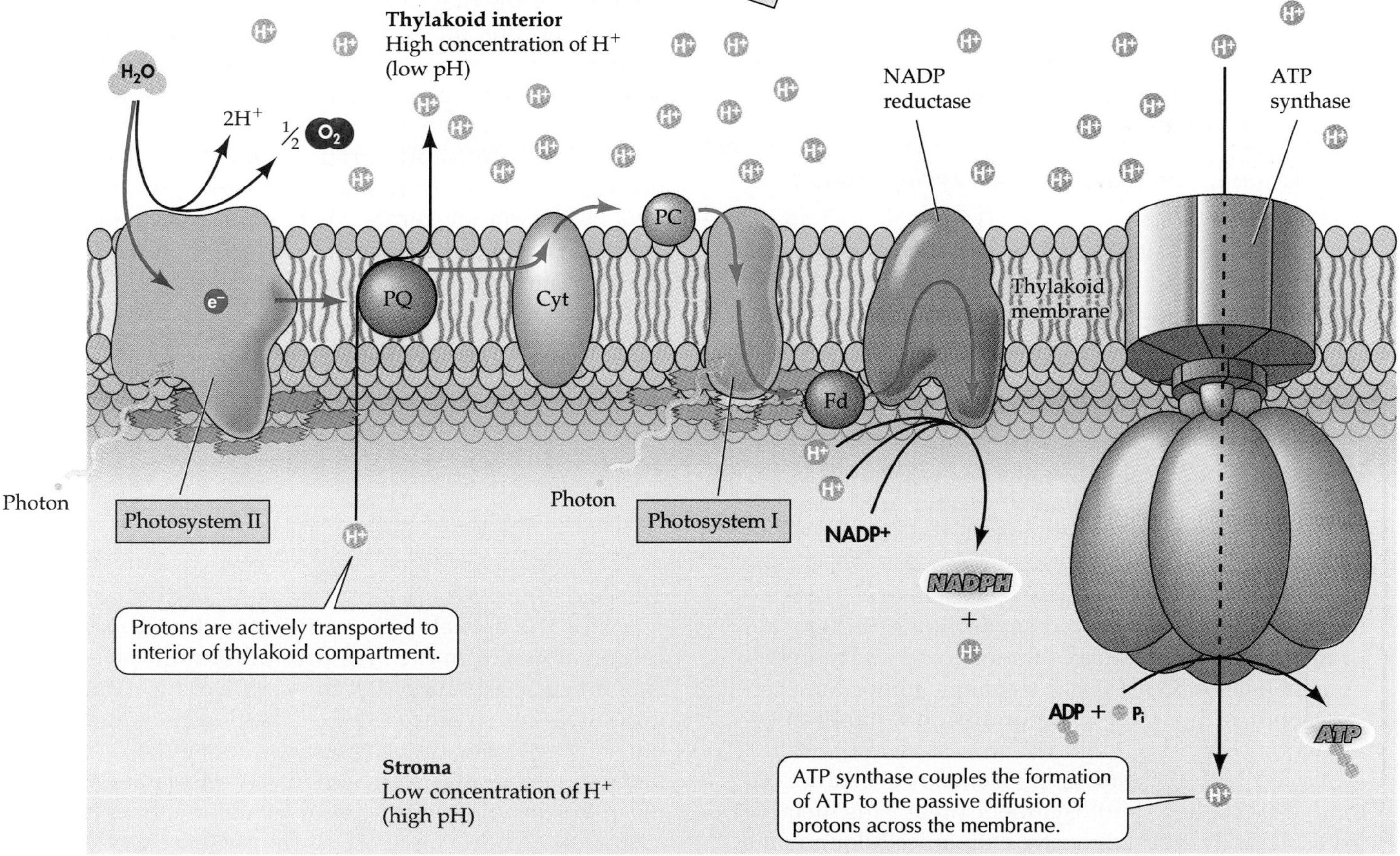

8.14 Chloroplasts Form ATP Chemiosmotically   Protons (H^+) pumped across the thylakoid membrane from the stroma during photophosphorylation make the interior of the thylakoid more acidic than the stroma. Driven by this pH difference, the protons then diffuse to the stroma through ATP synthase channels, which couple the energy of proton flow to the formation of ATP from ADP + P_i.

Light

Photophosphorylation
(Light reactions)

O_2

H_2O

ATP cycle

NADPH cycle

Calvin–Benson cycle
(Dark reactions)

CO_2

Glucose

Thylakoid interior

Thylakoid interior
High concentration of H^+
(low pH)

NADP reductase

ATP synthase

H_2O

$2H^+$ ½ O_2

PC

Thylakoid membrane

e^-

PQ   Cyt

Fd

Photon

Photosystem II

Photon

Photosystem I

$NADP^+$

NADPH
+
H^+

Protons are actively transported to interior of thylakoid compartment.

ADP + P_i

ATP

Stroma
Low concentration of H^+
(high pH)

ATP synthase couples the formation of ATP to the passive diffusion of protons across the membrane.

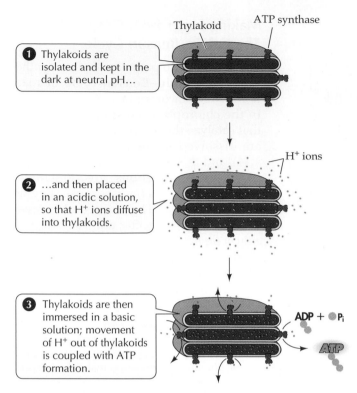

8.15 An Experiment Shows ATP Formation in the Dark
This experiment (described in the text) artificially induced ATP formation even in the absence of light energy, confirming the prediction of the chemiosmotic mechanism.

sults and critical thinking that were used to discover the sequence of reactions by which the gas CO_2 is incorporated into sugars.

Elucidation of the Calvin–Benson cycle required radioactive carbon

Real progress in understanding the dark reactions came only after World War II, when radioactive isotopes became available. A group of scientists at the University of California, Berkeley, led by Melvin Calvin and including Andrew Benson and James Bassham, wanted to identify the biochemical steps between the uptake of CO_2 and the appearance of the first complex carbohydrates in the chloroplast.

Accomplishing this goal depended on three advances in experimental methods: (1) the discovery and availability of a radioactive carbon isotope, ^{14}C; (2) the development of *paper partition chromatography*, a technique permitting the rapid separation of individual compounds from complex solutions; and (3) the development of *autoradiography*, a technique for locating colorless but radioactive compounds on a paper chromatogram.

Armed with these tools, the Berkeley group set out to investigate how photosynthetic organisms metabolize CO_2. They worked mostly with unicellular aquatic algae, such as the green alga *Chlorella*. Algae were grown in dense suspensions in a flattened flask (called a "lollipop" because of its shape) between two bright lights, ensuring a rapid rate of photosynthesis (Figure 8.16a). At the start of an experiment, a solution containing dissolved $^{14}CO_2$ was suddenly squirted into the lollipop. At a carefully measured time after this squirt, a sample of the culture was rapidly drained into a container of boiling ethanol.

The ethanol rapidly penetrated the algal cell walls, performing two functions: It killed the algae, stopping photosynthesis immediately, and it extracted the ^{14}C-containing intermediates of $^{14}CO_2$ metabolism from the algae (along with many other compounds). A drop of this ethanol extract was placed on filter paper for paper chromatography followed by autoradiography (Figure 8.16b). Many spots corresponding to radioactive substances appeared, indicating that many biochemical reactions had taken place during that short interval.

The first stable product of CO_2 fixation is the compound 3PG

The first reaction incorporating carbon dioxide into a larger compound is called **carbon fixation**. During 30 seconds of continuous exposure to $^{14}CO_2$, many different compounds were formed in the cells in Calvin's lollipop (see Figure 8.16b). But which of these compounds was the first to form?

To determine the carbon fixation reaction, the experiment was repeated several times, using ever-shorter exposures. Even after exposures of less than 2 seconds, half a dozen or more labeled compounds appeared in the autoradiographs. One, however, was produced most rapidly and in greatest abundance: **3PG** (3-phosphoglycerate, also called 3-phosphoglyceric acid), which we have already encountered as an intermediate in glycolysis (see Figure 7.6).

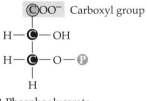

3-Phosphoglycerate

The compound 3PG is the first stable product of CO_2 fixation. The Berkeley group isolated the individual carbon atoms from the 3PG produced in the lollipop experiments and found that the carbon of the carboxyl group was much more intensely radioactive than the other two carbon atoms. (A single atom either is or is not radioactive. By "more intensely radioactive" we mean that in a population of molecules, the fraction of carboxyl carbons labeled with radioactivity was

(a)

The thin flask was filled with a suspension of algae and illuminated from both sides. After injection of $^{14}CO_2$, a sample was drained from the flask into boiling ethanol to kill the cells rapidly, stop enzymatic action, and extract molecules.

(b)

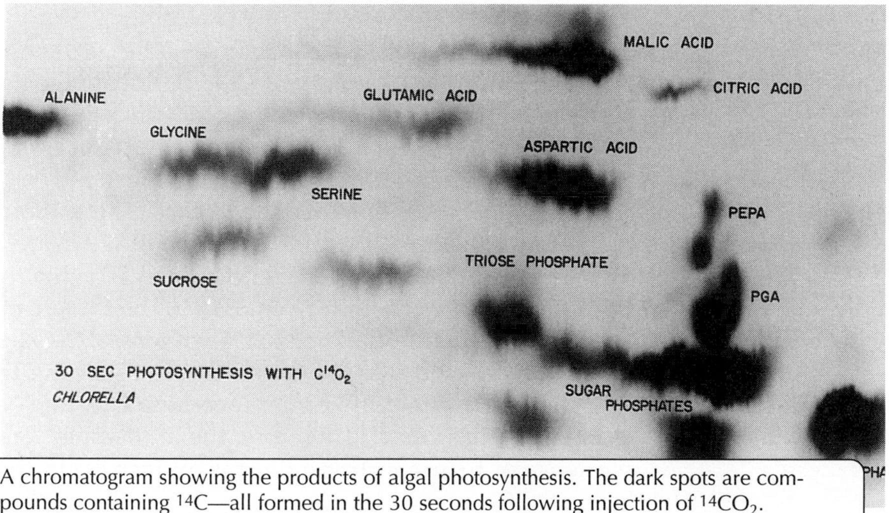

MALIC ACID

ALANINE
GLUTAMIC ACID   CITRIC ACID
GLYCINE
ASPARTIC ACID
SERINE
PEPA
SUCROSE   TRIOSE PHOSPHATE
PGA

30 SEC PHOTOSYNTHESIS WITH C$^{14}O_2$
CHLORELLA   SUGAR
PHOSPHATES

A chromatogram showing the products of algal photosynthesis. The dark spots are compounds containing ^{14}C—all formed in the 30 seconds following injection of $^{14}CO_2$.

8.16 A Lollipop and Its Products   *(a)* The "lollipop" apparatus used in early experiments on photosynthesis. *(b)* In this reproduction of an original chromatogram, the spot labeled PGA (an older notation) corresponds to the position of 3PG.

greater than the fractions of radioactive carbons in the other two carbon positions.)

The Berkeley group drew two important conclusions from this discovery. First, the heavy labeling showed that the carboxyl carbon is obtained directly from CO_2. Second, because radioactive carbon appears in the other carbon atoms of 3PG, they concluded that some kind of *cyclical* process is involved, a process by

which 3PG is made by adding CO_2 to another compound that is itself produced from photosynthetic 3PG.

The CO_2 acceptor is the compound RuBP

What is the compound, obtained from the metabolism of 3PG, that forms a covalent bond with CO_2 to make more 3PG? Given the structure of 3PG, it seemed reasonable to expect that the mysterious CO_2 acceptor would be a *two*-carbon compound that could react with CO_2 to become a *three*-carbon compound—3PG—with the CO_2 becoming a carboxyl group.

If this idea were correct, it should have been possible for the Berkeley group to find, on their chromatograms, a compound with only two carbon atoms, both of which were radioactive after a lollipop experiment. However, they did *not* find such a compound; thus the problem became more difficult. The "obvious" answer—that the CO_2 acceptor is a two-carbon compound—was wrong. Where would they go from here?

BASING A MODEL ON A CYCLE OF SUGARS. At this point in the investigation, concentrating on a photosynthetic *cycle* became useful. Consider a tentative cycle of the sort shown in Figure 8.17*a*: CO_2 reacts with X, the CO_2 acceptor, to produce 3PG. From the 3PG, photosynthetic organisms make products (things like glucose) and more X; this new X can react with another molecule of CO_2 and keep the cycle going. But what is X, and what are the other intermediates in the cycle?

It was observed that 3PG was the only acid phosphate produced in significant amount, whereas many kinds of *sugar* phosphates appeared on the chromatograms. On this basis, the Berkeley group guessed that the first thing to happen to 3PG is its conversion to a three-carbon sugar phosphate (glyceraldehyde 3-phosphate, which we will call G3P). Such a reaction (acid to aldehyde) is a reduction, and since reductions are highly endergonic, the Berkeley group proposed that this reaction would require ATP (Figure 8.17*b*).

They supported this proposal with two lollipop experiments, one in the light and one in the dark. When

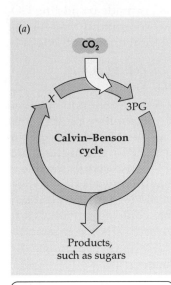

(a)

Products, such as sugars

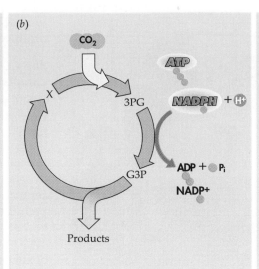

(b)

Products

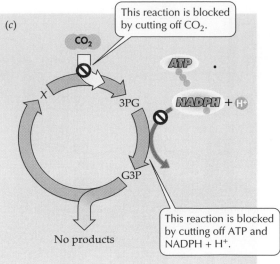

(c)

This reaction is blocked by cutting off CO_2.

No products

This reaction is blocked by cutting off ATP and NADPH + H^+.

After it became apparent that CO_2 combines with some molecule to form 3PG and that a cyclical process is involved, a pathway could be devised in which a molecule of compound X combines with CO_2 to form 3PG and in which further reactions regenerate X.

Ensuing speculation and experimentation led to the proposition that 3PG was reduced to a sugar phosphate, glyceraldehyde 3-phosphate (G3P); this endergonic reaction, a reduction, would require ATP, as shown here.

In the proposed model, a cutoff of CO_2 blocks the formation of 3PG from compound X and CO_2; a cutoff of ATP (by turning off the light) blocks the formation of G3P from 3PG.

8.17 Manipulating the Reactions of Photosynthesis
The Berkeley group manipulated the photosynthetic cycle in their laboratory by two simple means. First, turning off the light blocked the reaction 3PG → G3P. Second, the reaction X → 3PG was blocked by cutting off CO_2.

they supplied $^{14}CO_2$ to the algae in the lollipop, 3PG rapidly became radioactively labeled in both experiments, but G3P became radioactively labeled only in the light, showing that the reduction does require energy—presumably in the form of ATP generated in the light.

TWEAKING THE LIGHT AND CO_2 SUPPLIES. At this point an extremely clever suggestion was made: If we assume that Figure 8.17*b* accurately models what occurs in the chloroplast, then it should be possible to regulate this cycle in the laboratory by two simple means. First, turning off the light would specifically block the step from the acid, 3PG, to the aldehyde, G3P, because the necessary ATP and NADPH + H^+ can be produced only with an input of energy, and in photosynthesis the energy source is light. Therefore ATP should be made in the chloroplast only when the light is on. Second, the reaction from X to 3PG could easily be blocked by cutting off the supply of CO_2 to the lollipop (Figure 8.17*c*).

Assume that photosynthesis is proceeding at a steady pace, so the concentrations of 3PG and the CO_2

acceptor X in the cells are constant. The lights are on, of course, and there is plenty of CO_2. Suddenly the investigator turns off the lights, thus blocking the cycle as proposed in Figure 8.17*c*: How does the concentration of 3PG change immediately? How does the concentration of X change immediately? *Stop.* Think about this before you read on.

When the light is turned off, no more ATP is made. Without ATP, the reaction from 3PG to the three-carbon sugar G3P cannot take place. Therefore, 3PG is no longer being used up. However, there is nothing to stop the formation of 3PG from incoming CO_2 and X, as long as X is around. Therefore, the immediate consequence of turning off the light is an *increase* in the level of 3PG.

On the other hand, X continues to be used up, because its reaction with CO_2 does not depend on light or ATP, but the formation of X slows down because 3PG can no longer be reduced and ultimately form new X. Therefore, the concentration of X *decreases* (Figure 8.18). These are the changes we would expect to see *if* our model (see Figure 8.17*c*) were correct. Similar reasoning should convince you that when the CO_2 supply is cut off (with the lights on), the concentration of X rises and that of 3PG falls. With no CO_2 available, X is no longer used up and 3PG is no longer formed, but 3PG can still be reduced to G3P.

Having devised this model, the Berkeley group proceeded to study the effects of changes in light intensity and CO_2 supply on the concentrations of all the major radioactive compounds found in the lollipop experi-

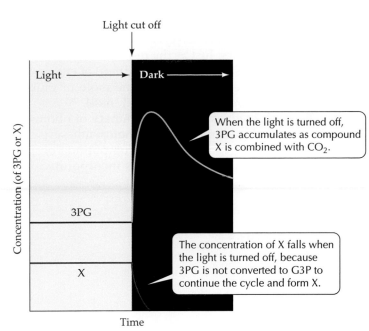

Light cut off

Light ⟶ Dark ⟶

When the light is turned off, 3PG accumulates as compound X is combined with CO_2.

3PG

The concentration of X falls when the light is turned off, because 3PG is not converted to G3P to continue the cycle and form X.

X

Time

8.18 Dark-Induced Changes Predicted by the Model
Light is necessary to provide the ATP for the endergonic reaction 3PG → G3P.

ments. The first thing they noticed was that only one compound exhibited the concentration changes proposed for 3PG—and that was 3PG itself. This result showed that their model was likely to be accurate.

But would any compound behave in the way predicted for the mysterious compound X, the CO_2 acceptor? Yes—and only one: a five-carbon sugar phosphate called **RuBP** (ribulose bisphosphate). It seemed, then, that instead of the originally proposed reaction (in which CO_2 was thought to react with a two-carbon sugar to form 3PG), there must be a reaction in which CO_2, with its single carbon, combines with the five-carbon RuBP to give two molecules of the three-carbon 3PG (Figure 8.19).

FINDING THE ENZYME THAT CATALYZES THE CAPTURE OF CO_2. The best way to prove that a proposed reaction takes place is to find an *enzyme* that catalyzes it. In this case, such an enzyme was soon discovered, and from more than one source. The enzyme, RuBP carboxylase, now commonly called **rubisco**, was found in the algae studied by the Berkeley group and also in the leaves of spinach. Of most importance, studies of spinach revealed that rubisco is found in only one part of the cell: the chloroplast—exactly where one should find an enzyme concerned with photosynthesis. The investigators concluded, then, that RuBP is the CO_2 acceptor, the previously unknown compound X.

Identifying intermediate reactions of the Calvin–Benson cycle

Having discovered the first product of CO_2 fixation (3PG) and the CO_2 acceptor (RuBP), the Berkeley group proceeded to work out the remaining reactions of the cycle. They found some relatively complicated steps between G3P and RuBP; among the intermediates are sugar phosphates with four, five, six, and seven carbon atoms. All the proposed intermediates have been found in chloroplasts, as have all the necessary enzymes.

The group also discovered that ATP is needed for an additional reaction in the Calvin–Benson cycle: the reaction producing RuBP (ribulose *bis*phosphate) from RuMP (ribulose *mono*phosphate). This additional ATP requirement makes it even easier to understand why turning off the light drastically reduces the concentration of RuBP (X in Figure 8.18).

Figure 8.20 summarizes the key features of the Calvin–Benson cycle. RuBP reacts with CO_2 to form 3PG. Then 3PG is reduced—in a reaction requiring ATP, as well as hydrogens provided by NADPH + H^+—to a three-carbon sugar phosphate (G3P). What follows this step is a complex sequence of reactions with two principal outcomes: the formation of more RuBP, and the release of products such as glucose. The

CO_2 + → Rubisco →

Carbon dioxide

The fate of the carbon atom in CO_2 is followed in red.

CH₂O P
|
C = O
|
H — C — OH
|
H — C — OH
|
CH₂O P

Ribulose bisphosphate (RuBP)

The enzyme rubisco catalyzes the reaction of CO_2 with RuBP.

Six-carbon skeleton of reaction intermediate

CH₂O P   COOH
| |
HO — C — H + H — C — OH
| |
COOH   CH₂O P

The reaction intermediate splits into two molecules of 3-phosphoglyceric acid (3PG).

8.19 RuBP Is the Carbon Dioxide Acceptor Ribulose bisphosphate (RuBP) is the CO_2-accepting compound X in Figures 8.17 and 8.18.

production of one molecule of glucose ($C_6H_{12}O_6$) requires the Calvin–Benson cycle to operate six times on six CO_2 molecules. Just as the respiration of one mole of glucose *yields* 686 kcal (2,867 kJ/mol) of energy (see Chapter 7), the production of one mole of glucose from CO_2 *requires* 686 kcal (2,867 kJ/mol).

Figure 8.21 gives a general summary of photosynthesis. The glucose produced in photosynthesis is subsequently used to make other compounds besides sugars. The carbon of glucose is incorporated into

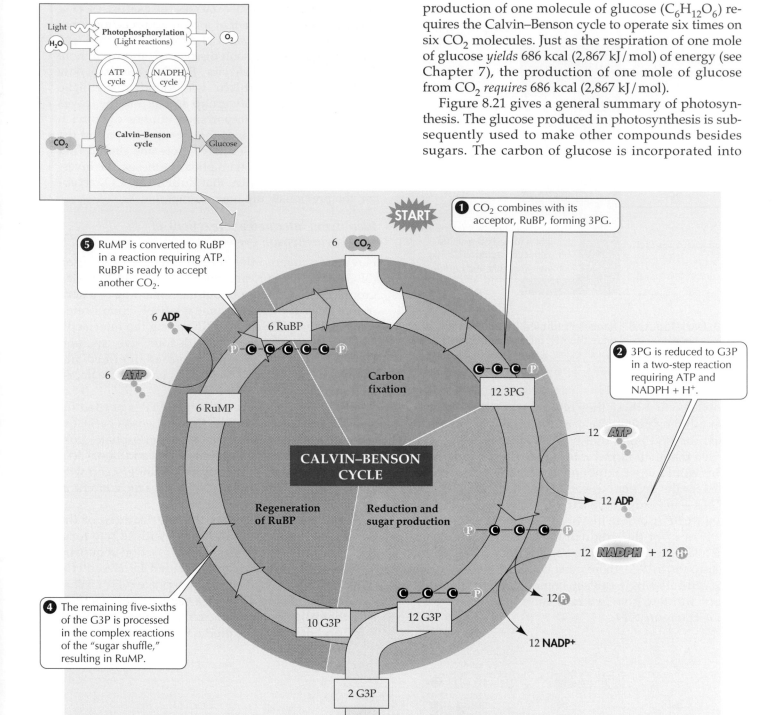

8.20 The Calvin–Benson Cycle   The Calvin–Benson cycle uses CO_2 and the ATP and NADPH + H$^+$ generated in the light reactions to produce glucose. This diagram shows only the key steps; the values given are those necessary to make one molecule of glucose, which requires six "turns" of the cycle.

amino acids, lipids, and the building blocks of the nucleic acids.

The products of the Calvin–Benson cycle are of crucial importance to the entire biosphere, for they serve as the food for all of life. Their covalent bonds represent the total energy yield from the harvesting of light by plants. Most of this stored energy is released by the plants themselves in their own glycolysis and cellular respiration. However, much plant matter ends up being consumed by animals. Glycolysis and cellular respiration in the animals then releases free energy from the plant matter for use in the animal cells.

Photorespiration and Its Evolutionary Consequences

The enzyme rubisco is the most abundant protein in the biosphere. Its properties are remarkably identical in all photosynthetic organisms, from bacteria to flowering plants. Its operation is the basis for the life of nearly

8.21 An Overview of the Photosynthetic Reactions

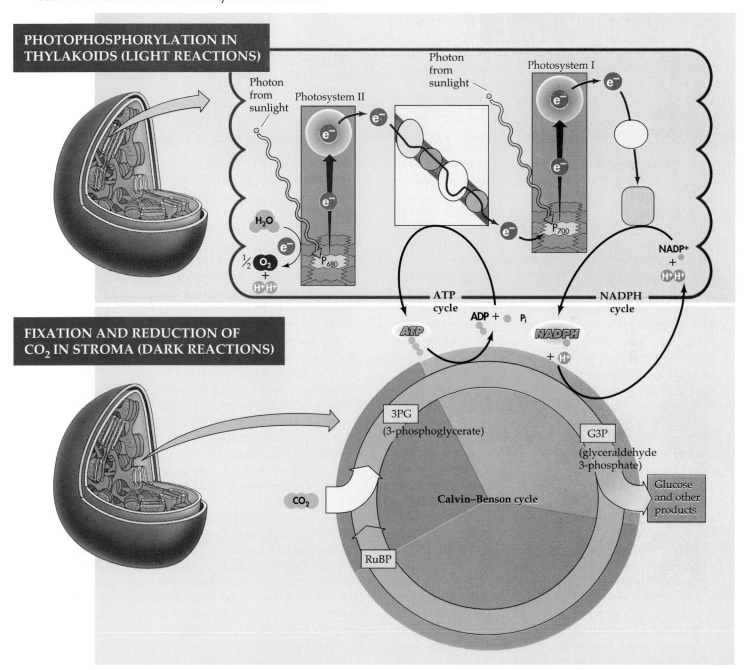

all heterotrophic organisms. However, some properties of this enzyme severely limit its effectiveness.

In the discussion that follows, we will identify and explore some of these limitations and see how evolution has constructed bypasses around them. First we'll look at photorespiration, and then we'll examine alternative pathways and plant anatomy that compensate for the limitations of rubisco and photorespiration.

In photorespiration, RuBP reacts with O_2

The substrate specificity of the enzyme rubisco is not limited to CO_2. Rubisco also functions as an oxidase, catalyzing the reaction of RuBP with O_2. (Indeed, the name *rubisco* stands for *ribulose bis*phosphate *carboxylase/oxygenase*.) Because this reaction requires light and takes up oxygen, it has been termed **photorespiration**.

The oxygenase function of rubisco is favored at higher temperatures (above 28°C) when CO_2 levels are low or O_2 levels are high. This temperature sensitivity of CO_2 fixation by rubisco will be important later in understanding how photosynthesis is accomplished under hot, dry conditions.

When RuBP and O_2 react, one of the products is glycolate, a two-carbon compound that leaves the chloroplasts and diffuses into membrane-enclosed organelles called *peroxisomes* (Figure 8.22). In peroxisomes, glycolate is oxidized by O_2 in a pathway whose product diffuses into mitochondria, where it is acted on and releases CO_2—undoing the carbon-fixing work of the Calvin–Benson cycle. Chloroplasts, peroxisomes, and mitochondria all play parts in photorespiration.

Photorespiration interferes with photosynthesis. In fact, it apparently reverses it—but without resulting in ATP formation as does cellular respiration. One estimate suggests that photorespiration reduces the rate of photosynthesis by 25 percent. This is a very significant detriment to plant growth and, put in human terms, crop productivity.

The role of photorespiration in the life of the plant is unknown. Perhaps it has no positive role and is merely wasteful. Perhaps it is an evolutionary "leftover" from the days when O_2 was still building up in Earth's atmosphere and hence was less concentrated. With this in mind, many scientists are attempting to develop a gene that codes for a form of rubisco that recognizes only CO_2 as its substrate, and to insert that gene into crop plants. (See Chapter 16 for a discussion of this recombinant-DNA technology.)

It seems odd that rubisco, so abundant in the living world, apparently functions less than optimally. Most types of plants photorespire away a substantial fraction of the CO_2 initially fixed in photosynthesis. But, as we are about to see, some plants have minimized photorespiration, thus maximizing the efficiency of their photosynthesis.

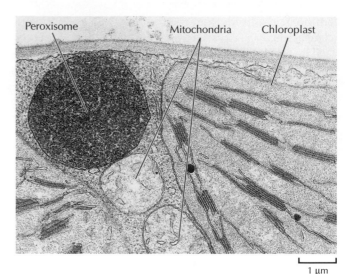

1 µm

8.22 Photorespiration: Chloroplasts, Peroxisomes, and Mitochondria Within the peroxisome, glycolate from chloroplasts is oxidized by O_2. The product diffuses into mitochondria, where it is acted on and releases CO_2.

Some plants have evolved systems to bypass photorespiration

The discoveries of the Berkeley group who elucidated the Calvin–Benson cycle led to the expectation that the exposure of a plant to both light and $^{14}CO_2$ would always lead to the appearance of the three-carbon compound 3-phospho[^{14}C]glycerate (3PG) as the first labeled product of CO_2 fixation. Thus scientists were surprised when they learned that in many plants, the first products of CO_2 fixation are four-carbon compounds. They named these plants C_4 **plants** in contrast to the C_3 plants, which make the three-carbon 3PG as their first products.

C_4 plants perform the normal Calvin–Benson cycle, but they have an additional early reaction that fixes CO_2 without losing carbon to photorespiration, greatly increasing the overall photosynthetic yield. Because this initial CO_2 fixation step functions even at low levels of CO_2 and high temperatures, C_4 plants very effectively optimize photosynthesis under conditions that inhibit the photosynthesis of C_3 plants.

C_4 plants live in dry environments exposed to high temperatures during daylight hours. Under these conditions, C_4 plants such as corn, sugar cane, and crabgrass maintain high rates of photosynthesis and growth, even though their stomata must close at times to limit water loss. In C_3 plants, which use only the Calvin–Benson cycle, any daytime closure of the stomata to limit water loss also limits the access of cells to CO_2 from the air, thus limiting their ability to photosynthesize. Why can C_4 plants tolerate a reduced CO_2 supply better than C_3 plants?

The leaves of C_4 plants contain the enzyme **PEP carboxylase** (phosphoenolpyruvate carboxylase), which catalyzes the reaction of PEP (phosphoenolpyruvate, a

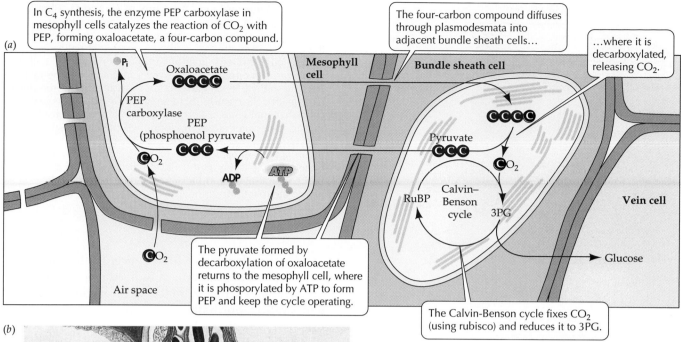

(a)

In C_4 synthesis, the enzyme PEP carboxylase in mesophyll cells catalyzes the reaction of CO_2 with PEP, forming oxaloacetate, a four-carbon compound.

The four-carbon compound diffuses through plasmodesmata into adjacent bundle sheath cells...

...where it is decarboxylated, releasing CO_2.

Oxaloacetate

Mesophyll cell

Bundle sheath cell

P_i

PEP carboxylase

PEP (phosphoenol pyruvate)

Pyruvate

CO_2

ADP

ATP

CO_2

RuBP

Calvin–Benson cycle

3PG

Vein cell

Glucose

The pyruvate formed by decarboxylation of oxaloacetate returns to the mesophyll cell, where it is phosporylated by ATP to form PEP and keep the cycle operating.

Air space

The Calvin-Benson cycle fixes CO_2 (using rubisco) and reduces it to 3PG.

(b)

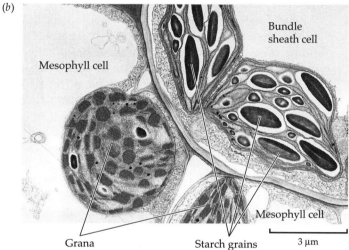

Bundle sheath cell

Mesophyll cell

Mesophyll cell

Grana

Starch grains

3 μm

8.23 C_4 Photosynthesis (*a*) The biochemistry of C_4 photosynthesis allows these plants to fix carbon and optimizes their fitness in hot, dry conditions. (*b*) In this electron micrograph, the chloroplasts in the mesophyll cells (left) have numerous grana and few starch grains. In contrast, the chloroplasts in the bundle sheath cell (right) have abundant starch grains, indicating that the enzymes of the Calvin–Benson cycle are actively forming 3PG, from which glucose and then startch are formed.

three-carbon acid) with CO_2 to yield the four-carbon compound *oxaloacetate* as the first product of CO_2 fixation (Figure 8.23). PEP carboxylase has a greater affinity for CO_2 than does rubisco, allowing C_4 plants to trap CO_2 more effectively than C_3 plants when stomatal closure depletes the CO_2 supply. And, because it lacks an oxygenase function, PEP carboxylase does not support photorespiration.

In C_4 plants, the two processes—initial CO_2 fixation and the Calvin–Benson cycle—take place in different cells. For the Calvin–Benson cycle to operate, oxaloacetate or another four-carbon compound derived from it diffuses from mesophyll cells to bundle sheath cells (see Figure 8.23).

The leaf anatomy of C_4 plants differs from that of C_3 plants

C_3 plants have only one type of cell capable of photosynthesis (Figure 8.24*a*). In contrast, the leaves of C_4 plants have two classes of photosynthetic cells, each with a distinctive type of chloroplast. The cells are arranged as shown in Figure 8.24*b*: a photosynthetic **mesophyll** layer surrounds an inner layer of photosynthetic **bundle sheath cells**. In C_4 plants, the four-carbon compound that is the product of initial CO_2 fixation diffuses out of the mesophyll cells and into the bundle sheath cells (see Figure 8.23).

In the bundle sheath cells, the four-carbon compound loses a carboxyl group to release CO_2, which is recaptured by rubisco and used in the Calvin–Benson cycle—the cycle that in C_3 plants takes place entirely within the mesophyll cells (see Figure 8.23). The chloroplasts in the bundle sheath cells of C_4 plants lack well-developed grana but typically have numerous, large starch grains, indicating that they (rather than the mesophyll cells) are the sites where sugars and starches form.

To summarize, in C_4 plants, the mesophyll cells pump CO_2 from a region where its concentration is low (the intercellular spaces within the leaf) to the

(a) **Arrangement of cells in a C₃ leaf**

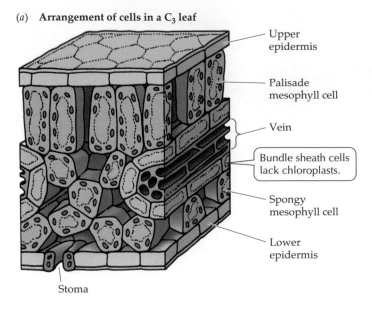

Upper epidermis

Palisade mesophyll cell

Vein

Bundle sheath cells lack chloroplasts.

Spongy mesophyll cell

Lower epidermis

Stoma

(b) **Arrangement of cells in a C₄ leaf**

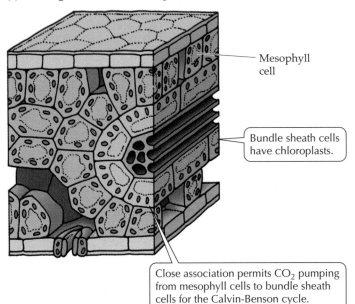

Mesophyll cell

Bundle sheath cells have chloroplasts.

Close association permits CO_2 pumping from mesophyll cells to bundle sheath cells for the Calvin-Benson cycle.

8.24 The Different Leaf Anatomy of C₃ and C₄ Plants

(a) In the leaf of a C₃ plant, the bundle sheath cells surrounding the vascular elements of a vein are relatively small; the upper part of the leaf is filled with upright palisade mesophyll cells; and loosely arranged spongy mesophyll cells allow gases to circulate in the lower layers within the leaf. Both mesophyll layers carry on photosynthesis, but the bundle sheath cells do not. (b) In a C₄ leaf, the bundle sheath cells are usually larger and contain prominent chloroplasts toward their outer edges; uniform mesophyll cells surround the entire vascular bundle. This arrangement facilitates the incorporation of carbon from CO_2 into four-carbon compounds by the mesophyll cells and the passage of these carbon-containing compounds to the bundle sheath cells, where the reactions of the Calvin–Benson cycle take place.

bundle sheath cells, so their concentration of CO_2 is high enough to maintain photosynthesis even on the hot, dry days when stomata are often closed and the temperature normally would favor the oxygenase function of rubisco. This system is effective because PEP carboxylase in the mesophyll cells can fix CO_2 at temperatures too high for rubisco to fix it effectively, and because PEP carboxylase, unlike rubisco, does not function as an oxygenase and hence its action does not lead to wasteful photorespiration. Table 8.1 compares C₃ and C₄ photosynthesis.

A related but distinguishable system called crassulacean acid metabolism (CAM) functions in certain plants that face frequent water shortages.

CAM plants also use PEP carboxylase

Some plants use PEP carboxylase to fix and accumulate CO_2 while conserving water during hot, dry daylight hours—including some succulents (water-storing plants of some families, such as the Crassulaceae), many cacti, pineapples, and several other kinds of flowering plants. These plants conserve water by keeping their stomata closed during the daylight hours, thus minimizing water loss by evaporation. How, then, can they perform photosynthesis? Their

TABLE 8.1   Comparison of Photosynthesis in C₃ and C₄ Plants

VARIABLE	C₃ PLANTS	C₄ PLANTS
Photorespiration	Extensive	Minimal
Perform Calvin–Benson cycle?	Yes	Yes
Primary CO_2 acceptor	RuBP	PEP
CO_2-fixing enzyme	Rubisco (RuBP carboxylase/oxygenase)	PEP carboxylase
First product of CO_2 fixation	3PG (3-carbon compound)	Oxaloacetate (4-carbon compound)
Affinity of carboxylase for CO_2	Moderate	High
Leaf anatomy: photosynthetic cells	Mesophyll	Mesophyll + bundle sheath
Classes of chloroplasts	One	Two

trick is to open their stomata at night and store CO_2 by a different mechanism.

The CO_2 metabolism of these plants is called **crassulacean acid metabolism**, or **CAM**, after the succulents in which it was first discovered. CAM is much like the metabolism of C_4 plants: CO_2 is initially fixed into four-carbon compounds that accumulate. In CAM plants, however, the two processes (initial CO_2 fixation and the Calvin–Benson cycle) are separated temporally (in *time*) rather than spatially (in *space*—in separate cells; Figure 8.25). And CAM plants lack the specialized cell relationships of C_4 plants.

In CAM plants, CO_2 is fixed initially in mesophyll cells during the *night*, when stomata are open and water loss is less of a problem. The products of CO_2 fixation accumulate in the vacuoles of mesophyll cells. When daylight arrives, the accumulated four-carbon compounds are shipped to the chloroplasts, where decarboxylation supplies the CO_2 for operation of the Calvin–Benson cycle, and the light reactions supply the necessary ATP and NADPH + H^+. We will discuss the stomatal behavior of CAM plants more thoroughly in Chapter 32.

Plants Perform Cellular Respiration

In plants, cellular respiration takes place both in the light and in the dark, whereas photosynthesis takes place strictly in the light. The site of glycolysis is the cytosol, that of respiration is the mitochondria, and that of photosynthesis is the chloroplasts. Thus photosynthesis and respiration can proceed simultaneously but in different organelles.

For a plant to live, it must photosynthesize more than it respires, giving it a net gain of carbon dioxide and energy from the environment. The excess of photosynthesis over respiration is great enough that plants—as well as photosynthetic bacteria and photosynthetic protists—can export food—and oxygen—to the animal kingdom and to all other nonphotosynthetic organisms.* Animals require both food and oxygen; they return carbon dioxide that plants may use in photosynthesis. Thus both carbon dioxide and oxygen have natural cycles.

Photosynthesis and respiration have important similarities. In eukaryotes, both processes reside in specialized organelles that have complex systems of inter-

*The exceptions are a few forms of archaea and bacteria that oxidize minerals deep in caves or on the ocean floor where light does not penetrate, and they support animal life in these isolated habitats.

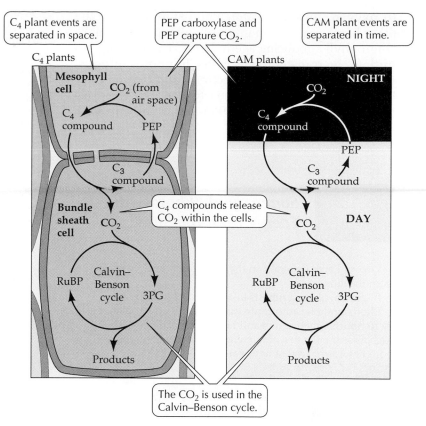

Sorghum stalks (C_4)

Sempervivum tectorum (CAM)

8.25 CAM and C_4 Plants Separate Two Sets of Reactions Differently Both plant types utilize four-carbon compounds whose production is separate from the Calvin–Benson cycle. The separation is spatial in C_4 plants, temporal in CAM plants.

nal membranes. ATP synthesis in both processes relies on the chemiosmotic mechanism, involving the pumping of protons across a membrane. Another key feature of both respiration and photosynthesis is electron transport, that is, the passing of electrons from carrier to carrier in a series of exergonic redox reactions.

In respiration, the carriers receive electrons from high-energy food molecules and pass them ultimately to oxygen, forming water. Photosynthesis, on the other hand, requires an input of light energy to make chlorophyll a reducing agent that is strong enough to initiate the transfer of electrons. In photosynthesis, water is the source of the electrons, and oxygen is released from water in a very early step. The electrons from water end up in NADPH and, finally, in food molecules.

Summary of "Photosynthesis: Energy from the Sun"

- Life on Earth depends on the absorption of light energy from the sun.
- In plants, photosynthesis takes place in chloroplasts.
- Photosynthetic autotrophs such as plants trap solar energy to nourish themselves and heterotrophs. Heterotrophs such as animals must obtain energy from autotrophs, or from heterotrophs that have eaten autotrophs.

Identifying Photosynthetic Reactants and Products

- Photosynthesizing plants take in CO_2, water, and light energy, producing O_2 and energy-rich, reduced carbon compounds. The overall reaction is $6 CO_2 + 12 H_2O + light \rightarrow C_6H_{12}O_6 + 6 O_2 + 6 H_2O$. The oxygen atoms in O_2 come from water, not from CO_2. **Review Figures 8.1, 8.2**

The Two Pathways of Photosynthesis: An Overview

- In the light-dependent reactions of photosynthesis, photophosphorylation produces ATP and reduces $NADP^+$ to $NADPH + H^+$. **Review Figure 8.3**
- ATP and $NADPH + H^+$ are needed for the reactions that fix and reduce CO_2 in the Calvin–Benson cycle, forming sugars. **Review Figure 8.3**

Properties of Light and Pigments

- Light energy comes in packets called photons, but it also has wavelike properties. **Review Figure 8.4**
- Pigments absorb light in the visible spectrum. **Review Figure 8.5**
- Absorption of a photon puts a pigment molecule in an excited state that has more energy than the ground state. **Review Figure 8.6**
- Each compound has a characteristic absorption spectrum. An action spectrum reveals the biological effectiveness of different wavelengths of light. **Review Figures 8.7, 8.8**
- Chlorophyll is essential for the photosynthesis of plants and algae. Chlorophyll and accessory pigments form antenna systems for absorption of light energy. If antenna systems funnel sufficient energy to a reactive chlorophyll molecule, it gives up an electron, reducing a carrier. **Review Figures 8.9, 8.10**
- An excited pigment molecule may lose its energy by fluorescence, by transferring it to another pigment molecule, or by transferring an electron to an oxidizing agent. **Review Figure 8.10**
- Energized electrons from excited chlorophyll are used to make ATP and $NADPH + H^+$.

Photophosphorylation and Reductions

- Noncyclic photophosphorylation uses two photosystems (I and II), producing ATP, $NADPH + H^+$, and O_2. Photosystem II uses P_{680} chlorophyll, from which light-excited electrons are passed to a redox chain that drives chemiosmotic ATP production. Light-driven oxidation of water releases O_2 and passes electrons from water to the P_{680} chlorophyll. Photosystem I passes electrons from P_{700} chlorophyll to another redox chain and then to $NADP^+$, forming $NADPH + H^+$. **Review Figure 8.11**
- Cyclic photophosphorylation uses P_{700} chlorophyll and produces only ATP. Its operation maintains the proper balance of ATP and $NADPH + H^+$ in the chloroplast. **Review Figures 8.12, 8.13**
- Chemiosmosis is the source of ATP in photophosphorylation. Electron transport pumps protons from the stroma into the thylakoids, establishing a proton-motive force. Diffusion of the protons back to the stroma via ATP synthase channels drives ATP formation from ADP and P_i. **Review Figures 8.14, 8.15**
- Cyclic photophosphorylation evolved before noncyclic photophosphorylation. Both are evolutionarily more ancient than cellular respiration, which requires the O_2 that noncyclic photophosphorylation added to the atmosphere.

Making Sugar from CO_2: The Calvin–Benson Cycle

- The Calvin–Benson cycle makes sugar from CO_2. The reactions of the cycle were elucidated by a combination of theorizing and experimentation using a radioactive carbon isotope, paper partition chromatography, and autoradiography. **Review Figures 8.16, 8.17, 8.18**
- The Calvin–Benson cycle consists of three phases: fixation of CO_2, reduction and sugar production, and regeneration of RuBP. RuBP is the initial CO_2 acceptor, and 3PG is the first stable product of CO_2 fixation. The enzyme rubisco catalyzes the reaction of CO_2 and RuBP to form 3PG. The remaining intermediates of the Calvin–Benson cycle include a variety of sugar phosphates with four, five, six, and seven carbon atoms, and some of the reactions use ATP and $NADPH + H^+$. **Review Figures 8.19, 8.20, 8.21**

Photorespiration and Its Evolutionary Consequences

- The enzyme rubisco can catalyze a reaction between O_2 and RuBP in addition to the reaction between CO_2 and RuBP. This consumption of O_2 is called photorespiration and significantly reduces the efficiency of photosynthesis. The reactions that constitute photorespiration are distributed over three organelles: chloroplasts, peroxisomes, and mitochondria.
- At high temperatures and low CO_2 concentrations, the oxygenase function of rubisco is favored, and the dominance of photorespiration would prevent plants from living in arid climates were it not for bypass reactions.
- C_4 plants bypass photorespiration with special chemical reactions and specialized leaf anatomy. In C_4 plants, PEP carboxylase in mesophyll chloroplasts initially fixes CO_2 in four-carbon acids that then diffuse into bundle sheath cells, where their decarboxylation produces locally high concentrations of CO_2 for rubisco operation and normal functioning of the Calvin–Benson cycle. **Review Figures 8.23, 8.24**
- CAM plants operate much like C_4 plants, but their initial CO_2 fixation by PEP carboxylase is temporally separated from the Calvin–Benson cycle, rather than spatially separated as in C_4 plants. **Review Figure 8.25**

Plants Perform Cellular Respiration

• Plants respire both in the light and in the dark but photosynthesize only in the light. To survive, a plant must photosynthesize more than it respires, giving it a net gain of reduced energy-rich compounds.

• Photosynthesis and respiration have several similarities. Both use systems of electron transport. Both rely on the chemiosmotic mechanism for the synthesis of ATP. However, photosynthesis traps light energy and converts it to chemical forms (sugars). Cellular respiration converts one form of chemical energy (sugar) to another (ATP).

Self-Quiz

1. Which statement about light is *not* true?
   a. Its velocity in a vacuum is constant.
   b. It is a form of energy.
   c. The energy of a photon is directly proportional to its wavelength.
   d. A photon of blue light has more energy than one of red light.
   e. Different colors correspond to different frequencies.

2. Which statement about light is true?
   a. An absorption spectrum is a plot of biological effectiveness versus wavelength.
   b. An absorption spectrum may be a good means of identifying a pigment.
   c. Light need not be absorbed to produce a biological effect.
   d. A given kind of molecule can occupy any energy level.
   e. A pigment loses energy as it absorbs a photon.

3. Which statement about chlorophylls is *not* true?
   a. They absorb light near both ends of the visible spectrum.
   b. They can accept energy from other pigments, such as carotenoids.
   c. Excited chlorophyll can either reduce another substance or fluoresce.
   d. Excited chlorophyll is an oxidizing agent.
   e. They contain magnesium.

4. In cyclic photophosphorylation
   a. oxygen gas is released.
   b. ATP is formed.
   c. water donates electrons and protons.
   d. NADPH + H$^+$ forms.
   e. CO_2 reacts with RuBP.

5. Which of the following does *not* happen in noncyclic photophosphorylation?
   a. Oxygen gas is released.
   b. ATP forms.
   c. Water donates electrons and protons.
   d. NADPH + H$^+$ forms.
   e. CO_2 reacts with RuBP.

6. In the chloroplast
   a. light leads to the pumping of protons out of the thylakoids.
   b. ATP forms when protons are pumped into the thylakoids.
   c. light causes the stroma to become more acidic than the thylakoids.
   d. protons return passively to the stroma through protein channels.
   e. proton pumping requires ATP.

7. Which statement about the Calvin–Benson cycle is *not* true?
   a. CO_2 reacts with RuBP to form 3PG.
   b. RuBP forms by the metabolism of 3PG.
   c. ATP and NADPH + H$^+$ form when 3PG is reduced.
   d. The concentration of 3PG rises if the light is switched off.
   e. Rubisco catalyzes the reaction of CO_2 and RuBP.

8. In C_4 photosynthesis
   a. 3PG is the first product of CO_2 fixation.
   b. rubisco catalyzes the first step in the pathway.
   c. four-carbon acids are formed by PEP carboxylase in bundle sheath cells.
   d. photosynthesis continues at lower CO_2 levels than in C_3 plants.
   e. CO_2 released from RuBP is transferred to PEP.

9. C_4 photosynthesis and crassulacean acid metabolism differ in that
   a. only C_4 photosynthesis uses PEP carboxylase.
   b. CO_2 is trapped by night in CAM plants and by day in C_4 plants.
   c. four-carbon acids are formed only in C_4 photosynthesis.
   d. only Crassulaceae commonly grow in dry or salty environments.
   e. only C_4 photosynthesis helps conserve water.

10. Photorespiration
    a. takes place only in C_4 plants.
    b. includes reactions carried out in peroxisomes.
    c. increases the yield of photosynthesis.
    d. is catalyzed by PEP carboxylase.
    e. is independent of light intensity.

Applying Concepts

1. Both photophosphorylation and the Calvin–Benson cycle stop when the light is turned off. Which specific reaction stops first? Which stops next? Continue answering the question "Which stops next?" until you have explained why both pathways have stopped.

2. In what principal ways are the reactions of photophosphorylation similar to the respiratory chain and oxidative phosphorylation discussed in Chapter 7? Differentiate between cyclic and noncyclic photophosphorylation in terms of (1) the products and (2) the source of electrons for the reduction of oxidized chlorophyll.

3. Draw a cartoon representation of noncyclic photophosphorylation using elements similar to those in Figure 8.13. (*Hint:* You have to add a second "photon hammer.")

4. The development of what three experimental techniques made it possible to elucidate the Calvin–Benson cycle? How were these techniques used in the investigation?

5. If water labeled with ^{18}O is added to a suspension of photosynthesizing chloroplasts, which of the following compounds will first become labeled with ^{18}O: ATP, NADPH, O_2, or 3PG? If water labeled with ^3H is added to a suspension of photosynthesizing chloroplasts, which of the same compounds will first become radioactive? If CO_2 labeled with ^{14}C is added to a suspension of photosynthesizing chloroplasts, which of those compounds will first become radioactive?

Readings

Alberts, B., D. Bray, J. Lewis, M. Raff, K. Roberts and J. D. Watson. 1994. *Molecular Biology of the Cell*, 3rd Edition. Garland Publishing, New York. Chapter 14, on energy conversion, contains a good discussion of photosynthesis.

Bazzazz, F. A. and E. D. Fajer. 1992. "Plant Life in a CO_2-Rich World." *Scientific American*, January. A comparison of C_3 and C_4 plants and their prospects.

Govindjee, and W. J. Coleman. 1990. "How Plants Make Oxygen." *Scientific American*, February. A description of the "clock" in photosystem II that splits water into oxygen gas, protons, and electrons.

Hall, D. O. and K. K. Rao. 1987. *Photosynthesis*, 4th Edition. Edward Arnold, New York. An intermediate-level treatment of all the major topics in photosynthesis and an excellent bibliography, all in 100 pages.

Stryer, L. 1995. *Biochemistry*, 4th Edition. W. H. Freeman, New York. Chapter 26 gives an advanced but clear treatment of topics in photosynthesis.

Voet, D. and J. G. Voet. 1995. *Biochemistry*, 2nd Edition. John Wiley & Sons, New York. Chapter 22 discusses photosynthesis.

Weinberg, C. J. and R. H. Williams. 1990. "Energy from the Sun." *Scientific American*, September. Photosynthesis and biomass technology, along with other solar-derived technologies such as wind and solar-thermal, are considered as sources of energy for industrial and other uses.

Youvan, D. C. and B. L. Marrs. 1987. "Molecular Mechanisms of Photosynthesis." *Scientific American*, June. A difficult but interesting article on events in the first fraction of a millisecond of photosynthesis in a bacterium. Part of the article is better read after reading Part Two of this book.

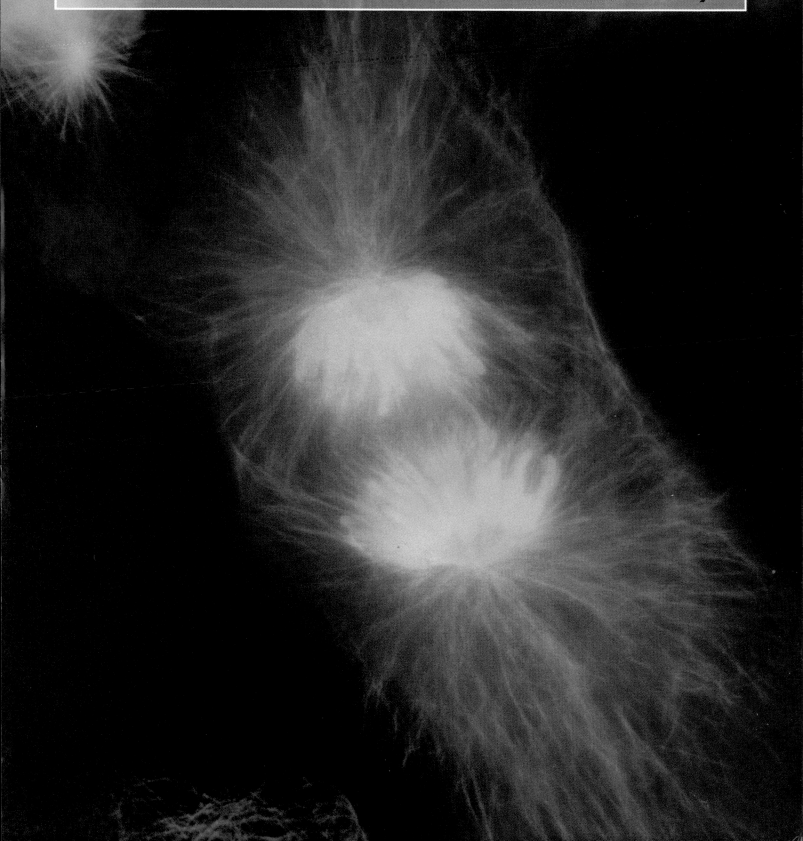

Part Two
Information and Heredity

Chapter 9

Chromosomes, the Cell Cycle, and Cell Division

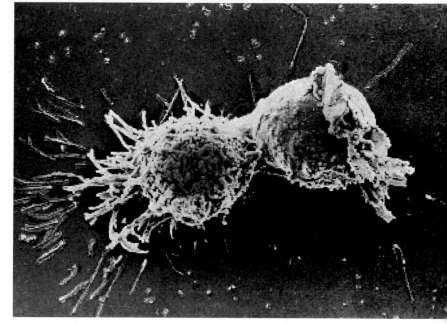

One Cell Becomes Two
The two human cells in this scanning electron micrograph are in the final stage of division and will soon separate from each other. The original nucleus has divided by a process called mitosis.

Of the more than 100 trillion (10^{14}) cells that make up an adult human, just the right cells must divide, at just the right time, to produce and maintain a healthy individual. Cell division is initiated by both internal and external signals, and these signals are highly regulated. When signals go wrong and cells divide inappropriately, a disorganized mass of cells—a tumor—may result. But in both tumor cells and normal cells, cell division consists of the same precise and elegant sequence of events. The nuclear DNA that controls cell function makes two copies of itself, and these copies are exactly distributed to the two new cells. The cytoplasm and its associated organelles are also parceled out to the two new cells. Finally, the two cells are separated by plasma membranes and, in some cases, cell walls.

Whereas unicellular organisms use cell division only as a means to reproduce, in multicellular organisms cell division also plays important roles in the growth and repair of tissues (Figure 9.1). The fact that a single cell, the fertilized egg, gives rise to a newborn baby with trillions of cells implies that development includes a lot of cell division. Less obvious but no less significant is the need to replace billions of blood cells and epithelial cells that die every day. In addition, a special type of reductive cell division is involved in forming the specialized reproductive cells—eggs and sperm—in complex eukaryotes.

In this chapter, after briefly discussing cell division in prokaryotes, we will describe the two types of eukaryotic cell division—mitosis and meiosis—and relate these to asexual and sexual reproduction.

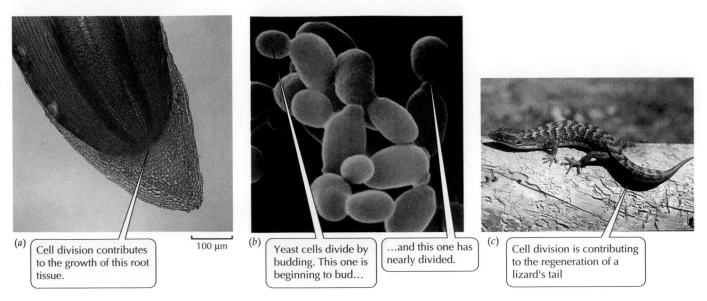

(a) Cell division contributes to the growth of this root tissue.

100 μm

(b) Yeast cells divide by budding. This one is beginning to bud…

…and this one has nearly divided.

(c) Cell division is contributing to the regeneration of a lizard's tail

9.1 Important Consequences of Cell Division Cell division is the basis for growth, reproduction, and regeneration.

Systems of Cell Division

Prokaryotes typically have a single chromosome

Prokaryotic cells, by definition, lack nuclei. Hence, they do not use the same mechanisms for cell division that eukaryotic cells use. Even so, when a prokaryote divides, by a process called **fission**, genetic information is distributed in an orderly fashion to its daughter cells.

A **chromosome** is a DNA molecule that contains the genetic information for an organism. Most prokaryotes have only one chromosome, a single long DNA molecule with proteins bound to it. In the bacterium *Escherichia coli*, the chromosome is a circular molecule of DNA about 1.6 million nm (1.6 mm) long. The bacterium itself is about 1 μm (1,000 nm) in diameter and about 4 μm long. Thus the space into which the long thread of DNA is packed in the bacterial nucleoid is very small relative to the length of the DNA molecule. It is thus not surprising that the molecule usually appears in electron micrographs as a hopeless tangle of fibers (Figure 9.2). The DNA molecule accomplishes some packing by folding on itself, and proteins bound to DNA contribute to this packing.

When prokaryotic cells are gently lysed (broken open) to release their contents, the chromosome sometimes becomes less tangled and spreads out. Several techniques have shown that the prokaryotic chromosome is a closed circle rather than the linear structure found in eukaryotes. Circular chromosomes are probably to be found in all prokaryotes, as well as in some viruses and in the chloroplasts and mitochondria of eukaryotic cells.

The prokaryotic chromosome is attached to the plasma membrane. When a new DNA molecule forms from the old one, it, too, attaches to the membrane. As new membrane material is added between the two sites of attachment, the two DNA molecules gradually separate. As the cell divides, new wall and membrane are inserted between the two chromosomes to form

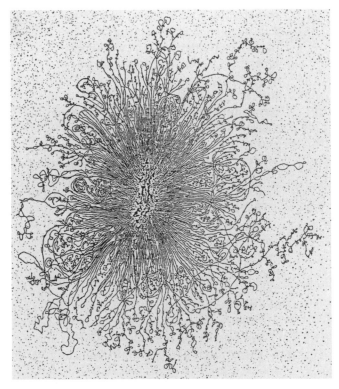

9.2 The Prokaryotic Chromosome Is a Circle The long, looping fibers of DNA from this cell of the bacterium *Escherichia coli* are all part of one continuous circular chromosome.

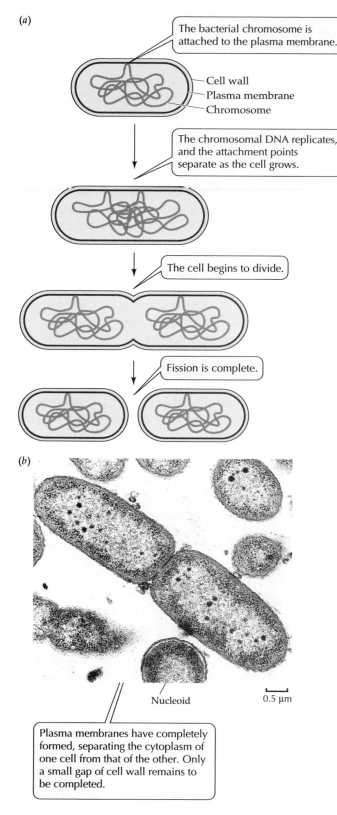

(a)

The bacterial chromosome is attached to the plasma membrane.

Cell wall
Plasma membrane
Chromosome

The chromosomal DNA replicates, and the attachment points separate as the cell grows.

The cell begins to divide.

Fission is complete.

(b)

Nucleoid

0.5 μm

Plasma membranes have completely formed, separating the cytoplasm of one cell from that of the other. Only a small gap of cell wall remains to be completed.

9.3 Prokaryotic Cell Division (a) The steps of cell division in prokaryotes. (b) These two cells of the bacterium *Pseudomonas aeruginosa* have almost completed fission. Each cell contains a complete chromosome, visible as the nucleoid in the center of the cell.

two daughter cells, each with an identical chromosome (Figure 9.3).

These three processes—DNA replication, chromosome separation, and cell division—occur in about 30 minutes total in an adequately nourished prokaryotic cell. This is a sharp contrast to the hours and even days it takes for eukaryotes to complete cell division.

Eukaryotic cells divide by mitosis or meiosis

The intricate mechanism mentioned at the beginning of this chapter that parcels out the duplicated copies of genetic material for cell division in eukaryotes is called **mitosis**. This replicating process produces identical nuclei for all the cells of an organism's adult body. A second mechanism for nuclear division, called *meiosis*, generates diversity among nuclei and plays a key role in sexual life cycles. (We will discuss meiosis in detail later in the chapter.)

The duplication of a eukaryotic cell typically consists of three steps: (1) the replication of the genetic material within the nucleus, (2) the packaging and separation of the genetic material into two new nuclei, and (3) the division of the cytoplasm. Between divisions—that is, for most of its life—a eukaryotic cell is in a condition called **interphase**.

What determines whether a cell in interphase will divide? How does mitosis lead to exact copies, and to diversity of products? Why do we need both exact copies and diverse products? Why do most organisms have sex in their life cycles? In the pages that follow we will describe the details of mitosis, meiosis, and interphase, as well as their consequences for heredity, development, and evolution.

Interphase and the Control of Cell Division

A cell lives and functions until it divides or dies—or, if it is a sex cell, until it fuses with another sex cell. Some, such as red blood cells, muscle cells, and nerve cells, lose the capacity to divide as they mature. Other cell types, such as cortical cells in plant stems, divide only rarely. Most cells, however, have some probability of dividing, and some are specialized for rapid division. For many kinds of cells we may speak of a **cell cycle** that has two phases: mitosis and interphase.

A given cell lives for one turn of the cell cycle and then becomes two cells. The cycle repeats again and again, a constant source of renewal. Even in tissues engaged in rapid growth, however, the cell cycle consists mainly of the time spent in interphase. Examination of any collection of cells, such as the tip of a root or a slice of liver, reveals that most of the cells are in interphase most of the time. Only a small percentage of the cells are in mitosis at any given moment. We can confirm this fact, in certain cultures of cells, by watching a single cell through its entire cycle.

In this section, we will describe the cell cycle events that occur during interphase, especially the "decision" to enter mitosis.

The subphases of interphase have different functions

Interphase consists of three subphases, identified as S, G1, and G2. The cell's DNA replicates during the **S phase** (the *S* stands for *synthesis*). The gap between the end of mitosis and the onset of S phase is called **G1**, or Gap 1. Another gap—**G2**—separates the end of the S phase and the beginning of mitosis. Mitosis itself is referred to as the **M phase** of the cell cycle (Figure 9.4). If a cell is not going to divide, it may remain in G1 for weeks or even years until it dies; the cell seemingly will not waste matter and energy replicating its genetic material in the S phase if it is not going to divide. (There are some exceptions in which cells that will not divide do synthesize DNA and are thus stuck in G2, but the continuation of G1 is the rule in the vast majority of nondividing cells.)

Although one key event—DNA replication—dominates and defines the S phase, important cell cycle events take place in the gap phases as well. G1 is quite variable in length; some rapidly dividing embryonic cells have dispensed with it entirely. The biochemical hallmark of a G1 cell is that it is preparing for S phase. It is at this time that the "decision" to enter another cell cycle is made. During G2, the cell makes preparations for mitosis, synthesizing components of the microtubules that form the spindle, for example.

Cyclins and other proteins signal events in the cell cycle

How are *appropriate* decisions to enter the S or M phases made? These transitions—from G1 to S and from G2 to M—require the activation of a protein complex called **cyclin-dependent kinase**, or **Cdk** (Figure 9.5). A *kinase* is an enzyme that catalyzes the transfer of a phosphate group from ATP to another molecule; such phosphate transfer is called *phosphorylation*. Cdk is a kinase that catalyzes the phosphorylation of certain amino acids in proteins. Phosphorylation by Cdk changes the three-dimensional structure of a targeted protein, sometimes simultaneously changing that protein's function. Active Cdk's catalyze the action of proteins that are important in initiating progress through the cell cycle.

The discovery that Cdk's induce mitosis is a beautiful example of research on different organisms and cells converging on a single mechanism. One group of scientists was studying immature sea urchin eggs, trying to find out how they are stimulated to divide and form mature eggs. A protein called maturation promoting factor was purified from the maturing eggs, which by itself prodded the immature eggs into division. At the same time, other scientists studying the cell cycle in yeasts, which are single-celled eukaryotes, found a strain that was stalled at the G1–S boundary. What this strain lacked was a Cdk, and it turned out that this Cdk was very similar to the sea urchin's maturation promoting factor. Similar Cdk's were soon found to control the G1–S transition in many other organisms, including humans.

But Cdk's are not active by themselves. They must be bound with a second type of protein, called **cyclins**. It is the cyclin–Cdk *complex* (in humans, cyclin D and Cdk4) that triggers protein kinase activity and the transition from G1 to S phase with the resulting DNA replication. Then the cyclin breaks down and the Cdk becomes inactive. Later in the cell cycle, another specific cyclin–Cdk partnership (cyclin B and Cdk2) takes over, activating a kinase activity that promotes the transition from G2 to chromosome condensation and mitosis.

What do cyclin–Cdk complexes target for phosphorylation? Not all such targets are known, but some are important for progression in the cell cycle. For ex-

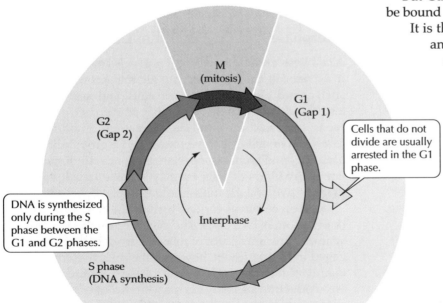

M
(mitosis)

G1
(Gap 1)

G2
(Gap 2)

Cells that do not divide are usually arrested in the G1 phase.

DNA is synthesized only during the S phase between the G1 and G2 phases.

Interphase

S phase
(DNA synthesis)

9.4 The Eukaryotic Cell Cycle   A cell's life history is made up of a short mitosis (purple) and a longer interphase (green). Interphase has three subphases (G1, S, and G2) in cells that divide.

9.5 Cyclin-Dependent Kinase and Cyclin Trigger Decisions in the Cell Cycle
A human cell makes the "decision" to enter the cell cycle during G1 when cyclin D binds to a cyclin-dependent kinase (Cdk4). There are four such cyclin–Cdk controls during the typical cell cycle in humans.

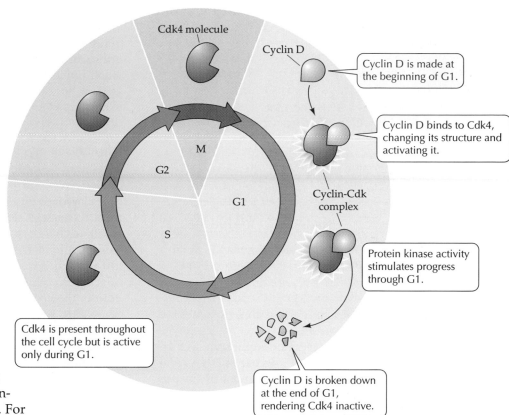

Cyclin D is made at the beginning of G1.

Cyclin D binds to Cdk4, changing its structure and activating it.

Protein kinase activity stimulates progress through G1.

Cyclin D is broken down at the end of G1, rendering Cdk4 inactive.

Cdk4 is present throughout the cell cycle but is active only during G1.

ample, the cyclin B–Cdk2 complex catalyzes the phosphorylation of certain target proteins, causing them to bind to DNA and initiate chromosome condensation. Phosphorylation of other target proteins results in the disaggregation of the nuclear envelope early in mitosis.

Since cancer results from inappropriate cell division, it is not surprising that the cyclin–Cdk controls are disrupted in cancer cells. For example, some fast-growing breast cancers have too much cyclin D, which overstimulates Cdk4 and cell division. Tumors of the parathyroid gland often make an entirely new cyclin that stimulates kinase activity in Cdk4. Finally, as we will describe in Chapter 17, a major protein in normal cells that prevents them from dividing is p53, which leads to inhibition of Cdk's. More than half of all human cancers have defective p53, resulting in the absence of cell cycle controls.

Mitotic inducers can stimulate cells to divide

Cyclin–Cdk complexes provide an *internal* control for progress through the cell cycle. But there are situations in the body where cells that are slowly cycling, or not cycling at all, must be stimulated to divide through *external* controls. When a person is cut and bleeds, specialized cell fragments called platelets gather at the wound and help initiate blood clotting. The platelets also produce and release a protein, *platelet-derived growth factor*, that diffuses to the adjacent cells in the skin and stimulates them to divide and heal the wound.

Other **growth factors** include *interleukins*, which are made by one type of white blood cell and promote cell division in other cells that are essential for the body's immune system defenses, and *erythropoietin*, made by the kidney to stimulate the division of bone marrow cells and the production of red blood cells. In addition, many hormones promote division in specific cells.

We will describe the different physiological roles of these external mitotic inducers in later chapters, but all growth factors act in similar ways. Growth factors bind to their target cells via specialized receptor proteins on the target cell surface. The specific binding event triggers events within the cell so that it begins a cell cycle. Cancer cells often cycle inappropriately because either they make their own growth factors or they no longer require growth factors to start cycling.

Eukaryotic Chromosomes

Most human cells other than eggs and sperm contain two full sets of genetic information, one from the mother and the other from the father. Eggs and sperm, however, contain only a single set; any particular egg or sperm in your body contains some information from your mother and some from your father. As in prokaryotes, the genetic information consists of molecules of DNA packaged as chromosomes.

However, eukaryotes have more than one chromosome, and during interphase these chromosomes reside within the cell's nucleus. Seen through a light microscope, the nucleus appears relatively featureless (except for the dark nucleolus) during most of the life of a cell (see Figure 4.8); the chromosomes cannot be seen. Mitosis has been defined historically as the stage

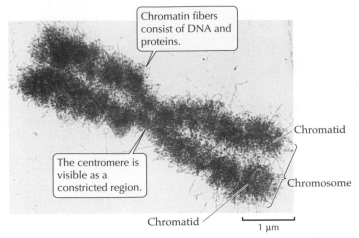

Chromatin fibers consist of DNA and proteins.

The centromere is visible as a constricted region.

Chromatid

Chromosome

Chromatid

1 μm

9.6 Chromosomes, Chromatids, and Chromatin A human chromosome, shown as the cell prepares to divide.

of the cell cycle at which condensed chromosomes become visible in the microscope.

The basic unit of the eukaryotic chromosome is a gigantic, linear, double-stranded molecule of DNA complexed with many proteins. During most of the eukaryotic cell cycle, each chromosome contains only one such double-stranded DNA molecule. However, when the DNA molecule duplicates, the chromosome consists of two joined **chromatids**, each made up of one double-stranded DNA molecule complexed with proteins (together known as *chromatin*; Figure 9.6). At the time chromosomes become visible in a microscope, the two chromatids are joined at a specific small region of the chromosome called the **centromere**. This single-centromere structure, whether it contains one or two DNA molecules, is properly called a *mitotic or meiotic chromosome*.

Chromatin consists of DNA and proteins

The complex of DNA and proteins in a eukaryotic chromosome is referred to as **chromatin**. The DNA carries the genetic information; the proteins organize the chromosome physically and regulate the activities of the DNA. By mass, the amount of chromosomal proteins is equivalent to that of DNA.

Chromatin changes form dramatically during mitosis and meiosis. During interphase, the chromatin is strung out so thinly that the chromosome cannot be seen as a defined body under the light microscope. But during most of mitosis and meiosis, the chromatin is so highly coiled and compacted that the chromosome appears as a dense, bulky object (see Figure 9.6).

This alternation of forms relates to the function of chromatin during different phases of the cell cycle. Before each mitosis the genetic material is duplicated. Remember that mitosis separates the duplicated genetic material into two new nuclei. This separation is

easier to accomplish if the DNA is neatly arranged in compact units rather than being tangled up like a plate of spaghetti. During interphase, however, the DNA must direct the growth and other activities of the cell. Such functions require that DNA be unwound and exposed so that it can interact with enzymes (see Chapters 12 and 14).

Chromatin proteins organize the DNA in chromosomes

Chromatin proteins associate closely with chromosomal DNA. Chromosomes contain large quantities of five classes of proteins called **histones**. Histones have a positive charge at the pH levels found in the cell. The positive charge is a result of their particular amino acid compositions. Histone molecules join together to produce complexes around which the DNA is wound (Figure 9.7). Eight histone molecules, two of each of four of the histone classes, unite to form a core or spool shaped so that the DNA molecule fits snugly in a coil around it. The other class of histone proteins (histone H1) appears to fit on the outside of the DNA, perhaps clamping it to the histone core. The resulting beadlike units, called **nucleosomes**, provide the major structure for packing the DNA in chromatin (Figure 9.8).

A chromatid has a single DNA molecule running through vast numbers of nucleosomes. The many nucleosomes of a mitotic chromatid may pack together and coil. During both mitosis and meiosis, the chromatin becomes ever more coiled and condensed, with further folding of the chromatin continuing up to the time at which chromosomes begin to move apart. Some diverse acidic proteins are also present in small quantities in chromosomes. Some of these proteins are involved in chromosome packaging; others are involved in the expression of DNA sequences.

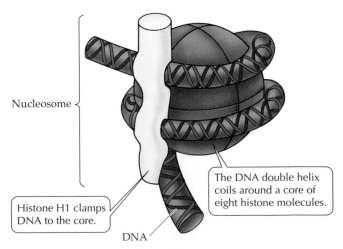

Nucleosome

Histone H1 clamps DNA to the core.

The DNA double helix coils around a core of eight histone molecules.

DNA

9.7 DNA Interacts with Histones to Form Nucleosomes In some cases, histone H1 may be on the inside of the core.

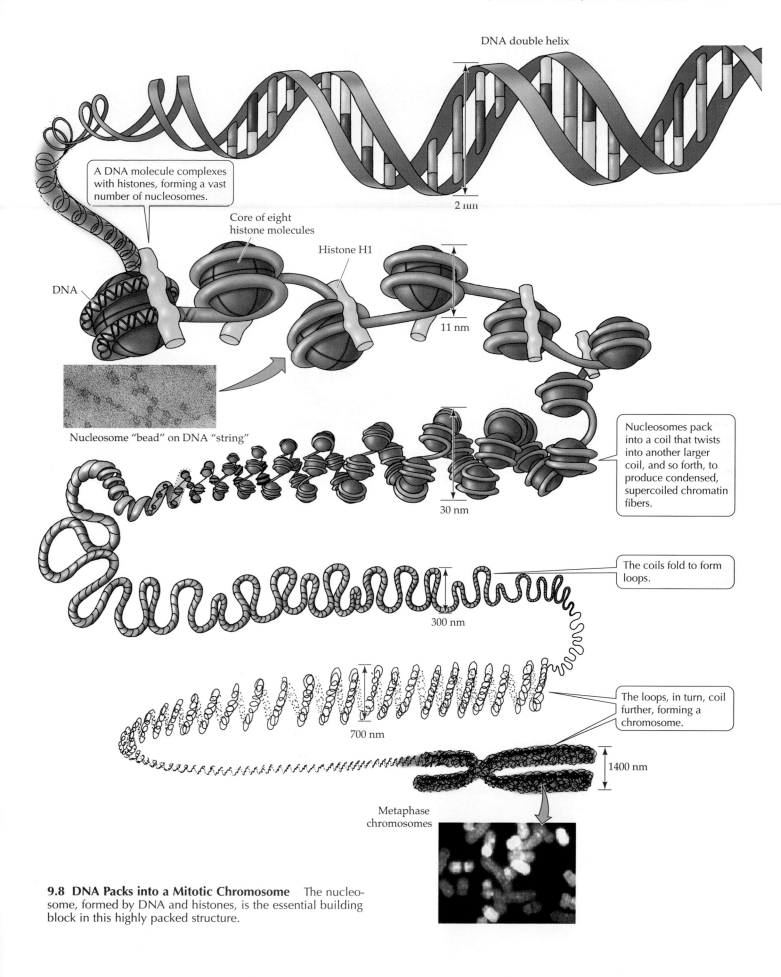

DNA double helix

A DNA molecule complexes with histones, forming a vast number of nucleosomes.

Core of eight histone molecules

Histone H1

2 nm

11 nm

DNA

Nucleosome "bead" on DNA "string"

Nucleosomes pack into a coil that twists into another larger coil, and so forth, to produce condensed, supercoiled chromatin fibers.

30 nm

The coils fold to form loops.

300 nm

The loops, in turn, coil further, forming a chromosome.

700 nm

1400 nm

Metaphase chromosomes

9.8 DNA Packs into a Mitotic Chromosome The nucleosome, formed by DNA and histones, is the essential building block in this highly packed structure.

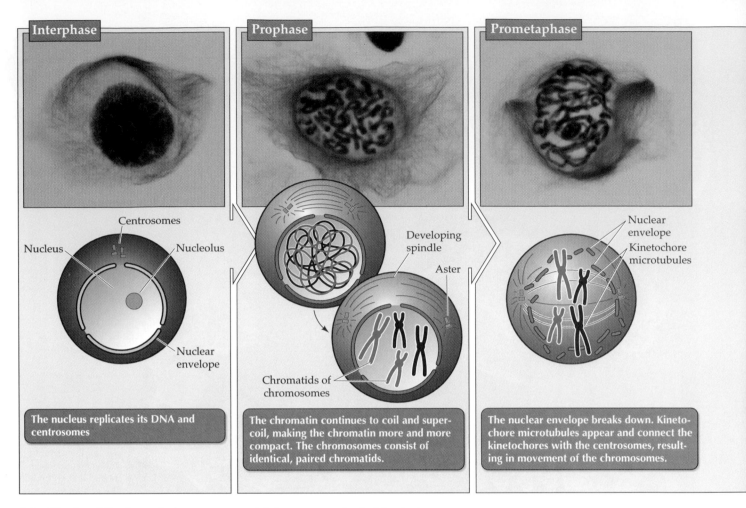

Interphase	Prophase	Prometaphase

Centrosomes

Nucleus

Nucleolus

Nuclear envelope

Developing spindle

Aster

Chromatids of chromosomes

Nuclear envelope

Kinetochore microtubules

The nucleus replicates its DNA and centrosomes

The chromatin continues to coil and super-coil, making the chromatin more and more compact. The chromosomes consist of identical, paired chromatids.

The nuclear envelope breaks down. Kineto-chore microtubules appear and connect the kinetochores with the centrosomes, result-ing in movement of the chromosomes.

9.9 Mitosis   Mitosis results in two new nuclei that are genetically identical to one another and to the nucleus from which they formed. These photomicrographs are of plant nuclei, which lack centrioles and asters. The diagrams are of corresponding phases in animal cells and introduce the structures not found in plants. In the micrographs, the red dye stains microtubules (and thus the spindle); the blue dye stains the chromosomes. In the diagrams, the chromosomes are stylized to emphasize the fates of the individual chromatids.

Although we know less about the organization of interphase chromatin than about that of mitotic chromatin, we know that interphase chromatin has nucleosomes that are spaced at the same intervals as in supercoiled chromatin (as shown in Figure 9.8). During interphase, DNA thus remains associated with histone molecules while it replicates and directs the synthesis of RNA. As we will show in Chapter 14, these protein–DNA interactions are important in regulating interphase activities.

Mitosis: Distributing Exact Copies of Genetic Information

Eukaryotic cells contain multiple chromosomes. Mitosis ensures the accurate distribution of these chromosomes to the daughter nuclei. During mitosis, a single nucleus gives rise to two nuclei that are genetically identical to each other and to the parent nucleus. In reality, mitosis is a continuous process in which each event flows smoothly into the next. For discussion, however, it is convenient to look at mitosis—the M phase of the cell cycle—as a series of separate events, or subphases: prophase, prometaphase, metaphase, anaphase, and telophase (Figure 9.9).

During interphase, the nucleus is between divisions. At this stage we see the nuclear envelope, the nucleoli, and a barely discernible tangle of chromatin. Once the decision to enter mitosis has been made, DNA replicates, and during G2 a pair of **centrosomes** forms from a single centrosome that lies near the nucleus. The centrosomes are regions of the cell that initiate the formation of microtubules and thus help orchestrate chromosomal movement. The regions are not

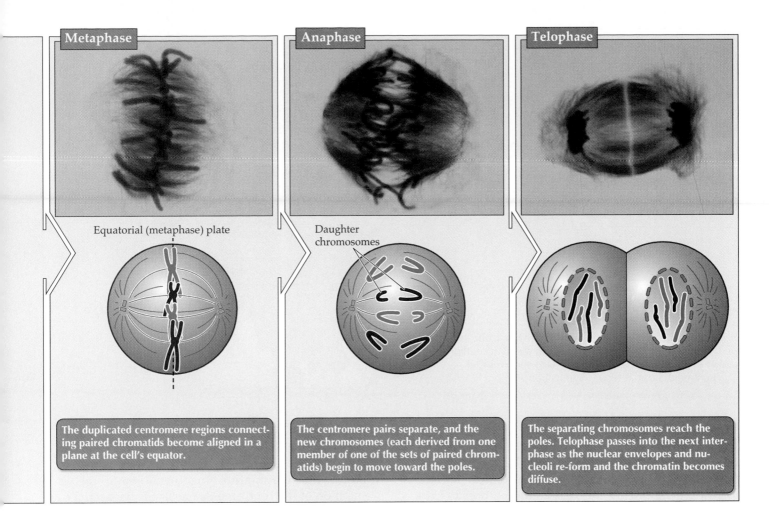

Metaphase	Anaphase	Telophase

Equatorial (metaphase) plate

Daughter chromosomes

The duplicated centromere regions connecting paired chromatids become aligned in a plane at the cell's equator.

The centromere pairs separate, and the new chromosomes (each derived from one member of one of the sets of paired chromatids) begin to move toward the poles.

The separating chromosomes reach the poles. Telophase passes into the next interphase as the nuclear envelopes and nucleoli re-form and the chromatin becomes diffuse.

enclosed by membranes and are not visible as discrete objects, but their positions are made evident by the arrangement of nearby microtubules.

In many organisms, each centrosome contains a pair of **centrioles**. However, the centrosomes of seed plants and some other organisms do not have centrioles. Where present, each centriole pair consists of one "parent" and one smaller "daughter" centriole at right angles to the parent centriole (see Figure 4.27).

The spindle forms during prophase

The appearance of the nucleus changes as the cell enters **prophase**—the beginning of mitosis (see Figure 9.9). The centrosomes move away from each other toward opposite ends, or *poles*, of the cell. Each centrosome then serves as a **mitotic center** that organizes microtubules. Some of the microtubules, called **polar microtubules**, run between the mitotic centers and make up the developing **spindle** (Figure 9.10).

The spindle is actually two *half spindles*: Each polar microtubule runs from one mitotic center to the middle of the spindle, where it overlaps with polar microtubules of the other half spindle. Polar microtubules are initially unstable, constantly forming and falling

apart until they contact polar microtubules from the other half spindle and become more stable. The spindle is responsible for the movement and distribution of chromosomes during mitosis and meiosis.

A prophase chromosome consists of two chromatids

The chromatin also changes during prophase. The extremely long, thin fibers take on a more orderly form as a result of coiling and compacting (see Figures 9.8 and 9.9). Under the light microscope, each prophase chromosome is seen to consist of two chromatids held tightly together over much of their length. The two chromatids of a single mitotic chromosome are identical in structure, chemistry, and the hereditary information they carry because of the way in which DNA replicates during the S phase that precedes mitosis.

Within the region of tight binding of the chromatids lies the centromere, which is where chromatids become associated with the microtubules of the spindle. Very late in prophase, specialized three-layered structures called **kinetochores** develop in the centromere region, one on each chromatid (see Figure 9.10). This is where the microtubules actually attach.

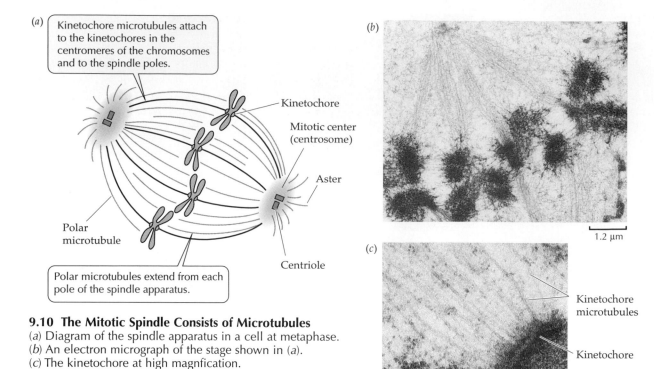

(a) Kinetochore microtubules attach to the kinetochores in the centromeres of the chromosomes and to the spindle poles.

Kinetochore

Mitotic center (centrosome)

Aster

Centriole

Polar microtubule

Polar microtubules extend from each pole of the spindle apparatus.

(b)

1.2 μm

(c)

Kinetochore microtubules

Kinetochore

0.6 μm

9.10 The Mitotic Spindle Consists of Microtubules
(a) Diagram of the spindle apparatus in a cell at metaphase.
(b) An electron micrograph of the stage shown in (a).
(c) The kinetochore at high magnfication.

Chromosome movements are highly organized

The somewhat condensed chromosomes start to move at the end of prophase, which is the beginning of **prometaphase** (see Figure 9.8). The nuclear lamina disintegrates and the nuclear envelope breaks into small vesicles, allowing the developing spindle to "invade" the nuclear region.

Some polar microtubules now attach to chromosomes at their kinetochores and so are called **kinetochore microtubules** (see Figure 9.10). The kinetochore of one chromatid is attached to microtubules coming from one pole, while the kinetochore of its sister chromatid is attached to microtubules emanating from the other pole. At the attachment points are proteins that act as "molecular motors"; these *motor proteins* have the ability to hydrolyze ATP to ADP and phosphate, thus releasing energy to move the chromosomes along microtubule "railroad tracks" toward the poles.

The movement of chromosomes toward the poles is counteracted by two factors. First, there seems to be a repulsive force that keeps the chromosomes in the middle of the spindle region. Second, the two chromatids are held together at their centromere. So chromosomes during prometaphase appear to move aimlessly back and forth between the centrosomes at the poles and the middle of the spindle. Gradually, the kinetochores approach this middle region, called the **equatorial plate** or **metaphase plate**, halfway between the poles (see Figure 9.9).

The cell is said to be in **metaphase** when all the kinetochores arrive at the equatorial plate. Metaphase lasts up to an hour and is the best time to see the sizes and shapes of chromosomes. At the end of metaphase the centromeres separate, possibly because a protein holding the chromatids together breaks down. Separation of the centromeres marks the beginning of **anaphase**, the phase of mitosis during which the two sister chromatids of each chromosome—now called *daughter chromosomes*, each containing one double-stranded DNA molecule—move to opposite ends of the spindle.

What propels this highly organized mass migration, which takes about 10 minutes, is not clear. Two things seem to move the chromosomes along. First, motor proteins propel them, and second, the kinetochore microtubules shorten from the poles, drawing the chromosomes toward them.

During anaphase the poles of the spindle are pushed farther apart, doubling the distance between them. The distance between poles increases because polar microtubules from opposite ends of the spindle have motors that cause them to slide, pushing the poles apart in much the same way that microtubules slide in cilia and flagella (see Chapter 4). This polar separation contributes to the separation of one set of daughter chromosomes from the other.

Nuclei re-form at the end of mitosis

When the chromosomes stop moving at the end of anaphase, the cell enters **telophase** (the final frame in Figure 9.9). Two nuclei with identical DNA, carrying identical sets of hereditary instructions, are at the opposite ends of the spindle, which begins to break down. The chromosomes begin to uncoil, continuing until they become the diffuse tangle of chromatin that is characteristic of interphase. The materials of nuclear envelopes and nucleoli, which were disaggregated during prophase, coalesce and form their respective structures. When these and other changes are complete, telophase—and mitosis—is at an end, and each of the daughter nuclei enters another interphase.

During interphase, the DNA duplicates and new chromatids form, so that each chromosome consists of two chromatids. The duplication of DNA is a major topic and will be discussed in Chapter 11. Centrioles, if present, replicate during interphase: The two paired centrioles first separate, and then each acts as a "parent" for the formation of a new "daughter" centriole at right angles to it.

Mitosis is beautifully precise. Its result is two nuclei that are *identical to each other* and to the parent nucleus in chromosomal makeup and hence in genetic constitution.

Cytokinesis: The Division of the Cytoplasm

Mitosis refers only to the division of the nucleus. The division of the cell's cytoplasm is accomplished by **cytokinesis**. Cytokinesis generally, but not always, follows mitosis. Animal cells usually divide by a furrowing of the membrane, as if an invisible thread were tightening between the two parts (Figure 9.11*a*). The "invisible" threads are microfilaments of actin and myosin located in a ring just beneath the plasma membrane. These two proteins interact to produce a contraction, just as they do in muscles, thus pinching the cell in two (see Figure 4.22*a*).

Plant cell cytoplasm divides differently, because plants have cell walls. As the spindle breaks down after mitosis, membranous vesicles derived from the Golgi apparatus appear in the equatorial region roughly midway between the two daughter nuclei of a dividing plant cell. With the help of microtubules, the vesicles fuse to form new plasma membrane and contribute their contents to a cell plate, which is the beginning of a new cell wall (Figure 9.11*b*).

Following cytokinesis, both daughter cells contain all the components of a complete cell. The precise distribution of chromosomes is ensured by mitosis. Organelles such as ribosomes, mitochondria, and chloroplasts need not be distributed equally between daughter cells, as long as some of each are present in both cells; accordingly, there is no mechanism with precision comparable to that of mitosis to provide for their equal allocation to daughter cells.

Mitosis: Asexual Reproduction and Genetic Constancy

The cell cycle repeats itself. By this process, a single cell can give rise to a vast number of others. The cell could be an entire organism reproducing with each cycle, or a cell that divides to produce a multicellular organism. The multicellular organism, in turn, may be able to reproduce itself by releasing one or more of its cells, derived from mitosis and cytokinesis, as a spore or by having a multicellular piece break away and grow on its own (Figure 9.12). The unicellular organ-

9.11 Cytokinesis Differs in Animal and Plant Cells
Plant cells form cell walls and thus must divide differently from animal cells. (*a*) A sea urchin egg has just completed cytokinesis at the end of the first division in its development into an embryo. (*b*) A dividing plant cell in late telophase.

(*a*)

The division furrow has completely separated the cytoplasm of one daughter cell from another, although their surfaces remain in contact.

Microvilli

170 μm

(*b*)

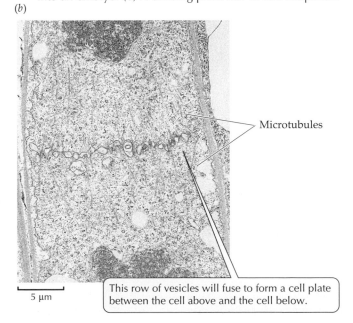

Microtubules

5 μm

This row of vesicles will fuse to form a cell plate between the cell above and the cell below.

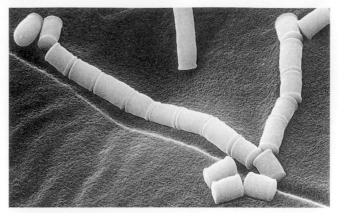

9.12 Asexual Reproduction These spool-shaped cells are asexual spores formed by a fungus. Each spore contains a nucleus produced by a mitotic division. A spore is the same genetically as the parent that fragmented to produce it.

ism and the multicellular organism reproducing by releasing cells provide examples of **asexual reproduction**, sometimes called vegetative reproduction. This mode of reproduction is based on mitotic division of the nucleus and, accordingly, produces offspring that

are genetically identical with the parent. Asexual reproduction is a rapid and effective means of making new individuals, and it is common in nature.

The uniformity of asexual reproduction, which leads to the production of a **clone** of genetically identical progeny, is very different from the situation in sexual reproduction. In sexual reproduction two parents, each contributing one cell, produce offspring that differ from each parent and from each other. This variety of genetic combinations results in a variety of offspring, some of whom may be better adapted to reproduce in a particular environment. Asexually reproducing organisms can also produce varied offspring. The cause of variation in asexual reproduction is primarily mutations, or changes, in the genetic material.

Meiosis: Sexual Reproduction and Diversity

Diversity is fostered by **sexual reproduction**, which combines genetic information from two different cells. Sexual life cycles are summarized in Figure 9.13. In the reproduction of most animal species, the two cells are contributed by two separate parents. Each parent provides a sex cell, or **gamete**. Each gamete is **haploid**,

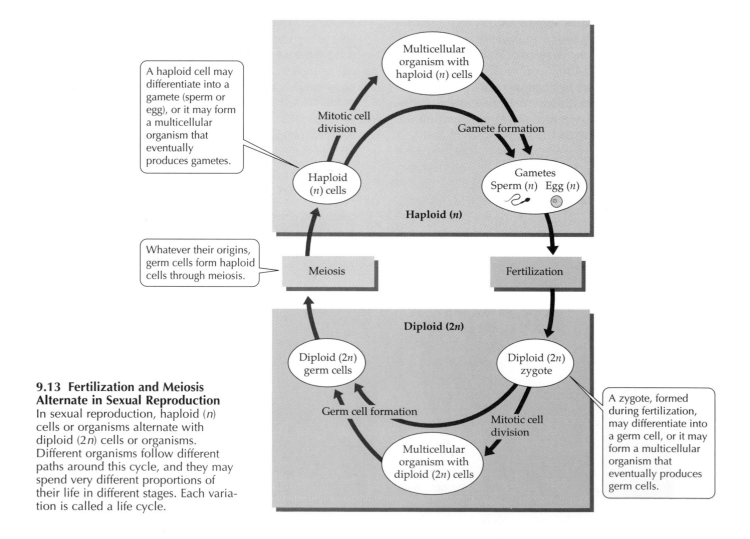

9.13 Fertilization and Meiosis Alternate in Sexual Reproduction
In sexual reproduction, haploid (*n*) cells or organisms alternate with diploid (2*n*) cells or organisms. Different organisms follow different paths around this cycle, and they may spend very different proportions of their life in different stages. Each variation is called a life cycle.

A haploid cell may differentiate into a gamete (sperm or egg), or it may form a multicellular organism that eventually produces gametes.

Whatever their origins, germ cells form haploid cells through meiosis.

Multicellular organism with haploid (*n*) cells

Mitotic cell division

Gamete formation

Haploid (*n*) cells

Gametes
Sperm (*n*)   Egg (*n*)

Haploid (*n*)

Meiosis

Fertilization

Diploid (2*n*)

Diploid (2*n*) germ cells

Diploid (2*n*) zygote

Germ cell formation

Mitotic cell division

Multicellular organism with diploid (2*n*) cells

A zygote, formed during fertilization, may differentiate into a germ cell, or it may form a multicellular organism that eventually produces germ cells.

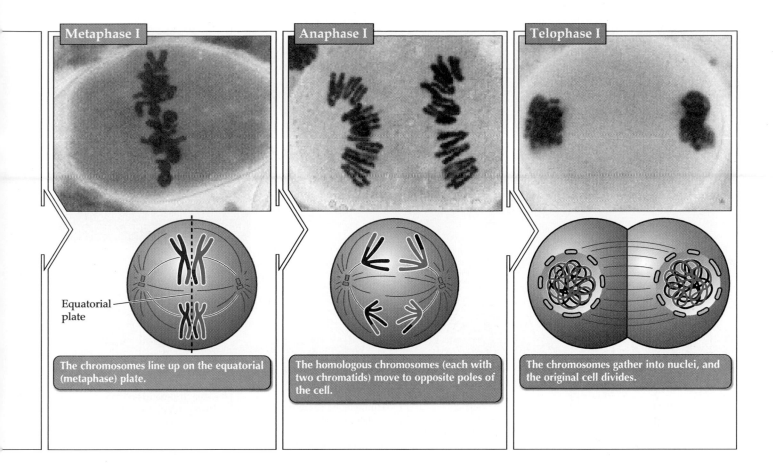

Metaphase I

Equatorial plate

The chromosomes line up on the equatorial (metaphase) plate.

Anaphase I

The homologous chromosomes (each with two chromatids) move to opposite poles of the cell.

Telophase I

The chromosomes gather into nuclei, and the original cell divides.

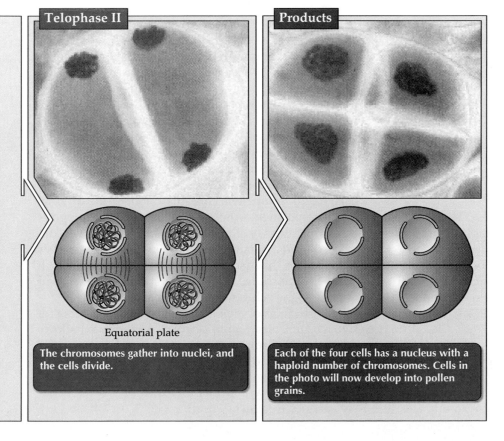

Telophase II

Equatorial plate

The chromosomes gather into nuclei, and the cells divide.

Products

Each of the four cells has a nucleus with a haploid number of chromosomes. Cells in the photo will now develop into pollen grains.

9.15 Meiosis In meiosis, two sets of chromosomes are divided among four cells, each of which then has half as many chromosomes as the original cell. These four haploid cells are the result of two successive nuclear divisions. The photomicrographs shown here are of meiosis in the male reproductive organ of a lily. As in Figure 9.9, the diagrams show corresponding phases in an animal.

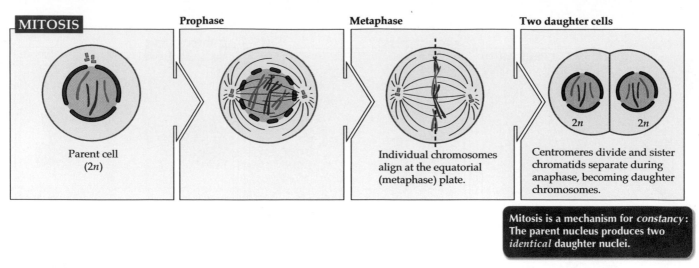

MITOSIS

	Prophase	Metaphase	Two daughter cells

Parent cell (2*n*)

Individual chromosomes align at the equatorial (metaphase) plate.

Centromeres divide and sister chromatids separate during anaphase, becoming daughter chromosomes.

Mitosis is a mechanism for *constancy*: The parent nucleus produces two *identical* daughter nuclei.

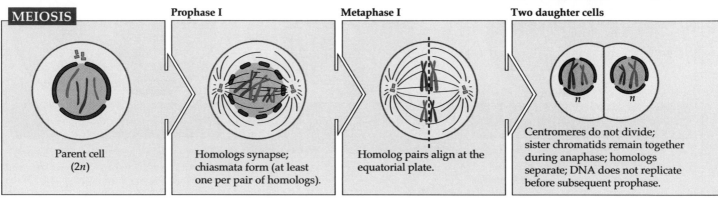

MEIOSIS

	Prophase I	Metaphase I	Two daughter cells

Parent cell (2*n*)

Homologs synapse; chiasmata form (at least one per pair of homologs).

Homolog pairs align at the equatorial plate.

Centromeres do not divide; sister chromatids remain together during anaphase; homologs separate; DNA does not replicate before subsequent prophase.

The two processes are also different. The three major differences between meiosis II and mitosis are:

- DNA replicates before mitosis but not before meiosis II.
- In mitosis, the sister chromatids that make up a given chromosome are identical; in meiosis II they differ over part of their length if they participated in crossing over during prophase of meiosis I.
- The number of chromosomes on the equatorial plate of each of the two nuclei in meiosis II is half the number in the single mitotic nucleus.

Figure 9.18 compares mitosis and meiosis. The result of meiosis is four nuclei; each nucleus is haploid and has a single full set of chromosomes that differs from other such sets in its exact genetic composition. The differences, to repeat a very important point, result from crossing over during prophase I and from the segregation of homologous chromosomes during anaphase I.

Meiosis leads to genetic diversity

What are the consequences of the synapsis and separation of homologous chromosomes during meiosis? In mitosis, each chromosome behaves independently of its homolog; its two chromatids are sent to opposite

poles at anaphase. If we start a mitotic division with *x* chromosomes, we end up with *x* chromosomes in each daughter nucleus, and each chromosome consists of one chromatid. In meiosis, things are very different.

In meiosis, synapsis organizes things so that chromosomes of maternal origin pair with the paternal homologs. Then the separation during meiotic anaphase I ensures that each pole receives one member from each pair of homologous chromosomes. (Remember that each chromosome still consists of two chromatids.) For example, at the end of meiosis I in humans, each daughter nucleus contains 23 out of the original 46 chromosomes—one member of each homologous pair. In this way, the chromosome number is decreased from diploid to haploid. Furthermore, meiosis I guarantees that each daughter nucleus gets one full set of chromosomes, for it must have one of each pair of homologous chromosomes.

The products of meiosis I are genetically diverse for two reasons. First, synapsis during prophase I allows the maternal chromosome to interact with the paternal one; if there is crossing over, the recombinant chromatids contain some genetic material from each chromosome. Second, which member of a pair of chromosomes goes to which daughter cell at anaphase I is a matter of pure chance. If there are two pairs of chromosomes in the diploid parent nucleus, a particular daughter nucleus could get paternal chromosome 1 and maternal chromosome 2, or paternal 2 and maternal 1, or both maternals, or both paternals. It all depends on the random way in which the homologous pairs line up at metaphase I.

Note that of the four possible chromosome combinations just described, two produce daughter nuclei that

9.18 Mitosis and Meiosis Compared Meiosis differs from mitosis by synapsis and by the failure of the centromeres to separate at the end of metaphase I.

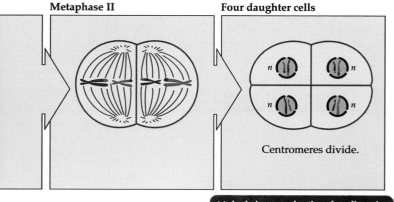

Metaphase II   Four daughter cells

Centromeres divide.

Meiosis is a mechanism for *diversity*: The parent nucleus produces four daughter nuclei, each *different* from the parent and from its sisters.

are the same as one of the parental types (except for any material exchanged by crossing over). The greater the number of chromosomes, the less probable that the original parental combinations will be reestablished. Most species of diploid organisms do, indeed, have more than two pairs. In humans, with 23 chromosome pairs, 2^{23} different combinations can be produced.

Meiotic Errors:
The Source of Chromosomal Disorders

A pair of homologous chromosomes may fail to separate during meiosis I, or sister chromatids may fail to separate during meiosis II or during mitosis. This phenomenon is called **nondisjunction**, and it results in the production of aneuploid cells (Figure 9.19). **Aneuploidy** is a condition in which one or more chromosomes or pieces of chromosomes are either lacking or are present in excess.

If, for example, the chromosome 21 pair fails to separate during the formation of a human egg (and thus both members of the pair go to one pole during anaphase I), the resulting egg contains either two copies of chromosome 21 or none at all. If an egg with two of these chromosomes is fertilized by a normal sperm, the resulting zygote and infant has three copies of the chromosome: He or she is **trisomic** for chromosome 21. As a result of carrying an extra chromosome 21, such a child demonstrates the symptoms of Down syndrome: impaired intelligence; characteristic abnormalities of the hands, tongue, and eyelids; and an increased susceptibility to cardiac abnormalities and diseases such as leukemia.

Other abnormal events can also lead to aneuploidy. In a process called *translocation*, a piece of a chromosome may break away and become attached to another chromosome. For example, a particular large part of one chromosome 21 may be translocated to another chromosome. Individuals who inherit this translocated piece along with two normal chromosomes 21 have Down syndrome.

Other human disorders result from particular chromosomal abnormalities. Sex chromosome aneuploidy causes disorders such as Turner syndrome and Klinefelter syndrome, discussed in Chapter 10 in connection with sex determination. Deletion of a portion of chromosome 5 results in cri du chat (French for "cat's cry") syndrome, so named because an afflicted infant's cry sounds like that of a cat. Symptoms of this syndrome include severe mental retardation.

Trisomies (and the corresponding monosomies) are surprisingly common in human zygotes, but most of the embryos that develop from such zygotes do not survive to birth. Trisomies for chromosomes 13, 15, and 18 greatly reduce the probability that an embryo will survive to birth, and virtually all infants who are born with such trisomies die before the age of 1 year.

Trisomies and monosomies for other chromosomes are lethal to the embryo. At least one-fifth of all recognized pregnancies spontaneously terminate during the first two months, largely because of such trisomies and monosomies. (The actual proportion of spontaneously terminated pregnancies is certainly higher, because the earliest ones often go unrecognized.)

Polyploids can have difficulty in cell division

Both diploid and haploid nuclei divide by mitosis. Multicellular diploid and multicellular haploid individuals develop from single-celled beginnings by mitotic divisions. Mitosis may proceed in diploid organisms even when a chromosome from one of the haploid sets is missing or when there is an extra copy of one of the chromosomes (as in Down syndrome).

Under some circumstances triploid ($3n$), tetraploid ($4n$), and higher-order polyploid nuclei form. Each of these *ploidy levels* represents an increase in the number

9.19 Nondisjunction during Meiosis I Leads to Aneuploidy
Nondisjunction can also occur during meiosis II.

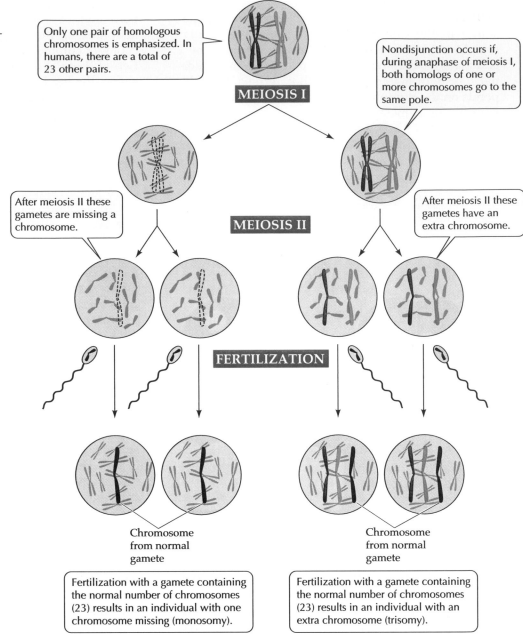

Only one pair of homologous chromosomes is emphasized. In humans, there are a total of 23 other pairs.

MEIOSIS I

Nondisjunction occurs if, during anaphase of meiosis I, both homologs of one or more chromosomes go to the same pole.

After meiosis II these gametes are missing a chromosome.

MEIOSIS II

After meiosis II these gametes have an extra chromosome.

FERTILIZATION

Chromosome from normal gamete

Chromosome from normal gamete

Fertilization with a gamete containing the normal number of chromosomes (23) results in an individual with one chromosome missing (monosomy).

Fertilization with a gamete containing the normal number of chromosomes (23) results in an individual with an extra chromosome (trisomy).

of complete sets of chromosomes present. If, through accident, the nucleus has one or more extra full sets of chromosomes—that is, if it is triploid, tetraploid, or of still higher ploidy—this abnormally high ploidy in itself does not prevent mitosis. In mitosis, each chromosome behaves independently of the others.

In meiosis, by contrast, chromosomes *synapse* to begin division. If even one chromosome has no homolog, anaphase I cannot send representatives of that chromosome to both poles. A diploid nucleus can undergo normal meiosis; a haploid one cannot. A tetraploid nucleus has an even number of each kind of chromosome, so each chromosome can pair with its homolog. But a triploid nucleus cannot undergo nor-

mal meiosis, because one-third of the chromosomes would lack partners.

This limitation has important consequences for the fertility of triploid, tetraploid, and other chromosomally unusual organisms that may be produced by plant breeding or by natural accidents. Modern bread wheat plants are hexaploids, the result of the accidental crossing of three different grasses, each having its own diploid set of 14 chromosomes.

Cell Death

As we mentioned at the start of the chapter, an essential role of cell division in complex eukaryotes is to re-

3. Which statement about the cell cycle is *not* true?
   a. It consists of mitosis and interphase.
   b. The cell's DNA replicates during G1.
   c. A cell can remain in G1 for weeks or much longer.
   d. Most proteins are formed throughout all subphases of interphase.
   e. Histones are synthesized primarily during S phase.

4. Which statement about mitosis is *not* true?
   a. A single nucleus gives rise to two identical daughter nuclei.
   b. The daughter nuclei are genetically identical to the parent nucleus.
   c. The centromeres separate at the onset of anaphase.
   d. Homologous chromosomes synapse in prophase.
   e. Mitotic centers organize the microtubules of the spindle fibers.

5. Which statement about cytokinesis is true?
   a. In animals, a cell plate forms.
   b. In plants, it is initiated by furrowing of the membrane.
   c. It generally immediately follows mitosis.
   d. In plant cells, actin and myosin play an important part.
   e. It is the division of the nucleus.

6. In sexual reproduction
   a. gametes are usually haploid.
   b. gametes are usually diploid.
   c. the zygote is usually haploid.
   d. the chromosome number is reduced during mitosis.
   e. spores are formed during fertilization.

7. In meiosis
   a. meiosis II reduces the chromosome number from diploid to haploid.
   b. DNA replicates between meiosis I and II.
   c. the chromatids that make up a chromosome in meiosis II are identical.
   d. each chromosome in prophase I consists of four chromatids.
   e. homologous chromosomes separate from one another in anaphase I.

8. In meiosis
   a. a single nucleus gives rise to two identical daughter nuclei.
   b. the daughter nuclei are genetically identical to the parent nucleus.
   c. the centromeres separate at the onset of anaphase I.
   d. homologous chromosomes synapse in prophase I.
   e. no spindle forms.

9. Which statement about aneuploidy is *not* true?
   a. It results from chromosomal nondisjunction.
   b. It does not happen in humans.
   c. An individual with an extra chromosome is trisomic.
   d. Trisomies are common in human zygotes.
   e. A piece of one chromosome may translocate to another chromosome.

10. In prokaryotes
    a. there are no meiotic divisions.
    b. mitosis proceeds as in eukaryotes.
    c. the genetic information is not carried in chromosomes.
    d. the chromosomes are identical to those of eukaryotes.
    e. cell division follows division of the nucleus.

Applying Concepts

1. How does a nucleus in the G2 phase of the cell cycle differ from one in the G1 phase?

2. What is a chromatid? When does a chromatid become a chromosome?

3. Compare and contrast mitosis (and subsequent cytokinesis) in animals and plants.

4. Suggest two ways in which, with the help of a microscope, one might determine the relative durations of the various phases of mitosis.

5. Contrast mitotic prophase and prophase I of meiosis. Contrast mitotic anaphase and anaphase I of meiosis.

6. Compare the sequence of events in the mitotic cell cycle with the sequence in programmed cell death.

Readings

Alberts, B., D. Bray, J. Lewis, M. Raff, K. Roberts and J. D. Watson. 1994. *Molecular Biology of the Cell*, 3rd Edition. Garland Publishing, New York. An outstanding book in which to pursue the topics of this chapter in greater detail.

Baserga, R. 1986 *The Biology of Cell Reproduction.* Harvard University Press, Cambridge, MA. A fine review of cell cycles in animal cells, with special emphasis on tumor cells.

The Cell Cycle. 1991. Cold Spring Harbor Symposia on Quantitative Biology, vol. 56. Cold Spring Harbor Laboratory, Cold Spring Harbor, NY. Summaries of research on cell cycle control by the scientists who did the work.

Cooper, G. M. 1997. *The Cell: A Molecular Approach.* ASM Press, Washington, DC, and Sinauer Associates, Sunderland, MA. A compact treatment of cell biology. Chapter 14 has a summary of the cell cycle.

Duke, R. C., D. M. Ojcius and J. D.-E. Young. 1996. "Cell Suicide in Health and Disease." *Scientific American*, December. A discussion of apoptosis.

Glover, D. M., C. Gonzalez and J. W. Raff. 1993. "The Centrosome." *Scientific American*, June. A description of the structure and function of the organelle that directs the assembly of the cytoskeleton and controls cell division.

Murray, A. and T. Hunt. 1993. *The Cell Cycle: An Introduction.* W. H. Freeman, New York. An excellent treatment of cell cycles, with historical background.

Nigg, E. 1993. "Targets of Cyclin-Dependent Protein Kinases." *Current Opinions in Cell Biology*, vol. 5, pages 187–193. This journal does what its title indicates: it gives up-to-date views of the field. The cell cycle is summarized each year.

Chapter 10

Transmission Genetics:
Mendel and Beyond

A Royal Family Stalked by Hemophilia
Although they did not suffer from the disease themselves, five of the women in this 1895 photograph of the British royal family carried the gene for the disease hemophilia, which afflicted several of their sons and grandsons. Because Victoria's son Edward VII was not affected and did not carry the gene, his descendants—including the present royal family of England—are free of the disease.

*I*n 1895, Queen Victoria and her numerous progeny sat for a family portrait. She and four other women in this photo shared a characteristic: All five carried a gene for the disease hemophilia, even though none of them had the disease. However, more than a dozen of Victoria's male descendants suffered from hemophilia, a failure of the blood to clot normally following an injury. All the family members with hemophilia were males; all the carriers—healthy individuals capable of transmitting the disease to their children—were women.

How can we explain this apparent correlation between sex and an inherited disease? How can we account for, and predict, patterns of inheritance in general? Experimental work in Mendelian genetics, begun in the nineteenth century and running through the first half of the twentieth century, enabled us to understand these patterns. The molecular details underlying the processing of inherited information have occupied us ever since the work of Mendelian genetics was completed.

The term "Mendelian genetics" refers to certain basic inheritance patterns and the origin of our concept of a particulate unit (the gene) responsible for those patterns. The term honors the Austrian monk Gregor Johann Mendel (1822–1884), the person who first made rigorous, quantitative observations of the patterns of inheritance and proposed plausible mechanisms to explain them. In organisms that reproduce sexually and have more than one chromosome (and orderly meiosis), many traits pass from parent to offspring in accord with these patterns.

216

In this chapter we will discuss how genes are transmitted from generation to generation of plants and animals, and show how many of the rules that govern genetics can be explained by the behavior of chromosomes during meiosis. We will also describe the interactions of genes with one another and with the environment, and the consequences of the fact that genes occupy specific positions in chromosomes.

Mendel's Work

Early genetic research did not use humans as subjects; there are obvious limitations to any such program. Even without considering ethical issues, we can readily see that the nine-month period of embryonic development, the comparatively small number of offspring each human produces, and the long time it takes to reach sexual maturity all argue against the use of our species for breeding experiments.

Much of the early genetic work was done with plants because of their economic importance; records show that people were deliberately cross-breeding date palm trees as early as 5,000 years ago. By the early 1800s, plant breeding was widespread, especially with ornamental flowers such as tulips. Half a century later, Gregor Mendel used the knowledge of plant reproduction to design and conduct experiments on inheritance between generations. Although his published results were neglected by scientists for 40 years, they ultimately became the foundation for the science of genetics.

Plant breeders control which plants mate

Plants are easily grown in large quantities, they often produce large numbers of offspring (in the form of seeds), and many have relatively short generation times. In many plant species, the same individuals have both male and female reproductive organs, in which case each plant may reproduce as a male, as a female, or as both. Best of all, it is often easy to control which individuals mate (Figure 10.1).

Some observations that Mendel found useful in his studies had been made in the late eighteenth century by a German botanist, Josef Gottlieb Kölreuter. Kölreuter attempted many crosses between plants, producing many **hybrids** (the offspring of genetically different parents) and learning a great deal about the biology of plant reproduction. He was the first to demonstrate that the two parents play equal roles in heredity, and he studied the offspring from **reciprocal**

10.1 A Controlled Cross between Two Plants   Mendel used the pea plant, *Pisum sativum*, in many of his experiments.

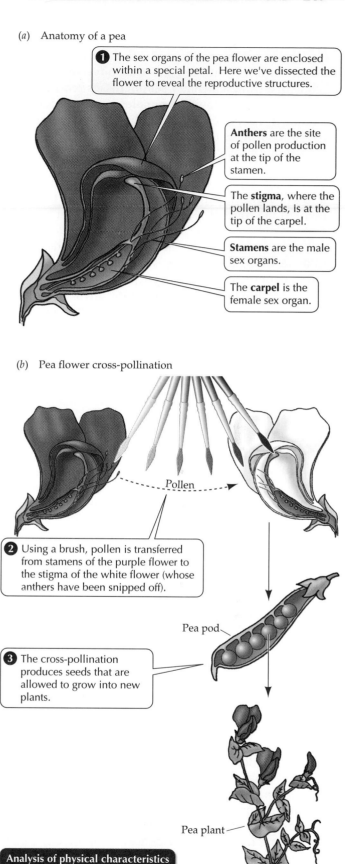

(a) Anatomy of a pea

1 The sex organs of the pea flower are enclosed within a special petal. Here we've dissected the flower to reveal the reproductive structures.

Anthers are the site of pollen production at the tip of the stamen.

The **stigma**, where the pollen lands, is at the tip of the carpel.

Stamens are the male sex organs.

The **carpel** is the female sex organ.

(b) Pea flower cross-pollination

Pollen

2 Using a brush, pollen is transferred from stamens of the purple flower to the stigma of the white flower (whose anthers have been snipped off).

Pea pod

3 The cross-pollination produces seeds that are allowed to grow into new plants.

Pea plant

Analysis of physical characteristics (see Figure 10.2) of the offspring shows evidence of hereditary transmission from both parents.

crosses—matings made in two directions. For example, in one set of crosses, males that have white flowers are crossed with females that have red flowers, while in a complementary set of crosses red-flowered males and white-flowered females are the parents. In Kölreuter's experience, such reciprocal crosses always gave identical results.

Although the concept of equal parental contributions of genetic determinants was an important discovery, the nature of these determinants remained unknown. Laws of inheritance proposed at the time favored the concept of *blending*. If a plant that had one form of a characteristic (say, red flowers) were crossed with one that had a different form of that characteristic (blue flowers), the offspring would be a blended combination of the two parents (purple flowers).

And following the blending motif, it was thought that once heritable elements were combined, they could not be separated again (like combined inks). The red and blue genetic determinants were thought to be forever blended into the new purple one. Then, about a century after Kölreuter completed his work, Mendel began his.

G. Mendel

Mendel's discoveries lay dormant for decades

When Mendel began his research, it was known that one female gamete combines with one male gamete to bring about fertilization. However, the role of the chromosomes as bearers of genetic information was unknown, and mitosis and meiosis were yet to be discovered.

Mendel himself was well qualified to make the big step forward. Although in 1850 he had failed an examination for a teaching certificate in natural science, he later undertook intensive studies in physics, chemistry, mathematics, and various aspects of biology at the University of Vienna. His work in physics and mathematics probably led to his applying experimental and quantitative methods to the study of heredity—and these were the key ingredients in his success.

Mendel worked out the basic principles of the heredity of plants and animals over a period of about nine years, the work culminating in a public lecture in 1865 and a detailed written account published in 1866. Mendel's paper on plant hybridization appeared in a journal that was received by 120 libraries, and he sent reprinted copies (of which he had obtained 40) to several distinguished scholars. However, his theory was not accepted. In fact, it was ignored.

Perhaps the chief difficulty was that the physical basis of his theory was not understood until the discovery of meiosis, some years later. Furthermore, the most prominent biologists of Mendel's time were not in the habit of thinking in mathematical terms, even the simple terms used by Mendel. Whatever the reasons, Mendel's pioneering paper had no discernible influence on the scientific world for more than 30 years.

Then, in 1900, Mendel's discoveries burst into prominence as a result of independent experiments by the Dutch Hugo de Vries, the German Karl Correns, and the Austrian Erich von Tschermak. Each of these scientists carried out crossing experiments and obtained quantitative data about the progeny; each published his principal findings in 1900; each cited Mendel's 1866 paper. At last the time was ripe for biologists to appreciate the significance of what these four geneticists had discovered—largely because meiosis had by then been described. That Mendel was able to make his discoveries *before* the discovery of meiosis was due in part to the methods of experimentation he used.

Mendel's Experiments and Laws of Inheritance

Mendel's work is a fine example of preparation, execution, and interpretation in the experimental method. Let's see how he approached each of these steps.

Mendel prepares to experiment

Mendel chose the garden pea for his studies because of its ease of cultivation, the feasibility of controlled pollination (see Figure 10.1), and the availability of varieties with differing traits. He controlled pollination, and thus fertilization, by manually moving pollen from one plant to another; thus he knew the parentage of the offspring in his experiments. If untouched, the peas Mendel studied naturally self-pollinate—that is, the female organs of flowers receive pollen from the male organs of the same flowers—and he made use of this natural phenomenon in some of his experiments.

Mendel began by examining varieties of peas in a search for heritable characters and traits suitable for study. A **character** is a feature such as flower color; a **trait** is a particular form of a character, such as white flowers; a **heritable** character trait is one that is passed from parent to offspring. Suitable characters for Mendel's experiments were those that had well-defined, contrasting alternatives, such as purple flowers versus white flowers, and that were *true-breeding*.

To be considered true-breeding, the observed trait must be the only form present in many generations. In other words, peas with white flowers, when crossed with one another, would have to give rise *only* to progeny with white flowers for many generations; tall plants bred to tall plants would have to produce *only* tall progeny.

Mendel isolated each of his true-breeding strains by repeated inbreeding (crossing of seemingly identical individuals, or of individuals with themselves) and se-

Character	Dominant trait	Recessive trait
Seed shape	Spherical	Wrinkled
Seed color	Yellow	Green
Flower color	Purple	White
Pod shape	Inflated	Constricted
Pod color	Green	Yellow
Flower position	Axial	Terminal
Stem height	Tall	Dwarf

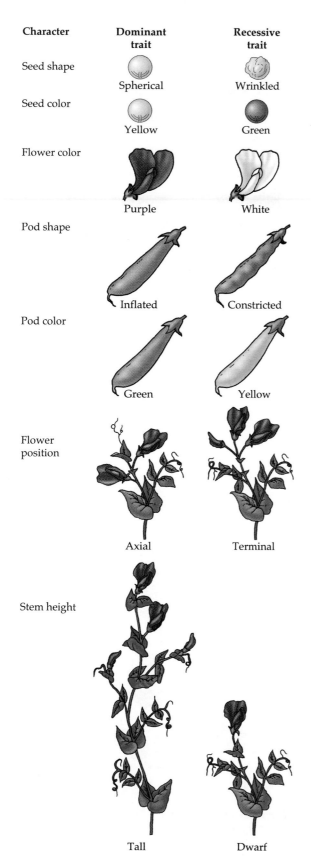

10.2 True-Breeding Traits Traits that are the only form present over many generations are considered to be true-breeding. Mendel's studies focused on seven contrasting pairs of such traits in *Pisum sativum*.

lection. For most of his work, Mendel concentrated on the seven pairs of contrasting traits shown in Figure 10.2. Before performing a given cross, he made sure that each potential parent was from a true-breeding strain—an essential point in his analysis of his experimental results.

Mendel then placed pollen he collected from one parental strain onto the stigma (female organ) of flowers of the other strain. The plants providing and receiving the pollen were the **parental generation**, designated **P**. In due course, seeds formed and were planted. The resulting new plants constituted the **first filial generation**, F_1. Mendel and his assistants examined each F_1 plant to see which traits it bore and then recorded the number of F_1 plants expressing each trait. In some experiments the F_1 plants were allowed to self-pollinate and produce a **second filial generation**, or F_2. Again, each F_2 plant was characterized and counted.

In sum, Mendel devised a well-organized plan of research, pursued it faithfully and carefully, recorded great amounts of quantitative data, and analyzed the numbers he recorded to explain the relative proportions of the different kinds of progeny. His 1866 paper stands to this day as a model of clarity. His results and the conclusions to which they led are the subject of the next few sections.

Mendel's Experiment 1 examined a monohybrid cross

"Experiment 1" in Mendel's paper included a monohybrid cross—one in which the parents were true-breeding for a given character but each displayed a different form of that character. He took pollen from plants of a true-breeding strain with wrinkled seeds and placed it on the stigmas of flowers of a true-breeding, spherical-seeded strain (Figure 10.3). He also performed the reciprocal cross, placing pollen from the spherical-seeded strain on the stigmas of flowers of the wrinkled-seeded strain.

In both cases, all the F_1 seeds that were produced were spherical—it was as if the wrinkled trait had disappeared completely. The following spring Mendel grew 253 F_1 plants from these spherical seeds, each of which was allowed to self-pollinate—this was the monohybrid cross—to produce F_2 seeds. In all, there were 7,324 F_2 seeds, of which 5,474 were spherical and 1,850 wrinkled (Figure 10.4).

Mendel observed that the spherical seed trait was **dominant**: In the F_1 generation, it was always expressed rather than the wrinkled seed trait, which he called **recessive**. In each of the other six pairs of traits studied by Mendel, one proved to be dominant over the other. When he crossed plants differing in one of these traits, only one of each pair of traits was evident in the F_1 generation. However, the trait that was not seen in the F_1 *reappeared* in the F_2.

10.3 Contrasting Traits In Experiment 1, Mendel studied the inheritance of seed shape. We know today that the wrinkled seeds possess an abnormal form of starch. Contrast their appearance with that of the spherical seeds below.

Of most importance, the ratio of the two traits in the F_2 generation was always the same—approximately 3:1; that is, three-fourths of the F_2 showed the dominant trait and one-fourth showed the recessive trait (Table 10.1). In Mendel's Experiment 1, the ratio was $5,474:1,850 = 2.96:1$. Reciprocal crosses in the parental generation gave similar outcomes in the F_2.

How Mendel interpreted his results

By themselves, the results from Experiment 1 disproved the widely held belief that inheritance is a blending phenomenon. According to the blending theory, Mendel's F_1 seeds should have had an appearance intermediate between those of the two parents—in other words, they should have been *slightly* wrinkled. Furthermore, the blending theory offered no explanation for the *reappearance* of the wrinkled trait in the F_2 seeds after its apparent absence in the F_1 seeds.

From his results Mendel proposed that the units responsible for the inheritance of specific traits are present as discrete units (particles) that occur in pairs and segregate (separate) from one another during the formation of gametes. In this particulate theory, the hereditary carriers retain their integrity in the presence of other units. This is a sharp contrast to blending, in which the carriers were believed to lose their identities when mixed together.

As he wrestled mathematically with his data, Mendel reached the conclusion that each pea has two units for each character, one from each parent. During the production of gametes, only one of the paired units for a given character is given to a gamete. Hence each gamete contains one unit, and the resulting zygote contains two, because it was produced by the fusion of two gametes. This conclusion is the core of Mendel's model of inheritance. Mendel's unit is now called a **gene**.

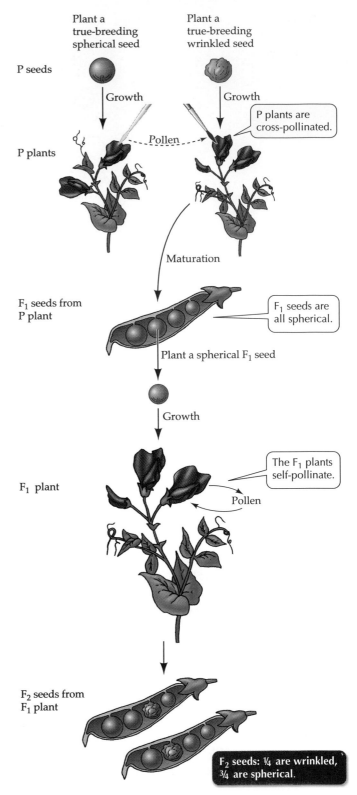

10.4 Mendel's Experiment 1 The pattern Mendel observed in the F_2 generation—$^1/_4$ of the seeds wrinkled, $^3/_4$ spherical—was the same no matter which variety contributed the pollen in the parental generation.

TABLE 10.1 Mendel's Results from Monohybrid Crosses

P			F$_2$			
DOMINANT	×	**RECESSIVE**	**DOMINANT**	**RECESSIVE**	**TOTAL**	**RATIO**
Spherical	×	Wrinkled seeds	5,474	1,850	7,324	2.96:1
Yellow	×	Green seeds	6,022	2,001	8,023	3.01:1
Purple	×	White flowers	705	224	929	3.15:1
Inflated	×	Constricted pods	882	299	1,181	2.95:1
Green	×	Yellow pods	428	152	580	2.82:1
Axial	×	Terminal flowers	651	207	858	3.14:1
Tall	×	Dwarf stems	787	277	1,064	2.84:1

Mendel reasoned that in Experiment 1, the spherical-seeded parent had a pair of genes of the same type, which we will call *S*, and the parent with wrinkled seeds had two *s* genes. The *SS* parent produced gametes each containing a single *S*, and the *ss* parent produced gametes each with a single *s*. Each member of the F$_1$ generation had an *S* from one parent and an *s* from the other; an F$_1$ could thus be described as *Ss*. We say that *S* is dominant over *s* because *s* is not evident when both forms of the gene are present.

The different forms of a gene (*S* and *s* in this case) are called **alleles**. Individuals that are true-breeding for a trait contain two copies of the same allele. For example, all the individuals in a population of a strain of true-breeding peas with wrinkled seeds must have the allele pair *ss*; if *S* were present, the plants would produce spherical seeds.

We say that individuals that produce wrinkled seeds are **homozygous** for the allele *s*, meaning that they have two copies of the same allele (*ss*). Some peas with spherical seeds—the ones with the genotype *SS*—are also homozygous. However, not all plants with spherical seeds have the *SS* genotype. Some spherical-seeded plants are **heterozygous**: They have two different alleles of the gene in question (in this case *Ss*).

To illustrate these terms with a more complex example, one in which there are three gene pairs, an individual with the genotype *AABbcc* is homozygous for the *A* and *C* genes—because it has two *A* alleles and two *c* alleles—but heterozygous for the *B* gene because it contains the *B* and *b* alleles. An individual that is homozygous for a character is sometimes called a *homozygote*; a *heterozygote* is heterozygous for the character in question.

The physical appearance of an organism is its **phenotype**. Mendel correctly supposed the phenotype to be the result of the **genotype**, or genetic constitution, of the organism showing the phenotype. In Experiment 1 we are dealing with two phenotypes (spherical seeds and wrinkled seeds). As we will see in the next section, the F$_2$ generation contains these two phenotypes and three genotypes: The wrinkled-seed phenotype is produced only by the genotype *ss*, whereas the spherical-seed phenotype may be produced by the genotypes *SS* or *Ss*.

Mendel's first law says that alleles segregate

How does Mendel's model of inheritance explain the composition of the F$_2$ generation in Experiment 1? Consider first the F$_1$, which has the spherical-seeded phenotype and the *Ss* genotype. According to the model, when any individual produces gametes, the alleles **segregate**, or separate, so each gamete receives only *one* member of the pair of alleles. This is Mendel's first law, the **law of segregation**. In Experiment 1, half the gametes produced by the F$_1$ contain the *S* allele and half the *s* allele. Years after Mendel's death, geneticists determined that the basis of segregation was the separation of homologous chromosomes during meiosis I (see Chapter 9).

During plant pollination, the random combination of these gametes produces the F$_2$ generation (Figure 10.5). Three different F$_2$ genotypes are possible: *SS*, *Ss* (which is the same thing as *sS*), and *ss*. Our quantitative way of looking at things may lead us to wonder what proportions of these genotypes we might expect to observe in the F$_2$ progeny. The expected frequencies of these three genotypes in our example may be determined by using the Punnett square, devised in 1905 by the British geneticist Reginald Crundall Punnett.

The **Punnett square** is a device to remind us to consider all possible combinations of gametes. The square looks like this:

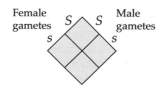

Female gametes *S* *S* Male gametes
s   *s*

It is a simple grid with all possible sperm genotypes shown across one side and all possible egg genotypes along another side. To complete the grid, fill each

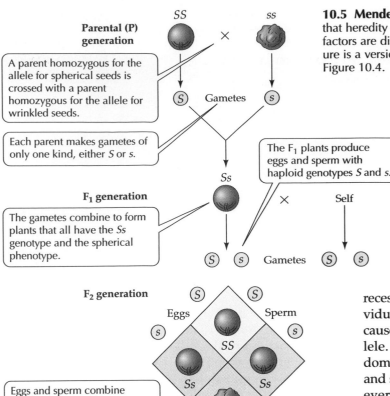

Parental (P) generation

SS × *ss*

A parent homozygous for the allele for spherical seeds is crossed with a parent homozygous for the allele for wrinkled seeds.

Gametes *S*   *s*

Each parent makes gametes of only one kind, either *S* or *s*.

The F₁ plants produce eggs and sperm with haploid genotypes *S* and *s*.

Ss

F₁ generation

The gametes combine to form plants that all have the *Ss* genotype and the spherical phenotype.

× Self

Gametes *S* *s*   *S* *s*

F₂ generation

Eggs   *s*   Sperm   *s*

SS   *Ss*   *Ss*

ss

Eggs and sperm combine randomly in four different ways to form F₂ plants, as this Punnett square shows.

Punnett square

10.5 Mendel's Explanation of Experiment 1 Mendel concluded that heredity depends on factors from each parent, and that these factors are discrete units that do not blend in the offspring. This figure is a version of Mendel's explanation of the experiment shown in Figure 10.4.

square with the corresponding sperm genotype and egg genotype, giving the diploid genotype of one member of the F₂ generation. For example, to fill the rightmost square, put in the *S* from the egg (female gamete) and the *s* from the sperm (male gamete), yielding *Ss*.

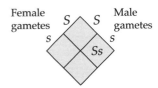

Female gametes   *S*   *S*   Male gametes
s   *s*
Ss

Examination of the Punnett square in Figure 10.5 reveals that self-pollination of the F₁ genotype *Ss* will give the three F₂ genotypes in the expected ratio 1 *SS*: 2 *Ss*:1 *ss*. Because *S* is dominant and *s* recessive, only two phenotypes result, in the ratio of 3 spherical (*SS* and *Ss*) to 1 wrinkled (*ss*), just as Mendel observed.

Mendel did not live to see his theory placed on a sound physical footing based on chromosomes and DNA. Genes are now known to be portions of the DNA molecules in chromosomes. More specifically, a gene is a portion of the DNA that resides at a particular site, called a **locus**, within the chromosome and that encodes a particular function. Mendel arrived at his law of segregation with no knowledge of chromosomes or meiosis, but today we can picture the alleles segregating as chromosomes separate into gametes in meiosis (Figure 10.6).

Mendel verified his hypothesis by performing a test cross

The **test cross** is a way to test whether a given individual showing a dominant trait is homozygous or heterozygous. In a test cross, the individual in question is crossed with an individual known to be homozygous for the recessive trait—an easy individual to identify because in order to have the recessive phenotype it must be homozygous; that is, it must contain alleles only for the recessive trait.

For the gene that we have been considering, the recessive homozygote for the test cross is *ss*. The individual being tested may be described initially as *S*– because we do not yet know the identity of the second allele. If the individual being tested is homozygous dominant (*SS*), all offspring of the test cross will be *Ss* and show the dominant trait (spherical seeds). If, however, the tested individual is heterozygous (*Ss*), then approximately half of the offspring of the test cross will show the dominant trait, but the other half will be homozygous for the recessive trait (Figure 10.7). These are exactly the results that are obtained; thus Mendel's model accurately predicts the results of such test crosses.

Mendel's second law says that alleles of different genes assort independently

What happens if two parents that differ at two or more loci are crossed? When a double heterozygote (for example, *AaBb*) makes gametes, do the alleles of maternal origin go together to one gamete and those of paternal origin to another gamete? Or does a single gamete receive some maternal and some paternal alleles? To answer these questions Mendel performed a series of **dihybrid crosses**, crosses made between parents that are identical double heterozygotes.

In these experiments Mendel began with peas that differed for two characters of the seeds: seed shape and seed color. One true-breeding strain produced only spherical, yellow seeds (*SSYY*) and the other strain produced only wrinkled, green ones (*ssyy*). A cross between these two strains produces an F₁ generation in which all the plants are *SsYy*. Because the *S* and *Y* alleles are dominant, these F₁ seeds would all be yellow and spherical.

Mendel continued this experiment to the next generation—the dihybrid cross. There are two ways in which these doubly heterozygous plants might produce gametes, as Mendel saw it. (Remember that he had never heard of chromosomes, let alone of meio-

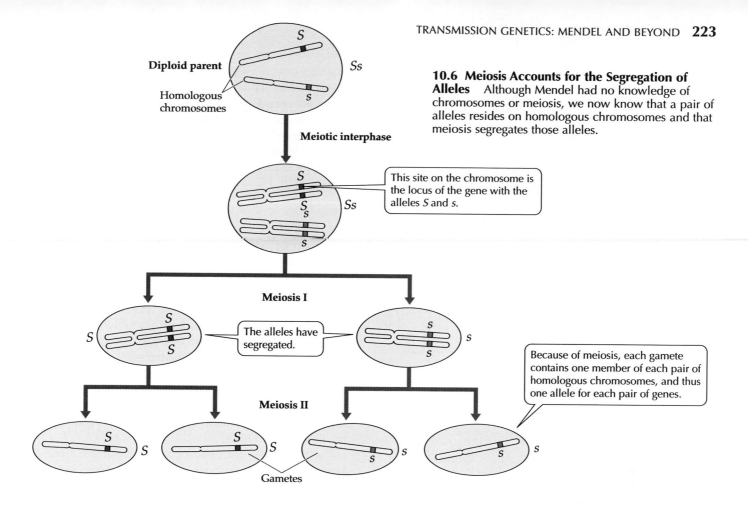

Diploid parent

Homologous chromosomes

Ss

Meiotic interphase

This site on the chromosome is the locus of the gene with the alleles *S* and *s*.

Ss

Meiosis I

The alleles have segregated.

Because of meiosis, each gamete contains one member of each pair of homologous chromosomes, and thus one allele for each pair of genes.

Meiosis II

Gametes

10.6 Meiosis Accounts for the Segregation of Alleles Although Mendel had no knowledge of chromosomes or meiosis, we now know that a pair of alleles resides on homologous chromosomes and that meiosis segregates those alleles.

sis.) First, if the alleles maintain the associations they had in the original parents (that is, if they are linked), then only two types of gametes would be produced (*SY* and *sy*); and the F_2 progeny resulting from self-pollination of the F_1 plants would consist of three times as many plants bearing spherical, yellow seeds as ones with wrinkled, green seeds. Were such results to be obtained, there would be no reason to suppose that seed shape and seed color were really regulated by two different genes, because spherical seeds would always be yellow, and wrinkled ones always green.

The second possibility is that the segregation of *S* from *s* is *independent* of the segregation of *Y* from *y* during the production of gametes (that is, they are un-linked). In this case, four kinds of gametes would be produced, and in equal numbers: *SY*, *Sy*, *sY*, and *sy*. When these gametes combined at random, they would produce an F_2 of nine different genotypes. The progeny can have any of three possible genotypes for shape (*SS*, *Ss*, or *ss*) and any of three genotypes for color (*YY*, *Yy*, or *yy*). The combined nine genotypes

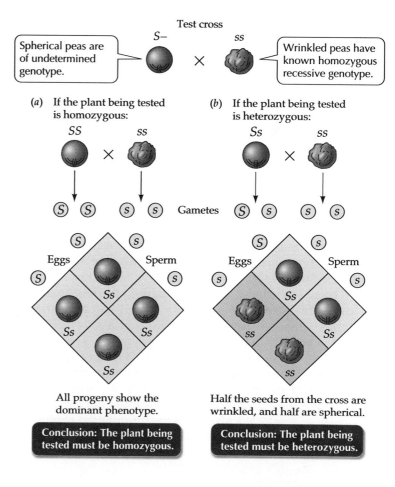

Test cross

Spherical peas are of undetermined genotype.

S—   ×   *ss*

Wrinkled peas have known homozygous recessive genotype.

(a) If the plant being tested is homozygous:

SS × *ss*

Gametes

Eggs   Sperm

All progeny show the dominant phenotype.

Conclusion: The plant being tested must be homozygous.

(b) If the plant being tested is heterozygous:

Ss × *ss*

Gametes

Eggs   Sperm

Half the seeds from the cross are wrinkled, and half are spherical.

Conclusion: The plant being tested must be heterozygous.

10.7 Homozygous or Heterozygous? A plant with a dominant phenotype may be homozygous or heterozygous. Its genotype can be determined by making a test cross, which means observing the phenotypes of progeny produced by crossing it with a homozygous recessive plant.

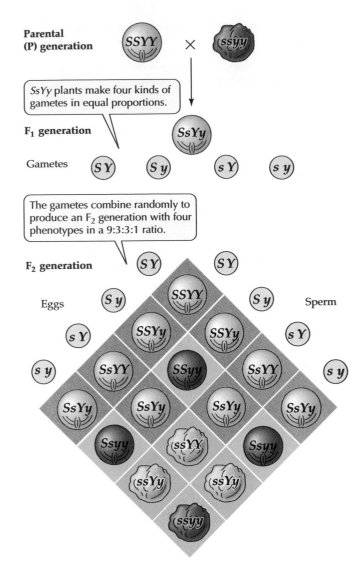

10.8 Independent Assortment The 16 possible combinations of gametes result in 9 different genotypes. Because S and Y are dominant over s and y, respectively, the 9 genotypes determine 4 phenotypes in the ratio of 9:3:3:1.

would produce just four phenotypes (spherical yellow, spherical green, wrinkled yellow, wrinkled green). By using a Punnett square, we can show that these four phenotypes would be expected to occur in a ratio of 9:3:3:1. (Figure 10.8)

Mendel's dihybrid crosses produced the results predicted by the second possibility. Four different phenotypes appeared in the F_2 in a ratio of about 9:3:3:1, rather than only the two parental types as predicted from the first possibility. The parental traits appeared in new combinations in two of the phenotypic classes (spherical green and wrinkled yellow). These new combinations are called **recombinant phenotypes**.

These results led Mendel to the formulation of what is now known as Mendel's second law: *Alleles of differ-*

ent genes assort independently of one another during gamete formation. This **law of independent assortment** is not as universal as the law of segregation, because it applies to genes that lie on separate chromosomes, but not necessarily to those that lie on the same chromosome. However, it is correct to say that *chromosomes* assort independently during the formation of gametes (Figure 10.9).

Punnett squares or probability calculations? You choose!

Many people find it easiest to solve genetics problems using probability calculations, perhaps because the basic underlying considerations are familiar. When we flip a coin, for example, we expect it has an equal probability of landing "heads" or "tails."

For a given toss, the probability of heads is independent of what happened in all the previous tosses. For a fair coin, a run of ten straight heads implies nothing about the next toss. No "law of averages" increases the likelihood that the next toss will come up tails, and no "momentum" makes an eleventh occurrence of heads any more likely. On the eleventh toss, the odds are still 50:50.

The basic conventions of probability are simple: If an event is absolutely certain to happen, its probability is 1. If it cannot happen, its probability is 0. Otherwise, its probability lies between 0 and 1. A coin toss results in heads approximately half the time, and the probability of heads is $\frac{1}{2}$—as is the probability of tails.

MULTIPLYING PROBABILITIES. If *two* coins (a penny and a dime, say) are tossed, each acts independently of the other. What, then, is the probability of both coins coming up heads? Half the time, the penny comes up heads; of that fraction, half the time the dime also comes up heads. Therefore, the joint probability of two heads is half of one-half, or $\frac{1}{2} \times \frac{1}{2} = \frac{1}{4}$. To find the joint probability of *independent* events, then, the general rule is to *multiply* the probabilities of the individual events (Figure 10.10).

THE MONOHYBRID CROSS. To apply a probabilistic approach to genetics problems, we need only deal with gamete formation and random fertilization. A homozygote can produce only one type of gamete, so, for example, an SS individual has a probability equal to 1 of producing gametes with the genotype S. The heterozygote Ss produces S gametes with a probability of $\frac{1}{2}$, and s gametes with a probability of $\frac{1}{2}$.

Consider the F_2 progeny of the cross in Figure 10.5. They are obtained by self-pollination of F_1 hybrids of genotype Ss. The probability that an F_2 plant has the genotype SS must be $\frac{1}{2} \times \frac{1}{2} = \frac{1}{4}$ because there is a 50:50 chance that the sperm will have the genotype S, and this chance is independent of the 50:50 chance that

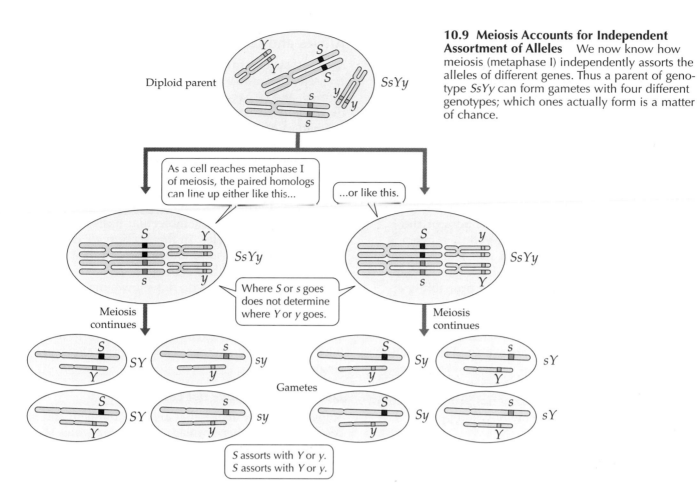

10.9 Meiosis Accounts for Independent Assortment of Alleles We now know how meiosis (metaphase I) independently assorts the alleles of different genes. Thus a parent of genotype *SsYy* can form gametes with four different genotypes; which ones actually form is a matter of chance.

the egg will have the genotype *S*. Similarly, the probability of *ss* offspring is ½ × ½ = ¼.

ADDING PROBABILITIES. The probability of getting *S* from the sperm and *s* from the egg is also ¼, but remember that the same genotype can also result from *s* in the sperm and *S* in the egg, with a probability of ¼. The probability of an event that can occur in two or more different ways is the sum of the individual probabilities of those ways. Thus the probability that an F₂ plant is a heterozygote is ¼ + ¼ = ½ (see Figure 10.10). The three genotypes are expected in the ratio ¼ *SS*: ½ *Ss*:¼ *ss*—hence the 1:2:1 ratio of genotypes and the 3:1 ratio of phenotypes seen in Figure 10.5.

10.10 Joint Probabilities of Independent Events Like two tosses of a coin, the segregation of alleles during sperm and egg formation are independent events. The probability (*P*) of any given combination of alleles from a sperm and an egg is obtained by multiplying the probabilities of each event; this is the probability of producing a homozygote. Since a heterozygote can be formed in two ways, the two probabilities are added together.

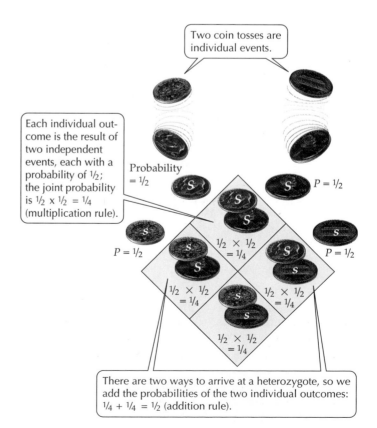

THE DIHYBRID CROSS. If F_1 plants heterozygous for two independent characters self-pollinate, the resulting F_2 plants express four phenotypes. The proportions of these phenotypes are easily determined by probabilities. Let's see how this works for the experiment of Figure 10.8.

The probability that a seed will be spherical is ¾, as we have just seen. By the same reasoning, the probability that a seed will be yellow is also ¾. The two characters are determined by separate genes and are independent of one another, so the joint probability that a seed will be both spherical and yellow is ¾ × ¾ = ⁹⁄₁₆. For the wrinkled, yellow members of the F_2 generation, the probability of being yellow is again ¾; the probability of being wrinkled is ½ × ½ = ¼. The joint probability that a seed will be both wrinkled and yellow, then, is ¾ × ¼ = ³⁄₁₆. The same probability applies, for similar reasons, to the spherical, green F_2 seeds. Finally, the probability that F_2 seeds will be both wrinkled and green must be ¼ × ¼ = ¹⁄₁₆. Looking at all four phenotypes, we see they are expected in the ratio of 9:3:3:1.

Probability calculations and Punnett squares give the same results. Learn to do genetics problems both ways, and then decide which method you prefer.

Alleles and Their Interactions

Mendel successfully interpreted the monohybrid crosses, test crosses, and dihybrid crosses he performed, and he gave us his laws of segregation and independent assortment. Decades later, genes became defined as chemical entities—DNA sequences—that are expressed as proteins.

Let's move on to the extensions to Mendelian genetics that have been developed by other workers, mostly in the early part of the twentieth century. Recall that alleles are alternative forms of the same gene. In the next chapter we'll see the molecular basis for the distinction between alleles. In this section we deal with how alleles relate to one another, some of their general properties, and how they arise.

In many cases, alleles do not show simple relationships between dominance and recessiveness. In others, a single allele may have multiple phenotypic effects when it is expressed. Existing alleles can form new alleles by mutation, so there can be many alleles for a single character.

Dominance is usually not complete

Some genes have alleles that are not dominant and recessive to each other. Instead, the heterozygotes show an intermediate phenotype—at first glance like that predicted by the old blending theory of inheritance.

For example, if a true-breeding red snapdragon is crossed with a true-breeding white one, all the F_1 flowers are pink. That this phenomenon can still be explained in terms of Mendelian genetics rather than of a blending theory is readily demonstrated by a further cross.

If one of these pink F_1 snapdragons is crossed with a true-breeding white one, the blending theory predicts that all the offspring would be a still-lighter pink. In fact, approximately ½ of the offspring are white and ½ the same pink as the original F_1. Suppose now that the F_1 pink snapdragons are self-pollinated. The resulting F_2 plants are distributed in a ratio of 1 red:2 pink:1 white (Figure 10.11). Clearly the hereditary particles—the genes—have *not* blended; in the F_2 they are readily sorted out in their original forms.

We can understand these results in terms of the Mendelian model. When a

Heterozygous snapdragons produce pink flowers because the allele for red flowers is incompletely dominant over the allele for white ones.

When true-breeding red and white parents cross, all plants in the F_1 generations are pink.

Parental (P) generation

rr White   *RR* Red

F_1 generation

Rr *Rr* *Rr* *Rr* *Rr*   *rr* White

F_2 generation

rr ¼ White   *Rr* ½ Pink   *RR* ¼ Red   *Rr* ½ Pink   *rr* ½ White

When F_1 plants self-pollinate, they produce F_2 offspring that are white, pink, and red in a ratio of 1:2:1.

A test cross confirms that pink snapdragons are heterozygous.

10.11 Incomplete Dominance Follows Mendel's Laws The study of incomplete dominance clearly disproves the idea of "blended" inheritance. In the F_2 generation, the genes are readily sorted out in their original forms.

heterozygous phenotype is intermediate, as in the snapdragon example, the gene is said to be governed by **incomplete dominance**. All we need to do in cases like this is recognize that the heterozygotes show a phenotype intermediate between those of the two homozygotes. Genes code for the production of specific proteins, many of which are enzymes. Different alleles at a locus code for alternative forms of a protein. When the protein is an enzyme, the different forms often have different degrees of catalytic activity.

In the snapdragon example, one allele codes for an enzyme that catalyzes a reaction leading to the formation of a red pigment in the flowers. The alternative allele codes for an altered protein that lacks catalytic activity for pigment production. Plants homozygous for this alternative allele cannot synthesize red pigment, and their flowers are white. Heterozygous plants, with only one allele for the functional enzyme, produce just enough red pigment that their flowers are pink. Homozygous plants that have two alleles for the functional enzyme produce more red pigment, resulting in red flowers.

There are more examples of incomplete dominance than of complete dominance. Thus an unusual feature of Mendel's report is that all seven of the examples he described (see Table 10.1) are characterized by complete dominance. For dominance to be complete, a single copy of the dominant allele must produce enough of its protein product to give the maximum phenotypic response.

For example, just one copy of the dominant allele T at one of the loci studied by Mendel leads to the production of enough of a growth-promoting chemical that the Tt heterozygotes are as tall as the homozygous dominant plants (TT)—the second copy of T causes no further growth of the stem. The homozygous recessive plants (tt) are much shorter because the allele t does not lead to the production of the growth promoter.

Sometimes two alleles at a locus produce two different phenotypes that *both* appear in heterozygotes. An example of this phenomenon, called **codominance**, is seen in the ABO blood group system in humans. One allele of the gene determines the A blood type, and another determines the B blood type. When both alleles are present, both are expressed, resulting in the AB blood type, which includes proteins produced by both alleles. We will return to other features of the ABO blood system shortly.

Some alleles have multiple phenotypic effects

When a single allele has more than one distinguishable phenotypic effect, we say that the allele is **pleiotropic**. A familiar example of pleiotropy is the allele responsible for the coloration pattern (light body, darker extremities) of Siamese cats, discussed later in

this chapter. The same allele is also responsible for the characteristic crossed eyes of Siamese cats. Although these effects appear to be unrelated, both result from the same protein produced under the influence of that allele.

New alleles arise by mutation

Why does a gene have different alleles? Different alleles exist because any gene is subject to **mutation**, which means that it can be changed to a *stable, heritable* new form. In other words, an allele can mutate to become a different allele.

One particular allele of a gene may be defined as the **wild type**, or standard, because it is present in most individuals in nature and gives rise to an expected trait or phenotype. Other forms of that same gene, often called **mutant** alleles, may produce a different phenotype. The wild-type and mutant alleles reside at the same locus and are inherited according to the rules set forth by Mendel. A genetic locus with a wild-type allele present less than 99 percent of the time (the rest of the alleles being mutant) is said to be **polymorphic** (from the Greek *poly*, "many," and *morph*, "form").

Mutation, to be discussed in detail in Chapter 12, is a random process; different copies of the same gene may be changed in different ways, depending on how and where the DNA changes. Genetic mutation is the raw material for the process of evolution—the subject of Part Three—and is the ultimate source of the vast biological diversity described in Part Four.

Many genes have multiple alleles

A group of individuals may have more than two alleles of a given gene. (Any one individual has only two alleles, of course—one from the mother and one from the father.) In fact, there are many examples of such **multiple alleles**. In the fruit fly *Drosophila melanogaster*, many alleles at one locus affect eye color by determining the amount of pigment produced (Figure 10.12). The exact color of the fly's eyes depends on which two alleles are inherited.

The ABO blood group system in humans is determined by a set of three alleles (I^A, I^B, and I^O) at one locus, which determines certain proteins (antigens) on the surface of red blood cells. Different combinations of these alleles in different people produce four different blood types, or phenotypes: A, B, AB, and O (Figure 10.13). Early attempts at blood transfusion—made before these blood types were understood—frequently killed the patient. Around 1900, however, the Austrian scientist Karl Landsteiner mixed blood cells and serum (blood from which cells have been removed) from different individuals. He found that only certain combinations of blood types are compatible. In other combinations, the red blood cells form clumps

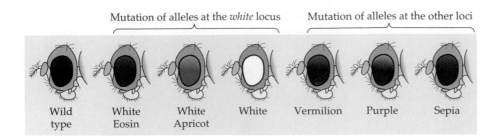

Mutation of alleles at the *white* locus   Mutation of alleles at the other loci

Wild type   White Eosin   White Apricot   White   Vermilion   Purple   Sepia

10.12 Multiple Alleles Govern Eye Color in *Drosophila* Wild-type flies have dark red eyes, but many alleles affect eye color. Mutations can produce a wide variety of eye colors by changing the amount of pigment produced.

because of the presence in the serum of specific proteins, called antibodies (see Chapter 18), that react with foreign, or "nonself," cells. This discovery led to our ability to administer compatible blood transfusions that do not kill the recipient.

The Rh factor, so named because it was first found in rhesus monkeys, is another substance on the surface of red blood cells. In most human populations, almost 100 percent of the individuals have the Rh factor, and their blood is said to be Rh⁺ (Rh-positive). Among Caucasians, however, only 83 percent are Rh⁺; the others lack the Rh factor, and their blood is called Rh⁻ (Rh-negative).

Like the ABO blood types, the Rh factor is genetically determined. A single locus with at least eight multiple alleles is responsible. Certain dominant alleles cause the production of the Rh factor; Rh⁻ individuals are homozygous recessives who lack these alleles. People who are Rh⁻ make antibodies that react with blood cells from Rh⁺ individuals. Because the blood systems of mother and newborn mix, if a woman who is Rh⁻ has an Rh⁺ child, the mother may make antibodies that react adversely with the blood of her newborn infant and can be fatal to the child (or, more commonly, to any subsequent Rh⁺ children she may have). Thus, pregnancies with Rh incompatibility are closely monitored.

Another system of multiple alleles is illustrated by the scallops in "Applying Concepts" Question 2 at the end of this chapter. The question of how differing alleles may be maintained in a population through time will be examined in Chapter 20.

Gene Interactions

Thus far we have treated the phenotype of an organism, with respect to a given character, as a simple result of its genotype, and we have implied that a single trait results from the alleles of a single gene. In fact, several genes may interact to determine a trait's phenotype. For example, height in people is determined by the actions of many genes, such as those that determine bone growth, hormones, and other aspects of development. Sometimes several genes act additively, so the phenotype can be predicted by how many of these genes are active. To complicate things further, the physical environment may interact with the genetic constitution of an individual in determining the phenotype. People's height, for example, is not determined only by their genes. Nutrition, one environmental factor, undoubtedly has a strong influence.

Some genes alter the effects of other genes

In **epistasis**, one gene alters the effect of another gene. For example, several genes determine coat color in mice. The wild-type color is agouti, a grayish pattern resulting from bands on the individual hairs. The dominant allele B determines that the

Blood type of cells	Genotype	Antibodies made by cells	Reaction to added antibodies	
			Anti-A	Anti-B
A	$I^A I^A$ or $I^A I^O$	Anti-B		
B	$I^B I^B$ or $I^B I^O$	Anti-A		
AB	$I^A I^B$	Neither anti-A nor anti-B		
O	$I^O I^O$	Both anti-A and anti-B		

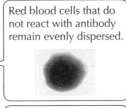

Red blood cells that do not react with antibody remain evenly dispersed.

Red blood cells that react with antibodies clump together (speckled appearance in the photograph).

10.13 ABO Blood Reactions Are Important in Transfusions Cells of blood types A, B, AB, and O were mixed with anti-A or anti-B antibodies. Note that anti-A reacts with A and AB cells but not with B or O. Which blood types do anti-B antibodies react with? As you look down the columns, note that each of the types, when mixed separately with anti-A and with anti-B, gives a unique pair of results; this is the basic method by which blood is typed.

Mice with genotype *aa* are albino regardless of their genotype for the other locus, because the *aa* genotype blocks all pigment production.

10.14 Genes May Interact Epistatically Epistasis occurs when one gene alters the effect of another gene. In these mice, the presence of the recessive genotype (*aa*) at one locus blocks pigment production, producing an albino mouse no matter what the genotype is at the second locus.

Mice with *bb* genotypes are black unless they are also *aa* (which makes them albino).

Mice that have at least one dominant allele at each locus are agouti.

hairs will have bands and thus that the color will be agouti, whereas the homozygous recessive genotype *bb* results in unbanded hairs and the mouse appears black. On another chromosome, a second locus affects an early step in the formation of hair pigments. The dominant allele *A* at this locus allows normal color development, but *aa* blocks all pigment production. Thus, *aa* mice are all-white albinos, irrespective of their genotype at the *B* locus (Figure 10.14).

If a mouse with genotype *AABB* (and thus the agouti phenotype) is crossed with an albino of genotype *aabb*, the F_1 is *AaBb* and of the agouti phenotype. If the F_1 mice are crossed with each other to produce an F_2, the epistasis of *aa* will result in an expected phenotypic ratio of 9 agouti:3 black:4 albino. Can you show why? The underlying ratio is the usual 9:3:3:1 for a dihybrid cross with unlinked genes, but be sure to look closely at each genotype and watch out for epistasis.

In another form of epistasis, two genes are mutually dependent: The expression of each depends on the alleles of the other. The epistatic action of such complementary genes may be explained as follows: Suppose the dominant gene *A* codes for enzyme A on the pathway for purple pigment in flowers and gene *B* codes for enzyme B:

colorless precursor $\xrightarrow{\text{enzyme A}}$ colorless intermediate $\xrightarrow{\text{enzyme B}}$ purple pigment

In order for the pigment to be produced, *both* reactions must take place. The alleles *a* and *b* code for nonfunc-

tional enzymes. If a plant is homozygous for either *a* or *b*, the corresponding reaction will not occur, no purple pigment will form, and the flowers will be white.

Polygenes mediate quantitative inheritance

Individual heritable characters are often found to be controlled by many genes, **polygenes**, of which each allele intensifies or diminishes the observed phenotype. As a result, variation in such characters is **continuous** rather than, as in the examples we have been considering, **discontinuous** (or discrete). Many characters that vary continuously—such as height and other aspects of size, or skin color—are under genetic control. Polygenes affecting a particular quantitative character are common on many different chromosomes.

Humans differ with respect to the amount of a dark pigment, melanin, in their skin (Figure 10.15). There is great variation in the amount of melanin among different people, but much of this variation is determined by alleles at just four (possibly three) loci. No alleles at these loci demonstrate dominance. Of course, skin color is not entirely determined by the genotype, since exposure to sunlight in fair-skinned people can cause the production of more melanin—that is, a suntan.

The environment affects gene action

Environmental variables such as light, temperature, and nutrition can sharply affect the translation of a genotype into a phenotype. A familiar example is the Siamese cat (Figure 10.16). This handsome animal normally has darker fur on its ears, nose, paws, and tail than on the rest of its body. The darkened parts have a lower temperature than the light parts.

A few simple experiments show that the Siamese cat has a genotype that results in dark fur, but only at temperatures below the general body temperature. If some dark fur is removed from the tail and the cat is kept at higher-than-usual temperatures, the new fur that grows in is light. Conversely, removal of light fur from the back, followed by local chilling of the area, causes the spot to fill with dark fur.

Genotype and environment interact to determine the phenotype of an organism. It is sometimes possible to determine the proportion of individuals in a group with a given genotype that actually show the expected phenotype. This proportion is called the **penetrance** of the genotype. The environment may also affect the **expressivity** of the genotype—that is, the degree to which it is expressed. For an example of environmental

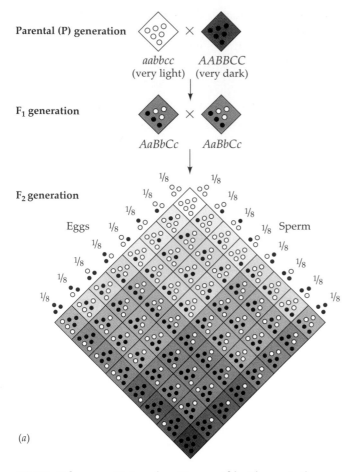

(a)

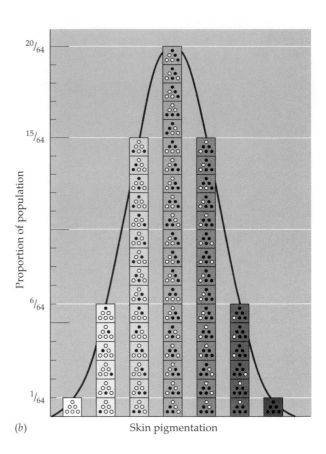

(b)   Skin pigmentation

10.15 Polygenes Determine Human Skin Pigmentation
A model of polygenic inheritance based on three genes. The
alleles *A*, *B*, and *C* contribute dark pigment to the skin, but
the alleles *a*, *b*, and *c* do not. The more *A*, *B*, and *C* alleles
an individual possesses, the darker that person's skin will be.
If both members of a couple have intermediate pigmentation
(*AaBbCc*, for example), they are unlikely to have children
with either very light or very dark skin.

effects on expressivity, consider how Siamese cats kept
indoors or outdoors in different climates might look.

Uncertainty over how much of the observed variation is due to the environment and how much to the
effects of the several polygenes complicates the analysis of quantitative inheritance. A useful approach that
avoids this difficulty is to study identical twins. Since
such twins are genetically identical, any differences
between them are attributed to environmental effects.

The phenotype of an organism depends on its total
genetic makeup and on its environment. Some of the
interactive effects will become more obvious when we
focus, in Chapter 14, on the regulation of gene expression.

10.16 The Environment Affects the Phenotype   This
Siamese cat has dark fur on its extremities, where the temperature is below the general body temperature.

Genes and Chromosomes

The recognition that genes occupy characteristic positions on chromosomes and thus are segregated by
meiosis enabled Mendel's successors to provide a physical explanation for his model of inheritance. It soon became apparent that the association of genes with chromosomes has other genetic consequences as well.

In this section we will address the following questions: What is the pattern of inheritance of genes that
occupy nearby loci on the same chromosome? How do

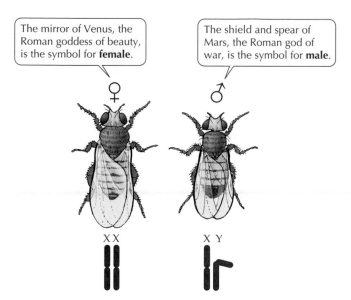

The mirror of Venus, the Roman goddess of beauty, is the symbol for **female**.

The shield and spear of Mars, the Roman god of war, is the symbol for **male**.

X X

X Y

10.17 *Drosophila melanogaster*, the Star of Morgan's Fly Room The fruit fly (whose Latin name means "vinegar-loving, dark-bodied fly") has a short generation time—a major reason for its widespread use as a laboratory organism in genetics experiments.

we determine the order of genes on a chromosome—and the distances between them? Why were all the carriers of hemophilia in Queen Victoria's family women, and why were all of her descendants who had hemophilia men?

The answers to these and many other genetic questions were worked out in studies of the fruit fly *Drosophila melanogaster* (Figure 10.17). Its small size, its ease of cultivation, and its short generation time made it an attractive experimental subject. Beginning in 1909, Thomas Hunt Morgan and his students established *Drosophila* as a highly useful laboratory organism in Columbia University's famous "fly room," where

10.18 Alleles That Don't Assort Independently
Morgan's studies showed that the genes for body color and wing size in *Drosophila* are linked, so their alleles do not assort independently. Linkage accounts for the departure of the actual phenotypes observed from the results predicted by Mendel's laws.

the group discovered the phenomena described in this section. *Drosophila* remains extremely important in studies of chromosome structure, population genetics, the genetics of development, and the genetics of behavior.

Linked genes are near each other on a chromosome

In the immediate aftermath of the rediscovery of Mendel's laws, the second law—independent assortment—was considered to be generally applicable. However, some investigators, including R. C. Punnett (the inventor of the square), began to observe strange deviations from the expected 9:3:3:1 ratio in some dihybrid crosses. T. H. Morgan, too, obtained data not in accord with Mendelian ratios and specifically not in accord with the law of independent assortment.

T. H. Morgan

Morgan crossed *Drosophila* of two known genotypes, $BbVgvg \times bbvgvg$, where B, the wild-type (gray) body, is dominant over b (black body), and Vg (wild-type wing) is dominant over vg (*vestigial*, a very small wing). Do you recognize this type of cross? It is a test cross (see Figure 10.7) for the two gene pairs. Morgan expected to see four phenotypes in a ratio of 1:1:1:1, but this was not what he observed. The body color gene and the wing size gene were not assorting independently; rather, they were for the most part inherited together (Figure 10.18).

These results became understandable to Morgan when he assumed that the two loci are on the *same*

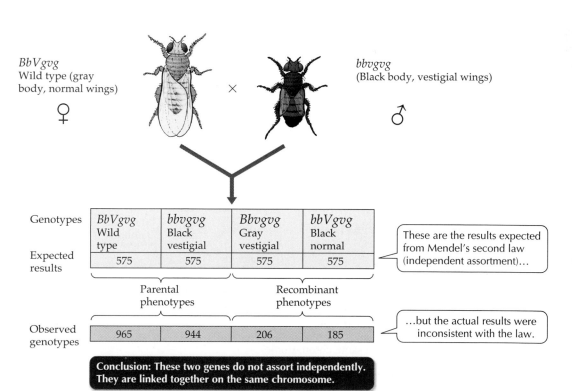

$BbVgvg$
Wild type (gray body, normal wings)
♀

×

$bbvgvg$
(Black body, vestigial wings)
♂

Genotypes	$BbVgvg$ Wild type	$bbvgvg$ Black vestigial	$Bbvgvg$ Gray vestigial	$bbVgvg$ Black normal
Expected results	575	575	575	575

These are the results expected from Mendel's second law (independent assortment)…

Parental phenotypes | Recombinant phenotypes

Observed genotypes	965	944	206	185

…but the actual results were inconsistent with the law.

Conclusion: These two genes do not assort independently. They are linked together on the same chromosome.

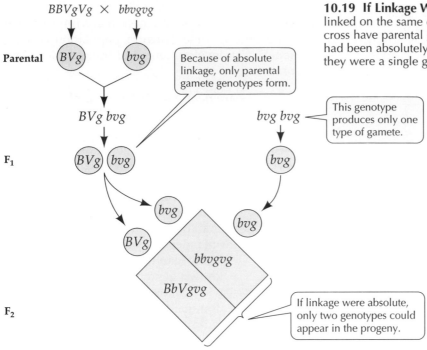

$BBVgVg \times bbvgvg$

Parental

Because of absolute linkage, only parental gamete genotypes form.

This genotype produces only one type of gamete.

F₁

F₂

If linkage were absolute, only two genotypes could appear in the progeny.

10.19 If Linkage Were Absolute   If two genes are absolutely linked on the same chromosome, all the F₂ offspring from a dihybrid cross have parental genotypes. If the genes in Morgan's experiment had been absolutely linked, they would have been inherited as if they were a single gene; such linkage is extremely rare.

Genes can be exchanged between chromatids

Complete linkage is extremely rare. If linkage were absolute, Mendel's second law (independent assortment of alleles of different loci) would apply only to loci on different chromosomes. What actually happens is more complex and therefore more interesting. The chromosome is not unbreakable, so **recombination** can occur. Genes at different loci on the same chromosome do sometimes separate from one another during meiosis.

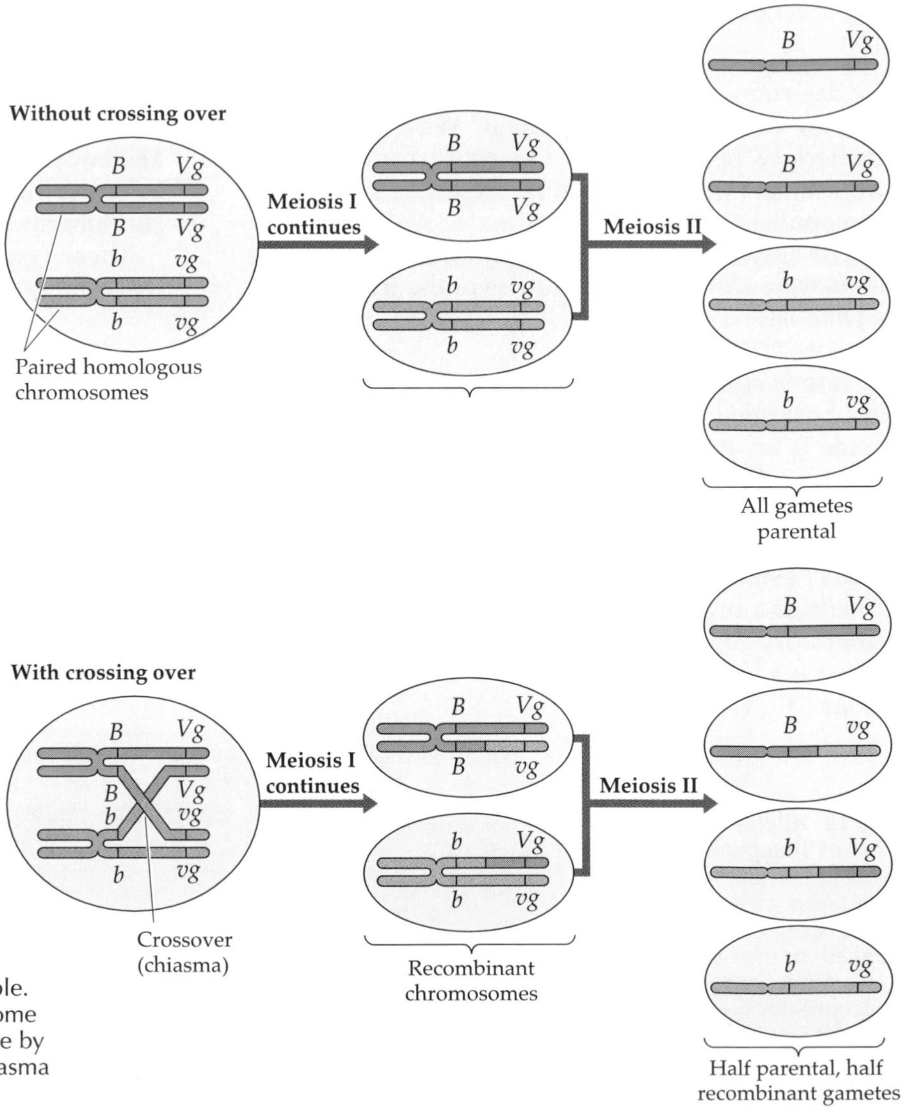

chromosome—that is, they are **linked**. After all, since the number of genes in a cell far exceeds the number of chromosomes, each chromosome must contain many genes. (Today we know that the human genome consists of perhaps 50,000 genes, distributed over 23 pairs of chromosomes.) The full set of loci on a given chromosome constitutes a *linkage group*. The number of linkage groups in a species equals the number of homologous chromosome pairs.

Suppose, now, that the *Bb* and *Vgvg* loci are on the same chromosome. If we assume that the linkage is absolute, we expect to see just *two* types of progeny from Morgan's cross (Figure 10.19). However, this is not always the case.

Without crossing over

Paired homologous chromosomes

Meiosis I continues

Meiosis II

All gametes parental

With crossing over

Crossover (chiasma)

Meiosis I continues

Recombinant chromosomes

Meiosis II

Half parental, half recombinant gametes

10.20 Crossing Over Results in Genetic Recombination   Chromosomes are breakable. Genes at different loci on the same chromosome can separate from one another and recombine by crossing over. Recombination occurs at a chiasma during prophase I of meiosis.

Genes on the same chromosome pair recombine by **crossing over**; that is, two homologous chromosomes physically exchange corresponding genetic segments during prophase I of meiosis (Figure 10.20). In other words, recombination occurs at a chiasma (see Figure 9.16) when homologous chromosomes are paired.

Recall that the DNA has duplicated by this stage, and each chromosome consists of two chromatids. The exchange event at any point along the length of the chromosome involves only two of the four chromatids, one from each member of the chromosome pair. The lengths of chromosome are exchanged reciprocally, so both chromatids involved in crossing over become recombinant (that is, each chromatid contains genes from both parents).

When crossing over takes place between linked genes, not all progeny of a cross are of parental types. Instead, recombinant offspring appear as well, and they appear in repeatable proportions called **recombinant frequencies**, which equal the number of recombinant progeny divided by the total number of progeny (Figure 10.21). Recombinant frequencies will be greater for loci that are far apart than for loci that are closer together. Recombination is a random event, much as if you were to close your eyes, pick up scissors, and try to cut a string held by a friend. You probably would not cut it if the string were very short, but if the string were long, the probability would be greater.

Geneticists make maps of eukaryotic chromosomes

If two loci are very close together on a chromosome, the odds for crossing over between them are small. In contrast, if two loci are far apart, crossing over could occur at many points. In 1911 Alfred Sturtevant, then an undergraduate student in T. H. Morgan's fly room, realized how that simple insight could be used to show where different genes lie on the chromosome in relation to one another. He suggested that the farther apart two genes are on a chromosome, the greater the likelihood that they will separate and recombine in meiosis.

The Morgan group had determined recombinant frequencies for many pairs of linked genes. Sturtevant used these recombinant frequencies to create genetic maps that indicated the arrangement of genes along the chromosome (Figure 10.22). Ever since Sturtevant demonstrated this important point, geneticists have mapped the chromosomes of eukaryotes, prokaryotes, and viruses, assigning distances in **map units**. A map unit corresponds to a recombinant frequency of 0.01; it is also referred to as a centimorgan (cM), in honor of the founder of the fly room. You, too, can work out a genetic map (Figure 10.23).

Sex is determined in different ways in different species

In Kölreuter's experience, and later in Mendel's, reciprocal crosses apparently always gave identical results.

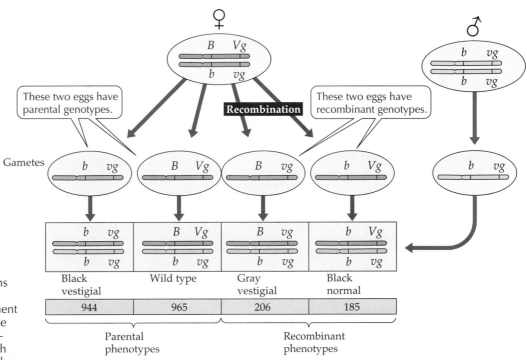

10.21 Recombinant Frequencies Crossing over between linked genes explains the nonparental genotypes Morgan found in the experiment described in Figure 10.18. The frequency of recombinant offspring can be calculated; such recombinant frequences will be larger for loci that are far apart than for those that are close together on the chromosome.

$$\text{RECOMBINANT FREQUENCY} = \frac{391 \text{ recombinants}}{2,300 \text{ total offspring}} = 0.17$$

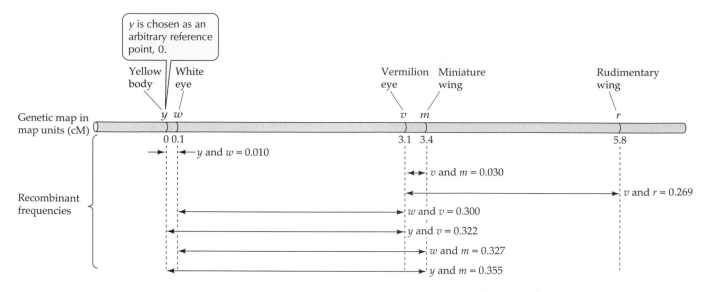

y is chosen as an arbitrary reference point, 0.

Yellow body — White eye — Vermilion eye — Miniature wing — Rudimentary wing

Genetic map in map units (cM)

y w   v m   r
0 0.1   3.1 3.4   5.8

Recombinant frequencies

y and w = 0.010
v and m = 0.030
v and r = 0.269
w and v = 0.300
y and v = 0.322
w and m = 0.327
y and m = 0.355

10.22 Steps toward a Genetic Map Because the chance of a recombinant genotype occurring increases the farther apart two loci fall, Sturtevant was able to derive this partial map of a *Drosophila* chromosome from the Morgan lab's data on the recombinant frequencies of five recessive traits. He assigned an arbitrary unit of distance—the map unit or centimorgan (cM)—equivalent to a recombinant frequency of 0.01.

At the outset, there appear to be several possible sequences (a-b-c, a-c-b, b-a-c), and we have no idea of the individual distances.

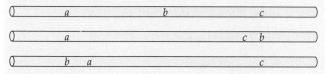

We make a cross *AABB* × *aabb*, and obtain an F$_1$ generation with a genotype *AaBb*. We test cross these *AaBb* individuals. How? By crossing them with *aabb*. Here are the genotypes of the first 1,000 progeny:

450 *AaBb*, 450 *aabb*, 50 *Aabb*, and 50 *aaBb*.

How far apart are the *a* and *b* genes? Well, what is the recombinant frequency? Which are the recombinant types, and which are the parental types?

Recombinant frequency (*a* to *b*) = (50 + 50)/1,000 = 0.1
So the map distance is
Map distance = 100 × recombinant frequency = 100 × 0.1 = 10 cM

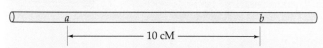

Now we make a cross *AACC* × *aacc*, obtain an F$_1$ generation, and test cross it, obtaining:

460 *AaCc*, 460 *aacc*, 40 *Aacc*, and 40 *aaCc*.

How far apart are the *a* and *c* genes?

Recombinant frequency (*a* to *c*) = (40 + 40)/1,000 = 0.08
Map distance = 100 × recombinant frequency = 100 × 0.08 = 8 cM

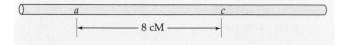

Now can you tell the order of the three genes on the chromosome? Why not? What else do you need to know?

We make a cross *BBCC* × *bbcc*, obtain an F$_1$ generation, and test cross it, obtaining:

490 *BbCc*, 490 *bbcc*, 10 *Bbcc*, and 10 *bbCc*.

Determine the map distance between *b* and *c*.

Recombinant frequency (*b* to *c*) = (10 + 10)/1,000 = 0.02
Map distance = 100 × recombinant frequency = 100 × 0.02 = 2 cM

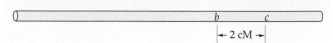

Which of the three genes is between the other two, then? Because *a* and *b* are the farthest apart, *c* must be between them.

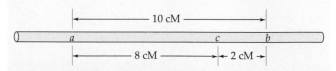

These numbers add up perfectly, but in most real cases they don't add up perfectly because of double crossovers.

10.23 Help Map These Genes We want to determine the order of three loci (*a*, *b*, and *c*) on a chromosome, as well as the map distances (in cM) between them. How do we determine a map distance?

The reason is that in diploid organisms, chromosomes come in pairs. One member of each chromosome pair derives from each parent; it does not matter, in general, whether a dominant allele was contributed by the mother or by the father. But this is not always the case. Sometimes the parental origin of a chromosome does matter. To understand the types of inheritance in which parental origin is important, we must consider the ways in which sex is determined in different species.

In corn (maize), a plant much studied by geneticists, every diploid adult has both male and female reproductive structures. These two types of tissue are genetically identical, just as roots and leaves are genetically identical. Plants such as maize and Mendel's pea plants, and animals such as earthworms, which produce both male and female gametes in the same organism, are said to be **monoecious** (from the Greek for "single house"). Other plants, such as date palms and oak trees, and most animals are **dioecious** ("two houses"), meaning that some of the individuals can produce only male gametes and the others can produce only female gametes.

In most dioecious organisms, sex is determined by differences in the chromosomes; but such determination operates in various different ways. For example, the sex of a honeybee (Figure 10.24*a*) depends on whether it develops from a fertilized or an unfertilized egg. A fertilized egg is diploid and gives rise to a female bee—either a worker or a queen, depending on the diet during larval life (again, note how the environment affects the phenotype). An unfertilized egg is haploid and gives rise to a male drone.

In many other animals, including ourselves, sex is determined by a single **sex chromosome** (or by a pair of them). Both males and females have two copies of each of the rest of the chromosomes, which are called **autosomes**.

For example, female grasshoppers have two **X chromosomes**, whereas males have only one. Females form eggs that contain one copy of each autosome and one X chromosome. The males form approximately equal amounts of two types of sperm: One type contains an X chromosome and one copy of each autosome; the other type contains only autosomes. Female grasshoppers are described as being XX (ignoring the autosomes) and males as XO (pronounced "ex-oh"). When an X-bearing sperm fertilizes an egg, the zygote is XX and develops into a female. When a sperm without an X fertilizes an egg, the zygote is XO and develops into a male. This chromosomal mechanism ensures that the two sexes are produced in approximately equal numbers.

As in grasshoppers, female mammals have two X chromosomes and males have one (Figure 10.24*b*). However, male mammals also have a sex chromosome that is not found in females: the **Y chromosome**. Females may be represented as XX and males as XY.

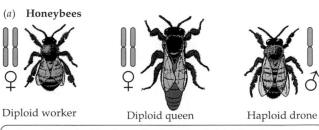

(*a*) **Honeybees**

♀ Diploid worker   ♀ Diploid queen   ♂ Haploid drone

In honeybees, fertilized eggs develop into diploid females and unfertilized eggs develop into haploid males.

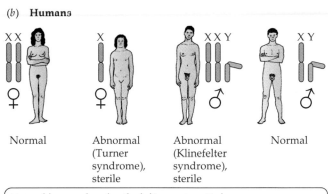

(*b*) **Humans**

XX ♀ Normal   X ♀ Abnormal (Turner syndrome), sterile   XXY ♂ Abnormal (Klinefelter syndrome), sterile   XY ♂ Normal

Normal human females (far left) carry two X chromosomes; normal males (far right) carry one X and one Y chromosome. Persons who have some other number of sex chromosomes may develop abnormally.

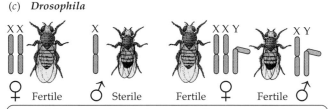

(*c*) *Drosophila*

XX ♀ Fertile   X ♂ Sterile   XXY ♀ Fertile   XY ♂ Fertile

Drosophila (fruit fly) females have two X chromosomes and may also have a Y chromosome; males have an X chromosome and, if they are fertile, a Y chromosome.

(*d*) **Chickens**

ZW ♀   ZZ ♂

In birds, the males carry two identical sex chromosomes (ZZ) and females have two differing ones (ZW).

10.24 Sex Is Determined in Different Ways For traits transmitted on the sex chromosomes, the parental origin of the gene is crucial, and varies from species to species.

The males produce two kinds of gametes: Each has a complete set of autosomes, but half the gametes carry an X chromosome and the rest carry a Y. When an X-bearing sperm fertilizes an egg, the resulting XX zygote is female; when a Y-bearing sperm fertilizes an egg, the XY zygote is male.

X and Y chromosomes have different functions

Some subtle but important differences show up clearly in mammals with abnormal sex chromosome constitutions. These conditions, resulting from nondisjunctions as described in Chapter 9, tell us something about the functions of the X and Y chromosomes. In humans, XO individuals sometimes appear. Human XO individuals are females who are physically moderately abnormal but mentally normal; usually they are also sterile. The XO condition in humans is called Turner syndrome. It is the only known case in which a human can live with only one member of a chromosome pair (here, the XY pair), although most XO conceptions terminate spontaneously early in development.

XXY individuals are also possible, a condition known as Klinefelter syndrome. They are sometimes taller than average, always sterile, and always male. Research with these individuals and related studies show that in humans the Y chromosome carries the genes that determine maleness.

The specific gene that determines maleness has been identified from observations of people with chromosome abnormalities. For example, some XY individuals who are women have been studied; a small portion of their Y chromosome was missing. In other cases, men who were genetically XX had a small part of the Y chromosome present, attached to another chromosome. The missing and present Y fragment in these two examples, respectively, contained the maleness-determining gene, which was identified as *SRY* (*sex-determining region* on the *Y* chromosome).

The *SRY* gene codes for a protein for *primary sex determination*. In the presence of functional SRY protein, the embryo develops sperm-producing testes. If SRY protein is absent, the primary sex determination is female: the presence of ovaries and eggs. *Secondary sex determination*, the outward manifestations of maleness and femaleness (such as body type, breast development, body hair, and voice) are not determined directly by the presence or absence of the Y chromosome. Rather, they are determined by the actions of hormones, such as testosterone and estrogen.

The Y chromosome functions differently in *Drosophila melanogaster* (Figure 10.24c). Superficially, *Drosophila* follows the same pattern of sex determination as mammals: Females are XX and males are XY. However, XO individuals are males (rather than females as in mammals) and almost always are indistinguishable from normal XY males except that they are sterile. XXY *Drosophila* are normal, fertile females. Thus, in *Drosophila*, sex is determined strictly by the ratio of X chromosomes to autosome sets. If there is one X chromosome for each set of autosomes, the individual is a female; if there is only one X chromosome

for the two sets of autosomes, the individual is a male. The Y chromosome plays no sex-determining role in *Drosophila*, but it is needed for male fertility.

In birds, moths, and butterflies, males are XX and females are XY. To avoid confusion, these forms are usually expressed as ZZ (male) and ZW (female) (Figure 10.24d). In these organisms, the female produces two types of gametes. Thus the egg determines the sex of the offspring, rather than the sperm as in humans and fruit flies.

Genes on sex chromosomes are inherited in special ways

How does the existence of sex chromosomes affect patterns of inheritance? In *Drosophila* and in humans, the Y chromosome carries few known genes, but a substantial number of genes affecting a great variety of characters are carried on the X chromosome. The result of this arrangement is an important deviation from the usual Mendelian ratios for the inheritance of genes located on the X chromosome. Any such gene is present in two copies in females, but in only one copy in males. Therefore, females may be heterozygous for genes that are on the X chromosome, but males will always be **hemizygous** for genes on the X chromosome—they will have only one of each.

Kölreuter's historic reciprocal crosses, mentioned at the beginning of this chapter, always gave the same outcome regardless of which parent displayed which trait. However, reciprocal crosses of parents that have different alleles on their sex chromosomes do not give identical results; this is a sharp deviation from the rules governing the inheritance of alleles on autosomes.

The first and still one of the best examples of **sex-linked inheritance**—inheritance of characters governed by loci on the sex chromosomes—is that of eye color in *Drosophila*. The wild-type eye color of these flies is red. In 1910, Morgan discovered a mutation that causes white eyes. He experimented by crossing flies of the wild-type and mutant phenotypes. His results demonstrated that the eye color locus is on the X chromosome.

When homozygous red-eyed females were crossed with (hemizygous) white-eyed males, all the sons and daughters had red eyes, because red is dominant over white and all the progeny had inherited a wild-type X chromosome from their mothers (Figure 10.25a). However, in the reciprocal cross, in which a white-eyed female was mated with a red-eyed male, all the sons were white-eyed and all the daughters red-eyed (Figure 10.25b).

The sons from the reciprocal cross inherited their only X chromosome from their white-eyed mother; the Y chromosome they inherited from their father does not carry the eye color locus. The daughters, on the

10.25 Eye Color Is a Sex-Linked Trait in *Drosophila*

Thomas Hunt Morgan demonstrated that the mutant allele that causes white eyes in *Drosophila* is carried on the X chromosome.

HUMANS BEINGS DISPLAY MANY SEX-LINKED CHARACTERS. The human X chromosome carries many genes. The alleles at these loci follow the same pattern of inheritance as those for white eyes in *Drosophila*. One human X chromosome gene, for example, has an allele that leads to red-green color blindness, a hereditary disorder. Red-green color blindness appears in individuals who are homozygous or hemizygous for a mutant recessive allele.

A color-blind man married to a homozygous normal woman will produce no color-blind children (Figure 10.26a). The sons inherit a single, normal X from their mother and will neither have the disorder nor transmit it to their children. The daughters get an X chromosome bearing a normal allele from their mother and one bearing the allele for color blindness from their father. Because color blindness is recessive, the daughters will not be color-blind (Figure 10.26b). However, they will be heterozygous *carriers*.

Female carriers of an X-linked trait will transmit the disorder to half their sons and the carrier role to half their daughters if the male parent has normal color vision. What parental genotypes would produce a color-blind female? Her father would have to be color-blind, and her mother either a carrier or herself color-blind. Because color blindness is rare, two such people are unlikely to meet and produce children. As a result, very few women are color-blind. This pattern of inheritance is the same as for hemophilia, the serious blood disorder discussed at the start of this chapter.

The small human Y chromosome carries very few genes. Among them are the maleness determinants, whose existence was suggested by the phenotypes of the XO and XXY individuals described earlier (see Figure 10.24b). The pattern of inheritance of Y-linked alleles should be easy for you to work out. Give it a try now. (Hint: Use a Punnett square to determine how the Y chromosome is inherited.)

Mendelian ratios are averages, not absolutes

You have been introduced to the basic phenotypic ratios observed for traits inherited following Mendelian patterns: 3:1, 1:1, 9:3:3:1. You will discover others as you do homework or test problems. It is essential to remember, however, that these ratios represent highest probabilities, not invariant rules.

The X–Y system of sex determination in our species results in roughly equal numbers of males and females in a substantial population, but you know that a given family of four children may not consist of two girls and two boys. It is not unusual for four children to be of the same sex; in fact, among families with four children, one in eight will have all boys or all girls. How do we know if this deviation is simply the result of chance?

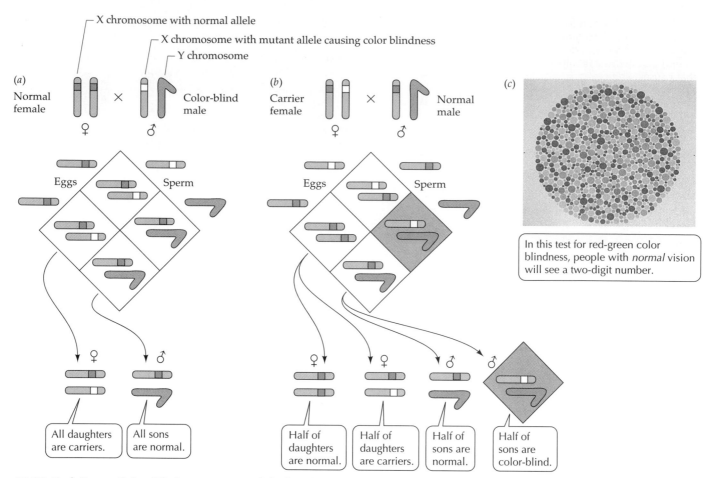

X chromosome with normal allele

X chromosome with mutant allele causing color blindness

Y chromosome

(a) Normal female × Color-blind male

Eggs Sperm

All daughters are carriers.

All sons are normal.

(b) Carrier female × Normal male

Eggs Sperm

Half of daughters are normal.

Half of daughters are carriers.

Half of sons are normal.

Half of sons are color-blind.

(c) In this test for red-green color blindness, people with *normal* vision will see a two-digit number.

10.26 Red-Green Color Blindness Is a Sex-Linked Trait in Humans The pattern of transmission of red-green color blindness is the same as that for hemophilia, discussed at the beginning of this chapter.

When we are trying to understand the genetic basis for the results of a cross, it is important to know how much deviation from a predicted ratio is reasonably expected as a result of normal chance. Statistical methods have been devised for determining whether the observed deviation from an expected ratio can be attributed to chance variation or whether the deviation is large enough to suggest that the observed ratio has a specific cause.

Statistical methods take several factors into consideration, but one of the most important is **sample size**. If we expect to find a 3:1 ratio between two phenotypes and we look at a sample of only 8 progeny, we should not be surprised to find 7 individuals of one phenotype and 1 of the other (rather than the expected 6 and 2). It would be surprising, however, in a sample of 80 to find 70 of one phenotype and 10 of the other—you would question whether that really represented a 3:1 ratio. In genetics experiments—and in quantitative biology in general—sample sizes should be large so that the data are easy to evaluate with confidence.

Cytoplasmic Inheritance

You have studied the basic patterns of Mendelian inheritance in terms of chromosomal behavior. Does all inheritance in eukaryotes conform to the Mendelian pattern?

Consider the four-o'clock plant shown in Figure 10.27. This particular four-o'clock shows three different patterns of chlorophyll distribution in three different parts of the shoot. One branch is all white; it has no chlorophyll. Another branch is entirely green. The rest of the shoot is variegated; that is, some patches of cells have chlorophyll and others have none. Each part of the plant has flowers, so we can do the following experiment.

Take pollen (which produces male gametes) from some flowers and transfer it to the female parts of flowers on another part of the plant, thus performing a cross. Table 10.2 shows the six possible crosses and their outcomes. Study the table for a moment before reading further and try to discern the basic pattern of inheritance. Do you see the surprising feature? Look again. What should become obvious is that the phenotype of the "father" (the parent producing the pollen) is irrelevant to the outcome. Only the "mother" (the

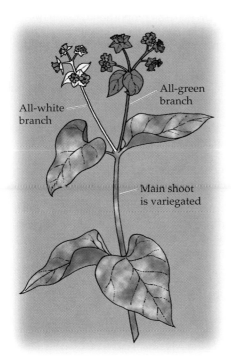

10.27 A Variegated Four-o'Clock with Green and White Branches   This is the plant used for the experiment summarized in Table 10.2. Pollen is transferred from a flower on one part of the plant to a flower on another part.

TABLE 10.2	Results of the Four-o'Clock Experiments	
PHENOTYPE OF BRANCH WITH FEMALE PARENT	**PHENOTYPE OF BRANCH WITH MALE PARENT**	**PHENOTYPE OF PROGENY**
White	White	White
White	Green	White
White	Variegated	White
Green	White	Green
Green	Green	Green
Green	Variegated	Green

parent producing the egg) seems to play a role in determining the phenotype of the offspring.

This pattern of inheritance, in which the progeny's phenotype is unaffected by the male parent, is found for particular characteristics in all sorts of eukaryotes and is referred to as **maternal inheritance**. How does it work?

The essence of Mendelian inheritance is that information carried on chromosomes is partitioned with great precision during meiosis, but eukaryotic cells have other self-reproducing entities besides the nuclear chromosomes. Chloroplasts and mitochondria carry some genetic information in small circular chromosomes (see Chapter 4). The DNA of these organelles is subject to mutation just as the DNA in the chromosomes of the nucleus is, so we may speak of alleles of nonnuclear genes.

Nonnuclear genes are not inherited in the same way as nuclear chromosomal genes, because meiosis does not segregate organelles precisely; cytokinesis does—randomly. Because the eggs of most species contain large amounts of cytoplasm and sperm contain almost no cytoplasm, the mitochondria in a zygote come from the cytoplasm of the female parent's egg, even though half the zygote's nuclear chromosomes come from the male parent.

In plant zygotes, all the chloroplasts come from the maternal cytoplasm. Hence any particle that is inherited through the cytoplasm is said to be maternally inherited. In such cases, reciprocal crosses give results that differ considerably from the predictions made by Mendel, as we saw in Table 10.2.

Chloroplasts and mitochondria are complex organelles, and only a minority of their functions are maternally inherited. For example, most components of the electron transport system are encoded by nuclear genes. Since mitochondria are essential to providing cellular energy, it is not surprising that inherited human diseases whose phenotypes show up as abnormal mitochondria often include serious muscle or nervous system abnormalities among their symptoms. Some of these genetic diseases are maternally inherited; others show typical Mendelian inheritance patterns.

Summary of "Transmission Genetics: Mendel and Beyond"

Mendel's Work
• Plant breeders can control which plants mate. Although it has long been known that both parent plants contribute equivalent determinants to the character traits of their offspring, before Mendel's time it was believed that, once they were brought together, these determinants blended and could never be separated. **Review Figure 10.1**
• Although Mendel's work was meticulous and well documented, his discoveries, reported in the 1860s, lay dormant until decades later, when (after meiosis was known) others rediscovered them.

Mendel's Experiments and Laws of Inheritance
• Mendel used garden pea plants for his studies because they were easily cultivated and crossed, and because they showed numerous characters (such as seed shape) with clearly different traits (spherical or wrinkled). **Review Figures 10.1, 10.2**
• In a monohybrid cross, the offspring showed only one of the two traits. Mendel proposed that the trait observed in the first generation (F_1) was dominant and the other was recessive. **Review Table 10.1**

• When the F_1 offspring were interbred, the resulting F_2 generation showed a 3:1 phenotypic ratio, with the recessive phenotype present in one-fourth of the offspring. This reappearance of the recessive phenotype refuted the blending hypothesis. **Review Figure 10.4 and Table 10.1**

• Because some alleles are dominant and some recessive, the same phenotype can result from different genotypes. Homozygous genotypes have two copies of the same allele; heterozygous genotypes have two different alleles. Heterozygous genotypes yield phenotypes that show the dominant trait.

• On the basis of many crosses with different characters, Mendel proposed his first law: that genetic determinants (now known as genes) are particulate, that there are two copies (alleles) of a gene in every parent, and that during gamete formation the two alleles for a character segregate from one another. **Review Figure 10.5**

• Geneticists who followed Mendel showed that genes are carried on chromosomes and that alleles are segregated during meiosis I. **Review Figure 10.6**

• Using a test cross, Mendel was able to determine whether a plant showing the dominant character was homozygous or heterozygous. The appearance of the recessive phenotype in half of the offspring of such a cross indicates that the dominant-appearing parent was heterozygous. **Review Figure 10.7**

• From studies of the simultaneous inheritance of two characters (seed shape and seed color, for example) on the same plants, Mendel concluded that alleles of different genes assort independently in these dihybrid crosses. **Review Figures 10.8, 10.9**

• We can predict the results of hybrid crosses either by using a Punnett square or by calculating probabilities. To determine the joint probability of independent events, multiply the individual probabilities. To determine the probability of an event that can occur in two or more different ways, add the individual probabilities. **Review Figure 10.10**

Alleles and Their Interactions

• Dominance is usually not complete, since both alleles in a heterozygous organism may be expressed in the phenotype. **Review Figure 10.11**

• Some alleles have multiple phenotypic effects. Alleles arise by mutation, and many genes (such as those that determine blood type) have multiple alleles. **Review Figures 10.12, 10.13**

Gene Interactions

• In epistasis, the products of different genes interact to produce a phenotype. **Review Figure 10.14**

• In some cases, the phenotype is the result of the additive effects of many genes (polygenes), and inheritance is quantitative. **Review Figure 10.15**

• Environmental variables such as temperature, nutrition, and light affect gene action.

Genes and Chromosomes

• Each chromosome carries many genes. Genes located near each other are said to be linked, and they are usually inherited together. **Review Figures 10.18, 10.19**

• Linked genes recombine by crossing over in prophase I of meiosis. The result is recombinant gametes, which have new combinations of linked genes because of the exchange. **Review Figures 10.20, 10.21**

• The distance between genes is proportional to the frequency of crossing over between them. Genetic maps are based on recombinant frequencies. **Review Figures 10.22, 10.23**

• Sex chromosomes carry genes that determine whether male or female gametes are produced. The specific functions of X and Y chromosomes differ depending on the species. **Review Figure 10.24**

• Genes on the sex chromosomes can be followed in crosses in fruit flies and mammals because the X chromosome carries many genes, but its homolog, the Y chromosome, has few. Males have only one allele for most X-linked genes, so rare alleles will show up phenotypically more often in males than in females. **Review Figures 10.25, 10.26**

Cytoplasmic Inheritance

• Cytoplasmic organelles such as chloroplasts and mitochondria have some genes.

• Cytoplasmic inheritance is generally by way of the egg (maternal), because male gametes contribute only their nucleus and no cytoplasm to the zygote at fertilization.

Self-Quiz

1. Which statement about Mendel's cross of *TT* peas with *tt* peas is *not* true?
   a. Each parent can produce only one type of gamete.
   b. F_1 individuals produce gametes of two types, each gamete being *T* or *t*.
   c. Three genotypes are observed in the F_2 generation.
   d. Three phenotypes are observed in the F_2 generation.
   e. This is an example of a monohybrid cross.

2. The phenotype of an individual
   a. depends at least in part on the genotype.
   b. is either homozygous or heterozygous.
   c. determines the genotype.
   d. is the genetic constitution of the organism.
   e. is either monohybrid or dihybrid.

3. Which statement about alleles is *not* true?
   a. They are different forms of the same gene.
   b. There may be several at one locus.
   c. One may be dominant over another.
   d. They may show incomplete dominance.
   e. They occupy different loci on the same chromosome.

4. Which statement about an individual that is homozygous for an allele is *not* true?
   a. Each of its cells possesses two copies of that allele.
   b. Each of its gametes contains one copy of that allele.
   c. It is true-breeding with respect to that allele.
   d. Its parents were necessarily homozygous for that allele.
   e. It can pass that allele to its offspring.

5. Which statement about a test cross is *not* true?
   a. It tests whether an unknown individual is homozygous or heterozygous.
   b. The test individual is crossed with a homozygous recessive individual.
   c. If the test individual is heterozygous, the progeny will have a 1:1 ratio.
   d. If the test individual is homozygous, the progeny will have a 3:1 ratio.
   e. Test cross results are consistent with Mendel's model of inheritance.

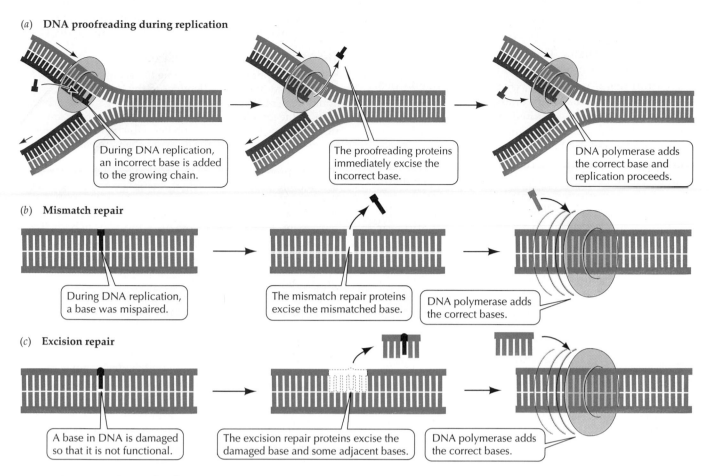

(a) **DNA proofreading during replication**

During DNA replication, an incorrect base is added to the growing chain.

The proofreading proteins immediately excise the incorrect base.

DNA polymerase adds the correct base and replication proceeds.

(b) **Mismatch repair**

During DNA replication, a base was mispaired.

The mismatch repair proteins excise the mismatched base.

DNA polymerase adds the correct bases.

(c) **Excision repair**

A base in DNA is damaged so that it is not functional.

The excision repair proteins excise the damaged base and some adjacent bases.

DNA polymerase adds the correct bases.

11.18 DNA Repair Mechanisms The proteins of DNA replication also play roles in the life-preserving repair mechanisms, helping to ensure the exact replication of template DNA.

When mismatch repair fails, DNA sequences are altered. One form of colon cancer arises in part from a failure of mismatch repair.

What if a DNA molecule becomes damaged during the life of a cell? Some cells live and play important roles for many years, even though their DNA is constantly at risk from hazards such as high-energy radiation, chemicals that induce mutations, and random spontaneous chemical reactions. Cells owe their lives to DNA repair mechanisms. For example, in **excision repair**, certain enzymes "inspect" the cell's DNA (Figure 11.18c). When they find mispaired bases, chemically modified bases, or points at which one strand has more bases than the other (with the result that one or more bases of one strand form an unpaired loop), these enzymes cut the defective strand. Another enzyme cuts away the bases adjacent to and including the offending base, and DNA polymerase and DNA ligase synthesize and seal up a new (usually correct) piece to replace the excised one.

Our dependence on this repair mechanism is underscored by our susceptibility to various diseases that arise from excision repair defects. One example is the skin disease xeroderma pigmentosum. People with this disease lack a mechanism that normally repairs damage caused by the ultraviolet radiation in sunlight. Without this mechanism, a person exposed to sunlight develops skin cancer.

DNA repair requires energy

What does it cost the cell to keep its DNA accurate and ensure that it replicates properly? At first glance, you might expect DNA polymerization to be fairly "neutral" energetically, because adding a new monomer to the chain requires the formation of a new phosphodiester bond but is supported by the hydrolysis of one of the high-energy bonds in the nucleoside triphosphate (see Figure 11.12). Overall, however, this reaction is slightly endergonic. But help is available in the form of the pyrophosphate ion released in the polymerization reaction. The enzyme pyrophosphatase cleaves the high-energy bond in the pyrophosphate. Coupling this reaction to the polymerization gives it a big boost.

Noncovalent bonds also play a major role in favoring DNA polymerization. Hydrogen bonds form between the complementarily paired bases, and other weak interactions form as the bases stack in the middle of the double helix. These bonds and interactions stabilize the DNA molecule and help drive the polymerization reaction. Thus DNA synthesis itself does not take a tremendous toll in energy.

DNA repair processes, however, are far from cheap energetically. Some are very inefficient. Nonetheless the cell deploys many DNA repair mechanisms, some overlapping in function with others. Why? Perhaps because the cell simply can't afford to leave its genetic information unprotected, regardless of the cost.

Normal, undamaged DNA is vital to life and its continuation. But what exactly is the function of DNA? What do genes do? What do they control? These are the topics of Chapter 12.

Summary of "DNA and Its Role in Heredity"

DNA: The Genetic Material

• In addition to circumstantial evidence (the location and quantity of DNA in the cell) that genes are made of DNA, two experiments provided convincing demonstration.
• In one experiment, DNA from a virulent strain of pneumococcus bacteria genetically transformed nonvirulent bacteria into virulent bacteria. **Review Figure 11.1**
• In the second set of experiments, labeled viruses were allowed to attach to host bacteria. The labeled protein coats of the viruses were shaken off the bacterial surface, but the labeled viral DNA entered the host cells, where it took control and caused the bacterial cells to produce hundreds of new viruses. **Review Figures 11.2, 11.3**

The Structure of DNA

• X ray crystallography showed that the DNA molecule is a helix. **Review Figure 11.4**
• Biochemical analysis revealed that DNA is composed of nucleotides, each containing one of four bases—adenine, cytosine, thymine, or guanine—and that the amount of adenine equals the amount of thymine and the amount of guanine equals the amount of cytosine. **Review Figure 11.5**
• Putting the accumulated data together, Watson and Crick proposed that DNA is a double-stranded helix in which the strands are antiparallel and the bases form opposite strands held together by hydrogen bonding. This model accounts for the genetic information, function, mutation, and replication of DNA. **Review Figures 11.6, 11.7**

DNA Replication

• Three possible models for DNA replication were hypothesized: semiconservative, conservative, and dispersive. **Review Figure 11.9**
• An experiment by Meselson and Stahl proved the replication of DNA to be semiconservative. Each parent strand acts as a template for the synthesis of a new strand; thus, the two replicated DNA helices contain one parent strand and one newly synthesized strand each. **Review Figures 11.10, 11.11**

The Mechanism of DNA Replication

• In DNA replication, the enzyme DNA polymerase catalyzes the addition of nucleotides to the 3′ end of each strand. The nucleotides are added according to base-pairing rules of the template strand of DNA. The substrates are deoxyribonucleoside triphosphates, which are hydrolyzed when they are added to the growing chain, releasing energy that fuels the synthesis of DNA. **Review Figure 11.12**
• Many proteins assist in DNA replication. DNA helicases unwind the double helix, and the template strands are stabilized by other proteins.
• Prokaryotes have a single origin of replication; eukaryotes have many. Replication in both cases proceeds in both directions from an origin of replication. **Review Figure 11.13**
• An RNA primase catalyzes the synthesis of short RNA primers, to which the nucleotides are added as the chain grows. **Review Figure 11.14**
• Using DNA polymerase, the leading strand grows continuously in the 5′-to-3′ direction until the replication of that section of DNA has been completed. Then the RNA primer is degraded and DNA added in its place. On the lagging strand, which grows in the other direction, DNA is still made in the 5′-to-3′ direction (toward the fork). But synthesis of the lagging strand is discontinuous: The DNA is added as short fragments to primers; then the polymerase skips toward the 5′ end to make the next fragment. **Review Figures 11.15, 11.16, 11.17**

DNA Proofreading and Repair

• The machinery of DNA replication makes about one error in 100,000 nucleotides added. These errors are repaired by three different mechanisms: proofreading, mismatch repair, and excision repair. DNA repair mechanisms lower the overall error rate of replication to about one base in 10 billion. **Review Figure 11.18**
• Although energetically costly and somewhat redundant, DNA repair is crucial to the survival of the cell.

Self-Quiz

1. Griffith's studies of *Streptococcus pneumoniae*
    a. showed that DNA is the genetic material of bacteria.
    b. showed that DNA is the genetic material of bacteriophages.
    c. demonstrated the phenomenon of bacterial transformation.
    d. proved that prokaryotes reproduce sexually.
    e. proved that protein is not the genetic material.

2. In the Hershey–Chase experiment
    a. DNA from parent bacteriophages appeared in progeny bacteriophages.
    b. most of the phage DNA never entered the bacteria.
    c. more than three-fourths of the phage protein appeared in progeny phages.
    d. DNA was labeled with radioactive sulfur.
    e. DNA formed the coat of the bacteriophages.

3. Which statement about complementary base pairing is *not* true?
    a. It plays a role in DNA replication.
    b. In DNA, T pairs with A.
    c. Purines pair with purines, and pyrimidines pair with pyrimidines.

d. In DNA, C pairs with G.

e. The base pairs are of equal length.

4. In semiconservative replication of DNA

 a. the original double helix remains intact and a new double helix forms.

 b. the strands of the double helix separate and act as templates for new strands.

 c. polymerization is catalyzed by RNA polymerase.

 d. polymerization is catalyzed by a double helical enzyme.

 e. DNA is synthesized from amino acids.

5. Which of the following does not occur during DNA replication?

 a. Unwinding of the parent double helix

 b. Formation of short pieces that are united by DNA ligase

 c. Complementary base pairing

 d. Use of a primer

 e. Polymerization in the 3'-to-5' direction

6. The primer used for DNA replication

 a. is a short strand of RNA added to the 3' end.

 b. is present only once on the leading strand.

 c. remains on the DNA after replication.

 d. ensures that there will be a free 5' end to which nucleotides can be added.

 e. is added to only one of the two template strands.

7. The 3' end of a DNA strand is defined as the place where

 a. the phosphate group is not bound to another nucleotide.

 b. both DNA strands end opposite each other.

 c. DNA polymerase binds to begin replication.

 d. there is a free —OH group at the 3' carbon of deoxyribose.

 e. three A residues are present.

8. The role of DNA ligase in DNA replication is to

 a. add more nucleotides to the growing chain one at a time.

 b. open up the two DNA strands to expose template strands.

 c. ligate base to sugar to phosphate in a nucleotide.

 d. bond Okazaki fragments to one another.

 e. remove incorrectly paired bases.

9. Incorrect bases that are added to DNA

 a. can be repaired by proofreading.

 b. cannot have been added by DNA polymerases, since these enzymes make no errors.

 c. do not result in mispairing.

 d. are replaced along with adjacent nucleotides.

 e. are methylated.

10. The following events occur in excision repair of DNA. What is their proper order?

 1 Base-paired DNA is made complementary to the template.

 2 Damaged bases are recognized.

 3 DNA ligase seals the new strand to existing DNA.

 4 Part of a single strand is excised.

 a. 1234 *b.* 2134 *c.* 2413

 d. 3421 *e.* 4231

Applying Concepts

1. Outline a series of experiments using radioactive isotopes to show that bacterial DNA and not protein enters the host cell and is responsible for bacterial transformation.

2. Suppose that Meselson and Stahl had continued their experiment on DNA replication for another ten bacterial generations. Would there still have been any $^{14}N–^{15}N$ hybrid DNA present? Would it still have appeared in the centrifuge tube? Explain.

3. If DNA replication were conservative rather than semiconservative, what results would Meselson and Stahl have observed? Diagram the results using the conventions of Figure 11.11.

4. Using the following information, calculate the number of origins of DNA replication on a human chromosome: DNA polymerase adds nucleotides at 3,000 base pairs per minute in one direction; replication is bidirectional; S phase lasts 300 minutes; there are 120 million base pairs per chromosome. With a typical chromosome 3 μm long, how many origins are there per micrometer?

5. The drug dideoxycytidine (used to treat certain viral infections) is a nucleotide made with 2',3'-dideoxyribose. This sugar lacks —OH groups at both the 2' and the 3' positions. Explain why this drug would stop the growth of a DNA chain if added to the DNA.

Readings

Felsenfeld, G. 1985. "DNA." *Scientific American*, October. A well-illustrated description of DNA structure and function.

Griffiths, A. J. F., J. H. Miller, D. T. Suzuki, R. C. Lewontin and W. M. Gelbart. 1996. *An Introduction to Genetic Analysis*, 6th Edition. W. H. Freeman, New York. An excellent textbook of modern genetics. Chapters 11, 12, and 13 are particularly relevant.

Judson, H. F. 1996. *The Eighth Day of Creation: Makers of the Revolution in Biology*, Expanded Edition. CSHL Press, Plainview, NY. A sparkling history of molecular biology, with the best available description of the events surrounding the discovery of the structure of DNA.

Modrich, P. 1994. "Mismatch Repair, Genetic Stability, and Cancer." *Science*, vol. 266, pages 1959–1960. This brief article is from the *Science* issue recognizing DNA repair as the "Molecule of the Year."

Radman, M. and R. Wagner. 1988. "The High Fidelity of DNA Duplication." *Scientific American*, August. A description of how error avoidance and error correction work that poses the question, Why don't they work even better?

Sancar, A. 1994. "Mechanisms of DNA Excision Repair." *Science*, vol. 266, pages 1954–1956. This brief article is from the *Science* issue recognizing DNA repair as the "Molecule of the Year."

Stent, G. S. and R. Calendar. 1978. *Molecular Genetics*, 2nd Edition. W. H. Freeman, New York. A brilliant technical and historical introduction to molecular genetics and the role of DNA.

Upton, A. C. 1982. "The Biological Effects of Low-Level Ionizing Radiation." *Scientific American*, February. An explanation of how radiation leads to mutations.

Watson, J. D. 1968. *The Double Helix*. Atheneum, New York. A captivating and, to some, infuriating book in which Watson describes the events leading to the discovery of DNA structure.

Chapter 12

From DNA to Protein: Genotype to Phenotype

One Member of the Translation Team
Transfer RNA molecules recognize the genetic message encoded in the nucleotide sequence of messenger RNA and simultaneously carry specific amino acids, enabling them to translate the language of DNA into the language of proteins.

*F*lying alone in your corporate jet, you make a crash landing on an uncharted island in the North Sea. The batteries in your radio are dying, but you manage to send off a frantic message. A ham radio operator on the northern German coast, who does not know English, hears your message and writes it down phonetically. He gives it to an English-speaking friend, who deciphers the words and converts them into German. A message is sent to a German ship, and you are rescued.

This story and its happy ending illustrate the main themes of this chapter, which is concerned with what DNA *does* (Chapter 11 covered what DNA *is*). The information content of a gene consists of its base sequence, just as your frantic message consisted of the words you spoke. In the cell, the information in DNA is *transcribed* into information in RNA, as your spoken words were transcribed into phonetic English. A remarkable team of proteins and RNAs, some of which are constituents of ribosomes, *translates* the information from a sequence of bases in the RNA into a sequence of amino acids in an enzyme or other protein, just as the ham radio operator's friend converted your English into German. The enzyme produces the final effect—catalysis of a chemical reaction—just as the German translation of your message led to effective action.

Of course, both our adventure story and the cellular processing of DNA depend on accurate functioning of all components of the system. The original message itself may be erroneous—DNA is subject to mutation, and you might use a wrong word in your haste. Errors can also arise in transcription or translation.

In this chapter we will focus first on how it became established that the gene in DNA expresses itself as the phenotype in a polypeptide. Next we will describe how this expression occurs: First, the information content of DNA is converted to mRNA (transcription); then mRNA travels to the ribosome, where its information determines the order of specific amino acids in a polypeptide (translation). Finally, we will describe the genetic concept of mutation in more precise terms than we have up to now, defining alleles in terms of alterations in the nucleotide sequences of DNA.

Genes and the Synthesis of Polypeptides

There are many steps between genotype and phenotype. Genes cannot, all by themselves, directly produce a phenotypic result such as a particular eye color, a specific seed shape, or a cleft chin, any more than a compact disk can play a symphony without the help of a CD player.

With the gene defined as DNA, the first step in relating it to its phenotype is to define phenotypes in chemical terms. This chemical definition of phenotypes was developed actually before the discovery of DNA as the genetic material. Using organisms as diverse as humans and bread molds, scientists studied the chemical differences between organisms carrying wild-type and mutant alleles and found that the major phenotypic differences were in specific proteins.

Some hereditary diseases feature defective enzymes

The first hints as to how genes are expressed came early in the twentieth century from the work of the English physician Archibald Garrod. Alkaptonuria is a hereditary disease in which the patient's urine turns black when exposed to air. Garrod recognized that this symptom showed that the biochemistry of the affected individual was different from that of other people. He suggested in 1908 that alkaptonuria and some other hereditary diseases are consequences of "inborn errors of metabolism."

Garrod proposed that what makes the urine dark is a defect in an enzyme that metabolizes the amino acid tyrosine. He studied the pattern of inheritance of alkaptonuria and reasoned that it affects individuals who are homozygous for the recessive allele of a particular gene (see Chapter 10), which in normal individuals codes for active enzyme. His studies and proposal explicitly linked genotype and phenotype by means of enzymes.

However, like Mendel's explanation of inheritance in the garden pea, Garrod's ideas were too advanced for their time and sat almost unappreciated for more than 30 years. We will return to Garrod's work and human genetic diseases in Chapter 17.

The one-gene, one-polypeptide hypothesis

A series of experiments performed by George W. Beadle and Edward L. Tatum at the California Institute of Technology in the 1940s confirmed and extended Garrod's ideas. Beadle and Tatum experimented with the bread mold *Neurospora crassa*. The nuclei in the mass of the mold are haploid (*n*), as are the reproductive spores. This fact is important because it means that even recessive mutant alleles are easy to detect in experiments.

Neurospora can be grown on a simple, completely defined medium (that is, one in which all the ingredients are known) that contains inorganic ions, a simple source of nitrogen (such as ammonium chloride), an organic source of energy and carbon (such as glucose), and a single vitamin (biotin). From this minimal medium, the enzymes of wild-type *Neurospora* can catalyze the metabolic reactions needed to make all the chemical constituents of its cells.

Beadle and Tatum hypothesized that mutations might lead to altered enzymes that could no longer do their jobs. In that case, mutants of *Neurospora* might be found that could not make certain compounds they needed; such mutants would grow only on media to which those compounds were added. We call mutants of this type **auxotrophs** ("increased eaters"), in contrast to the wild-type **prototrophs** ("original eaters") that constituted the original *Neurospora* population. Whereas prototrophs can grow on minimal medium, auxotrophs require specific additional nutrients, such as a particular amino acid, a vitamin, or a purine or pyrimidine.

Beadle and Tatum isolated a number of auxotrophic mutant strains of *Neurospora*. These auxotrophs did not grow on the minimal medium that supported the growth of the wild-type strain. But they did grow on a complete medium, to which all supplements (amino acids, nucleotides, vitamins, and so on) had been added. For each strain, Beadle and Tatum were able to find a single compound that, when added to the minimal medium, supported the growth of that mutant. This result supported the idea that mutations have simple effects—and, perhaps, the idea that each mutation causes a defect in only one enzyme in the metabolic pathway leading to the synthesis of the required nutrient.

One group of auxotrophs could grow on minimal medium supplemented with the amino acid arginine. They were classified as *arg* mutants. Mapping studies established that some of the *arg* mutations were at different loci on a chromosome or were on different chromosomes. Beadle and Tatum concluded from this observation that different genes can participate in governing a single biosynthetic pathway—in this case, the pathway leading to arginine.

They grew 15 different *arg* mutants in the presence of various suspected intermediates in the synthetic metabolic pathway for arginine:

$$X \rightarrow \text{ornithine} \rightarrow \text{citrulline} \rightarrow \text{arginine}$$

Some of the mutants were able to grow on different intermediates, as well as on arginine-supplemented medium (Figure 12.1). For example, some mutants could grow on either arginine or citrulline, and some could grow on arginine, citrulline, or ornithine.

Beadle and Tatum concluded that the mutants that grew on arginine- or citrulline-supplemented medium but not on ornithine-supplemented medium were deficient in an enzyme that catalyzes the conversion of ornithine to citrulline. Similarly, these investigators were able to relate each of the other *arg* mutants to a particular enzyme. Subsequent analysis of cell extracts of the various mutant strains for the relevant enzymes confirmed their hypotheses.

This work led Beadle and Tatum to formulate the one-gene, one-enzyme hypothesis. According to this hypothesis, the function of a gene is to control the production of a single, specific enzyme. This proposal strongly influenced the subsequent development of the sciences of genetics and molecular biology. Garrod had pointed in the same direction more than three decades earlier, but only after Beadle and Tatum's research were other scientists prepared to act on the suggestion.

Many enzymes are composed of more than one polypeptide chain (that is, they have quaternary structure). Each chain is specified by its own separate gene. Thus, it is more correct to speak of a **one-gene, one-polypeptide hypothesis**: The function of a gene is to control the production of a single, specific polypeptide.

Much later, it was discovered that some genes code for forms of RNA that do not become translated into polypeptides, and still other genes are sequences of DNA involved in controlling which DNA sequences are expressed. But these discoveries did not invalidate the relations between other genes and polypeptides.

DNA, RNA, and the Flow of Information

Now we turn our attention to the mechanisms by which a gene expresses itself as a polypeptide. The first mechanism, transcription, transcribes the information of DNA (the gene) into corresponding information in an RNA sequence. The second mechanism, translation, translates this RNA information into an appropriate amino acid sequence in a polypeptide.

RNA differs from DNA

To understand the transcription and translation of genetic information, you need to know about RNA. **RNA** (ribonucleic acid) is a polynucleotide similar to DNA (see Figure 3.23) but different in three ways.

1. RNA generally consists of only one polynucleotide strand (thus Chargaff's equalities, $G = C$ and $A = T$ [see Figure 11.5], are true only for DNA and not for RNA).
2. The sugar molecule found in ribonucleotides is ribose rather than the deoxyribose found in DNA.
3. Although three of the nitrogenous bases (adenine, guanine, and cytosine) in ribonucleotides are identical to the bases in deoxyribonucleotides, the fourth base in RNA is uracil (U), which is similar to thymine but lacks the methyl ($-CH_3$) group.

RNA can base-pair with single-stranded DNA, and this pairing obeys the AT, UA, and GC hydrogen-bonding rules. RNA can also fold over and base-pair within its own sequence, as we will see with tRNA later in this chapter.

Information flows in one direction when genes are expressed

Francis Crick proposed what he called the **central dogma** of molecular biology. The central dogma is, simply, that DNA codes for the production of RNA (transcription), RNA codes for the production of protein (translation), and protein does *not* code for the production of protein, RNA, or DNA (Figure 12.2). In Crick's words, "once 'information' has passed into protein *it cannot get out again.*"

Crick contributed two key ideas to the development of the central dogma. The first solved a difficult problem: How could one explain the relationship between a specific nucleotide sequence (in DNA) and a specific amino acid sequence (in protein), since the nucleotides of DNA do not attach to amino acids? Crick made a clever suggestion: He proposed that an adapter molecule carries a specific amino acid at one end and recognizes a sequence of nucleotides with another region.

In due course, other molecular biologists found and characterized these adapter molecules, called transfer RNAs, or **tRNAs**. Because they recognize the genetic message (a series of nucleotides) and simultaneously carry specific amino acids, tRNAs can translate the language of DNA into the language of proteins.

Crick's second major contribution to the central dogma addressed another problem: How does the genetic information get from the nucleus to the cytoplasm? (Most of the DNA of a eukaryotic cell is confined to the nucleus, but proteins are synthesized in the cytoplasm.) Crick, together with the South African geneticist Sydney Brenner and the French molecular

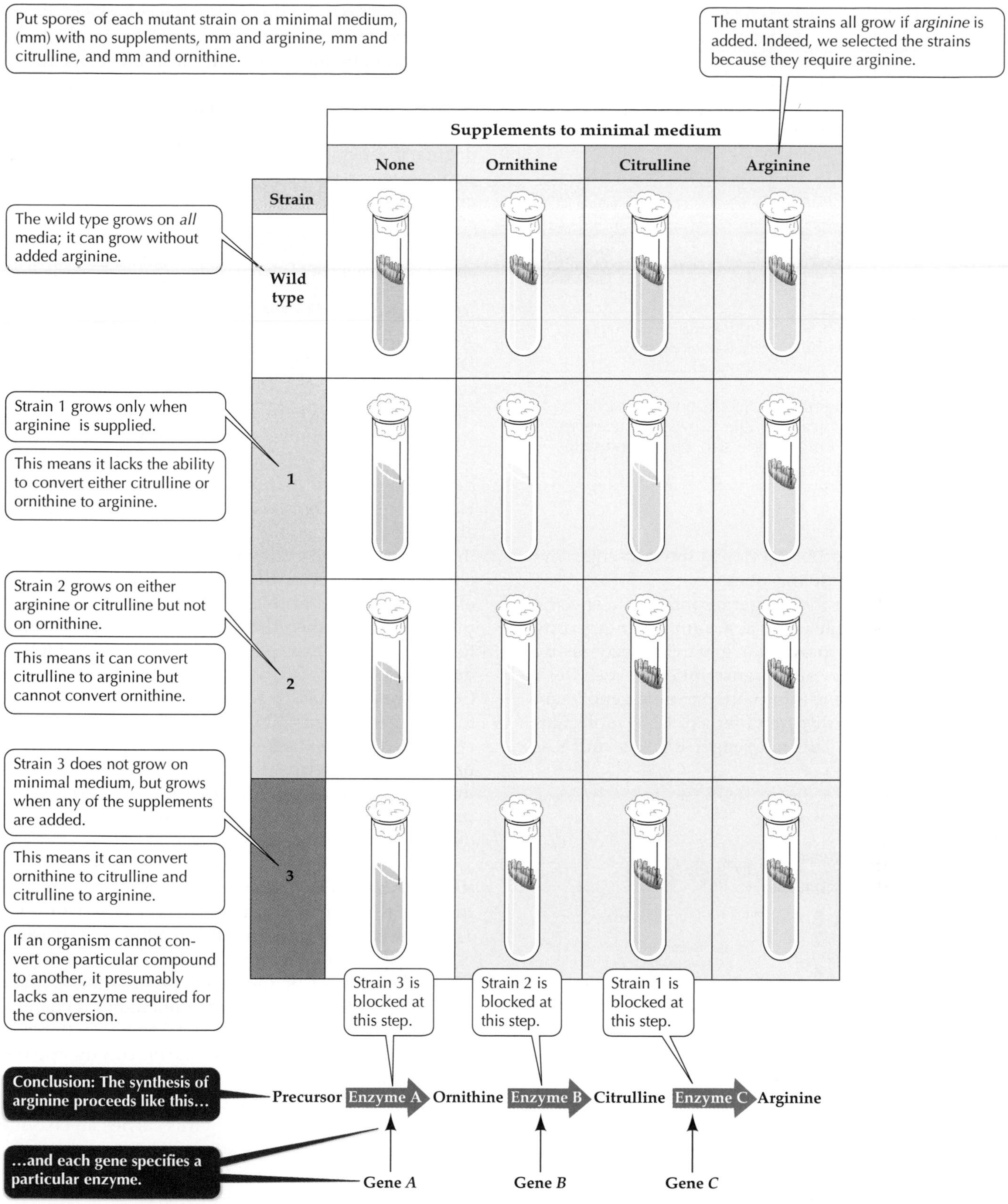

Put spores of each mutant strain on a minimal medium, (mm) with no supplements, mm and arginine, mm and citrulline, and mm and ornithine.

The mutant strains all grow if *arginine* is added. Indeed, we selected the strains because they require arginine.

	Supplements to minimal medium			
Strain	None	Ornithine	Citrulline	Arginine
Wild type				
1				
2				
3				

The wild type grows on *all* media; it can grow without added arginine.

Strain 1 grows only when arginine is supplied.

This means it lacks the ability to convert either citrulline or ornithine to arginine.

Strain 2 grows on either arginine or citrulline but not on ornithine.

This means it can convert citrulline to arginine but cannot convert ornithine.

Strain 3 does not grow on minimal medium, but grows when any of the supplements are added.

This means it can convert ornithine to citrulline and citrulline to arginine.

If an organism cannot convert one particular compound to another, it presumably lacks an enzyme required for the conversion.

Strain 3 is blocked at this step.

Strain 2 is blocked at this step.

Strain 1 is blocked at this step.

Conclusion: The synthesis of arginine proceeds like this...

...and each gene specifies a particular enzyme.

Precursor → Enzyme A → Ornithine → Enzyme B → Citrulline → Enzyme C → Arginine

Gene *A*   Gene *B*   Gene *C*

12.1 Genes and Enzymes Beadle and Tatum studied several nutritional mutants of *Neurospora,* as shown here. Wild-type, prototrophic strains grow on the minimal medium, but different auxotrophic mutants required the addition of different nutrients in order to grow; step through the figure to follow the reasoning that upheld the "one-gene, one-enzyme" hypothesis.

(a)

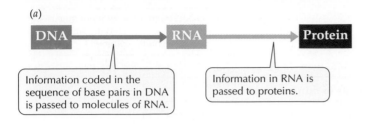

Information coded in the sequence of base pairs in DNA is passed to molecules of RNA.

Information in RNA is passed to proteins.

(b)

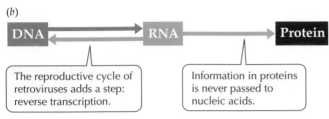

The reproductive cycle of retroviruses adds a step: reverse transcription.

Information in proteins is never passed to nucleic acids.

12.2 The Central Dogma *(a)* Information flows from DNA to proteins, as indicated by the arrows. *(b)* The reproductive cycle of retroviruses adds a step, reverse transcription, to the central dogma.

biologist François Jacob, developed the messenger hypothesis in response to this question.

According to the messenger hypothesis, a specific type of RNA molecule forms as a complementary copy of one strand of a particular gene. The process by which RNA forms is called **transcription**. If each RNA molecule contains the information from a gene, there should be as many different kinds of RNA molecules as there are genes. This messenger RNA, or **mRNA**,

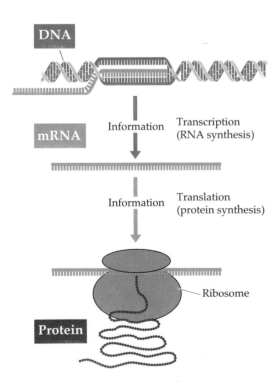

then travels from the nucleus to the cytoplasm. In the cytoplasm, mRNA serves as a template on which the tRNA adapters line up so that the amino acids are in the proper sequence for a growing polypeptide chain—a process called **translation** (Figure 12.3).

Summarizing the main features of the central dogma, the messenger hypothesis, and the adapter hypothesis, we may say that *a given gene is transcribed to produce a messenger RNA (mRNA) complementary to one of the DNA strands,* and that *transfer RNA (tRNA) molecules translate the sequence of bases in the mRNA into the appropriate sequence of amino acids.*

RNA viruses modify the central dogma

According to the central dogma of molecular biology, DNA codes for RNA and RNA codes for protein. All cellular organisms have DNA as their hereditary material. Only among viruses (which are not cellular) are variations on the central dogma found.

Many viruses, such as the tobacco mosaic virus, have RNA rather than DNA as their nucleic acid. Heinz Fraenkel-Conrat of the University of California at Berkeley separated the protein and RNA fractions of the tobacco mosaic virus and then recombined them to obtain active virus particles. When he took RNA from one mutant strain of this virus and combined it with protein from another, the resulting viruses replicated to produce more virus particles like the first (the RNA-donating) strain. With this experiment Fraenkel-Conrat showed that RNA is the genetic material of the tobacco mosaic virus. RNA itself is the template for the synthesis of the next generation of viral RNA and viral proteins. In this virus, DNA is left out of the flow of information (which is normally from DNA to RNA to protein). A more radical variation on the central dogma is seen in the retroviruses.

MAKING DNA FROM RNA: RETROVIRUSES. Rous sarcoma virus is an RNA virus that causes a cancer in chickens. The virus enters a chicken cell and subsequently causes the cell to make a DNA "transcript" of the viral RNA, the reverse of the usual process (see Figure 12.2*b*). The afflicted cell does not burst, but it changes permanently in shape, metabolism, and growth habit. The new DNA becomes part of the hereditary apparatus of the infected chicken cell.

In 1964, Howard Temin of the University of Wisconsin hypothesized that the virus carries an enzyme

12.3 From Gene to Phenotype in Prokaryotes This figure summarizes the messenger hypothesis as it appears in prokaryotes. In eukaryotes the process is more complex in that transcription forms a pre-mRNA in the nucleus. Pre-mRNA must then be processed to form the mature mRNA that is translated in the cytoplasm (see Chapter 14).

for the manufacture of DNA, using viral RNA as the information template. Five years later he and David Baltimore simultaneously but independently discovered the enzyme, which was named **reverse transcriptase** because it transcribes DNA from RNA rather than RNA from DNA. The DNA copy of the viral RNA can then use cellular machinery to make more viral RNA. Viruses that employ reverse transcriptase are known as *retroviruses*.

The central dogma requires slight modification to account for the flow of information in retroviruses and their hosts. However, the stipulation that information does not flow from protein back to the nucleic acids still holds.

Transcription: DNA-Directed RNA Synthesis

Transcription, the formation of a specific RNA under the control of a specific DNA, requires the enzyme **RNA polymerase**. It also requires the appropriate ribonucleoside triphosphates (ATP, GTP, CTP, and UTP) and the DNA template. In a given region of DNA, such as a gene, only *one* of the strands—the **template strand**—is transcribed. The other, complementary DNA strand remains untranscribed. For different genes in the same DNA molecule, different strands may be transcribed. That is, the strand that is the complementary strand in one gene may be the template strand in another.

Not only mRNA is produced by transcription. The same process is responsible for the synthesis of tRNA and ribosomal RNA (**rRNA**), which constitutes a major fraction of the ribosome. Like mRNA, these other forms of ribonucleic acid are encoded by specific genes. In prokaryotes, most of the DNA acts as a template for the production of mRNA, tRNA, or rRNA. The situation in eukaryotes is more complicated, as will be explained later in this chapter and in more detail in Chapter 14.

In DNA replication, the two strands of the parent molecule unwind, and each strand becomes paired with a new strand. In transcription, DNA partly unwinds so that it may serve as a template for RNA synthesis. As the RNA transcript forms, it peels away, allowing the DNA that has already been transcribed to rewind into the double helix (Figure 12.4).

Transcription can be divided into three distinct processes: initiation, elongation, and termination. Let's consider each of these in turn.

Initiation of transcription requires a promoter and an RNA polymerase

RNA polymerase has a relatively weak attraction for any DNA sequence, but it binds very tightly and is effective at beginning transcription only at special sequences on DNA called **promoters**. These sequences can be identified in several ways. For example, DNA can be chopped up into short stretches and presented to RNA polymerase in the test tube; sequences to which RNA polymerase binds tightly are promoters. Or, a promoter region can be detected by a change in its DNA sequence; in some cases a single base-pair change results in a promoter that can no longer bind tightly to the RNA polymerase.

There is one promoter for each gene (or, in prokaryotes, each set of genes) to be transcribed into mRNA. Promoters serve as punctuation marks, telling the RNA polymerase where to start and which strand of DNA to read. A promoter, being a specific sequence in the DNA and reading in a particular direction, orients the RNA polymerase and thus "aims" it at the correct strand to use as a template. Part of each promoter is the initiation site, where transcription begins. Farther toward the 3' end of the promoter lie groups of nucleotides that help the RNA polymerase bind. RNA polymerase moves in a 3'-to-5' direction along the template strand (see Figure 12.4).

Not all promoters are identical. One promoter may bind RNA polymerase very effectively and therefore trigger frequent transcription of its gene; in other words, it competes effectively for the available RNA polymerase. Another promoter may bind the polymerase poorly, and its genes will rarely be transcribed. The efficiency of the promoter sets a limit on how often each gene can be transcribed. An enzyme that is needed in large amounts is encoded by a gene whose promoter is efficient, but the synthesis of an enzyme that is needed only in tiny amounts is controlled by an inefficient promoter. We will consider prokaryotic promoters in more detail in Chapter 13 and eukaryotic promoters in Chapter 14.

Not all RNA polymerases are identical. Prokaryotes have a single RNA polymerase that produces mRNA, tRNA, and rRNA. Eukaryotes have three different RNA polymerases with distinct roles. Of these, RNA polymerase II is responsible for mRNA production.

Initiation of transcription in eukaryotes differs from that in prokaryotes in another respect. As we will see in slightly more detail in Chapter 14, eukaryotic RNA polymerases cannot bind the promoter and start the process of transcription until other proteins bind to sites in the promoter and thus prepare a docking site for the RNA polymerase. For now, recognize that the requirement for these proteins affords a way to regulate the transcription of particular genes, augmenting the differences already imposed by the varying efficiency of promoters.

Transcription can proceed as soon as RNA polymerase binds to the promoter and starts unwinding the DNA strands at the initiation site. The next stage is elongation of the RNA transcript, in which additional ribonucleotides are added to the growing chain.

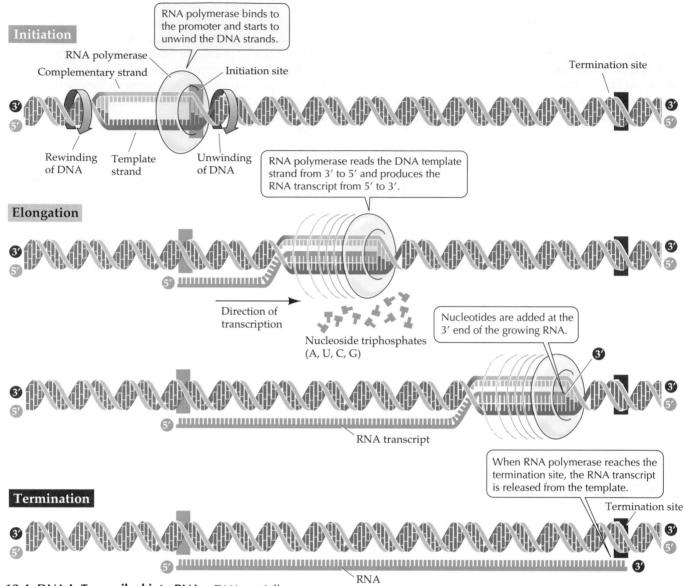

Initiation

RNA polymerase
Complementary strand

RNA polymerase binds to the promoter and starts to unwind the DNA strands.

Initiation site

Termination site

Rewinding of DNA

Template strand

Unwinding of DNA

Elongation

RNA polymerase reads the DNA template strand from 3' to 5' and produces the RNA transcript from 5' to 3'.

Direction of transcription

Nucleoside triphosphates (A, U, C, G)

Nucleotides are added at the 3' end of the growing RNA.

RNA transcript

When RNA polymerase reaches the termination site, the RNA transcript is released from the template.

Termination site

Termination

Termination site

RNA

12.4 DNA Is Transcribed into RNA DNA partially unwinds to serve as a template for RNA synthesis. The RNA transcript forms and then peels away, allowing the DNA that has already been transcribed to rewind into a double helix. Three distinct processes—initiation, elongation, and termination—comprise DNA transcription. The RNA polymerase is much larger in reality than indicated here, covering about 50 base pairs.

RNA polymerases elongate the transcript

RNA polymerase unwinds the DNA about 20 base pairs at a time and reads the template strand of DNA in the 3'-to-5' direction (see Figure 12.4). Like DNA polymerases, RNA polymerases add new nucleotides to the 3' end of the growing strand. That is, the new RNA grows from its own 5' end to its 3' end. The RNA transcript is thus antiparallel to the DNA template strand.

As the RNA polymerase moves along the template strand, unwinding the next stretch of DNA, it rewinds the stretch of DNA that it just processed. Transcription, like DNA replication, is a lot of work, requiring a lot of energy. As with replication, transcription draws on energy released by both the removal and the breakdown of the pyrophosphate group from each added nucleotide.

Unlike DNA polymerases, RNA polymerases do not inspect and correct their work. Transcription errors occur at a rate of one mistake for every 10^4 to 10^5 bases.

Transcription terminates at particular base sequences

What tells RNA polymerase to stop adding nucleotides to a growing transcript? Just as initiation sites specify the start of transcription, particular base sequences in the DNA specify its termination. The mechanisms of termination are complex and of more than one kind. For some genes the newly formed transcript simply falls away from the DNA template and the RNA polymerase. For others, a helper protein pulls the transcript away.

In prokaryotes the translation of mRNA often begins (at the 5′ end of the mRNA) before transcription of the mRNA molecule is complete. In eukaryotes the situation is more complicated. First, there is a spatial separation of transcription (in the nucleus) and translation (in the cytoplasm). Second, the first product of transcription is a pre-mRNA that is longer than the final mRNA and must undergo considerable processing before it becomes the mRNA and can be translated. The reasons for this processing, and its mechanisms, will be discussed in Chapter 14.

The Genetic Code

You can think of the genetic information transcribed in an mRNA molecule as a series of three-letter "words." Each sequence of three nucleotides (the "letters") along the chain specifies a particular amino acid. The three-letter "word" is called a **codon**. That codon is complementary to the corresponding codon in the DNA molecule from which it was transcribed.

The complete genetic code is shown in Figure 12.5. Notice that there are many more RNA codons than there are different amino acids in proteins. Combinations of the four "letters" (the bases) give 64 (4^3) different three-letter codons, yet these determine only 20 amino acids. AUG, which codes for methionine, is also the **start codon**, the initiation signal for translation. Three of the codons (UAA, UAG, UGA) are **stop codons**, or chain terminators; when the translation machinery reaches one of these codons, translation stops and the polypeptide is released from the translation complex.

After describing the properties of the genetic code, we will examine

some of the scientific thinking and experimentation that went into deciphering it.

The genetic code is degenerate but not ambiguous

After the start and stop codons, the remaining 60 codons are far more than enough to code for the other 19 amino acids—and indeed there are repeats. Thus we say that the code is **degenerate**; that is, an amino acid may be represented by more than one codon. The degeneracy is not evenly divided among the amino acids. For example, methionine and tryptophan are represented by only one codon each, whereas leucine is represented by six different codons (see Figure 12.5).

The term "degeneracy" should not be confused with "ambiguity." To say that the code was ambiguous would mean that a single codon could specify either of two (or more) different amino acids; there would be doubt whether to put in, say, leucine or something else. The genetic code is not ambiguous. Degeneracy in the code means that there is more than one clear way to say, "Put leucine here." In other words, a given amino acid may be encoded by more than one codon, but a codon can code for only one amino acid. But just as people in different places prefer different ways of saying the same thing—"Good-bye!" "See you!" "Ciao!" and "So long!" have the same meaning—different organisms prefer one or others of the degenerate codons. These preferences are important in genetic engineering (see Chapter 16).

The code appears to be relatively universal, applying to all the species on our planet. Thus the code must be an ancient one that has been maintained intact throughout the evolution of living things. Exceptions are known: Within mitochondria and chloroplasts the code differs slightly from that in

12.5 The Universal Genetic Code
Genetic information is encoded in mRNA in three-letter units—codons—made up of the bases uracil (U), cytosine (C), adenine (A), and guanine (G). To decode a codon, find its first letter in the left column, then read across the top to its second letter, then read down the right column to its third letter. The amino acid the codon specifies is given in the corresponding row. For example, AUG codes for methionine, and GUA codes for valine.

prokaryotes and elsewhere in eukaryotic cells; in one group of protists, UAA and UAG code for glutamine rather than functioning as stop codons. The significance of these differences is not yet clear. What is clear is that the exceptions are few and slight.

You should remember that the codons in Figure 12.5 are mRNA codons. The master codons on the DNA strand that was transcribed to produce the mRNA are complementary and antiparallel to these codons. Thus, for example, AAA in the template DNA strand corresponds to phenylalanine (which is coded for by the mRNA codon UUU), and CCA in the DNA template corresponds to tryptophan (which is coded for by the mRNA codon UGG). If the last example surprised you, note that we normally list the base sequences of nucleic acids in a 5'-to-3' order.

Does this code really work? Convincing evidence comes from experiments in which artificial DNA of a known base sequence is introduced into prokaryotes and the prokaryotes are induced to produce the specific protein encoded by that DNA (see Chapter 16). We can now program bacteria to synthesize proteins that no organism has ever made before.

Biologists broke the genetic code by translating artificial messengers

Molecular biologists broke the code in which genetic information is stored in the early 1960s. The problem seemed difficult: How could more than 20 "code words" be written with an "alphabet" consisting of only four "letters"? How, in other words, could four bases code for 20 or so different amino acids?

The idea that the code was a triplet code, based on three-letter codons, was considered likely. With only four letters (A, G, C, U), a one-letter code clearly could not unambiguously encode 20 amino acids; it could encode only four of them. A two-letter code could contain only $4 \times 4 = 16$ codons—still not enough. But a triplet code could contain up to $4 \times 4 \times 4 = 64$ codons.

Marshall W. Nirenberg and J. H. Matthaei, at the National Institutes of Health, made the first "decoding" breakthrough in 1961 when they realized that they could use a very simple artificial polynucleotide instead of a complex, natural mRNA as a messenger. They could then identify the polypeptide that the artificial messenger encoded.

Nirenberg had prepared an artificial mRNA in which all the bases were uracil: poly U. When poly U was added to a reaction mixture containing all the ingredients necessary for cell-free protein synthesis (ribosomes, amino acids, activating enzymes, tRNAs, and other factors), a polypeptide formed. This polypeptide contained only one kind of amino acid: phenylalanine (Phe). Poly U coded for poly Phe! Accordingly, UUU appeared to be the mRNA code word—the codon—for phenylalanine. Following up

on this success, Nirenberg and Matthaei soon showed that CCC codes for proline and AAA for lysine. (Poly G presented some chemical problems and was not tested initially.) UUU, CCC, and AAA were three of the easiest codons; different approaches were required to work out the rest.

Other scientists later found that simple "mRNAs," only three nucleotides long and each amounting to a codon, can bind to ribosomes and that the resulting complex can then cause the binding of the corresponding charged tRNA. Thus, for example, simple UUU causes the tRNA charged with phenylalanine to bind to the ribosome. After this discovery, complete deciphering of the code book was relatively simple. To find the "translation" of a codon, Nirenberg could use a sample of that codon as an artificial mRNA and see which amino acid became bound.

The Key Players in Translation

Prokaryotic mRNAs are ready to be translated as they peel away from the DNA template strand, and they degrade within minutes. Eukaryotic pre-mRNAs are processed extensively before they become translatable mRNAs, and these mRNAs continue to be functional for many minutes to several hours. In both cases, however, translation is rapid, taking only a few minutes to make a polypeptide of hundreds of amino acids. This complex process occurs at the ribosome, which binds to mRNA, carrying the genetic code from DNA. Before translation begins, however, each amino acid becomes attached to its specific tRNA.

Transfer RNAs carry specific amino acids

Before we examine translation, let's see how a codon is related to the amino acid for which it codes. As predicted by Crick, the codon and the amino acid are related by way of an adapter—a specific type of tRNA. For each of the 20 amino acids, there is at least one specific tRNA molecule.

A tRNA molecule is small, consisting of only about 75 to 80 nucleotides (Figure 12.6). At the 3' end of every tRNA molecule is a site to which the amino acid attaches. At about the midpoint is a group of three bases, called the **anticodon**, that constitutes the point of contact with mRNA. At contact, the tRNA and mRNA are antiparallel to each other. Each tRNA species has a unique anticodon, allowing it to unite by complementary base pairing with a particular codon. Complementary base pairing is what enables translation to be so specific.

Recall that 61 different codons encode the 20 amino acids in proteins. Does this mean that the cell must produce 61 different tRNA species, each with a different anticodon? No. The cell gets by with about two-thirds that number of tRNA species, because the speci-

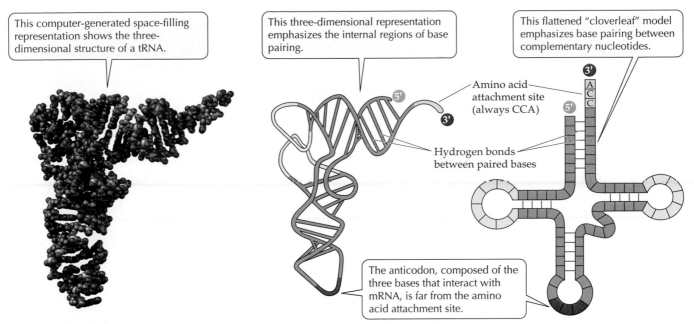

This computer-generated space-filling representation shows the three-dimensional structure of a tRNA.

This three-dimensional representation emphasizes the internal regions of base pairing.

This flattened "cloverleaf" model emphasizes base pairing between complementary nucleotides.

Amino acid attachment site (always CCA)

Hydrogen bonds between paired bases

The anticodon, composed of the three bases that interact with mRNA, is far from the amino acid attachment site.

12.6 Transfer RNA: Crick's Adapter The tRNA molecules carry amino acids, associate with mRNA molecules, and interact with ribosomes. There is at least one specific tRNA molecule for each of the amino acids.

ficity for the base at the 3′ end of the codon (and 5′ end of the anticodon) is relaxed. This phenomenon, called *wobble*, allows the alanine codons GCA, GCC, and GCU all to be recognized by the same tRNA. Wobble is allowed in some matches but not in others; of most importance, it does not allow the genetic code to be ambiguous!

The three-dimensional shape of tRNAs (see Figure 12.6) allows them to combine specifically with binding sites on ribosomes. The structure of tRNA molecules relates clearly to their functions: They carry amino acids, associate with mRNA molecules, and interact with ribosomes.

Activating enzymes link the right tRNAs and amino acids

How does a tRNA molecule combine with the correct amino acid? A family of **activating enzymes**, known more formally as aminoacyl-tRNA synthetases, accomplishes this task (Figure 12.7). Each activating enzyme is specific for one amino acid and for one tRNA. The enzyme has a three-part active site that recognizes three smaller molecules: a specific amino acid, ATP, and a specific tRNA.

The enzyme reacts first with a molecule of amino acid and a molecule of ATP, producing a high-energy amino acid–AMP (adenosine monophosphate, which remains bound to the enzyme). The high energy results from the breaking of the bonds in the ATP—the high-energy bond between AMP and the terminal py-

rophosphate group (see Figure 6.7), and then the high-energy bond in the pyrophosphate (PP_i). The energy is conserved in the bond between the amino acid (AA) and AMP:

$$\text{enzyme} + \text{ATP} + \text{AA} \rightarrow \text{enzyme—AMP—AA} + PP_i$$

The enzyme then catalyzes a shifting of the amino acid from the AMP to the 3′-terminal nucleotide of the tRNA:

$$\text{enzyme—AMP—AA} + \text{tRNA} \rightarrow$$
$$\text{enzyme} + \text{AMP} + \text{tRNA—AA}$$

The activating enzyme finally releases this charged tRNA (tRNA with its attached amino acid) and can then charge another tRNA molecule. The bond between the amino acid and tRNA is a high-energy bond; it provides the energy for the synthesis of a peptide bond joining adjacent amino acids.

A clever experiment showed the importance of the specificity of the attachment of tRNA to its amino acid—a specificity that has been called the "second genetic code." The amino acid cysteine, already properly attached to its tRNA, was chemically modified to become a different amino acid, alanine. Which component—the amino acid or the tRNA—would be recognized when the hybrid charged tRNA was put into a protein-synthesizing system? The answer was, the latter: Everywhere in the synthesized protein where cysteine was supposed to be, alanine appeared instead. The cysteine-specific tRNA delivered its cargo (alanine) to every address where cysteine was called for.

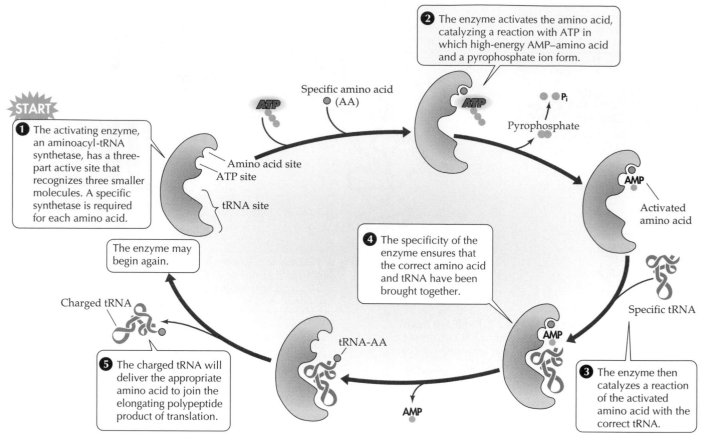

START

❶ The activating enzyme, an aminoacyl-tRNA synthetase, has a three-part active site that recognizes three smaller molecules. A specific synthetase is required for each amino acid.

Amino acid site
ATP site
tRNA site

Specific amino acid (AA)

ATP

❷ The enzyme activates the amino acid, catalyzing a reaction with ATP in which high-energy AMP–amino acid and a pyrophosphate ion form.

ATP

P_i

Pyrophosphate

AMP

Activated amino acid

❹ The specificity of the enzyme ensures that the correct amino acid and tRNA have been brought together.

Specific tRNA

AMP

❸ The enzyme then catalyzes a reaction of the activated amino acid with the correct tRNA.

The enzyme may begin again.

Charged tRNA

tRNA-AA

AMP

❺ The charged tRNA will deliver the appropriate amino acid to join the elongating polypeptide product of translation.

12.7 Charging a tRNA Molecule Each activating enzyme must make the correct association of an amino acid and its tRNA; the enzyme is the essential link between nucleic acid "language" and protein "language."

The ribosome is the staging area for translation

Ribosomes are required for translation of the genetic information into a polypeptide chain. Although ribosomes are the smallest cellular organelles, their mass of several million daltons makes them large in comparison with the charged tRNAs.

Each ribosome consists of two subunits, a large one and a small one (Figure 12.8). In eukaryotes, the large subunit consists of three different molecules of rRNA (ribosomal RNA) and about 45 different protein molecules, arranged in a precise pattern. The small subunit in eukaryotes consists of one rRNA molecule and 33 different protein molecules. These different proteins and RNAs are held together by ionic and hydrophobic forces, not covalent bonds. If these forces are dis-

rupted by detergents, for example, the proteins and rRNAs separate from each other. When the detergent is removed, the entire complex structure *self-assembles*. This is like separating the pieces of a jigsaw puzzle and having them fit together again without human hands to guide them.

The ribosomes of prokaryotes are somewhat smaller than those of eukaryotes. Mitochondria and chloroplasts also contain ribosomes, some of which are even smaller than those of prokaryotes. When not active in the translation of mRNA, the ribosomes exist as sepa-

12.8 Ribosome Structure Each ribosome consists of a large and a small subunit, which separate when they are not in use.

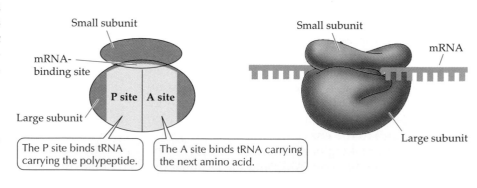

Small subunit

mRNA-binding site

P site | A site

Large subunit

The P site binds tRNA carrying the polypeptide.

The A site binds tRNA carrying the next amino acid.

Small subunit

mRNA

Large subunit

rated subunits. Each ribosome has two tRNA-binding sites (P and A) that participate in translation. The ribosome also binds to the mRNA that it is translating.

A given ribosome is not specifically adapted to produce just one kind of protein. A ribosome can combine with any mRNA and all tRNAs and thus can be used to make different polypeptide products. The mRNA contains the information that specifies the polypeptide sequence. The ribosome is simply the molecular machine that accomplishes the task. Its structure enables it to hold the mRNA and tRNAs in the right positions, thus allowing the growing polypeptide to be assembled efficiently.

Translation: RNA-Directed Polypeptide Synthesis

We have been working our way through the steps by which the sequence of bases in the template strand of a DNA molecule specifies the sequence of amino acids in a protein (see Figure 12.3). We are now at the last step: translation, the RNA-directed assembly of a protein. Like transcription, translation occurs in three steps: initiation, elongation, and termination. Most of these reactions are catalyzed by ribosomal proteins. A polypeptide contains information for its three-dimensional shape, as well as for its ultimate cellular destination.

Translation begins with an initiation complex

The translation of mRNA begins with the formation of an initiation complex, which consists of a charged tRNA bearing the first amino acid and a small ribosomal subunit, both bound to the starting point on the mRNA chain (Figure 12.9). The small ribosomal unit binds to a recognition sequence on the mRNA. Recall that the start codon in the genetic code is AUG. Thus the first amino acid is methionine (see Figure 12.5). The anticodon of a methionine-charged tRNA binds to the appropriate point on the mRNA by complementary base pairing with AUG, the initiation codon. (Not all proteins have methionine as their N-terminal amino acid. In many cases the initiator methionine is removed by an enzyme.)

After the methionine-charged tRNA has bound to the mRNA, the large subunit of the ribosome joins the complex. The ribosome has two tRNA-binding sites: the A site (which accepts a tRNA molecule bearing one amino acid) and the P site (which carries a tRNA molecule bearing a growing polypeptide chain). The first charged tRNA, bearing methionine, now lies in the P site of the ribosome, and the A site is aligned with the second codon.

How are all these ingredients—mRNA, two ribosomal subunits, and methionine-charged tRNA—put together properly? A group of proteins called **initiation factors** help direct the process, using GTP as an energy supply.

The polypeptide elongates from the N terminus

During translation the ribosome moves along the mRNA in the 5′-to-3′ direction (Figure 12.10). A charged tRNA whose anticodon is complementary to the second codon enters the open A site. The large subunit then catalyzes the formation of a peptide bond between the amino acid on the P site and the amino acid on the A site in such a way that the first amino acid is the N terminus of the new protein, while the second amino acid remains attached to its tRNA by its carboxyl group (—COOH).

In 1992, Harry Noller and his colleagues at the University of California at Santa Cruz found that if they removed almost all the proteins in the large ribosomal subunit, it still catalyzed peptide bond formation. But if the rRNA was destroyed, so was the catalytic activity. Part of rRNA in the large subunits interacts with the end of the charged tRNA where the amino acid is attached. Thus rRNA appears to be the catalyst for peptide bond formation.

The idea that RNA—instead of the usual protein—can act as a catalyst, or **ribozyme**, is not unprecedented. For example, we

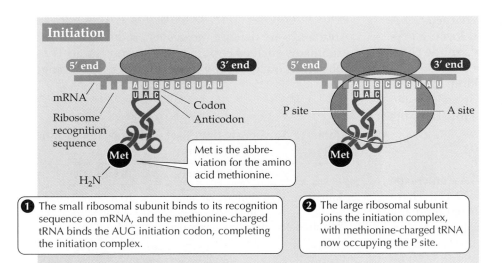

1 The small ribosomal subunit binds to its recognition sequence on mRNA, and the methionine-charged tRNA binds the AUG initiation codon, completing the initiation complex.

2 The large ribosomal subunit joins the initiation complex, with methionine-charged tRNA now occupying the P site.

12.9 The Initiation of Translation
Translation begins with the formation of an initiation complex.

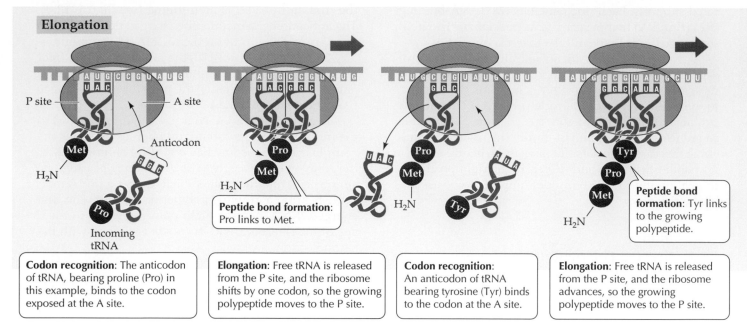

Elongation

Codon recognition: The anticodon of tRNA, bearing proline (Pro) in this example, binds to the codon exposed at the A site.

Elongation: Free tRNA is released from the P site, and the ribosome shifts by one codon, so the growing polypeptide moves to the P site.

Codon recognition: An anticodon of tRNA bearing tyrosine (Tyr) binds to the codon at the A site.

Elongation: Free tRNA is released from the P site, and the ribosome advances, so the growing polypeptide moves to the P site.

12.10 Translation: The Elongation Stage The polypeptide chain elongates as the mRNA is translated.

will describe catalytic RNAs that act in mRNA splicing in Chapter 14. Because of its base pairing, RNA can fold into three-dimensional shapes (see tRNA, Figure 12.6) and bind substrates, just as protein-based enzymes do.

Elongation continues and the polypeptide grows

After the first tRNA releases its amino acid, it dissociates from the complex, returning to the cytosol to become charged with another amino acid of the same kind. The second tRNA, now bearing a *dipeptide*, shifts to the P site of the ribosome, which moves along the mRNA by another triplet codon. Energy for this movement comes from the hydrolysis of another molecule of GTP.

The process continues: (1) The next charged tRNA enters the open A site; (2) its amino acid forms a peptide bond, picking up the growing polypeptide chain from the tRNA in the P site; and (3) the entire tRNA–polypeptide complex, along with its codon, moves to the newly vacated P site. All these steps are assisted by proteins called elongation factors. How does the cycle end?

A release factor terminates translation

When a stop codon—UAA, UAG, or UGA—enters the A site, translation terminates (Figure 12.11). These codons encode no amino acids, nor do they bind any tRNA. Rather, they bind a protein **release factor**, which causes a water molecule instead of an amino acid to attach to the forming protein.

The newly completed protein thereupon separates from the ribosome. Its C terminus is the last amino acid to join the chain. Its N terminus, at least initially, is methionine, as a consequence of the AUG start codon. Table 12.1 summarizes the initiation and termination of transcription and translation.

Regulation of Translation

As in any factory, the machinery of translation can work at varying rates. For example, externally applied chemicals such as some antibiotics can stop translation. The presence of more than one ribosome on an mRNA can speed up protein synthesis. And the endoplasmic reticulum (ER) can be used to segregate a protein as it is being made.

Some antibiotics work by inhibiting translation

Antibiotics are defense molecules produced by microorganisms such as certain bacteria and fungi. These substances often destroy other microbes, which might

Table 12.1 Signals That Start and Stop Transcription and Translation

	TRANSCRIPTION	TRANSLATION
Initiation	Promoter sequence in DNA	AUG start codon in mRNA
Termination	Terminator sequence in DNA	UAA, UAG, or UGA stop codon in mRNA

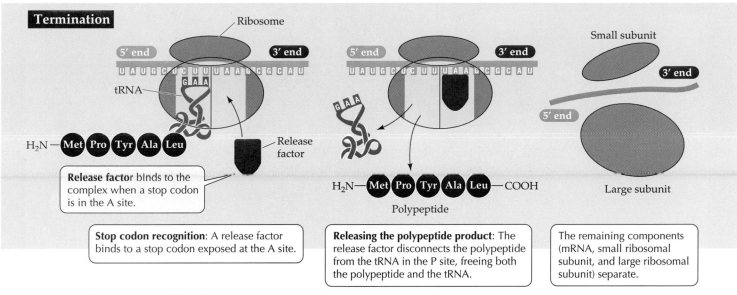

12.11 The Termination of Translation Translation terminates when the ribosome encounters a stop signal on the mRNA.

compete with the defender for nutrients. Since the 1940s, scientists have isolated increasing numbers of antibiotics, and physicians use them to treat a great variety of infectious diseases, ranging from bacterial meningitis to pneumonia to gonorrhea.

The key to antibiotic action is specificity: An antibiotic must work to destroy the microbial invader but not harm the human host. One way in which antibacterials accomplish this task is to block the synthesis of the bacterial cell wall, something that is essential to the microbe but that is not part of human biochemistry. Penicillin works in this fashion.

Another way is to inhibit bacterial protein synthesis. Recall that the bacterial ribosome is smaller and has a different collection of proteins than the eukaryotic ribosome has. Some antibiotics bind only to bacterial ribosomal proteins that are important in protein synthesis (Table 12.2). So, without the ability to make proteins, the bacterial invaders die and the infection is stemmed.

Polysome formation increases the rate of protein synthesis

Several ribosomes can work simultaneously at translating a single mRNA molecule to produce multiple molecules of the protein at the same time. As soon as the first ribosome has moved far enough from the initiation point, a second initiation complex can form, then a third, and so on. The assemblage of a thread of mRNA with its beadlike ribosomes and their growing polypeptide chains is called a polyribosome, or **polysome** (Figure 12.12).

A polysome is like a cafeteria line, where patrons follow each other, adding items to their trays. The person at the start has a little food (initiation); the person at the end has a complete meal (completed protein). Cells that are actively synthesizing proteins contain large numbers of polysomes and fewer free ribosomes or ribosomal subunits.

A signal sequence leads a protein through the ER

As a polypeptide chain forms on the ribosome, it spontaneously folds into its three-dimensional shape. As described in Chapter 3, this shape is determined by the particular order of amino acids that make up the protein, as well as factors such as their polarity and charge. Ultimately, this shape allows the polypeptide to interact with other molecules in the cell, such as a substrate if it acts as an enzyme. In addition to this structural information, the amino acid sequence contains information indicating where in the cell the polypeptide belongs.

As you learned in Chapter 4, an important difference between prokaryotes and eukaryotes is that eu-

Table 12.2	Antibiotics That Inhibit Bacterial Protein Synthesis
ANTIBIOTIC	**STEP INHIBITED**
Chloromycetin	Formation of peptide bonds
Erythromycin	Translocation of mRNA along ribosome
Neomycin	Interactions between tRNA and mRNA
Streptomycin	Initiation of translation
Tetracycline	Binding of tRNA to ribosome

(a)

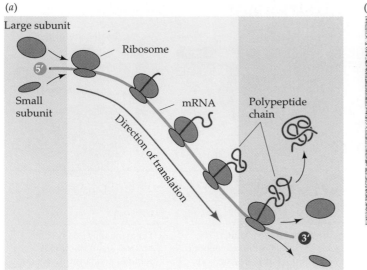

(b)

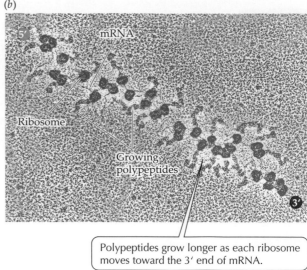

Polypeptides grow longer as each ribosome moves toward the 3′ end of mRNA.

12.12 A Polysome   (a) A polysome consists of ribosomes and their growing polypeptide chains moving in single file along an mRNA molecule. (b) An electron microscope gave us this detailed view of a polysome.

karyotic cells have many individual compartments. Different compartments need different proteins. Are proteins synthesized where they are needed, or are they transported from a synthesis site? How are particular proteins targeted to the correct site—electron transport chain components to the mitochondria, histones to the nucleus, and so forth?

Proteins that are to remain soluble within the cell are synthesized on "free" ribosomes—that is, ribosomes that are not attached to the endoplasmic reticulum (ER). Proteins that are to become parts of membranes, or are to be exported from the cell, or are to end up in lysosomes or peroxisomes are synthesized on ribosomes of the rough ER. All protein synthesis, however, *begins* on free ribosomes. The first few amino acids of a polypeptide chain determine whether production of the protein will be completed on the rough ER or on free ribosomes.

If a specific sequence of amino acids, the **signal sequence**, is present at the beginning of the chain, the finished product will be a membrane protein or a protein destined for export. The signal sequence attaches to a *signal recognition particle* composed of protein and RNA (Figure 12.13). This attachment blocks further protein synthesis until the ribosome can become attached to a specific receptor protein in the membrane of the ER. The receptor protein becomes a channel through which the growing polypeptide is extruded, either into the membrane itself or into the interior of the ER, as synthesis continues.

An enzyme within the ER interior then removes the signal sequence from the new protein, which ends up either built into the ER membrane or retained within the ER rather than in the cytosol. From the ER the newly formed protein can be transported to its appropriate location—to other cellular compartments or to the outside of the cell—without mixing with other molecules in the cytoplasm (see Figure 12.13). Signals, consisting of amino acid sequences or sugars added in the ER and the Golgi apparatus, determine the cellular destination of a protein, much as postal zip codes direct mail.

Mutations: Heritable Changes in Genes

Accurate DNA replication, transcription, and translation all depend on the reliable pairing of complementary bases. Errors occur, though infrequently, in all three processes. In particular, errors in the DNA replication during the production of the gametes produce mutations. **Mutations** are heritable changes in genetic information.

Minute changes in the genetic material often lead to easily observable changes in the phenotype. Some effects of mutation in humans are readily detectable—dwarfism, for instance, or the presence of more than five fingers on each hand. A mutant genotype in a microorganism may be obvious if, for example, it results in a change in nutritional requirements, as we described for *Neurospora* (see Figure 12.1).

However, other mutations may be unobservable. In humans, for example, a particular mutation drastically lowers the level of an enzyme called glucose 6-phosphate dehydrogenase that is present in many tissues, including red blood cells. The red blood cells of a person carrying the mutant gene are abnormally sensitive

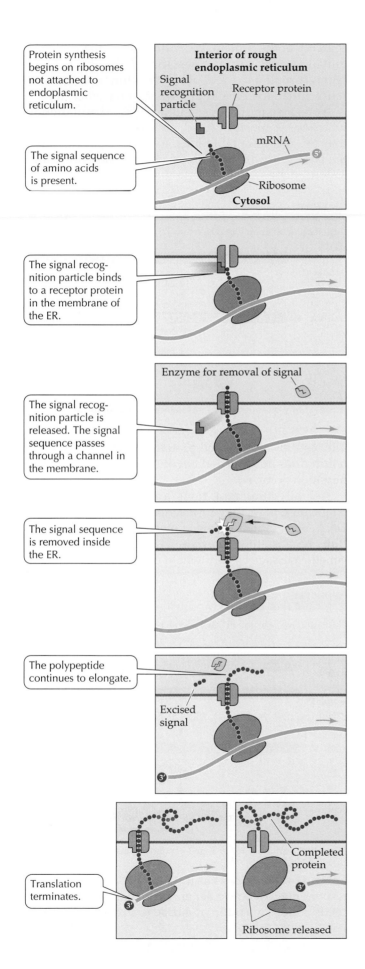

Protein synthesis begins on ribosomes not attached to endoplasmic reticulum.

The signal sequence of amino acids is present.

Interior of rough endoplasmic reticulum

Signal recognition particle

Receptor protein

mRNA

Ribosome

Cytosol

The signal recognition particle binds to a receptor protein in the membrane of the ER.

The signal recognition particle is released. The signal sequence passes through a channel in the membrane.

Enzyme for removal of signal

The signal sequence is removed inside the ER.

The polypeptide continues to elongate.

Excised signal

Translation terminates.

Completed protein

Ribosome released

12.13 A Signal Sequence Moves a Polypeptide into the ER When a signal sequence of amino acids is present at the beginning of the chain, the polypeptide will be taken into the endoplasmic reticulum and elongation completed there. The finished protein is thus segregated from the cytosol.

to an antimalarial drug called primaquine; when such people are treated with this drug, their red blood cells rupture, causing serious medical problems. People with the normal allele have no such problem. Before the drug came into use, no one was aware that such a mutation existed. Similarly, distinguishing a mutant bacterium from a normal bacterium may require sophisticated chemical methods, not just visual inspection.

Some mutations cause their phenotypes only under certain restrictive conditions and are not detectable under other, permissive conditions. We call organisms carrying such mutations *conditional mutants*. Many conditional mutants are temperature-sensitive, unable to grow at a particular restrictive temperature, such as 37°C, but able to grow normally at a lower, permissive temperature, such as 30°C. The mutant allele in such an organism may code for an enzyme with an unstable tertiary structure that is altered at the restrictive temperature.

All mutations are alterations in the nucleotide sequence in DNA. We divide mutations into two categories: point mutations and chromosomal mutations. **Point mutations** are mutations of single genes: One allele becomes another because of small alterations in the sequence or number of nucleotides—even as small as the substitution of one nucleotide for another. **Chromosomal mutations** are more extensive alterations. They may change the position or direction of a DNA segment without actually removing any genetic information, or they may cause a segment of DNA to be irretrievably lost. Both point mutations and chromosomal mutations are heritable.

Point mutations may be silent, missense, nonsense, or frame-shift

Many point mutations consist of the substitution of one base for another in the DNA and hence in the mRNA. Because of the degeneracy of the genetic code, some of these mutations result in no change in amino acids after the altered mRNA is translated; for this reason they are called **silent mutations**.

For example, four mRNA codons code for the amino acid proline: CCA, CCC, CCU, and CCA (see Figure 12.5). If the template strand of DNA for this particular region has the sequence CGG, it will be transcribed to CCG in mRNA and proline-charged *tRNA will bind at the ribosome*. But if there is a mutation in the DNA such that the triplet in the template DNA reads GGA, the mRNA codon will be CCU—and the tRNA that binds will still bring proline:

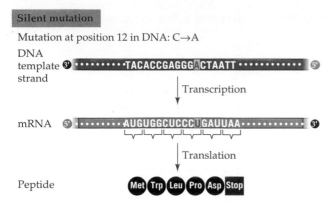

Silent mutation

Mutation at position 12 in DNA: C→A

DNA template strand 3′ ···········TACACCGAGGG**A**CTAATT··········· 5′

↓ Transcription

mRNA 5′ ···········AUGUGGCUCCC**U**GAUUAA··········· 3′

↓ Translation

Peptide Met Trp Leu Pro Asp Stop

Result: No change in amino acid sequence

Silent mutations are quite common and account for genetic diversity that is not expressed as phenotypic differences.

In contrast to silent mutations, some base substitution mutations may change the genetic message such that one amino acid substitutes for another in the protein. This is a **missense mutation**:

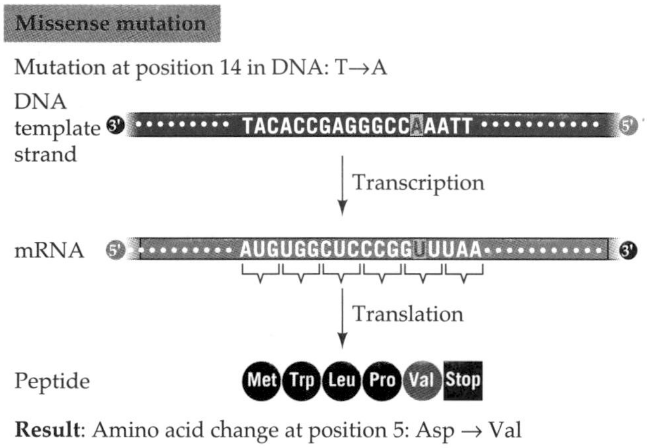

Missense mutation

Mutation at position 14 in DNA: T→A

DNA template strand 3′ ···········TACACCGAGGGCC**A**AATT··········· 5′

↓ Transcription

mRNA 5′ ···········AUGUGGCUCCCGG**U**UUAA··········· 3′

↓ Translation

Peptide Met Trp Leu Pro Val Stop

Result: Amino acid change at position 5: Asp → Val

A specific example of such a mutation is the sickle allele for human β-globin. Sickle-cell anemia is the consequence of a recessive allele that, when homozygous, results in defective red blood cells. Where oxygen is abundant, as in the lungs, the cells are normal in structure and function. But at the low oxygen levels characteristic of working muscles, the red blood cells collapse into the shape of a sickle (Figure 12.14).

The disease results from a defect in hemoglobin, a protein that carries oxygen. One of the polypeptides in hemoglobin differs by one amino acid between normal and sickle-cell hemoglobin. A missense mutation such as this may sometimes cause the protein not to function, but often the effect is only to reduce the functional

efficiency of the protein. Individuals carrying missense mutations may survive, even though the affected protein is essential to life. Through evolution, some missense mutations even improve functional efficiency.

Nonsense mutations, another type of mutation in which bases are substituted, are more often disruptive than are missense mutations. In a nonsense mutation, the base substitution causes a chain terminator (stop) codon, such as UAG, to form in the mRNA product.

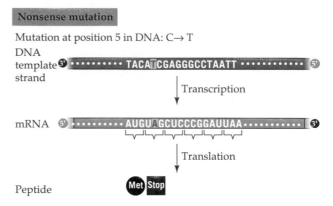

Nonsense mutation

Mutation at position 5 in DNA: C→ T

DNA template strand 3′ ···········TACA**T**CGAGGGCCTAATT··········· 5′

↓ Transcription

mRNA 5′ ···········AUGU**A**GCUCCCGGAUUAA··········· 3′

↓ Translation

Peptide Met Stop

Result: Only one amino acid translated; no protein made

The result is a shortened protein product, since translation does not proceed beyond the point where the mutation occurred.

Not all point mutations are base substitutions. Single base pairs may be inserted into or deleted from DNA. Such mutations are known as **frame-shift mutations** because they interfere with the decoding of the genetic message by throwing it out of register:

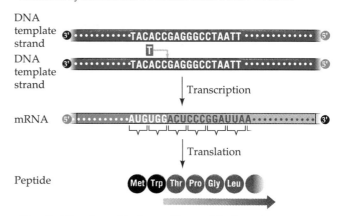

Frame-shift mutation

Mutation by insertion of T between bases 6 and 7 in DNA

DNA template strand 3′ ···········TACACCGAGGGCCTAATT··········· 5′
                                    T
DNA template strand 3′ ···········TACACCGAGGGCCTAATT··········· 5′

↓ Transcription

mRNA 5′ ···········AUGUGG**A**CUCCCGGAUUAA··········· 3′

↓ Translation

Peptide Met Trp Thr Pro Gly Leu

Result: All amino acids changed beyond the insertion

Think again of codons as three-letter words, each corresponding to a particular amino acid. Translation proceeds codon by codon; if a base is added to the mes-

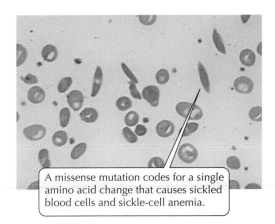

A missense mutation codes for a single amino acid change that causes sickled blood cells and sickle-cell anemia.

12.14 Sickled Blood Cells   This misshapen red blood cell is caused by a mutation that substitutes an incorrect amino acid in one of the two polypeptides of hemoglobin.

sage or subtracted from it, translation proceeds perfectly until it comes to the one-base insertion or deletion. From that point on, the three-letter words in the message are one letter out of register. In other words, such mutations shift the "reading frame" of the genetic message. Frame-shift mutations almost always lead to the production of completely nonfunctional proteins.

Chromosomal mutations are extensive changes

Genetic strands can break and rejoin, grossly disrupting the sequence of genetic information. There are four types of such chromosomal mutations: deletions, duplications, inversions, and translocations (Figure 12.15).

Deletions remove part of the genetic material (Figure 12.15*a*). Like frame-shift point mutations, they cause death unless they affect unnecessary genes or are masked by the presence, in the same cell, of normal copies of the deleted genes. It is easy to imagine one mechanism that could produce deletions: A DNA molecule might break at two points and the two end pieces might rejoin, leaving out the DNA between the breaks.

Another mechanism by which deletion mutations might arise would lead simultaneously to the production of a second kind of chromosomal mutation: a **duplication** (Figure 12.15*b*). Duplication would arise if homologous chromosomes broke at

12.15 Chromosomal Mutations   Chromosomes may break during replication, and parts of chromosomes may then rejoin incorrectly. Letters on the colored chromosomes distinguish segments and identify consequences of duplications, deletions, inversions, and reciprocal translocations.

different positions and then reconnected to the wrong partners. One of the two molecules produced by this mechanism would lack a segment of DNA (it would have a deletion), and the other would have two tandem copies (a duplication) of the information that was deleted from the first.

Breaking and rejoining can also lead to **inversion**—the removal of a segment of DNA and its reinsertion into the same location, but "flipped" end for end so that it runs in the opposite direction (Figure 12.15*c*). If an inversion includes part of a segment of DNA that codes for a protein, the resulting protein will be drastically altered and almost certainly nonfunctional.

The fourth type of chromosomal mutation, called **translocation**, results when a segment of DNA breaks, moves from a chromosome, and is inserted into a different chromosome. Translocations may be reciprocal, as in Figure 12.15*d*, or nonreciprocal, as the mutation involving duplication and deletion in Figure 12.15*b* illustrates. Translocations can make synapsis in meiosis difficult and thus sometimes lead to aneuploidy (the lack or excess of chromosomes; see Chapter 9).

Some chemicals induce mutations and cancer

Mutagens are agents that cause mutations. Among the chemical mutagens are *base analogs*—purines or pyrimidines that are not found in natural DNA but are enough like the natural bases that they can be incorporated into DNA. Base analogs are mutagenic presumably because they are more likely than natural DNA bases to mispair. For example, 5-bromouracil is very

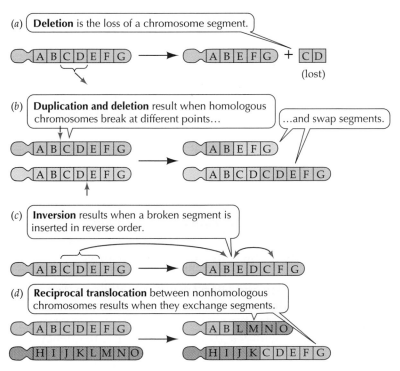

(*a*) **Deletion** is the loss of a chromosome segment.

(*b*) **Duplication and deletion** result when homologous chromosomes break at different points… …and swap segments.

(*c*) **Inversion** results when a broken segment is inserted in reverse order.

(*d*) **Reciprocal translocation** between nonhomologous chromosomes results when they exchange segments.

similar to thymine, and it is easily incorporated into DNA in place of thymine:

Thymine (Base)   5-Bromouracil (Base analog)

But 5-bromouracil is much more likely than thymine to engage in an abnormal pairing with guanine, and it therefore induces mutations of AT to GC and GC to AT. Thus 5-bromouracil is a potent chemical mutagen.

Cancerous cells have lost control of their cell cycle (see Chapter 9). Biomedical scientists have traced many kinds of cancers to mutations in particular genes, as we will see in Chapter 17. In addition, many substances present in polluted air, tobacco smoke, and foods we eat are known to be chemical **carcinogens**— cancer producers. Many known carcinogens either are mutagens or are converted to mutagens by enzymes in the endoplasmic reticulum.

The most widely used method to determine whether a substance is carcinogenic is called the **Ames test**, named for its inventor Bruce Ames and based on the idea that liver cells convert carcinogens to mutagens. To perform this test, we combine the suspected carcinogenic compound with a suspension of ground-up liver cells (the source of modifying enzymes) and mutant bacterial cells that cannot grow in the absence of, say, a particular amino acid. We then look for the appearance of bacteria that can grow in the absence of that amino acid (Figure 12.16). Such bacteria must result from a mutation that reverses the original mutation; thus their presence indicates that the compound added to the suspension is mutagenic. There is a correlation between activity in the Ames test and tumor-forming activity: If the Ames test indicates that a substance is mutagenic, that substance may also be carcinogenic. Thus we can test many compounds for carcinogenic activity without unnecessarily sacrificing the lives of large numbers of rodents or other test animals.

Mutations are the raw material of evolution

Without mutation, there would be no evolution. As we will see in Part Three, mutation does not drive evolution, but it provides the genetic diversity on which natural selection and other agents of evolution act.

All mutations are rare events, but mutation frequencies vary from organism to organism and from gene to gene within a given organism. The frequency of mutation is usually much lower than one mutation per 10^4 genes per DNA duplication, and sometimes the frequency is as low as one mutation per 10^9 genes per duplication. Most mutations are point mutations in which one nucleotide is substituted for another during the synthesis of a new DNA strand.

Most mutations harm the organism that carries them, and some are neutral (they have no effect on the organism's ability to survive or produce offspring). Once in a while, however, a mutation improves an organism's adaptation to its environment or becomes favorable when environmental variables change. Duplication mutations may be the source of "extra" genes. Most of the complex creatures living on Earth have more DNA and therefore more genes than the simpler creatures do. Humans, for example, have 1,000 times more genetic material than prokaryotes have.

How do new genes arise? If whole genes were sometimes duplicated by the mechanism described in the previous section, the bearer of the duplication

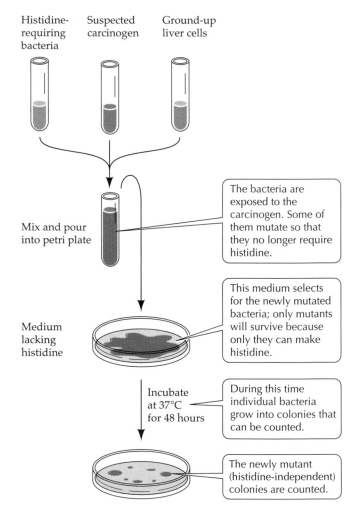

12.16 The Ames Test The Ames test to determine the ability of a suspected carcinogen to cause mutations works because liver cells convert carcinogens to mutagens.

would have a surplus of genetic information that might be turned to good use. Subsequent mutations in one of the two copies of the gene might not have an adverse effect on survival, because the other copy of the gene would continue to produce functional protein. The extra gene might mutate over and over again without ill effect because its function would be fulfilled by the original copy.

If the random accumulation of mutations in the extra gene led to the production of a useful protein (for example, an enzyme with an altered specificity for the substrates it binds, allowing it to catalyze different—but related—reactions), natural selection would tend to perpetuate the existence of this new gene. New copies of genes also arise through the activity of transposable elements, which are discussed in Chapters 13 and 14.

Summary of "From DNA to Protein: Genotype to Phenotype"

Genes and the Synthesis of Polypeptides

• Genes are made up of DNA and are expressed in the phenotype as polypeptides (proteins).

• Certain genetic diseases in humans were initially found to be caused by abnormal alleles expressed as defective enzymes. This discovery led to the one-gene, one-polypeptide hypothesis.

• Experiments with the bread mold *Neurospora* confirmed the one-gene, one-polypeptide hypothesis. Mutant strains of this haploid mold were found to be blocked, each at a specific enzymatic step, along a biochemical pathway. **Review Figure 12.1**

DNA, RNA, and the Flow of Information

• A gene is expressed in two steps: First, DNA is transcribed to RNA; then RNA is translated into protein. RNA differs from DNA in three ways: It is single-stranded, its sugar molecule is ribose rather than deoxyribose, and its fourth base is uracil rather than thymine.

• The central dogma of molecular biology is DNA → RNA → protein. In retroviruses, the rule for transcription is reversed: RNA → DNA. Other RNA viruses exclude DNA altogether, going directly from RNA to protein. **Review Figures 12.2, 12.3**

Transcription: DNA-Directed RNA Synthesis

• RNA is transcribed from a DNA template strand after the bases of DNA are exposed by unwinding of the double helix.

• In a given region of DNA, only one of the two strands (the template strand) can act as a template for transcription.

• RNA polymerase catalyzes transcription from the template strand of DNA.

• The initiation of transcription requires that RNA polymerase recognize and bind tightly to a special promoter sequence on DNA. RNA is synthesized (elongates) in a 5'-to-3' direction, reading in an antiparallel (3'-to-5') way from the template DNA. Special sequences and protein helpers terminate transcription. **Review Figure 12.4**

The Genetic Code

• The genetic code is present in the RNA transcripts (mRNA) that contain information for protein synthesis.

• The code is read sequentially, as nonoverlapping triplets of nucleotides (codons). Since there are four bases, there are 64 possible codons.

• One codon indicates the start of translation (protein synthesis) and codes for methionine. Three codons indicate the end of translation. The other 60 codons code only for particular amino acids.

• Since there are only 20 different amino acids, the genetic code is degenerate; that is, there is more than one codon for many amino acids. But the code is not ambiguous: A single codon does not specify more than one amino acid. **Review Figure 12.5**

The Key Players in Translation

• In prokaryotes, translation begins before the mRNA is completed. In eukaryotes, transcription occurs in the nucleus and translation occurs in the cytoplasm.

• Translation requires three special components to function properly: tRNAs, activating enzymes, and ribosomes.

• In translation, amino acids are linked in an order specified by the triplet codons in mRNA. This task is achieved by an adapter, transfer RNA (tRNA), which binds the correct amino acid and has an anticodon complementary to the mRNA codon. **Review Figure 12.6**

• The aminoacyl-tRNA synthetases, a family of activating enzymes, attach specific amino acids to their appropriate tRNAs. Each amino acid–identifying codon has a specific enzyme. **Review Figure 12.7**

• The mRNA with its DNA-directed codons meets the tRNAs with their amino acids at the ribosome, which contains most of the molecules that catalyze events in translation. The ribosome has two tRNA-binding sites: The P site binds tRNA containing the growing polypeptide; the A site binds the tRNA carrying the next amino acid. **Review Figure 12.8**

Translation: RNA-Directed Polypeptide Synthesis

• An initiation complex consisting of an amino acid–charged tRNA and a small ribosomal subunit bound to mRNA triggers the beginning of translation. **Review Figure 12.9**

• Polypeptides grow from the N terminus toward the C terminus. The ribosome moves along the mRNA one triplet codon at a time. **Review Figure 12.10**

• The presence of a stop codon in the A site of the ribosome causes translation to terminate. **Review Figure 12.11**

Regulation of Translation

• Some antibiotics work by blocking events in translation. **Review Table 12.2**

• In a polysome, more than one ribosome moves along the mRNA at one time. **Review Figure 12.12**

• As polypeptides are made, their amino acid sequences contain the information to fold them into three-dimensional shapes. Amino acid sequences also contain information on the cellular destination of proteins. All protein synthesis begins on free ribosomes in the cytoplasm. However, if translation produces a signal sequence of amino acids, the ribosomal complex binds to the outside of the ER. The signal sequence routes proteins into the lumen of the endoplasmic reticulum, which is the first destination of proteins headed to membrane-enclosed organelles. **Review Figure 12.13**

Mutations: Heritable Changes in Genes

• Mutations in DNA are often expressed as abnormal proteins. However, the result may not be easily observable phenotypic changes. Some mutations appear only under certain conditions, such as exposure to a certain environmental agent (such as a drug) or condition (such as temperature).
• Point mutations (silent, missense, nonsense, or frame-shift) result from alterations in single base pairs of DNA. **Review Page 276**
• Chromosomal mutations (deletions, duplications, inversions, or translocations) involve large regions of a chromosome. **Review Figure 12.15**
• Some chemicals induce mutations and may also cause cancer. These substances may chemically alter bases or substitute for them. The Ames test uses bacteria and liver enzymes to rapidly screen for mutagens (and hence carcinogens). **Review Figure 12.16**
• Mutation is essential to evolution: It is the source of the genetic variation on which natural selection and other evolutionary agents act.

Self-Quiz

1. Which of the following is *not* a difference between RNA and DNA?
   a. RNA has uracil and DNA has thymine.
   b. RNA has ribose and DNA has deoxyribose.
   c. RNA has 5 bases and DNA has 4.
   d. RNA is a single polynucleotide strand and DNA is a double strand.
   e. RNA is relatively smaller than human chromosomal DNA.

2. A *Neurospora* mutant deficient in the enzyme that catalyzes the conversion of citrulline to arginine (see the pathway of arginine synthesis in Figure 12.1) would grow on
   a. minimal medium.
   b. minimal medium + ornithine.
   c. minimal medium + citrulline.
   d. minimal medium + arginine.
   e. both *b* and *c*.

3. A region of DNA template strand has the sequence 3'-ATTCGC-5'. What is the sequence of RNA transcribed from this DNA?
   a. 3'-AUUCGC-5'
   b. 3'-TAAGCG-5'
   c. 5'-UAAGCG-3'
   d. 5'-AUUCGC-3'
   e. 5'-ATTCGC-3'

4. Which of the following is *not* a type of chromosome mutation?
   a. duplication
   b. inversion
   c. deletion
   d. frame-shift
   e. translocation

5. At a certain location in a gene, the template strand of DNA has the sequence GAA. A mutation alters the triplet to GAG. This type of mutation is called
   a. silent.
   b. missense.
   c. nonsense.
   d. frame-shift.
   e. translocation.

6. Transcription
   a. produces only mRNA.
   b. requires ribosomes.
   c. requires tRNAs.
   d. produces RNA growing from the 5' end to the 3' end.
   e. takes place only in eukaryotes.

7. Which statement about translation is *not* true?
   a. It is RNA-directed polypeptide synthesis.
   b. An mRNA molecule can be translated by only one ribosome at a time.
   c. The same genetic code operates in all organisms.
   d. Any ribosome can be used in the translation of any mRNA.
   e. There are both start and stop codons.

8. Which statement is *not* true?
   a. Transfer RNA functions in translation.
   b. Ribosomal RNA functions in translation.
   c. RNAs are produced in transcription.
   d. Messenger RNAs are produced on ribosomes.
   e. DNA codes for mRNA, tRNA, and rRNA.

9. The genetic code
   a. is different for prokaryotes and eukaryotes.
   b. has changed during the course of recent evolution.
   c. has 64 codons that code for amino acids.
   d. is degenerate.
   e. is ambiguous.

10. A mutation that results in the codon UAG where there had been UGG
    a. is a nonsense mutation.
    b. is a missense mutation.
    c. is a frame-shift mutation.
    d. is a large-scale mutation.
    e. is unlikely to have a significant effect.

Applying Concepts

1. The genetic code is described as degenerate. What does this mean? How is it possible that a point mutation, consisting of the replacement of a single nitrogenous base in DNA by a different base, might not result in an error in protein production?

2. Har Gobind Khorana, at the University of Wisconsin, synthesized artificial mRNAs such as poly CA (CACA-CA…) and poly CAA (CAACAACAA…). He found that poly CA codes for a polypeptide consisting of threonine (Thr) and histidine (His), in alternation (His–Thr–His–Thr…). There are two possible codons in poly CA, CAC and ACA. One of these must code for histidine and the other for threonine—but which is which? The answer comes from results with poly CAA, which produces three different polypeptides: poly Thr, poly Gln (glutamine), and poly Asn (asparagine). (An artificial messenger can be read, inefficiently, beginning at any point in the chain; there is no specific initiator region.) Thus poly CAA can be read as a polymer of CAA, of ACA, or of AAC. Compare the results of the poly CA and poly CAA experiments, and determine which codon codes for threonine and which for histidine.

3. Look back at Question 2. Using the genetic code (Figure 12.5) as a guide, deduce what results Khorana would have obtained had he used poly UG and poly UGG as artificial messengers. In fact, very few such artificial messengers would have given useful results. For an example of what could happen, consider poly CG and poly CGG. If poly C were the messenger, a mixed polypeptide of arginine and alanine (Arg–Ala–Ala–Arg . . .) would be obtained; poly CGG would give three polypeptides: poly Arg, poly Ala, and poly Gly (glycine). Can any codons be determined from only these data? Explain.

4. Errors in transcription occur about 100,000 times as often as do errors in DNA replication. Why can this high rate be tolerated in RNA synthesis but not in DNA synthesis?

Readings

Cooper, G. M. 1997. *The Cell: A Molecular Approach*. ASM Press, Washington, DC, and Sinauer Associates, Sunderland, MA. Chapter 7 is a concise, up-to-date treatment of protein synthesis.

Griffiths, A. J. F., J. H. Miller, D. T. Suzuki, R. C. Lewontin and W. M. Gelbart. 1996. *An Introduction to Genetic Analysis*, 6th Edition. W. H. Freeman, New York. Genetics texts, and this is one of the best, have descriptions of how the one-gene, one-polypeptide hypothesis was developed, as well as the use of genetics in deciphering the genetic code.

Hill, W. E., E. A. Dahlberg, R. Garrett, P. B. Moore, D. Schlesinger and J. R. Warner. 1990. *The Ribosome: Structure, Function and Evolution*. American Society for Microbiology, Washington, DC. A detailed summary of knowledge of this organelle and its role in protein synthesis.

Judson, H. F. 1996. *The Eighth Day of Creation: Makers of the Revolution in Biology*, Expanded Edition. CSHL Press, Plainview, NY. A sparkling history of molecular biology, including the discovery of messenger RNA.

Lodish, H., D. Baltimore, A. Berk, S. L. Zipursky, P. Matsudaira and J. Darnell. 1995. *Molecular Cell Biology*, 3rd Edition. Scientific American Books, New York. Superb, detailed chapters on protein synthesis.

Saks, M. E., J. R. Sampson and J. N. Abelson. 1995. "The Transfer RNA Identity Problem." *Science*, vol. 263, pages 191–197. A look at how the correct amino acid becomes attached to the correct tRNA.

Stent, G. S. and R. Calendar. 1978. *Molecular Genetics*, 2nd Edition. W. H. Freeman, New York. A brilliant technical and historical introduction to molecular genetics and the role of DNA.

Upton, A. C. 1982. "The Biological Effects of Low-Level Ionizing Radiation." *Scientific American*, February. A discussion of how radiation leads to mutations.

Watson, J. D., M. Gilman, J. Witkowski and M. Zoller. 1992. *Recombinant DNA*, 2nd Edition. W. H. Freeman, New York. A superbly written outline of molecular biology, including protein synthesis.

Chapter 13

The Genetics of Viruses and Prokaryotes

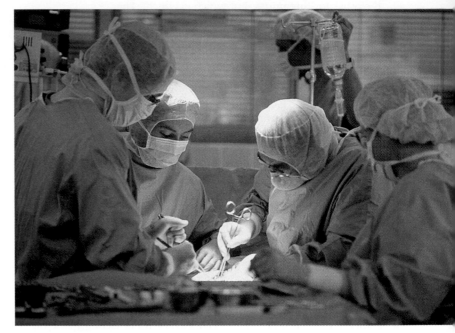

Are There Uninvited Guests Here?
A masked team performs surgery on a patient. But have harmful, drug-resistant bacteria invaded the surgical suite?

A member of her university's cross-country team, Janet entered the hospital just after final exams for some long-delayed surgery on a tendon in her knee. The tendon repair went well, but she left the hospital with something new: Bacteria called *Pseudomonas aeruginosa* had infected the surgical wounds. The antibiotics typically used to kill these bacteria did not work. She ended up back in the hospital two weeks later, where she received intensive antibiotic therapy and ultimately recovered.

Janet developed what is called a *nosocomial infection*—an infection acquired as a result of a hospital stay. Why would a hospital, which we think of as a place to get better, sometimes—in fact for about 10% of all patients—be a place where we get sick? Of course, the stresses of Janet's surgery could have reduced her immunity to the bacteria that are common everywhere in our environment; but, increasingly, the heavy use of antibiotics in hospitals makes them breeding grounds for bacteria that have genes for resistance to these antibiotics.

What's going on here? How have bacteria acquired antibiotic resistance so rapidly, and how do they pass that acquired resistance along to other bacteria? To answer these questions, we will discuss remarkable pieces of DNA known as R factors. But first we need to introduce the ways in which viruses and prokaryotes undergo genetic recombination and how their genetic information is organized.

Viruses are not cells but intracellular parasites that can reproduce only within living cells. As we saw in Chapter 11, the molecular biology of bacterial viruses has been important in determining the principles of molecular biology. Viruses have also been useful in genetic research. In this chapter we will examine the structures, classification, reproduction, and genetics of viruses.

Bacteria and other prokaryotes are a lot more complex than viruses, but still are relatively simple when compared to multicellular eukaryotes. Their genetics, too, has been extensively studied. The elegance of transcriptional controls in prokaryotes has provided valuable lessons for the more complicated situation in eukaryotes. In this chapter we will explore reproduction, mating, and genetic recombination in bacteria. We will see how bacteria acquire genes from other prokaryotes, from viruses, and from naked nucleic acids.

Using Prokaryotes and Viruses for Genetic Experiments

Prokaryotes such as *Escherichia coli* and viruses such as bacteriophages have often been easier to study than eukaryotes. What are the advantages of working with prokaryotes and viruses?

First, it is easier to work with small amounts of DNA than with large amounts. A typical bacterium contains about 1/1,000 as much DNA as a single human cell, and a typical bacterial virus contains about 1/100 as much DNA as a bacterium. Second, data on large numbers of organisms can be obtained easily from prokaryotes, but not from most eukaryotes. A single milliliter of medium can contain more than 10^9 *E. coli* cells or 10^{11} bacteriophages. In addition, most prokaryotes grow rapidly. A culture of *E. coli* can be grown under conditions that allow cells to double every 20 minutes. By contrast, 10^9 mice would cost more than 10^9 dollars and would require a cage that would cover about 3 square miles, and growth of a generation of mice takes about 3 months instead of 20 minutes. Third, prokaryotes and viruses are usually haploid, facilitating genetic analyses.

The ease of growing and handling bacteria and their viruses permitted the explosion of genetics and molecular biology that began shortly after the mid-twentieth century (you read about some of these discoveries in Chapters 11 and 12). The relative biological simplicity of bacteria and bacteriophages contributed immeasurably to the discoveries about the genetic material, the replication of DNA, and the mechanisms of gene expression. Later these bacteria and bacteriophages were the first subjects of recombinant-DNA technology (see Chapter 16).

Questions of interest to all biologists continue to be studied in prokaryotes, and prokaryotes continue to be important tools for biotechnology and for research on eukaryotes.

Viruses: Reproduction and Recombination

When Oswald Avery and his colleagues showed that DNA is the hereditary material of bacteria, the finding did not create a sensation. The results of the Hershey–Chase experiment, on the other hand, which supported the hypothesis that DNA is the hereditary material of certain viruses, captured the imagination of scientists, including James Watson and Francis Crick. What made the difference? In the decade between those two discoveries, scientists had shown that bacteria and viruses have genes, and that they undergo genetic recombination. These discoveries all increased the value of using bacteria and viruses for genetic study.

Although there are many kinds of viruses, they are usually composed of just nucleic acid and a few proteins, and they have relatively simple means of infecting their targeted host cells. Some viruses—the best studied are certain bacteriophages—can infect a cell but postpone reproduction until conditions are favorable. In these cases, we will discover how viral genetic information is inserted into the host chromosome. Finally, we will describe the simplest infective agents: *viroids*, which are made up only of genetic material, and *prions*, which apparently have no nucleic acid but are made entirely of protein.

Scientists studied viruses before they could see them

Most viruses are much smaller than most bacteria (Table 13.1). Viruses have become well understood only within the last half century, but the first step on this path of discovery was taken by the Russian botanist Dmitri Ivanovsky in 1892. He was trying to find the cause of *tobacco mosaic disease*, which results in the destruction of photosynthetic tissues and can devastate a tobacco crop. Ivanovsky passed an infectious extract of diseased tobacco leaves through a fine porcelain filter, a technique that had been used previously by physicians and veterinarians to isolate disease-causing bacteria, which would stay on the filter.

To Ivanovsky's surprise, the disease agent in this case did not stick to the filter: It passed through, and the liquid filtrate still caused tobacco mosaic disease. Clearly, the agent was smaller than a bacterium. Other

TABLE 13.1 Common Sizes of Microorganisms

MICROORGANISM	TYPE	TYPICAL SIZE RANGE (μm^3)
Protists	Eukaryote	5,000–50,000
Photosynthetic bacteria	Prokaryote	5–50
Spirochetes	Prokaryote	0.1–2
Mycoplasmas	Prokaryote	0.01–0.1
Poxviruses	Virus	0.01
Influenza virus	Virus	0.0005
Poliovirus	Virus	0.00001

workers soon showed that similar filterable agents, so tiny that they cannot be seen under the light microscope, cause several plant and animal diseases. And alcohol, which kills bacteria, does not destroy the ability of these tiny agents to cause disease.

In 1935, Wendell Stanley, working at what is now Rockefeller University, was the first to succeed in crystallizing viruses. The crystalline viral preparation became infectious again when it was dissolved. It was soon shown that crystallized viral preparations consist of protein and nucleic acid. Finally, direct observation of viruses with electron microscopes showed clearly how much they differ from bacteria and other organisms.

Viruses reproduce only with the help of living cells

Unlike the organisms that make up the six taxonomic kingdoms of the living world, viruses are **acellular**; that is, they are not cells and do not consist of cells. Unlike cellular creatures, viruses do not metabolize energy—they neither produce ATP nor conduct fermentation, cellular respiration, or photosynthesis.

Whole viruses never arise directly from preexisting viruses. They are *obligate intracellular parasites*; that is, they develop and reproduce only within the cells of specific hosts. The cells of animals, plants, fungi, protists, and prokaryotes (both bacteria and archaea) serve as hosts to viruses. When they reproduce, viruses usually destroy the host cells, releasing progeny viruses that then seek new hosts. Many diseases of humans, animals, and plants are caused by viruses.

Because they lack the distinctive cell wall and ribosomal biochemistry of bacteria, viruses are not affected by antibiotics. Few specific drugs affect viruses.

The best way a person can stem a viral infection is to rely on the immune system (we will describe how this works in Chapter 18). The best way to prevent the spread of a viral infection in a population is to contain it. If there are no nearby susceptible hosts, the viral infection will run its course and the particles will die because they have no place to reproduce. This was the strategy used to eliminate smallpox, which used to be a worldwide scourge and is now extinct. Currently an effort is under way to do the same for polio.

Viruses outside of host cells exist as individual particles called **virions** (Figure 13.1). The virion, the basic unit of a virus, consists of a central core of either DNA or RNA (but not both) surrounded by a **capsid**, or coat, which is composed of one or more proteins. The way in which these proteins are assembled gives the virion a characteristic shape. In addition, many animal viruses have a lipid and protein membrane acquired from host cell membranes in the course of viral reproduction and release. Many bacterial viruses (bacteriophages) have specialized "tails" made of protein. The complex architecture of HIV, the AIDS virus, is shown in Figure 13.2.

There are many kinds of viruses

A common way to classify viruses separates them by whether they have DNA or RNA and then by whether their nucleic acid is single- or double-stranded. Some of the RNA viruses have more than one molecule of RNA, and the DNA of one virus family is circular. Further levels of classification depend on factors such as the overall shape of the virus and the symmetry of the capsid.

Most capsids may be categorized as **helical** (coiled like a spring; Figure 13.1*a*), **icosahedral** (a regular solid

(*a*)

(*b*)

(*c*)

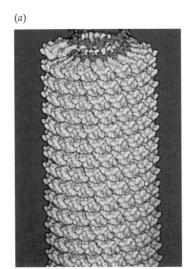
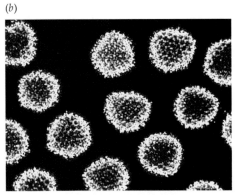
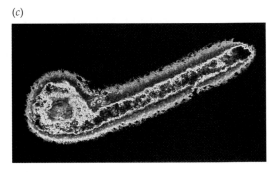

13.1 Virions Come in Various Shapes (*a*) A computer model of the tobacco mosaic virus, a plant virus. The model shows only about one-seventh of this long virus, which consists of an inner helix of RNA covered with a helical array of protein molecules. (*b*) Many animal viruses, such as the adenovirus modeled here, have an icosahedral (20-sided) capsid as an outer shell. Inside the shell is a spherical mass of proteins and DNA. (*c*) Not all virions are regularly shaped. Wormlike virions of the Ebola virus infect humans, causing internal hemorrhaging that is usually fatal.

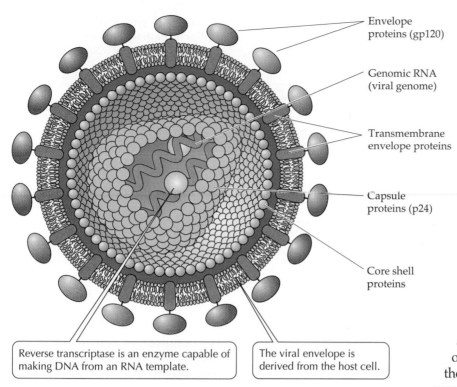

Envelope proteins (gp120)

Genomic RNA (viral genome)

Transmembrane envelope proteins

Capsule proteins (p24)

Core shell proteins

Reverse transcriptase is an enzyme capable of making DNA from an RNA template.

The viral envelope is derived from the host cell.

13.2 The Structure of HIV, the AIDS Virus HIV has a core containing two molecules of genomic RNA (it is diploid) and various proteins, including the enzyme reverse transcriptase. Another protein (shown in blue) surrounds the core. The surface of the virus is complex.

with 20 faces; Figure 13.1*b*), or **binal** (having a polyhedral, or many-faced, head and a helical tail; see the structure of the T2 bacteriophage in Figure 11.2). Another level of classification is based on the presence or absence of a membranous envelope around the virion; still further subdivision is based on capsid size.

The first level of virus classification is based on the type of host. Let's see how reproductive cycles and other properties vary among the major groups of viruses: bacterial, animal, and plant viruses.

Bacteriophages reproduce by either a lytic cycle or a lysogenic cycle

Viruses that parasitize bacteria are known as **bacteriophages**. Bacteriophages recognize their hosts by means of a specific interaction between the proteins of the bacteriophage and the molecules of the host bacteria's cell wall. The virions, which also must penetrate cell walls, are often equipped with tail assemblies that inject their nucleic acid through the cell wall into the host bacterium or archaeon. After the phage has injected its nucleic acid into the host, one of two things happens, depending on the kind of phage.

We saw one type of viral reproductive cycle when we studied the Hershey–Chase experiment (see Figure 11.3). That was the **lytic cycle**, in which the infected bacterium lyses (bursts), releasing progeny phage. In the **lysogenic cycle** the infected bacterium does not lyse, but instead harbors the viral nucleic acid for many rounds of bacterial cell division before one or more of the progeny lyse.

THE LYTIC CYCLE. Phages that reproduce only by the lytic cycle are called **virulent viruses**. The phage attacks a prokaryotic cell by specific combination of part of the capsid with a receptor protein on the cell surface (Figure 13.3). The phage injects its nucleic acid into the host cell, where the phage nucleic acid takes over the host's synthetic machinery, making it copy the phage nucleic acid and produce the phage proteins. In one clever mechanism for this process, the phage genetic material has a promoter sequence that attracts the host's RNA polymerase. One of the proteins made by the translation of phage mRNAs destroys the host cell's own DNA, thus removing a potential competitor for the viral nucleic acid.

As synthesis proceeds, viral parts accumulate within the infected cell. The new nucleic acid molecules are loaded into the "head" proteins, and the other proteins (tails, tail fibers) join the complex. At this point the host cell is teeming with new phages. One of the phage genes encodes an enzyme that attacks the host cell wall, causing lysis and the release of many new infectious phages. The whole process—from attachment and infection to lysis of the host—takes about half an hour.

THE LYSOGENIC CYCLE. Phage infection does not always result in lysis of the bacteria. Some phages seem to disappear from the culture, leaving the bacteria immune to further attack by the same strain of phage. In such cultures, however, a few free phages are always present. Bacteria harboring phages that are not lytic are called **lysogenic**, and the phages are called **temperate viruses**. When lysogenic bacteria are combined with other bacteria that are sensitive to the phages, they cause the sensitive cells to lyse.

Experiments revealed that the lysogenic bacteria contain a noninfective entity called a prophage. A **prophage** is a molecule of phage DNA that has been integrated into the bacterial chromosome (Figure 13.4). As part of the bacterial genome, the prophage can remain quiet within bacteria through many cell divisions. However, an occasional lysogenic cell can be in-

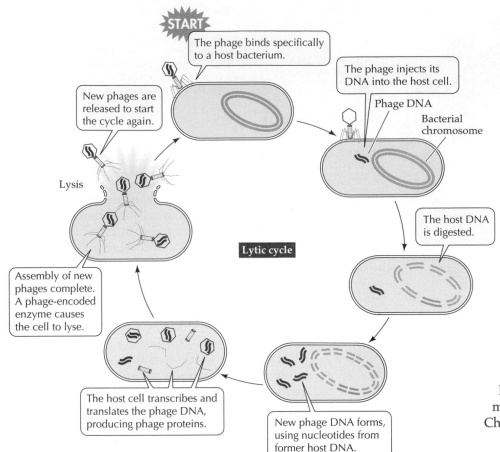

START

The phage binds specifically to a host bacterium.

The phage injects its DNA into the host cell.

Phage DNA

Bacterial chromosome

New phages are released to start the cycle again.

Lysis

The host DNA is digested.

Lytic cycle

Assembly of new phages complete. A phage-encoded enzyme causes the cell to lyse.

The host cell transcribes and translates the phage DNA, producing phage proteins.

New phage DNA forms, using nucleotides from former host DNA.

13.3 The Lytic Cycle of a Virulent Bacteriophage In the lytic cycle, infection by viral DNA leads directly to the multiplication of the virus and lysis of the host bacterial cell.

phages to a culture of *E. coli* to produce substantial numbers of phages of both parental types.

However, when *E. coli* were infected with both phage strains, not only the parental types appeared, but also many phages of genotypes h^+r^+ and hr—that is, recombinant phages (Figure 13.5). Thus, the *map distance* between the two genetic markers could be calculated in a way very similar to the method of mapping in eukaryotes (see Chapter 10):

$$\text{Map distance} = \frac{\begin{array}{c}\text{number of}\\\text{recombinant phages}\end{array}}{\begin{array}{c}\text{total number}\\\text{of phages}\end{array}}$$

$$= \frac{(h^+r^+)+(hr)}{(h^+r^+)+(hr)+(h^+r)+(hr^+)}$$

As more genes in such phage crosses were studied, a map began to take form. In due course, it was established that the phage has a single chromosome.

duced to lyse, releasing a large number of free phages, which can then infect other uninfected bacteria and renew the reproductive cycle.

This *"genetic switch"* from the lysogenic to the lytic cycle (and vice versa) is very useful to the phage, whose purpose is to reproduce as many offspring as possible. When the host cell of an infecting phage is growing slowly and is low on energy, the phage becomes lysogenic. Then, when the host's health is restored to a level that allows maximal phage reproduction, the prophage is released from its dormant state and the lytic cycle proceeds. In the laboratory, ultraviolet radiation is a potent inducer of the switch to the lytic cycle.

Bacteriophage genes can recombine

When investigators simultaneously infected *E. coli* with *two* different mutant strains of the bacteriophage T2, they observed genetic recombination. In one such experiment, one of the phage strains was genotypically h^+r and the other was hr^+. (We need not worry here about what the phenotypes were; just note that h^+ and h are alleles at one locus and r^+ and r are alleles at another locus.) We would expect the addition of these

Animal viruses have diverse reproductive cycles

Among the more common viral diseases of humans are the common cold and influenza. Almost all vertebrates are susceptible to viral infections, but among invertebrates, such infections are common only in arthropods (a group that includes insects, crustaceans, and some other animals). A group of viruses called **arboviruses** (short for "*arthropod-borne viruses*") causes serious diseases, such as encephalitis, in humans and other mammals. Arboviruses are transmitted to the mammalian host through a bite (certain arboviruses are carried by mosquitoes, for example). Although carried within the arthropod host's cells, arboviruses apparently do not affect that host severely; they affect only the bitten and infected mammal.

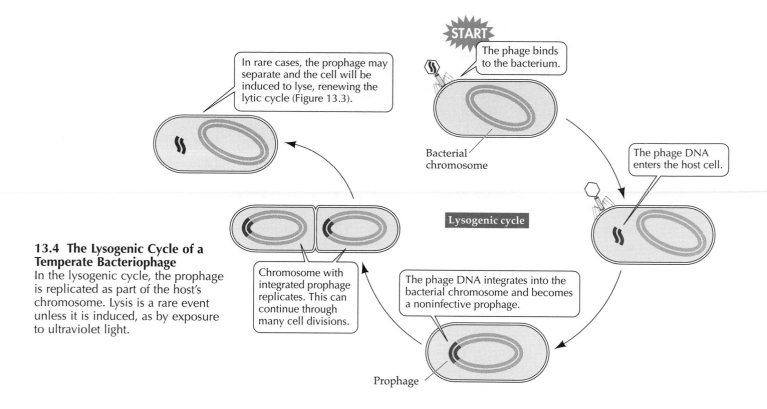

13.4 The Lysogenic Cycle of a Temperate Bacteriophage
In the lysogenic cycle, the prophage is replicated as part of the host's chromosome. Lysis is a rare event unless it is induced, as by exposure to ultraviolet light.

START

The phage binds to the bacterium.

Bacterial chromosome

The phage DNA enters the host cell.

In rare cases, the prophage may separate and the cell will be induced to lyse, renewing the lytic cycle (Figure 13.3).

Lysogenic cycle

Chromosome with integrated prophage replicates. This can continue through many cell divisions.

The phage DNA integrates into the bacterial chromosome and becomes a noninfective prophage.

Prophage

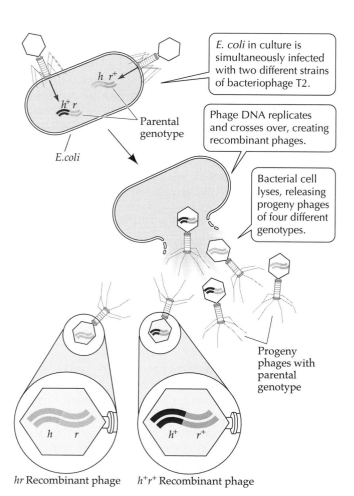

E. coli in culture is simultaneously infected with two different strains of bacteriophage T2.

Parental genotype

E.coli

Phage DNA replicates and crosses over, creating recombinant phages.

Bacterial cell lyses, releasing progeny phages of four different genotypes.

Progeny phages with parental genotype

hr Recombinant phage

h⁺r⁺ Recombinant phage

Animal viruses are very diverse. Table 13.2 shows only the major levels of animal virus classification and some examples.

Animal viruses begin the process of infection by attaching to the plasma membrane of the host cell. The viruses themselves may have membranes, which derive from the host cell that produced them. Some, such as herpesviruses, obtain their membranes from the host's nuclear envelope. Others, such as HIV (the AIDS virus), obtain their membranes from the host plasma membrane.

Animal viruses without membranes are taken up by endocytosis, which traps them within a membranous vesicle inside the host cell. Then the membrane of the vesicle breaks down, releasing the virion, and the host cell digests the protein capsid, liberating the viral nucleic acid, which takes charge. This general pattern of events is common, but the details vary sharply among the different types of animal viruses.

In the case of DNA viruses with membranes, viral membranes are studded with glycoproteins that bind to receptors on the host's plasma membrane. The host

13.5 Genetic Recombination in Bacteriophages If a bacterial culture is exposed simultaneously to a sufficient concentration of bacteriophages of two different strains, viruses of both strains may infect the same bacterium. This—a phage cross—results in recombinant progeny as well as parental types.

TABLE 13.2 A Classification Scheme for Some Animal Viruses

	NUCLEIC ACID			VIRION		
VIRUS GROUP	MOL. WT. (millions)	TYPE	NUMBER OF STRANDS	SHAPE	SIZE (nm)	NOTES
Families of viruses that affect both vertebrates and other hosts						
Poxviridae	160–200	DNA	2	Brick-shaped	$300 \times 240 \times 100$	Poxviruses
Parvoviridae	1.2–1.8	DNA	1	Icosahedral	20	Hosts include rats and insects
Reoviridae	15	RNA	2	Icosahedral	50–80	Vertebrate, insect, and plant hosts
Rhabdoviridae	4	RNA	1	Bullet-shaped	175×70	Rabies, vesicular stomatitis
Families of viruses that affect only vertebrates						
Herpetoviridae	100–200	DNA	2	Icosahedral	150	Herpes
Adenoviridae	20–29	DNA	2	Icosahedral	70–80	Adenovirus
Papovaviridae	3–5	DNA	2	Icosahedral	45–55	Papillomas
Retroviridae	10–12	RNA	1	Spherical	100–200	Tumor viruses
Paramyxoviridae	7	RNA	1	Spherical	100–300	Measles, Newcastle disease
Orthomyxoviridae	5	RNA	1	Spherical	80–120	Influenza
Togaviridae	4	RNA	1	Spherical	40–60	Rubella, hog cholera, arboviruses
Coronaviridae	?	RNA	1	Spherical	80–120	Common cold
Arenaviridae	3.5	RNA	1	Spherical	85–120	Lassa fever
Picornaviridae	2.6–2.8	RNA	1	Icosahedral	20–30	Digestive and respiratory diseases
Bunyaviridae	6	RNA	1	Spherical	90–100	Encephalitis

and viral membranes fuse, releasing the rest of the virion into the cell. The host cell then replicates the viral nucleic acid and synthesizes new capsid protein as directed by the viral nucleic acid. New capsids and new viral nucleic acid combine spontaneously, and in due course, the host cell releases the new virions (Figure 13.6). Such viruses usually escape from the host cell by budding through virus-modified areas of the plasma membrane. During this process the completed virions acquire a membrane similar to that of the host cell.

Retroviruses such as HIV have a much more complex reproductive cycle (Figure 13.7). A major feature of that cycle, the reverse transcription of retroviral RNA to produce a DNA provirus, was the notable exception to Francis Crick's "central dogma" of molecular biology (see Figure 12.2). The resulting DNA transcript is integrated into a host chromosome, where it may reside permanently, occasionally being expressed to produce new virions. We discuss other aspects of HIV and AIDS in Chapter 18.

As noted in Chapter 12, there are RNA viruses other than the retroviruses. These other RNA viruses have a simpler reproductive cycle, lacking the reverse transcription step (step 3 in Figure 13.7). In cases such as poliovirus, the viral RNA serves as the mRNA for virally coded proteins, which the host cell synthesizes following infection. Some of these proteins are the coat and packaging proteins for the virus; others are involved in viral RNA replication. In other RNA viruses, such as the ones that cause influenza, the replication enzymes are part of the virus particle.

Many plant viruses spread with the help of vectors

Viral diseases of flowering plants are very common. After reproduction, plant viruses must pass through a cell wall and through the host plasma membrane. Most plant viruses accomplish this penetration through their association with **vectors**—intermediate carriers of disease from one organism to another. Infection of a plant usually results from attack by a virion-laden insect vector.

The insect vector uses its proboscis (snout) to penetrate the cell wall, allowing the virions to move via the proboscis from the insect into the plant. Plant viruses, such as tobacco mosaic virus, can be introduced artificially without insect vectors: First a leaf or other plant part is bruised; then it is exposed to a suspension of virions. Viral infections may also be inherited vegetatively or sexually, in which cases a vector is not needed.

13.6 The Reproductive Cycle of an Animal Virus with a Membrane The infected host cell replicates the viral genome and synthesizes capsid proteins, which combine spontaneously. The new viral membrane is aquired as the virus leaves the cell.

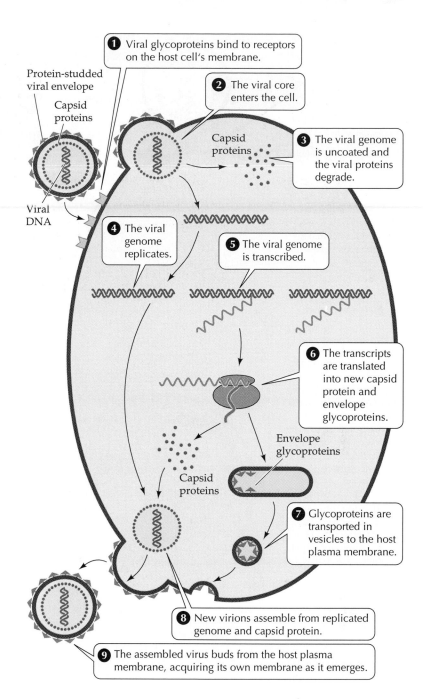

Protein-studded viral envelope
Capsid proteins
Viral DNA

1 Viral glycoproteins bind to receptors on the host cell's membrane.

2 The viral core enters the cell.

Capsid proteins

3 The viral genome is uncoated and the viral proteins degrade.

4 The viral genome replicates.

5 The viral genome is transcribed.

6 The transcripts are translated into new capsid protein and envelope glycoproteins.

Envelope glycoproteins

Capsid proteins

7 Glycoproteins are transported in vesicles to the host plasma membrane.

8 New virions assemble from replicated genome and capsid protein.

9 The assembled virus buds from the host plasma membrane, acquiring its own membrane as it emerges.

Once inside a plant cell, the virus reproduces and now must spread to other cells in the plant. Within an organ such as a leaf, the virus spreads through the plasmodesmata, the cytoplasmic connections between cells (see Chapter 4). However, some modification is needed. Because the viruses are too large to go through these channels, special proteins bind to them and help change their shape to squeeze the viruses through the pores.

Viroids are infectious agents consisting entirely of RNA

Pure viral nucleic acids can produce viral infections under laboratory conditions, although inefficiently. Might there be infectious agents in nature that consist of nucleic acid without a protein capsid? In 1971, Theodore Diener of the U.S. Department of Agriculture reported the isolation of agents of this type, called viroids. **Viroids** are circular, single-stranded RNA molecules consisting of a few hundred nucleotides. They are one-thousandth the size of the smallest viruses. These RNAs are most abundant in the nuclei of infected cells.

Viroids have been found only in plants, where they produce a variety of diseases. Two mechanisms are known by which viroids may be transmitted from plant to plant. Like tobacco mosaic virus, viroids can be transmitted *horizontally*, from one plant to another, if a bruised infected plant contacts an injured uninfected one. They can also be transmitted *vertically*, from parent to offspring. When a pollen grain or an ovule produced by an infected plant contains viroids, these infect the daughter plant produced by fertilization.

There is no evidence that viroids are translated to synthesize proteins, and it is not known how they cause disease. Viroids are replicated by the enzymes of their plant hosts. Similarities in base sequences between viroids and certain nontranslated sequences (introns) of plant genes suggest that viroids evolved from introns. This conclusion is supported by the fact that viroids, although made of RNA, are catalytically active in the way that some introns are (see Chapter 14).

Prions may be infectious proteins

If an RNA molecule by itself can be infectious, what about a protein? A class of protein fibrils, called **prions**, appear to cause certain degenerative diseases of mammalian central nervous systems. These fibrils consist entirely of protein, with no nucleic acid component. Prions appear to cause scrapie, a disease of sheep and goats. Prions have also been identified in connection with "mad cow disease," which is thought to have been spread to humans when people ate contaminated beef from cows that had eaten the remains of infected sheep. Prions are involved in other, similar

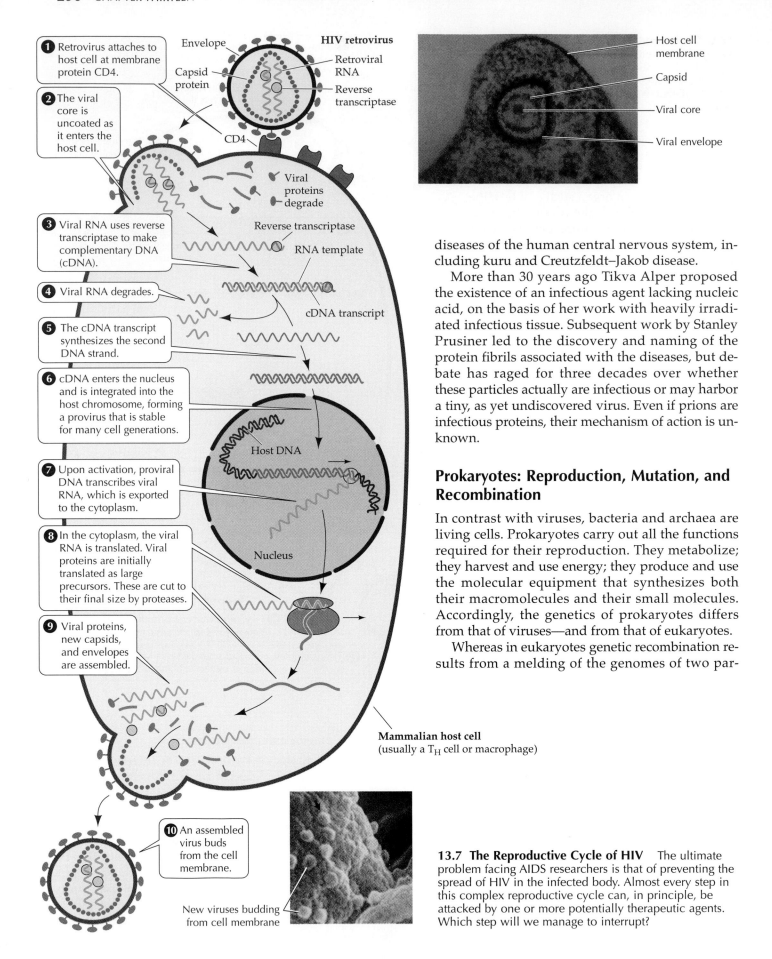

1 Retrovirus attaches to host cell at membrane protein CD4.

2 The viral core is uncoated as it enters the host cell.

3 Viral RNA uses reverse transcriptase to make complementary DNA (cDNA).

4 Viral RNA degrades.

5 The cDNA transcript synthesizes the second DNA strand.

6 cDNA enters the nucleus and is integrated into the host chromosome, forming a provirus that is stable for many cell generations.

7 Upon activation, proviral DNA transcribes viral RNA, which is exported to the cytoplasm.

8 In the cytoplasm, the viral RNA is translated. Viral proteins are initially translated as large precursors. These are cut to their final size by proteases.

9 Viral proteins, new capsids, and envelopes are assembled.

10 An assembled virus buds from the cell membrane.

Envelope

HIV retrovirus

Capsid protein

Retroviral RNA

Reverse transcriptase

CD4

Viral proteins degrade

Reverse transcriptase

RNA template

cDNA transcript

Host DNA

Nucleus

Mammalian host cell (usually a T_H cell or macrophage)

New viruses budding from cell membrane

Host cell membrane

Capsid

Viral core

Viral envelope

diseases of the human central nervous system, including kuru and Creutzfeldt–Jakob disease.

More than 30 years ago Tikva Alper proposed the existence of an infectious agent lacking nucleic acid, on the basis of her work with heavily irradiated infectious tissue. Subsequent work by Stanley Prusiner led to the discovery and naming of the protein fibrils associated with the diseases, but debate has raged for three decades over whether these particles actually are infectious or may harbor a tiny, as yet undiscovered virus. Even if prions are infectious proteins, their mechanism of action is unknown.

Prokaryotes: Reproduction, Mutation, and Recombination

In contrast with viruses, bacteria and archaea are living cells. Prokaryotes carry out all the functions required for their reproduction. They metabolize; they harvest and use energy; they produce and use the molecular equipment that synthesizes both their macromolecules and their small molecules. Accordingly, the genetics of prokaryotes differs from that of viruses—and from that of eukaryotes.

Whereas in eukaryotes genetic recombination results from a melding of the genomes of two par-

13.7 The Reproductive Cycle of HIV   The ultimate problem facing AIDS researchers is that of preventing the spread of HIV in the infected body. Almost every step in this complex reproductive cycle can, in principle, be attacked by one or more potentially therapeutic agents. Which step will we manage to interrupt?

ents, in prokaryotes it results from the interaction of the genome of a parent cell with a much smaller sample of genes from another cell. There are three ways in which the smaller sample is introduced into the parent cell: conjugation, transformation, and transduction.

In *conjugation*, or "bacterial sex," two bacteria come close to each other and one produces a hollow tube through which some DNA can pass between the cells. In *transformation*, which was described in Chapter 11, a fragment of DNA from one cell is taken up from the environment by another cell. In *transduction*, a bacteriophage actually carries bacterial DNA from one bacterial cell to another. Small extra pieces of DNA, called *plasmids*, are present in some bacteria. These plasmids are involved in recombination, as well as in the antibiotic resistance we described at the beginning of this chapter. We will discuss each mode of genetic recombination in more detail in this section.

The reproduction of prokaryotes gives rise to clones

All prokaryotes reproduce by the division of single cells into two identical offspring (see Figure 9.3). A single cell gives rise to a **clone**—a population of genetically identical individuals. As long as conditions remain favorable, a population of *E. coli*, for example, can double every 20 minutes.

Pure cultures of *E. coli* can be grown on the surface of a solid medium that contain a sugar, minerals, a nitrogen source such as ammonium chloride (NH_4Cl), and a solidifying agent such as agar (Figure 13.8). If the number of cells spread on the surface is small, each cell will give rise to a small, rapidly growing colony. If an extremely large number of cells is poured onto the solid medium, their growth will produce one continuous colony—a bacterial "lawn." Bacteria can also be grown on a liquid nutrient medium. We'll see examples of all these techniques in this chapter.

Prokaryotic genes mutate

We can do an experiment to demonstrate that prokaryotic mutants arise. First we mix a large sample of the bacterium *E. coli* with a suspension of a suitable bacteriophage and pour the mixture over solid growth medium in a wide, shallow dish called a petri plate. Wherever the virus finds a bacterial cell, it attaches to it, infects it, and eventually causes it to burst, killing the bacterium and releasing many new viruses. These viruses, in turn, attack neighboring cells. Soon circular clearings, called **plaques**, begin to appear in the lawn wherever the viruses have killed bacteria (Figure 13.9). The plaques, which are caused by the virus-induced bursting, or lysis, of the bacteria, grow larger and larger.

However, if you scan several such petri plates, here and there you will find a bacterial colony growing in the midst of a plaque, in spite of the surrounding hordes of viruses. Each of these colonies has arisen from a mutant bacterium that is resistant to the virus. We know that resistance is a heritable trait because the mutant bacteria give rise to colonies of cells that are similarly resistant to the virus.

The preceding description of the growth of bacteria and bacteriophages in the laboratory and the consequence of a certain mutation illustrates evolution on a small scale. In nature, *E. coli* constantly mutate to a virus-resistant form at a low rate. The mutants normally do not take over the entire population but exist in low frequency as members of the bacterial population. However, when the environment favors one genotype in a population over others, the proportions of the different genotypes in the population change. In the case of *E. coli*, the virus kills the wild type but not

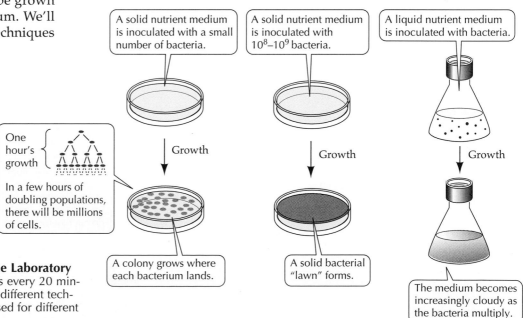

A solid nutrient medium is inoculated with a small number of bacteria.

A solid nutrient medium is inoculated with 10^8–10^9 bacteria.

A liquid nutrient medium is inoculated with bacteria.

One hour's growth

In a few hours of doubling populations, there will be millions of cells.

Growth

Growth

Growth

A colony grows where each bacterium lands.

A solid bacterial "lawn" forms.

The medium becomes increasingly cloudy as the bacteria multiply.

13.8 Growing Bacteria in the Laboratory
A population of *E. coli* doubles every 20 minutes in laboratory culture. The different techniques of culture shown are used for different applications.

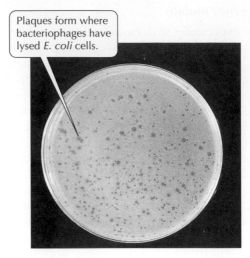

Plaques form where bacteriophages have lysed *E. coli* cells.

13.9 Bacteriophages Clear Bacterial Lawns The dark circles are clear plaques in an opaque lawn of *E. coli*.

13.10 New Prototrophic Colonies Appear After growing together, a mixture of complementary auxotrophic strains contains a few cells that can give rise to new prototrophic colonies. This experiment proved that genetic recombination takes place in prokaryotes.

the resistant strain, so the resistant strain soon predominates.

Some bacteria conjugate, recombining their genes

The existence and heritability of mutations in bacteria attracted the attention of geneticists to these microbes. But if there were no form of exchange of genetic information between individuals, bacteria would not be useful for genetic analysis. Luckily, in 1946 Joshua Lederberg and Edward Tatum demonstrated that such exchanges do occur; in *E. coli*, however, genetic recombination is a rare event.

Lederberg and Tatum used two nutrient-requiring, or *auxotrophic*, strains of *E. coli* as parents. Like the *Neurospora* in Figure 12.1, these strains will not grow on normal media, but require supplementation with a substance they cannot synthesize because of an enzyme defect. *E. coli* strain 1 requires the amino acid methionine and the vitamin biotin for growth, and its genotype is symbolized as *met⁻bio⁻*. Strain 2 requires neither of these substances but could not grow without the amino acids threonine and leucine. Considering all four factors, we note that strain 1 is *met⁻bio⁻thr⁺leu⁺* and strain 2 is *met⁺bio⁺thr⁻leu⁻*.

These two mutant strains were mixed and cultured together for several hours on a medium supplemented with methionine, biotin, threonine, and leucine, so that both strains could grow. The bacteria were then removed from the medium by centrifugation, washed,

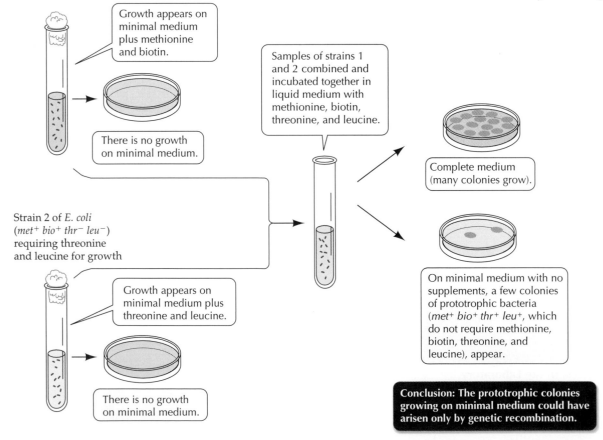

Strain 1 of *E. coli* (*met⁻ bio⁻ thr⁺ leu⁺*) requires methionine and biotin for growth

Growth appears on minimal medium plus methionine and biotin.

There is no growth on minimal medium.

Strain 2 of *E. coli* (*met⁺ bio⁺ thr⁻ leu⁻*) requiring threonine and leucine for growth

Growth appears on minimal medium plus threonine and leucine.

There is no growth on minimal medium.

Samples of strains 1 and 2 combined and incubated together in liquid medium with methionine, biotin, threonine, and leucine.

Complete medium (many colonies grow).

On minimal medium with no supplements, a few colonies of prototrophic bacteria (*met⁺ bio⁺ thr⁺ leu⁺*, which do not require methionine, biotin, threonine, and leucine), appear.

Conclusion: The prototrophic colonies growing on minimal medium could have arisen only by genetic recombination.

and transferred to minimal medium, which lacked all four supplements. Neither parent strain could grow on this medium, because of their nutritional requirements. However, a few colonies *did* appear on the plates. Because they grew in the minimal medium, these colonies must have consisted of bacteria that were *met⁺bio⁺thr⁺leu⁺*; that is, they were prototrophic (Figure 13.10). These colonies appeared at a rate of approximately 1 for every 10 million cells put on the plates (1 / 10⁷).

How might these prototrophic colonies have arisen? Lederberg and Tatum were able to rule out mutation, and other investigators ruled out bacterial transformation (see Chapter 11). A third possibility is that the bacteria had **conjugated** in pairs, allowing their genetic material to mix and recombine to produce prototrophic colonies from cells containing *met⁺* and *bio⁺* alleles from strain 2 and *thr⁺* and *leu⁺* alleles from strain 1 (see Figure 13.10).

Later experiments showed that conjugation had indeed occurred, and that the two cells of differing genotype mated. One bacterial cell—the *recipient*—received DNA from the other cell—the *donor*—that included the two wild-type alleles for the loci in the recipient. Recombination then created a genotype with four wild-type alleles. The physical contact required for conjugation could be observed under the electron microscope (Figure 13.11).

What sort of a process brings about the recombination of genes after bacteria conjugate? We will learn about this shortly, but first we want to examine what distinguishes a recipient cell from a donor cell in conjugation.

Male bacteria have a fertility plasmid

The transfer of genetic material during conjugation in *E. coli* is a one-way process from a donor to a recipient.

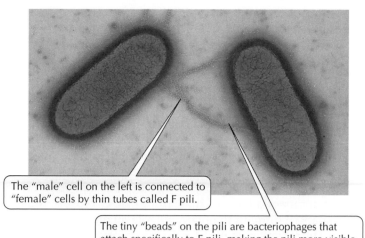

The "male" cell on the left is connected to "female" cells by thin tubes called F pili.

The tiny "beads" on the pili are bacteriophages that attach specifically to F pili, making the pili more visible.

13.11 Bacterial Conjugation After cells are joined by F pili, they are drawn into closer contact, and DNA is transferred from one cell to the other.

English microbiologist William Hayes characterized many strains and found that each strain is either a recipient or a donor, which can also be called female and male, respectively.

In many cases, the female bacterium becomes male after conjugation—in bacteria, maleness is an infectious venereal disease! Maleness in bacteria is defined by the presence of a fertility factor, called the **F plasmid**. A **plasmid** is a small, circular piece of DNA that is separate from the main chromosome. The F plasmid contains about 25 genes. Males, which possess the plasmid, are F⁺; females, lacking the plasmid, are F⁻. In a cross of F⁺ × F⁻, the F plasmid replicates and a copy is transferred to the female, thus rendering the female F⁺, while the original male remains F⁺ (Figure 13.12).

The F plasmid can replicate itself and persist in the cell population as if it were a second chromosome independent of the normal bacterial chromosome. Genes on the F plasmid direct the formation of long, thin structures called **F pili** (singular *pilus*, "hair") projecting from the surface of the male bac-

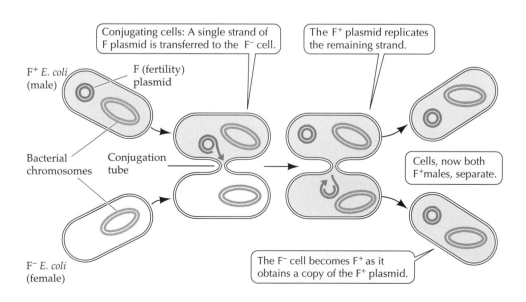

Conjugating cells: A single strand of F plasmid is transferred to the F⁻ cell.

The F⁺ plasmid replicates the remaining strand.

F⁺ *E. coli* (male)

F (fertility) plasmid

Bacterial chromosomes

Conjugation tube

Cells, now both F⁺ males, separate.

The F⁻ cell becomes F⁺ as it obtains a copy of the F⁺ plasmid.

F⁻ *E. coli* (female)

13.12 Transfer of the F Plasmid in *E. coli* During conjugation, the F⁻ recipient cell receives a copy of the F plasmid—an extra piece of DNA—from the donor cell by way of a connecting tube (the conjugation tube) and becomes F⁺.

terium. The ends of these F pili attach to the surface of female cells (see Figure 13.11). Although an F pilus makes the initial contact, a subsequent contact allows DNA to be transferred.

Genes from the male integrate into the female's chromosome

The discovery that genetic recombination could follow conjugation opened the possibility of mapping the genetic material of bacteria by determining recombinant frequencies (see Chapter 10). However, early attempts at mapping were complicated by the fact that very few recombinant offspring arose from F+ × F− crosses, thus making it difficult to obtain reliable quantitative data.

The situation changed with the discovery of certain mutant male strains that gave more abundant recombinant offspring. These strains were called **Hfr mutants** (for "*h*igh *f*requency of *r*ecombination"). Hfr males, unlike ordinary F+ males, do not generally transfer their F plasmid to the female. Furthermore, only certain genes are transferred with high frequency—they transfer other genes no more frequently than ordinary F+ males do. We know now that in Hfr strains the F plasmid is actually integrated into the bacterial chromosome.

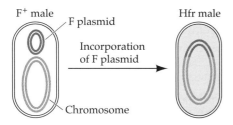

Work in 1955 by the French biologists Elie Wollman and François Jacob explained these observations. Wollman and Jacob showed that the genes from an Hfr male enter the female in sequence. In their most dramatic experiments, they used the technique of **interrupted mating**.

Wollman and Jacob allowed a brief time for the male and female bacteria to attach and transfer their DNA; then they separated the two cells and observed what happened. They mixed Hfr and F− *E. coli* bacteria at high concentration to initiate conjugation; at various times thereafter

they diluted samples of the mixture and agitated them in a kitchen blender for 2 minutes. Such agitation separates conjugating bacteria but does not damage them.

The number of Hfr genes passed to the female depended on the length of time allowed before conjugation was interrupted—the longer the conjugation, the more genes were transferred (Figure 13.13*b*). The genes always entered in the same order from any particular Hfr strain. The Hfr mutant almost never transferred the F plasmid.

Jacob and Wollman recognized immediately that this interrupted mating technique provided a simple way to map the chromosome. They prepared different mutant strains and crossed pairs of strains; then they

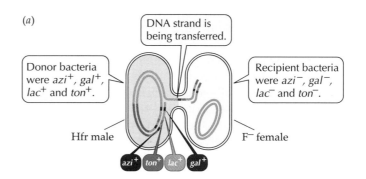

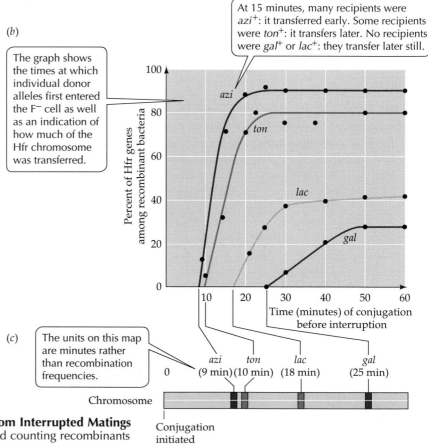

13.13 Creating a Chromosome Map of *E. coli* from Interrupted Matings By interrupting conjugating cells at various times and counting recombinants from the matings, we can prepare chromosome maps.

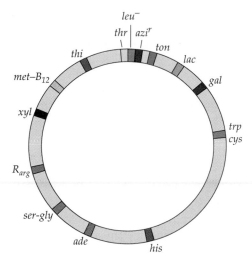

13.14 The *E. coli* Chromosome This map of the *E. coli* chromosome summarizes data from interrupted-conjugation experiments with many Hfr strains (see Figure 13.13). This early version of the map shows only a few genes, but well over half of the estimated 3,000 *E. coli* genes have been mapped. The three-letter abbreviation for each gene is derived from its mutant phenotype: *leucine*-requiring, *azide*-resistant, and so on.

interrupted successive samples from the crosses. Because the genes are transferred in a particular sequence, the length of mating time required before a particular gene is transferred and thus available to appear in recombinant progeny is a measure of its location on the chromosome (Figure 13.13*c*).

Different Hfr strains had different orders for gene transfer. Strain 1 might transfer in the order *azi-ton-lac-gal*, while strain 2 transferred in the order *lac-gal-azi-ton* and strain 3 in the order *ton-lac-gal-azi*. These and other observations led Jacob and Wollman to the following conclusions: (1) The *E. coli* chromosome is circular (Figure 13.14), (2) Hfr males have the F plasmid incorporated into their chromosome, (3) the location where the F plasmid is integrated in the chromosome varies, giving rise to different Hfr strains (thus the different orders of gene transfer), (4) the inserted F plasmid marks the point at which the chromosome "opens" as conjugation begins (see Figure 13.15), and (5) one end—always the same one—of the opened chromosome leads the way into the female. The piece of chromosome continues to move through the conjugation tube until mating is interrupted naturally or otherwise. At the extreme end of the opened chromosome lies the portion of the F plasmid that determines maleness; this will be the last portion of the donor chromosome to be transferred.

What moves from the Hfr to the F⁻ is just one strand of the double-stranded Hfr DNA. Transfer is initiated when one strand within the sequence of the F plasmid is cut once and opens (is "nicked"; Figure

13.15). The 5′ end of the nicked strand begins to unravel from the chromosome and moves to the F⁻ cell. Meanwhile, the transferred strand is replaced in the Hfr cell by DNA synthesis at the 3′ end of the nick, using the intact circular strand as the template. Thus the male still contains a complete double-stranded set of DNA sequences, even after donating some DNA to the F⁻ cell.

As it enters the F⁻ cell, the nicked DNA strand functions as a template, becoming double-stranded. Genes on this piece of DNA will give rise to recombinant bacteria only if the genes become integrated into the F⁻ chromosome by crossing over (Figure 13.16). About half of the transferred Hfr genes become integrated in this way. The DNA containing the other, unincorporated genes does not replicate when the cell divides, and it is eventually destroyed.

Prokaryotic conjugation can be summarized as follows: A female cell (F⁻) is contacted by an F pilus growing on a male cell (either F⁺ or Hfr), and the male cell transfers some DNA to the female. An F⁺ cell transfers the F plasmid, turning the F⁻ cell into F⁺. An Hfr cell transfers DNA other than the F plasmid. The transferred DNA that integrates into the F⁻ chromosome becomes part of the genome of the F⁻ cell.

As we suggested earlier, there are other, quite different ways to transfer genes from one bacterium to

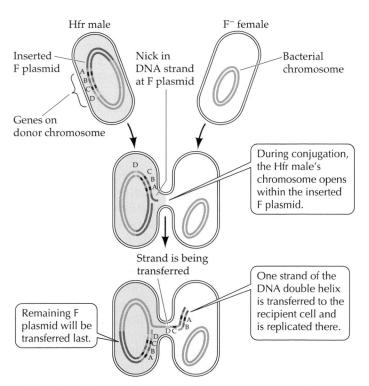

13.15 The Behavior of Hfr Males Because most of the F plasmid is the last DNA to be transferred, the recipient cell usually does not become a male, since the complete chromosome is rarely transferred.

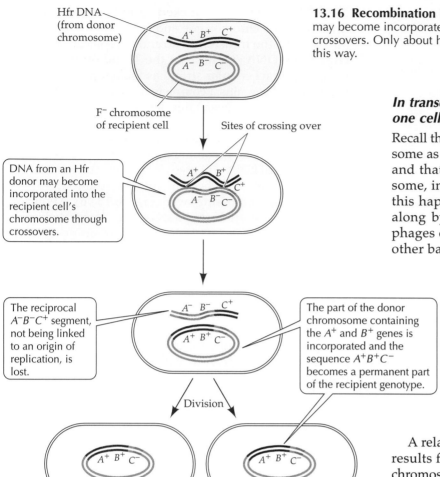

Hfr DNA
(from donor
chromosome)

F⁻ chromosome
of recipient cell

Sites of crossing over

DNA from an Hfr
donor may become
incorporated into the
recipient cell's
chromosome through
crossovers.

The reciprocal
$A^-B^-C^+$ segment,
not being linked
to an origin of
replication, is
lost.

The part of the donor
chromosome containing
the A^+ and B^+ genes is
incorporated and the
sequence $A^+B^+C^-$
becomes a permanent part
of the recipient genotype.

Division

13.16 Recombination Following Conjugation DNA from an Hfr donor may become incorporated into the recipient cell's chromosome through crossovers. Only about half the transferred Hfr genes become integrated in this way.

another, that also result in genetic recombination. Next, we will examine transformation.

In transformation, cells pick up genes from their environment

Frederick Griffith obtained the first evidence for the transfer of prokaryotic genes more than 70 years ago when he discovered **transformation** (see Figure 11.1). Recall that living bacteria of a nonvirulent strain of *Pneumococcus* picked up heritable information from dead virulent pneumococci. We now know that DNA had leaked from the dead cells and was taken up as free DNA by the living cells, within the body of a living mouse.

Once transforming DNA is inside the host cells, an event very similar to the recombination of Hfr and F⁻ occurs, and new genes can be spliced into the host chromosome. Bacteria can also take up and express DNA from a solution in a test tube. As we'll see in Chapter 16, on genetic engineering, we can transform bacteria with recombinant DNA, inducing cells to make proteins they have never seen before.

In transduction, viruses carry genes from one cell to another

Recall that a temperate phage can integrate its chromosome as a prophage into a host bacterial chromosome, and that the prophage can escape from the chromosome, initiating a lytic cycle (see Figure 13.4). When this happens, bacterial genes are occasionally taken along by the departing phage DNA. The resulting phages can then introduce these bacterial genes into other bacteria that they infect, and the genes may be randomly integrated into the chromosomes of the new hosts. This phenomenon is called **specialized transduction** (here "transduce" means "transfer") (Figure 13.17a). Transducing phages, which carry bacterial genes, cause newly infected bacteria to become lysogenic. In specialized transduction, only the chromosomal genes that are adjacent to the site of integration of the prophage may be taken along with the phage DNA.

A related phenomenon, **generalized transduction**, results from the incorporation of part of the *bacterial* chromosome, *without* the prophage DNA, into a phage coat (Figure 13.17b). The resulting particle, even though it lacks any phage genes, can infect another bacterium, injecting the piece of DNA from its former host. A bacterium thus infected with a piece of foreign bacterial DNA does *not* become lysogenic, nor does it form new phages and burst as in the lytic cycle (see Figure 13.3). It simply contains extrachromosomal bacterial DNA, as if it had conjugated with an Hfr cell.

If crossing over takes place between the host chromosome and the transduced DNA, the transfer of genes is completed. In contrast to specialized transduction, generalized transduction can move any small part of the bacterial chromosome. There is no limitation on which bacterial genes might become enclosed in a phage coat.

The phage particle is big enough to house several adjacent bacterial genes. Generalized transduction is thus another powerful tool for mapping the bacterial chromosome—with viral assistance. If two genes are transduced together, they must be close together on the chromosome. See Question 2 in "Applying Concepts" for an example of mapping by generalized transduction.

Transduction is also an important method for genetic engineering. As you will see in Chapter 16, viruses are used as vectors to transfer DNA into a cell, which becomes genetically transformed.

13.17 Transduction Phages may carry bacterial DNA from cell to cell.

Resistance factors are plasmids carrying harmful genes

The F plasmid and viral prophages are examples of episomes. **Episomes** are nonessential genetic elements that can exist in either of two states: independently replicating within a cell, or integrated into the main chromosome. Episomes cannot arise by mutation; they must be obtained by infection from outside the bacterium. The infection can come from a virus or from another bacterium. As we have seen, episomes may be used as vehicles for transferring genes from one bacterium to another.

Plasmids, free circles of DNA such as the F plasmid we encountered earlier (see Figure 13.12), are also nonessential genetic elements; they usually are not incorporated into the bacterial chromosome. (An episome is simply a plasmid that has the possibility of becoming part of the chromosome.) Every plasmid is a replicon, which means that it contains a sequence where DNA replication starts, and so is capable of independent replication.

Resistance factors, or **R factors**, are plasmids that carry genes for resistance to antibiotics. R factors first came to the attention of biologists in 1957 during an epidemic in Japan, when it was discovered that some strains of the *Shigella* bacterium, which causes dysentery, were resistant to several antibiotics. Researchers found that resistance to the entire spectrum of antibiotics could be transferred by conjugation even when no genes on the main chromosome were transferred. Also, F⁻ cells could serve as donors, indicating that the genes for antibiotic resistance were not carried by the F plasmid.

Eventually it was shown that the genes for antibiotic resistance are carried on plasmids, but not the F plasmid. Each of these plasmids (the R factor) carries one or more genes conferring resistance to particular antibiotics, as well as genes that code for proteins involved in the transfer of DNA to a recipient bacterium. As far as biologists can determine, R factors appeared long before antibiotics were discovered, but they seem to have become more abundant in modern times. Can you propose a hypothesis to explain why R factors might be more widespread now than they were in the past?

R factors pose a serious threat to hospital patients, as we noted at the beginning of this chapter. The heavy use of antibiotics in the hospital environment selects for bacterial strains bearing the R factors, and the unfortunate infected patient becomes home to bacterial strains that can't be knocked out by antibiotics.

(a) **Specialized transduction**

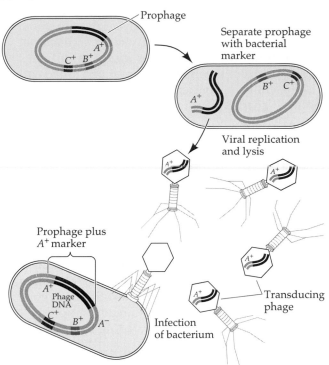

An A^- bacterial cell becomes A^+ when a transducing phage introduces the A^+ gene to the recipient cell when the phage DNA inserts into the host chromosome.

(b) **Generalized transduction**

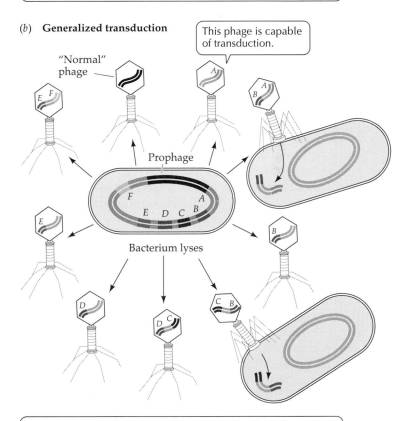

Rarely during phage production, parts of the bacterial chromosome are incorporated into phage particles. This DNA may be injected into a recipient bacterial cell and may become part of the recipient's chromosome.

R factors also pose a threat to people in the general clinical setting if antibiotics are used inappropriately. You probably have gone to see a physician with a sore throat, which can have either a viral or a bacterial cause. The best way to know is for the doctor to take a small sample from the inflamed throat and try to culture and identify any bacteria that are present. But you cannot wait another day. Impatient, you ask the doctor to give you something to cure the disease. She prescribes an antibiotic, which you take. The sore throat gradually gets better, and you think that the antibiotic did the job.

But suppose the infection was viral? In this case, the antibiotic did nothing to combat the disease, which just ran its normal course. However, the antibiotic may have done something harmful: By killing many normal bacteria in your body, it may have exerted a selection for bacteria harboring R factors. These bacteria reproduce in the presence of the antibiotic, and soon could be quite numerous. Then the next time you got a *bacterial* infection, there would be a ready supply of R factors to be transferred to the invading bacteria and antibiotics would be ineffective.

Transposable elements move genes among plasmids and chromosomes

As we have seen, plasmids, viruses, and even phage coats (in the case of transduction) can transport genes from one bacterial cell to another. Another type of "gene transport" within the individual cell relies on segments of chromosomal or plasmid DNA called **transposable elements**. Copies of transposable elements can be inserted at new locations in the same chromosome, or into another chromosome. Insertion often produces multiple physiological effects by disrupting the genes into which the transposable elements are inserted (Figure 13.18a).

The first transposable elements to be discovered in prokaryotes were large pieces of DNA, typically 1,000 to 2,000 base pairs long, found in many places in the *E. coli* chromosome. In one mechanism of transposition, the sequence of a transposable element can replicate independently of the rest of the chromosome. The copy then inserts itself at other, seemingly random places in the chromosome. The genes encoding the enzymes necessary for this insertion are found within the transposable element itself. Some other transposable elements are cut from their original sites and inserted elsewhere without replication. Many transposable elements discovered later were longer (about 5,000 base pairs) and carried one or more additional genes. These longer elements with additional genes are called **transposons** (Figure 13.18b).

What do transposons and other transposable elements have to do with the genetics of prokaryotes—or with hospitals? Transposable elements have contributed

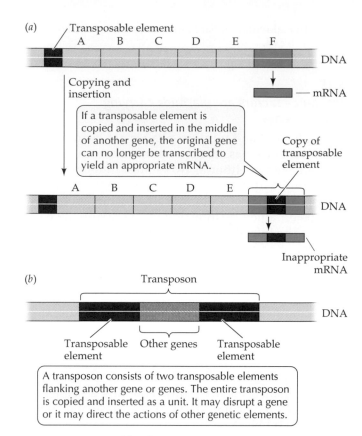

If a transposable element is copied and inserted in the middle of another gene, the original gene can no longer be transcribed to yield an appropriate mRNA.

A transposon consists of two transposable elements flanking another gene or genes. The entire transposon is copied and inserted as a unit. It may disrupt a gene or it may direct the actions of other genetic elements.

13.18 Transposable Elements and Transposons
(a) Transposable elements are segments of DNA that can be inserted at new locations, either on the same chromosome or on a different chromosome. (b) Transposons consist of transposable elements and other genes.

to the evolution of plasmids. R plasmids originally gained their genes for antibiotic resistance through the activity of transposable elements; one piece of evidence for this conclusion is that each resistance gene in an R plasmid is part of a transposon. Transposons on the F plasmid and on the bacterial chromosome interact to direct the insertion of the F plasmid into the chromosome in the formation of an Hfr male.

We will have more to say about transposable elements in the next chapter, which is concerned with the genetics of eukaryotes.

Regulation of Gene Expression in Prokaryotes

We have now seen that prokaryotes have genes, that prokaryotic genes mutate, that they are exchanged among cells by three basic mechanisms, that they may be parts of chromosomes or plasmids, and that they may move among these DNA molecules. In preceding chapters we learned how genes replicate and how they are expressed. But are all prokaryotic genes expressed all the time? If not, why not? And if not, how then does the prokaryotic cell regulate gene expression?

13.19 An Inducer Stimulates the Synthesis of an Enzyme

It is most efficient for a cell to produce an enzyme only when it is needed; some enzymes are induced by the presence of the substance they act upon (for example, the induction of β-galactosidase by the presence of lactose).

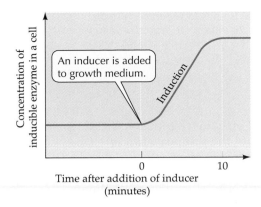

An inducer is added to growth medium.

Induction

Concentration of inducible enzyme in a cell

Time after addition of inducer (minutes)

Regulation of transcription conserves energy

To express all the genes of a cell all the time would be inefficient in terms of energy. Why should a cell use valuable ATP and other resources to produce an enzyme when that enzyme is not needed? In fact, it does not. Prokaryotes produce some enzymes only when their substrates are present in the environment, and they turn off the synthesis of others when the products of those enzymes build to excessive concentrations.

As a normal inhabitant of the human intestine, *E. coli* has to adjust to sudden changes in its chemical environment. Its host may present it with one foodstuff one hour and another the next. For example, these bacteria may suddenly be deluged with milk, the main carbohydrate of which is the sugar lactose. Lactose is a β-galactoside—a disaccharide containing galactose β-linked to glucose (see Chapter 3). Before lactose can be of any use to the bacteria, it must first be taken into their cells by a membrane transport carrier called β-galactoside permease. Then it must be hydrolyzed to glucose and galactose by the enzyme β-galactosidase. A third protein, the enzyme thiogalactoside transacetylase, is also required for lactose metabolism.

When *E. coli* is grown on a medium that does not contain lactose or other β-galactosides, the levels of all three of these proteins within the bacterial cell are extremely low—the cell does not waste energy and material making the unneeded proteins. If, however, the environment changes such that lactose is the predominant sugar and very little glucose is present, the synthesis of all three of these proteins begins promptly

and they increase in abundance. There are only two molecules of β-galactosidase in an *E. coli* cell when glucose is in the medium. But when it is absent, lactose can induce the synthesis of 3,000 molecules of β-galactosidase per cell! Regulation of enzyme synthesis by the genes that code for them thus promotes efficiency in the cell.

Compounds that stimulate the synthesis of an enzyme (as does lactose in this example) are called **inducers** (Figure 13.19). The enzymes that are produced are called **inducible enzymes**, whereas enzymes that are made all the time at a constant rate are called **constitutive enzymes**. If lactose is removed from *E. coli*'s environment, synthesis of the three enzymes stops almost immediately. The enzyme molecules that have already formed do not disappear; they are merely diluted during subsequent growth and reproduction until their concentration falls to the original low level within each bacterium.

We have now seen two basic ways to regulate the rates of metabolic pathways. In Chapter 6, we described allosteric regulation of enzyme *activity* (the rate of enzyme-catalyzed reactions); this mechanism allows rapid, fine-tuning of metabolism. Regulation of gene expression—that is, regulation of the *concentration* of enzyme—is slower but produces a greater savings of energy. Figure 13.20 compares these two modes of regulation.

Transcriptional regulation uses some familiar tools and some new tools

The blueprints for the synthesis of the three proteins that process lac-

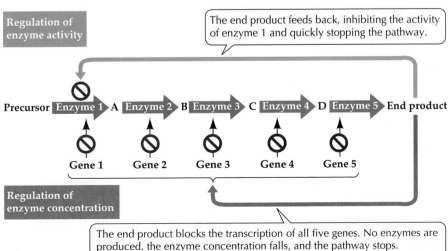

Regulation of enzyme activity

The end product feeds back, inhibiting the activity of enzyme 1 and quickly stopping the pathway.

Precursor ▸ Enzyme 1 ▸ A ▸ Enzyme 2 ▸ B ▸ Enzyme 3 ▸ C ▸ Enzyme 4 ▸ D ▸ Enzyme 5 ▸ End product

Gene 1   Gene 2   Gene 3   Gene 4   Gene 5

Regulation of enzyme concentration

The end product blocks the transcription of all five genes. No enzymes are produced, the enzyme concentration falls, and the pathway stops.

13.20 Two Ways to Regulate a Metabolic Pathway

Feedback from the end product can block enzyme activity, or it can stop the transcription of genes that code for the enzymes.

13.21 Repressor Bound to Operator Blocks Transcription Portions of the repressor bind to the major and minor grooves in the DNA helix, preventing transcription.

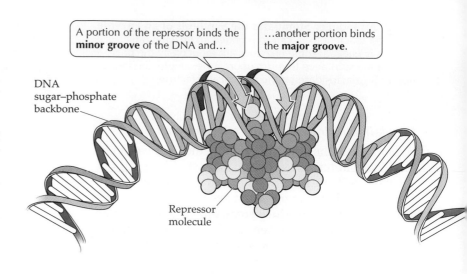

> A portion of the repressor binds the **minor groove** of the DNA and...

> ...another portion binds the **major groove**.

DNA sugar–phosphate backbone

Repressor molecule

tose are called **structural genes**, indicating that they specify the primary structure (the amino acid sequence) of a protein molecule. In other words, structural genes can be transcribed into mRNA. When Jacob, Wollman, and Monod mapped the particular structural genes coding for proteins involved in the metabolism of lactose, they discovered that all three lie close together in a region that covers about 1 percent of the *E. coli* chromosome.

It is no coincidence that these three genes lie next to one another. The information from them is transcribed into a single, continuous molecule of mRNA. Because this particular messenger governs the synthesis of all three lactose-metabolizing enzymes, either all or none of the enzymes are made, depending on whether their common message—their mRNA—is present in the cell.

The three genes share a single promoter. Recall from Chapter 12 that a promoter is a site on DNA where RNA polymerase binds to initiate transcription. The promoter for these three structural genes is very efficient so that the maximum rate of mRNA synthesis can be high, but there must also be a way to shut down mRNA synthesis when the enzymes are not needed. Such flexibility requires some new tools.

Operons are units of transcription in prokaryotes

Prokaryotes shut down transcription by placing an obstacle between the promoter and its structural genes. A short stretch of DNA called the **operator** can bind very tightly a special type of protein molecule, a **repressor**, thereby creating the obstacle. When the repressor protein is bound to the operator region DNA, it blocks the transcription of mRNA (Figure 13.21). When the repressor is not attached to the operator, messenger synthesis proceeds rapidly.

The whole unit, consisting of closely linked structural genes and the stretches of DNA that control their transcription, is called an **operon** (Figure 13.22). An operon always consists of a promoter, an operator, and one or more structural genes. The promoter and operator are binding sites on DNA and are not transcribed.

13.22 The *lac* Operon of *E. coli* and Its Regulator The *lac* operon is a segment of DNA that includes a promoter, an operator, and the three structural genes that code for the *lac*tose-metabolizing enzymes.

E. coli has three different ways of controlling the transcription of operons. Two depend on interactions of repressor with the operator, and the third depends on interactions of other proteins with the promoter. Let's consider each of the three control systems in turn.

Operator–repressor control that induces transcription: The lac operon

The operon that controls and contains the genes for the three *lac*tose-metabolizing enzymes is called the ***lac* operon** (see Figure 13.22). As we just learned, RNA polymerase can bind to the promoter, and a repressor can bind to the operator. How is the operon controlled?

The key lies in the repressor and its binding to the operator. The repressor is a protein that has two binding sites: one for the operator and the other for small molecules called *inducers*. Inducers of the *lac* operon are molecules of lactose and certain other β-galactosides. Binding of the inducer changes the shape of the repressor (by allosteric modification; see Chapter 6). This change in shape prevents the repressor from binding to the operator (Figure 13.23). At this point,

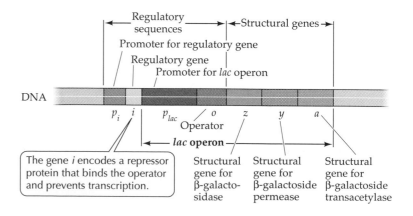

Regulatory sequences | Structural genes

Promoter for regulatory gene

Regulatory gene

Promoter for *lac* operon

DNA

p_i i p_{lac} o z y a

Operator

lac operon

The gene *i* encodes a repressor protein that binds the operator and prevents transcription.

Structural gene for β-galactosidase

Structural gene for β-galactoside permease

Structural gene for β-galactoside transacetylase

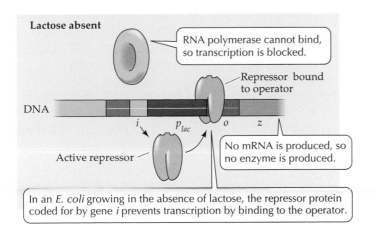

Lactose absent

RNA polymerase cannot bind, so transcription is blocked.

Repressor bound to operator

DNA

i p_{lac} o z

Active repressor

No mRNA is produced, so no enzyme is produced.

In an *E. coli* growing in the absence of lactose, the repressor protein coded for by gene *i* prevents transcription by binding to the operator.

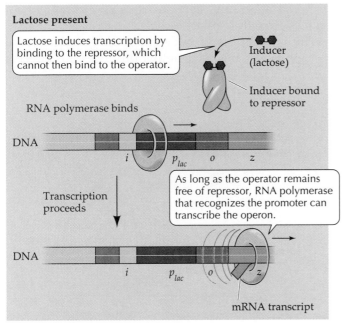

Lactose present

Lactose induces transcription by binding to the repressor, which cannot then bind to the operator.

Inducer (lactose)

Inducer bound to repressor

RNA polymerase binds

DNA

i p_{lac} o z

Transcription proceeds

As long as the operator remains free of repressor, RNA polymerase that recognizes the promoter can transcribe the operon.

DNA

i p_{lac} o z

mRNA transcript

13.23 The *lac* Operon: Transcription Induced by the Removal of a Repressor   Lactose (the inducer) leads to enzyme production by preventing the repressor protein (which would have stopped transcription) from binding to the operator.

RNA polymerase can bind to the promoter and start transcribing the structural genes of the *lac* operon. The mRNA transcribed from these genes is translated on ribosomes, synthesizing the three proteins required for metabolizing lactose.

What happens if the concentration of lactose drops? As the lactose concentration decreases, the inducer (lactose) molecules separate from the repressor. Free of lactose molecules, the repressor returns to its original shape and binds to the operator, and transcription of the *lac* operon stops. Translation stops soon thereafter, because the mRNA that is already present breaks down quickly. Thus, the inducer regulates the binding of the repressor to the operator.

Repressor proteins are coded by **regulatory genes**. The regulatory gene that codes for the repressor of the *lac* operon is called the *i* (inducibility) gene. The *i* gene happens to lie close to the operon that it controls, but some other regulatory genes are distant from their operons. Like all other genes, the *i* gene itself has a promoter, which can be designated p_i. Because this promoter does not bind RNA polymerase very effectively, only enough mRNA to synthesize about ten molecules of repressor protein per cell per generation is produced. This quantity of the repressor is enough to regulate the operon effectively—to produce more would be a waste of energy. There is no operator between p_i and the *i* gene. Therefore, the repressor of the *lac* operon is constitutive; that is, it is made at a constant rate that is not subject to environmental control.

Let's review the important features of inducible systems such as the *lac* operon. In the absence of inducer the *lac* operon is turned *off*. Adding inducer turns the operon *on*. Control is exerted by a regulatory protein—the repressor—that turns the operon *off*. Some genes, such as *i*, produce proteins whose sole function is to regulate the expression of other genes; certain other DNA sequences (namely, operators and promoters) do not code for proteins but are binding sites for regulatory proteins.

Operator–repressor control that represses transcription: The trp operon

We have seen that *E. coli* benefits from having an inducible system for lactose metabolism. Only when lactose is present does the system switch on. Equally valuable to a bacterium is the ability to switch *off* the synthesis of certain enzymes in response to the excessive accumulation of their products. For example, if the amino acid tryptophan, an essential constituent of proteins, is present in ample concentration, it is advantageous to stop making the enzymes for tryptophan synthesis. When the formation of an enzyme can be turned off in response to such a biochemical cue, synthesis of the enzyme is said to be **repressible**.

Monod realized that repressible systems, such as the *trp* operon for *tryp*tophan synthesis, could work by mechanisms similar to those of inducible systems, such as the *lac* operon. In repressible systems, the repressor cannot shut off its operon unless it first binds to a **corepressor**, which may be either the metabolite itself (tryptophan in this case) or an analog of it (Figure 13.24). If the metabolite is absent, the operon is transcribed at a maximum rate. If the metabolite is present, the operon is turned off.

The difference between inducible and repressible systems is small but significant. In inducible systems, a substance in the environment (the inducer) interacts with the regulatory-gene product (the repressor), rendering it *incapable* of binding to the operator and thus

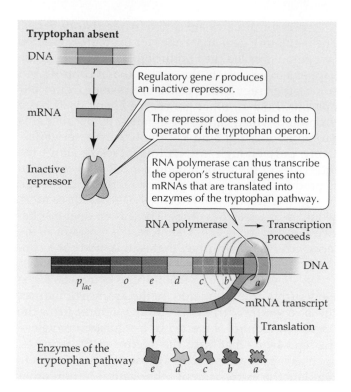

Tryptophan absent

Regulatory gene *r* produces an inactive repressor.

The repressor does not bind to the operator of the tryptophan operon.

RNA polymerase can thus transcribe the operon's structural genes into mRNAs that are translated into enzymes of the tryptophan pathway.

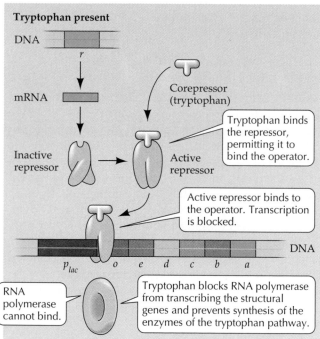

Tryptophan present

Tryptophan binds the repressor, permitting it to bind the operator.

Active repressor binds to the operator. Transcription is blocked.

RNA polymerase cannot bind.

Tryptophan blocks RNA polymerase from transcribing the structural genes and prevents synthesis of the enzymes of the tryptophan pathway.

13.24 The *trp* Operon: Transcription Repressed by the Binding of a Repressor Because tryptophan activates an otherwise inactive repressor, it is called a corepressor.

incapable of blocking transcription. In repressible systems, a substance in the cell (the corepressor) interacts with the regulatory-gene product to make it *capable* of binding to the operator and blocking transcription. Although the effects of the substances are exactly opposite, the systems as a whole are strikingly similar.

In both the inducible lactose system and the repressible tryptophan system, the regulatory molecule functions by binding the operator. Next we'll consider an example of control by binding the *promoter*.

Protein synthesis can be controlled by increasing promoter efficiency

A prokaryotic cell has the means to increase the transcription of certain relevant genes when it needs a new energy source. *E. coli* prefers to get its energy from glucose in the environment. When glucose is unavailable, *E. coli* must get energy from another source, such as lactose or certain other sugars or even amino acids. The alternative energy source must be catabolized (broken down) by reactions requiring catabolic enzymes. Specific regulatory molecules enhance the transcription of operons that contain the genes for these enzymes. The mechanism that produces this effect is

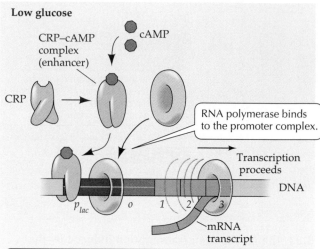

Low glucose

RNA polymerase binds to the promoter complex.

When supplies of glucose are low, a receptor protein (CRP) and cAMP form a complex that binds to the promoter and activates it, allowing transcription of structural genes that encode enzymes for catabolizing the alternative energy source.

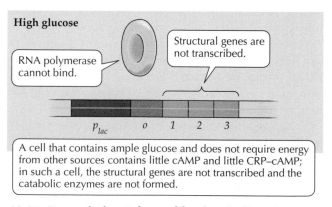

High glucose

RNA polymerase cannot bind.

Structural genes are not transcribed.

A cell that contains ample glucose and does not require energy from other sources contains little cAMP and little CRP–cAMP; in such a cell, the structural genes are not transcribed and the catabolic enzymes are not formed.

13.25 Transcription Enhanced by the Binding of CRP at the Promoter Site The structural genes of this operon encode enzymes that break down a food source other than glucose.

entirely different from the two operator–repressor mechanisms just discussed, which turn the operon on or off. This type of mechanism makes the promoter function more efficiently.

Suppose that a bacterial cell lacks a glucose supply but has access to another food that can be catabolized to yield energy. In operons containing genes for catabolic enzymes, the promoters bind RNA polymerase in a series of steps (Figure 13.25). First, a special protein (abbreviated **CRP**, for *c*AMP *r*eceptor *p*rotein) binds the low-molecular-weight compound adenosine 3′, 5′-cyclic monophosphate, better known as cyclic AMP or cAMP. Next, the CRP–cAMP complex binds close to the binding site of the RNA polymerase and enhances the binding of the polymerase to the promoter 50-fold. The promoters of the *lac* operon and many other genes responsible for sugar metabolism are activated in this way.

When glucose becomes abundant in the medium, alternative food molecules do not need to be broken down, so the cell diminishes or ceases synthesizing the enzymes that catabolize these alternative sources. Glucose decreases synthesis of these enzymes—a phenomenon called *catabolite repression*—by lowering the cellular concentration of cAMP.

The *lac* and *trp* systems—the two operator–repressor systems—are examples of *negative* control of transcription because the regulatory molecule (the repressor) in each case *prevents* transcription. The promoter system is an example of *positive* control of transcription because the regulatory molecule (the CRP–cAMP complex) *enhances* transcription.

As we will see in Chapter 13, the regulation of gene expression in eukaryotes is somewhat different than in prokaryotes.

Summary of "The Genetics of Viruses and Prokaryotes"

Using Prokaryotes and Viruses for Genetic Experiments

- Bacteria and viruses are useful to the study of genetics and molecular biology because they contain much less DNA than complex eukaryotes have, they grow and reproduce rapidly, often in less than an hour, and they are haploid, which means their genetics is simpler.

Viruses: Reproduction and Recombination

- Viruses were discovered as disease-causing agents small enough to pass through a filter that retains bacteria. They consist of a nucleic acid genome that codes for a few proteins, and a protein coat. Some viruses also have a lipid membrane that is derived from host membranes.

- Viruses are obligate intracellular parasites, needing the biochemical machinery of living cells to reproduce.
- There are many types of viruses, classified by their size, shape, and genetic material (RNA or DNA), or host (plant, animal, protist, fungus, or prokaryote). **Review Figure 13.1**
- Bacteriophages infect bacteria. In the lytic cycle, about 30 minutes after phage infection the host cell breaks open, releasing many new phage particles. Some phages can undergo a lysogenic cycle, in which their DNA is inserted into the host chromosome and replicates for generations. When conditions are appropriate, the lysogenic DNA exits the host chromosome and enters a lytic cycle. **Review Figures 13.3, 13.4**
- If two phages infect the same host cell simultaneously, their DNAs can come into close proximity and genetic recombination can occur. The resulting data make it possible to generate a map of the phage chromosome. **Review Figure 13.5**
- Most of the many types of RNA and DNA viruses that infect animals cause diseases. Some animal viruses are surrounded by membranes. Retroviruses, such as HIV, have RNA and reproduce their genomes through a DNA intermediate. Other RNA viruses use their RNA as mRNA to code for enzymes and replicate their genomes without using DNA. **Review Figures 13.6, 13.7, and Table 13.2**
- Many plant viruses are spread by other organisms, such as insects.
- Viroids are made only of RNA and infect plants, where they are replicated by the plant's enzymes.
- Prions appear to be infective proteins and cause diseases in animals, including humans.

Prokaryotes: Reproduction, Mutation, and Recombination

- When bacteria divide, they form clones of identical cells that can be observed as colonies when grown on solid media. **Review Figure 13.8**
- Prokaryotic genes can mutate, as shown by experiments in which prokaryotes demonstrate resistance to bacteriophages. **Review Figure 13.9**
- A bacterium can transfer its genes to another bacterium by conjugation, transformation, or transduction.
- In conjugation, a bacterium attaches to another bacterium and passes a partial copy of its DNA to the adjacent cell. **Review Figures 13.10, 13.11**
- The F plasmid codes for the conjugation tube. A bacterium harboring F is male (F⁺). Without F, the cell is female (F⁻). During conjugation, an F⁺ cell can transfer a copy of its F plasmid to an F⁻ cell, converting it to an F⁺ cell. **Review Figure 13.12**
- If the F plasmid integrates into the main chromosome (as in Hfr males), it can prompt the transfer of part of that chromosome into a recipient cell, where the extra DNA can undergo genetic recombination with the host chromosome, transferring new genes to the host chromosome permanently. The order of transfer reflects the positions of different markers on the circular bacterial chromosome. **Review Figures 13.13, 13.14, 13.15, 13.16**
- In transformation, genes are transferred between prokaryotes when fragments of bacterial DNA are taken up by a cell from the medium. These genetic fragments may recombine with the host chromosome, thereby permanently adding new genes.

- In transduction, phage particles carry bacterial DNA from one bacterium to another. This transfer can include a few specific bacterial genes attached to a lysogenic phage genome when it leaves the bacterial chromosome. Or it can involve fragments of bacterial DNA that are packaged by themselves into phage particles. In both cases, the new bacterial DNA can recombine with the chromosome of the recipient bacteria. **Review Figure 13.17**
- Plasmids are independently replicating, extrachromosomal DNA elements that usually are not incorporated into the main chromosome. Bacterial R factors are plasmids that carry genes for antibiotic resistance, as well as genes coding for interbacterial transfer by conjugation. R factors are a serious public health threat.
- Transposable elements are movable stretches of DNA that can jump from one place to another on the bacterial chromosome—either by actually moving or by making a new copy, which inserts at a new location. **Review Figure 13.18**

Regulation of Gene Expression in Prokaryotes

- In bacteria, constitutive genes are constantly expressed, and their products are essential to the cells at all times. The expression of other genes is regulated; their products are made only when they are needed. Genes are regulated by inducers (to turn on expression) or corepressors (to turn off expression). **Review Figures 13.19, 13.20, 13.24**
- The functionally related bacterial genes contained in an operon are transcribed into a single mRNA, so they are under the same regulatory control. **Review Figure 13.22**
- The expression of prokaryotic genes is regulated by three different mechanisms: inducible operator–repressor systems, repressible operator–repressor systems, and enhancement systems that increase the efficiency of a promoter.
- The *lac* operon is an example of an inducible system whose proteins allow bacteria to use lactose. When glucose is the energy source, very few of the *lac* operon proteins are present in the cell. But if glucose is absent, lactose can induce the syntheses of these proteins. **Review Figure 13.19**
- A promoter controls RNA polymerase–catalyzed transcription of the entire *lac* operon. Between the promoter and the structural genes lies a stretch of DNA called the operator. When glucose is present and lactose is absent, a repressor protein binds tightly to the operator, preventing RNA polymerase from binding to the promoter. Thus the operon is transcriptionally inactive.
- When glucose is absent and lactose is present, the latter acts as an inducer by binding to the repressor. This changes the repressor's shape so that it no longer recognizes the operator. With the operator unoccupied, RNA polymerase binds to the promoter, and transcription occurs to produce the proteins for lactose utilization. **Review Figures 13.23, 13.24, 13.25**
- The *trp* operon is a repressible system in which the presence of the end product of a biochemical pathway, tryptophan, represses the syntheses of enzymes involved in its synthesis. Tryptophan acts as a corepressor, binding to an inactive repressor protein, giving it a strong affinity for the operator. Where the operator is occupied with repressor, transcription of the structural genes coding for the tryptophan biosynthesis pathway is blocked. **Review Figure 13.24**
- The efficiency of RNA polymerase can be increased by regulation of the level of cyclic AMP, which binds to a protein, CRP. The altered CRP then binds to a site near the promoter of a target gene, enhancing the binding of RNA polymerase and hence transcription. **Review Figure 13.25**

Self-Quiz

1. In bacterial conjugation
   a. each cell donates DNA to the other.
   b. a bacteriophage carries DNA between bacterial cells.
   c. one partner possesses a fertility plasmid.
   d. the two parent bacteria merge like sperm and egg.
   e. all the progeny are recombinant.

2. Which statement about the bacterial fertility factor is *not* true?
   a. It is a plasmid.
   b. It confers "maleness" on the cell in which it resides.
   c. It can be transferred to a female cell, making the female cell male.
   d. It has thin projections called F pili.
   e. It can become part of the bacterial chromosome.

3. Hfr mutants
   a. are female bacteria that are highly efficient recipients of genes.
   b. rarely transfer all the genes on the chromosome.
   c. keep their F plasmid separate from the chromosome at all times.
   d. are unable to conjugate with other bacteria.
   e. transfer genes in random order.

4. Lysogenic bacteria
   a. lack a prophage.
   b. are accompanied by free phages when growing in culture.
   c. lyse immediately.
   d. cannot by induced to enter a lytic cycle.
   e. are susceptible to further attack by the same strain of phage.

5. Which statement about transduction is *not* true?
   a. The viral DNA is an episome.
   b. Transduction is a useful tool for mapping a bacterial chromosome.
   c. In specialized transduction, the newly infected cell becomes lysogenic.
   d. Transduction can result in genetic recombination.
   e. To carry bacterial genes, the viral coat must contain viral DNA.

6. Plasmids
   a. are circular protein molecules.
   b. are required by bacteria.
   c. are tiny bacteria.
   d. may confer resistance to antibiotics.
   e. are a form of transposable element.

7. Which statement about transposable elements is *not* true?
   a. They can be copied to another DNA molecule.
   b. They can be copied to the same DNA molecule.
   c. They are typically 100 to 500 base pairs long.
   d. They may be part of a plasmid.
   e. They encode the enzyme transposase.

8. In an inducible operon
   a. an outside agent switches on enzyme synthesis.
   b. a corepressor unites with the repressor.
   c. an inducer affects the rate at which repressor is made.
   d. the regulatory gene lacks a promoter.
   e. the control mechanism is positive.

9. The promoter is
   a. the region that binds the repressor.
   b. the region that binds RNA polymerase.

c. the gene that codes for the repressor.

d. a structural gene.

e. an operon.

10. The CRP–cAMP system

    a. produces many catabolites.

    b. requires ribosomes.

    c. operates by an operator–repressor mechanism.

    d. is an example of positive control of transcription.

    e. relies on operators.

Applying Concepts

1. Viruses sometimes carry DNA from one cell to another by transduction. Sometimes a segment of bacterial DNA is incorporated into a phage protein coat without any phage DNA. These particles can infect a new host. Would the new host become lysogenic if the phage originally came from a lysogenic host? Why or why not?

2. Genetic markers A^+, B^+, and C^+ are in one strain of *E. coli* that is used for transduction into a second strain: $A^-B^-C^-$. After transduction, recombinants are selected for in the recipient bacteria. The frequencies of single recombinants, which are positive (+) for only one of A, B, or C, are all 1 percent. The frequencies of double transductants are as follows: $A^+ B^+ = 10^{-6}$ percent; $A^+C^+ = 10^{-6}$ percent; $B^+C^+ = 10^{-3}$ percent. Explain these data.

3. You are provided with two strains of *E. coli*. One, an Hfr strain that is sensitive to streptomycin, carries the alleles A^+, B^+, and C^+. The other is an F^- strain that is resistant to streptomycin and carries the alleles A^-, B^-, and C^-. You mix the two cultures. After 20, 30, and 40 minutes you take samples of the mixed culture and swirl them vigorously in a blender. Next you add streptomycin to the swirled cultures. You examine surviving bacteria to determine their phenotypes. Some of the bacteria from the 20-minute sample are B^+; in the 30-minute sample there are both B^+ and C^+ bacteria; but A^+ bacteria are found only in the 40-minute sample. What can you say about the arrangement of the A, B, and C loci on the *E. coli* chromosome? Explain your answer fully.

4. You have isolated three strains of *E. coli*, which you name 1, 2, and 3. You attempt to cross these strains, and you find that recombinant progeny are obtained when 1 and 2 are mixed or when 2 and 3 are mixed, but not when 1 and 3 are mixed. By diluting a suspension of strain 2 and plating it out on solid medium, you isolate some separate clones. You find that almost all these clones can conjugate with strain 1 to produce recombinant offspring. One of the clones derived from strain 2, however, lacks the ability to conjugate with strain 1. Characterize strains 1, 2, and 3 and the nonconjugating clone of strain 2 in terms of the fertility (F) plasmid.

5. In the lactose (*lac*) operon of *E. coli*, repressor molecules are encoded by the regulatory gene. The repressor molecules are made in very small quantities and at a constant rate per cell. Would you surmise that the promoter for these repressor molecules is efficient or inefficient? Is synthesis of the repressor constitutive, or is it under environmental control?

6. A key characteristic of a repressible enzyme system is that the repressor molecule must react with a corepressor (typically, the end product of a pathway) before it can combine with the operator of an operon to shut the operon off. How is this different from an inducible enzyme system?

Readings

Griffiths, A. J. F., J. H. Miller, D. T. Suzuki, R. C. Lewontin and W. M. Gelbart. 1996. *An Introduction to Genetic Analysis*, 6th Edition. W. H. Freeman, New York. A revision of one of the best genetics textbooks, with excellent coverage of prokaryotic genetics.

Judson, H. F. 1996. *The Eighth Day of Creation: Makers of the Revolution in Biology*, Expanded Edition. CSHL Press, Plainview, NY. A constantly fascinating history of molecular biology, with much attention to the regulation of gene expression. For a lay audience.

Lodish, H., D. Baltimore, A. Berk, S. L. Zipursky, P. Matsudaira and J. Darnell. 1995. *Molecular Cell Biology*, 3rd Edition. Scientific American Books, New York. A comprehensive yet comprehensible summary of gene control.

Matthews, K. S. 1996. "The Whole Lactose Repressor." *Science*, vol. 271, pages 1245–1246. This nontechnical summary and the subsequent, technical article by Lewis (pages 1247–1254) illustrate the current sense of the structure and allosteric changes of the *lac* repressor.

Nomura, M. 1984. "The Control of Ribosome Synthesis." *Scientific American*, February. A discussion of how ribosomes are assembled and the roles of operons in regulating ribosome production in bacteria.

Ptashne, M. 1992. *A Genetic Switch*, 2nd Edition. Cell Press and Blackwell Scientific Publications, Cambridge, MA. An in-depth look at the life cycles of a single bacteriophage type, lambda, that reveals elegant control mechanisms.

Watson, J. D., M. Gilman, J. Wikowski and M. Zoller. 1992. *Recombinant DNA*, 2nd Edition. Scientific American Books, New York. Superbly written descriptions of prokaryotic and viral molecular genetics.

Chapter 14

The Eukaryotic Genome and Its Expression

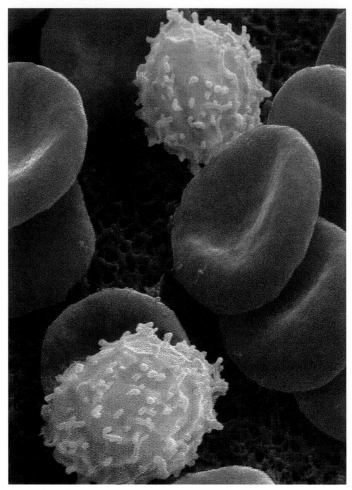

Two Cells, Two Different Protein Products
The red blood cells—erythrocytes—contain abundant hemoglobin, while the white blood cells synthesize proteins of the immune system.

When Tom was diagnosed with acute leukemia—cancer of the blood cells—his initial treatment included chemotherapy, in which combinations of powerful antimitotic drugs were administered to kill the dividing cancer cells that were rapidly spreading through his body. But the dosages his physicians prescribed were not up to the task, and the cells continued to spread. Higher dosages of these drugs would be lethal, as they would kill not only the cancer cells, but healthy and essential dividing cells—such as the stem cells in bone marrow that divide by the hundreds of millions to form blood cells. Without these stem cells, Tom's body would no longer produce red blood cells with their vital oxygen-carrying protein hemoglobin, nor would he be able to produce white blood cells, which make the proteins of the immune system that can eliminate infectious diseases as well as some tumors.

Tom's doctors tried a new approach. They extracted some of his bone marrow and removed the cancer cells from it, then stored the marrow in a refrigerator. Then they gave Tom extremely high doses of the chemotherapeutic drugs, which killed the cancer cells. Finally, the stored bone marrow was replaced into Tom's body. The stem cells began to divide, and after a few weeks they were differentiating into adequate populations of normal red and white blood cells. Tom's leukemia had disappeared.

The success of Tom's bone marrow transplant depended on many things, but the principle behind it is based on the specificity of gene expression during cell differentiation. What are the genetic mechanisms that ensure that healthy

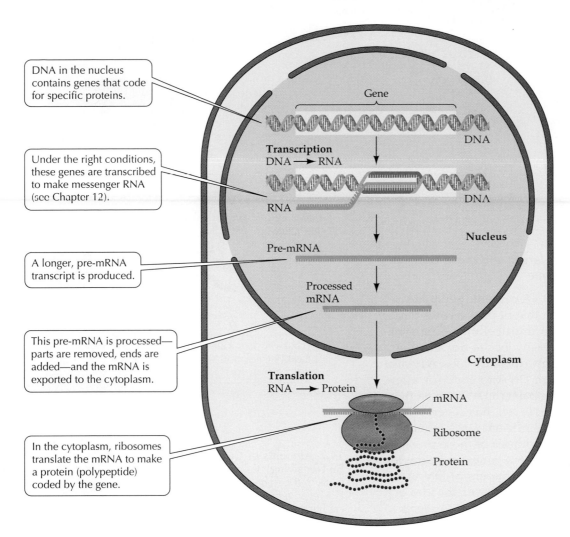

DNA in the nucleus contains genes that code for specific proteins.

Under the right conditions, these genes are transcribed to make messenger RNA (see Chapter 12).

A longer, pre-mRNA transcript is produced.

This pre-mRNA is processed—parts are removed, ends are added—and the mRNA is exported to the cytoplasm.

In the cytoplasm, ribosomes translate the mRNA to make a protein (polypeptide) coded by the gene.

Gene

DNA

Transcription
DNA → RNA

RNA

DNA

Nucleus

Pre-mRNA

Processed mRNA

Cytoplasm

Translation
RNA → Protein

mRNA

Ribosome

Protein

14.1 Eukaryotic mRNA Is Processed in the Nucleus and Exported to the Cytoplasm Compare this "road map" to Figure 12.3.

red blood cells will have hemoglobin, and that white blood cells are able to create the vital antibody proteins of the immune system? What features of the DNA sequences of eukaryotic genes determine these mechanisms, and how do they differ from the genes that code for proteins in prokaryotes?

In this chapter, you will see that although the eukaryotic genetic material, DNA, is the same as that of prokaryotes, its *organization* in the chromosomes is often very different. Eukaryotic chromosomes contain vast amounts of repetitive sequences. Some of these sequences have vital roles, such as the genes for ribosomal and transfer RNAs, and others have unknown functions. Of special interest are the extreme ends of chromosomal DNA, the telomeres, which play a role in DNA replication and chromosome integrity.

The transcription and later tailoring of RNAs are more complicated processes in eukaryotes than in prokaryotes. A large pre-mRNA is transcribed from each gene (Figure 14.1), and this molecule is extensively modified; even internal stretches of nucleotides are removed before the mRNA is exported from the nucleus to the cytoplasm.

Finally, we consider the fascinating molecular machinery that allows the precise regulation of gene expression needed for a eukaryote to develop and function. In contrast to the operons of prokaryotes, related eukaryotic genes are often scattered throughout the genome. Many elegant mechanisms control the selective transcription and translation of eukaryotic genes.

The Eukaryotic Genome

As biologists unraveled the intricacies of gene structure and expression in prokaryotes, they tried to generalize their findings by saying, "What's true for *E. coli* is also true for elephants." Although much of prokaryotic biochemistry does apply to eukaryotes, the old saying has its limitations (Table 14.1). For example, eukaryotes have much more DNA in each cell than prokaryotes have.

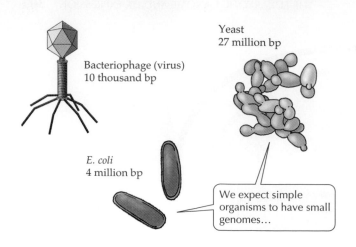

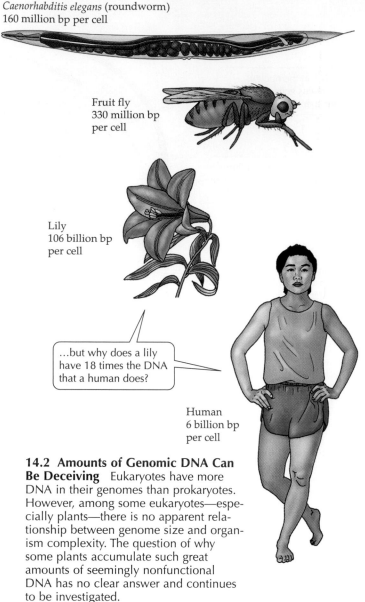

Bacteriophage (virus)
10 thousand bp

E. coli
4 million bp

Yeast
27 million bp

We expect simple organisms to have small genomes…

Caenorhabditis elegans (roundworm)
160 million bp per cell

Fruit fly
330 million bp per cell

Lily
106 billion bp per cell

…but why does a lily have 18 times the DNA that a human does?

Human
6 billion bp per cell

The eukaryotic genome is larger and more complex than the prokaryotic genome

The fact that the haploid DNA content (genome size) of eukaryotes is larger than that of prokaryotes might be expected, given that in multicellular organisms there are many cell types, many jobs to do, and many proteins—coded for by DNA—to do the jobs. A typical virus contains only enough DNA to code for a few proteins—about 10,000 base pairs (bp). The most thoroughly studied prokaryote, *E. coli*, has several thousand different proteins and sufficient DNA (about 4.7 million bp) to regulate the synthesis of those proteins. Humans have considerably more genes and complex controls; nearly 6 billion bp (2 meters of DNA) are crammed into each human cell. However, the idea of a more complex organism needing more DNA seems to break down with some plants. For example, the lily (which produces beautiful flowers each spring but produces fewer proteins than a human does) has 18 times more DNA than humans have (Figure 14.2).

Unlike prokaryotic DNA, most eukaryotic DNA does not code for proteins. Instead, interspersed

14.2 Amounts of Genomic DNA Can Be Deceiving Eukaryotes have more DNA in their genomes than prokaryotes. However, among some eukaryotes—especially plants—there is no apparent relationship between genome size and organism complexity. The question of why some plants accumulate such great amounts of seemingly nonfunctional DNA has no clear answer and continues to be investigated.

TABLE 14.1 A Comparison of Prokaryotic and Eukaryotic Genes and Genomes

CHARACTERISTIC	PROKARYOTES	EUKARYOTES
Genome size (base pairs)	10^4–10^7	10^8–10^{11}
Repeated sequences	Few	Many
Noncoding DNA within coding sequences	Rare	Common
Transcription and translation separated in cell	No	Yes
DNA segregated within a nucleus	No	Yes
DNA bound to proteins	Some	Extensive
Promoter	Yes	Yes
Enhancer/silencer	Rare	Common
Capping and tailing of mRNA	No	Yes
RNA splicing required	Rare	Common
Number of chromosomes in genome	One	Many

throughout the eukaryotic genome are various kinds of repeated DNA sequences. Even the coding regions of genes contain sequences that do not end up in mature mRNA. Some noncoding DNA maintains chromosomal integrity at the ends (telomeres), and some helps control gene expression. But the presence of much of this noncoding DNA remains an enigma.

In contrast to the single main DNA molecule of most prokaryotes, the eukaryotic genome is partitioned into several separate molecules of DNA, or chromosomes. In humans, each chromosome contains a double helix with 20 million to 100 million bp. This separation of the DNA into different volumes of the genomic encyclopedia requires that each chromosome have at a minimum three defining DNA sequences: recognition sequences for the DNA replication machinery, a *centromere region* that holds the replicated sequences together before mitosis, and a *telomeric sequence* at the ends of the chromosome.

In eukaryotes, the nuclear envelope separates DNA and its transcription (inside the nucleus) from the cytoplasmic site where mRNA is translated into protein. This separation allows for many points of control in the synthesis, processing, and export of mRNA to the cytoplasm. The organization of the nuclear eukaryotic genome is fundamentally about regulation: Great complexity requires a great deal of regulation, and this fact is evident in the many processes associated with the expression of the eukaryotic genome.

In addition, most eukaryotic DNA is not even fully exposed to the nuclear environment. Instead, it is extensively packaged by proteins into nucleosomes, fibers, and ultimately chromosomes (Figure 14.3). This extensive compaction is a means for restricting access of the RNA synthesis machinery to the DNA, as well as a way to segregate replicated DNAs during mitosis and meiosis.

Like the genes of prokaryotes that code for proteins, eukaryotic genes have noncoding flanking sequences that control their transcription. These include the *promoter* region, where RNA polymerase ultimately binds to begin transcription. Of equal importance in eukaryotes, but rare in prokaryotes, is a second set of controlling DNA sequences, the *enhancers* and *silencers*. Enhancers and silencers are often located quite distant from the promoter and act by binding proteins that then stimulate or inhibit transcription.

Within a protein-coding gene (that is, between the DNA bases coding for start and stop codons) are base sequences that interrupt the coding region and do not code for amino acids. These noncoding DNA sequences within protein-coding genes present a special problem: How do cells ensure that these noncoding regions do not end up in the mature mRNA that exits the nucleus?

The answer lies in an elaborate cutting and splicing mechanism within the nucleus that, after transcrip-

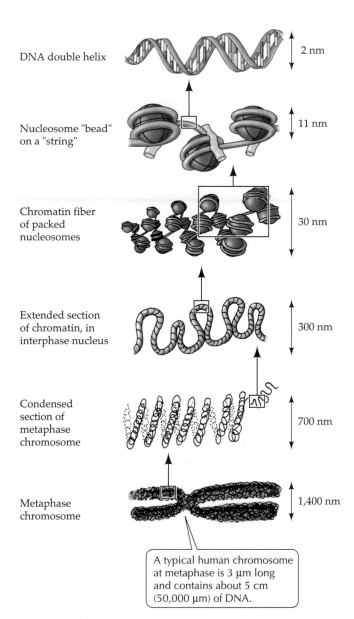

DNA double helix — 2 nm

Nucleosome "bead" on a "string" — 11 nm

Chromatin fiber of packed nucleosomes — 30 nm

Extended section of chromatin, in interphase nucleus — 300 nm

Condensed section of metaphase chromosome — 700 nm

Metaphase chromosome — 1,400 nm

A typical human chromosome at metaphase is 3 μm long and contains about 5 cm (50,000 μm) of DNA.

14.3 DNA is Packed into Chromosomes Proteins are essential in winding and folding DNA into a very compact structure.

tion, modifies an initial transcript called **pre-mRNA**. The noncoding sequences are cut out of the pre-mRNA and the coding regions are spliced together. Thus, in contrast to the "what is transcribed is what is translated" scheme of most prokaryotic genes, the mature mRNA at the eukaryotic ribosome is a modified and much smaller molecule than the one initially made in the nucleus.

Hybridization is a tool for genome and gene analysis

Investigations of eukaryotic genome and gene sequence organization (referred to as gene structure) have been greatly aided by **nucleic acid hybridization**.

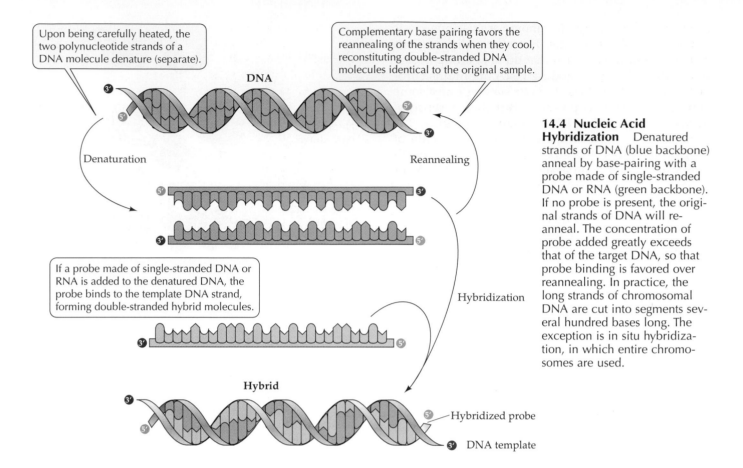

Upon being carefully heated, the two polynucleotide strands of a DNA molecule denature (separate).

Complementary base pairing favors the reannealing of the strands when they cool, reconstituting double-stranded DNA molecules identical to the original sample.

DNA

Denaturation

Reannealing

If a probe made of single-stranded DNA or RNA is added to the denatured DNA, the probe binds to the template DNA strand, forming double-stranded hybrid molecules.

Hybridization

Hybrid

Hybridized probe

DNA template

14.4 Nucleic Acid Hybridization Denatured strands of DNA (blue backbone) anneal by base-pairing with a probe made of single-stranded DNA or RNA (green backbone). If no probe is present, the original strands of DNA will reanneal. The concentration of probe added greatly exceeds that of the target DNA, so that probe binding is favored over reannealing. In practice, the long strands of chromosomal DNA are cut into segments several hundred bases long. The exception is in situ hybridization, in which entire chromosomes are used.

This technique depends on the association, through complementary base pairing, of single-stranded nucleic acids (Figure 14.4). It begins with separation of the strands of DNA. If DNA is carefully heated, the hydrogen bonds between base pairs break and the two strands of each double helix separate. This process is termed *DNA denaturation*. If this single-stranded mixture is allowed to cool, the two complementary strands eventually line up beside one another and re-form hydrogen bonds by the AT and GC base-pairing rules. This process, called *DNA reannealing*, has been very useful in research on eukaryotic genomes.

Suppose that instead of being allowed to reanneal, the single strands of denatured DNA are mixed with a high concentration of a second single-stranded nucleic acid, with a sequence complementary to one of the immobilized DNA strands. This second molecule is given time to find and anneal to its target sequence. The second molecule is called a **probe** because it is "probing" the target DNA, "looking" for its complementary sequence.

A probe may be DNA or RNA. If the probe is labeled with nucleotides that are radioactive or fluorescent, the formation of a base-paired hybrid between the probe and target can be identified by the experimenter. Hybridization has been invaluable in the ex-amination of eukaryotic gene structures, as well as in mapping genes (see Chapters 16 and 17). It has also been essential to revealing several types of repetitive DNA in the eukaryotic genome.

Highly repetitive sequences contribute to chromosome structure and spindle attachment

If the DNA of a prokaryote is broken into fragments of about 1,000 bp and these fragments are denatured and reannealed in solution, each single-stranded region takes a long time to find its complementary partner. But when the same experiment is done with eukaryotic DNA, although some parts of the genome do reanneal slowly, others reanneal much faster.

Why do these rates differ in different sequences of the eukaryotic genome? The answer is that in eukaryotic DNA there are multiple copies of some, but not all, DNA sequences. If a particular single strand of DNA has, say, a few hundred complementary partners with which it can anneal, it will be able to find a partner much more rapidly than one for which only a single acceptable partner exists.

By measuring rates of reannealing, researchers discovered that there are three classes of eukaryotic DNA: single-copy sequences ("slow" reannealers), highly repetitive sequences ("fast" reannealers), and moderately repetitive sequences ("moderate" reannealers). The class that reanneals the slowest consists of single-copy sequences—genes that, like most prokaryotic genes, have only one copy in each genome. Single-copy sequences code for most enzymes and structural proteins in eukaryotes. Some noncoding single-copy sequences form long spacers between genes on a chromosome.

The class of DNA that reanneals the fastest consists of **highly repetitive DNA** sequences. These sequences typically are simple, 5 to 50 bp long, and repeated up to millions of times. Typically, there are several such blocks of repeated sequences in the genome. For example, in the guinea pig the sequence CCCTAA* is repeated at least 10,000 times at the centromere region of each chromosome. At the centromere, the highly repetitive DNA may contribute to the functional integrity of the chromosomes and the attachment of spindle fibers.

In most cases, however, the location of these simple sequences shows no apparent logic, and their role is unclear. What is clear is that they are usually not transcribed, since no RNAs that contain these tandemly repeated sequences are found. Humans have at least ten simple, highly repetitive sequence types.

Telomeres are repetitive sequences at the ends of chromosomes

There are several types of **moderately repetitive DNA** sequences in the eukaryotic genome. One type is important in maintaining the ends of chromosomes when DNA is replicated. Recall from Chapter 11 that replication proceeds differently on the two strands of a DNA molecule. Both new strands form in the 5'-to-3' direction, but one strand (the leading strand) grows continuously from one end to the other, while the other (the lagging strand) grows as a series of short Okazaki fragments (see Figure 11.18).

With the circular prokaryotic chromosome, as both DNA strands grow around the chromosome, production of a complete series of Okazaki fragments is not a problem. There is always some DNA at the 5' end of a primer, ready to replace the primer after it is removed. But things are more complicated in the eukaryotic chromosome. The leading strand can grow without interruption to the very end. How does the last Okazaki fragment for the end of the lagging strand form? Replication must begin with an RNA primer at the 5' end of the forming strand, but there is nothing beyond the primer in the 5' direction to replace the RNA. So the new chromosome formed after DNA replication lacks a bit of double-stranded DNA at each end. The ends of the chromosome have been clipped off.

In many eukaryotes, there are moderately repetitive sequences at the ends of chromosomes called **telomeres**; in humans the sequence is TTAGGG and it is repeated about 2,500 times (Figure 14.5a). The need for these repeats can be shown experimentally by putting fragments of DNA from a human into a yeast cell. The fragments will be stably maintained in their new home only if they have the telomeric repeats at their ends. Otherwise, they rapidly break down.

When human cells are removed from the body and put in a nutritious medium in the laboratory, they will grow and divide. But each chromosome can lose about 50 to 200 bp of telomeric DNA after each round of DNA replication and cell division. This shortening compromises the stability of the chromosomes, and after 20 to 30 cell divisions they are unable to take part in cell division. The same thing happens in the body, and explains in part why cells do not last the entire lifetime of the organism.

Yet constantly dividing cells, such as bone marrow cells and germ line cells, manage to maintain their moderately repetitive telomeric DNA. An enzyme, appropriately called telomerase, prevents the loss of this DNA by catalyzing the addition of any lost telomeric sequences (Figure 14.5b). Telomerase is made up not only of proteins but also of an RNA sequence that acts as the template for the telomeric sequence addition.

Considerable interest has been generated by the finding that telomerase is expressed in more than 90 percent of human cancers. Telomerase may be an important factor in the ability of cancer cells to divide continuously. Since most normal cells do not have this activity, telomerase is an attractive target for drugs designed to attack tumors specifically.

Some moderately repetitive sequences are transcribed

Some moderately repetitive DNA sequences code for tRNA and rRNA, which are used in protein synthesis. These RNAs are constantly being made, but even the maximum rate of transcription of single-copy sequences would be inadequate to supply the large amounts of these molecules needed by most cells; hence there are multiple copies of the DNA sequences coding for them. Since these moderately repetitive sequences are transcribed into RNA, they are properly termed "genes," and we can speak of rRNA genes and tRNA genes.

As an example, in mammals there are four sizes of rRNA in the ribosome—the 18S, 5.8S, 28S, and 5S rRNAs.[†] The 18S, 5.8S, and 28S rRNAs are transcribed

---

*When a DNA sequence such as CCCTAA is written, the complementary bases are assumed.

[†]The term "S" refers to the movement of a molecule in a centrifuge: In general, larger molecules have a higher S value.

14.5 Telomeres and Telomerase (a) The loss of repeat sequences from the telomere leads to cell death. (b) In cells that divide continuously, the enzyme telomerase prevents the loss of repeat sequences.

from a repeated unit of DNA that, as a single precursor, is twice the size of the three ultimate products (Figure 14.6). Several posttranscriptional steps cut this precursor into its final three rRNAs and discard the nonuseful, or "spacer," RNA. The DNA coding for these RNAs is moderately repetitive in humans: A total of 280 copies of the unit are located in clusters on five different chromosomes.

Other moderately repetitive sequences in mammals are not clustered, but instead are scattered throughout the genome. These DNAs usually are not transcribed and usually are short, about 300 bp long. In humans, half of these DNAs are of a single type, called the *Alu* family (because they have a sequence in them that is recognized by a nuclease enzyme, Alu I). There are 300,000 copies of the *Alu* family in the genome, and they may act as multiple origins for DNA replication.

Surprising those who long believed in genetic stability, some moderately repetitive DNA sequences move about the genome.

(a)

3' GGGATTGGGATTGGGATTGGGATTGGGATTGGGATT 5'
5' CCCTAACCCTAACCCTAACCCTAACCCTAA 3'

> Human telomeres have about 2,500 repeats of this sequence.

3' GGGATTGGGATTGGGATTGGGATTGGGATT 5'
5' CCCTAACCCTAACCCTAACCCTAA 3'

> Normally, DNA replicates from a primer in a 5'-to-3' direction. Because there is no primer at the extreme 5' end of a chromosome, there is a gap in replication, leading to shortening of the chromosome after each round of replication. Chromosome shortening leads in turn to cell death.

(b)

3' GGGATTGGGATTGGGATTGGGATTGGGATTGGGATT 5'
5' CCCTAACCCTAACCCTAACCCTAA 3'

CCCUAA
RNA
Telomerase

> An RNA in telomerase acts as a template for DNA. This adds the telomeric sequence to the end of the chromosome.

3' GGGATTGGGATTGGGATTGGGATTGGGATTGGGATT 5'
5' CCCTAACCCTAACCCTAACCCTAACCCTAA 3'

> The original length of the chromosomal DNA has been restored. Note the gap where the primer for DNA replication has been removed.

Transposable elements move about the genome

Most of the remaining scattered moderately repetitive DNA is not stably integrated into the genome. Instead, these DNA sequences can move from place to place in the genome and are called **transposable DNA**. There are four main types of transposable elements, or transposons, in eukaryotes (Table 14.2).

SINEs (short *in*terspersed *e*lements) are up to 300 bp long and are transcribed but not translated. LINEs (*l*ong *in*terspersed *e*lements) are up to 7,000 bp long, and some are transcribed and translated into proteins. Both of these elements are present in more than 100,000 copies and move about the genome in a distinctive way: They make an RNA copy, which acts as a template for the new DNA, which then inserts at a new location in the genome.

This mechanism is also employed by the third type of movable DNA, the retrotransposons, which are rare in mammals but are more common in yeasts and animals other than mammals. The genetic organization of viral retrotransposons resembles that of retroviruses such as HIV, but these segments lack the genes for protein coats and thus cannot produce viruses.

> The rRNA coding unit is repeated many times (280 in humans).

DNA

13,000 bp
Transcribed region

30,000 bp
Nontranscribed spacer region

DNA

18S 5.8S 28S

RNA primary transcript

18S 5.8S 28S

> Processing steps splice out the spacers.

18S 5.8S 28S

Three rRNAs are the result.

14.6 A Moderately Repetitive Sequence Codes for rRNAs This rRNA gene, along with its nontranscribed spacer, is repeated 280 times in the human genome.

TABLE 14.2 Transposable Elements in Eukaryotes

TYPE OF ELEMENT	DESCRIPTION
DNA transposons	DNA segments that move or are copied at new sites
Retrotransposons	DNA segments that are copied by means of an RNA intermediate
Viral retrotransposons	Retrotransposons that possess long terminal repeats
Nonviral retrotransposons	Retrotransposons that lack long terminal repeats
LINEs	Long interspersed elements
SINEs	Short interspersed elements

The fourth type of moderately repetitive transposable element is the DNA transposon. Like its counterpart in prokaryotes, the eukaryotic DNA transposon does not use an RNA intermediate but actually moves to a new spot in the genome without replicating (Figure 14.7).

What role do these repeated sequences play in the cell? There are few answers. The best answer so far seems to be that transposons are cellular parasites that simply replicate themselves. But these replications can lead to the insertion of a transposon at a new location, and this event has important consequences. For example, insertion of a transposon into the coding region of a gene causes a mutation because of the addition of new base pairs.

If this process takes place in the germ line, a gamete with a new mutation results. If the process takes place in a somatic cell, cancer may result. If a transposon replicates not just itself but also an adjacent gene, the result may be a gene duplication. Clearly, transposition stirs the genetic pot in the eukaryotic genome and thus contributes to genetic variability.

In Chapter 4, we described the endosymbiosis theory of the origin of chloroplasts and mitochondria, which proposes that these organelles are the descendants of once free-living prokaryotes. Transposable elements may have played a role in this process. In living forms, although the organelles have some DNA, the nucleus has most of the genes that encode the organelle proteins. If the organelles were once independent, they must originally have had all of these genes. How did the genes move to the nucleus? The answer may lie in DNA transposition. Genes once in the organelle may have moved to the nucleus by well-known molecular events that still occur today. The current DNA in the organelles may be the remnants of more complete genomes.

The Structures of Protein-Coding Genes

Like their prokaryotic counterparts, protein-coding genes in eukaryotes are generally single-copy DNA sequences. But there are two distinctive characteristics of the eukaryotic genes that we will examine. First, they contain noncoding internal sequences, and second, they form gene families with structurally and functionally related cousins in the genome.

Protein-coding genes contain noncoding internal and flanking sequences

Eukaryotic proteins are usually encoded by single-copy genes. Preceding the beginning of the coding re-

14.7 Transposons and Transposition
At the end of each transposable element in DNA is an inverted repeat sequence that helps in the transposition process. If the transposon inserts within a gene, it disrupts the coding sequence and an abnormal protein can result.

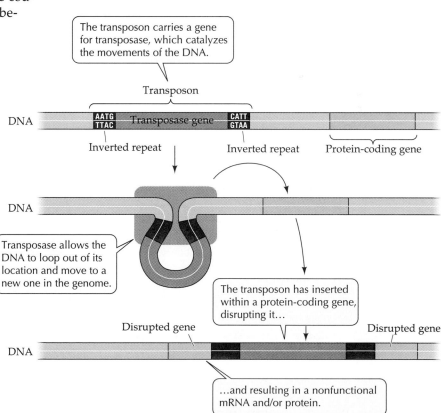

gion is a **promoter**, where RNA polymerase begins the transcription process. Unlike the prokaryotic enzyme, the eukaryotic RNA polymerase does not itself recognize the promoter sequence, as we'll see later. At the other end of the gene, after the coding region, is a DNA sequence appropriately called the **terminator**, which RNA polymerase recognizes as the end point for transcription (Figure 14.8).

Both the promoter and terminator sequences are parts of the gene that do not become transcribed into RNA. Base sequences called **introns** (intervening sequences within the gene) also do not end up in mRNA. Although the introns are transcribed into RNA, they are *spliced* out within the nucleus. The locations of these noncoding base pairs can be determined by comparing the base sequences of a gene (DNA) with those of its final mRNA.

Remarkably, these extra base pairs are within the coding region of the gene. One or more introns are present in most eukaryotic genes (see Figure 14.8). These intron sequences appear in the primary tran-

script of RNA, the pre-mRNA within the nucleus, but by the time the mature mRNA exits the organelle, the introns have been removed (the pre-mRNA has been cut) and the dangling ends of the mRNA have been reconnected (spliced). The parts of the gene present in mRNA are called **exons**.

Although direct sequencing of the DNA that codes for an mRNA is the easiest way to map the locations of introns within a gene, nucleic acid hybridization is the method that originally revealed the existence of introns in protein-coding genes (Figure 14.9). Biologists denatured the DNA coding for one of the globin proteins that make up hemoglobin and added globin mRNA. From the resulting hybridization they expected to obtain a linear matchup of the mature mRNA to the globin-coding DNA.

They got their wish, in part: There were indeed stretches of RNA–DNA hybridization. But there were also some double-stranded, looped structures. These loops were the introns, stretches of DNA that did not have complementary bases on the mRNA. Later evidence showed that hybridization to the gene using the initial RNA transcript was complete and the introns were indeed transcribed. Somewhere on the path from transcript to mRNA the introns had been removed and the exons had been spliced together. We will examine this splicing process later in the chapter.

To summarize, most (but not all) vertebrate genes and many other eukaryotic genes, and even a few prokaryotic ones, contain introns. Introns interrupt but do not scramble the DNA sequence that codes for a polypeptide chain. The base sequence of the exons, taken in order, is exactly complementary to that of the mature mRNA product. The introns, therefore, separate a gene's protein-coding region into distinct parts. In some cases, these parts code for different functional regions, or domains, of the protein. For example, the

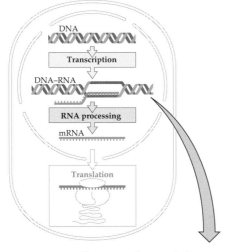

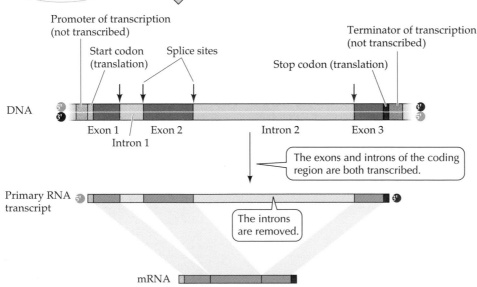

Promoter of transcription (not transcribed)

Start codon (translation)

Splice sites

Terminator of transcription (not transcribed)

Stop codon (translation)

DNA

Exon 1 Exon 2 Intron 2 Exon 3

Intron 1

The exons and introns of the coding region are both transcribed.

The introns are removed.

Primary RNA transcript

mRNA

14.8 The Structure and Transcription of a Member of the β-Globin Gene Family
This entire gene is about 1,600 bp long. The coding region (blue) has 441 base pairs (triplets coding for 146 amino acids plus a triplet stop codon). Noncoding sequences of DNA—introns—are initially transcribed between codons 30 and 31 (130 bp long) and 104 and 105 (850 bp long) but are spliced out of the final gene product.

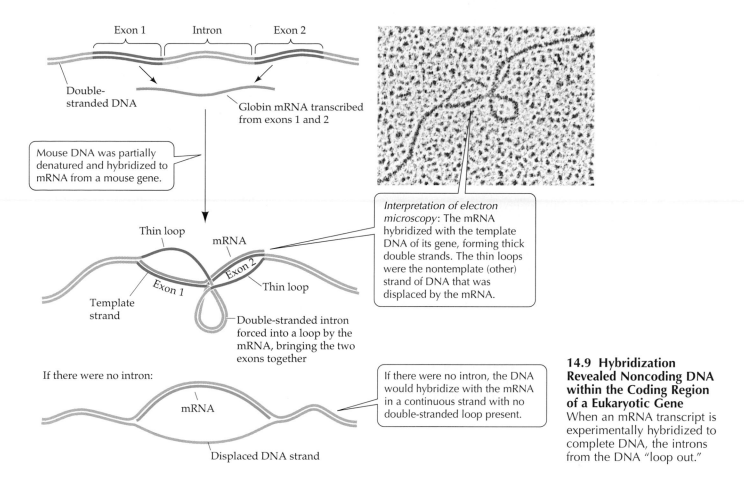

Exon 1 Intron Exon 2

Double-stranded DNA

Globin mRNA transcribed from exons 1 and 2

Mouse DNA was partially denatured and hybridized to mRNA from a mouse gene.

Thin loop

mRNA

Exon 2

Exon 1

Thin loop

Template strand

Double-stranded intron forced into a loop by the mRNA, bringing the two exons together

Interpretation of electron microscopy: The mRNA hybridized with the template DNA of its gene, forming thick double strands. The thin loops were the nontemplate (other) strand of DNA that was displaced by the mRNA.

If there were no intron:

mRNA

If there were no intron, the DNA would hybridize with the mRNA in a continuous strand with no double-stranded loop present.

Displaced DNA strand

14.9 Hybridization Revealed Noncoding DNA within the Coding Region of a Eukaryotic Gene
When an mRNA transcript is experimentally hybridized to complete DNA, the introns from the DNA "loop out."

globin proteins that make up hemoglobin have two domains: one for binding to heme and another for binding to the other globin chains. These two domains are coded for by different exons in the globin gene.

Many eukaryotic genes are members of gene families

About half of all eukaryotic protein-coding genes are present in only one copy in the haploid genome. The rest have multiple copies. Often inexact, nonfunctional copies of a particular gene, called *pseudogenes*, are located nearby. The duplicates may have arisen by an abnormal event in chromosomal crossing over during meiosis or by retrotransposition. A set of duplicated or related genes is called a **gene family**. Some families, such as the β-globins that are part of hemoglobin, contain only a few members; other families, such as the immunoglobulins that make up antibodies, have hundreds of members.

Like the members of any family, the DNA sequences in a gene family are usually different from each other to a certain extent. As long as one member retains the original DNA sequence and thus codes for the proper protein, the other members can mutate slightly, extensively, or not at all. The availability of extra genes is important for "experiments" in evolu-

tion: If the mutated gene is useful, it will be selected for in succeeding generations. If the gene is a total loss (a pseudogene), the functional copy is still there to save the day.

A good example of gene families found in vertebrates is the gene family for the globins. As mentioned earlier, these are the proteins found in hemoglobin and also in myoglobin (an oxygen-binding protein present in muscle). The globin genes probably all arose from a single common ancestor gene long ago. In humans, there are three functional members of the alpha (α) globin family and five in the beta (β) globin family (Figure 14.10). Each hemoglobin contains the heme pigment held inside four globin polypeptides, two identical α-globins and two identical β-globins.

During human development, different members of the globin gene family are expressed at different times and in different tissues (Figure 14.11). This differential gene expression has great physiological significance. For example, the form of hemoglobin found in the fetus ($\alpha_2\gamma_2$) binds O_2 more tightly than adult hemoglobin ($\alpha_2\beta_2$) does. (Note that both γ and β are members of the β-globin family.) This specialized form of hemoglobin ensures that in the placenta, where maternal and fetal bloods come near each other, O_2 will be transferred from the mother to the developing child's

14.10 Gene Familes The human α- and β-globin gene families are located on different chromosomes. Each family is organized into a cluster of genes separated by "spacer" DNA. The nonfunctional pseudogenes are indicated by the Greek letter psi (ψ).

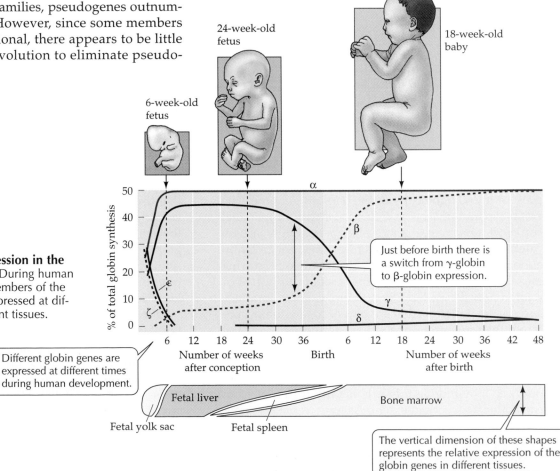

Spacer DNA is noncoding DNA between gene family members.

β-Globin gene cluster

ε   Gγ   Aγ   ψβ1   δ   β

Pseudogenes

α-Globin gene cluster

ζ2   ψζ1   ψα1   α2   α1

Pseudogenes are family members that do not code for functional mRNAs or proteins.

circulation. Just before birth, the synthesis of fetal hemoglobin in the liver stops, and the adult form takes over in bone marrow cells. The precise developmental regulation of transcription of this gene family is controlled at many levels, which we will discuss later.

In addition to genes that encode proteins, the globin family includes nonfunctional genes called **pseudogenes**, designated with the Greek letter psi (ψ). These pseudogenes are the "black sheep" of any gene family: They are experiments in evolution that went wrong.

The DNA sequence of a pseudogene may not differ vastly from that of other family members. It may just lack a promoter, for example, and thus cannot be transcribed. Or it may lack the recognition sites for the removal of introns and thus will be transcribed into pre-mRNA but not correctly processed into a useful mRNA. In some gene families, pseudogenes outnumber functional genes. However, since some members of the family are functional, there appears to be little selective pressure in evolution to eliminate pseudogenes.

RNA Processing

Unlike the situation in prokaryotic genes, the primary RNA transcript (pre-mRNA) of a eukaryotic gene is not the same as the mature mRNA. To produce the mRNA, the primary transcript is processed by the ad-

14.11 Differential Expression in the Globin Gene Families During human development, different members of the globin gene family are expressed at different times and in different tissues.

24-week-old fetus

18-week-old baby

6-week-old fetus

α

β

Just before birth there is a switch from γ-globin to β-globin expression.

ε

ζ

γ

δ

% of total globin synthesis

50
40
30
20
10
0

6   12   18   24   30   36   6   12   18   24   30   36   42   48

Number of weeks after conception   Birth   Number of weeks after birth

Different globin genes are expressed at different times during human development.

Fetal liver   Bone marrow

Fetal yolk sac   Fetal spleen

The vertical dimension of these shapes represents the relative expression of the globin genes in different tissues.

14.12 Processing the Ends of Eukaryotic mRNA The modifications at both ends— "caps" and "tails"— are important for mRNA function.

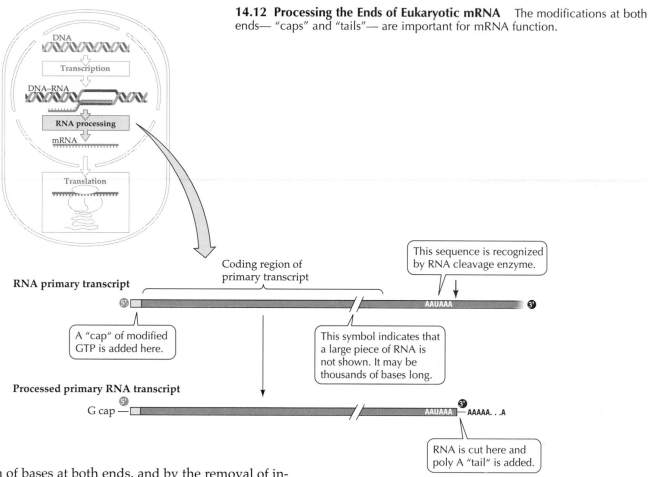

RNA primary transcript

Coding region of primary transcript

This sequence is recognized by RNA cleavage enzyme.

AAUAAA

A "cap" of modified GTP is added here.

This symbol indicates that a large piece of RNA is not shown. It may be thousands of bases long.

Processed primary RNA transcript

G cap

AAUAAA — AAAAA...A

RNA is cut here and poly A "tail" is added.

dition of bases at both ends, and by the removal of introns and joining of exons.

The primary transcript of a protein-coding gene is modified at both ends

Two early steps in the processing of mRNA are the addition of a "cap" at the 5' end and the addition of a poly A "tail" at the 3' end (Figure 14.12). The cap is a chemically modified molecule of guanosine triphosphate (GTP). It apparently facilitates the binding of mRNA to the ribosome for translation and protects the mRNA from breaking down.

Following the last codon of most eukaryotic mRNAs, but not at the end of the pre-mRNA, is the sequence AAUAAA. This sequence acts as a signal for an enzyme to cut the pre-mRNA. Immediately after cleavage, another enzyme adds 100 to 300 residues of adenine (poly A) to the 3' end of the mRNA. This "tail" may assist in export of the mRNA from the nucleus.

Splicing removes introns from the primary transcript

The next step in the processing of eukaryotic mRNA within the nucleus is the removal of the introns. If these RNA regions were not removed, a nonfunctional mRNA producing an improper amino acid sequence in the protein would result. The process called RNA

splicing removes introns and splices the exons together (see Figure 14.8).

After the primary transcript is made, it is quickly bound to several **small nuclear ribonucleoprotein particles** (**snRNPs**, commonly pronounced "snurps"), which begin the splicing process. How do these particles recognize the mRNA? The answer lies in the base sequences at the junctions between exons and introns and in the sequences of the RNAs in the snRNPs.

At the boundaries between introns and exons there are **consensus sequences**—short stretches of DNA that appear, with little variation, in many different genes. The RNA in one of the snRNPs (called U1) has a stretch of bases complementary to the consensus 5' exon–intron boundary. Another snRNP (U2) binds near the 3' intron–exon boundary. In both cases, the two RNAs—snRNP and pre-mRNA—bind by base pairing (Figure 14.13). Still other snRNPs bind sequences within the intron itself, forming a large RNA–protein complex called the **spliceosome**. The spliceosome uses ATP energy to cut the RNA, releases the introns, and joins the exons to produce mature mRNA (Figure 14.14).

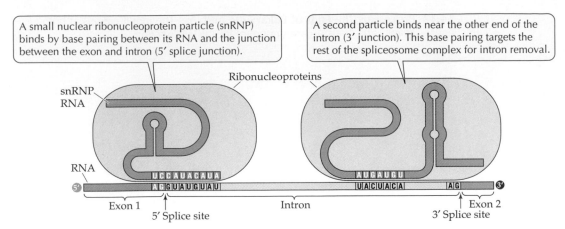

A small nuclear ribonucleoprotein particle (snRNP) binds by base pairing between its RNA and the junction between the exon and intron (5' splice junction).

A second particle binds near the other end of the intron (3' junction). This base pairing targets the rest of the spliceosome complex for intron removal.

Ribonucleoproteins

snRNP RNA

RNA

UCCAUACAUA

AUGAUGU

A G GUAUGUAU

UACUACA

A G

Exon 1

5' Splice site

Intron

3' Splice site

Exon 2

14.13 Recognition of Exon–Intron Junctions for Splicing
Two large RNA–protein complexes line up the splicing machinery.

Molecular studies of human diseases have been valuable tools in the investigation of consensus sequences and splicing machinery. Beta thalassemia is a human genetic disease inherited as an autosomal recessive trait in which people make an inadequate amount of the β-globin polypeptide that is part of hemoglobin. These people suffer from severe anemia because they have an inadequate supply of red blood cells. In some cases, the genetic mutation occurs at the consensus sequence at the boundary between an intron and an exon. Consequently, the mRNA cannot be spliced correctly, and nonfunctional globin mRNA is made.

This is an excellent example of the use of mutations in determining a cause-and-effect relationship in biology. In the logic of science, merely linking two phenomena (for example, consensus sequences and splicing) does not prove that one is necessary for the other. In an experiment, the scientist alters one phenomenon (for example, the base sequence at the consensus region) to see whether the other event (for example, splicing) occurs. In thalassemia, nature has done the experiment for us.

People who have certain connective-tissue diseases make antibodies against their own proteins (autoimmunity). In the rheumatic disease lupus, for example, antibodies are made against the U1 snRNP used in mRNA splicing. Addition of these antibodies to the spliceosome inhibits splicing. Although it is not clear whether the antibodies actually cause this inhibition in the patient, they have proved invaluable as a tool for extracting and studying the splicing complex.

After the processing events are completed in the nucleus, the mRNA exits the organelle, apparently through the nuclear pores (see Figure 4.8). A receptor at the nuclear pore recognizes the processed mRNA (or a protein bound to it). Unprocessed or incompletely processed primary transcripts remain in the nucleus.

Transcriptional Control

In a multicellular organism with specialized cells and tissues, every cell has every gene for that organism. For development to proceed normally and for each cell to acquire and maintain its proper function, certain specific proteins must be synthesized at just the right times and in just the right cells. Thus, the expression of eukaryotic genes is precisely regulated. The methods for this regulation are varied (Figure 14.15).

In some cases, gene regulation depends on changes in the DNA itself: Genes are rearranged on the chromosome or even selectively replicated to give more templates to transcribe. In other cases, changes in the proteins that bind to DNA can make genes more or less available for transcription. Both the transcription and the processing of pre-mRNA can be controlled. Transport of the mRNA into the cytoplasm and its stability in the new location can also be controlled. Once the mRNA is in the cytoplasm, its translation into protein can be regulated. Finally, once the protein itself is made, its structure can be modified, thereby affecting its activity.

In this section, we will describe several ways in which cells control the transcription of specific genes. First, we will examine how a high degree of selectivity in gene transcription can be achieved by specific activation and inhibition of proteins that bind to DNA. Then we will see how the overall structure and compactness of the protein–DNA complex that makes up chromatin within the nucleus often influences whether transcription can occur. In addition, the locations of genes can be changed to make them more accessible to the transcriptional apparatus or genes can be selectively replicated to make more templates for mRNA production.

14.14 The Spliceosome, an RNA Splicing Machine

After the snRNPs bind to initiate the process (see Figure 14.13), other proteins come to the complex and form the spliceosome.

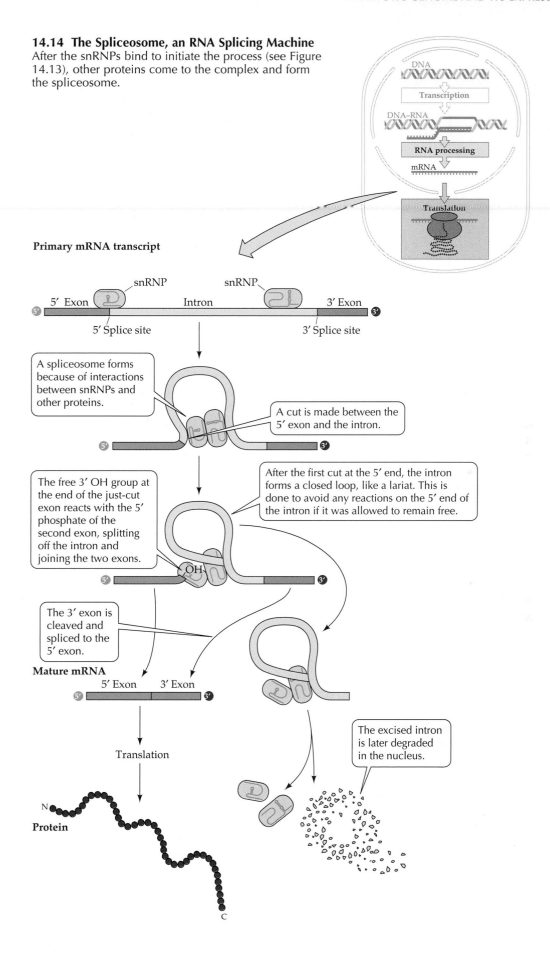

Primary mRNA transcript

snRNP   snRNP

5′ Exon   Intron   3′ Exon

5′ Splice site   3′ Splice site

A spliceosome forms because of interactions between snRNPs and other proteins.

A cut is made between the 5′ exon and the intron.

The free 3′ OH group at the end of the just-cut exon reacts with the 5′ phosphate of the second exon, splitting off the intron and joining the two exons.

After the first cut at the 5′ end, the intron forms a closed loop, like a lariat. This is done to avoid any reactions on the 5′ end of the intron if it was allowed to remain free.

OH

The 3′ exon is cleaved and spliced to the 5′ exon.

Mature mRNA

5′ Exon   3′ Exon

The excised intron is later degraded in the nucleus.

Translation

N

Protein

C

DNA

Transcription

DNA–RNA

RNA processing

mRNA

Translation

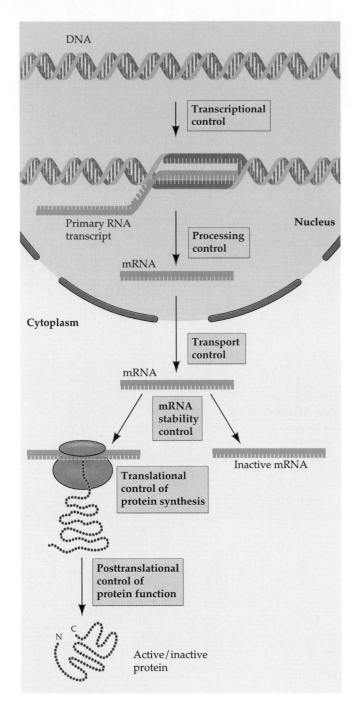

DNA

Transcriptional control

Primary RNA transcript

Nucleus

Processing control

mRNA

Cytoplasm

Transport control

mRNA

mRNA stability control

Inactive mRNA

Translational control of protein synthesis

Posttranslational control of protein function

N C

Active/inactive protein

14.15 Potential Places for the Regulation of Gene Expression Genes are rearranged on the chromosome or even selectively replicated to give more templates to transcribe. In other cases, changes in the proteins that bind to DNA can make genes more or less available for transcription. And, of course, both transcription of pre-mRNA and its processing are controllable. Once the mRNA is in the cytoplasm, its translation into protein can be regulated. Finally, once the protein itself is made, modifications to its structure can occur which affect its activity.

Specific genes can be selectively transcribed

Brain and liver cells in a mouse have some proteins in common and some that are distinctive for each cell type. Yet both cells have the same DNA sequences and, therefore, the same genes. Are the differences in protein content due to differential transcription of genes, or is it that all the genes are transcribed in both cell types and a posttranscriptional mechanism (splicing, export of mRNA to the cytoplasm, translation of the mRNA, or protein longevity) is responsible for the differences in proteins?

The two alternatives—transcriptional or posttranscriptional control—can be distinguished by examination of the actual RNA sequences made within the nucleus of each cell type. Such analysis indicates that the major mechanism is differential gene transcription. Both brain and liver cells transcribe "housekeeping" genes, such as those for glycolysis enzymes and ribosomal RNAs. But liver cells transcribe some genes for liver-specific proteins, and brain cells transcribe some genes for brain-specific proteins. And neither cell type transcribes the special genes for proteins that are characteristic of muscle, blood, bone, and the other cell types in the body.

CONTRASTING EUKARYOTES AND PROKARYOTES. Unlike prokaryotes, in which related genes are transcribed as a unit in operons and under coordinate control, eukaryotes tend to have solitary genes. Thus, to regulate several genes at once requires common control elements in each gene. For example, steroid hormones such as estrogen stimulate the transcription of several different genes because each of these widely separated genes has, near its promoter, a specific sequence in common called the *hormone response element.*

In contrast to the single RNA polymerase in bacteria, eukaryotes have three different RNA polymerases. Each eukaryotic polymerase catalyzes the transcription of a specific type of gene. Only one (RNA polymerase II) transcribes protein-coding genes to mRNA. The other two transcribe the DNA that codes for rRNA (polymerase I) and for tRNA and small nuclear RNPs such as U1 (polymerase III).

The diversity of eukaryotic polymerases is reflected in the diversity of promoters, which tend to be much more variable than are prokaryotic promoters. In addition, most eukaryotic genes have enhancer and silencer elements (which we will discuss shortly) that can be quite distant from the protein-coding sequence and that control its rate of transcription. Whether a eukaryotic gene is transcribed depends on the sum total of the effects of all of these elements; thus there are many points of possible control.

Finally, the transcriptional apparatus is very different in eukaryotes than in prokaryotes, in which a sin-

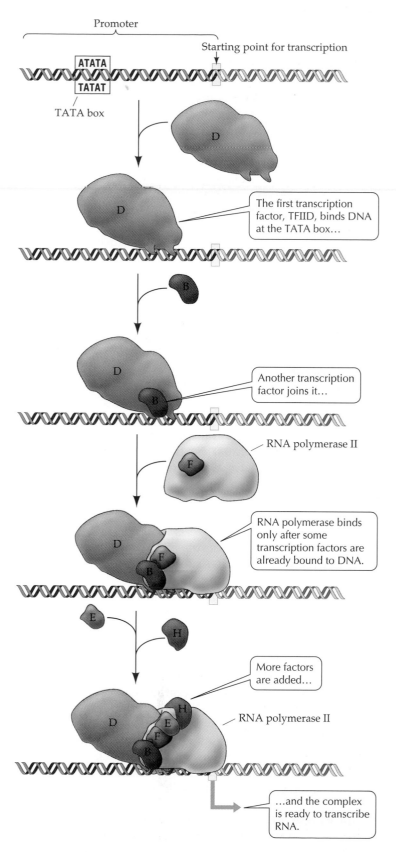

14.16 The Initiation of Transcription on Eukaryotic DNA
Each transcription factor binds to specific regions on DNA. Except for TFIID, which also binds to the TATA box, each factor also has binding sites only for the other proteins.

gle peptide subunit can cause RNA polymerase to recognize the promoter. In eukaryotes, many proteins are involved at the initiation stage of transcription. We will confine our subsequent discussion to RNA polymerase II, which catalyzes the transcription of most protein-coding genes. The mechanisms for the other two polymerases are similar.

TRANSCRIPTION FACTORS. Recall from Chapter 13 that the *promoter* is a sequence of DNA near the 5' end of the the coding region where RNA polymerase begins transcription. A prokaryotic promoter has two essential regions: One, about 40 bp 5' to the initiation point of transcription, is the sequence recognized by RNA polymerase. The second, nearer to the initiation point, is rich in AT base pairs (it is called the TATA box) and is the site at which DNA begins to denature so that its templates can be exposed. In eukaryotes, there is a TATA box about 25 bp away from the initiation site for transcription, and one or two recognition sequences about 50 to 70 bp 5' from the TATA box.

Eukaryotic RNA polymerase II by itself cannot simply bind to the DNA at the promoter and initiate transcription. Rather, it binds and acts only after various regulatory proteins, called **transcription factors**, have assembled on the chromosome (Figure 14.16). First, the protein TFIID ("TF" stands for *transcription factor*) binds to the TATA box. The binding event changes the shapes of both the protein and the DNA, presenting a new surface that attracts the binding of other transcription factors. RNA polymerase II does not bind until several other proteins have already bound to the complex.

Some sequences, such as the TATA box, are common to the promoters of many genes and are recognized by transcription factors found in all the cells of an organism. Other sequences in promoters may be specific to only a few genes and are recognized by transcription factors found only in certain tissues. These specific transcription factors play an important role in differentiation, the specialization of cells during development.

How do these transcription factors deal with the histone proteins present in nucleosomes? The nucleosomal proteins could block the binding of transcription factors and inhibit the initiation of transcription, but several mechanisms appear to be at work to prevent this inhibition. In some cases, the transcription factors simply bind to DNA just after it is replicated and before the histones have had a chance to bind. In other cases, proteins called nucleosome disruptors open up the nucleosome complex and allow the transcription factors to bind.

REGULATORS, ENHANCERS, AND SILENCERS. In addition to the initiation complex of transcription factors, two

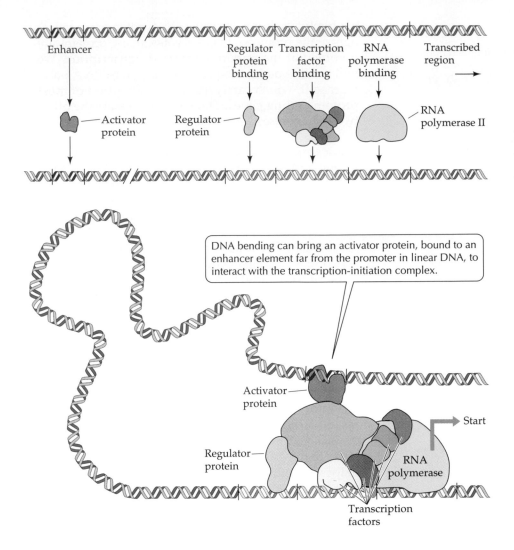

Enhancer — Activator protein

Regulator protein binding — Regulator protein

Transcription factor binding

RNA polymerase binding — RNA polymerase II

Transcribed region →

DNA bending can bring an activator protein, bound to an enhancer element far from the promoter in linear DNA, to interact with the transcription-initiation complex.

Activator protein

Regulator protein

RNA polymerase

Start

Transcription factors

14.17 The Roles of Transcription Factors, Regulators, and Activators

The actions of many proteins determines whether and where RNA polymerase will transcribe DNA.

bination of all the factors is what determines the maximum rate of transcription. In the immature red blood cells of bone marrow, which make a large amount of the protein β-globin, the transcription of globin genes is stimulated by the binding of seven inducers and six activators. By contrast, in white blood cells in the same bone marrow, these regulatory proteins are not made and they do not bind to their sites adjacent to the β-globin genes; consequently these genes are hardly transcribed at all.

COORDINATING THE EXPRESSION OF GENES. In eukaryotes, how do cells coordinate the regulation of several genes whose transcription must be turned on at the same time? In prokaryotes, where related genes are linked together in an operon, coordination is clear: A single regulatory system can regulate several adjacent genes. But in eukaryotes, the several genes whose regulation requires coordination may be on different chromosomes.

In this case, regulation can be achieved if the various genes all have the same controlling sequences near them, which bind to the same activators and regulators. One of the many examples of this phenomenon is provided by the *stress proteins* (or heat shock proteins), which are made when eukaryotic cells are exposed to an elevated temperature. Under conditions of stress, various scattered genes become transcriptionally active to make the various stress proteins. Each of these genes has a specific regulatory sequence near its promoter called the heat shock element. Binding of a specific protein to this element causes the stimulation of RNA synthesis.

other regions of DNA bind proteins that activate RNA polymerase. The recently discovered **regulator** regions are clustered just upstream of the promoter. Various different proteins (in the β-globin gene, seven) may bind here. Their net effect is to bind to the adjacent transcription factor complex and activate it. Much farther away—up to 20,000 bp away—are the **enhancer** regions. Enhancer regions also bind *activator* proteins that strongly stimulate the transcription complex. How they can exert this influence from a distance is not clear. In one model, the DNA bends—it is known to do so—so that the activator is in contact with the RNA polymerase complex (Figure 14.17). Finally, there are negative regulatory regions on DNA called **silencers**, which have the reverse effect of enhancers. Silencers turn off transcription by binding to proteins appropriately called *repressors*.

How do these various proteins and DNA sequences—transcription factors, activators, repressors, regulators, enhancers, and silencers—regulate transcription? Apparently, all genes in most tissues can transcribe a small amount of RNA. But the right com-

THE BINDING OF PROTEINS TO DNA. A key to transcriptional control in eukaryotes is that transcription factors, activators, and repressors all bind to specific DNA sequences. How do they recognize and bind to DNA? There are four common structural themes for protein structures that bind to DNA. These themes are

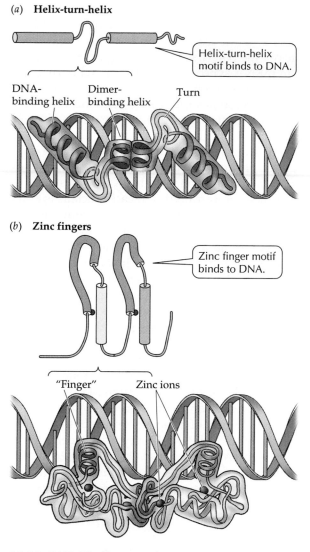

(a) **Helix-turn-helix**

Helix-turn-helix motif binds to DNA.

DNA-binding helix

Dimer-binding helix

Turn

(b) **Zinc fingers**

Zinc finger motif binds to DNA.

"Finger"

Zinc ions

14.18 DNA-Binding Proteins Most transcription factors and other DNA-binding proteins have one of these four structural regions (motifs) by which they bind to DNA.

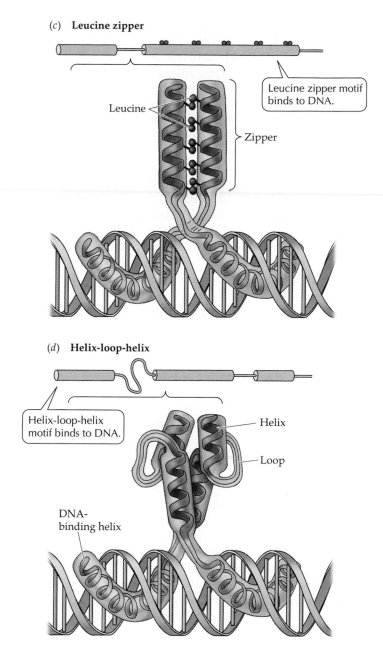

(c) **Leucine zipper**

Leucine zipper motif binds to DNA.

Leucine

Zipper

(d) **Helix-loop-helix**

Helix-loop-helix motif binds to DNA.

Helix

Loop

DNA-binding helix

called *motifs* and consist of combinations of structures and special components. The *helix-turn-helix* motif (Figure 14.18*a*) occurs in many transcription factors that stimulate specific genes during development. This motif appears in the proteins that activate genes involved in development of the embryo (homeobox proteins; see Chapter 15) and in the proteins that regulate development of the immune and central nervous systems.

The *zinc finger* motif (Figure 14.18*b*) occurs in transcription factors, most notably the steroid hormone receptors. Steroid hormones, such as estrogen, can enter cells though the lipid membrane because they are nonpolar. The hormones then bind to receptor proteins in

the cytoplasm, and the complex enters the nucleus. There, the complex binds to the hormone response elements near promoters to activate transcription.

The *leucine zipper* motif (Figure 14.18*c*) occurs in many DNA-binding proteins—for example, the inducer AP-1, which binds near promoters of genes involved in mammalian cell growth and division. Overactivity of AP-1 has been linked to several types of cancer.

The *helix-loop-helix* motif (Figure 14.18*d*) occurs in the activator proteins that bind to enhancers for the immunoglobulin genes that synthesize antibodies, as well as in the inducers involved in muscle protein synthesis.

Genes can be inactivated by chromatin structure

The packaging of DNA by nuclear proteins (see Figure 14.3) can make DNA physically inaccessible to RNA polymerase and the rest of the transcription apparatus, much like the binding of repressor to the operator in the prokaryotic *lac* operon prevents transcription. As mitosis or meiosis ends, chromosomes partly uncoil (see Chapter 9). Two kinds of chromatin can be distinguished by staining of the interphase nucleus: euchromatin and heterochromatin. *Euchromatin* is diffuse and stains lightly, and is transcribed into mRNA. *Heterochromatin* stains densely and is generally not transcribed; any genes that it contains are thus inactivated. Heterochromatin contains much of the highly repeated DNA.

How chromatin inactivation controls gene expression is best understood for the X and Y chromosomes of mammals. The normal female has two X chromosomes, the normal male an X and a Y. The Y chromosome has few genes that are also present on the X, and the Y is largely transcriptionally inactive in most cells. So there is a great difference between females and males in the "dosage" of X chromosome genes.

In other words, each female cell has two copies of the genes on the X chromosome and therefore has the potential of producing twice as much protein product of these genes as a male has. When the gene involved is on an autosome, the result is usually lethal and the embryo fails to develop. How do both sexes develop, when one of them obviously has an "extra" (or one too few, depending on your viewpoint) chromosome?

The answer was found in 1961 independently by Mary Lyon, Liane Russell, and Ernest Beutler. They suggested that one of the X chromosomes in each cell of an XX female is transcriptionally inactivated early in the life of the embryo and remains inactive, as do all the cells arising from it. In a given cell, the "choice" of which X in the pair of Xs to inactivate is usually random. Recall that one of the Xs in a female came from her father and one from her mother. Thus, in one embryonic cell the paternal X might be the one remaining active in mRNA synthesis, but in its neighboring cell, the maternal X might be active.

This suggestion is supported by genetic, biochemical, and cytological evidence. Interphase cells of XX females have a single, stainable nuclear body called a **Barr body** after its discoverer, Murray Barr (Figure 14.19). This clump of heterochromatin, which is not present in males, is the inactivated X chromosome.

The number of Barr bodies in each nucleus is equal to the number of X chromosomes minus one (the one represents the X chromosome that remains transcriptionally active). So a female with the normal two X chromosomes will have one Barr body, one with three X's has two, an XXXX female has three, and an XXY male has one. We may infer that interphase cells of each person, male or female, have a single *active* X chromosome, making the dosage of the expressed X chromosomes genes constant in both sexes.

The mechanism of *X inactivation* is chromosome condensation that makes the DNA sequences physically unavailable for the transcription machinery. This process involves chromosomal proteins. One method may be the addition of a methyl group to the 5' position of cytosine on DNA. Methylation seems to be most prevalent in transcriptionally inactive genes. For example, most of the DNA of the inactive X chromosome has many cytosines methylated, while few of them on the active X are methylated. Methylated DNA appears to bind certain specific chromosomal proteins, and these may be responsible for heterochromatin formation. However, this has not yet been proved.

The otherwise inactive X chromosome has one gene that is lightly methylated and *is* transcriptionally active. This gene is called *XIST* (for *X* *i*nactivation *s*pecific *t*ranscript) and is heavily methylated and *not* transcribed from the other, "active" X chromosome in a female. The RNA transcribed from *XIST* does not leave the nucleus and is not an mRNA. Instead, it appears to bind to the X chromosome that transcribes it, and this binding somehow leads to a spreading of inactivation along the chromosome.

Each Barr body is the condensed, inactive member of a pair of X chromosomes in the cell. The other X is not condensed and is active in transcription.

14.19 Barr Bodies in the Nuclei of Female Cells The number of Barr bodies per nucleus is equal to the number of X chromosomes minus one. Thus males (XY) have no Barr body, whereas females (XX) have one.

A DNA sequence can move to a new location to activate transcription

In some instances gene expression is regulated by movement of a gene to a new location on the chromosome. One example is in the yeast *Saccharomyces cerevisiae*. The haploid single cells of this fungus exist in one of two mating types, *a* and α, which fuse to form a diploid zygote. Although all cells have an allele for each of these types, the allele that is expressed determines the mating type. In some yeasts, the mating type changes with almost every cell division cycle. How does it change so rapidly?

The yeast cell keeps the two different alleles (coding for type α and type *a*) at separate locations on the chromosome, away from a third site, the MAT locus, which is the site of expression. These two alleles are usually transcriptionally silent because a repressor protein binds to them. However, when a copy of the α or *a* allele inserts at the MAT region, the gene for proteins of the appropriate mating type is transcribed.

A change in mating type requires three steps: First, a new DNA copy of the nonexpressed allele is made (if the cell is now α, the new copy will be the *a* allele). Second, the current occupant of the MAT region (in this case, the α DNA) is removed by a nuclease. Third, the new allele (*a*) is inserted at the MAT region and transcribed. The *a* proteins are now made, and the mating type is changed.

DNA rearrangement is also important in producing the highly variable proteins that make up the human repertoire of antibodies. We will return to this subject in Chapter 18.

Selective gene amplification results in more templates for transcription

Another way for a cell to make more of a gene product than another cell does is to have more copies of the appropriate gene and to transcribe them all. The process of creating more copies of a specific gene in order to increase transcription is called **gene amplification**.

As described earlier, the genes that code for three of the four human ribosomal RNAs are linked together in a unit, and this unit is repeated several hundred times in the genome to provide more templates for rRNA synthesis (rRNA is the most abundant kind of RNA in the cell). In some instances, however, even this moderate repetition is not enough to satisfy the demands of the cell. For example, the mature eggs of frogs and fish have up to 1 trillion ribosomes. These ribosomes are used for the massive protein synthesis that follows fertilization. How can a single cell end up with so many ribosomes (and so much rRNA)? Adding to the problem is the fact that the cell that differentiates into the egg would take 50 years to make 1 trillion ribosomes if it transcribed its rRNA genes at peak efficiency.

The egg cell solves this problem by selectively amplifying its rRNA gene clusters until there are more than a million copies of a cluster that was originally present in fewer than 1,000 copies. In fact, this gene complex goes from being 0.2 percent of the total genome DNA to being 68 percent. These million copies transcribing at maximum rate (Figure 14.20) are just enough to make the necessary trillion ribosomes in a few days.

The mechanism for selective overreplication of a single gene is not clearly understood, but it is important. As Chapter 17 will show, in some cancers, a cancer-causing gene called an oncogene becomes amplified. Also, in some tumors treated with a drug that targets a

14.20 Transcription from Multiple Genes for rRNA
Elongating strands of rRNA transcripts form arrowhead-shaped regions, each centered on a strand of DNA that codes for rRNA.

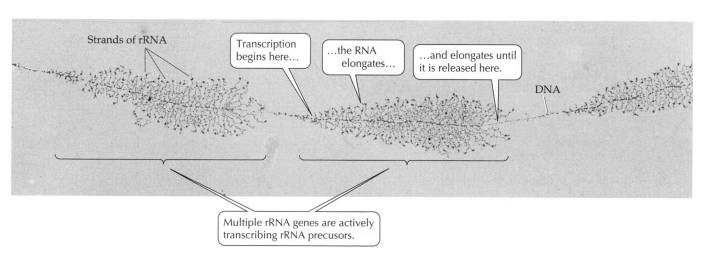

single protein, amplification of the gene for the target protein leads to an excess of that protein and the cell becomes resistant to the prescribed dose of the drug.

Posttranscriptional Control

There are many ways to control the appearance of mature mRNA even after a precursor has been transcribed. For example, the exons of the pre-mRNA can be recombined in different ways by alternate splicing. The result is that different proteins are synthesized, depending on which exons are used to make up the final mRNA. Or, the longevity of mRNA in the cytoplasm can be regulated. The longer an mRNA exists in the cytoplasm, the more of its coded protein can be made.

Different mRNAs can be made from the same gene by alternate splicing

Most primary transcripts have several introns (see Figure 14.8). The splicing mechanism recognizes the boundaries between exons and introns. What would happen if the β-globin pre-mRNA, which has two introns, was spliced from the start of the first intron to the end of the second? Not only the two introns but also the middle exon would be spliced out. An entirely new protein (certainly not a β-globin) would be made and the functions of normal β-globin would be lost.

Although alternate splicing is not common in normal RNA processing, it can be a deliberate mechanism to generate a family of different proteins from a single gene. For example, a single primary transcript for the structural protein tropomyosin can be alternately spliced to give five different mRNAs and five different proteins in five different tissues: skeletal muscle, smooth muscle, fibroblast, liver, and brain (Figure 14.21). The same mechanism is involved in generating the wide variety of antibodies in the immune system.

The stability of mRNA can be regulated

As the genetic material, DNA must remain stable, and there are elaborate repair mechanisms if it becomes damaged. RNA, on the other hand, has no such repair system. After it arrives in the cytoplasm, mRNA is subject to breakdown catalyzed by ribonucleases, which exist both in the cytoplasm and in lysosomes. But not all eukaryotic mRNAs have the same life span. The differences in how long mRNAs last make possible the control of protein synthesis by the differential stabilities of mRNAs.

Tubulin is a protein that polymerizes to form microtubules, a component of the cytoskeleton. When a large pool of free tubulin is available in the cytoplasm, there is no particular need for the cell to make more of it. Under these conditions some tubulin molecules bind to tubulin mRNA, and this binding makes the mRNA especially susceptible to breakdown, and less tubulin is made. Other examples illustrate the same mechanism—that the less time an mRNA stays in the cytoplasm, the less it can be translated into protein.

Translational and Posttranslational Control

Just as proteins can control the synthesis of mRNA by binding to DNA, they can also control the translation

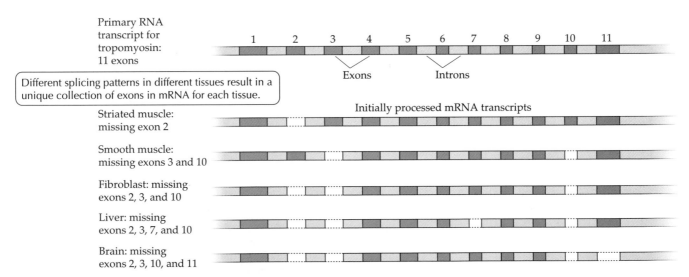

14.21 Alternate Splicing Results in Different mRNAs and Proteins In mammals, the protein tropomyosin is coded for by a gene that has 11 exons. Different tissues splice tropomyosin pre-mRNA differently, resulting in 5 different proteins. For example, in striated (skeletal) muscle, exon 2 (along with its flanking introns) is spliced out, resulting in a protein that is made up of amino acids coded for by exons 1 and 3–11.

of mRNA by binding to mRNA in the cytoplasm. This mode of control is especially important in long-lived mRNAs. A cell must not continue to make proteins that it does not need. For example, mammalian cells respond to certain stimuli by making cytokines, which stimulate specific cells to divide (see Chapter 9). If the mRNA for a cytokine is still in the cytoplasm and available for translation long after the cytokine is needed, it will be made and released inappropriately. This might cause a target cell population to divide inappropriately, forming a tumor.

Let's examine the role of translational repressors in protein synthesis and the ways in which the activity and lifetime of a protein can be regulated after it is made.

The translation of mRNA can be controlled

One way to control translation is by the capping mechanism on mRNA. As already noted, mRNA is capped at its 5′ end by a modified guanosine residue (see Figure 14.12). Messenger RNAs that have unmodified caps are not translated. For example, stored mRNA in the oocyte of the tobacco hornworm moth has the guanosine added to its end, but the G is not modified. Hence, this stored mRNA is not translated. However, after fertilization, the cap is modified, allowing the mRNA to be translated to produce proteins needed for early embryogenesis.

Free iron ions (Fe^{2+}) within a mammalian cell are bound by a storage protein, *ferritin*. When iron is in excess, ferritin synthesis rises dramatically. Yet the amount of ferritin mRNA remains constant. This increase in ferritin synthesis is due to an increased rate of mRNA translation. When the iron level in the cell is low, a *translational repressor* protein binds to ferritin mRNA and prevents its translation by blocking its attachment to the ribosome. When iron levels rise, the excess iron binds to the repressor and alters the three-dimensional structure of the translational repressor, causing it to detach from the mRNA, and translation proceeds.

Translational control also acts in the synthesis of hemoglobin. As we described earlier, hemoglobin consists of four polypeptide chains and a nonprotein pigment, heme. If heme synthesis does not equal globin synthesis, some polypeptide chains will stay free in the cell, waiting for their heme partner. One way that the balance is maintained is that excess heme in the cell stimulates the rate of translation of globin mRNA by removing a block of the initiation of translation at the ribosome.

Protein function and lifetime can be regulated after translation

We have considered how gene expression may be regulated by the control of transcription, of RNA process-

ing, and of translation. However, the story does not end here, because most proteins are modified after translation. Some of these changes are permanent, such as the addition of sugars (glycosylation) or the removal of a signal sequence after a protein has crossed a membrane (see Chapter 11). Other changes are reversible, such as the addition of phosphate groups (phosphorylation), which we will discuss further in later chapters.

An important way to regulate the action of a gene product (protein) in a cell is to regulate its lifetime in the cell. Proteins involved in cell division (e.g., the cyclins) are hydrolyzed at the right moment to time the sequence of events. Transcriptional inducers must be present only when they are needed, lest the affected genes be always "on." Proteins identified for breakdown are often linked to a small protein called *ubiquitin* (so called because it is ubiquitous, or widespread). Ubiquitin forms short chains on the targeted protein, and these chains attract a large complex of proteases that catalyze the breakdown of the protein.

Even single-celled eukaryotes such as yeasts have many of the complex mechanisms of gene regulation that we have described in this chapter. In multicellular organisms, from worms to wheat, these mechanisms must also coordinate the activities of different types of cells and tissues. Gene regulation is about the expression of genotype into phenotype. Nowhere is this more dramatic than in the unfolding of development from a fertilized egg to an adult organism. It is not surprising that many genes are expressed and then silenced during this process. We turn to these phenomena in the next chapter.

Summary of "The Eukaryotic Genome and Its Expression"

The Eukaryotic Genome

• Although eukaryotes have more DNA in their genomes than prokaryotes, in some cases there is no apparent relationship between genome size and organism complexity. **Review Figure 14.2**

• Unlike prokaryotic DNA, eukaryotic DNA in interphase cells is separated from the cytoplasm by being contained within a nucleus.

• Eukaryotic DNA is extensively packaged by proteins. **Review Figure 14.3**

• Nucleic acid hybridization is an important technique for analyzing eukaryotic genes. **Review Figure 14.4**

• Nucleic acid hybridization reveals that there are three classes of DNA. The fastest-reannealing DNA represents short sequences repeated up to millions of times in the genome. This highly repeated DNA is usually not tran-

scribed and may assist in spindle attachment to chromosomes during mitosis. DNA that reanneals more slowly consists of moderately repetitive sequences. The slowest-reannealing DNA consists of single-copy sequences.
• Some telomeric DNA may be lost during each DNA replication, eventually leading to chromosome instability and cell death. The enzyme telomerase catalyzes the restoration of the lost telomeric DNA. Most somatic cells lack telomerase and thus have limited life spans. Cancer cells are able to divide continuously because they express telomerase. **Review Figure 14.5**
• Some moderately repetitive DNA sequences, such as the sequences coding for rRNAs, are transcribed. **Review Figure 14.6**
• Some moderately repetitive DNA sequences are transposable, or able to move about the genome. **Review Figure 14.7**

The Structures of Protein-Coding Genes
• Eukaryotic protein-coding genes are usually present in only one copy per haploid genome.
• A typical protein-coding gene has noncoding internal sequences (introns) and flanking sequences that are involved in the machinery for transcription and translation. **Review Figures 14.8, 14.9**
• Some eukaryotic genes form families with related genes that have similar sequences and code for similar proteins. These related proteins may be made at different times and in different tissues. Some sequences in gene families are pseudogenes, which code for nonfunctional mRNAs or proteins. **Review Figures 14.10**
• Differential expression of different genes in the globin family ensures important physiological changes during human development. **Review Figure 14.11**

RNA Processing
• After transcription, the precursor to mature mRNA is covalently altered by the addition of a 5′ cap of a modified guanosine and a 3′ poly A tail. **Review Figure 14.12**
• The introns are removed from the mRNA precursor by the spliceosome, a complex of RNAs and proteins. **Review Figures 14.13, 14.14**

Transcriptional Control
• Eukaryotic gene expression can be controlled at the transcriptional, posttranscriptional, translational, and posttranslational levels. **Review Figure 14.15**
• The major method of control of eukaryotic gene expression is by selective transcription resulting from specific proteins binding to regulatory regions on DNA.
• A series of proteins must bind to the promoter before RNA polymerase can bind. Whether RNA polymerase will initiate transcription also depends on the binding of regulatory proteins and activator proteins (which are bound by enhancers and stimulate transcription), and repressor proteins (which are bound by silencers and inhibit transcription). **Review Figures 14.16, 14.17**
• Regulatory, enhancer, and silencer proteins bind to DNA by recognizing specific sequences. There are four domains, or regions, that are used in these DNA-binding proteins. **Review Figure 14.18**
• The clumping of chromatin by proteins bound to DNA can block transcription by making DNA physically unavailable to RNA polymerase. This clumping results in hetero-

chromatin, as in X inactivation in female mammals. Specific DNA modifications may be involved in this process. **Review Figure 14.19**
• In some cases, such as the change from one mating type to another in yeast, the movement of a gene to a new location on a chromosome may alter the ability of the gene to be transcribed.
• Some genes become selectively amplified, and the extra copies result in increased transcription. **Review Figure 14.20**

Posttranscriptional Control
• Because eukaryotic genes have several exons, alternate splicing can be used to produce different proteins. **Review Figure 14.21**
• The longevity of mRNA in the cytoplasm can be regulated by the binding of proteins.

Translational and Posttranslational Control
• Translational repressors can inhibit the translation of mRNA.
• The function and stability of a completed protein can be regulated by chemical modifications.

Self-Quiz

1. Eukaryotic protein-coding genes differ from their prokaryotic counterparts in that only eukaryotic genes
   a. are double-stranded.
   b. are present in only a single copy.
   c. contain introns.
   d. have a promoter.
   e. transcribe mRNA.

2. Which statement about nucleic acid hybridization is *not* true?
   a. Part of the process is complementary base pairing.
   b. A DNA strand can hybridize with another DNA strand.
   c. An RNA strand can hybridize with a DNA strand.
   d. A polypeptide can hybridize with a DNA strand.
   e. Double-stranded DNA is denatured at high temperatures.

3. Which statement about repetitive DNA is true?
   a. Highly repetitive DNA reanneals most slowly of the three classes of DNA.
   b. Highly repetitive DNA is usually transcribed rapidly.
   c. Much highly repetitive DNA is at the centromeres.
   d. Single-copy DNA is rare in eukaryotes.
   e. Transposable elements are single-copy genes.

4. Which of the following does *not* occur after mRNA is transcribed?
   a. binding of RNA polymerase II to the promoter
   b. capping of the 5′ end
   c. addition of a poly A tail to the 3′ end
   d. splicing out of the introns
   e. transport to the ribosome

5. Which statement about RNA splicing is *not* true?
   a. It removes introns.
   b. It is performed by small nuclear ribonucleoprotein particles (snRNPs).
   c. It always removes the same introns.
   d. It is directed by consensus sequences.
   e. It shortens the RNA molecule.

6. Telomeres are
   a. present in equal length in all cells.
   b. removed by telomerase.
   c. essential for the stability of chromosomes.
   d. located at the ends and middle of eukaryotic chromosomes.
   e. caused by errors in DNA replication.

7. Which statement about selective gene transcription in eukaryotes is *not* true?
   a. Different classes of RNA polymerase transcribe different parts of the genome.
   b. Transcription requires transcription factors.
   c. Genes are transcribed in groups called operons.
   d. Both positive and negative regulation occur.
   e. Many proteins bind at the promoter.

8. Heterochromatin
   a. contains more DNA than does euchromatin.
   b. is transcriptionally inactive.
   c. is responsible for all negative transcriptional control.
   d. clumps the X chromosome in human males.
   e. occurs only during mitosis.

9. Translational control
   a. is not observed in eukaryotes.
   b. is a slower form of regulation than transcriptional control.
   c. can be achieved by only one mechanism.
   d. requires that mRNA be uncapped.
   e. ensures that heme synthesis equals globin synthesis.

10. Which of the following are *not* used in transcriptional regulation in eukaryotes?
    a. enhancers
    b. silencers
    c. transcription factors
    d. RNA polymerase subunits
    e. promoters

Applying Concepts

1. In rats, a gene 1,440 bp long codes for an enzyme made up of 192 amino acid units. Discuss this apparent discrepancy. How long would the initial and final mRNA transcripts be?

2. The activity of the enzyme dihydrofolate reductase (DHFR) is high in some tumor cells. This activity makes the cells resistant to the anticancer drug methotrexate, which targets DHFR. Assuming that you had the complementary DNA for the gene that encodes DHFR, how would you show whether this increased activity was due to increased transcription of the single-copy DHFR gene or to amplification of the gene?

3. Describe the steps in the production of a mature, translatable mRNA from a eukaryotic gene that contains introns. Compare this to the situation in prokaryotes (see Chapter 13).

4. A protein-coding gene has three introns. How many different proteins can be made from alternate splicing of the pre-mRNA transcribed from this gene?

5. Most somatic cells in a mammal do not express telomerase. Yet the germ line cells that produce gametes by meiosis do express this enzyme. Explain.

Readings

Baeuerle, P. A. 1995. *Inducible Gene Expression.* Birkhauser, Boston, MA. A survey of the many ways that eukaryotic genes are regulated.

Cech, T. R. 1986. "RNA as an Enzyme." *Scientific American,* November. A description of the exciting discovery of RNA catalysts, some of which are involved in RNA splicing.

Elgin, S. (Ed.). 1995. *Chromatin Structure and Gene Expression.* IRL Press, Oxford. A description of how chromosomal proteins influence transcription.

Gesteland, R. F. and J. F. Atkins (Eds.). 1993. *The RNA World.* Cold Spring Harbor Laboratory Press, Cold Spring Harbor, NY. A collection of papers from experts on the occurrence and significance of RNA.

Grunstein, M. 1992. "Histones as Regulators of Genes." *Scientific American,* October. An account showing that histones not only organize DNA, but also regulate transcription.

Hershey, J. W. B., M. Matthews and N. Sonenberg (Eds.). 1996. *Translational Control.* Cold Spring Harbor Laboratory, Plainview, NY. An authoritative and extensive description of the control of translation, with an emphasis on eukaryotes.

Lewin, B. 1997. *Genes VI.* Oxford University Press, New York. This molecular biology text has excellent summaries of gene structure and transcription.

Rennie, J. 1993. "DNA's New Twists." *Scientific American,* March. A discussion of transposable elements and other features.

Rhodes, D. and A. Klug. 1993. "Zinc Fingers." *Scientific American,* February. A description of how zinc finger regions in proteins help them bind to DNA and regulate transcription.

Chapter 15

Development: Differential Gene Expression

New Organs from Small Changes
The fruit fly in the bottom picture has a genetic mutation that altered the pattern of its development and produced a complete and functional second set of wings.

At one time, each of the *Drosophila melanogaster* in these photographs was a fertilized egg, a single cell. This cell contained the entire genome for the fly, with complete instructions for the production of a mature organism. In a matter of days, these instructions were carried out as a series of genes was expressed at just the right times and in just the right places. In addition, some cells produced growth factors and other signals, which interacted precisely with their target cells. Through a combination of cell division, cell movement, and cell differentiation (permanent changes in cell structure and function), the two eggs developed first into small, worm-like larvae and then into the handsome flies you see here.

These flies are obviously very different from each other. The normal one has a single pair of large wings. The other has two pairs of wings, both pairs properly constructed and fully functional. How did that happen? The fly with extra wings is a mutant—it has point mutations in just two genes. How can such a tiny genetic alteration produce such a dramatic phenotypic result? How does any multicellular organism develop its characteristic shape and functions? In this chapter we will consider how an organism's genome directs the development of an adult, with its many different cell types and complex body pattern, from a single cell.

Throughout this chapter we will describe how biologists have studied development in invertebrates, such as the fruit fly and nematode worm, as well as "simpler" vertebrates such as the frog. Two major conclusions have come out of these studies: First, all types of somatic cells—all the cells except for gametes—in an organism retain all of the genes that were present in the fertilized egg. Cell differentiation does not result from a loss of DNA. Instead (and this is the second major theme of developmental biology) cellular actions during development and cell differentiation result from the differential expression of genes. During development, the various mechanisms described in Chapter 14, such as transcriptional and translational control, as well as intercellular signaling, work to produce a complex organism.

It is very appropriate to start (and end) this chapter with the fruit fly, for the mechanisms of development revealed by studying the fly turn out to be very similar to those of more complex eukaryotes, including humans.

The Processes of Development

Development is a process of progressive change during which an organism takes on the successive forms that characterize its life cycle (Figure 15.1). In its early stages of development a plant or animal is called an **embryo**. Sometimes the embryo is contained within a protective structure such as a seed coat, an eggshell, or a uterus. An embryo does not photosynthesize or actively feed; instead it obtains its food from its mother directly or indirectly (by way of nutrients stored in the egg, for example). Later stages, such as a human fetus, may precede the birth of the new, independent, organism. Development continues throughout an organism's life, ceasing only when the organism dies.

Growth (increase in size) is an important part of development, continuing throughout the individual's life in some species but reaching a more or less stable end point in others. A multicellular body cannot grow—indeed, it cannot arise from a single cell—without **cell division**. Repeated mitotic divisions generate the multicellular body. In plants, unless the daughter cells become longer (expand) after they form, the embryo does not grow; thus in plant development, cell expansion begins with the first cell divisions. In animal development, on the other hand, such **cell expansion** is often slow to begin: The animal embryo may consist of thousands of cells before it becomes larger than the fertilized egg.

Development has two major components in addition to cell division and cell expansion: differentiation and morphogenesis. The remainder of the chapter will focus on these processes.

Differentiation is the generation of cellular specificity; that is, differentiation defines the specific structure and function of a cell. Mitosis, as we have seen, produces daughter nuclei that are chromosomally and genetically identical to the nucleus that divides to pro-

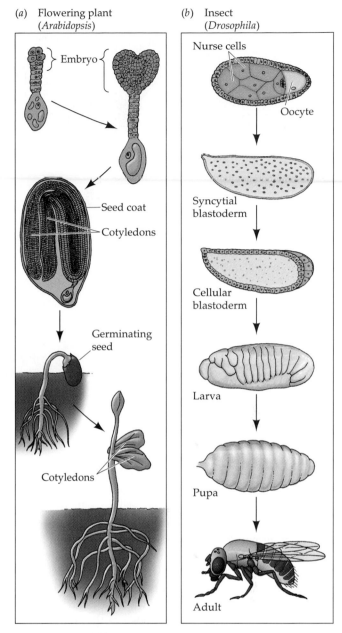

(a) Flowering plant (*Arabidopsis*)

Embryo

Seed coat

Cotyledons

Germinating seed

Cotyledons

(b) Insect (*Drosophila*)

Nurse cells

Oocyte

Syncytial blastoderm

Cellular blastoderm

Larva

Pupa

Adult

15.1 Stages of Development Stages from embryo to adult shown for a plant and an animal. Cell division and expansion, growth, cell differentiation, and the creation of the organs and tissues of the adult body are all part of the complex process.

duce them. However, the cells of an animal or plant body are not all identical in structure or function. The human body, with its approximately 100 trillion (10^{14}) cells, consists of about 200 functionally distinct cell types—for example, muscle cells, blood cells, and nerve cells. When the embryo consists of only a few cells, each cell has the potential to develop in many different ways. As development proceeds, the possibilities available to individual cells gradually narrow, until each cell's *fate* is fully determined and the cell has differentiated.

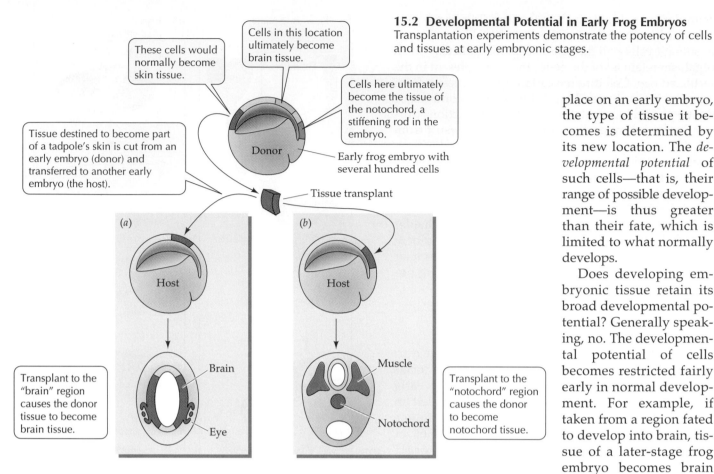

15.2 Developmental Potential in Early Frog Embryos
Transplantation experiments demonstrate the potency of cells and tissues at early embryonic stages.

These cells would normally become skin tissue.

Cells in this location ultimately become brain tissue.

Cells here ultimately become the tissue of the notochord, a stiffening rod in the embryo.

Tissue destined to become part of a tadpole's skin is cut from an early embryo (donor) and transferred to another early embryo (the host).

Donor

Early frog embryo with several hundred cells

Tissue transplant

(a) Host

(b) Host

Transplant to the "brain" region causes the donor tissue to become brain tissue.

Brain

Eye

Muscle

Notochord

Transplant to the "notochord" region causes the donor to become notochord tissue.

place on an early embryo, the type of tissue it becomes is determined by its new location. The *developmental potential* of such cells—that is, their range of possible development—is thus greater than their fate, which is limited to what normally develops.

Does developing embryonic tissue retain its broad developmental potential? Generally speaking, no. The developmental potential of cells becomes restricted fairly early in normal development. For example, if taken from a region fated to develop into brain, tissue of a later-stage frog embryo becomes brain tissue even if transplanted to parts of an early-stage embryo destined to become other structures. The tissue of the later-stage embryo is thus said to be *determined*: Its fate has been sealed, regardless of its surroundings. By contrast, the younger transplant tissue in Figure 15.2 has not yet become determined.

Determination precedes differentiation

Determination, the commitment of a cell to a particular fate, is a process influenced by the cellular environment and the cellular contents acting on the cell's genome. Determination is followed by differentiation, the actual changes in biochemistry, structure, and function that result in cells of different types. *Determination* is not something that is visible under the microscope—cells do not change appearance when they become determined. *Differentiation* often involves a change in appearance as well as function. Determination precedes differentiation. Determination is a commitment; the final realization of this commitment is differentiation.

The Role of Differential Gene Expression in Differentiation

Differentiated cells are recognizably different from each other, sometimes dramatically so. Certain cells in our

Whereas differentiation gives rise to cells of different kinds, **morphogenesis** (literally, "creation of form") gives rise to the shape of the multicellular body and its organs. Morphogenesis results from *pattern formation*, the organization of differentiated tissues into specific structures. (We will discuss pattern formation in detail later in the chapter.) The organized division and expansion of motionless cells are the major tools available for building the plant body form. In animals, cell movements are very important in morphogenesis, as is the programmed death of certain cells.

In plants and animals alike, differentiation and morphogenesis result ultimately from the regulated activities of genes and their products.

Cells have fewer and fewer options as development proceeds

The technique of marking specific cells of early embryos with stains and observing which cells of older embryos contain the stain enabled biologists to determine which adult structures develop from certain parts of the early embryo. For instance, the shaded area of the frog embryo shown in Figure 15.2 has the **fate** of becoming (that is, it is destined to become) part of the skin of the tadpole larva if left in place. However, if we cut out a piece from this region and transplant it to another hair follicles continuously produce keratin, the protein of hair, nails, feathers, and porcupine quills. Blood cells do not produce keratin, nor do other kinds of cells in our bodies. In the hair follicle cells, the gene that encodes keratin is transcribed; in most other cells in our body that gene is not transcribed. During plant development, some cells activate a gene that encodes an enzyme that catalyzes the formation of suberin, the substance that gives cork its characteristic waxy feel. Activation of that gene is a key step in the differentiation of those cells. Generalizing from these observations, we may say that *differentiation results from differential gene expression*—that is, from the differential regulation of transcription, posttranscriptional events such as mRNA splicing, and translation (see Chapter 14).

Because the cells of a multicellular organism arise by mitotic divisions of a single-celled fertilized egg, or zygote, most of them are genetically identical. In the absence of mutation, all the cells in an organism have the same hereditary makeup; yet the adult organism is composed of many distinct types of cells. This apparent contradiction results from regulation of the expression of various parts of the genome.

Because the zygote has the ability to give rise to every type of cell in the adult body, we say it is **totipotent**. Its genetic "library" is complete, containing instructions for all the structures and functions that will arise throughout the life cycle. Later in the development of animals (and probably to a lesser extent in plants), the cellular descendants of the zygote lose their totipotency and become determined—that is, committed to form only certain parts of the embryo. Determined cells differentiate into specific types of cells such as neurons in the nervous system or muscle cells. When a cell achieves the fate for which it was determined, it is said to have differentiated.

The mechanisms of differentiation relate primarily to changes in the transcription and translation of genetic information.

Differentiation usually does not include a permanent change in the genome

An early explanation of the mechanisms of differentiation stated that the cell nucleus undergoes irreversible genetic changes during development. It was suggested that chromosomal material is lost, or that some of it is irreversibly inactivated.

Differentiation is irreversible in certain types of cells. Examples include the mammalian red blood cell, which loses its nucleus during development, and the tracheid, a water-conducting cell in vascular plants. Tracheid development culminates in the death of the cell, leaving only the pitted cell walls that formed while the cell was alive (see Chapter 31). In these two extreme cases, the irreversibility of differentiation can be explained by the absence of a nucleus.

Generalizing about mature cells that retain functional nuclei is more difficult. We tend to think of plant differentiation as reversible and of animal differentiation as irreversible, but this is not a hard-and-fast rule. A lobster can regenerate a missing claw, but a cat cannot regenerate a missing paw. Why is differentiation reversible in some cells but not in others? At some stage of development do changes within the nucleus permanently commit a cell to specialization?

At the Institute of Cancer Research in Philadelphia in the 1950s, Robert Briggs and Thomas J. King performed experiments to see whether all the genetic material of an organism is preserved in potentially active form, or if some of it is permanently inactivated or lost during normal development. To find out whether the nuclei of early frog embryos had lost the ability to do what the totipotent zygote nucleus could do, Briggs and King carried out a series of meticulous transplants. First they removed the nucleus from an unfertilized egg (thus forming what is called an *enucleated* egg). Then, with a very fine glass tube, they punctured a cell in an early embryo and drew up part of its contents, including a nucleus, which they then injected into the enucleated egg.

More than 80 percent of these nuclear transplant operations resulted in the formation, from the egg and its new nucleus, of a normal early embryo. Of these embryos, more than half developed into normal tadpoles and, ultimately, normal adult frogs. These experiments showed that no information is lost from the nuclei of cells as they pass through the early stages of embryonic development. On the other hand, Briggs and King found that when the donor nuclei were derived from older embryonic stages, fewer tadpoles developed (Figure 15.3).

The Briggs and King experiments demonstrated that every cell in the early frog embryo has all of the genes necessary to produce an adult frog. The same is probably true for humans, leading to a practical application in human genetic testing. An eight-celled human embryo can be isolated in the laboratory and a single cell removed to determine if a harmful allele is present in homozygous form. Each remaining cell, being totipotent, can be stimulated to divide and act as a zygote.

Briggs and King's work was carried further by John B. Gurdon at Oxford University, who performed similar transplants on a different species of frog, using nuclei from more advanced embryos and even swimming tadpoles. These nuclei, transplanted into an egg whose nucleus had been inactivated, could produce all the types of cells present in embryos, tadpoles, and, in some cases, adult frogs.

Gurdon's results confirmed the totipotency of the nucleus of a differentiated cell. Clearly, nuclei change in their activities during differentiation, but the changes need not be irreversible. The environment

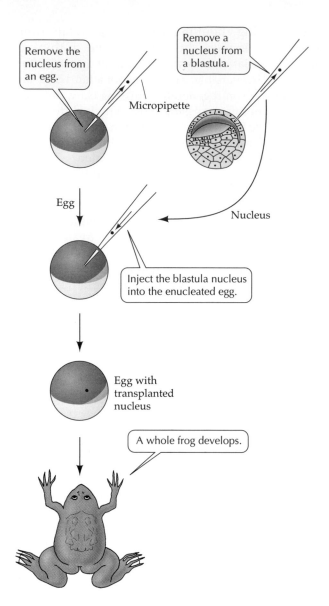

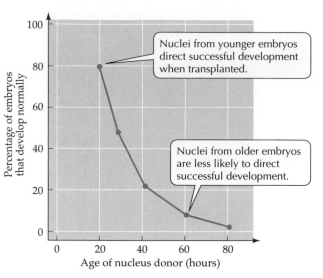

15.3 Nuclei Lose Differentiation Potential with Age
Nuclear transplant experiments on frogs by Briggs and King showed that the potential for cell differentiation declines with the age of the embryo.

around the nucleus—the cytoplasm—exerts a great influence over which nuclear genes are expressed. Thus a nucleus inside a cell in the tadpole's intestine is surrounded by "intestinal" signals, but it retains the ability to act as a fertilized egg nucleus if it is put into the cytoplasm of a fertilized egg.

A dramatic example of totipotency was reported in 1997, when Ian Wilmut and his colleagues at a biotechnology company in Scotland used a modification of Gurdon's nuclear transplant procedure to clone sheep (Figure 15.4). Previous attempts to produce mammals by this method had worked only if the donor nucleus was part of an entire cell and if the donor cell was from an early embryo. Using adult donor cells, as Gurdon did with frogs, resulted in chromosomal abnormalities and embryonic death. Apparently, when mammalian donor cells were in the G2 phase of the cell cycle and were fused with egg cytoplasm also in

G2, some extra DNA replication took place that created havoc with the cell cycle in the egg when it attempted to divide.

Wilmut took differentiated cells from a ewe's udder and starved them of nutrients for a week, thus halting the cells in G1. After one of these cells was fused with an enucleated egg from a different ewe (fusion of the donor cell and enucleated egg was achieved by electrical stimulation), mitotic inducers in the egg cytoplasm (see Chapter 9) were able to stimulate the donor nucleus to enter S phase and the rest of the cell cycle proceeded normally. After several cell divisions, the early embryo was transplanted into the womb of a surrogate mother. Of 272 successful attempts to fuse adult cells with enucleated eggs, one lamb, Dolly, survived to be born. Dolly was genetically identical to the ewe from whose udder the donor nucleus had been obtained.

The purpose of Wilmut's experiment was to clone sheep that have been genetically programmed to produce products such as pharmaceuticals in their milk (see Chapter 16). The cloning procedure could make multiple, identical copies of sheep that are reliable producers of a drug such as α-antitrypsin, which is used to treat people with cystic fibrosis. Not surprisingly, Wilmut's work touched off a storm of controversy, since Dolly was the first mammal to be born through cloning.

An example of nuclear totipotency gone awry occurs in a human tumor called a teratocarcinoma. Here, a differentiated cell "dedifferentiates" and divides, forming a tumor, as occurs in most cancers. But some

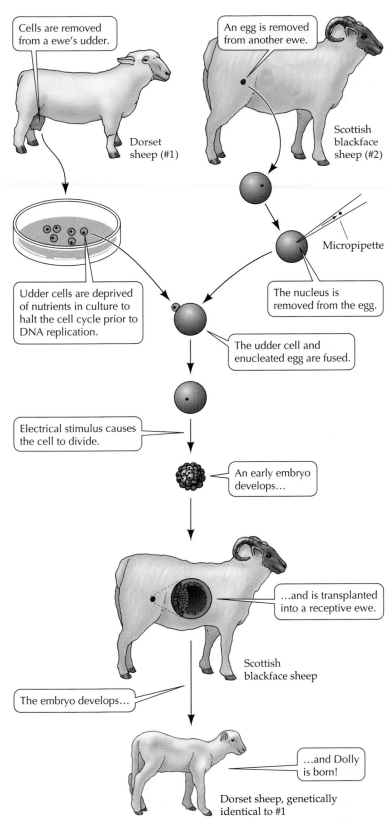

Cells are removed from a ewe's udder.

Dorset sheep (#1)

An egg is removed from another ewe.

Scottish blackface sheep (#2)

Udder cells are deprived of nutrients in culture to halt the cell cycle prior to DNA replication.

Micropipette

The nucleus is removed from the egg.

The udder cell and enucleated egg are fused.

Electrical stimulus causes the cell to divide.

An early embryo develops…

…and is transplanted into a receptive ewe.

Scottish blackface sheep

The embryo develops…

…and Dolly is born!

Dorset sheep, genetically identical to #1

15.4 Cloning a Mammal Dolly, a cloned sheep resulting from this experiment, has the same genes as the ewe that donated the udder cells.

cells in the tumor redifferentiate to form specialized tissue arrangements. So the tumor can be a single ball of cells inside the abdomen, with some of them forming kidney tubes, others hair, and still others teeth! How this occurs is not clear.

For a clear demonstration that genes are not lost from differentiated cells, we consider next a phenomenon observed in developing insects.

Transdetermination shows that developing insects retain their genes

The overall pattern of development in butterflies, moths, and many other insects is probably familiar to you. From a fertilized egg there develops a creeping *larva* that feeds voraciously, growing through a series of molts, in which the external coat is shed to permit the body to grow. Some specialized cells, arranged in clusters called **imaginal discs**, remain undifferentiated throughout larval growth but later give rise to the tissues of the adult, such as wings, antennae, and legs. The larva eventually stops feeding and then surrounds itself with a cocoon and transforms into a *pupa*. In the pupa, tremendous changes take place. Some larval cells die, others are reprogrammed to make different products that are characteristic of the adult, and the imaginal discs differentiate into new adult structures (Figure 15.5). Such a major transformation between larva and adult is referred to as **complete metamorphosis**.

The fates of the imaginal discs of insects are determined long before metamorphosis. If transplanted from one larva to another, an imaginal disc still develops into the same type of organ (for example, a wing or an antenna) it would have if left undisturbed. In addition, that organ forms wherever the imaginal disc for it is placed in the host body.

If transplanted into an *adult* insect, an imaginal disc remains undifferentiated (because the hormonal signal for its development is lacking), but the cells of the disc continue to divide within the new host. Later these

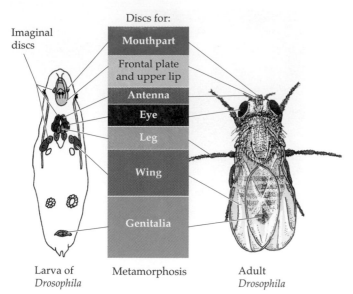

Discs for:

Imaginal discs

Mouthpart
Frontal plate and upper lip
Antenna
Eye
Leg
Wing
Genitalia

Larva of *Drosophila*   Metamorphosis   Adult *Drosophila*

15.5 Metamorphosis in Fruit Flies   In metamorphosis, most of the larval tissues die and are resorbed, providing building blocks for subsequent development. The remaining larval tissues are specialized imaginal discs, which differentiate to form the organs of the adult insect.

transplanted disc cells can be transplanted to other adult insects or to larvae. If returned to a larva, the disc cells almost always develop into the adult organ for which they were originally determined. Occasionally, however, an imaginal disc will **transdetermine** during a series of transplants; that is, it will develop into an organ other than that normally expected. Transdetermination shows that imaginal discs have not lost the genes that they normally do not express, since a disc *can*, under the right circumstances, express these genes and produce a different organ.

Many plant cells remain totipotent

A food storage cell in a carrot root faces a dark future. It is not destined to photosynthesize or to give rise to new carrot plants. However, if we isolate that cell from the root, maintain it in a suitable nutrient medium, and provide it with appropriate chemical cues, we can fool the cell into changing its behavior. In effect, the cell "thinks" it is a fertilized egg. It divides and gives rise to a typical carrot embryo and, eventually, a complete plant.

The ability to clone an entire carrot plant from a differentiated root cell indicates that the cell contains the entire carrot genetic library and that it can express the appropriate genes in the right sequence. Many cells from other plant species show similar behavior in the laboratory, and this ability to generate a whole plant from a single cell has been invaluable for genetically altering plants in biotechnology (see Chapter 16).

Genes are differentially expressed in cell differentiation

All of these experiments—nuclear transplants in frogs and sheep, imaginal disc manipulations in fruit flies, and plant cell cloning—point to the conclusion of genome constancy in all somatic cells of an organism. Molecular biology experiments have provided even more convincing evidence. For example, the gene for β-globin, one of the protein components of hemoglobin, is present and expressed in red blood cells as they form in the bone marrow of mammals. Is the same gene also present—but unexpressed—in nerve cells in the brain, which do not make the protein?

Nucleic acid hybridization (see Figure 14.4) can provide an answer. A probe for the β-globin gene can be applied to DNA from both immature red blood cells (recall that mature red blood cells lose their nuclei) and brain cells. In both cases, the probe finds its complement, showing that the β-globin gene is present in both types of cells.

On the other hand, if the probe is applied to cellular mRNAs rather than cellular DNA, it finds β-globin mRNA only in the red blood cells, and not in the brain cells. This result shows that the gene is expressed in only one of the two tissues. Many similar experiments have shown convincingly that differentiated cells lose none of the genes that were present in the fertilized egg.

What molecular program leads to this differential gene expression? One well-studied system is the conversion of undifferentiated muscle precursor cells, the myoblasts, into the large, multinucleated cells that make up skeletal muscle fibers (see Chapter 44). The key event that starts this conversion is the expression of *MyoD1* (*Myo*blast *D*etermining Gene *1*). The protein product of this gene is a transcription factor (MyoD1) with a helix-loop-helix domain (see Figure 14.18) that not only binds to promoters of the muscle-determining genes to stimulate their transcription, but also acts at the gene promoter of *MyoD1* to keep its levels high in the cells and in the offspring resulting from cell divisions.

Strong evidence for the controlling role of MyoD1 comes from experiments in which *MyoD1* mRNA is injected into the precursors of other cell types. For example, if *MyoD1* mRNA is put into fat cell precursors, they become reprogrammed to become muscle cells. Genes such as *MyoD1*, which code for proteins that direct fundamental decisions in development, often by regulating genes on other chromosomes, are called **selector genes**. These genes usually code for transcription factors. We will describe other selector genes, such as the homeotic genes of *Drosophila*, later in this chapter.

The Role of Cytoplasmic Segregation in Cell Determination

What initially stimulates the *MyoD1* promoter to begin transcription is unclear, but a chemical signal clearly is involved. In general, two overall mechanisms have been found to cause such signals to stimulate their target cells. In one mechanism—*cytoplasmic segregation*—a factor within eggs, zygotes, or precursor cells is unequally distributed such that it ends up in some cells or regions of cells and not others. In the second mechanism—*induction*—a factor is actively produced and secreted to induce the target cells to differentiate.

First we will consider cytoplasmic segregation, beginning with its role in distinguishing one end of an organism from the other.

Eggs, zygotes, and embryos develop polarity

Polarity—the difference of one end from the other—is obvious in development. Our heads are distinct from our feet, and the distal ends of our arms (wrists and fingers) differ from the proximal ends (shoulders). An animal's polarity develops early, even in the egg itself. Yolk may be distributed asymmetrically in the egg and the embryo, and other chemical substances may be confined to specific parts of the cell or may be more concentrated at one pole than at the other.

In some animals, the original polar distribution of materials in the egg's cytoplasm changes as a result of fertilization. As division proceeds, the resulting cells contain unequal amounts of the materials that were not distributed uniformly in the zygote. As we learned from the work of Briggs and King, Gurdon, and Wilmut, cell nuclei do not always undergo irreversible changes during early development; thus we can explain some embryological events on the basis of the *cytoplasmic* differences in cells.

Even a structure as apparently simple as a sea urchin egg has polarity. A striking difference between cells can be demonstrated very early in embryonic development. The Swedish biologist Sven Hörstadius showed in the 1930s that the development of sea urchin embryos that have been divided in half at the eight-cell stage depends on how they are separated.

If the embryo is split into "left" and "right" halves, with each half containing cells from both the upper and the lower half, normal-shaped but dwarf larvae develop. If, however, the cut separates the upper four cells from the lower four, the result is different. The upper four cells develop into an abnormal early embryo with large cilia at one end that cannot form a larva. The lower four cells develop into a small, somewhat misshapen larva with an oversized gut (Figure 15.6). For fully normal development, factors from both the upper and lower parts of the embryo are neces-

sary. In similar experiments on eggs, Hörstadius showed that material is distributed unequally between upper and lower parts already in the unfertilized egg.

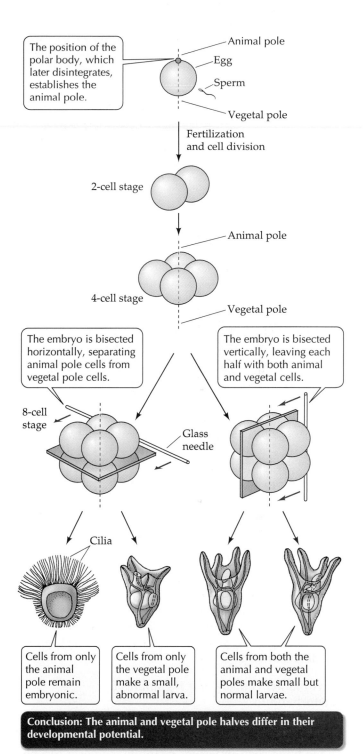

Conclusion: The animal and vegetal pole halves differ in their developmental potential.

15.6 Early Asymmetry in the Embryo Experiments by Sven Hörstadius showed that the upper (animal pole) and lower (vegetal pole) halves of very early sea urchin embryos differ in their developmental potential, and that cells from both halves are necessary to produce a normal larva.

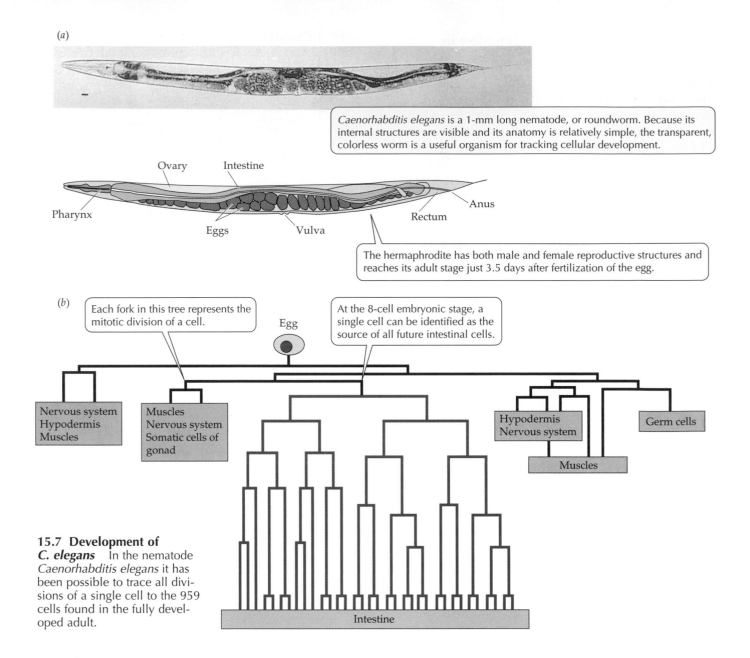

(a)

Caenorhabditis elegans is a 1-mm long nematode, or roundworm. Because its internal structures are visible and its anatomy is relatively simple, the transparent, colorless worm is a useful organism for tracking cellular development.

Ovary   Intestine

Pharynx

Eggs   Vulva   Rectum   Anus

The hermaphrodite has both male and female reproductive structures and reaches its adult stage just 3.5 days after fertilization of the egg.

(b)

Each fork in this tree represents the mitotic division of a cell.

Egg

At the 8-cell embryonic stage, a single cell can be identified as the source of all future intestinal cells.

Nervous system
Hypodermis
Muscles

Muscles
Nervous system
Somatic cells of gonad

Hypodermis
Nervous system

Germ cells

Muscles

Intestine

15.7 Development of C. elegans In the nematode *Caenorhabditis elegans* it has been possible to trace all divisions of a single cell to the 959 cells found in the fully developed adult.

These and other experiments are among many that established that the unequal distribution of materials in the egg cytoplasm plays a role in directing embryonic development. Such materials are called **cytoplasmic determinants**.

Microfilaments distribute P granules in Caenorhabditis

The tiny nematode (roundworm) *Caenorhabditis elegans* (Figure 15.7) lives in the soil, where it feeds on bacteria. As we have seen in earlier chapters, bacteria can be grown in the laboratory on petri plates containing medium in agar. *C. elegans* roundworms are grown on such cultures of bacteria. The entire process of development from the egg to larva takes about 12 hours at 25°C and is easily observed using a low-magnification dis-

secting microscope, because the body covering is transparent. The facts that *C. elegans* is easy to culture, develops rapidly, and is easily observed have not surprisingly made this worm a favorite organism of developmental biologists. Indeed, its entire genome is being sequenced.

Because the development of *C. elegans* does not vary, it has been possible to construct a cellular "tree" that describes the origin of each of the 959 somatic cells of the adult form. One way to do this is to inject a marker dye into an embryonic cell: That cell and its descendants will all be labeled. But in some cases, the worm itself "marks" the cells that will differentiate along a certain pathway.

One pathway in which cells are marked is the *germ line*, which consists of cells that can form gametes. Particles called **P granules** appear to be cytoplasmic determinants for this line of differentiation. The positions of P granules in the zygote and embryo are determined by the action of microfilaments. Before the zygote divides, the P granules collect at the posterior end of the cell (thus the term P—for "pole"). Thus, all the granules appear in only one of the first two daughter cells (Figure 15.8). The P granules continue to be precisely distributed during the early cell divisions, ending up in only those cells—the germ cells—that will eventually give rise to eggs and sperm. It is uncertain whether P granules are cytoplasmic determinants themselves; they may simply be distributed together with something else that is the "real" cytoplasmic determinant. Their composition gives no obvious clue, since they are made up of many proteins and RNA.

Having seen some examples of determination by cytoplasmic segregation, next we will examine induction—the second general mechanism of determination—in which some tissues induce the determination of other tissues.

The Role of Embryonic Induction in Cell Determination

Experimental work has clearly established that the fates of particular tissues are determined by interactions with other specific tissues in the embryo. In developing animal embryos there are many such instances of induction, in which one tissue causes an adjacent tissue to develop in a different manner. We will describe two examples of such induction: one in the developing vertebrate eye, and the other in a developing reproductive structure in the nematode *C. elegans*.

Tissues direct the development of their neighbors by secreting inducers

The development of the lens in the vertebrate eye is a classic example of induction. In a frog embryo, the developing forebrain bulges out at both sides to form the *optic vesicles*, which expand until they come in contact with the cells at the surface of the head (Figure 15.9). The surface tissue in the region of contact with the optic vesicles thickens, forming a *lens placode*. The lens placode bends inward, folds over on itself, and ultimately detaches from the surface to produce a structure that will develop into the lens.

If the growing optic vesicle is cut away before it contacts the surface cells, no lens forms in the head region from which the optic vesicle has been removed. An impermeable barrier placed between the optic vesicle and the surface cells also prevents the lens from forming. These observations suggest that surface tissue begins to develop into a lens when it receives a signal—an **inducer**—from its contact with an optic vesicle.

Distribution of nuclei | Distribution of P granules

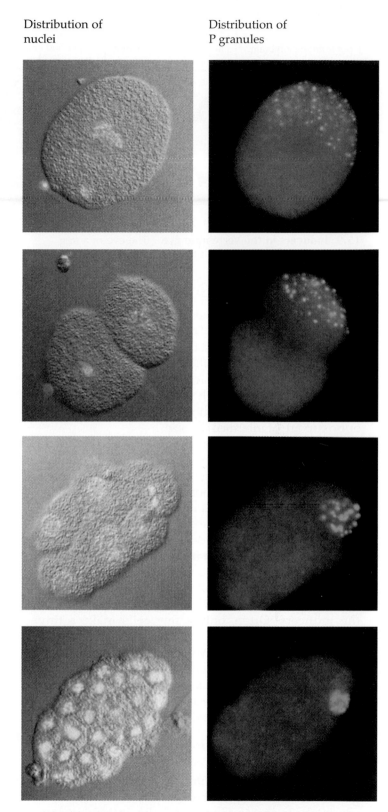

15.8 Distribution of P Granules in *C. elegans* These micrographs show a developing embryo of *Caenorhabditis elegans*. Whereas the P granules (bright spots) move to the posterior end of the embryo and are eventually confined to the single cell that gives rise to gametes, the nuclei (stained blue) of the same embryo are distributed evenly among its cells.

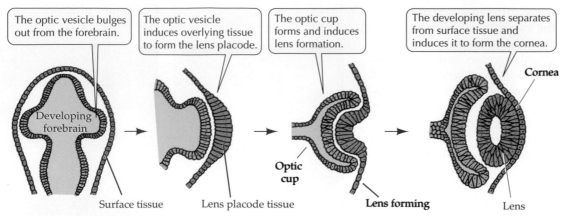

The optic vesicle bulges out from the forebrain.

The optic vesicle induces overlying tissue to form the lens placode.

The optic cup forms and induces lens formation.

The developing lens separates from surface tissue and induces it to form the cornea.

Developing forebrain

Cornea

Optic cup

Lens forming

Surface tissue   Lens placode tissue

Lens

15.9 Inducers in the Vertebrate Eye   The eye of a frog develops as inducers take their turns.

The interaction of tissues in eye development is a two-way street: There is a "dialogue" between the developing optic vesicle and the surface tissue. The developing lens determines the size of the optic cup that forms from the optic vesicle. If head surface tissue from a species of frog with small eyes is grafted over the optic vesicle of one with large eyes, both lens and optic cup are of intermediate size.

The developing lens also induces the surface tissue over it to develop into a cornea, a specialized layer that allows light to pass through and enter the eye. Thus a chain of inductive interactions participates in development of the parts required to make an eye. Induction triggers a sequence of gene expression in the responding cells. Tissues do not induce themselves; rather, different tissues interact and induce each other. We will return to embryonic induction in Chapter 40.

One of the most difficult problems in developmental biology has been identifying the chemical nature of the inducers. Often, only a few cells make a tiny amount of the substance. In some cases, specific diffusible proteins may be involved; the inducer that acts earliest in frog embryos appears to be a growth factor (see Chapter 40). In other cases, however, insoluble extracellular materials such as collagen and other proteins may be involved in induction.

Even single cells can induce changes in their neighbors

The tiny worm *Caenorhabditis elegans* is an excellent organism for studying the mechanisms of induction because, as we saw in Figure 15.7, the entire development of *C. elegans* takes only a few days and is clearly visible. The hermaphroditic form contains both male and female reproductive organs and lays eggs through a pore called the *vulva* on the ventral surface.

During development, a single cell, called the *anchor cell*, induces the vulva to form. If the anchor cell is de-stroyed by laser surgery, no vulva forms. The anchor cell controls the fates of six cells on the animal's ventral surface. Each of these cells has three possible fates. By the manipulation of two genetic switches, a given cell becomes a primary vulval precursor, a secondary vulval precursor, or simply part of the worm's surface, an epidermal cell (Figure 15.10).

The anchor cell produces an inducer that diffuses out of the cell and interacts with adjacent cells. Cells that receive enough of the inducer become vulval precursors; cells slightly farther from the anchor cell become epidermis. The first switch, controlled by the inducer from the anchor cell, determines whether a cell takes the "track" toward becoming part of the vulva or the pathway toward becoming epidermis.

The cell closest to the anchor cell, having received the most inducer, becomes the primary vulval precursor and produces its own inducer, which acts on the two neighboring cells and directs them to become secondary vulval precursors. Thus the primary vulval precursor cell controls a second switch, determining whether a vulval precursor will take the primary track or the secondary track. The two inducers control the activation or inactivation of specific genes in the responding cells.

There is an important lesson to draw from this example. Much of development is controlled by switches that allow a cell to proceed down one of two alternative tracks. One challenge for the developmental biologist is to find these switches and determine how they work. In the case of vertebrates, some progress has been made.

As we will describe in Chapter 40, a key event early in embryology is the induction of differentiation of a layer of cells called the mesoderm. In this case, several proteins have been identified as having inducing properties. Three of these proteins are growth factors, which act by binding to receptors on recipient cells and causing a signal transduction cascade of events, ultimately leading to altered gene transcription. Another inducer is a transcription factor, the vertebrate version of a fruit fly protein called wingless. This

15.10 Two Gene "Switches" in *C. elegans*
Two secreted proteins act as the primary and secondary inducers. The gene activation patterns triggered by these switches determine cell fate.

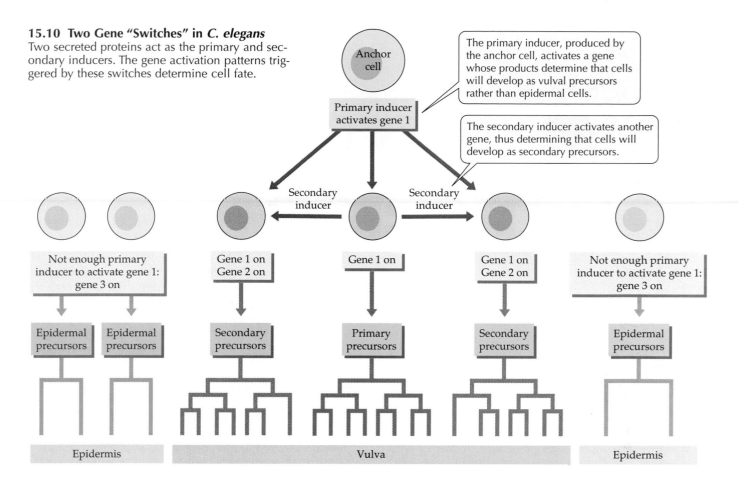

The primary inducer, produced by the anchor cell, activates a gene whose products determine that cells will develop as vulval precursors rather than epidermal cells.

The secondary inducer activates another gene, thus determining that cells will develop as secondary precursors.

"vertebrate version" of a fruit fly protein is yet another example of the conservation of developmental mechanisms through evolutionary time.

The Role of Pattern Formation in Organ Development

A highly active area of current research in developmental biology is the study of **pattern formation**, the spatial organization of a tissue or organism. Pattern formation is inextricably linked to *morphogenesis*, the appearance of body form. The differentiation of cells is beginning to be understood in terms of molecular events, but how do molecular events contribute to the organization of multitudes of cells into specific body parts, such as a leaf, a flower, a shoulder blade, or a tear duct? There are several different processes, including apoptosis, cell adhesion, the establishment of chemical gradients, and cell movements.

Animal development results in part from the movement of cells, as we saw in the example of induction in the developing frog eye (see Figure 15.9). Plant development is restricted in this regard. The cell wall anchors a plant cell in place, preventing movement. Let's first see how a particular form can develop when cells cannot move around.

Directed cell division and cell expansion form the plant body pattern

Although a plant cell cannot move, it may be able to elongate preferentially in one direction. In addition, the direction in which a cell divides is often regulated genetically. Cytoskeletal elements play determining roles in both of these events. The strongest elements in the cell wall are cellulose microfibrils whose orientation is determined by microtubules of the cytoskeleton. The orientation of the cellulose microfibrils, in turn, determines the direction of cell elongation (Figure 15.11*a*). Other microtubules form a preprophase band that determines the plane of cell division (Figure 15.11*b*).

Some cells are programmed to die

As described in Chapter 9, *apoptosis* is programmed cell death, a series of events caused by the expression of certain genes (see Figure 9.22). Some of these "death genes" have been pinpointed, and there are related ones in organisms as diverse as worms and humans.

Apoptosis is vital to the normal development of all animals. For example, the nematode *C. elegans* produces precisely 1,030 somatic cells when it develops from a fertilized egg to an adult (see Figure 15.7). But 131 of these cells die. The sequential expression of two

(a) Expansion

Cellulose microfibrils encircle the cell in a specific orientation and constrain cell expansion.

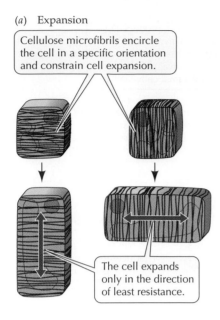

The cell expands only in the direction of least resistance.

(b) Division

The preprophase bands of cytoplasmic microtubules determine the orientation of cell division.

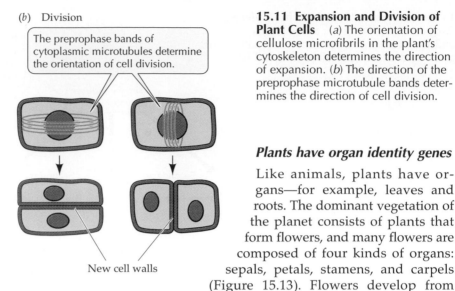

New cell walls

15.11 Expansion and Division of Plant Cells (a) The orientation of cellulose microfibrils in the plant's cytoskeleton determines the direction of expansion. (b) The direction of the preprophase microtubule bands determines the direction of cell division.

Plants have organ identity genes

Like animals, plants have organs—for example, leaves and roots. The dominant vegetation of the planet consists of plants that form flowers, and many flowers are composed of four kinds of organs: sepals, petals, stamens, and carpels (Figure 15.13). Flowers develop from groups of cells in the shape of domes at growing points on the stem. How does the dome give rise, in short order, to *whorls*—groups of organs that encircle a central axis—of four different organs? The answer involves the activities of a group of genes.

A group of **organ identity genes** work in concert to specify the successive whorls. We recognize the presence of organ identity genes because mutations in these genes lead to major alterations in flower structure. Table 15.1 describes the phenotypes of some of these mutations. Analyses of these mutant genes and their products are leading us to a preliminary understanding of how normal flowers develop. The developmental genetics of flowers is best understood for *Arabidopsis thaliana* (see Figure 15.1a) and snapdragons.*

*Not surprisingly, the gene products of organ identity genes are DNA-binding proteins. A single gene, *leafy*, appears to be important in the initiation of flower development in many species.

genes called *ced-4* and *ced-3* (for *c*ell *d*eath) appears to control this process. In the nervous system, for example, there are 302 nerve cells that come from 405 precursors; thus 103 cells undergo apoptosis. If either *ced-3* or *ced-4* is nonfunctional, all 405 cells form neurons and organizational chaos results. A third gene, *ced-9*, acts as an inhibitor of apoptosis: that is, its protein blocks the function of the *ced-4* gene. So, where cell death is required, *ced-3* and *ced-4* are active and *ced-9* is inactive, and where cell death does not occur, the reverse is true.

Remarkably, a similar system of cell death genes acts in humans. During early development, human hands and feet look like tiny paddles—the fingers and toes are linked by webbing. Between days 41 and 56 of development, cells in the webbing die, freeing the individual fingers and toes (Figure 15.12). The gene (*caspase*) that stimulates this apoptosis is similar in DNA sequence to *ced-3*, and a second gene (*bcl-2*) that inhibits apoptosis is similar to *ced-9*. So humans and worms, two creatures separated by more than 600 million years of evolutionary time, have similar genes controlling programmed cell death.

Apoptosis is essential to pattern formation in animal development and plays other roles in your life. The lens of your eye consists of the specialized remains of cells that have undergone apoptosis. The dead cells that form the outermost layer of your skin and those from the uterine wall that are lost during menstruation have also undergone apoptosis. The white blood cells live only a few months in the circulation, then undergo apoptosis. In a form of cancer called follicular large-cell lymphoma, these white blood cells do not die but continue to divide. The reason for this malfunction is that a genetic mutation has caused the overexpression of *bcl-2*, the gene that inhibits cell death.

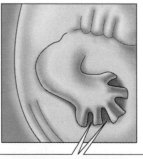

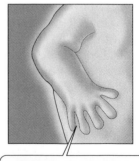

41 days after fertilization: Genes for programmed cell death are expressed only in the tissue between the digits.

56 days after fertilization: Apoptosis is complete. Cells of the digits have absorbed the remains of the dead cells.

15.12 Apoptosis Removes the Webbing between Fingers Early in the second month of human development, the webbing connecting the fingers is removed by apoptosis.

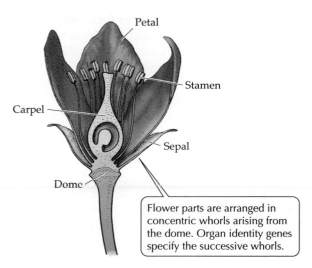

Flower parts are arranged in concentric whorls arising from the dome. Organ identity genes specify the successive whorls.

15.13 The Organs of a Flower The sepal is the outermost whorl, enclosing the petals, then the stamen, and finally the carpel. Mutations in organ identity genes alter the flower's structure.

TABLE 15.1 Organ Identity Mutations in Flower Development

GENOTYPE	PHENOTYPE			
	WHORL 1	WHORL 2	WHORL 3	WHORL 4
Wild type	Sepals	Petals	Stamens	Carpels
apetala 2 mutant	Carpels	Stamens	Stamens	Carpels
apetala 3 mutant	Sepals	Sepals	Carpels	Carpels
agamous mutant	Sepals	Petals	Petals	Sepals

In addition to being fascinating to biologists, these organ identity genes have caught the attention of agricultural scientists. The foods that constitute much of the human diet are seeds and fruits, such as the grains rice, wheat, and corn. Seeds and fruits form from the female reproductive organs on the flower. Thus, genetically manipulating the number of these organs on a particular plant could increase the amount of grain a crop could produce. More carpels mean more seeds—that is, a larger crop.

Plants and animals use positional information

Certain cells in both plants and animals appear to "know" where they are with respect to the body as a whole; this spatial sense is called **positional information**. In plants, the pattern of development of two major types of conducting tissue—one for water and minerals and the other for the products of photosynthesis—suggested long ago that distance from the body surface may play a role in their formation.

Cells destined to become water conductors are farther from the body surface than are those destined to become photosynthate conductors. Thus those destined to become water conductors are exposed to lower concentrations of O_2 and higher concentrations of CO_2, and these differences may help determine which genes are expressed in which parts of the stem and root. Recently it has been suggested that the cells on the surface of the stem secrete a protein or other signal that is more concentrated close to the surface than deeper in the stem. Other signals may diffuse from the stem tip and root tip, establishing positional information along the plant's axis. These ideas are still speculative.

Morphogens provide positional information in the developing limbs of animals

More concrete evidence is available concerning positional information in animal embryos than in plants. In the 1960s and 1970s, the English developmental biologist Lewis Wolpert developed a theory of positional information, based on gradients of morphogens in developing limb buds of chick embryos. A **morphogen** is a substance, produced in one place, that diffuses and produces a concentration gradient, with the result that different cells are exposed to different concentrations of the morphogen and thus develop along different lines.

A morphogen concentration gradient results in the development of a chick wing from a wing bud, which is a bulge on the surface of a 3-day-old embryo. Like any other three-dimensional object, a wing can be described in terms of three perpendicular axes. The *anteroposterior axis* of the wing is the axis that corresponds to the axis of the body running from the head to the tail. The *proximodistal axis* runs from the base of the limb to its tip, and the *dorsoventral axis* from the top of the wing to the undersurface. Each axis has a corresponding type of positional information. Here we will consider just the anteroposterior axis.

Pattern formation along the anteroposterior axis can be modified experimentally in ways that suggest that it is controlled at least in part by a particular part of the wing bud, called the *zone of polarizing activity*, or **ZPA**, that lies on the posterior margin of the bud. Different parts of the limb appear to develop normally at *specific distances from the ZPA*. This hypothesis is supported by the results of delicate surgery and grafting experiments such as those depicted in Figure 15.14, in which a ZPA from one bud is grafted onto another bud that still has its own ZPA.

The donor ZPA can be placed in different positions on different hosts. If the extra ZPA is grafted on the anterior margin, opposite from the host ZPA, the anteroposterior axis of the wing tip is duplicated, and two complete mirror images form (Figure 15.14*b*). If the extra ZPA is placed somewhat closer to the host ZPA, incomplete mirror images appear (Figure 15.14*c*): One digit is missing from each of the units, as if the

(a)

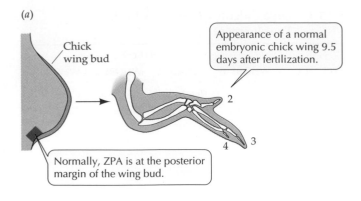

Chick wing bud

Appearance of a normal embryonic chick wing 9.5 days after fertilization.

Normally, ZPA is at the posterior margin of the wing bud.

(b)

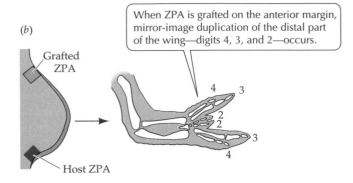

When ZPA is grafted on the anterior margin, mirror-image duplication of the distal part of the wing—digits 4, 3, and 2—occurs.

Grafted ZPA

Host ZPA

(c)

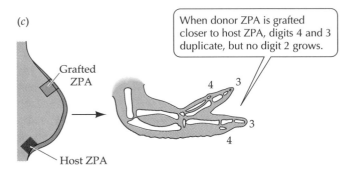

When donor ZPA is grafted closer to host ZPA, digits 4 and 3 duplicate, but no digit 2 grows.

Grafted ZPA

Host ZPA

(d)

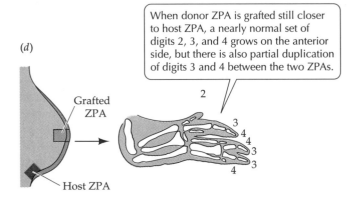

When donor ZPA is grafted still closer to host ZPA, a nearly normal set of digits 2, 3, and 4 grows on the anterior side, but there is also partial duplication of digits 3 and 4 between the two ZPAs.

Grafted ZPA

Host ZPA

Conclusion: Different parts of the limb develop at specific distances from the ZPA.

15.14 The ZPA Provides Positional Information In each experiment, the zone of polarizing activity from a donor wing bud was grafted onto a host wing bud that still had its own ZPA.

missing units are of a type that can form only if they are more than some minimum distance from a ZPA. In the third case (Figure 15.14d), with the two ZPAs close together, there is room for some duplication, but also enough room on the anterior side for a complete, nearly normal unit to form.

How might the ZPA produce these effects? The answers have been slow in coming and are still incomplete. However, the explanation includes at least two elements: a gene and a small molecule. The gene is called *sonic hedgehog* (for reasons not worth explaining here). This gene normally is expressed in the limb bud only where ZPA activity is greatest. If *sonic hedgehog* is inserted into other cells and caused to be expressed, those cells show ZPA activity when grafted into a developing limb bud. The sonic hedgehog protein appears to be a morphogen, secreted from the cells and forming a gradient in neighboring tissue.

The small molecule that plays a role in ZPA activity is retinoic acid, a derivative of vitamin A. Retinoic acid can replace the ZPA in transplant experiments. For example, if some retinoic acid inside a porous bead is placed in the anterior region, it induces a duplication of the digits. So which comes first—retinoic acid or sonic hedgehog? The answer appears to be the former. In the bead experiments, the vitamin derivative induces cells in the ZPA to produce sonic hedgehog, which is the true stimulus for cell differentiation.

Sonic hedgehog protein is also an important morphogen in mammals, and is especially active during the differentiation of parts of the nervous system. Mice that lack the gene for this protein die around the time of birth as a result of severe malformations of the brain. These abnormalities are very similar to a human disorder called holoprosencephaly, which is the cause of miscarriage pregnancies of one out of every 250 human fetuses. At least some cases of this devastating disease are caused by a mutation in the gene that codes for the human version of sonic hedgehog protein.

The Role of Differential Gene Expression in Establishing Body Segmentation

Another experimental subject that developmental biologists have used to study pattern formation is the *Drosophila* fruit fly. Insects (and many other animals) develop a highly modular body composed of different types of modules, called *segments*. Complex interactions of different sets of genes underlie the pattern formation of segmented bodies.

Unlike the body segments of segmented worms such as earthworms, the segments of the *Drosophila* body are different from one another. The *Drosophila* adult has a head (several fused segments), three different thoracic segments, eight abdominal segments, and a terminal segment at the posterior end. Thirteen seemingly identical segments in the *Drosophila* larva correspond to these specialized adult segments. Several types of genes are expressed sequentially in the embryo to define these segments. The first step in this process is to establish the polarity of the embryo.

Maternal effect genes determine polarity in Drosophila

In *Drosophila* eggs and larvae, polarity is based on the distribution of more than a dozen morphogens, of which some are mRNAs and some are proteins. These morphogens are products of specific **maternal effect genes** in the mother and are distributed to the eggs, often in a nonuniform manner. The cytoskeleton (especially microtubules) is essential to this process. Maternal effect genes produce effects on the embryo regardless of the genotype of the father. They determine the dorsoventral (back–belly) and anteroposterior (head–tail) axes of the embryo.

The fact that morphogens specify these axes has been established by the results of experiments in which cytoplasm was transferred from one egg to an-

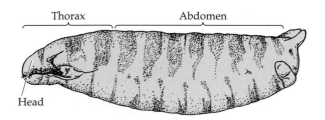

A normal larva produced by a wild-type female has normal body parts.

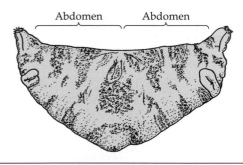

A larva produced by a female homozygous for a mutant allele of *bicoid*, one of the maternal effect genes, has no head or thorax. It consists of two hind ends.

15.15 Polarity Gone Wild The anteroposterior axis of *Drosophila* larvae arises from the interaction of several morphogens. If one of them is missing, the other predominates—in this case, the one that forms the abdomen.

other. Females homozygous for a particular mutation of the maternal effect gene *bicoid* produce larvae with no head and no thorax (Figure 15.15). However, if eggs of homozygous mutant *bicoid* females are inoculated at the anterior end with cytoplasm from the anterior region of a wild-type egg, the treated eggs develop into normal larvae—with heads developing from the part of the egg that receives the wild-type cytoplasm. On the other hand, removal of 5 percent or more of the cytoplasm from the anterior of a wild-type egg results in an abnormal larva that looks like a *bicoid* mutant larva.

Another maternal effect gene, *nanos*, plays a comparable role in the development of the posterior end of the larva. Eggs from homozygous mutant *nanos* females develop into larvae with missing abdominal segments. Injecting the *nanos* eggs with cytoplasm from the posterior region of wild-type eggs allows normal development. In wild-type larvae, the overall framework of anteroposterior and dorsoventral axes is laid down by the activity of the maternal effect genes. Their gene products are made by cells that surround and nurture the developing egg and are localized at certain specific regions of the egg as it forms.

After the axes of the embryo are determined, the next step in the segmentation process is to determine the larval segments.

Segmentation and homeotic genes act after the maternal effect genes

The number, boundaries, and polarity of the larval segments are determined by proteins encoded by the **segmentation genes**. The maternal effect genes set the segmentation genes in motion. Three classes of segmentation genes participate, one after the other, to regulate finer and finer details of the segmentation pattern (Figure 15.16).

First, **gap genes** organize large areas along the anteroposterior axis. Mutations in gap genes result in gaps in the body plan—the omission of several larval segments. Second, **pair rule genes** divide the embryo into units of two segments each. Mutations in pair rule genes result in embryos missing every second segment. Third, **segment polarity genes** determine the boundaries and anteroposterior organization of the segments. Mutations in segment polarity genes result in segments in which posterior structures are replaced by reversed (mirror-image) anterior structures.

Finally, after the basic pattern of segmentation has been established by the segmentation genes, differences between the segments are mediated by the activities of **homeotic genes**. These genes are expressed in different combinations along the length of the body and tell each segment what to become. Homeotic genes are analogous to the organ identity genes of plants, which are sometimes called homeotic-like genes.

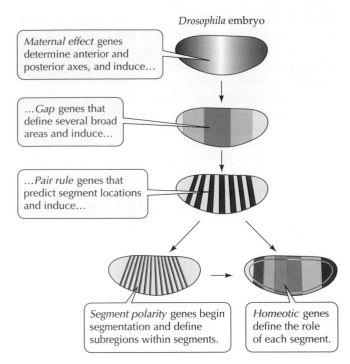

Maternal effect genes determine anterior and posterior axes, and induce…

…Gap genes that define several broad areas and induce…

…Pair rule genes that predict segment locations and induce…

Drosophila embryo

Segment polarity genes begin segmentation and define subregions within segments.

Homeotic genes define the role of each segment.

15.16 A Gene Cascade Controls Pattern Formation in the *Drosophila* Embryo Gap, pair rule, and segment polarity genes are collectively referred to as the segmentation genes. The shading shows the locations of the gene products, most of which are DNA-binding proteins.

To see how the maternal effect, segmentation, and homeotic genes interact, let's watch them "build" a *Drosophila* larva step by step, beginning with the unfertilized egg.

Drosophila *development results from a transcriptionally controlled cascade*

One of the most striking and important observations about development in *Drosophila*—and in other animals—is that it results from a sequence of changes, each change triggering the next. The sequence, or *cascade*, is controlled at transcription.

In general, unfertilized eggs are storehouses of mRNAs, which are made prior to fertilization to support protein synthesis during the early stages of embryo development. Indeed, early embryos do not carry out transcription. After several cell divisions, mRNA production resumes, forming the mRNAs needed for later development.

15.17 Bicoid and Nanos Protein Gradients Provide Positional Information Translation of mRNAs at the ends of the *Drosophila* larva leads to gradients of the morphogen products, which in turn control the expression of the gap genes.

Some of these prefabricated mRNAs in the egg provide positional information. Before the egg is fertilized, mRNA for the bicoid protein is localized at the end destined to become the anterior end of the animal. After the egg is fertilized and laid, and nuclear divisions begin, the *bicoid* mRNA is translated, forming bicoid protein that diffuses away from the anterior end, establishing a gradient of the protein (Figure 15.17). Another morphogen, the nanos protein, diffuses from the posterior end, forming a gradient in the other direction. Thus, each nucleus in the developing embryo is exposed to a different concentration of bicoid protein and to a different ratio of bicoid protein to nanos protein. What do these morphogens do?

The two morphogens regulate the expression of the gap genes, although in different ways. Bicoid protein affects transcription, while nanos affects translation. The high concentrations of bicoid protein in the anterior portion of the egg turn on a gap gene called *hunchback*, while simultaneously turning off another gap gene, *Krüppel*. Nanos at the posterior end reduces the translation of *hunchback*, so the difference in concentration of gap gene products at the two ends is established. The pattern of gap gene activity resulting from the actions of these morphogens is shown in Figure 15.18.

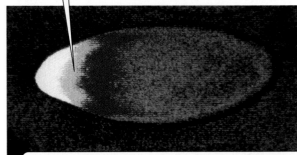

The concentration of bicoid protein is highest at the embryo's anterior end (bright yellow in this photograph).

The color of the gradient moves from orange to red as bicoid concentration decreases into the dark blue posterior end.

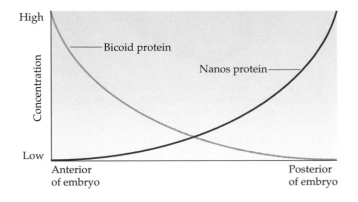

High

Low

Concentration

Bicoid protein

Nanos protein

Anterior of embryo

Posterior of embryo

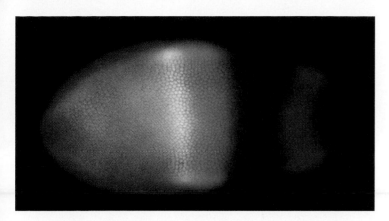

15.18 Gap Genes in Action Interactions of proteins encoded by gap genes define domains of the larval body in *Drosophila*. In this larva, hunchback (orange) and Krüppel (green) proteins overlap, forming a boundary (yellow) between two domains.

The proteins encoded by the gap genes are another set of transcription factors that control the expression of the pair rule genes. Many pair rule genes in turn encode transcription factors that control the expression of the segment polarity genes, giving rise to a complex, striped pattern (see Figure 15.16) that foreshadows the segmented body plan of *Drosophila*.

By this point, each nucleus of the embryo—it is still a single, multinucleated cell, or *syncytium*—is exposed to a distinctive set of transcription factors. The segmented body pattern of the larva is established even before any sign of segmentation is visible. When the segments do appear, they are not all identical, because the homeotic genes give the different segments their different structural and functional properties. Each homeotic gene is expressed over a characteristic portion of the embryo. The homeotic genes of *Drosophila* are arranged on the same chromosome—in the same order as the order in which they are expressed from the anterior to the posterior end of the larva.

Let's turn now to the homeotic genes and how their mutation can alter the course of development.

Homeotic mutations produce large-scale effects

Our present understanding of the genetics of pattern formation began with the discovery of dramatic mutations in *Drosophila* called homeotic mutations in which normal body parts are formed in inappropriate segments. Two bizarre examples are the *Antennapedia* mutant, in which legs grow in the place of antennae, and the *ophthalmoptera* mutant, in which wings grow in the place of eyes.

Homeotic genes fall into a few clusters on a chromosome. Of these, the best characterized is the **bithorax complex**. The eight or more genes of the bithorax complex control development of the abdomen and posterior thorax of the fly. The mutant fly shown at the beginning of this chapter resulted from mutations in the bithorax complex. Development of the head and anterior thorax is controlled by another homeotic gene cluster, the **Antennapedia complex**. The functions of the two complexes interact substantially.

If the entire bithorax complex is deleted, the larva produced, although highly abnormal, still has the normal number of segments. Thus the bithorax complex does not determine the number of segments. Rather, the segmentation genes determine the number and polarity of segments. During normal development, the homeotic genes act later than the segmentation genes and give each segment its distinctive character. For this reason, they are also called selector genes.

Homeobox-containing genes encode transcription factors

In the early 1980s Walter Gehring and his associates William McGinnis and Michael Levine, working in Switzerland, and, independently, Thomas Kaufman and Matthew Scott at Indiana University, undertook a study of the Antennapedia complex using the techniques of recombinant DNA technology.

These investigators set out to isolate and clone the *Antennapedia* (*Antp*) gene, a member of the Antennapedia complex, from *Drosophila*. As part of this study, they prepared a DNA that could hybridize to the *Antp* gene. To their surprise, the *Antp* DNA hybridized with both the *Antp* gene and a nearby segmentation gene (the *fushi tarazu* gene, abbreviated *ftz*, from the Japanese for "too few segments"). The *Antp* and *ftz* genes must have DNA sequences of close similarity, because part of each gene hybridizes with the same DNA.

Further hybridization studies demonstrated that this particular shared stretch of DNA is also found in the *bicoid* gene of the bithorax complex, in some other parts of the *Drosophila* genome, and in genes in other insect species. In fact, this important sequence of 180 base pairs of DNA, called the **homeobox**, has now been shown to be part of numerous genes of many animals and plants. The homeotic genes of animals contain the homeobox, but the organ identity genes of plants do not.

Mice and humans (the two best-studied mammals) have clusters of homeobox-containing genes. These genes are expressed in particular segments of the animal, just as in the fruit fly. Thirty-eight genes are divided into four clusters, each located on a different chromosome (Figure 15.19). As in *Drosophila*, these homeobox genes are arranged in the same order on each chromosome as they are expressed from anterior to posterior of the developing animal.

What is the function of the homeobox, which is present in almost all eukaryotes? The homeobox sequence

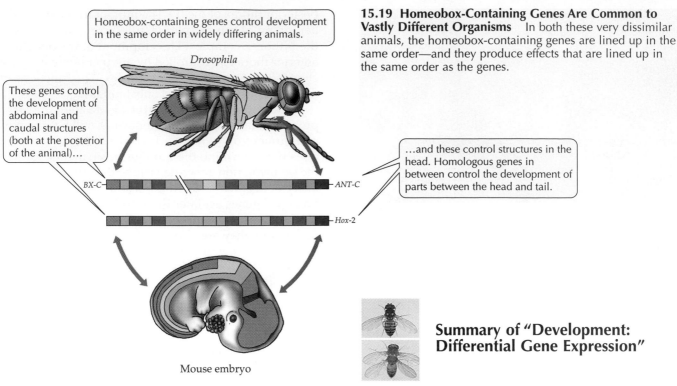

Homeobox-containing genes control development in the same order in widely differing animals.

Drosophila

These genes control the development of abdominal and caudal structures (both at the posterior of the animal)…

BX-C

ANT-C

…and these control structures in the head. Homologous genes in between control the development of parts between the head and tail.

Hox-2

Mouse embryo

15.19 Homeobox-Containing Genes Are Common to Vastly Different Organisms In both these very dissimilar animals, the homeobox-containing genes are lined up in the same order—and they produce effects that are lined up in the same order as the genes.

codes for a region of 60 amino acids—the **homeo-domain**—that is part of some proteins. Some of these proteins, the transcription factors, return to the nucleus and bind to DNA, regulating the transcription of other genes. A computerized search of published sequences of DNA from numerous species revealed a similarity between the homeobox and parts of certain regulatory genes in yeast—genes that produce proteins that also bind to specific DNA sequences. Some genes with homeoboxes are expressed only at certain times and in certain tissues as development proceeds, as would be expected if these proteins regulate development.

What are we to make of the presence of a homeobox in species as diverse as humans, fruit flies, frogs, nematodes, and tomatoes—and of its presence in several genes in the same organism? Its presence is consistent with the hypothesis that a single gene in an ancient organism was the evolutionary progenitor of what is now a widespread controlling system for development (see Chapter 23). One of the most astounding findings of recent developmental biology research has been this conservation of genes involved in developmental pathways.

As the DNA of more and more organisms is sequenced (see Chapter 17), more such similarities are emerging. Although there are certainly major differences in the end products of human and fruit fly development, the basic genetic mechanisms appear quite similar.

Summary of "Development: Differential Gene Expression"

The Processes of Development

• A multicellular organism develops through embryonic stages and eventually into an adult. Development continues until death. **Review Figure 15.1**
• Growth results from a combination of cell division and cell expansion.
• Differentiation produces specialized cell types. The overall form of the multicellular organism is the result of morphogenesis, resulting from pattern formation.
• In many organisms, the fates of the earliest embryonic cells have usually not yet been decided. These early embryonic cells may develop into different tissues if transplanted to other parts of an embryo. **Review Figure 15.2**
• As the embryo develops, its cells gradually become determined—committed to developing into particular parts of the embryo and into adult structures. Following determination, cells eventually differentiate into their final, often specialized, forms.

The Role of Differential Gene Expression in Differentiation

• The zygote is totipotent; it contains the entire genetic constitution of the organism and is capable of forming all adult tissues.
• Two lines of evidence show that differentiation does not involve permanent changes in the genome. First, nuclear transplant and cloning experiments show that the nucleus of a differentiated cell retains the ability to act like a zygote nucleus and stimulate the production of an entire organism. Second, molecular biological investigations have shown directly that all cells contain all genes for that organism, but that only certain genes are expressed in a given tissue. **Review Figures 15.3, 15.4**
• In insect metamorphosis, adult body parts arise from imaginal discs that normally have precise fates. Under certain circumstances imaginal discs may transdetermine and

develop into quite different parts of the body, showing that determination does not entail the loss of genes. Instead, cells become differentiated by the differential expression of genes. **Review Figure 15.5.**

The Role of Cytoplasmic Segregation in Cell Determination

• Unequal distribution of cytoplasmic determinants in the egg, zygote, or embryo leads to patterns of cell determination in normal development. Experimentally altering this distribution can alter gene expression and produce abnormal or nonfunctional organisms. **Review Figure 15.6**

• The nematode *Caenorhabditis elegans* provides a striking example of cytoplasmic segregation. The hermaphroditic form of this worm consists of 959 cells that develop from the fertilized egg by a precise pattern of cell divisions and other events. The gametes arise from embryonic cells containing granules that are probably cytoplasmic determinants. **Review Figures 15.7, 15.8**

The Role of Embryonic Induction in Cell Determination

• Some embryonic animal tissues direct the development of their neighbors by secreting inducers.

• Induction is often reciprocal: One tissue induces a neighbor to change, and the neighbor, in turn, induces the first tissue to change, as in eye formation in vertebrate embryos. **Review Figure 15.9**

• Induction in *C. elegans* can be very precise, with individual cells producing specific effects in just two or three neighboring cells. **Review Figure 15.10**

The Role of Pattern Formation in Organ Development

• Pattern formation triggers the sequence of cell divisions, cell expansions, cell movements, and programmed cell deaths that constitutes organ development and morphogenesis.

• Plant cells do not move, but cytoskeletal elements direct cell division and expansion, which form the plant body pattern. **Review Figure 15.11**

• In plants and animals, programmed cell death (apoptosis) is important in development. Some genes whose protein products regulate apoptosis have been identified. **Review Figure 15.12**

• Plants have organ identity genes that interact to cause the formation of sepals, petals, stamens, and carpels. Mutation of these genes causes undifferentiated cells to form a different organ. **Review Figure 15.13 and Table 15.1**

• Both plants and animals use positional information as a basis for pattern formation.

• Positional information is well understood in chick limb formation. Gradients of morphogens are established in the embryo. Cells at different distances from the zone of polarizing activity are exposed to different concentrations of the morphogens and thus respond differently. Some morphogens are transcription factors. **Review Figure 15.14**

The Role of Differential Gene Expression in Establishing Body Segmentation

• The fruit fly *Drosophila melanogaster* has provided much information about the development of body segmentation; some of this information applies to mice and other mammals.

• The first genes to act in determining *Drosophila* segmentation are maternal effect genes, such as *bicoid* and *nanos*, which encode morphogens that form gradients in the egg. These morphogens act on segmentation genes to define the anteroposterior organization of the embryo. **Review Figures 15.15, 15.16, 15.17**

• Segmentation develops as the result of a transcriptionally controlled cascade, the product of one gene promoting or repressing the expression of another gene. There are three kinds of segmentation genes, each responsible for a different step in segmentation. Gap genes organize large areas along the anteroposterior axis, pair rule genes divide the axis into pairs of segments, and segment polarity genes see to it that each segment has an appropriate anteroposterior axis. **Review Figure 15.16**

• The bicoid and nanos proteins act as a transcription factor and translation regulator, respectively, to control the level of expression of gap genes. Gap genes encode transcription factors that regulate the expression of pair rule genes. The products of the pair rule genes are transcription factors that regulate the segment polarity genes. **Review Figures 15.16, 15.17, 15.18**

• Activation of the segmentation genes leads to the activation of the appropriate homeotic genes in different segments. The homeotic genes define the functional characteristics of the segments. **Review Figure 15.16**

• Mutations in homeotic genes often have bizarre effects, causing structures to form in inappropriate parts of the body. Homeotic genes contain the homeobox, which encodes an amino acid sequence that is part of many transcription factors. The homeobox is found in key genes of distantly related species; thus numerous regulatory mechanisms may trace back to a single evolutionary precursor. **Review Figure 15.19**

Self-Quiz

1. Which statement about determination is true?
   - *a.* Differentiation precedes determination.
   - *b.* All cells are determined after two cell divisions in most organisms.
   - *c.* A determined cell will keep its determination no matter where it is placed in an embryo.
   - *d.* A cell changes its appearance when it becomes determined.
   - *e.* A differentiated cell has the same pattern of transcription as a determined cell.

2. The Briggs and King, Gurdon, and Wilmut experiments showed that
   - *a.* all nuclei of an organism have the same genes.
   - *b.* nuclei of embryonic cells can be totipotent.
   - *c.* nuclei of differentiated cells have different genes than zygote nuclei have.
   - *d.* differentiation is fully reversible in all cells of a frog.
   - *e.* differentiation involves permanent changes in the genome.

3. If an imaginal disc for a wing is transplanted into an adult fruit fly, and then put into a larva, it will
   - *a.* always form a wing.
   - *b.* divide and remain undifferentiated.
   - *c.* undergo programmed cell death.
   - *d.* sometimes form an organ other than a wing.
   - *e.* form an eye.

4. A major difference between early human and fruit fly embryology is that only in the latter
   a. does cytokinesis *not* occur.
   b. are polarity genes not expressed.
   c. is the fertilized egg totipotent.
   d. do nuclei become determined before differentiation.
   e. is *sonic hedgehog* expressed.

5. Which statement about cytoplasmic determinants in *Drosophila* is *not* true?
   a. They specify the dorsoventral and anteroposterior axes of the embryo.
   b. Their positions in the embryo are determined by microfilament action.
   c. They are products of specific genes in the mother insect.
   d. They often produce striking effects in larvae.
   e. They have been studied by the transfer of cytoplasm from egg to egg.

6. In fruit flies, the following genes are used to determine segment polarity: (k) gap genes; (l) homeotic genes; (m) maternal effect genes; (n) pair rule genes. In what order are these genes expressed during development?
   a. klmn
   b. lknm
   c. mknl
   d. nkml
   e. nmkl

7. Which statement about embryonic induction is *not* true?
   a. One tissue induces an adjacent tissue to develop in a certain way.
   b. It triggers a sequence of gene expression in target cells.
   c. It may be either instructive or permissive.
   d. A tissue may induce itself.
   e. The chemical identification of specific inducers has been difficult.

8. In the process of body segmentation in *Drosophila* larvae,
   a. the first steps are specified by homeotic genes.
   b. mutations in pair rule genes result in embryos missing every second segment.
   c. mutations in gap genes result in the insertion of extra segments.
   d. segment polarity genes determine the dorsoventral axes of segments.
   e. segmentation is the same as in earthworms.

9. Homeotic mutations
   a. are often so severe that they can be studied only in larvae.
   b. cause subtle changes in the forms of larvae or adults.
   c. occur only in prokaryotes.
   d. do not affect the animal's DNA.
   e. are confined to the zone of polarizing activity.

10. Which statement about the homeobox is *not* true?
    a. It is transcribed and translated.
    b. It is found only in animals.
    c. Some proteins containing the homeodomain bind to DNA.
    d. It is a stretch of DNA shared by many genes.
    e. Its activities often relate to body segmentation.

Applying Concepts

1. Molecular biologists can insert genes attached to high-level promoters into cells (see Chapter 16). What would happen if the following were inserted and overexpressed? Explain your answers.
   a. *ced-9* in embryonic nerve cell precursors in *C. elegans*
   b. *MyoD1* in undifferentiated myoblasts
   c. *sonic hedgehog* in a chick limb bud
   d. *nanos* at the anterior end of the *Drosophila* embryo

2. A powerful method to test for the function of a gene in development is to generate a "knockout" organism, in which the gene in question is inactivated. What do you think would happen in each of the following?
   a. knockout *C. elegans* for *ced-9*
   b. knockout *Drosophila* for *nanos*
   c. knockout *C. elegans* for P granules

3. Look at the chart of organ identity mutations in Table 15.1. What pattern do you perceive in the results of these mutations, and what might this pattern mean?

4. During development, the potential of a tissue becomes ever more limited, until, in the normal course of events, the potential is the same as the original prospective fate. On the basis of what you have learned in this chapter and in Chapter 14, discuss possible mechanisms for the progressive limitation of the potential.

5. How were biologists able to obtain such a complete accounting of all the cells in the roundworm *Caenorhabditis elegans*? Why can't we reason directly from studies of *C. elegans* to comparable problems in our own species?

Readings

Ameisen, J. C. 1996. "The Origin of Programmed Cell Death." *Science*, vol. 272, pages 1278–1279. Interesting speculation based on the observation of programmed cell death in four unicellular eukaryotes.

Duke, R. C., D. M. Ojcius and J. D.-E. Young. 1996. "Cell Suicide in Health and Disease." *Scientific American*, December. A discussion of the mechanisms and functions of apoptosis.

Gehring, W. J. 1985. "The Molecular Basis of Development." *Scientific American*, October. A clear account of homeotic mutations and the homeobox.

Gilbert, S. F. 1997. *Developmental Biology*, 5th Edition. Sinauer Associates, Sunderland, MA. An exceptionally well balanced treatment of developmental biology, covering both molecular and cellular concepts and embryology. Gives a feeling for the history of the discipline.

Holliday, R. 1989. "A Different Kind of Inheritance." *Scientific American*, June. A description of how patterns of gene activity are transmitted from one cell generation to the next.

Hynes, R. O. 1986. "Fibronectins." *Scientific American*, June. A look at proteins that guide migrating cells during development and their possible role in cancer.

Lawrence, P. A. 1992. *The Making of a Fly: The Genetics of Animal Design*. Blackwell Scientific, Oxford. A detailed, advanced treatment of the *Drosophila* material covered in this chapter.

Nüsslein-Volhard, C. 1996. "Gradients That Organize Embryo Development." *Scientific American*, August. Experiments conducted by the Nobel laureate author, her colleagues, and others to investigate morphogen gradients in *Drosophila* embryos.

Chapter 16

Recombinant DNA and Biotechnology

Making Medicine
This flask of genetically engineered mammalian cells is the starting point for the biotechnological production of a useful product, in this case a protein targeted to cancer cells.

John is at home preparing dinner when one side of his face starts to twitch. He tries to call out to his wife, but his speech is slurred. Like several million people each year, John is having a stroke because a blood clot has blocked a major artery in his brain, depriving nerve cells of essential oxygen. Normally, the body's clot-dissolving mechanism would be activated slowly and blood flow might eventually resume. But if John's brain waits for this slow process, it will be deprived of oxygen for so long that essential nerve cells will die. Instead, his emergency room physicians inject directly onto the clot a substance that initiates the dissolving process, a protein called tissue plasminogen activator. The clot quickly dissolves, restoring blood flow to the brain. John leaves the hospital a day later, his face and speech entirely normal.

In a small village in Africa, a one-year-old boy, Kwame, is exposed to *Neisseria meningitidis*. Three years earlier, his sister, at the time also one year old, had been infected by this bacterium. She developed meningitis, with the bacteria infecting her central nervous system. Within a day, a stiff neck and high fever led to seizures; she was dead a week later. Fortunately, her brother is receiving an antimeningitis vaccine, but in a very unusual way. The bananas that Kwame eats have been genetically engineered to express the gene for the vaccine protein. The vaccine protein successfully protects him, and he does not develop the disease.

The gene for human tissue plasminogen activator has been inserted into bacteria, which are used as factories for the production of large quantities of this rare protein. Genes encoding vaccine proteins have been inserted into the ba-

351

nana genome, and the fruits are being tested for their effectiveness as oral vaccines in children with limited access to health care personnel. These are but two of many examples of the uses of **DNA technology**, the ability of humans to manipulate genes almost at will. DNA technology has revolutionized much of experimental biology. Its use has been important in most of the recent advances in our understanding of how eukaryotic genes are regulated and organized. It is also changing agriculture, medicine, and other areas of the chemical industry, as well as forensics and the battle against environmental pollution.

In this chapter, we consider the basic techniques of DNA manipulation and some of its applications. Although many of these techniques have been called revolutionary, most of them come from the knowledge of DNA transcription and translation that we described in earlier chapters. After a specific sequence of DNA is isolated from cells or made in the chemistry laboratory, it can be introduced into almost any prokaryotic or eukaryotic cell. Such genetic transformation, often across species lines, has provided an invaluable tool for studying molecular physiology.

We will also see how the new gene in a host cell can be coaxed into making its protein product, as is the case with tissue plasminogen activator and with the meningitis vaccine protein. Finally, a process called the polymerase chain reaction (PCR) allows DNA to replicate in a test tube—a capability that has wide applications.

Cleaving and Rejoining DNA

Scientists have realized that the chemical reactions used in living cells for one purpose may be applied in the laboratory for other, novel purposes. Recombinant DNA technology, the manipulation and combination of DNA molecules from different sources, is based on this realization, and on an understanding of the properties of certain enzymes and of DNA itself. In this section we will identify the numerous naturally occurring enzymes that cleave DNA, help it replicate, and repair it. Many of these enzymes have been isolated and purified, and are now used in the laboratory to manipulate and combine DNA. Then we will see how fragments of DNA can be separated and covalently linked to other fragments.

As we saw in previous chapters, the nucleic acid base-pairing rules underlie many fundamental processes of molecular biology. The mechanisms of DNA replication, transcription, and translation rely on complementary base pairing. Similarly, all the key techniques in recombinant DNA technology—sequencing, rejoining, amplifying, and locating DNA fragments—make use of the complementary base pairing of A with T (or U) and of G with C.

Restriction endonucleases cleave DNA at specific sequences

All organisms must have mechanisms to deal with their enemies. As we saw in Chapter 13, bacteria are attacked by viruses called bacteriophages that inject their genetic material into the host bacterial cell and disrupt its operations. Eventually the phage genes may be replicated by the enzyme systems of the host, produce new viruses, and kill the host. Some bacteria defend themselves against such invasions by producing enzymes called **restriction endonucleases** that catalyze the cleavage of double-stranded DNA molecules—such as those injected by phages—into smaller, noninfectious fragments (Figure 16.1). The bonds cut are between the 3' hydroxyl of one nucleotide and the 5' phosphate of the next one.

There are many such restriction enzymes, each of which cleaves DNA at a specific site defined by a sequence of bases called a **recognition site** or **restriction site**. The DNA of the host bacterial cell is not cleaved by its own restriction enzymes, because of specific modifying enzymes called methylases that add methyl (—CH_3) groups to certain bases at the restriction sites when host DNA is being replicated. The methylation of the host's bases makes the recognition sequence unrecognizable to the endonuclease, thus preventing cleavage of the host DNA. But the unmethylated phage DNA is efficiently cleaved.

A specific sequence of bases defines each recognition site. For example, the enzyme *Eco*RI (named after

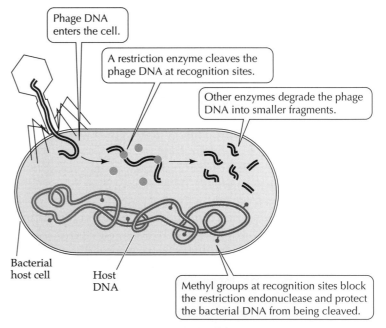

Phage DNA enters the cell.

A restriction enzyme cleaves the phage DNA at recognition sites.

Other enzymes degrade the phage DNA into smaller fragments.

Bacterial host cell

Host DNA

Methyl groups at recognition sites block the restriction endonuclease and protect the bacterial DNA from being cleaved.

16.1 Bacteria Fight Invading Viruses with Restriction Enzymes Bacteria produce restriction enzymes that cleave and degrade phage DNA. Other enzymes protect the bacteria's own DNA from being cleaved.

its source, a strain of the bacterium *E. coli*) cuts DNA only where it encounters the following paired sequence in the DNA double helix:

$$5' \ldots \text{GAATTC} \ldots 3'$$
$$3' \ldots \text{CTTAAG} \ldots 5'$$

Notice that this sequence reads the same in the 5'-to-3' direction on both strands. It is palindromic, like the word "mom," in the sense that the double-stranded "word" is the same in both directions. *Eco*RI has two identical subunits that cleave each strand between the G and A.

This recognition sequence occurs on average about once in 4,000 base pairs in a typical genome—or about once per four prokaryotic genes. So *Eco*RI can chop a large piece of DNA into smaller pieces containing, on average, just a few genes. For small genomes such as those of viruses that have only a few thousand base pairs, cleavage may result in a few fragments. For a huge eukaryotic chromosome with tens of millions of base pairs, the number of fragments is very large.

Of course, "on average" does not mean that the enzyme cuts all stretches of DNA at regular intervals. The *Eco*RI recognition sequence does not occur even once in the 40,000 base pairs of the genome of a phage called T7—a fact that is crucial to the survival of this virus, since its host is *E. coli*. Fortunately for the *E. coli* that makes *Eco*RI, the DNA of other phages does contain the recognition sequence. The ability to cleave the phage DNA at such sequences prevents the *E. coli* from being overrun by phages.

Hundreds of restriction enzymes have been purified from various microorganisms. In the test tube, different restriction enzymes that recognize different recognition sequences may cut the same sample of DNA. Thus, cutting a sample of DNA in many different, specific places is an easy task, and restriction enzymes can be used as "knives" for genetic "surgery."

Gel electrophoresis identifies the sizes of DNA fragments

After a sample of DNA has been digested with a restriction enzyme, the DNA is in fragments, each of which is bounded at its ends by the recognition sequence. As we noted, these fragments are not all the same size, and this property provides a way to separate the fragments from each other. Fragments are separated to determine the number and sizes of fragments produced, or to identify and purify an individual fragment.

The best way to separate DNA fragments is by **gel electrophoresis** (Figure 16.2). Because of its phosphate groups, DNA is negatively charged at

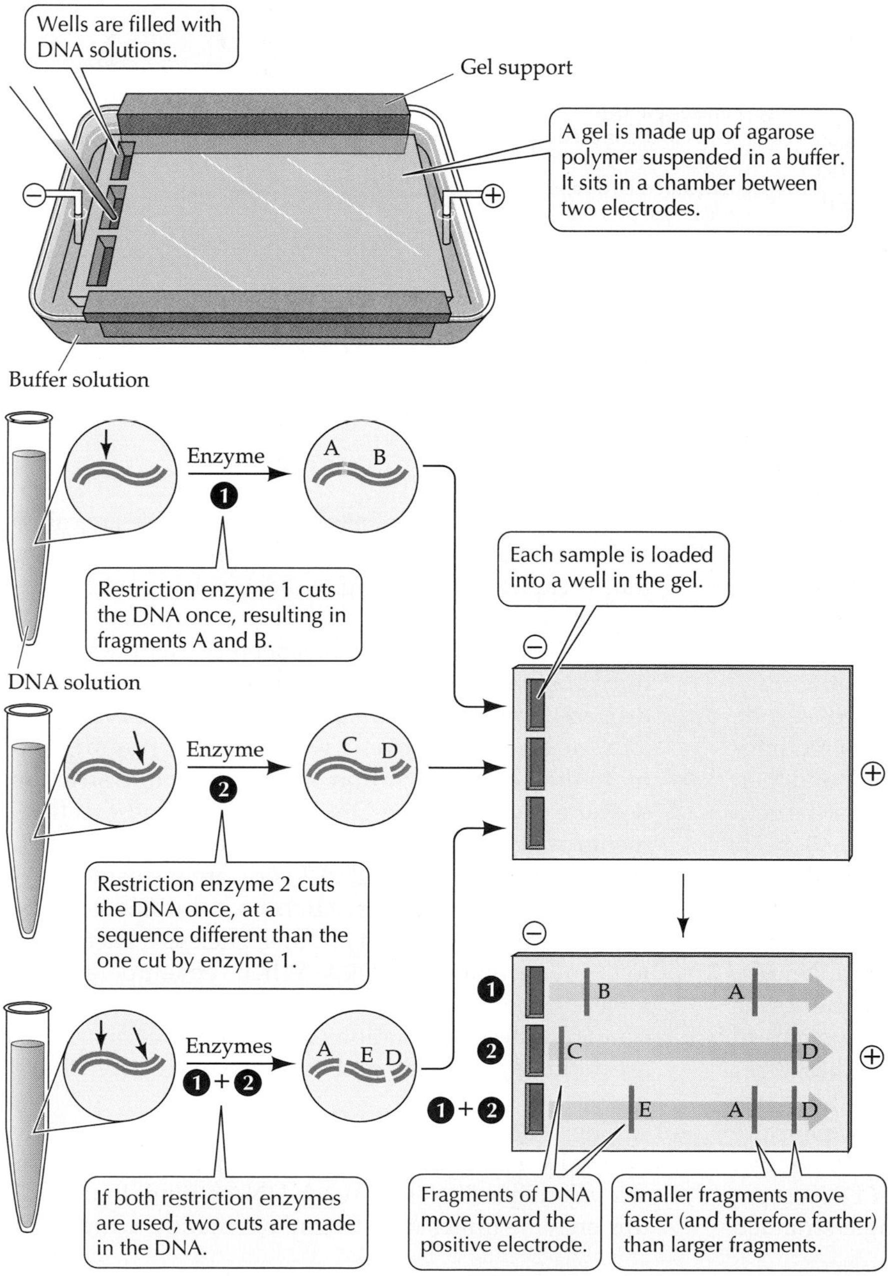

Wells are filled with DNA solutions.

Gel support

A gel is made up of agarose polymer suspended in a buffer. It sits in a chamber between two electrodes.

Buffer solution

DNA solution

Enzyme ❶ → A B

Restriction enzyme 1 cuts the DNA once, resulting in fragments A and B.

Enzyme ❷ → C D

Restriction enzyme 2 cuts the DNA once, at a sequence different than the one cut by enzyme 1.

Enzymes ❶+❷ → A E D

If both restriction enzymes are used, two cuts are made in the DNA.

Each sample is loaded into a well in the gel.

❶ B A
❷ C D
❶+❷ E A D

Fragments of DNA move toward the positive electrode.

Smaller fragments move faster (and therefore farther) than larger fragments.

16.2 Separating Fragments of DNA by Gel Electrophoresis A mixture of DNA fragments is placed in a gel and an electric field is applied across the gel. The negatively charged DNA moves toward the positive end of the field, with smaller molecules moving faster than larger ones. When the electric field is shut off, the separate fragments can be analyzed.

(a)

(b)

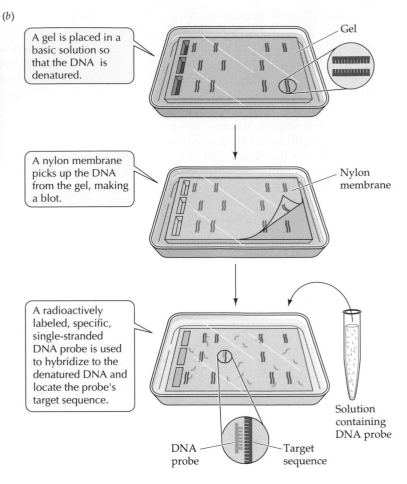

A gel is placed in a basic solution so that the DNA is denatured.

Gel

A nylon membrane picks up the DNA from the gel, making a blot.

Nylon membrane

A radioactively labeled, specific, single-stranded DNA probe is used to hybridize to the denatured DNA and locate the probe's target sequence.

Solution containing DNA probe

DNA probe

Target sequence

16.3 Analyzing DNA Fragments   (a) Identifying the DNA fragments on an electrophoresis gel. (b) Technique for using a hybridization probe to identify a specific DNA fragment.

neutral pH. In gel electrophoresis, a mixture of fragments is placed in a porous gel and an electric field (with positive and negative ends) is applied across the gel. Because opposite charges attract, the DNA moves toward the positive end of the field. Since the porous gel acts as a sieve, the smaller molecules move faster than the larger ones. After a fixed time and while all fragments are still on the gel, the electric power is shut off and the separated fragments can be examined or removed individually.

Different DNA samples may be "run" on a gel side by side in different "lanes." DNA fragments of known molecular size (in base pairs) are often run in a lane on each gel to provide a size reference. We can visualize the separated DNA fragments by staining them with a dye that fluoresces under ultraviolet light. Or, we can identify a specific DNA sequence by denaturing the DNA in the gel, affixing the denatured DNA to a nylon membrane to make a "blot" of the gel, and hybridizing the fragments with a specific single-stranded DNA probe (Figure 16.3). The gel region containing a desired fragment can be removed when the gel is sliced, and then the pure DNA fragment can be removed from the gel.

Recombinant DNA can be made in the test tube

An important property of some restriction enzymes is that they make staggered cuts in DNA rather than cutting both strands at a single base pair. For example, *Eco*RI cuts DNA within the recognition sequence, as shown at the top of Figure 16.4. After the two cuts in the opposing strands are made, they are held together only by the hydrogen bonding between four base pairs. The hydrogen bonds of these few base pairs are too weak to persist at warm temperatures (above room temperature), so the two strands of DNA separate, or denature. As a result, there are single-stranded

"tails" at the location of each cut. These tails are called *sticky ends* because they have a specific base sequence that is capable of binding by base pairing with complementary sticky ends.

After a DNA molecule has been cut with a restriction enzyme, the complementary sticky ends can hydrogen-bond to one another. The original ends may rejoin, or an end may pair with another fragment. If more than one recognition site for a given restriction enzyme is present in a DNA sample, numerous fragments can be made, all with the same sequence at their sticky ends. Because all *Eco*RI ends are the same, fragments from one source, such as a human, can be joined to fragments from another, such as a bacterium, to create recombinant DNA. When the temperature is lowered, the fragments anneal (come together) at random, but these associations are unstable because they are held together by only four pairs of hydrogen bonds.

The associated sticky ends can be permanently united by a second enzyme, **DNA ligase**, which forms a covalent bond to "seal" each DNA strand. In the cell, this enzyme unites new fragments made during DNA

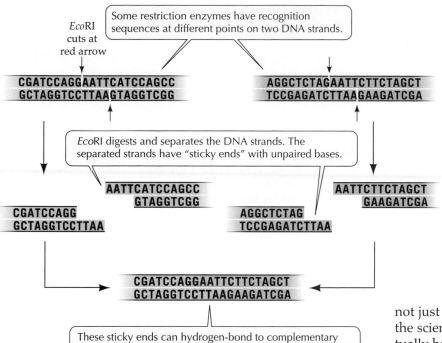

16.4 Cutting and Splicing DNA Some restriction enzymes (*Eco*RI is shown here) make staggered cuts in DNA from two different sources (shown here in blue and green). At warm temperatures, the two DNA strands will separate (denature), leaving "sticky ends" that can recombine with complementary fragments when the temperature is lowered.

replication and mends breaks in DNA (see Chapter 11). In the laboratory, DNA ligase can be used to seal breaks in DNA generated by reannealing the fragments from restriction enzyme digestion.

With these two enzyme tools—restriction endonucleases and DNA ligases—scientists can cut and rejoin different DNA molecules to form recombinant DNA (see Figure 16.4). This simple concept has changed the directions of biological science. For example, a piece of nonbacterial DNA can be inserted into a plasmid, a small circular chromosome often present in bacteria, provided that both the plasmid and the nonbacterial DNAs have the same recognition sequence.

The nonbacterial DNA is cleaved at many places, producing fragments with sticky ends. The circular plasmid DNA with one restriction site is cut, transforming it into a linear molecule with sticky ends. The sticky ends of the nonbacterial DNA hydrogen-bond to the sticky ends of the plasmid DNA. If ligase is then added, its activity produces a circular plasmid containing the nonbacterial DNA.

Many restriction enzymes do not produce sticky ends. Instead, they cut both DNA strands at the same base pair within the recognition sequence, making

"blunt" ends. Chemical methods have been developed to ensure the ligation of a DNA fragment into blunt ends of a target DNA.

Cloning Genes

The goal of recombinant DNA work is to manipulate host cells to produce many copies of a particular gene, either for purposes of analysis or to produce its protein product in quantity. In this section, we will discuss host selection, the entry of DNA into cells, and genetic markers. The choice of host cell, prokaryotic or eukaryotic, is important. Once the host species is selected, the DNA of interest is mixed with the cells and, under specific conditions, can enter some of them.

Because all the host cells proliferate—not just the few that receive the recombinant DNA—the scientist must be able to determine which cells actually have the targeted DNA sequence. One common method of identifying recombinant DNA cells is to tag the inserted sequence with genetic markers whose phenotypes are easily observed. For example, the inserted sequence might carry a marker gene for resistance to an antibiotic; thus any cells that don't die when exposed to the antibiotic must contain the new DNA.

Genes can be cloned in prokaryotic or eukaryotic cells

The initial successes of recombinant DNA technology were achieved with bacteria as hosts. As noted in preceding chapters, bacterial cells are easily grown and manipulated in the laboratory. Much of their molecular biology is known, especially for certain bacteria, such as *E. coli*, and numerous genetic markers can be used to select for cells harboring the manipulated DNA. In some important ways, however, bacteria are not ideal organisms for studying and expressing eukaryotic genes. Bacteria lack the splicing complex to excise introns from the initial RNA transcript of eukaryotic genes.

Many eukaryotic proteins are extensively modified after translation by reactions such as glycosylation and phosphorylation (see Chapter 14). Often these modifications are essential for the activity of the protein. Unfortunately, prokaryotes usually lack the machinery to perform these eukaryotic modifications. Finally, in some instances the addition of a new gene and its expression in a eukaryote are the point of the experiment. That is, the aim is to produce a *transgenic organism*. In these cases, the host for the new DNA may be a mouse, a wheat plant, yeast, or a human, to name a few examples.

Yeasts, such as *Saccharomyces*, the baker's and brewer's yeasts, are common eukaryotic hosts for recombinant DNA studies. Advantages of using yeasts include rapid cell division (a life cycle completed in 2 to 8 hours), ease of growth in the laboratory, and a relatively small genome size (about 20 million base pairs)—several times larger than that of *E. coli*, yet 1/300 the size of the mammalian one.

Plant cells can be used as hosts, especially if the desired result is a transgenic plant. The property that makes plant cells good hosts is their *totipotency*—that is, the ability of a differentiated cell to act like a fertilized egg and produce an entire new organism. Isolated plant cells grown in culture can take up recombinant DNA, and by manipulation of the growth medium, these transgenic cells can be induced to form an entire new plant. This plant can then be reproduced naturally in the field and will carry and express the gene carried on the recombinant DNA.

Vectors can carry new DNA into host cells

In natural environments, DNA released from one bacterium can sometimes be taken up by another bacterium and genetically transform that bacterium (see Chapter 10), but this is not common. The challenge of inserting new DNA into a cell is not just entry, but the replication of the molecule in the host cell as it divides. DNA polymerase, the enzyme that catalyzes replication, does not bind to just any sequence of DNA and begin the replication. Rather, like any DNA-binding protein, it recognizes a specific sequence, the *origin of replication* (see Chapter 11).

There are two general ways in which the newly introduced DNA can be part of a *replicon*, or replication unit. First, it can insert into the host chromosome after entering the cell. Although this is often a random event, it is nevertheless a common method of integrating a new gene into the host cell. Alternatively, the new DNA can enter the host cell as part of a carrier DNA that already has the appropriate origin of replication.

This carrier DNA, targeted at the host cell, is called a **vector**. In addition to its ability to replicate independently in the host cell, a vector must have two other properties. First, it must have sequences that allow the new DNA to be added to it—a recognition sequence for a restriction enzyme. Thus the vector must be able to form recombinant DNA. Second, the vector should have a marker that will announce to the scientist its presence in the host cell. Typically, this marker is a gene that codes for a protein whose phenotype is easily detected, such as resistance to a drug. Another useful property for a vector is its ease of isolation and manipulation, which usually is reflected by its small size in comparison to host chromosomes.

PLASMIDS AS VECTORS. The properties of plasmids make them ideal vectors for genes in bacteria. Each plasmid is a naturally occurring bacterial chromosome, with an origin of DNA replication. The plasmid is small, usually 2,000 to 6,000 base pairs, as compared to the main *E. coli* chromosome, which has more than 4 million base pairs. Because it is so small, a plasmid not surprisingly has only single sites for various restriction enzymes (Figure 16.5a). The fact that there is only one restriction site for a given enzyme is essential because it allows for insertion of new DNA at only that location (see Figure 16.4).

Many plasmids contain genes for enzymes that confer antibiotic resistance. This characteristic provides the marker for a host cell carrying the plasmid. It is relatively easy to determine if a colony of bacteria is resistant to an antibiotic (see Chapter 13). A final useful property of plasmids is the capacity to replicate independently of the host chromosome, often many times more than the host. It is not uncommon for a

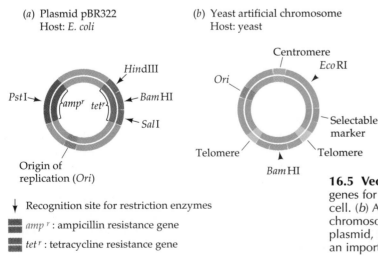

(a) Plasmid pBR322
Host: *E. coli*

*Hin*dIII
*Pst*I
amp^r *tet*^{r}
*Bam*HI
*Sal*I
Origin of replication (*Ori*)

(b) Yeast artificial chromosome
Host: yeast

Centromere
*Eco*RI
Ori
Selectable marker
Telomere
Telomere
*Bam*HI

↓ Recognition site for restriction enzymes

▬ *amp*^{r}: ampicillin resistance gene

▬ *tet*^{r}: tetracycline resistance gene

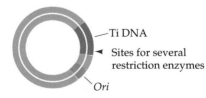

(c) Ti plasmid
Hosts: *Agrobacterium tumefaciens* (plasmid) and infected plants (Ti DNA)

Ti DNA
Sites for several restriction enzymes
Ori

16.5 Vectors for Carrying DNA into Cells (a) A plasmid with genes for antibiotic resistance can be incorporated into an *E. coli* cell. (b) A DNA molecule synthesized in the laboratory becomes a chromosome that can carry its inserted DNA into yeasts. (c) The Ti plasmid, isolated from the bacterium *Agrobacterium tumefaciens*, is an important vector for inserting DNA into many types of plants.

bacterial cell with a single main chromosome to have hundreds of copies of a recombinant plasmid.

VIRUSES AS VECTORS. Constraints to plasmid replication limit the size of new DNA that can be spliced into a plasmid to about 5,000 base pairs. Although a prokaryotic gene may be this small, 5,000 base pairs is much smaller than most eukaryotic genes, with their introns and extensive flanking sequences that are important in gene expression. So, a vector that accommodates larger DNA inserts is needed.

Both prokaryotic and eukaryotic viruses are often used as vectors for eukaryotic DNA. Bacteriophage lambda, which infects *E. coli*, has a DNA genome of more than 45,000 base pairs. If the genes that cause the host cell to die and lyse—about 20,000 base pairs—are eliminated, the virus can still infect a host cell and inject its DNA. The deleted 20,000 base pairs can be replaced with DNA from another organism, thereby creating recombinant viral DNA.

Because viruses infect cells naturally, they offer a great advantage as vectors over plasmids, which often require artificial means to coax them to enter cells. As we will see in Chapter 17, viruses are important vectors for delivering new genes to people in gene therapy.

ARTIFICIAL CHROMOSOMES AS VECTORS. Bacterial plasmids are not good vectors for yeast hosts, because prokaryotic and eukaryotic DNA sequences use different origins of replication. Thus a recombinant bacterial plasmid will not replicate in yeast. To remedy this problem, scientists have created in the laboratory a "minimalist chromosome" called the **yeast artificial chromosome**, or **YAC** (Figure 16.5*b*).

This DNA molecule has not only the yeast origin of replication, but sequences for the yeast centromere and telomere as well, making it a true chromosome. With artificially synthesized single recognition sites for restriction enzymes and useful marker genes for yeast (nutritional requirements), YACs are only about 10,000 base pairs in size but can accommodate 50,000 to 1.5 million base pairs of inserted DNA.

PLASMID VECTORS FOR PLANTS. An important vector for carrying new DNA into many types of plants is a plasmid that is found in *Agrobacterium tumefaciens*. This bacterium lives in the soil and causes a plant disease called crown gall, which is characterized by the presence of growths, or tumors, in the plant. *A. tumefaciens* contains a plasmid, called Ti (for *tumor-inducing*) (Figure 16.5*c*).

Part of the Ti plasmid is T DNA, a transposon that produces copies of itself in the chromosomes of infected plant cells. The T DNA has recognition sequences for restriction enzymes, and new DNA can be spliced into the T DNA region. When the T DNA is thus replaced, the plasmid no longer produces tumors, but the transposon, with the new DNA, is inserted into the host cell's chromosomes. The plant cell can then be grown in culture or induced to form a new, transgenic, plant.

There are many ways to insert recombinant DNA into host cells

Although some vectors, such as naturally infecting viruses, can enter host cells directly, most vectors require help to enter host cells. A major problem for DNA entry is that the exterior surface of the plasma membrane, with its phospholipid head groups, is negatively charged, and so is DNA. The resulting charge repulsion can be alleviated if the exterior of the cells and the DNA are neutralized with Ca^{2+} (calcium) salts. The salts reduce the charge effect, and the plasma membrane becomes permeable to DNA. In this way, almost any cell, prokaryotic or eukaryotic, can be transfected by taking up a DNA molecule from its environment. In plants and fungi, the cell walls must first be removed by hydrolysis with fungal enzymes; the resulting separated plant cells lacking cell walls are called protoplasts.

In addition to the "naked" DNA approach, DNA can be introduced into host cells by a variety of mechanical methods (Figure 16.6). In one, called *particle bombardment*, tiny high-velocity particles of tungsten or gold are coated with DNA and then shot into host cells. This "gene gun" approach must be undertaken with great care to prevent the cell contents from being damaged. In a second method, *electroporation*, cells are exposed to rapid pulses of high-voltage current. This treatment temporarily renders the plasma membrane permeable to DNA in the surrounding medium. A third mechanical way to insert DNA into cells is to inject it with a very fine pipette. This method is especially useful on large cells such as eggs. Finally, DNA can be coated in ways that allow it to pass through the plasma membrane. For example, it can be encased in liposomes, bubbles of lipid that fuse with the membranes of the host cell.

Genetic markers identify host cells that contain recombinant DNA

Following interaction with an appropriate vector, a population of host bacterial cells is heterogeneous, since only a small percentage of the cells have actually taken up the vector. Also, since the reaction of cutting the plasmid and inserting the new DNA to make recombinant DNA is far from perfect, only a few of the plasmids that have moved into the host cells actually contain the new DNA sequence. How can we select only the cells that contain the plasmid with the desired target DNA?

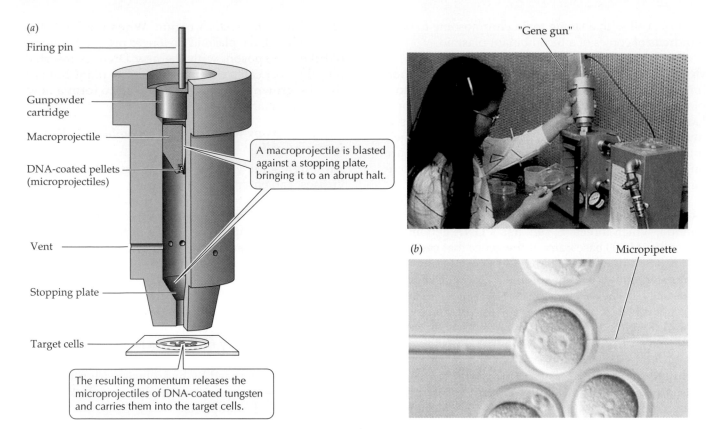

(a)

Firing pin

Gunpowder cartridge

Macroprojectile

DNA-coated pellets (microprojectiles)

A macroprojectile is blasted against a stopping plate, bringing it to an abrupt halt.

Vent

Stopping plate

Target cells

The resulting momentum releases the microprojectiles of DNA-coated tungsten and carries them into the target cells.

"Gene gun"

(b)

Micropipette

16.6 Methods of Mechanically Inserting DNA into Cells
(a) The "gene gun." A macroprojectile is blasted against a stopping plate, bringing it to an abrupt halt. The resulting momentum releases the microprojectiles of DNA-coated tungsten and carries them into the target cells. (b) A less equipment-intensive method is to inject DNA into a cell using a micropipette.

The experiment we are about to describe illustrates an elegant, commonly used approach to this problem. In this example, we use *E. coli* bacteria as hosts and the plasmid vector pBR322 (see Figure 16.5a), which carries the genes for resistance to the antibiotics ampicillin and tetracycline. When the plasmid is incubated with the restriction enzyme *Bam*HI, it encounters its recognition sequence, GGATCC, only once, at a site within the gene for tetracycline resistance.

If foreign DNA is inserted into this restriction site, the presence of these "extra" base pairs within the tetracycline resistance gene inactivates it. So plasmids containing the inserted DNA will carry an intact gene for ampicillin resistance but *not* an intact gene for tetracycline resistance. This is the key to selection of host bacteria that have the recombinant plasmid (Figure 16.7).

The cutting and splicing reaction results in three types of DNA, all of which can be taken up by the host bacteria, which normally are susceptible to killing by both ampicillin and tetracycline. The recombinant plas-

mid—the one we want—turns out to be the rarest type of DNA. Its uptake confers on host *E. coli* only resistance to ampicillin. More common are bacteria that take up pBR322 plasmid that has sealed back up on itself and retains intact genes for resistance to both antibiotics.

Still more common are bacteria that take up the targeted DNA sequence alone; since it is not part of a replicon, it does not survive as the bacteria divide. These host cells will remain susceptible to both antibiotics, as will the vast majority (more than 99.9 percent) of cells that take up no DNA. So the unique drug resistance phenotype of the cells with recombinant DNA marks them in a way that can be detected by simply altering the medium surrounding the cells with ampicillin and/or tetracycline.

In addition to genes for antibiotic resistance, several other marker genes are used to detect recombinant DNA in host cells. Scientists have created several artificial vectors in the laboratory that include sites for restriction enzymes within the *lac* operon (see Chapter 13). When this gene is inactivated by the insertion of foreign DNA, the vector no longer carries this operon's function into the host cell. Other "reporter genes" that have been used in vectors include the gene for luciferase, the enzyme that causes fireflies to glow in the dark when supplied with substrate. Green fluorescent protein, which does not require a substrate to glow, is now widely used (Figure 16.8).

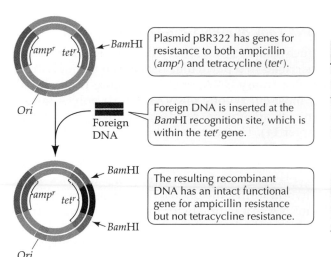

Plasmid pBR322 has genes for resistance to both ampicillin (*amp*^r) and tetracycline (*tet*^r).

Foreign DNA is inserted at the *Bam*HI recognition site, which is within the *tet*^r gene.

The resulting recombinant DNA has an intact functional gene for ampicillin resistance but not tetracycline resistance.

Detection of Recombinant DNA in *E. coli*		
DNA TAKEN UP BY *AMP*^S AND *TET*^S *E. COLI*	PHENOTYPE FOR AMPICILLIN	PHENOTYPE FOR TETRACYCLINE
None	Sensitive	Sensitive
Foreign DNA only	Sensitive	Sensitive
pBR322 plasmid	Resistant	Resistant
pBR322 recombinant plasmid	Resistant	Sensitive

16.7 Marking Recombinant DNA by Inactivating a Gene Scientists manipulate marker genes within plasmids so they will know which host cells have incorporated the recombinant genes. The bacteria in this experiment, which were "marked" by inactivating the gene for tetracycline resistance, could display any of the phenotype combinations indicated in the table; assuming we wish to select only those that have taken up the pBR322 *recombinant* plasmid, we can do so by altering antibiotic ingredients in the medium surrounding the cells.

After DNA uptake (or not), host cells are usually first grown on a solid medium for selection. If the concentration of cells dispersed on the solid medium is low, each cell will divide and grow into a distinct colony (see Chapter 13). The colonies that contain recombinant DNA can be picked off the medium and then grown in large amounts in liquid culture. The power of recombinant DNA technology to amplify a

gene is indicated by the fact that a 1-liter culture of bacteria harboring the human β-globin gene in pBR322 plasmid has as many copies of the gene as the sum total of all the cells in a typical human being.

Sources of Genes for Cloning

There are three principal sources of the genes or DNA fragments used in recombinant DNA work. One source is random pieces of chromosomes inserted into vectors; these DNA–vector units are maintained as gene libraries. The second source is complementary DNA, obtained by reverse transcription from specific mRNAs. The third source is DNA synthesized by organic chemists in the laboratory. The specific fragments can be deliberately modified to create mutations or to revert a mutant sequence back to wild type. We will elaborate on these sources in the following sections.

Genomic libraries contain pieces of a genome

The 23 human chromosomes (or 24, in a male) are a library that contain all the genes of the species. Each chromosome, or "volume" in the library, contains on average 100 million base pairs of DNA. Such a huge molecule is not very useful for studying genome organization or for isolating a specific gene. Thus each chromosome must be broken into smaller pieces, and each piece is then analyzed. These smaller fragments still represent a library; however, the information is now in many more volumes than 23. Each DNA fragment can be inserted into a vector, which can then be taken up by a host cell (Figure 16.9). Each host cell colony, then, harbors a single fragment of human DNA.

Using plasmids, which are able to insert only a few thousand base pairs of foreign DNA, about a million separate fragments are required to make a library of the human genome. For phage lambda, which can

16.8 A Reporter Gene Announces the Presence of a Vector in Eukaryotic Cells These cells have taken up a vector that expresses a gene producing green fluorescent protein.

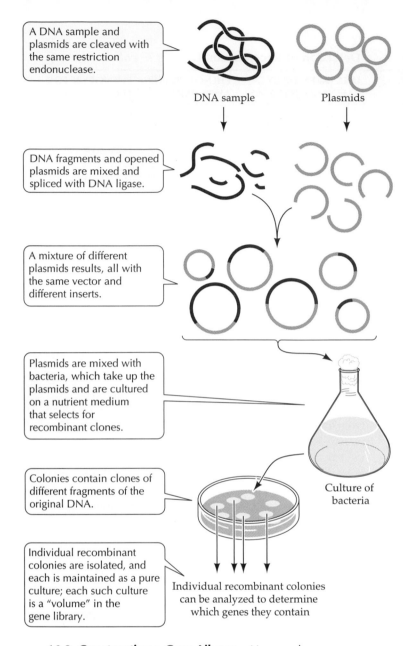

A DNA sample and plasmids are cleaved with the same restriction endonuclease.

DNA sample

Plasmids

DNA fragments and opened plasmids are mixed and spliced with DNA ligase.

A mixture of different plasmids results, all with the same vector and different inserts.

Plasmids are mixed with bacteria, which take up the plasmids and are cultured on a nutrient medium that selects for recombinant clones.

Colonies contain clones of different fragments of the original DNA.

Culture of bacteria

Individual recombinant colonies are isolated, and each is maintained as a pure culture; each such culture is a "volume" in the gene library.

Individual recombinant colonies can be analyzed to determine which genes they contain

16.9 Constructing a Gene Library Human chromosomes are broken up into fragments of DNA that are inserted into vectors (plasmids are shown here) and taken up by host bacterial cells, each of which harbors a single fragment of the human DNA. The information in these cell colonies constitutes a gene library.

carry four times as much DNA as a plasmid, the number of colonies for the library is reduced to about 250,000. Although this seems like a large number, a single growth plate can hold up to 80,000 phage colonies and is easily screened for the presence of a particular DNA sequence by denaturing the phage DNA and applying a particular probe for hybridization.

A DNA copy of mRNA can be made

A much smaller DNA library, that includes only genes transcribed in a particular tissue, can be made from complementary DNA, or cDNA (Figure 16.10). Recall that most eukaryotic mRNAs have a poly A tail—a string of adenosine residues at their 3′ end (see Chapter 14). The first step in cDNA production is to extract mRNA from a tissue and allow it to hybridize with a molecule (called oligo dT—the "d" indicates *d*eoxyribose) consisting of a string of T residues. After the hybrid forms, the oligo dT serves as a primer and the mRNA as a template for the enzyme reverse transcriptase, which synthesizes DNA from RNA templates. A cDNA strand complementary to mRNA is made; after it is removed from the template RNA, this DNA can be manipulated by cloning.

A collection of cDNAs from a particular tissue at a certain time in the lifetime of the organism is called a cDNA library. Because mRNAs are often present in small amounts and such libraries can be cloned, cDNA libraries have been invaluable in comparisons of gene expression in different tissues at different stages of development. For example, their use has shown that up to one-third of all the genes of an animal are expressed only during prenatal development. Complementary DNA is also a good starting point for the cloning of eukaryotic genes. It is especially useful for genes that are present in few copies and rarely expressed.

DNA can be synthesized chemically in the laboratory

When we know the amino acid sequence of a protein, we can obtain the DNA that codes for it by simply making it in the laboratory, using organic chemistry techniques. DNA synthesis has even been automated, and at many institutions a special service laboratory can make short to medium-length sequences overnight for any number of investigators.

How do we design a synthetic gene? Using the genetic code (see Figure 12.5) and the known amino acid sequence, we can figure out the appropriate base sequence for the gene. With this sequence as a starting point, we can add other sequences such as codons for translation initiation and termination, and flanking sequences for transcription initiation, termination, and regulation. Of course, these noncoding DNA sequences must be the ones actually recognized by the host cell if the synthetic gene is to be transcribed. It does no good to have a prokaryotic promoter sequence near a gene if that gene is to be inserted into a yeast for expression. Codon usage is also important: Many amino acids have more than one codon, and different organisms stress the use of different synonymous codons.

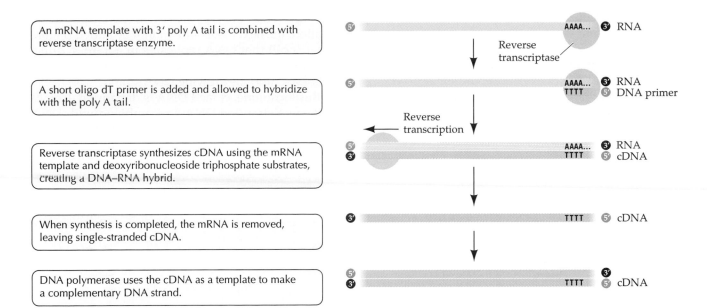

An mRNA template with 3' poly A tail is combined with reverse transcriptase enzyme.

A short oligo dT primer is added and allowed to hybridize with the poly A tail.

Reverse transcriptase synthesizes cDNA using the mRNA template and deoxyribonucleoside triphosphate substrates, creating a DNA–RNA hybrid.

When synthesis is completed, the mRNA is removed, leaving single-stranded cDNA.

DNA polymerase uses the cDNA as a template to make a complementary DNA strand.

16.10 Synthesizing Complementary DNA Gene libraries that include only genes transcribed in a particular tissue at a particular time can be made from complementary DNA. cDNA synthesis is especially useful for identifying genes that are present only in a few copies, and is often a starting point for gene cloning.

DNA can be mutated in the laboratory

Mutations that occur in nature have been important in proving cause-and-effect relationships in biology. For example, in Chapter 14 we learned that some people with the disease beta thalassemia have a mutation at the splice site for intron removal and so cannot make proper β-globin mRNA. This example shows the importance of the splice site consensus sequence.

Recombinant DNA technology has allowed us to ask the "What if?" questions without having to look for mutations in nature. Because synthetic DNA can be made in any sequence desired, we can clone such DNA to create specific mutations and then see what happens when the mutant DNA expresses itself in the host cell. Additions, deletions, and base-pair substitutions are all possible on isolated or synthetic DNA.

These mutagenesis techniques have allowed scientists to bypass the search for naturally occurring mutant strains, leading to many cause-and-effect proofs. For example, it was proposed that the signal sequence at the beginning of a secreted protein is essential to its insertion across endoplasmic reticulum membranes. So, a gene coding for such a protein, with the codons for the signal sequence deleted, was made. Sure enough, when this gene was expressed in yeast cells, the protein did not cross the ER membrane.

Mutagenesis has also begun to be useful in the design of specific drugs. The advent of a new branch of biology called computational biology has led to so-phisticated studies of the three-dimensional shapes and chemical properties of enzymes, substrates, and their possible regulators. Attempts are being made to devise rules to predict the tertiary structure of a protein from its primary structure. For example, the three-dimensional design of a polypeptide that regulates an enzyme might be proposed. It could then be made and induced to mutate to test the relationship between structure and activity.

Some Additional Tools for DNA Manipulation

Biological methods are not the only ways that DNA can be managed in the laboratory. In this section, we examine two important enzyme-based techniques that are invaluable, and both can be automated. First, the nucleotide sequence of DNA can be determined, and second, a DNA sequence can be replicated many times by the polymerase chain reaction. (Not surprisingly, both of these techniques earned Nobel prizes for their discoverers.) Finally, we will describe the use of antisense RNA to block the translation of specific mRNAs.

The nucleotide sequence of DNA can be determined

Cloned DNA fragments can be isolated and their nucleotide sequences determined. This ability is important in distinguishing true protein-coding genes from any surrounding "spacer" DNA, in order to study gene structure and identify the base-pair changes in mutations. Although there are several methods for sequencing, a simple one invented by Frederick Sanger has been automated and is being used in the Human Genome Project, an undertaking in which the entire 3 billion bases of the human chromosome set are being sequenced.

The method for sequencing relies on the mechanism for DNA replication. The nucleoside triphosphates (NTPs) that are the normal substrates for DNA replication use the sugar 2-deoxyribose; thus they are referred to as dNTPs. If instead of deoxyribose, the sugar used in the NTPs is 2,3-dideoxyribose, the resulting nucleoside triphosphates (ddNTPs) will be picked up by DNA polymerase and added to the growing DNA chain.

However, because ddNTPs lack a hydroxyl group at the 3' position, the next nucleotide cannot be added (Figure 16.11). (Recall that DNA replicates in a 5'-to-3' direction.) Thus synthesis stops at the place where ddNTP is incorporated into the growing DNA strand.

In this technique, cloned DNA is denatured. Copies of the single-stranded DNA are combined with DNA polymerase (to synthesize the complementary strand),

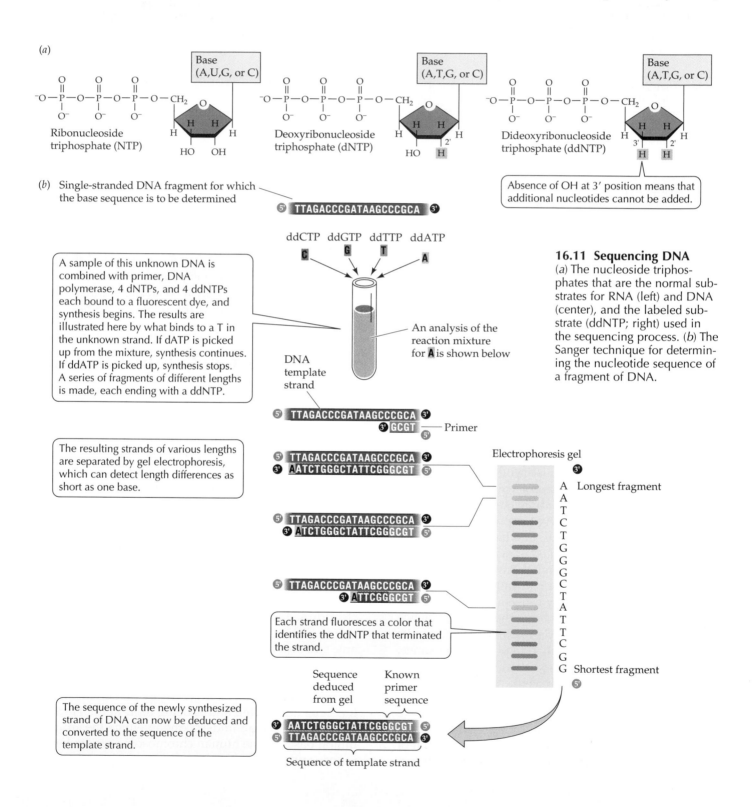

(a)

Ribonucleoside triphosphate (NTP)

Deoxyribonucleoside triphosphate (dNTP)

Dideoxyribonucleoside triphosphate (ddNTP)

Absence of OH at 3' position means that additional nucleotides cannot be added.

(b) Single-stranded DNA fragment for which the base sequence is to be determined

5' TTAGACCCGATAAGCCCGCA 3'

ddCTP   ddGTP   ddTTP   ddATP
  C   G   T   A

A sample of this unknown DNA is combined with primer, DNA polymerase, 4 dNTPs, and 4 ddNTPs each bound to a fluorescent dye, and synthesis begins. The results are illustrated here by what binds to a T in the unknown strand. If dATP is picked up from the mixture, synthesis continues. If ddATP is picked up, synthesis stops. A series of fragments of different lengths is made, each ending with a ddNTP.

An analysis of the reaction mixture for **A** is shown below

DNA template strand

16.11 Sequencing DNA
(a) The nucleoside triphosphates that are the normal substrates for RNA (left) and DNA (center), and the labeled substrate (ddNTP; right) used in the sequencing process. (b) The Sanger technique for determining the nucleotide sequence of a fragment of DNA.

5' TTAGACCCGATAAGCCCGCA 3'
                3' GCGT 5' — Primer

The resulting strands of various lengths are separated by gel electrophoresis, which can detect length differences as short as one base.

5' TTAGACCCGATAAGCCCGCA 3'
3' AATCTGGGCTATTCGGGCGT 5'

Electrophoresis gel

5' TTAGACCCGATAAGCCCGCA 3'
   3' ATCTGGGCTATTCGGGCGT 5'

A Longest fragment
A
T
C
T
G
G
C
T
A
T
T
C
G
G Shortest fragment

5' TTAGACCCGATAAGCCCGCA 3'
         3' ATTCGGGCGT 5'

Each strand fluoresces a color that identifies the ddNTP that terminated the strand.

Sequence deduced from gel   Known primer sequence

The sequence of the newly synthesized strand of DNA can now be deduced and converted to the sequence of the template strand.

3' AATCTGGGCTATTCGGGCGT 5'
5' TTAGACCCGATAAGCCCGCA 3'

Sequence of template strand

primers, the four dNTPs (dATP, dGTP, dCTP and dTTP), and small amounts of the four ddNTPs, each of them with a fluorescent "tag" that emits a different color of light. The reaction mixture soon contains a DNA mixture made up of the cloned DNA single strands (whose sequence will be determined) and shorter complementary strands. The latter, each ending with a ddNTP, are of varying lengths.

For example, each time a T is reached on the template strand, the growing complementary strand adds either dATP or ddATP. If dATP is added, the strand grows until a ddNTP stops it. If ddATP is added, chain growth terminates at that point. After DNA replication has been allowed to proceed for a while in a test tube, the numerous short fragments are denatured from the template and separated by gel electrophoresis (see Figure 16.11). During the electrophoresis run, the strands pass through a laser beam that excites the fluorescent tags. The light emitted is then detected, and the information—that is, which ddNTP is at the end of a strand of which length—is fed into a computer, which processes it and prints out the sequence.

Increasingly powerful analytical tools are being developed to analyze DNA sequences. In the computer, these sequences may be scanned for protein-coding regions, promoters, start and stop codons, and intron–exon boundaries. This type of analysis has been an es-

sential step in the isolation of genes. Alternatively, sequences can be electronically "deposited" in a computerized data bank and compared to other known DNA sequences. Such comparisons have yielded many surprises, such as the presence of common protein-coding regions in regulatory genes for development in organisms ranging from fruit flies to humans (see Chapter 15).

The polymerase chain reaction makes multiple copies of DNA in the test tube

Cloning in a host organism is not the only way to obtain many copies of a particular DNA sequence for analysis. In fact, for DNA molecules whose sequences are at least partly known, cloning in cells has been largely replaced by the **polymerase chain reaction** (PCR) in test tubes. PCR is not very complicated: A short region of DNA, such as a gene, is copied many times in the test tube by DNA polymerase.

PCR is a cyclic process in which the following sequence of steps is repeated over and over again (Figure 16.12). Double-stranded DNA is denatured by heat into single strands, short primers for DNA replication are added to the 3' end of the single DNA template strands, and DNA polymerase catalyzes the production of complementary new strands. A single cycle, taking a few minutes, doubles the amount of DNA

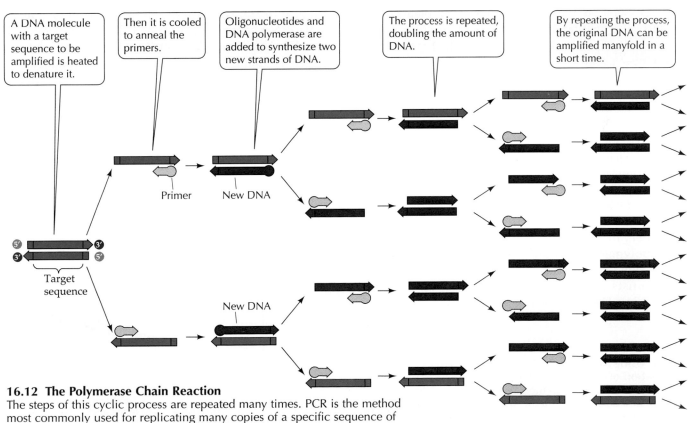

16.12 The Polymerase Chain Reaction
The steps of this cyclic process are repeated many times. PCR is the method most commonly used for replicating many copies of a specific sequence of DNA; this is referred to as amplifying the sequence.

and leaves the new strands in the double-stranded state. Repeating the cycle many times can theoretically lead to a geometric increase in the number of copies of the DNA sequence.

The PCR technique requires that the base sequences at the 3' end of the target DNA be known so that a complementary primer, usually 15 to 20 bases long, can be made in the laboratory. Because of the uniqueness of DNA sequences, usually the two primers will bind to only one region of DNA in the organism's genome. For example, the two primers

$$5'\text{--AGACTCAGAGAGAACCC--}3'$$

and

$$3'\text{--GGGGCACCAGAAACTTA--}5'$$

will bind to a region of human DNA that surrounds the β-globin gene and no other region. Thus, in a mixture of fragments that contains the entire 3 billion base pairs of the human genome, these two primers can pinpoint the β-globin gene, which represents just 1 millionth of the human genome. This specificity in the face of the incredible diversity of target DNAs is a key to the power of PCR.

One potential problem with PCR involves its temperature requirements. To denature the DNA during each cycle, it must be heated to more than 90°C. Then it must be cooled to about 55°C to allow the primer to hydrogen-bond to the single strands of template DNA. Because the heating step destroys most DNA polymerases, the PCR method would not be feasible because new polymerase must be added during each cycle after denaturation—an expensive and laborious proposition.

This problem has been solved by nature: In the hot springs at Yellowstone National Park, as well as other locations, there live bacteria called, appropriately, *Thermus aquaticus*. The means by which these organisms survive temperatures up to 95°C was investigated by bacteriologist Thomas Brock and his colleagues. They discovered that *T. aquaticus* has an entire metabolic machinery that is heat-resistant, including DNA polymerase that does not denature at this high temperature.

Scientists pondering the problem of amplifying DNA by PCR read Brock's basic research articles and got a clever idea: Why not use *T. aquaticus* DNA polymerase in the PCR reaction? It would not be denatured and thus would not have to be added during each cycle. The idea worked, and earned biochemist Kerry Mullis a Nobel prize. PCR has had enormous impact. Some of its most striking applications will be described later in this chapter and in Chapter 17.

Antisense RNA and ribozymes prevent specific gene expression

The base-pairing rules not only can be used to make genes; they can be employed to stop the translation of mRNA. As is often the case, this is an example of scientists imitating nature. In normal cells, a rare method of the control of gene expression is the production of an RNA molecule that is complementary to mRNA. This complementary molecule is called **antisense RNA** because it binds by base pairing to the "sense" bases that make up the coding sequence on mRNA. The formation of the double-stranded RNA hybrid prevents tRNA from binding to the mRNA, and the hybrid tends to be broken down in the cytoplasm. So, although the gene continues to be transcribed, this is an effective method of preventing translation—the synthesis of a protein from its mRNA.

In the laboratory, after determining the sequence of a gene and its mRNA, scientists can add antisense RNA to the cell to prevent translation of the mRNA (Figure 16.13). The antisense RNA can be added as itself—RNA can be taken up by cells in the same way that DNA is introduced—or it can be made in the cell by transcription from a DNA molecule introduced as a part of a vector. This technique is especially useful if a tissue-specific promoter is used to prime antisense transcription, so that its expression occurs only in a targeted tissue. In either case, translation of the targeted gene is prevented. An even more effective way to ensure that antisense RNA works is to couple the

16.13 Using Antisense RNA to Block Translation of an mRNA Once a gene's sequence is determined in the laboratory, it may be desirable to prevent the synthesis of its protein. This can be done with antisense RNA that is complementary to the mRNA.

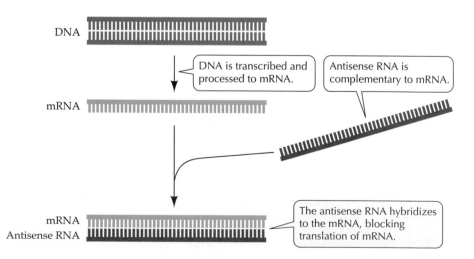

DNA

DNA is transcribed and processed to mRNA.

Antisense RNA is complementary to mRNA.

mRNA

mRNA
Antisense RNA

The antisense RNA hybridizes to the mRNA, blocking translation of mRNA.

antisense sequence to a special RNA sequence, a **ribozyme**, that catalyzes the cutting of its target RNA.

Antisense RNA (with or without a ribozyme) has been widely used to test cause-and-effect relationships. For example, it can be used to repress the synthesis of a specific protein—something that would otherwise be very difficult. Using antisense RNA, the synthesis of a protein essential for the growth of cancer cells has been blocked, and the cells have reverted to a normal state.

Biotechnology: Applications of DNA Manipulation

Biotechnology is the use of microbial, plant, and animal cells to produce materials useful to people. These products include foods, medicines, and chemicals. Some of them have been made biologically for a long time. For example, using yeasts to brew beer and wine dates back at least 8,000 years in human history, and using bacterial cultures to make cheese and yogurt is a technique many centuries old.

For a long time, people were not aware of the cellular bases of these biochemical transformations. About 100 years ago, thanks largely to Pasteur's work, it became clear that bacteria, yeasts, and other microbes could be harnessed as biological converters to make defined products. Alexander Fleming's discovery that the mold *Penicillium* makes the antibiotic penicillin led to the large-scale commercial culture of microbes to produce vast quantities of antibiotics, as well as other useful chemicals. Today, microbes are harnessed to make much of the industrial-grade alcohol, glycerol, butyric acid, and citric acid that are used by themselves or in manufacturing other products.

In the past, the list of such products was limited to those that were naturally made by the microbes. The many products that eukaryotic cells make, such as hormones and certain enzymes, had to be extracted from those complex organisms. Yields were low, and purification was difficult and costly. All this has changed with the advent of gene cloning. The insertion of almost any gene into bacteria or yeasts, along with methods to induce the genes to make their products, have turned these microbes into versatile factories for important products.

In this section we will describe how this is done and what results have been achieved in the production of pharmaceuticals. With increasing knowledge of the controls of gene activity, as well as new methods to insert genes into eukaryotic cells, it is possible to genetically transform complex organisms. Genetic transformation has been especially successful in agriculture, and a beginning at human gene therapy has been made (see Chapter 17). Finally, we describe some of the many uses of PCR.

Expression vectors turn cells into factories

If a eukaryotic gene is inserted into a plasmid such as pBR322 (see Figure 16.5a) and cloned in *E. coli*, little if any of the gene product of the eukaryotic gene will be made by the host cell. The reason is that the eukaryotic gene lacks the bacterial promoter for RNA polymerase binding, the terminator for transcription, and a special sequence on mRNA for ribosome binding. All of these are necessary for the eukaryotic gene to be expressed and its products synthesized in the bacterial cells.

Expression vectors have all the characteristics of typical vectors, as well as the extra sequences needed for the foreign gene to be expressed in the host cell. For bacteria, these additional sequences include the promoter, the transcription terminator, and the sequence for ribosome binding (Figure 16.14). For eukaryotes, also included might be the poly A addition site, transcription factor binding sites, and enhancers. Once these sequences are placed at the appropriate location on the vector, the gene will be expressed.

An expression vector can be refined in various ways. For example, a specific promoter can be used that responds to hormonal stimulation so that the foreign gene can be induced to transcribe its mRNA only when the scientist adds the hormone. Hormonal stimulation might also activate the promoter so that transcription and protein production occur at very high rates—a goal of obvious importance in the manufacture of an industrial product. Another refinement is to add a DNA sequence to the gene that leads either to the packaging of the expressed protein or to its secretion outside of the cell. Again, this is a great convenience for purification of the product.

Medically useful proteins can be made by DNA technology

Many medically useful products have been made by recombinant DNA technology (Table 16.1), and hundreds more are in various stages of development. We will focus on three of these products (the first of which was introduced at the beginning of this chapter) to illustrate exactly how scientists have developed pharmaceutically important products by recombinant DNA methods.

TISSUE PLASMINOGEN ACTIVATOR TO DISSOLVE BLOOD CLOTS. In most people, when a wound begins bleeding, a blood clot soon forms to stop the flow. Later, as the wound heals, the clot dissolves. How does the blood perform these conflicting functions at the right times? Mammalian blood has an enzyme called *plasmin* that catalyzes the dissolution of the clotting proteins. But plasmin is not always active; if it were, a blood clot would dissolve as soon as it formed! Instead, plasmin is "stored" in the blood in an inactive form called *plasminogen*. The conversion of plasminogen to plasmin is

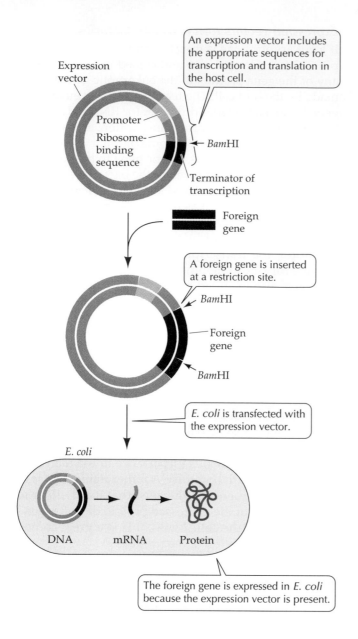

Expression vector

An expression vector includes the appropriate sequences for transcription and translation in the host cell.

Promoter

Ribosome-binding sequence

BamHI

Terminator of transcription

Foreign gene

A foreign gene is inserted at a restriction site.

BamHI

Foreign gene

BamHI

E. coli is transfected with the expression vector.

E. coli

DNA   mRNA   Protein

The foreign gene is expressed in E. coli because the expression vector is present.

16.14 An Expression Vector Allows a Foreign Gene to Be Expressed An inserted eukaryotic gene may not be expressed in *E. coli* because the eukaryotic gene lacks the necessary bacterial sequences for promotion, termination, and ribosome binding. Expression vectors contain these additional sequences, enabling the eukaryotic protein to be synthesized in the prokaryotic cell.

activated by an enzyme appropriately called *tissue plasminogen activator* (TPA), which is produced by cells lining the blood vessels. Thus, the reaction is

$$\text{plasminogen} \xrightarrow{\text{TPA}} \text{plasmin}$$
$$\text{(inactive)} \qquad\qquad \text{(active)}$$

Heart attacks and strokes are caused by blood clots that form in important blood vessels leading to the heart or the brain, respectively. During the 1970s, a bacterial enzyme, streptokinase, was found to stimulate the quick dissolution of clots in some patients with these afflictions. Treating people with this enzyme saved lives, but there were side effects. The drug was a protein foreign to the body, so people's immune systems reacted against it. Of more importance, the drug sometimes prevented clotting throughout the circulatory system, leading to an almost hemophilia-like condition in some people.

The discovery of TPA and its isolation from human tissues led to the hope of a human protein that would not provoke an immune reaction but would specifically bind at the clot, as TPA does. But the amounts of TPA available from human tissues were tiny, certainly not enough to inject at the site of a clot in a patient in the emergency room.

Recombinant DNA technology solved the problem. An antibody against human TPA was used to isolate

TABLE 16.1 Some Medically Useful Products of Biotechnology

PRODUCT	USE
Brain-derived neurotropic factor	Stimulates regrowth of brain tissue in patients with Lou Gehrig's disease
Colony-stimulating factor	Stimulates production of white blood cells in patients with AIDS
Erythropoietin	Prevents anemia in patients undergoing kidney dialysis
Factor VIII	Replaces clotting factor missing in patients with hemophilia A
Growth hormone	Replaces missing hormone in people of short stature
Insulin	Stimulates glucose uptake from blood in some people with diabetes
Platelet-derived growth factor	Stimulates wound healing
Tissue plasminogen activator	Dissolves blood clots after heart attacks and strokes
Vaccine proteins: hepatitis B, herpes, influenza, Lyme disease, meningitis, pertussis, etc.	Prevent and treat infectious diseases

the mRNA for TPA from cell extracts (Figure 16.15). The antibody precipitated not only finished TPA in the cytoplasm, but also TPA still being made on the ribosome but sufficiently folded for antibody recognition. Once TPA mRNA was available, it was used to make a cDNA copy, which was then inserted into an expression vector and introduced into *E. coli*. The transfected bacteria made the protein in quantity, and it soon became available commercially. This protein has had considerable success in dissolving blood clots in both heart attack and, especially, stroke patients.

ERYTHROPOIETIN TO REDUCE ANEMIA. Another widely used protein made through recombinant DNA methods is *erythropoietin* (EPO). The kidneys produce this hormone, which travels through the blood to bone marrow, where it stimulates the division of stem cells to produce red blood cells. People who have suffered kidney failure often require a procedure called kidney dialysis to remove toxins from the organ. However, because dialysis also removes EPO, these patients can become severely anemic (depleted of red blood cells).

As with TPA, the amounts of EPO that can be obtained from healthy people to give to these patients are extremely small, but once again, biotechnology has come to the rescue. The gene for EPO was isolated and cloned in an expression vector in bacteria. Large amounts of the protein are produced by the bacteria, and it is now given to tens of thousands of kidney dialysis patients, with great success at reducing anemia.

HUMAN INSULIN TO STIMULATE GLUCOSE UPTAKE. One of the first important medications made by recombinant DNA methods was human insulin. This hormone, normally made by the pancreas gland, stimulates cells to take up glucose from the blood. People who have certain forms of diabetes mellitus have a deficiency of pancreatic insulin. Injections of the hormone can compensate for this deficiency.

In the past, the injected insulin was obtained from the pancreases of cows and pigs, which caused two problems. First, animal insulin is laborious to purify; second, it is slightly different in its amino acid sequence from human insulin. Some diabetics' immune systems detect these differences and react against the foreign protein.

The ideal solution is to use human insulin, but until the advent of recombinant DNA technology, it was

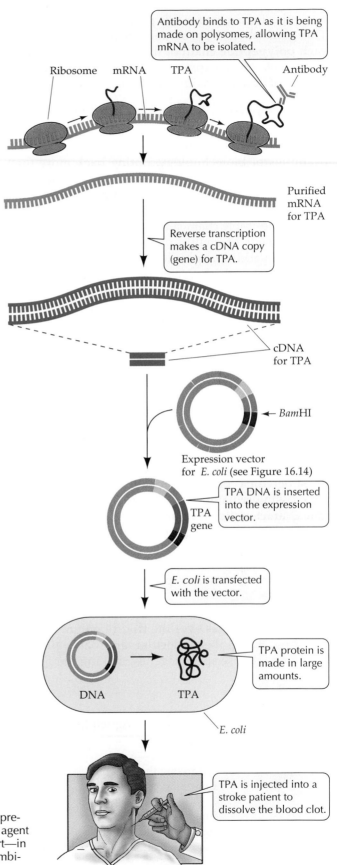

16.15 Tissue Plasminogen Activator: From Protein to Gene to Pharmaceutical TPA is a naturally occurring human protein that prevents blood from clotting. Its isolation and use as a pharmaceutical agent in treating patients suffering from blood clotting in the brain or heart—in other words, strokes and heart attacks—was made possible by recombinant DNA technology.

available only in minuscule amounts. Since insulin is made up of only 51 amino acids, scientists were able to synthesize a gene for this protein in the laboratory. This gene (there were actually two of them, one for each polypeptide chain of the protein) was inserted into *E. coli* via an expression vector, making feasible the use of human insulin in diabetics.

DNA manipulation is changing agriculture

The cultivation of plants and husbanding of animals that constitute agriculture give us the world's oldest examples of biotechnology, dating back more than 8,000 years in human history. Over the centuries, people have adapted crops and farm animals to their needs. Through cultivation and breeding (artificial selection), desirable characteristics such as ease of cooking the seeds or quality of the meat have been improved. In addition, people have selected crops with desirable growth characteristics, such as a reliable ripening season and resistance to diseases.

Until recently, the most common way to improve crop plants was to select varieties that existed in nature through mutational variation and that had the desired genotypes and phenotypes. The advent of genetics in the past century led to its application to plant and animal breeding. A crop plant with a desirable gene could be identified, and through deliberate crosses the single gene or, more usually, many genes could be introduced into a widely used variety of that crop.

Despite spectacular successes, such as the breeding of "supercrops" of wheat, rice, and corn, such deliberate crossing remains a hit-or-miss affair. Many desirable characteristics are complex in their genetics, and it is hard to predict accurately the results of a cross. Moreover, traditional crop plant breeding takes a long time: Many plants can reproduce only once or twice a year—a far cry from the rapid reproduction of bacteria or fruit flies.

Modern biotechnology based on recombinant DNA has two advantages over the traditional methods of breeding. First, the molecular approach allows a breeder to choose specific genes, making the process more precise and less likely to fail as a result of the incorporation of unforeseen genes. The ability to work with cells in the laboratory and then regenerate a whole plant by cloning makes the process much faster than the years needed for traditional breeding. The second advantage—and it is truly an amazing one—is that the molecular methods allow breeders to introduce any gene from any organism into a plant or animal species. This, along with deliberate mutagenesis, expands the range of possible new characteristics to an almost limitless horizon.

As with medicine, there are many examples of molecular biotechnology applied to agriculture (Table 16.2), ranging from improving the nutritional properties of crops, to using animals as gene product factories (transgenic goats are making human TPA in their milk), to using edible crops to make oral vaccines (as we saw at the beginning of the chapter). We will focus on a few examples to show the approaches used.

PLANTS THAT MAKE THEIR OWN INSECTICIDES. Humans are not the only species that consume crop plants. Plants are subject to infections by viruses, bacteria, and fungi, but probably the most important pests are insects. From the locusts of biblical (and modern) times to the cotton boll weevil, insects have continually plagued the crops people grow.

The development of insecticides has improved the situation somewhat, but insecticides have their problems. Most, such as the organophosphates, are relatively nonspecific, killing not only the pest in the field but beneficial insects in the ecosystem as well. Some even have toxic effects on other organisms, including people. What's more, insecticides are applied to the surface of the plants and tend to blow away to adjacent areas, where they may have unforeseen effects.

Some bacteria have solved their pest problem by producing proteins that kill insect larvae that eat the bacteria. For example, there are dozens of strains of *Bacillus thuringiensis*, each of which produces a protein toxic to the insect larvae that plague these bacteria. The toxicity of this protein is 80,000 times that of the usual commercial insecticides. When the hapless larva

---

TABLE 16.2 Agricultural Biotechnologies under Development

PROBLEM	TECHNOLOGY/GENES
Improving the environmental adaptations of plants	Drought tolerance; salt tolerance
Controlling crop pests	Herbicide tolerance; resistance to viruses, bacteria, fungi, and insects
Improving breeding	Male sterility for hybrid seeds
Improving nutritional traits	High-lysine seeds
Improving crops after harvest	Delay of fruit ripening; high-solids tomatoes; sweeter vegetables
Using plants as bioreactors	Plastics, oils, and drugs produced in plants

16.16 Some Transgenic Plants Make Their Own Insecticides The corn plant on the left has been eaten by earworms. The one on the right is healthy because it has been genetically transformed to make a natural insecticide, the toxin of *Bacillus thuringiensis*.

eats the bacteria, the toxin becomes activated, binding specifically to the insect gut to produce holes. The insect starves to death.

Dried formulations of *B. thuringiensis* have been sold for decades as a safe, biodegradable insecticide. But biodegradation is their limitation, because it means that the dried bacteria must be applied repeatedly during the growing season. A better approach would be to have the crop plants make the toxin themselves.

The toxin genes from different strains of *B. thuringiensis* have been isolated and cloned. They have been extensively modified by the addition of plant promoters and terminators, plant poly A signals, plant codon preferences, and plant controlling elements on DNA. Following introduction of the modified genes into cells in the laboratory using the Ti plasmid vector (see Figure 16.5c), transgenic plants have been grown and tested for insect resistance in the field. So far, transgenic tomato, corn, potato, and cotton crops have been successfully shown to have considerable resistance to their insect predators (Figure 16.16).

TOMATOES THAT RIPEN SLOWLY. Most people agree that for taste, tomatoes bought at the market are vastly inferior to the ones we can pick right off the vine. The problem is that tomatoes develop their full flavor while they ripen on the plant during a period of 6 weeks, so if they are picked early so that they are firm enough to be transported easily from farm to market, they ripen rapidly on the way and end up being bland. Could the ripening process be slowed down after picking?

Plant biochemists have identified an enzyme called *polygalacturonase* (PG) that is important in the ripening

process because it hydrolyzes components of the tomato cell walls. To specifically reduce PG synthesis, they used an antisense strategy (see Figure 16.13). They isolated the gene for PG from tomato plants and cloned it. They then used this DNA to make a gene that would code for an RNA complementary to PG mRNA. Then they attached the antisense gene to an active promoter in a Ti plasmid vector in the laboratory and inserted it into tomato protoplasts.

The transgenic tomatoes that resulted expressed the antisense RNA in their fruits as they developed, thus blocking PG production during the 6 weeks on the plant when ripening would normally occur. Fruit from these transgenic plants can be picked while immature, ships easily without ripening, and has a much longer and more flavorful shelf life than do conventional tomatoes.

CROPS THAT ARE RESISTANT TO HERBICIDES. Glyphosate (known by the trade name Roundup) is a widely used and effective weed killer, or herbicide. It works only on plants, by inhibiting an enzyme system in the chloroplast that is involved in the synthesis of amino acids. Glyphosate is truly a "miracle herbicide," killing 76 of the world's 78 worst weeds that grow in fields and rob crop plants of needed water and nutrients from the soil.

Unfortunately, it also kills crop plants, so great care must be taken with its use. In fact, it is best used to rid a field of weeds before the crop plant starts to grow. But as any gardener knows, when the crop begins to grow the weeds reappear. So it would be advantageous if the crop were not affected by the herbicide. Then, the herbicide could be applied at will to the field any time and would kill only the weeds.

Fortunately, some soil bacteria have mutated to develop an enzyme that breaks down glyphosate. Scientists have isolated the gene for this enzyme, cloned it, and added plant sequences for transcription, translation, and targeting to the chloroplast. The gene has been inserted into cotton and soybean plants and these transgenic crops are resistant to glyphosate, permitting the herbicide to be used more effectively.

DNA fingerprinting uses the polymerase chain reaction

"Everyone is unique." This old saying applies not only to human behavior but also to the human genome. Mutations, genetics, and recombination through sexual reproduction ensure that each member of a species (except identical twins) has a unique DNA sequence. An individual can be definitively characterized ("fingerprinted") by his or her DNA base sequence.

The ideal way to distinguish an individual from all the other people on Earth would be to describe his or her entire genomic DNA sequence. But since the human genome contains more than 3 billion nucleotides, this idea is clearly not practical. Instead, scientists have looked for genes that are highly polymorphic—that is, genes that have multiple alleles in the human population and are therefore different in different individuals. One easily analyzed genetic system consists of the short moderately repeated DNA sequences that occur side by side in the chromosomes.

These repeat patterns are inherited. For example, an individual might inherit a chromosome 15 with the short block repeated six times from her mother, and the same block repeated two times from her father. These repeats are easily detectable if they lie between two recognition sites for a restriction enzyme. If the DNA from this individual is isolated and cut with the restriction enzyme, she will have two different-sized fragments of DNA between the sites: one larger (the one from the mother) and the other smaller (the one from the father). These patterns are easily seen by use of gel electrophoresis (Figure 16.17). With several different repeated sequences (as many as eight are used, each with numerous alleles) an individual's unique pattern becomes apparent.

Typically, these methods require 1 μg of DNA, or the DNA content of about 100,000 human cells, but this amount is not always available. The power of PCR (see Figure 16.12) permits the amplification of the DNA from a single cell to produce in a few hours the necessary 1 μg for restriction and gel analysis.

In forensics, DNA fingerprints are used to help prove the innocence or guilt of a suspect. For example, in a rape case, DNA can be extracted from dried semen or hair left by the attacker. After PCR and restriction analyses, this extracted DNA can be compared to DNA from a suspect. So far, this method has been used to prove innocence (the DNA patterns are different) more often than guilt (the DNA patterns are the same). The reason is that it is easy to exclude someone on the basis of the tests, but two people could have the same patterns, since what is being tested is just a small sample of the genome, so proving that a suspect is guilty cannot rest on DNA fingerprinting alone.

In addition to this highly publicized use, there are many other applications of PCR-based DNA fingerprinting. In 1992, there were 52 California condors, looked after by the San Diego and Los Angeles zoos. Scientists made DNA fingerprints of all these birds so that the geneticists at the zoos can mate unrelated individuals that have many allelic differences in order to increase genetic variation in the offspring of this endangered species. A similar program is under way for the threatened Galapagos tortoises (see Chapter 55).

Plant scientists have found in nature or produced by breeding programs thousands of varieties of crops such as rice, wheat, corn, and grapes. Many of these varieties are kept in cold storage in "seed banks."

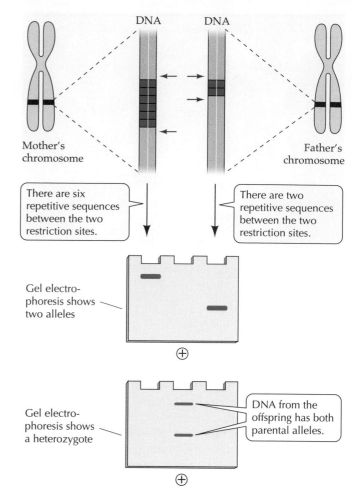

16.17 DNA Fingerprinting   As many as eight different repeated gene sequences can be used to make a DNA fingerprint.

Samples of these plants are being DNA-fingerprinted to determine which varieties are genetically the same and which are the most diverse, as a guide to breeding programs.

A related use of PCR is in the diagnosis of infections. In this case, the test is whether the DNA of an infectious agent is present in blood or a tissue sample. Specifically, the question is whether a primer strand that matches the pathogen's DNA and is used for PCR will be amplified on DNA from the patient. Because so little of the target is needed and the primers can be made to bind only to a specific viral or bacterial genome, the PCR-based test is extremely sensitive. If an organism is present in small amounts, PCR testing will detect it.

Finally, the isolation and characterization of genes for various human diseases, such as sickle-cell anemia and cystic fibrosis, has made PCR-based genetic testing a reality. We will discuss this subject in depth in the next chapter.

Summary of "Recombinant DNA and Biotechnology"

Cleaving and Rejoining DNA

- Knowledge of DNA transcription, translation, and replication has been used to create recombinant DNA molecules, made up of sequences from different organisms.
- Restriction enzymes, which are made by microbes as a defense mechanism, bind to DNA at specific sequences and cut it. **Review Figure 16.1**
- DNA fragments generated from cleavage by restriction enzymes can be separated by size using gel electrophoresis. The sequences of these fragments can be further identifed by hybridization with a probe. **Review Figures 16.2, 16.3**
- Many restriction enzymes make staggered cuts in the two strands of DNA, creating "sticky ends," with unpaired bases. The sticky ends can be used to create recombinant DNA if DNA molecules from different species are cut with the same restriction enzyme and then mixed in the test tube. **Review Figure 16.4**

Cloning Genes

- Bacteria, yeasts, and cultured plant cells are commonly used as hosts for recombinant DNA experiments.
- Newly introduced DNA must be part of a replicon if it is to be propagated in host cells. One way to make sure that the introduced DNA is part of a replicon is to introduce it as part of a carrier DNA, or vector DNA, that has a replicon.
- There are specialized vectors for bacteria, yeasts, and plant cells. These vectors must contain a replicon, recognition sequences for restriction enzymes, and genetic markers to identify their presence in the host. **Review Figure 16.5**
- Naked DNA may be introduced by chemical or mechanical means. In this case, the DNA must integrate into the host DNA. **Review Figure 16.6**
- When vectors carrying recombinant DNA are incubated with host cells, nutritional, antibiotic resistance, or fluorescent markers can be used to identify which cells contain the vector. **Review Figures 16.7, 16.8**

Sources of Genes for Cloning

- The cutting of DNA by a restriction enzyme produces many fragments that can be individually and randomly put into a vector and inserted into a host to create a genomic library.
- A smaller library of cDNAs can be made from the mRNAs produced in a tissue or organism at a certain moment. In this case, mRNA is extracted and used to create DNA (cDNA) by reverse transcription. This cDNA is then used to make a library. **Review Figure 16.10**
- A third source of DNA for cloning is DNA made by chemists in the laboratory. Organic chemistry methods can be used to create specific, mutated DNA sequences.

Some Additional Tools for DNA Manipulation

- Cloned DNA fragments can be isolated so that their nucleotide sequences can be determined.

- Several hundred bases in DNA can be sequenced at a time. One method utilizes dideoxyribonucleotides, which cause the chain to terminate, to create fragments that are then separated by gel electrophoresis and analyzed by computer. **Review Figure 16.11**
- The polymerase chain reaction makes multiple copies of DNA in the test tube, using a heat-resistant DNA polymerase. If the sequences of the ends of DNA are known, primers can be made complementary to them. These primers can be extended in the test tube as part of the polymerase chain reaction. PCR can amplify a target DNA sequence manyfold in a few hours. **Review Figure 16.12**
- Preparation of an antisense RNA complementary to a specific mRNA can prevent translation when the antisense RNA hybridizes to the mRNA. A vector expressing antisense RNA can prevent translation continuously. **Review Figure 16.13**

Biotechnology: Applications of DNA Manipulation

- The ability to clone genes has made possible many new applications of biotechnology, such as the large-scale production of eukaryotic gene products.
- For a vector carrying a gene of interest to be expressed in a host cell, the gene must be adjacent to appropriate sequences for its transcription and translation in the host cell. **Review Figure 16.14**
- Recombinant DNA and expression vectors have been used to make medically useful proteins that would otherwise have been difficult to obtain in necessary quantities. **Review Figure 16.15 and Table 16.1**
- Because plant cells can be cloned to produce adult plants, the introduction of new genes into plants via vectors has been more rapid than with animal cells. The result is crop plants that carry new, useful genes. **Review Figure 16.16 and Table 16.2**
- Because the DNA of an individual or species is unique, the polymerase chain reaction can be used to identify an organism from a small sample of its cells—that is, to create a DNA fingerprint. **Review Figure 16.17**

Self-Quiz

1. Restriction endonucleases
   a. play no role in bacteria.
   b. cleave DNA at highly specific recognition sequences.
   c. are inserted into bacteria by bacteriophages.
   d. are made only by eukaryotic cells.
   e. add methyl groups to specific DNA sequences.

2. Sticky ends
   a. are double-stranded ends of DNA fragments.
   b. are identical for all restriction enzymes.
   c. are removed by restriction enzymes.
   d. are complementary to other sticky ends generated by the same restriction enzyme.
   e. are hundreds of bases long.

3. In gel electrophoresis,
   a. DNA fragments are separated on the basis of size.
   b. DNA does not have an electric charge.
   c. DNA fragments cannot be removed from the gel.
   d. the electric field separates positively charged DNA fragments from negatively charged DNA fragments.
   e. the DNA fragments are naturally fluorescent.

4. Possession of which feature is *not* desirable in a vector for gene cloning?
   a. An origin of DNA replication
   b. Genetic markers for the presence of the vector
   c. Multiple recognition sites for the restriction enzyme to be used
   d. One recognition site each for many different restriction enzymes
   e. Genes other than the target for cloning

5. DNA can be introduced into any cell by
   a. injection.
   b. being complexed with calcium salts.
   c. being placed along with the cell into a particle gun.
   d. gel electrophoresis.
   e. being heated to be denatured.

6. Complementary DNA (cDNA)
   a. is produced from ribonucleoside triphosphates.
   b. is produced by reverse transcription.
   c. is the "other strand" of single-stranded DNA.
   d. requires no template for its synthesis.
   e. cannot be placed into a vector, since it has the opposite base sequence of the vector DNA.

7. In a genomic library of frog DNA in *E. coli* bacteria,
   a. all bacterial cells have the same sequences of frog DNA.
   b. all bacterial cells have different sequences of DNA.
   c. each bacterial cell has a random fragment of frog DNA.
   d. each bacterial cell has many fragments of frog DNA.
   e. the frog DNA is transcribed into mRNA in the bacterial cells.

8. An expression vector requires all of the following, except
   a. genes for ribosomal RNA.
   b. a selectable genetic marker.
   c. a promoter of transcription.
   d. an origin of DNA replication.
   e. restriction enzyme recognition sites.

9. The polymerase chain reaction
   a. is a method for sequencing DNA.
   b. is used to transcribe specific genes.
   c. amplifies specific DNA sequences.
   d. does not require primers for DNA replication.
   e. uses a DNA polymerase that denatures at 55°C.

10. In DNA fingerprinting,
    a. a positive identification can be made.
    b. a gel blot is all that is required.
    c. multiple restriction digests generate unique fragments.
    d. the polymerase chain reaction amplifies finger DNA.
    e. the variability of repeated sequences between two restriction sites is evaluated.

Applying Concepts

1. Compare PCR and cloning as methods to amplify a gene. What are the requirements, benefits, and drawbacks of each method?

2. As specifically as you can, outline the steps you would take to (a) insert and express the gene for a new, nutritious seed protein in wheat; (b) insert and express a gene for a human enzyme in sheep's milk.

3. The *E. coli* plasmid pSCI carries genes for resistance to the antibiotics tetracycline and kanamycin. The *tet*[r] gene has a single restriction site for the enzyme *Hind*III. Both the plasmid and the gene for corn glutein protein are cleaved with *Hind*III and incubated to create recombinant DNA. The reaction mixture is then incubated with *E. coli* that are sensitive to both antibiotics. What would be the characteristics, with respect to antibiotic sensitivity or resistance, of colonies of *E. coli* containing, in addition to its own genome: (a) no new DNA; (b) native pSCI DNA; (c) recombinant pSCI DNA; (d) corn DNA only? How would you detect these colonies?

4. Using a DNA synthesizer, you make a fragment of DNA with the proposed sequence 3'-ATTGTCCTCTGA-5'. Using the sequencing protocol described in Figure 16.11, how would you verify that this is the correct sequence? Describe the reactions and gel patterns that you would observe.

Readings

Chrispeels, M. J. and D. Sadava. 1994. *Plants, Genes and Agriculture.* Jones and Bartlett, Boston, MA. A comprehensive summary of the role of biotechnology in agriculture.

Gilbert, W. 1991. "Toward a Paradigm Shift in Biology." *Nature,* January 10, volume 349, page 99. A discussion of the impact of DNA technology on pure research in biology.

Lewin, B. 1997. *Genes VI.* Oxford University Press, Oxford. This text contains excellent outlines of gene libraries, cDNA cloning, and other specialized methods.

Lodish, H., D. Baltimore, A. Berk, S. L. Zipursky, P. Matsudaira and J. Darnell, 1995. *Molecular Cell Biology,* 3rd Edition. Scientific American Books, New York. Excellent chapters on recombinant DNA methods and the polymerase chain reaction.

Sambrook, J., E. Fritsch and T. Maniatis. 1989. *Molecular Cloning: A Laboratory Manual,* 2nd Edition. This is the book on the lab benches of most molecular biologists.

Chapter 17

Molecular Biology and Medicine

Dr. Asbjørn Følling
Trained as both a physician and a chemist, this Norwegian medical doctor (center) is shown at a 1962 ceremony where he was honored for his research establishing the genetic and biochemical nature of phenylketonuria. Følling's work led to effective treatment of the condition; today the idea of correcting the genetic defect that causes it is being contemplated.

The mother brought her two children to Dr. Asbjørn Følling in 1934 as a last resort. Ever since their births, she had watched the conditions of her 6-year-old daughter and 4-year old son deteriorate. Now both were severely mentally retarded. So far, all of the doctors who had examined the children had expressed sympathy but could do nothing. The mother had noticed a peculiar smell clinging to her children and she had heard that Dr. Følling was trained as both a chemist and a physician. Could he help? It turned out he couldn't, because their retardation was irreversible. But while examining these children, Dr. Følling made a major discovery.

As part of his examination, Dr. Følling tested the children's urine by adding a brown solution of ferric chloride to look for ketones, which are often excreted by diabetics. In normal people, this solution stays brown, but in diabetics it turns purple. To his surprise, the urine of these children turned the solution dark green. He had never seen this color before in the urine test, and it had not been described in any of his reference books. At first, he suspected the children were taking a medication that ended up in the urine and reacted with ferric chloride. So he asked the mother to refrain from giving her children any medications for a week and then to bring him two new urine samples. Once again, the samples tested green. Clearly, a substance unique to the bodies of these two children was responsible for the strange color.

373

Here's where Følling's chemistry training served him well. Using analytical chemistry, he purified the substance from the children's urine and tentatively identified it. Then, he used organic chemistry to make the substance in the lab and proved it to be identical to the one in the urine. This substance was phenylpyruvic acid. On the basis of the similarity between phenylpyruvic acid and the amino acid phenylalanine, Følling hypothesized that these children were unable to metabolize phenylalanine and that the excess was being converted to phenylpyruvic acid. He soon found other mentally retarded people who excreted this substance, and among the first ten were three pairs of siblings. Følling observed that the parents of these children were mentally normal and did not excrete phenylpyruvic acid. All of these observations fit the idea of an autosomally recessive inherited condition.

Dr. Følling had discovered the genetic disease *phenylketonuria*. But it was not the first such disease to be described in biochemical terms. In 1909, Dr. Archibald Garrod had found the cause of *alkaptonuria*—an inherited disorder in which the patients' urine turns black—to be on the same biochemical pathway as the cause of phenylketonuria. Garrod coined the term "inborn errors of metabolism" as a general description of diseases for which genetics and biochemistry are clearly linked.

Later the phenotypes of both phenylketonuria and alkaptonuria were identified as abnormalities in specific enzymes. Today the causes of hundreds of such single-gene, single-enzyme diseases are known. In some cases, these discoveries have led to the design of specific therapies and ways to screen for the abnormal proteins in people who do not overtly show the disease.

As we will see in this chapter, more precision in describing these abnormalities at the DNA level has come from molecular biology. Even cancer, it turns out, is caused in most cases by abnormalities in genes. The rise of "molecular medicine" is most dramatically shown by undertakings such as gene therapy and the Human Genome Project, which we will discuss at the end of the chapter.

Protein as Phenotype

The different types of a person's cells are distinguished by their unique proteins. For example, hair follicles make the hair protein keratin, and red blood cells contain hemoglobin. As we saw in Chapters 11 and 12, genetic mutations are often expressed phenotypically as proteins that differ from the normal, wild type. Many genetic diseases also have specific proteins that differ from the normal proteins.

In this first section of the chapter, we will identify and discuss the kinds of abnormal proteins that can result from inheritance of an abnormal allele or its origin by mutation. Then we will consider the role of the environment and patterns of inheritance resulting from autosomal recessives, autosomal dominants, X linkages, and chromosome abnormalities.

Many genetic diseases result from abnormal or missing proteins

Proteins have many roles in cells, and the genes that code for these proteins can be mutated to cause genetic diseases. Enzymes, carriers such as hemoglobin, receptors, transport proteins, and structural proteins have all been implicated in genetic diseases.

ENZYMES. Følling's patients with phenylketonuria (PKU)—the inability to metabolize phenylalanine and related compounds—had a dramatically unusual phenotype in that they were mentally retarded. In addition, they had a musty odor, resulting from phenylpyruvic acid, and they had lighter-colored hair and skin than their relatives. Although Følling made his discovery in 1934, not until 1957 was this complex phenotype traced back to its primary phenotype: a single abnormal protein.

As Følling had predicted, phenylalanine hydroxylase, the enzyme that catalyzes the conversion of dietary phenylalanine to tyrosine, was not active in these patients' livers (Figure 17.1). Lack of this conversion led to excess phenylalanine in the blood and explained the accumulation of phenylpyruvic acid, which is the product of an offshoot pathway activated because of the accumulation of phenylalanine.

Later, the sequences of the 451 amino acids that constitute phenylalanine hydroxylase in normal people were compared with those in individuals suffering from PKU, in many cases revealing only a single difference: Instead of the arginine at position 408 in the long polypeptide chain, many people with PKU have tryptophan. Once again, the ideas of

$$\text{one gene} \rightarrow \text{one enzyme}$$

and

$$\text{one mutation} \rightarrow \text{one amino acid change}$$

hold true in human diseases as they do in studies of so many other organisms.

How does the abnormality in PKU lead to its multitude of clinical symptoms? Since the pigment melanin is made from tyrosine, which is not synthesized but obtained mostly in the patients' diet, lighter skin and hair color seem reasonable to expect. But the mental retardation is much more difficult to explain. This symptom could result from the accumulation of the substrate (phenylalanine), due to the missing enzyme activity; alternatively, it could be due to a relative lack of the product (tyrosine), or to the accumulation of phenylpyruvic acid. While knowledge of liver biochemistry has led to specification of the enzyme abnormality, knowledge of brain biochemistry is more

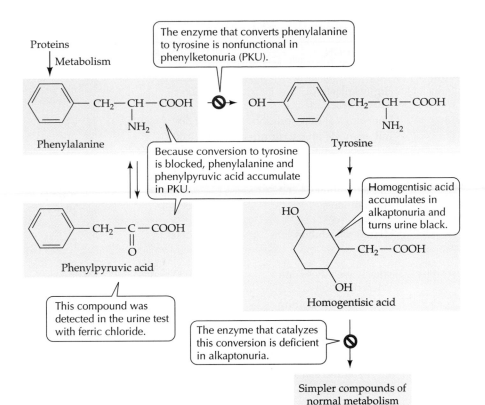

17.1 One Gene, One Enzyme in Humans
Phenylketonuria and alkaptonuria are caused by abnormalities in specific enzymes. Knowing the causes of such single-gene, single-enzyme metabolic diseases can aid in the development of screening tests and therapeutic regimens.

primitive. The exact cause of mental retardation in PKU remains elusive, but as we will see later in this chapter, it can be prevented.

Hundreds of human genetic diseases that result from enzyme abnormalities have been discovered, many of which lead to mental retardation and premature death (Table 17.1). Most of these diseases are rare; PKU, for example, shows up in one newborn infant out of every 12,000. But this is just the tip of the mutational iceberg. Undoubtedly, some mutations result in altered proteins that have no obvious clinical effects.

For example, there could be many amino acid changes in phenylalanine hydroxylase that do not affect its catalytic activity. Analysis of the same protein in different people often shows variations that have no functional significance. In fact, at least 30 percent of all proteins have detectable amino acid differences among individuals. If one protein variant exists in less than 99 percent of a population (that is, if the protein has another variant at least 2 percent of the time), the protein is said to be **polymorphic**. The key point is that polymorphism does not necessarily mean disease.

HEMOGLOBIN. The first human genetic disease for which an amino acid abnormality was tracked down as

the cause was not PKU. It was the blood disease *sickle-cell anemia*, which most often afflicts people whose ancestors came from the Tropics or from the Mediterranean. Among African-Americans, about 1 in 655 are homozygous for the sickle allele and have the disease. The abnormal allele produces an abnormal globin protein that leads to sickled red blood cells (see Figure 12.14). These cells tend to block narrow blood capillaries, especially when the O_2 concentration is low, and the result is tissue damage.

Human hemoglobin is a protein with quaternary structure, containing four polypeptide chains and the pigment heme. Hemoglobin has two α and two β chains. In sickle-cell anemia, one of the 146 amino acids in the β chain is abnormal: At position 6, the normal glutamic acid has been replaced by valine. This replacement changes the charge of the protein (glutamic acid is negatively charged and valine is neutral; see Table 3.1), and the protein can form long aggregates in the red blood cells. The result is anemia, a deficiency of red blood cells.

Because hemoglobin is easy to isolate and study, its variations in the human population have been extensively documented (Figure 17.2). Hundreds of single amino acid alterations in β-globin have been reported. Some of these polymorphisms are even at the same amino acid position. For example, at the same position

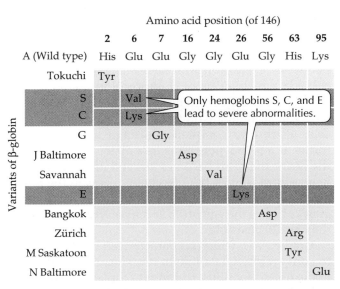

17.2 Hemoglobin Polymorphism Only three of the many variants of hemoglobin lead to clinical abnormalities.

TABLE 17.1 **Some Human Genetic Disorders**

DISORDER	DESCRIPTION	NEWBORN FREQUENCY
AUTOSOMAL DOMINANT CONDITIONS		
Familial hypercholesterolemia	High cholesterol; early heart attacks	1 in 500
Huntington's disease	Neurological deterioration	1 in 2,500
Marfan syndrome	Tall, spindly; artery ruptures	1 in 20,000
AUTOSOMAL RECESSIVE CONDITIONS		
Sickle-cell anemia	Blocked capillaries; organ damage	1 in 655 (African-Americans)
Cystic fibrosis	Thick mucus; infections	1 in 2,500
Phenylketonuria	Mental retardation	1 in 12,000
X-LINKED CONDITIONS		
Duchenne's muscular dystrophy	Muscle weakness and deterioration	1 in 3,000 males
Hemophilia A	Lack of blood clotting	1 in 10,000 males
CHROMOSOME ABNORMALITIES		
Trisomy (extra) 21 (Down syndrome)	Mental retardation	1 in 600
XXY (Klinefelter syndrome)	Short stature; sterility	1 in 700 males
XO (Turner syndrome)	Sterility	1 in 1,500 females
XYY syndrome	Tall stature; acne	1 in 800 males
Fragile-X syndrome	Mental retardation	1 in 1,250 males; 1 in 2,000 females

that is mutated in sickle-cell anemia, the normal glutamic acid may be replaced by lysine in *hemoglobin C disease*. In this case, the anemia is usually not severe. Many hemoglobin variants, although obviously altering the primary phenotype, have no effect on the protein's function and thus no clinical phenotype. This is fortunate, since about 5 percent of all humans are carriers for one of these variants.

RECEPTORS AND TRANSPORT PROTEINS. Some of the most common human genetic diseases show their primary phenotype as altered membrane proteins. About one person in 500 is born with *familial hypercholesterolemia* (FH), in which levels of cholesterol in the blood are several times higher than normal. The excess cholesterol can accumulate on the inner walls of blood vessels, leading to complete blockage if a blood clot forms. If a blood clot forms in a major vessel serving the heart, the heart becomes starved of oxygen and a heart attack results. If a blood clot forms in the brain, the result is a stroke. People with FH often die of heart attacks before the age of 45, and in severe cases, before they are 20 years old.

Unlike PKU, which is characterized by the inability to convert phenylalanine to tyrosine, the problem in FH is not an inability to convert cholesterol to other products. People with FH have all the machinery to metabolize cholesterol, primarily in liver cells. The problem is that they are unable to transport the cholesterol into the cells that use it.

Cholesterol travels in the bloodstream in protein-containing particles called lipoproteins. One type of lipoprotein, *low-density lipoprotein*, carries cholesterol to the liver cells. After binding to a specific receptor on the membrane of the liver cell, the lipoprotein is taken up by the cell by endocytosis and then delivers its cholesterol "baggage" to the interior of the cell. People with FH lack a functional version of the receptor protein. Of the 840 amino acids that make up the receptor, often only one is abnormal in FH, but this is enough to change its structure so that it cannot bind to the lipoprotein and thus the lipoprotein (carrying the cholesterol) cannot enter the liver cell (Figure 17.3a).

In Caucasians, about one baby in 2,500 is born with *cystic fibrosis*. The clinical phenotype of this genetic disease is an unusually thick and dry mucus that lines organs such as the tubes that serve the respiratory system (see Chapter 45). This dryness prevents cilia on the surfaces of the epithelial cells from working efficiently to clear out bacteria and fungal spores that people take in with every breath. The results are recurrent and serious infections, as well as liver, pancreatic, and digestive failures. Patients often die in their 20s or 30s.

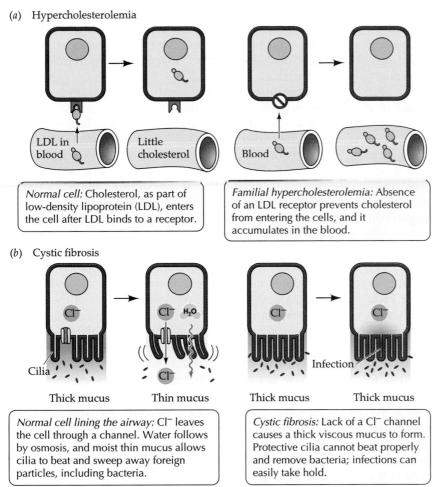

(a) Hypercholesterolemia

LDL in blood

Little cholesterol

Blood

Normal cell: Cholesterol, as part of low-density lipoprotein (LDL), enters the cell after LDL binds to a receptor.

Familial hypercholesterolemia: Absence of an LDL receptor prevents cholesterol from entering the cells, and it accumulates in the blood.

(b) Cystic fibrosis

Cl⁻

Cl⁻   H₂O

Cl⁻

Cl⁻

Cilia

Infection

Thick mucus

Thin mucus

Thick mucus

Thick mucus

Normal cell lining the airway: Cl⁻ leaves the cell through a channel. Water follows by osmosis, and moist thin mucus allows cilia to beat and sweep away foreign particles, including bacteria.

Cystic fibrosis: Lack of a Cl⁻ channel causes a thick viscous mucus to form. Protective cilia cannot beat properly and remove bacteria; infections can easily take hold.

17.3 Genetic Diseases of Membrane Proteins
The left two panels illustrate normal cell function, while the right panels show the abnormalities caused by (*a*) hypercholesterolemia and (*b*) cystic fibrosis.

working copy of dystrophin, so their muscles also do not work.

A rarer genetic disease that is caused by a change in a structural protein is *osteogenesis imperfecta*, or brittle-bone disease, which occurs in one in 10,000 births. The culprit here is collagen, a protein that is essential in bone formation (see Chapter 44). Depending on the nature of the amino acid changes, the symptoms may be fairly minor— for example, skin, ligament, and tendon ailments. In more severe cases, bones are markedly malformed.

Coagulation proteins are involved in the clotting of blood at a wound. Inactive clotting proteins are always present in blood plasma (the noncellular portion of blood), but they become active only at a wound. People afflicted by the genetic disease *hemophilia* lack one of the coagulation proteins. Some people with this disease risk death from even minor cuts, since they cannot stop bleeding.

The reason for the thick mucus in patients with cystic fibrosis is a defective membrane protein, the *chloride transporter* (Figure 17.3*b*). In normal cells, this membrane channel opens after ATP hydrolysis to release Cl⁻ to the outside of the cell. The imbalance of ions (more are now on the outside of the cell) causes water to leave the cell by osmosis, resulting in a moist mucus. A single amino acid change in the transporter renders it nonfunctional, which leads to a dry mucus and the consequent clinical problems.

STRUCTURAL PROTEINS. About one boy in 3,000 is born with *Duchenne's muscular dystrophy*, in which the problem is not an enzyme or receptor but a protein involved in biological structure. People with this disease show progressively weaker muscles and are wheelchair-bound by their teenage years. Afflicted patients usually die in their 20s, when the muscles that serve their respiratory system fail. Normal people have a protein in their skeletal muscles called dystrophin that may bind the major muscle protein actin to the plasma membrane (see Chapter 44). Patients with Duchenne's muscular dystrophy do not have a

Most diseases are caused by both heredity and environment

Human diseases for which clinical phenotypes can be traced to a single altered protein may number in the thousands, and they are dramatic evidence of the one gene, one polypeptide idea. But taken together, these diseases have a frequency of about 1 percent in the total population.

Far more common are diseases that are *multifactorial*; that is, they are caused by many genes and proteins interacting with the environment. Although we tend to call humans either normal (wild type) or abnormal (mutant), the sum total of our genes is what determines which of us who eats a high-fat diet will die of a heart attack or which of us exposed to infectious bacteria actually come down with a disease. Estimates suggest that up to 60 percent of all people have diseases that are genetically influenced.

Human genetic diseases have several patterns of inheritance

PKU, sickle-cell anemia, and cystic fibrosis are inherited as **autosomal recessives**. This genetic term refers to the clinical phenotype: Typically, both parents are normal, heterozygous carriers of the abnormal allele. They have a 25 percent (one in four) chance of having an affected (homozygous recessive) boy or girl. Because of this low probability and the fact that many families in Western societies now have fewer than four children, it is unusual for more than one child in a family to have an autosomal recessive disease.

In the cells that produce the altered protein in people who have these recessive diseases—that is, liver cells for PKU, immature red blood cells for sickle-cell anemia, and epithelial cells for cystic fibrosis—only the nonfunctional, mutant, protein is made. In these homozygotes, a biochemical pathway or important cell function is disrupted, and disease results.

Not unexpectedly, heterozygotes, with one normal and one mutant allele, often have only 50 percent of the normal level of functional protein. For example, people who are heterozygous for the allele for PKU have half the number of active molecules of phenylalanine hydroxylase in their liver cells as individuals who carry two normal alleles for this enzyme. The cells compensate for this deficiency simply by having the active enzymes in the heterozygote do twice the work, and normal function is achieved.

Familial hypercholesterolemia and osteogenesis imperfecta are inherited as **autosomal dominants**. The presence of only one mutant allele is enough to produce the adverse clinical outcome, and direct transmission of the disease from parent to offspring is the rule. In people who are heterozygous for familial hypercholesterolemia, having half the number of functional receptors for low-density lipoprotein on the surface of liver cells is simply not enough to clear the cholesterol from the blood.

Osteogenesis imperfecta shows a *dominant negative* inheritance. Collagens are composed of three polypeptide chains, and when one of them is mutant, it prevents the other two from working properly, and the whole structure is disrupted. The expression of this abnormal collagen in the heterozygote is so harmful that it is often lethal.

The most prevalent types of hemophilia and Duchenne's muscular dystrophy are inherited as **X-linked** alleles. As we described in Chapter 10, a man always passes his X chromosome on to his daughters and not his sons. So if a father carries a mutant allele on the X chromosome, all of his daughters must inherit it, and none of his sons will inherit the mutant allele. On the other hand, a mother who is a carrier of the mutant allele can pass on the mutant X to both her sons and her daughters.

Both hemophilia and Duchenne's muscular dystrophy are inherited as recessives. Thus the son who inherits a mutant allele from his mother will have the disease, while a daughter will be a heterozygous carrier. Since until recently, few males with these diseases lived to reproduce, the most common pattern of inheritance has been from carrier mother to offspring, and these diseases are much more common in males than in females.

Chromosome abnormalities also cause human diseases. Such abnormalities include an excess or loss of one or two chromosomes (*aneuploidy*), loss of a piece of a chromosome (*deletion*), and transfer of a piece of one chromosome to another chromosome (*translocation*). About one newborn in 200 is born with a chromosome abnormality. While some of them are inherited, many are the result of meiotic problems such as nondisjunction (see Chapter 9).

Many zygotes that have abnormal chromosomes do not survive development and are spontaneously aborted. Of the 20 percent of pregnancies that are spontaneously aborted during the first 3 months of human development, an estimated half of them have chromosomal aberrations. For example, more than 90 percent of human zygotes that have only one X chromosome and no Y (Turner syndrome) do not live beyond the fourth prenatal month.

Why is the addition or deletion of chromosomal material so harmful to the developing human? This is the same question we ask in trying to explain X chromosome inactivation (see Chapter 14). The reasons are not yet clear. In translocations, the shuffling of chromosomal material to a new location causes a dramatic change in gene expression.

The most common form of inherited mental retardation is *fragile-X syndrome* (Figure 17.4). About one male in 1,500 and one female in 2,000 are affected. Near the tip of the abnormal X chromosome is a constriction that tends to break during preparation for microscopy, giving the name for this syndrome. Although the basic pattern of inheritance is that of an X-linked recessive trait, there are departures from this behavior. Not all people with the fragile-X abnormality are mentally retarded, as we will describe later.

Mutations and Human Diseases

The isolation and description of the precise nature of human mutations has proceeded rapidly since the development of molecular biology techniques (see Chapter 16). When the primary phenotypic expression was known, as in the case of abnormal hemoglobins, cloning the gene was straightforward, although time-consuming.

In other instances, such as Duchenne's muscular dystrophy, a chromosome deletion was associated with the disease, and this deletion pointed the way to the missing gene. In still other cases, such as cystic fibrosis, only a subtle molecular marker was available, leading investigators to the gene. In both of the latter

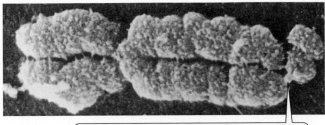

The constriction at the lower tip of this chromosome is the location of the fragile-X abnormality.

17.4 A Fragile-X Chromosome at Metaphase The chromosomal abnormality that causes the mental retardation symptomatic of fragile-X syndrome shows up physically as a constriction.

examples, the primary phenotype, the defective protein, was unknown when the gene was isolated; only then was the protein found.

In the discussions that follow, we will examine how mRNA, chromosome deletions, and DNA markers can be useful in identifying both mutant genes and abnormal proteins for genetic diseases such as muscular dystrophy and cystic fibrosis. We close this discussion by considering triplet repeats in several genetic diseases.

The logical way to isolate a gene is from its protein

As mentioned earlier, the primary phenotype for sickle-cell anemia was described in the 1950s as a single amino acid change in β-globin. On the basis of the clinical picture of sickled red blood cells, β-globin was certainly the right protein to examine. By the 1970s, β-globin mRNA could be isolated from immature red blood cells, which transcribe the globins as their major gene product. Then, a cDNA copy could be made and used to probe a human DNA library to isolate the entire β-globin gene (see Chapter 16), including introns and flanking sequences (Figure 17.5a). DNA sequencing could then be used to compare the normal gene with the gene from patients with sickle-cell anemia; sure enough, a single mutation was found.

Sometimes we can make an educated guess as to which protein is altered in a genetic disease. This was the case for *Marfan syndrome*. People with this dominantly inherited disorder often die suddenly when

17.5 Strategies for Isolating Human Genes (a) Once the seqeunce for the normal β-globin gene is established from the isolated protein, it can be compared to the gene sequence of patients with sickle-cell anemia. (b) When an abnormality is caused by a missing gene, as in Duchenne's muscular dystrophy, researchers compare the affected chromosome with a normal chromosome and isolate the DNA that is missing, then determine the protein for which this DNA codes.

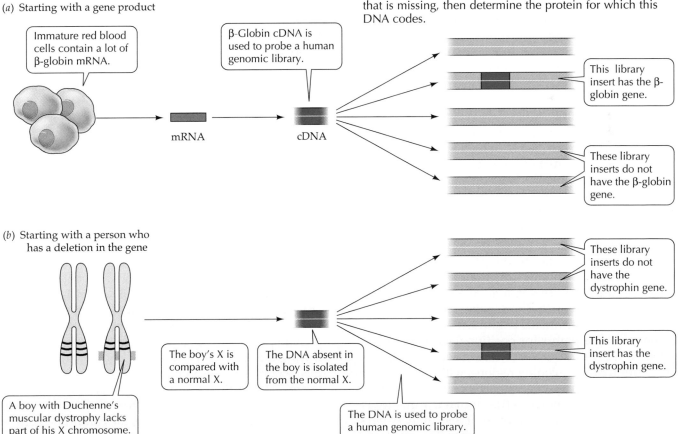

(a) Starting with a gene product

Immature red blood cells contain a lot of β-globin mRNA.

mRNA

β-Globin cDNA is used to probe a human genomic library.

cDNA

This library insert has the β-globin gene.

These library inserts do not have the β-globin gene.

(b) Starting with a person who has a deletion in the gene

A boy with Duchenne's muscular dystrophy lacks part of his X chromosome.

The boy's X is compared with a normal X.

The DNA absent in the boy is isolated from the normal X.

The DNA is used to probe a human genomic library.

These library inserts do not have the dystrophin gene.

This library insert has the dystrophin gene.

their heart beats vigorously, and, in trying to expand with the increased pressure from the blood, a major blood vessel ruptures. Until 1985, searching for a vessel wall protein in patients afflicted with Marfan syndrome that differed from the same protein in unaffected people proved fruitless. All the major isolated proteins seemed identical in the two groups of people. Then, electron microscopy revealed a new filamentous protein, called *fibrillin*, in the aorta wall. Fibrillin was purified, and in much the same way as β-globin, its gene was isolated and turned out to be mutated in people with Marfan syndrome.

Chromosome deletions can lead to gene and then protein isolation

The inheritance pattern of Duchenne's muscular dystrophy is consistent with an X-linked recessive trait. But until the late 1980s, neither the abnormal protein involved nor its gene had been described. This failure to describe either the protein or the gene was not from lack of effort: Almost every muscle protein that could be isolated at the time had been, and comparisons between affected people and normal muscles showed no differences.

Then several boys were found to have the disease because of a small deletion in their X chromosome. Comparison of the deleted and normal X chromosomes made possible isolation of the gene that was missing in the boys (Figure 17.5*b*). An important but not unexpected lesson that emerged from this research was that many different mutations can give rise to the Duchenne's muscular dystrophy phenotype. Some are deletions, and others are point mutations at intron splice sites.

DNA markers can point the way to important genes

In cases in which no candidate protein or visible chromosome deletion is available to help scientists in isolating a gene responsible for a disease, a technique called *positional cloning* has been invaluable. To understand this method, imagine an astronaut looking down from space, trying to find her son on a park bench on Chicago's North Shore. Unable to spot the boy with her naked eye, the astronaut picks out landmarks that will lead her to the park. She recognizes the shape of North America, then moves to Lake Michigan, the Sears Tower, and so on. Once she has zeroed in on the North Shore park, she can use advanced optical instruments to find her son.

The reference points for positional cloning are genetic markers on the DNA. These markers can be located within protein-coding regions, within introns, or in spacer DNA between genes. The only requirement is that they be polymorphic (recall that a sequence is polymorphic if one form is present less than 99 percent of the time—that is, if there is more than one form of the sequence).

As we described in Chapter 16, restriction enzymes cut DNA molecules at specific recognition sequences. On a particular human chromosome, a given restriction enzyme may make hundreds of cuts, producing many DNA fragments that can be probed on gel electrophoresis. For example, the enzyme *Eco*RI cuts DNA at

$$5' \dots \text{GAATTC} \dots 3'$$

Suppose this recognition site exists in a stretch of human chromosome 7. The enzyme will cut this stretch once and make two fragments of DNA. Now suppose in some people this sequence is mutated as follows:

$$5' \dots \text{GAGTTC} \dots 3'$$

This sequence will not be recognized by the restriction enzyme; thus it will remain intact and yield one larger fragment of DNA. The existence of such DNA differences is called a **restriction fragment length polymorphism**, or **RFLP** (Figure 17.6). An RFLP band pattern is inherited in a Mendelian fashion and can be followed through a pedigree. More than 1,000 such markers have been described for the human genome.

Genetic markers can be used as landmarks to find genes of interest if they too are polymorphic. The key to this method is the well-known observation from genetics that if two genes are located near each other on the same chromosome, they are passed on together from parent to offspring. The same holds true for any pair of genetic markers.

So, in order to narrow down the location of a gene on the human genome, a scientist must find a marker and gene that are always inherited together. The gene might be the one that causes cystic fibrosis, for example, and the marker might be the absence of the *Eco*RI site on the chromosome noted earlier. Family medical histories are taken and pedigrees are constructed. If the DNA marker and genetic disease are inherited together, they must be near each other on the chromosome. Unfortunately, "near each other" might be as much as a million base pairs away. This process of locating a marker and gene that are inherited together is thus similar to the astronaut focusing on Chicago: The landmarks lead to only an approximate location.

How can the gene be isolated? Many sophisticated methods are available. For example, the neighborhood around the RFLP can be screened for further RFLPs involving other restriction enzymes. With luck, one of them might be linked to the disease-causing gene, narrowing the search further. Then, DNA fragments from this region can be used to probe for sequences that are expressed and therefore encode protein. In the case of cystic fibrosis, a cDNA library from epithelial cells of sweat glands was used. Finally, the candidate gene is sequenced from normal people and from people who have the disease in question. If appropriate mutations

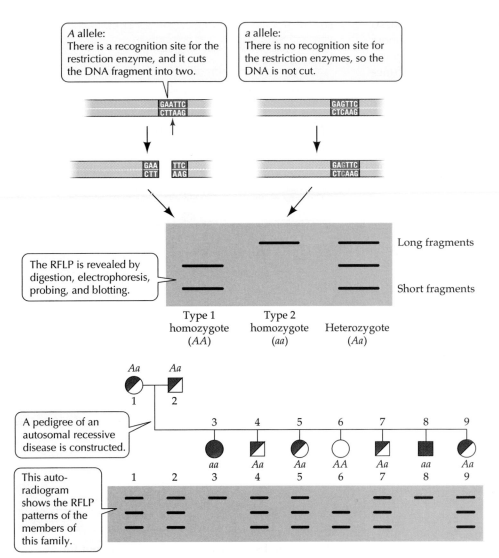

17.6 RFLP Mapping Restriction fragment length polymorphisms are differences in DNA sequences that serve as genetic markers. More than 1,000 such markers have been described for the human genome.

are found, the gene of interest has been isolated.

Since the isolation in 1989 of the gene responsible in mutant form for cystic fibrosis, positional cloning has been used to isolate dozens of important human genes, including those involved in fragile-X syndrome, Huntington's disease, and hereditary breast cancer.

Human gene mutations come in many sizes

The isolation of genes responsible for hereditary diseases has led to spectacular advances in the understanding of human biology. Before the genes, and then their proteins, were isolated for Duchenne's muscular dystrophy and for cystic fibrosis, proteins such as dystrophin and the chloride transporter had never been described. This identification of mutant genes has opened up new vistas in our understanding of how the human body works. Much more such information will come from the Human Genome Project, which we describe later in this chapter.

What mutations, then, give rise to human genetic diseases? Analyses of DNA have shown that the causes are both micro, at the level of point mutations, and macro, at the level of deletions, insertions, and a special case called expanding triplets. A single disease can have one of several micro and/or macro causes.

Phenylketonuria and sickle-cell anemia are caused by point mutations in the genes coding for the relevant proteins (Table 17.2). In addition, some of the variants of the β-globin gene cause disease, but others do not. Single base-pair mutations that alter a protein's function are usually important to the protein's three-dimensional structure—at the active site of an enzyme, for example.

Some mutations lead to not much of a protein at all. For example, some people with cystic fibrosis have a "nonsense" mutation such that a codon for an amino acid near the beginning of the

---

TABLE 17.2 Comparison of Two Genetic Diseases Caused by Point Mutations

VARIABLE	SICKLE-CELL ANEMIA	PHENYLKETONURIA
Protein in phenotype	β-globin	Phenylalanine hydroxylase
Length of chain	146 amino acids	451 amino acids
Normal protein	Glutamic acid at position 6	Arginine at position 408
Disease protein	Valine at position 6	Tryptophan at position 408
Length of gene	1512 base pairs	2448 base pairs
Normal allele	CGG at codon 6	GAA at codon 408
Disease allele	TGG at codon 6	GTA at codon 408

long protein chain has been changed to a stop codon, so protein translation stops at that point and a very short peptide is made. As we noted in Chapter 12, other point mutations affect RNA processing, leading to nonfunctional mRNA and a lack of protein synthesis.

DNA sequencing has revealed that mutations occur most often at certain base pairs in human DNA. These "hot spots" are often located where cytosine residues have been methylated to 5-methylcytosine (see Chapter 14). The explanation of this mutation phenomenon has to do with the natural instability of the bases in DNA.

Either spontaneously, or with chemical prodding, cytosine residues can lose their amino group and form uracil (Figure 17.7a). But the cell nucleus has a repair system that recognizes this uracil as being inappropriate for DNA: After all, uracil occurs only in RNA! So, the uracil is removed and cytosine replaced.

The fate of 5-methylcytosine that loses its amino group is rather different, since the result of that loss is thymine, a natural base for DNA. The uracil repair system ignores this thymine (Figure 17.7b). However, since the GC pair is now a mismatched GT pair, a different type of repair system comes in and tries to fix the mismatch. Half the time, the repair system matches a new C to the G, but the other half of the time, it matches a new A to the T, resulting in a mutation.

Larger mutations involve many base pairs of DNA. For example, earlier we described deletions in the X chromosome involving the dystrophin gene in Duchenne's muscular dystrophy. Some of these cover only part of the gene, leading to a partially complete protein and a mild form of the muscle disease. Others cover all of the gene and thus the protein is missing from muscle, resulting in the severe form of the disease. Still other deletions involve millions of base pairs, and cover not only the dystrophin gene but adjacent genes as well; the result may be several diseases simultaneously.

Insertions that cause disease can result from retrotransposition (see Chapter 14). For example, in some cases of hemophilia, a moderately repetitive LINE (long interspersed element) sequence has been inserted within the protein-coding region of the clotting-factor gene. This insertion causes the gene to code for a larger and nonfunctional protein.

Triplet repeats show the fragility of some human genes

About one-fifth of all males that have a fragile-X chromosome are phenotypically normal, as are most of their daughters. But many of those daughters' sons are mentally retarded. In a family in which the fragile-X syndrome appears, later generations tend to show earlier onset and more severe symptoms of the

(a)

When cytosine loses its amino group, uracil is formed.

Cytosine → Uracil

A DNA repair system removes this abnormal base and replaces it with cytosine.

GGATCACTC / CCTAGTGAG → GGAT U ACTC / CCTA G TGAG → GGATCACTC / CCTAGTGAG

(b)

When 5-methylcytosine loses its amino group, thymine results. Since this is a normal DNA base, it is not repaired.

5-Methylcytosine → Thymine

When DNA replicates, half the daughter DNA is mutant...

GGATCACTC / CCTAGTGAG → GGAT T ACTC / CCTA G TGAG

Replication

50% → GGATTACTC / CCTAATGAG

50% → GGATCACTC / CCTAGTGAG

...and half is normal.

17.7 5-Methylcytosine in DNA Is a Hot Spot for Mutagenesis Cytosine can lose an amino group either spontaneously or because of exposure to certain chemical mutagens. The abnormality is usually repaired unless the cytosine residue has been methylated to 5-methylcytosine, in which case a mutation is likely to occur.

disease. It is almost as if the abnormal allele itself is changing—and getting worse. And that's exactly what's happening.

The gene responsible for fragile-X syndrome contains a repeated triplet of CGG at a certain point. In normal people, this triplet is repeated 6 to 54 times (average 29). In the alleles of mentally retarded people with fragile-X syndrome, the CGG region is repeated 200 to 1,300 times. In the "premutated" males who show no symptoms but who are likely to have affected offspring, the repeats are fewer—52 to 200 times. These repeats become more numerous as the daughters of these males pass the chromosome on to their children (Figure 17.8).

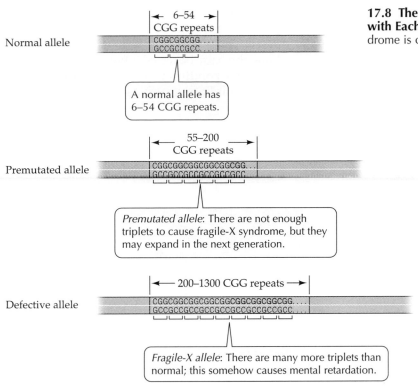

Normal allele

← 6–54 →
CGG repeats

CGGCGGCGG....
GCCGCCGCC....

A normal allele has 6–54 CGG repeats.

Premutated allele

← 55–200 →
CGG repeats

CGGCGGCGGCGGCGGCGG...
GCCGCCGCCGCCGCCGCC...

Premutated allele: There are not enough triplets to cause fragile-X syndrome, but they may expand in the next generation.

Defective allele

← 200–1300 CGG repeats →

CGGCGGCGGCGGCGGCGGCGGCGGCGG....
GCCGCCGCCGCCGCCGCCGCCGCCGCC....

Fragile-X allele: There are many more triplets than normal; this somehow causes mental retardation.

17.8 The CGG Repeat in the Fragile-X Gene Expands with Each Generation The genetic defect in fragile-X syndrome is caused by excessive repetitions of the CGG triplet.

posed to, a certain disease. It can be applied at many times and for many purposes: Prenatal testing can identify an embryo or fetus with a disease so that either a medical intervention can be applied or decisions about continuing the pregnancy can be considered. Newborn babies are tested so that proper medical intervention can be initiated. Asympto-matic people who have a relative with a genetic disease are tested to determine if they are carriers.

The goal of any screening is not just to provide information; it is to provide information that can be used to reduce an individual's burden resulting from the disease. In the broader arena of society, genetic screening poses ethical questions concerning our collective responsibility for people with genetic disorders.

Expanding triplet repeats have been found in other diseases, such as myotonic dystrophy (involving repeated CTG triplets) and Huntington's disease (in which CAG is repeated). Many non-disease-causing genes also appear to have these repeats, and because they are so common, they are assumed to play an important role. How they expand is not known. They may be found within the protein-coding region or outside of it.

Detecting Human Genetic Variations

Determination of the precise molecular phenotype and genotype for human genetic diseases has had three consequences. First, knowing what goes wrong is a way to find out what goes right: Mutants illuminate normal cell physiology. Second, knowing the cause of a disease at the biochemical level can lead to specific biochemical treatments and, potentially, cures. We return to some of these later in this chapter. Third, knowledge of molecular phenotypes and genotypes has led to ways of making precise diagnoses for many of these diseases, often before symptoms first appear, thus making medical intervention possible. DNA screening permits the identification of specific genotypes at any time of life and in any cell.

Genetic screening is the application of a test to identify people in a population who have, or are predis-

Screening for abnormal phenotypes makes use of protein expression

Screening for phenylketonuria in newborns is legally mandatory in many countries, including all of the United States and Canada. The reason for screening in the first days of life is that babies who are homozygous for this genetic disease are born with a normal phenotype, because excess phenylalanine in their blood before birth diffuses across the placenta to the mother's circulation. Since the mother is almost always heterozygous and therefore has adequate phenylalanine hydroxylase activity, the excess phenylalanine of the fetus is metabolized by its mother. Thus at birth the baby has not yet accumulated abnormal levels of phenylalanine.

After birth, however, the situation is different. The baby begins to consume protein-rich food (milk) and to break down some of its own proteins. In the process, phenylalanine enters the blood, and without the mother's enzyme to help, a baby with PKU, lacking its own phenylalanine hydroxylase activity, accumulates phenylalanine. The level of phenylalanine in its blood rises, and after a few days the level may be ten times higher than normal. Within days, the developing brain begins being damaged, and as Dr. Følling saw, untreated PKU patients are profoundly mentally retarded.

If PKU is detected early, it can be treated with a special diet (described later in this chapter). Thus detection is imperative. At first, physicians used Følling's ferric chloride test for phenylpyruvic acid in the urine. Unfortunately, babies with PKU do not start excreting large quantities of this substance until they are 4 to 8 weeks old, which can be too late to prevent brain damage. In 1963, Robert Guthrie described a simple screening test for PKU in newborns that today is used almost universally (Figure 17.9).

The Guthrie test uses a biological assay for blood phenylalanine. First, a drop of blood is collected from the baby's heel and placed on a piece of blotting paper. This paper is sent to the laboratory, where *Bacillus subtilis* bacteria are exposed to the dried blood spot. A small amount of a phenylalanine-like substance, 2-thienylalanine, is added to the growth medium. If there is no phenylalanine in the blood spot, the analog blocks bacterial growth.

A level of phenylalanine that is typical of normal individuals promotes a moderate amount of growth; more phenylalanine promotes more growth, and so on. Since the bacteria grow in a halo away from the dried spot, the diameter of the halo offers a simple estimate of the level of phenylalanine in the blood spot. This elegant application of bacterial physiology can be automated so that the screening laboratory can process many samples in a day.

If an infant tests positive for PKU in this screening, he or she must be retested using a more accurate chemical measurement for phenylalanine. If this test also shows a very high level in the blood, dietary intervention is begun. The whole process—the newborn genetic screen, confirmatory test, diagnosis, and initiation of treatment—are completed by the end of the second week of life. Since the screening test is inexpensive ($1) and since babies with PKU who receive early medical intervention are practically normal, the benefit is significant. Indeed, the benefit-to-cost ratio of newborn screening for PKU is one reason that legislators so readily make screening a legal requirement.

Even so, the screening test for PKU has come up against the concern with costs in contemporary medical care. Although the test itself is not costly, obtaining a blood sample may be. There is increasing financial pressure for mothers and their babies to leave the hospital less than a day after birth. Blood phenylalanine levels in some infants with PKU may rise slowly; they may not be significantly above normal levels during the first 2 days of life. So a blood sample taken before

the mother and child leave the hospital may not yet show a positive result on the screening test, even if the baby does have PKU. For this reason, it is recommended that the baby be tested a few days later.

Various other conditions are tested in newborns. Some of these conditions, such as galactosemia (an in-

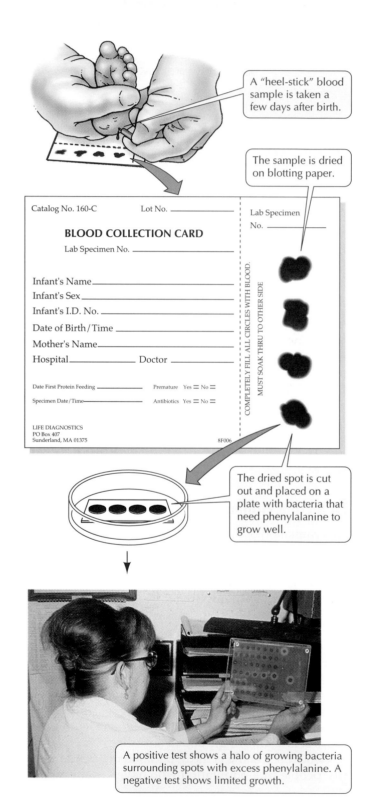

A "heel-stick" blood sample is taken a few days after birth.

The sample is dried on blotting paper.

The dried spot is cut out and placed on a plate with bacteria that need phenylalanine to grow well.

A positive test shows a halo of growing bacteria surrounding spots with excess phenylalanine. A negative test shows limited growth.

17.9 Genetic Screening of Newborns for Phenylketonuria A simple lab test devised by Robert Guthrie in 1963 is used today to screen newborns for phenylketonuria. Early detection means the symptoms of the condition can be prevented by following a therapeutic diet.

herited inability to break down milk sugar) have bacterial tests. Others, such as sickle-cell anemia, require differentiating between proteins with different charges (recall that normal hemoglobin is more negatively charged than is sickle hemoglobin). Still others use sophisticated immunoassays, such as the screening test for hypothyroidism. All of these tests provide valuable genetic information to the parents and clinical direction for the physicians to help the affected individuals.

There are several ways to screen for abnormal genes

Screening tests based on enzyme activity or protein structure, such as those for PKU and sickle-cell anemia, have limitations. They must be performed on tissues in which the relevant gene is expressed. For example, the blood level of phenylalanine is an indirect measure of phenylalanine hydroxylase activity in the liver, and hemoglobin electrophoresis shows the presence of sickle β-globin. But what if blood is difficult to test, as in a fetus? What about diseases that are expressed only in liver or brain and are not reflected in blood? What about proteins that are expressed under cellular controls, such that low activity might be the result of a simple dietary factor? Finally, since tissues in heterozygotes often compensate for having just one functional gene by raising the activity of the remaining active proteins to near normal levels, heterozygote testing is very difficult.

These problems are overcome by DNA testing, which is the most accurate way to test for an abnormal gene. With the description of the mutations in human genetic diseases (for example, see Table 17.2), it has become possible to directly examine any cell in the body at any time during the life span for gene mutations. However, these methods work best for diseases caused by one or a few mutations in the population. If there are dozens of possible gene mutations, simple tests such as the ones we will describe are much less informative.

The polymerase chain reaction (PCR) allows testing of the DNA of even one cell. For example, suppose a couple is heterozygous for the cystic fibrosis allele, have had a child with the disease, and want a normal child. If the woman is treated with the appropriate hormones, she can be induced to "superovulate," releasing several eggs. Following release, her eggs have completed the first meiotic division, with one large cell (the oocyte) and a small cell (the polar body) still attached to one another. In a heterozygote, one of these cells will have the normal allele and the other will have the allele for cystic fibrosis.

In polar body diagnosis, the polar body is removed and its DNA amplified by PCR. Then, a genetic test can be done to see which allele is present. If the polar body has the allele for cystic fibrosis, the oocyte must have the normal allele. The oocyte is then fertilized in the test tube with the husband's sperm, allowed to develop for a few days, and implanted in the mother's womb. The genetic diagnosis has been confirmed in all cases, and the result is normal children.

Polar body diagnosis is perhaps the extreme in DNA testing before birth. More typical are analyses of fetal cells after implantation in the womb. Fetal cells can be analyzed at about the tenth week of pregnancy (by chorionic villus sampling) or the thirteenth to seventeenth weeks (by amniocentesis). These two sampling methods are described in Chapter 39. In either case, only a few fetal cells are required for PCR and then genetic testing.

Newborns can also be screened for genes. The blood spots used for screening for PKU and other disorders contain enough of the baby's blood cells to permit extraction of the DNA, its amplification by PCR, and then genetic testing. Pilot studies are under way for testing for sickle-cell anemia and cystic fibrosis, and other genes will surely follow.

DNA testing is also now widely used to test adults for heterozygosity. For example, a sister or female cousin of a boy with Duchenne's muscular dystrophy may want to know if she is a carrier for the X chromosome that contains the dystrophin gene mutation. Similarly, the relatives of children with cystic fibrosis can determine their carrier status via DNA testing.

More problematic are genetic tests that show a predisposition to a disease. For example, as we will see later in this chapter, mutations in certain genes are associated with increased risk for certain types of cancer. If we can identify whether or not an individual carries such a mutation, who should have access to this information? DNA testing carries with it the potential for abuse. For example, given that there is a DNA marker for the Y chromosome, would it be ethical to test for the sex of an embryo or fetus and then choose to terminate the pregnancy on the basis of this information?

Of the numerous methods of genetic analysis, two are the most widespread. We will describe their use to detect the mutation in the DNA coding for β-globin, which results in sickle-cell anemia. The first method, *allele-specific cleavage*, employs differences between the normal and sickle alleles in the β-globin gene with respect to a restriction enzyme recognition site. Around position 6 in the normal gene is the sequence 5'...CCTGAGGAG...3'. This sequence is recognized by the restriction enzyme *Mst*II, which cleaves DNA at 5'...CCTNAGG...3', where "N" is any base. In the sickle mutation (see Table 17.2), the DNA sequence is changed to 5'...CCTGTGGAG...3'. The single basepair alteration, while it results in the clinical phenotype of sickle-cell anemia, also makes this sequence unrecognizable by *Mst*II. So, when *Mst*II fails to make the cut in the mutant gene, gel electrophoresis detects

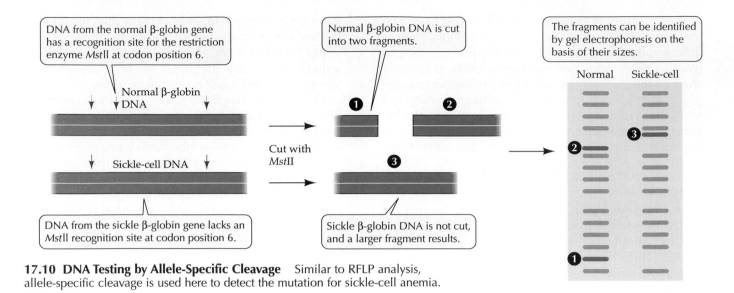

17.10 DNA Testing by Allele-Specific Cleavage Similar to RFLP analysis, allele-specific cleavage is used here to detect the mutation for sickle-cell anemia.

a larger DNA fragment (Figure 17.10). This analysis is similar to RFLP investigations (see Figure 17.6).

The second way to detect mutations is by *allele-specific oligonucleotide hybridization*. In this case, oligonucleotides are synthesized in the lab that will hybridize either to the denatured normal β-globin DNA sequence around position 6 or to the sickle mutant sequence. Usually, a probe of at least a dozen bases is needed to form a stable double helix with the target DNA, which may be fixed on a piece of filter paper. If the probe is labeled with radioactivity or with a colored or fluorescent substrate, hybridization is readily detected (Figure 17.11). This method is easier and faster than allele-specific cleavage.

Cancer: A Disease of Genetic Changes

Perhaps no malady affecting people in the developed world instills more fear than cancer. One in three Americans will have some form of cancer in their lifetime, and at present, one in four will die of it. With a million new cases and half a million deaths in the United States annually, cancer ranks second only to heart disease as a killer. Cancer was less common a century ago; then, as now in many poor regions of the world, people died of infectious diseases and did not live long enough to get cancer. Cancer is a disease of the later years of life; children are less frequently afflicted.

In 1970, the U.S. government declared "war on cancer." Although progress has been made, it is obvious from the statistics and from the people affected that the war is far from won. What has happened, however, is that the groundwork for winning the war has been laid during the past several decades. A tremendous amount of information on cancer cells—their growth and spread and their molecular changes—has been obtained by scientists and physicians mobilized to attack this challenging problem. Perhaps the most remarkable discovery is that cancer is a disease caused primarily by genetic changes.

In this discussion of cancer, we will examine the genetic changes in cancer cells and the distinctions between benign and malignant tumors. We will identify the viral causes of some cancers, and consider the roles of oncogenes and tumor suppressor genes. With these foundations, we will see why multiple genetic changes are often necessary to produce cancers.

Cancer cells differ from their normal cell counterparts

Cancer cells differ from the normal cells from which they originate in two major ways. First, a cancerous cell loses the control over cell division that exists in most tissues. Most cells in the body divide only if they are exposed to outside influences, such as growth factors and hormones. Cancer cells do not respond to these controls and instead divide more or less continuously, ultimately forming tumors (large masses of cells). By the time a physician can feel a tumor or see one on an X ray or CAT scan, it already has millions of cells.

Benign tumors resemble the tissue they came from, grow slowly, and remain localized where they develop. For example, a lipoma is a benign tumor of fat cells that may arise in the armpit and just stay there. Benign tumors are usually not a problem, but they must be removed if they impinge on an important organ, such as the brain.

Malignant tumors, on the other hand, are dedifferentiated and do not look like their parent tissue at all. A flat, specialized lung epithelial cell in a highly or-

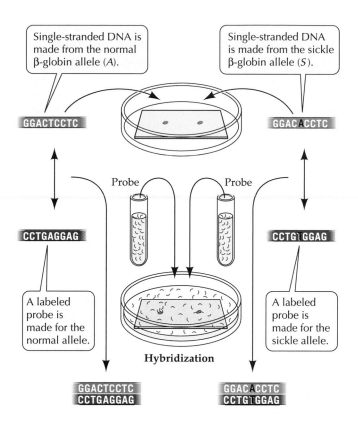

Single-stranded DNA is made from the normal β-globin allele (*A*).

GGACTCCTC

Single-stranded DNA is made from the sickle β-globin allele (*S*).

GGACACCTC

Probe   Probe

CCTGAGGAG

CCTGTGGAG

A labeled probe is made for the normal allele.

A labeled probe is made for the sickle allele.

Hybridization

GGACTCCTC
CCTGAGGAG

GGACACCTC
CCTGTGGAG

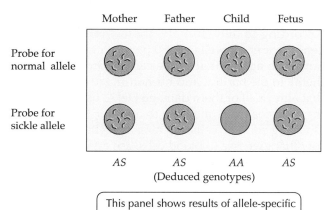

	Mother	Father	Child	Fetus
Probe for normal allele				
Probe for sickle allele				
	AS	*AS*	*AA*	*AS*

(Deduced genotypes)

This panel shows results of allele-specific hybridizations for a family.

17.11 DNA Testing by Allele-Specific Oligonucleotide Hybridization The hybridization process for this family reveals that three of them are carriers of the sickle-cell allele; the first child, however, inherited two normal alleles and is neither affected by the disease nor a carrier.

rounds them by actively secreting digestive enzymes to "eat" their way toward a blood vessel. Then some of the cancer cells enter either the bloodstream or the lymphatic system (Figure 17.12). Their journey through these vessels is perilous—imagine a small clump of cells going through the heart as it beats—and few cells survive, perhaps one in 10,000. When by chance cancer cells arrive at a suitable organ for further growth, they express cell surface proteins that allow them to bind to the new host tissue. This binding allows the tumor cells to grow away from the vessel, and they proceed to invade the host tissue. A malignant tumor also secretes chemical factors that cause blood vessels to grow to it and supply it with oxygen and nutrients.

Different forms of cancer affect different parts of the body. About 85 percent of all human tumors are *carcinomas*—cancers that arise in surface tissues such as the skin and linings (epithelial cells) of organs. Lung cancer, breast cancer, colon cancer, and liver cancer are all carcinomas. The first three account for more than half of all cancer deaths in Europe and North America; liver cancer is more common in Africa and Asia. *Sarcomas* are cancers of tissues such as bone, blood vessels, and muscle. *Leukemias* and *lymphomas* affect the cells that give rise to blood cells.

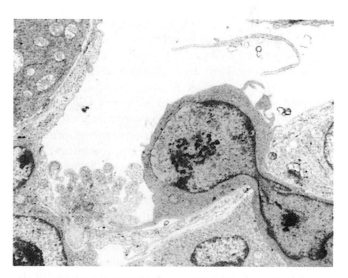

17.12 The Spread of Cancer A cancer cell squeezes into a small blood vessel through the vessel's wall. The cancer cell is then carried through the blood and, if it survives the journey, it may spread into other tissue.

dered arrangement in the wall of the lung may turn into an undistinguished, round lung cancer cell, clumps of which may break off. Malignant cells often have irregular structures, such as variable sizes and shapes of nuclei. Many malignant cells express the gene for telomerase (see Chapter 14) and thus do not shorten the ends of their chromosomes after each DNA replication.

The second, and most fearsome, characteristic of cancer cells that distinguishes them from normal cells is their ability to invade surrounding tissues and spread to other parts of the body. This spreading of cancer, called **metastasis**, occurs in several stages. First, the cancer cells extend into the tissue that sur-

TABLE 17.3   Human Cancers Known to Be Caused by Viruses

CANCER	ASSOCIATED VIRUS
Liver cancer	Hepatitis B virus
Lymphoma, nasopharyngeal cancer	Epstein–Barr virus
T cell leukemia	Human T cell leukemia virus
Anogenital cancers	Papilloma (wart) virus
Kaposi sarcoma	Herpes simplex virus

Some cancers are caused by viruses

Viruses cause many human diseases, and Peyton Rous's discovery in 1910 that a sarcoma in chickens is caused by a virus that is transmitted from one bird to another spawned an intensive search for cancer-causing viruses in humans. At least five types of human cancer are probably caused by viruses (Table 17.3).

Hepatitis B, a liver disease that affects people all over the world, is caused by a virus that contaminates blood or is carried from mother to child during birth. The viral infection can be long-lasting and may flare up numerous times. This virus is associated with liver cancer, especially in Asia and Africa, where millions of people are infected. But it does not act to cause cancer by itself: Some gene mutations also appear necessary in the tumor cells of Asians and Africans afflicted with the disease (although apparently not in Europeans and North Americans).

Similarly, Epstein–Barr virus, the agent that causes mononucleosis, is associated with certain types of lymphoma in Africa and Asia, but usually not in North America and Europe. Again, additional genetic events appear to occur in people who develop tumors. A rare form of leukemia is caused by HTLV-I, the human T cell leukemia virus. This virus is especially frequent in certain areas of Japan. Kaposi sarcoma, an otherwise rare tumor that is common in AIDS patients, is caused by a variant of herpes simplex virus. This virus is probably easily fought off by the immune systems of healthy people, but the crippled immune system of these patients cannot remove the virus, so it causes cancer.

An important group of virus-caused cancers in North Americans and Europeans consists of the various anogenital cancers caused by papillomaviruses. The genital and anal warts that these viruses cause often develop into tumors. These viruses seem to be able to act on their own, not needing mutations in the host tissue for tumors to arise. Sexual transmission of these papillomaviruses is unfortunately widespread.

Most cancers are caused by genetic mutations

Worldwide, no more than 15 percent of all cancers may be caused by viruses. What causes the other 85 percent? Because most cancers develop in older people, it is reasonable to assume that one must live long enough for a series of events to accumulate to finally produce the malignant cell. This assumption turns out to be correct, and the events are genetic mutations. But these are usually not the mutations found in genetic diseases that are present in germ line cells and passed on to offspring. Instead, the mutations in cancer cells are usually **somatic mutations**, which alter the genes of nonsex cells.

DNA can become damaged in many ways. Spontaneous mutations arise because of chemical changes in the nucleotides. For example, the conversion of cytosine to uracil, and of 5-methylcytosine to thymine (see Figure 17.7) occur with some frequency in DNA, as do errors in DNA replication. In addition, certain substances called *carcinogens* cause cancer. Familiar carcinogens include chemicals that are present in tobacco smoke and salted meats, ultraviolet light from the sun, and ionizing radiation from sources of radioactivity.

Less familiar but just as harmful are thousands of chemicals present naturally in the foods people eat. According to one estimate, these "natural" carcinogens account for well over 80 percent of the human exposure to agents that cause cancer. But the common theme in the natural and human-made carcinogens is that almost all of them damage DNA, usually by causing changes from one base to another (Figure 17.13). Dividing somatic cells, such as the cells in bone marrow that give rise to blood cells, are especially susceptible to genetic damage.

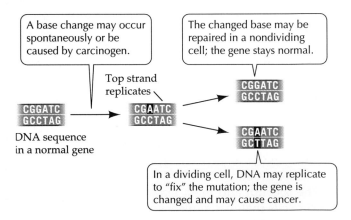

17.13 Dividing Cells Are Especially Susceptible to Genetic Damage and Cancer   A base change is more likely to be repaired in a nondividing cell.

17.14 Proto-Oncogene Products Stimulate Cell Division
Mutations can affect any of the several ways proto-oncogenes stimulate cell division, thus causing cancer.

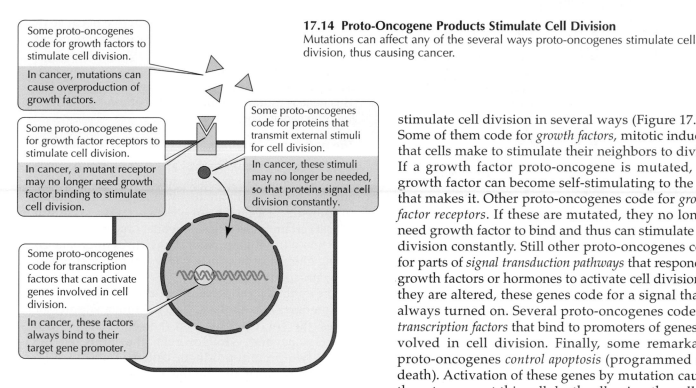

Some proto-oncogenes code for growth factors to stimulate cell division.

In cancer, mutations can cause overproduction of growth factors.

Some proto-oncogenes code for growth factor receptors to stimulate cell division.

In cancer, a mutant receptor may no longer need growth factor binding to stimulate cell division.

Some proto-oncogenes code for proteins that transmit external stimuli for cell division.

In cancer, these stimuli may no longer be needed, so that proteins signal cell division constantly.

Some proto-oncogenes code for transcription factors that can activate genes involved in cell division.

In cancer, these factors always bind to their target gene promoter.

Two kinds of genes are changed in many cancers

The changes in the control of cell division that lie at the heart of cancer can be likened to the control of the speed of an automobile. To make a car move, two things must happen: The gas pedal must be pressed and the brake must be released. In the human genome, there are both **oncogenes**, which act as the gas pedal to stimulate cell division, and **tumor suppressor genes**, which put the brake on to inhibit it.

Normal, differentiated cells typically do not divide, so their oncogenes are turned off and their tumor suppressor genes are turned on. Somatic mutations cause these sets of genes to do the reverse: turn the oncogenes on and the tumor suppressors off.

ONCOGENES. The first hint that oncogenes (from the Greek *onco-*, "mass") were necessary for cells to become cancerous came with identification of the virally induced cancers. In many cases, these viruses infect their host cells and bring in a new gene, one that acts to stimulate cell division by being actively expressed in the viral genome. It soon became apparent that these oncogenes had counterparts in the genomes of the host cells, but some of these cellular genes, called *proto-oncogenes*, were not actively transcribed in differentiated, nondividing cells. So the search for genes that are damaged by carcinogens quickly zeroed in on the proto-oncogenes.

The cellular proto-oncogenes are attractive targets for carcinogenesis, since they have the capacity to stimulate cell division in several ways (Figure 17.14). Some of them code for *growth factors*, mitotic inducers that cells make to stimulate their neighbors to divide. If a growth factor proto-oncogene is mutated, the growth factor can become self-stimulating to the cell that makes it. Other proto-oncogenes code for *growth factor receptors*. If these are mutated, they no longer need growth factor to bind and thus can stimulate cell division constantly. Still other proto-oncogenes code for parts of *signal transduction pathways* that respond to growth factors or hormones to activate cell division. If they are altered, these genes code for a signal that is always turned on. Several proto-oncogenes code for *transcription factors* that bind to promoters of genes involved in cell division. Finally, some remarkable proto-oncogenes *control apoptosis* (programmed cell death). Activation of these genes by mutation causes them to prevent this cell death, allowing the cells to continue dividing.

The same types of mutations that we observed in human genetic diseases activate cellular proto-oncogenes. Some proto-oncogenes are activated by point mutations; others by chromosome changes such as translocations; still others by gene amplification. Whatever the mechanism, the result is the same: The "gas pedal" for cell division is pressed.

TUMOR SUPPRESSOR GENES. About 10 percent of all cancer is clearly inherited. Often the inherited form of the cancer is clinically similar to the noninherited kind that comes later in life, called the *sporadic form*. The major differences are that the inherited form strikes a person much earlier in life and usually shows up as multiple tumors.

In 1971, Alfred Knudson used these observations to predict that for a tumor to occur, a tumor suppressor gene must be inactivated. But unlike oncogenes, where one mutated form is all that is needed for activation, the full inactivation of a tumor suppressor requires that both alleles be turned off, which requires two mutational events. For people with sporadic cancer, it takes a long time for both genes in a single cell to mutate. But in inherited cancer, people are born with one mutant allele for the tumor suppressor and need just one more event for full inactivation of the "brakes" (Figure 17.15).

Various tumor suppressor genes have been isolated and confirm Knudson's "two-hit" hypothesis. Some of these tumor suppressor genes are involved in inher-

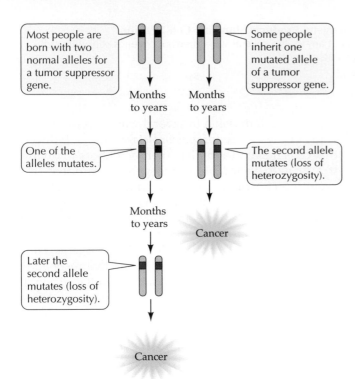

17.15 The "Two-Hit" Hypothesis for Cancer   Although a single mutation can activate a proto-oncogene, *two* mutations are needed to inactivate a tumor suppressor gene. An inherited predisposition occurs in people born with one allele already mutated.

ited forms of rare childhood cancers such as retinoblastoma (a tumor of the eye) and Wilms' tumor of the kidney, as well as in inherited breast and prostate cancers. In all instances, both copies of the tumor suppressor gene must be inactivated for the tumor to develop.

People born with one inactivated copy are prone to early cancer and are much more likely to get the disease than are those born with two normal copies of the gene. A good example is breast cancer. The 9 percent of women who inherit one mutated copy of the gene *BRCA1* have a 60 percent chance of having breast cancer by age 50 and an 82 percent chance of developing it by age 70. The comparable figures for women who inherit two normal copies of the gene are 2 percent and 7 percent, respectively.

How do tumor suppressor genes act in the cell? Not surprisingly, they appear to control progress through the cell cycle. The protein encoded by *Rb*, a gene that was first described for its contribution to retinoblastoma, is active during G1. In the active form it binds some transcription factors that are necessary for progress to S phase and the rest of the cycle. In nondividing cells, *Rb* remains active, preventing cell division until the proper growth factor signals are present. When the Rb protein is inactive because of mutation, the cell cycle moves forward independent of growth factors.

Another widespread tumor suppressor gene, *p53*, has a protein product that also stops cells during G1. This gene is mutated in many types of cancers, including lung cancer and colon cancer.

The pathway from normal to cancerous is complex

The "gas pedal" and "brake" analogies for oncogenes and tumor suppressor genes, respectively, are elegant but simplified. There are many oncogenes and tumor suppressor genes, some of which act only in certain cells at certain times. The anogenital cancers caused by infection with human papillomaviruses offer a fascinating glimpse at how oncogenes and tumor suppressor genes interact.

The epithelial cells that are infected make the protein products of both the *Rb* and *p53* tumor suppressor genes, thereby ensuring that the cell cycle will not progress beyond G1. However, the viral genome makes two protein products, called E7 and E6 ("E" stands for the *e*arly time in the infection cycle that the proteins are made). Remarkably, the viral E7 protein binds to the Rb protein and inactivates it, and the E6 protein similarly binds to and inactivates the p53 protein. With both tumor suppressors inactive, cell division can proceed.

Because cancers progress to full malignancy slowly, it is possible to describe the oncogenes and tumor suppressor genes at each stage in great molecular detail. Such a description has been developed for colon cancer (Figure 17.16). At least three tumor suppressor genes and one oncogene must be mutated in sequence for a cell to become metastatic. Although the occurrence of all these events in a single cell might appear unlikely, remember that the colon has millions of cells, that the cells giving rise to epithelial cells are constantly dividing, and that these changes take place over many years of exposure to natural and human-made carcinogens and to spontaneous mutations.

The characterization of the molecular changes in tumor cells has opened up the possibility of genetic diagnoses and screening, as is done for genetic diseases. Many tumors are now commonly diagnosed in part by oligonucleotide-specific probes for oncogene and/or tumor suppressor gene alterations. For hereditary cancers, it is possible to detect early in life whether an individual has inherited a mutated tumor suppressor gene, making possible preventive action. For example, a person who inherits mutated copies of the tumor suppressor genes involved in colon cancer normally would have a high probability of developing this tumor by age 40. Surgical removal of the colon would prevent this metastatic tumor from arising.

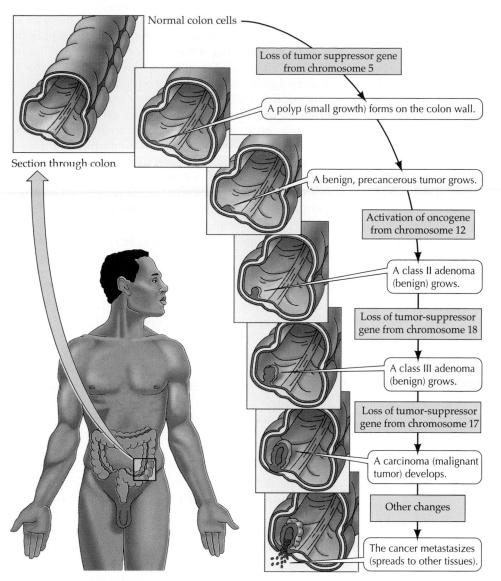

17.16 Multiple Mutations Transform a Normal Colon Epithelial Cell into a Cancerous Cell

Within the figure:
- Normal colon cells
- Section through colon
- Loss of tumor suppressor gene from chromosome 5
- A polyp (small growth) forms on the colon wall.
- A benign, precancerous tumor grows.
- Activation of oncogene from chromosome 12
- A class II adenoma (benign) grows.
- Loss of tumor-suppressor gene from chromosome 18
- A class III adenoma (benign) grows.
- Loss of tumor-suppressor gene from chromosome 17
- A carcinoma (malignant tumor) develops.
- Other changes
- The cancer metastasizes (spreads to other tissues).

Treating Genetic Diseases

Most treatments of genetic diseases try to alleviate the symptoms that affect the patient. But to effectively treat diseases caused by genes—whether they affect all cells, as in inherited disorders such as PKU, or only somatic cells, as in lung cancer—physicians must be able to diagnose the disease accurately, must know how the disease works at the biochemical level, and must be able to intervene early, before the disease ravages or kills the individual. As we have seen, basic research has provided the knowledge for accurate diagnostic tests, as well as a beginning at understanding pathogenesis (the cause of diseases) at the molecular level. Physicians are now applying this knowledge to treat genetic diseases. In this section, we will see that these treatments range from specifically modifying the mutant phenotype, such as supplying a missing product of a defective enzyme, to supplying the normal version of a mutant gene.

One approach is to modify the phenotype

After a newborn baby is diagnosed with PKU, therapy is aimed at *restricting the substrate* of the deficient enzyme. In this case, the substrate is phenylalanine, which is not synthesized by the infant but comes mostly from its diet. So the infant is immediately put on a special diet that contains only enough of this amino acid for immediate use.

Lofenelac, a milk-based product that is low in phenylalanine, is fed to these infants just like formula. Later, certain fruits, vegetables, cereals, and noodles low in phenylalanine can be added to the diet. Meat, fish, eggs, dairy products, and bread, which contain high amounts of phenylalanine, must be avoided, especially during childhood, when brain development is most rapid. The artificial sweetener aspartame must also be avoided, because it is made of two amino acids, one of which is phenylalanine.

People with PKU are generally advised to stay on a low-phenylalanine diet for life. Although maintaining these dietary restrictions may be difficult, it is effective. Numerous follow-up studies since newborn screening was initiated have shown that people with PKU who stay on the diet are no different from the rest of the population in terms of mental ability. This is an impressive achievement in public health, given the extent of mental retardation in untreated patients.

Another way to modify a disease phenotype is by *metabolic inhibitors*. As we described earlier, people with familial hypercholesterolemia accumulate dangerous levels of cholesterol in their blood. Not only are these people unable to metabolize dietary cholesterol, but they also synthesize a lot of it. One effective treatment for people with this disease is the drug mevinolin, which blocks the patients' own cholesterol synthesis. Patients who receive this drug need only worry about cholesterol in their diet and not about what their cells are making.

Metabolic inhibitors also form the basis of cancer therapy with radiation and drugs. The strategy is to kill rapidly dividing cells, since rapid cell division is the hallmark of malignancy. Radiation, which can be focused on the tumor, damages DNA, breaks chromosomes, and generally kills cells by forming highly reactive chemicals called free radicals. If the tumor is localized, all of it may be killed. The same holds true of the most common way to treat cancers: surgical removal.

Unfortunately, by the time they are diagnosed, many tumors have already begun to metastasize. Treatment of metastatic cancers involves attempting to control them at many sites in the body. Many drugs have been designed to kill the dividing cells (Figure 17.17), but most of these drugs are given in the blood and thus also damage other, noncancerous, dividing cells in the body.

Given the broad scope of drug treatment, it is not surprising that people undergoing chemotherapy suffer side effects such as loss of hair (the skin epithelium constantly divides to replace the cells that die), digestive upsets (gut epithelial cells also divide constantly), and depletion of blood cells (bone marrow stem cells). The effective dose of these highly toxic drugs for treating the cancer is often just below the dose that would kill the patient, so they must be used with utmost care. Usually they can control the spread of cancer, but not cure it.

An obvious way to treat a disease phenotype in which a functional protein is missing is to *supply the missing protein*. This is the basis of treatment of hemophilia A, in which the missing blood clotting factor is supplied in pure form. The production of human clotting protein by DNA technology (see Chapter 16) is critically important here, because it allows a pure protein to be given instead of blood transfusions of crude blood products, which could be contaminated. Blood contamination was a problem in the early years of the AIDS epidemic, when testing for HIV in blood was not yet possible; as a result, many people with hemophilia developed AIDS because they were transfused with HIV-contaminated blood.

Unfortunately, the phenotypes of many diseases caused by abnormalities in genes are very complex. Simple interventions, such as some of those we have described, do not work for most such diseases. Indeed, a recent survey showed that current therapies for 351 diseases caused by single gene mutations improved the patients' life span by only 15 percent.

For the many polygenic disorders such as cancers, "magic bullet" specific therapies are unlikely. Knowledge of the precise genetic errors often far outstrips knowledge of clinical mechanisms. Is it possible to treat the root cause of these disorders at the DNA level?

Gene therapy offers the hope of specific therapy

Perhaps the most obvious thing to do when a cell lacks a functional allele is to provide one. Such **gene therapy** is under intensive investigation for diseases ranging from the rare inherited disorders caused by single mutations, to cancer, AIDS, and atherosclerosis. Like genetic transformation in other organisms (see Chapter 13), gene therapy in humans seeks to insert a new gene that will be expressed in the host. Thus, the new DNA is often attached to a promoter that will be active in the human cells.

Presenting the DNA for cellular uptake follows the methods used in biotechnology: Ca^{2+} complexes, liposomes, and viral vectors are used to enable uptake of the "good gene" into the human cells. Physicians who practice this "molecular medicine" are confronted by the challenges of genetic engineering: efficient uptake; effective vectors; appropriate expression and processing of mRNA and protein; and selection within the body for the cells that contain the recombinant DNA.

5-Fluorouracil blocks the synthesis of nucleotides.

Etoposide prevents DNA from unwinding, blocking DNA replication and transcription.

DNA

Arabinocytosine blocks DNA replication.

Transcription

Adriamycin blocks transcription to inhibit protein synthesis.

mRNA

Translation

Vincristine and Taxol block the mitotic spindle microtubules from functioning.

Cell division proteins

17.17 Drug Strategies for Killing Cancer Cells
These medications kill dividing cancer cells. Unfortunately, most of them also affect noncancerous dividing cells.

17.18 Gene Therapy   New genes have been added to somatic cells taken from a patient's body, then returned to express normal alleles.

Which human cells should be the targets of gene therapy? The best idea for a genetic disease would be to replace the nonfunctional gene with a functional one in every cell of the body. But vectors to do this are simply not available, and delivery to every cell poses a formidable challenge. Until recently, the major attempts at gene therapy have been *ex vivo*. That is, the scientists and physicians have taken cells from the patient's body, added the new gene to the cells in the laboratory, and then returned the cells to the patient in the hope that the correct gene product would be made (Figure 17.18). A widely publicized example of this approach was the introduction of a functional gene for the enzyme adenosine deaminase into the white blood cells of a girl with a genetic deficiency of this enzyme. Unfortunately, these were mature white blood cells, and although they survived for a time in the girl and made the enzyme for some therapeutic benefit, they eventually died, as is the normal fate of such cells. It would be more effective to insert the functional gene into *stem cells*, the bone marrow cells that constantly divide. This is a major thrust of many current clinical experiments on gene therapy.

The other method of gene therapy is to insert the gene directly into cells in the body of the patient, using a vector or complex. This *in vivo* approach is being attempted for various types of cancer. For example, lung cancers are accessible if the DNA or vector is given as an aerosol through the respiratory system. Vectors expressing DNA for functional tumor suppressor genes that are mutated in the tumors, as well as vectors expressing antisense RNAs against oncogene mRNAs, have been successfully introduced in this way to patients with lung cancer, with some clinical improvement.

The Human Genome Project

In 1984 the U.S. government sponsored a conference to examine the problem of detecting DNA damage in people exposed to low levels of radiation, such as Japanese who had survived the atomic bombs 39 years earlier. The scientists attending this conference quickly realized that monitoring the human genome for changes would also be useful in evaluating environmental mutagens. But new, more efficient and sensitive technologies would be needed in order to be able to sequence the entire human genome. In 1986, Renato Dulbecco, who won the Nobel prize for his pioneering work on cancer

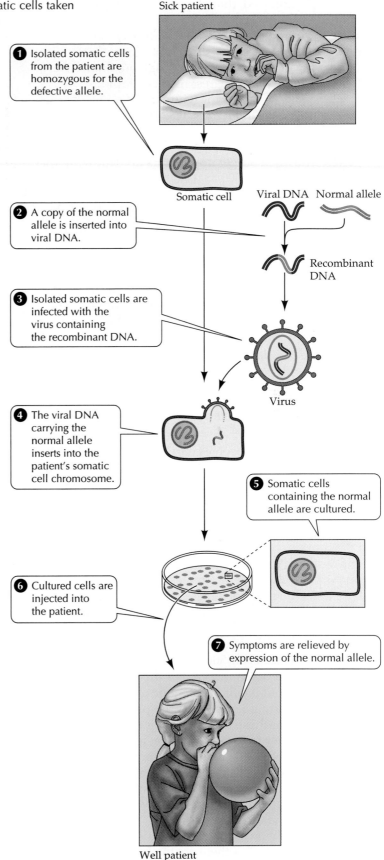

Sick patient

1 Isolated somatic cells from the patient are homozygous for the defective allele.

Somatic cell   Viral DNA   Normal allele

2 A copy of the normal allele is inserted into viral DNA.

Recombinant DNA

3 Isolated somatic cells are infected with the virus containing the recombinant DNA.

Virus

4 The viral DNA carrying the normal allele inserts into the patient's somatic cell chromosome.

5 Somatic cells containing the normal allele are cultured.

6 Cultured cells are injected into the patient.

7 Symptoms are relieved by expression of the normal allele.

Well patient

viruses, suggested that determining the entire sequence of human DNA could be a boon to cancer research. He proposed that the scientific community be mobilized to the task. By 1991, ambitious efforts were under way in the United States, England, France, Italy, Canada, and Japan. Progress has been rapid.

The major goal of the Human Genome Project (HGP) is to determine the nucleotide sequence of all the DNA of an individual. Of course, genetic polymorphisms ensure that every person is different. But some generalities will surely apply to us all. Knowing the sequences of the 50,000 to 100,000 human genes, most of which are not known today, could unlock the secrets of the polygenic disorders that plague humankind, as well as provide a wealth of basic information on who we are at the molecular level.

In this section, we will examine HGP approaches to its challenge and some of the benefits of knowing the entire sequence of the human genome. To reach this goal, the HGP has established three intermediate goals: (1) to develop ways of mapping the genome at increasingly fine levels of precision, (2) to test new techniques by determining the genome sequences of some "simple" organisms, and (3) to study some of the ethical issues that the knowledge will raise.

Mapping and sequencing entire genomes uses new molecular methods

Each human chromosome consists of a single double-stranded molecule of DNA, and because of their differing sizes (see Figure 9.14), the chromosomes can be separated from each other. So it is possible to carefully isolate the DNA of human chromosome 4, for example. The straightforward approach to sequencing the DNA of this chromosome would be to start at one end and simply do it all. Unfortunately, this approach does not work. The DNA of a molecule that is 50 million base pairs long cannot be sequenced in that form; only 500 base pairs at a time can be sequenced. (See Figure 16.11 for a review of DNA sequencing.)

In the HGP, then, chromosomal DNA is first cut into fragments, each 500 base pairs long, and then each fragment is sequenced. For the human genome, which has about 3 billion base pairs, there are about 6 million such fragments. Then the problem becomes one of putting these millions of pieces of the jigsaw puzzle together to form a long, linear molecule. This problem can be overcome by breaking up the DNA in several ways into "sub-jigsaws" that overlap, and using *mapped markers* to align the overlapping fragments. The data can be stored in powerful computers for analysis and ordering.

What are the mapped markers along the chromosome? The crudest of them are the **bands** that appear when chromosomes are stained with certain dyes (see Figure 9.14). DNA in a chromosome is approximately

evenly split among the 20 or so dark and light bands, so each band represents about 5 million base pairs.

Banding is useful to localize a gene or other DNA fragment by *fluorescence in situ hybridization* (FISH), in which a probe is hybridized to metaphase chromosomes whose DNA has been denatured (Figure 17.19). In interphase nuclei, the extended nature of the chromosomes can allow FISH to localize DNA sequences that are separated by only about 100,000 base pairs. By careful manipulation of separated chromosomes under the microscope, a single band can be dissected out and then further analyzed by fragmentation of its DNA.

In **physical mapping** of the chromosome, landmarks are ordered and the distances between them are determined (Figure 17.20). The result can be compared to a road map showing towns with the mileage separating them. The "towns" are DNA markers and the "mileage" is in base pairs. The simplest of the markers are the recognition sites for restriction enzymes. Since there are hundreds of these sequence–enzyme pairs, there are hundreds of ways to cut the DNA and then generate a restriction map. Physical mapping has been useful in generating maps for relatively small DNAs of thousands of base pairs, such as individual chromosome regions or the genomes of viruses.

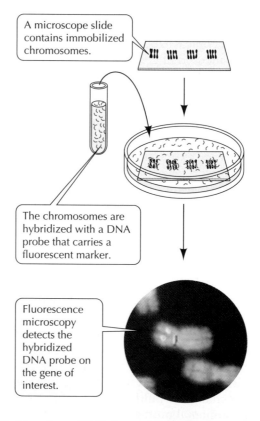

A microscope slide contains immobilized chromosomes.

The chromosomes are hybridized with a DNA probe that carries a fluorescent marker.

Fluorescence microscopy detects the hybridized DNA probe on the gene of interest.

17.19 Mapping a DNA Sequence by Fluorescence *in situ* Hybridization In the FISH technique, banding patterns are used to locate a gene using a fluorescent probe. (The red band is a marker for the centromere of the X chromosome, further specifying the location of the gene.)

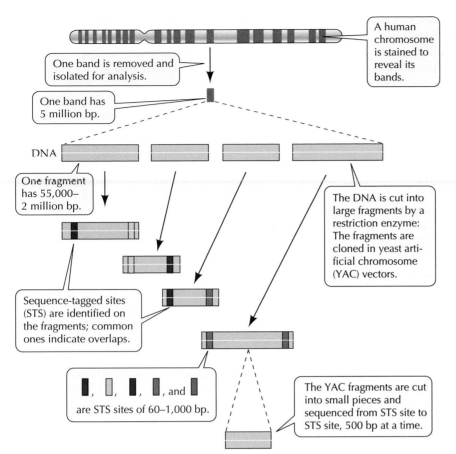

17.20 Physical Mapping of Short Stretches of DNA Mapping a single band on a chromosome arranges the DNA fragments for sequencing.

A human chromosome is stained to reveal its bands.

One band is removed and isolated for analysis.

One band has 5 million bp.

DNA

One fragment has 55,000–2 million bp.

The DNA is cut into large fragments by a restriction enzyme: The fragments are cloned in yeast artificial chromosome (YAC) vectors.

Sequence-tagged sites (STS) are identified on the fragments; common ones indicate overlaps.

, , , , and
are STS sites of 60–1,000 bp.

The YAC fragments are cut into small pieces and sequenced from STS site to STS site, 500 bp at a time.

base pairs long. These pieces are then sequenced.

In addition to physical maps, **genetic maps** of human chromosomes are being generated. Genetic maps also consist of signposts along the chromosome, but in this case they are DNA sequences whose locations are determined genetically. Many DNA sequences are polymorphic. If particular polymorphisms of two genes are always inherited together, they must be closely linked on the chromosome. Such linked DNA sequences might be a disease-causing gene and a particular RFLP. Short, side-by-side repeats of very simple sequences (such as CA) appear to occur in fixed numbers. For example, one sequence might be repeated five times (CACACACACA), another 36 times, and so on throughout the genome and are thus quite useful as genetic linkage markers.

Some restriction enzymes recognize 8 to 12 base pairs in DNA, not just the usual 4 to 6 base pairs. A DNA molecule with several million base pairs will have relatively few of these larger sites, and the enzyme thus will generate a small number of relatively large fragments. These large fragments can be put into yeast artificial chromosome (YAC) vectors (see Figure 16.5*b*) and made into a library. How are the books (fragments) in this library arranged?

A powerful approach to arranging the fragments involves the development of *sequence-tagged sites* (STSs)—unique stretches of DNA, 60 to 100 base pairs long, whose sequences are known. An STS can be obtained by simply taking a random piece of genomic DNA and seeing if it is unique to a location in the genome by FISH. If it is, the STS is now a marker for that particular chromosomal location. About 20,000 STSs have been mapped on human chromosomes, meaning that each is about 100,000 base pairs (or just a few genes) away from the next STS.

To arrange DNA fragments of a map, libraries made from different restriction enzyme cuts are compared. If two large fragments of DNA cut with different enzymes have the same STS, they must overlap. The individual fragments can be cut again—this time with more common restriction enzymes—into pieces 500

Genetic maps are useful to the HGP in two ways. First, as we described earlier, linkage studies with markers have been very important in tracking down disease-causing genes by positional cloning. Second, the genetic and physical maps can complement each other—one providing new markers for the other.

Sequencing technologies for the final, 500-base-pair DNA fragments are being improved. The use of ultra-thin gels to speed the separation of DNA pieces, automation, and computer programs to process the huge amounts of information have reduced the cost and increased the efficiency of the sequencing process.

The genomes of several organisms have been sequenced

Although the "grunt work" of mapping and sequencing DNAs is not very exciting, the results for organisms whose complete genomes have been sequenced, as well as preliminary data from humans, have been spectacular. Complete genomic sequences have been determined for 141 viruses, 51 organelles, two bacteria, one archaean, and yeast, a single-celled eukaryote. We will describe some of the revelations from the sequences of a bacterium and from yeast.

The only host for the bacterium *Haemophilus influenzae* is humans, where it lives in the upper respiratory tract and can cause either ear infections or, more seriously, meningitis in children. Its 1,830,137 base pairs are in a single circular chromosome. In addition to its origin of DNA replication and genes coding for ribosomal and tRNAs, this bacterial chromosome has 1,743 regions containing amino acid codons along with the transcriptional (promoter) and translational (start and stop codons) information for protein synthesis. Only 1,007 (58 percent) of these have predicted amino acid sequences that correspond to known proteins. So there are probably many proteins from this bacterium to be discovered.

Of the genes and proteins with assigned roles, most confirm a century of biochemical description of bacterial enzymatic pathways. For example, there are genes for the entire pathways of glycolysis, fermentation, and electron transport. Some of the gene sequences for unknown proteins may code for membrane proteins, possibly involved in active transport. Another finding is that highly infective strains of this bacterium have genes coding for surface proteins that attach them to the respiratory tract, while noninfective strains lack these genes.

Not unexpectedly, the yeast genome is much more complex. It has 16 chromosomes and more than 12,068,000 base pairs. More than 600 scientists around the world collaborated in mapping and sequencing the yeast genome. When they began, they knew of about 1,000 yeast genes coding for RNA or protein. Now there are apparently 5,885! This means that there is a protein-coding gene every 2,000 base pairs in the genome, and 70 percent of the entire DNA is taken up by coding regions. The estimated 100,000 genes in the human genome, in contrast, take up less than 2 percent of all human DNA.

It is now possible to estimate what fractions of the yeast genome code for specific metabolic roles. Apparently 11 percent of the proteins are for general metabolism, 3 percent for energy production and storage, 3 percent for DNA replication and repair, and 13 percent for protein synthesis—there are more than 200 transcription factors! There are also many structural proteins, and, of course, many proteins whose functions are as yet unknown.

Comparisons with human genes indicate that many yeast genes have homologous sequences—similar but not identical—in humans. Since it is possible by molecular biology to selectively inactivate individual yeast genes, the resulting phenotypic changes will be relevant to the roles of these genes in humans.

Preliminary information reveals much about the human genome

Human DNA mapping has developed quickly, and large-scale sequencing is under way. Instead of taking the approach of the scientists studying yeast and bac-teria—waiting for the complete genomic sequence before isolating protein-coding genes—molecular biologists have tried to get at the protein-coding genes first.

Since these regions are transcribed into mRNA, the approach has been to sequence cDNA. Actually, the entirety of each cDNA has not been sequenced from libraries of different tissues at different stages of life. Instead, only short yet unique regions, called *expressed-sequence tags*, have been developed to act as chromosome markers (STSs). These tags can be used to find the sequences of entire protein-coding genes.

Of the 30,000 human genes isolated and sequenced from this cDNA–STS approach, many have yeast homologs, so we can speculate as to their functions. About 40 percent of them appear to be genes for basic metabolism, about 22 percent are for protein synthesis, and 12 percent are involved in physiological signaling. Identification of many of these genes will no doubt lead to a better understanding of human biology and its pathologies. But these genes represent at most only half of the protein-coding regions of the human genome.

As the HGP progresses, more and more human genes are being identified, either by disease-related positional cloning or by cDNA analysis. Once a gene is sequenced from one person, it can be sequenced in others, enabling the discovery of common polymorphisms. Computerized databases of polymorphisms can then be matched to databases of diseases, to see if people with a certain gene variant have a certain disease. For example, an allele of the gene for apolipoprotein E is strongly correlated with increased risk for Alzheimer's disease.

The technologies developed for the HGP are already being applied to human diagnostics. Automation of PCR and DNA sequencing are making possible the diagnosis of many diseases caused by genes, from prenatal detection of cystic fibrosis to biopsy of developing tumors. "DNA-on-a-chip" technologies may allow the simultaneous amplification and detection of human variants by placing a drop of blood or other tissue fluid onto a computer chip–like structure made of oligonucleotides instead of silicon.

What will be the uses of genetic information?

After the primary genetic defect that causes cystic fibrosis was discovered, many people predicted a "tidal wave" of genetic testing for heterozygous carriers. Everyone would want the test, it was said—especially the relatives of patients with the genetic disease. But the tidal wave has not developed. To find out why, a team of psychologists, ethicists, and geneticists interviewed 20,000 people in the United States. What the researchers found surprised them: Most people are simply not very interested in their genetic makeup, unless they have a close relative with the disease and are involved in a pregnancy.

Some other people, however, might be very interested in genetic testing. People who test positive for genetic abnormalities, from hypercholesterolemia to cancer, might be denied employment or health insurance. The many linkages of genetic abnormalities to behavioral characteristics, ranging from manic depression to schizophrenia, has led to the potential for screening and then social manipulations of those at risk. Consequently, many legislative bodies are considering and passing bills that prohibit genetic discrimination. The HGP has set aside more than 5 percent of its budget for investigations into the ethical, legal, and social issues of its findings. Such an approach is unusual in scientific history, where the pattern has been to invent technology first and ask social questions later.

Although the HGP has the potential for social disruption, it also has enormous possible benefits. As the ultimate extension of molecular medicine, it will lead not only to the identification of the genes that are altered in disease, but also to targeted treatments, and possibly cures. It will also tell us a lot about ourselves. Comparing the DNA sequences of different people and other organisms will shed new light on how humans got to be where they are, genetically. And by focusing on polymorphisms—the differences between us—the HGP will help physicians understand the complex diseases that result from the interactions of many genes and the environment.

Summary of "Molecular Biology and Medicine"

Protein as Phenotype

• The idea of the one gene, one polypeptide relationship, developed with prokaryotic systems, also applies to human genetic diseases. In many human genetic diseases (for example, phenylketonuria), an enzyme is missing or inactive. **Review Figure 17.1 and Tables 17.1, 17.2**
• Point mutations in the genes that encode the protein components of hemoglobin either lead to clinical abnormalities such as sickle-cell anemia or are relatively benign. **Review Figure 17.2**
• Some diseases are caused by mutations that affect structural proteins; examples include Duchenne's muscular dystrophy, osteoporosis imperfecta, and hemophilia.
• The genes that code for receptors and membrane transport proteins can be mutated and cause diseases such as familial hypercholesterolemia and cystic fibrosis. **Review Figure 17.3**
• Few human diseases are caused by a single gene mutation. Most are caused by the interactions of many genes and proteins with the environment.

• Human genetic diseases show different patterns of inheritance. Phenylketonuria, sickle-cell anemia, and cystic fibrosis, for example, are inherited as autosomal recessives; familial hypercholesterolemia and osteoporosis imperfecta as autosomal dominants; hemophilia and Duchenne's muscular dystrophy as X-linked conditions; and fragile-X syndrome as a chromosome abnormality.

Mutations and Human Diseases

• Molecular biology techniques have made possible the isolation of many genes responsible for human genetic diseases. One method is to isolate the mRNA for the protein in question and then use the mRNA to isolate the gene from a genomic library. Another method is to compare DNA from a patient that lacks a piece of a chromosome with DNA from a person who does not show this deletion to isolate the missing DNA. **Review Figure 17.5**
• In positional cloning, DNA sequence landmarks are used as guides to point the way to a gene. These landmarks might be restriction fragment length polymorphisms, which are linked to a mutant gene. **Review Figure 17.6**
• Human mutations range from single point mutations to large deletions. Some of the most common mutations occur where the modifed base 5-methylcytosine is converted to thymine. **Review Figure 17.7 and Table 17.2**
• The fragile-X chromosome is inherited, and the allele's effects on mental retardation worsen with each generation. This effect is caused by a triplet repeat that tends to expand with each new generation. **Review Figure 17.8**

Detecting Human Genetic Variations

• Genetic screening detects human gene mutations. Some protein abnormalities can be detected by simple tests, such as detection of excess substrate or lack of product. **Review Figure 17.9**
• The advantage of testing DNA for mutations directly is that any cell can be tested at any time in the life of the organism.
• There are two methods of DNA testing: allele-specific cleavage and allele-specific oligonucleotide hybridization. **Review Figures 17.10, 17.11**

Cancer: A Disease of Genetic Changes

• Most cancers are caused by genetic changes.
• Tumors may be benign, which grow to a certain extent and then stop, or malignant, spreading through organs and to other places in the body.
• The most common cancers occur in dividing cells such as epithelia.
• At least five types of human cancers are caused by viruses, which account for about 15 percent of all cancers. **Review Table 17.3**
• Eighty-five percent of human cancers are caused not by viruses, but by genetic mutations of somatic cells. Normal cells contain proto-oncogenes, which when mutated can become oncogenes and cause cancer by stimulating cell division or preventing cell death. **Review Figures 17.13, 17.14**
• About 10 percent of all cancer is inherited as a result of the mutation of tumor suppressor genes, which normally act to slow down the cell cycle. For cancer to develop, both alleles of a tumor suppressor gene must be mutated. In inherited cancer, an individual inherits one mutant allele and then somatic mutation occurs in the second one. In sporadic can-

cer, two normal alleles are inherited, so two mutational events must occur in the same somatic cell. **Review Figure 17.15**
• Mutations must activate several oncogenes and inactivate several tumor suppressor genes for a cell to produce a malignant tumor. **Review Figure 17.16**

Treating Genetic Diseases
• Most genetic diseases are treated symptomatically. However, as more knowledge is accumulated, specific treatments are being devised.
• One treatment approach is to modify the phenotype—for example, by manipulating the diet or providing specific inhibitors to prevent the accumulation of a harmful substrate, or by supplying a missing protein. **Review Figure 17.17**
• In gene therapy, a mutant gene is replaced with a normal gene. Either the affected cells can be removed, the new gene added, and the cells returned to the body, or the new gene can be inserted via a vector directly into the patient. **Review Figure 17.18**

The Human Genome Project
• The aim of the Human Genome Project is to determine the entire DNA sequence of a human, which will require sequencing many 500-base-pair fragments and then putting the sequences together.
• The broadest genomic maps are chromosome bands. Individual sequences may be mapped on the bands by hybridization. **Review Figure 17.19**
• Various short DNA sequences have been mapped on chromosomes, physically or genetically, to act as guideposts for fragments of DNA. The various fragments may be ordered, cut with restriction enzymes, and sequenced. **Review Figure 17.20**
• The genome of the bacterium *Haemophilus* contains 1.8 million base pairs and has been sequenced. Many new proteins have been found encoded by this genome. The sequence of the yeast genome, which contains more than 12 million base pairs, reveals almost 6,000 protein-coding genes, of which only 1,000 are currently known. The various roles of the yeast genes can be inferred from their sequence.
• Sequencing has identified more than 30,000 human genes, some of which are responsible for diseases. As more genes relevant to human health are described, concerns about how such information is used are growing.

Self-Test

1. Phenylketonuria is an example of a genetic disease in which
   a. a single enzyme is not functional.
   b. inheritance is sex-linked.
   c. two parents without the disease cannot have a child with the disease.
   d. mental retardation always occurs, regardless of treatment.
   e. a transport protein does not work properly.

2. Mutations of the gene for β-globin
   a. are usually lethal.
   b. occur only at amino acid position 6.
   c. number in the hundreds.
   d. always result in sickling of red blood cells.
   e. can always be detected by gel electrophoresis.

3. Multifactorial diseases
   a. are less common than single-gene diseases.
   b. involve the interaction of many genes with the environment.
   c. affect less than 1 percent of humans.
   d. involve the interactions of several mRNAs.
   e. are exemplified by sickle-cell anemia.

4. In fragile-X syndrome,
   a. females are affected more severely than males.
   b. a short sequence of DNA is repeated many times to create the fragile site.
   c. both the X and Y chromosomes tend to break when prepared for microscopy.
   d. all people who carry the gene that causes the syndrome are mentally retarded.
   e. the basic pattern of inheritance is autosomal dominant.

5. Which of the following is *not* a practical way to isolate a human gene that mutates to cause a disease?
   a. Use an antibody to isolate the mRNA for the protein involved, make a cDNA copy, and then use this cDNA to probe a genomic library.
   b. Compare the DNA of a person with the disease (who has a deleted chromosome) with the DNA from a person without the disease to see if the latter has extra sequences, which contain the target gene.
   c. Use DNA markers that are closely linked to the mutant gene and then isolate the gene by molecular methods.
   d. Isolate DNAs from people with and without the disease and insert each into cloning vectors to see what proteins are made.
   e. Sequence DNAs from people with and without the genetic mutation to determine the differences.

6. Mutational "hot spots" in human DNA
   a. always occur in genes that are transcribed.
   b. are common at cytosines that have been modified to 5-methylcytosine.
   c. involve long stretches on nucleotides.
   d. occur where there are triplet repeats, such as CTG.
   e. are very rare in genes that code for proteins.

7. Newborn genetic screening for PKU
   a. is very expensive.
   b. detects phenylketones in urine.
   c. has not led to the prevention of mental retardation resulting from this disorder.
   d. must be done during the first day of an infant's life.
   e. uses bacterial growth to detect excess phenylalanine in blood.

8. Genetic diagnosis by DNA testing
   a. detects only mutant and not normal alleles.
   b. can be done only on eggs or sperm.
   c. involves hybridization to rRNA.
   d. utilizes restriction enzymes and a polymorphic site.
   e. cannot be done with PCR.

9. Most human cancers
   a. are caused by viruses.
   b. are in blood cells or their precursors.
   c. involve mutations of somatic cells.
   d. spread through solid tissues rather than by the blood or lymphatic system.
   e. are inherited.

10. Current treatments for genetic diseases include all of the following *except*
    a. restricting a dietary substrate.
    b. replacing the mutated gene in all cells.
    c. alleviating the patient's symptoms.
    d. inhibiting the function of a harmful metabolite.
    e. supplying a protein that is missing.

Applying Concepts

1. Compare the roles of proto-oncogenes and tumor suppressor genes in normal cells. How do these genes and their functions change in tumor cells? Propose targets for cancer therapy involving these gene products.

2. In the past, it was common for people with phenylketonuria (PKU) who were placed on a low-phenylalanine diet after birth to be allowed to return to a normal diet during their teenage years. Although the levels of phenylalanine in their blood were high, their brains were thought to be beyond the stage of being harmed. If a woman with PKU becomes pregnant, however, a problem arises. Typically, the fetus is heterozygous but is unable at early stages of development to metabolize the high levels of phenylalanine that arrive from the mother's blood. Why is the fetus heterozygous? What do you think would happen to the fetus during this "maternal PKU" situation? What would be your advice to a woman with PKU who wants to have a child?

3. Cystic fibrosis is an autosomal recessive disease in which thick mucus is produced in the lungs and airway. The gene responsible for this disease codes for a protein composed of 1,480 amino acids. In most patients with cystic fibrosis, the protein has 1,479 amino acids: A phenylalanine is missing at position 508. A baby is born with cystic fibrosis. He has an older brother. How would you test the DNA of the brother to determine if he is a carrier for cystic fibrosis? How would you design a gene therapy protocol to "cure" the cells in the lung and airway?

4. The Human Genome Project aims to sequence an entire human genome. A related endeavor, the Human Genome *Diversity* Project, aims to sequence genomes from different groups of people from around the world by collecting hair and blood samples. What would be the value of such information? What concerns do you think are being raised by the people whose DNAs are being analyzed?

Readings

Anas, G. and S. Elias. 1992. "Gene Mapping: Using Law and Ethics as Guides." Oxford University Press, New York. A discussion of the social, legal, and ethical impacts of molecular genetic medicine.

"Cancer: Special Issue." 1996. *Scientific American*, September. A series of articles, covering topics ranging from causes to cures.

Mange, E. J. and A. P. Mange. 1994. *Basic Human Genetics*. Sinauer Associates, Sunderland, MA. A brief yet comprehensive text.

McKusick, V. A. 1994. *Mendelian Inheritance in Man: A Catalog of Human Genes and Genetic Disorders*, 11th Edition. Johns Hopkins University Press, Baltimore. The definitive work on human genetic variations.

"Molecular Medicine." 1994–1997. *New England Journal of Medicine*. An excellent series of articles on all aspects of molecular biology, including methods.

Pollock, R. 1995. "Signs of Life." Houghton-Mifflin, New York. A popular book on the molecular revolution in medicine, written by a distinguished scientist.

Scriver, C., A. Beaudet, W. Sly and D. Valle. 1996. *The Metabolic and Molecular Bases of Inherited Disease*. McGraw-Hill, New York. The definitive work on human biochemical genetics.

Chapter 18

Natural Defenses against Disease

Feel a Sneeze Coming On?
This artificially colored scanning electron micrograph shows highly magnified household dust. The unique shapes of these structures cause some of them to provoke reactions in the immune system.

*O*uch! A splinter enters your finger. It smarts, and you know what's going to happen next. If you don't remove the splinter and cleanse the wound immediately, the tissue around the splinter will redden, heat up, swell, and start to throb. These four responses are the symptoms of inflammation, an effective mechanism our bodies use to limit the spread of infection by disease-producing organisms (pathogens) on the splinter. Our environment teems with pathogens such as viruses, bacteria, protists, and fungi. We are challenged by them daily. How do we survive?

You have probably been ill with the flu (influenza) at least a few times in your life. And you have almost certainly encountered the influenza virus since the last time you were actually ill with the flu. Here's just part of what happened in your recent encounter with the newest flu strain. In a lymph node, cells called macrophages phagocytosed some of the flu viruses and partly digested them. The macrophages displayed viral pieces on their plasma membranes, where specialized white blood cells called T cells recognized the pieces and were activated to divide and differentiate further. Many of the descendants of the activated T cells differentiated to attack your virus-infected cells, preventing viral reproduction and thus saving you from another case of the flu. Other descendants of the activated T cells persist in your body today as "memory cells" prepared to rapidly defend you the next time this variety of flu virus strikes.

These defensive events require the participation of many kinds of proteins. Some cellular proteins function as specific receptors, some as markers identify-

400

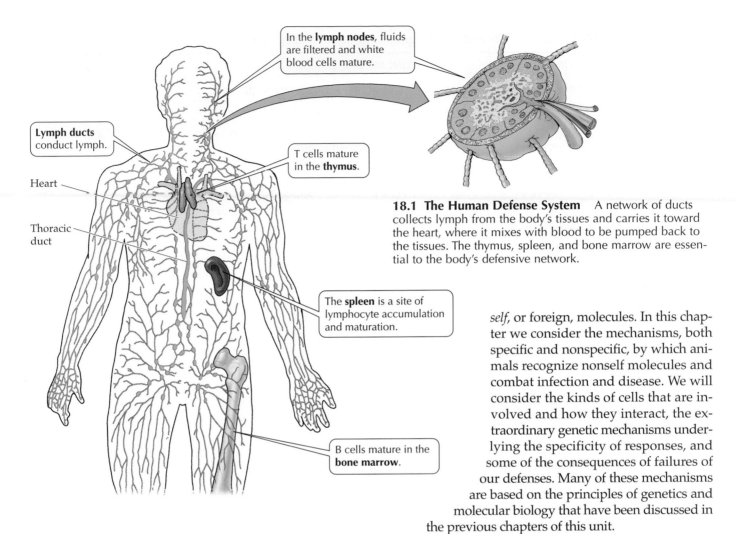

In the **lymph nodes**, fluids are filtered and white blood cells mature.

Lymph ducts conduct lymph.

Heart

Thoracic duct

T cells mature in the **thymus**.

The **spleen** is a site of lymphocyte accumulation and maturation.

B cells mature in the **bone marrow**.

18.1 The Human Defense System   A network of ducts collects lymph from the body's tissues and carries it toward the heart, where it mixes with blood to be pumped back to the tissues. The thymus, spleen, and bone marrow are essential to the body's defensive network.

ing your cells, some as signals triggering events in the macrophages and T cells, and others as weapons leading to the breakdown of the infected cells.

Not all pathogens are greeted by the same set of defenses, but all do elicit a response—usually one that successfully wards off disease. Inflammation is one of several *nonspecific* defenses that we employ. Your response to the flu virus is an example of *specific* defenses, which can distinguish between one type of virus and another.

Our defensive responses typically involve interactions between different kinds of defender cells. They often result in the establishment of a memory of the invading pathogen that can last for years. Most of the responses are highly specific, tailored to meet very specific challenges. Each of us seems able to deal specifically with millions of different kinds of challenges. Genetic diversity underlies these specific defenses, as well as the differences between the individuals in a species—differences that lead, for example, to the rejection of tissues grafted from other individuals.

Animal defense systems are based on the distinction between *self*—the animal's own molecules—and *non-*

self, or foreign, molecules. In this chapter we consider the mechanisms, both specific and nonspecific, by which animals recognize nonself molecules and combat infection and disease. We will consider the kinds of cells that are involved and how they interact, the extraordinary genetic mechanisms underlying the specificity of responses, and some of the consequences of failures of our defenses. Many of these mechanisms are based on the principles of genetics and molecular biology that have been discussed in the previous chapters of this unit.

Defensive Cells and Proteins

Components of our defense system (also called the **immune system**) are dispersed throughout the body and interact with all the other tissues and organs of the body. The thymus, bone marrow, spleen, and certain other lymphoid tissues are essential parts of our defense system (Figure 18.1), but central to their functioning are the blood and lymph.

Blood and lymph are fluid tissues that consist of water, dissolved solutes, and cells. About 60 percent of the **blood** volume is the yellowish solution called plasma. The remainder consists of red blood cells, white blood cells, and platelets (cell fragments essential to clotting) (Table 18.1). The red blood cells are normally confined to the closed circulatory system consisting of the heart, arteries, capillaries, and veins, but the other blood cells are also found in the lymph.

The **lymph** is a fluid that is derived from the blood and other tissues. It accumulates in spaces outside the blood vessels of the circulatory system and contains many of the components of blood, except red blood

TABLE 18.1   Cells and Cell Fragments in Blood and Lymph

TYPE OF CELL	FUNCTION
Red blood cells (erythrocytes)	Transport oxygen and carbon dioxide
White blood cells (leukocytes)	Defend against pathogens
Basophils	Release histamine; may promote the development of T cells
Phagocytes	Destroy nonself materials
Eosinophils	Kill antibody-coated parasites
Neutrophils	Phagocytize antibody-coated pathogens
Monocytes	Develop into macrophages
Macrophages	Engulf and digest microorganisms; activate T cells
Lymphocytes	Have many roles in the immune system
B cells	Differentiate to form antibody-producing cells and memory cells
T cells	Kill virus-infected cells; regulate activities of other white blood cells
Natural killer cells	Attack and lyse virus-infected or cancerous body cells
Platelets (cell fragments without nuclei)	Initiate blood clotting

cells. From the spaces around body cells, the lymph moves slowly into tiny lymph capillaries and then into larger vessels that join together, forming one large lymph duct that joins a major vein (the superior vena cava) near the heart.

By this system of vessels, the lymphatic fluid is eventually returned to the blood and the circulatory system. At many sites along the lymph vessels are small roundish structures called **lymph nodes** that contain a variety of white blood cells involved in nonspecific and specific defenses. Lymph nodes filter the blood and present foreign materials to the defensive cells.

In this first section of the chapter, we will introduce the variety of white blood cells, their defensive roles, and the secreted protein signals that coordinate their activities.

White blood cells play many defensive roles

One milliliter of blood typically contains about 5 billion red blood cells and 7 million larger white blood cells (Figure 18.2). White blood cells have nuclei and are colorless. They can receive signals that direct them to leave the circulatory system by squeezing through junctions between the cells that form the walls of blood capillaries. In response to invading pathogens, the number of white blood cells in the blood and lymph may rise sharply, providing medical professionals with a useful clue for detecting an infection.

The most abundant types of white blood cells are the **lymphocytes** (see Table 18.1). A healthy person has about a trillion lymphocytes, a number similar to the total number of cells in the brain. Two groups of lymphocytes, the **B cells** and **T cells**, together with spe-

cialized cells that arise from them, are the important cells of the immune system that detects specific stimuli and responds in a defensive way to them. B cells and T cells are found in the blood and lymphoid tissues (see Figure 18.1).

Both B and T cells originate from cells in the bone marrow. The precursors of T cells migrate via the blood

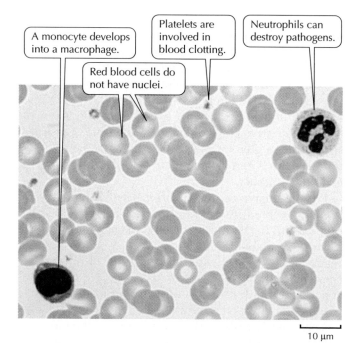

A monocyte develops into a macrophage.

Red blood cells do not have nuclei.

Platelets are involved in blood clotting.

Neutrophils can destroy pathogens.

10 μm

18.2 Blood Cells Two white blood cells are shown among the many red blood cells in this micrograph. Table 18.1 defines the many types of white blood cells. Platelets are cell fragments.

to the thymus, where they become mature T cells. The B cells leave the bone marrow and circulate through the blood and lymph vessels, passing through the lymph nodes and spleen. B and T cells look the same under the light microscope, but they have quite different functions in immune responses, which are methods of reacting against nonself or altered-self substances. Discussion of these functions and their mechanisms will occupy most of this chapter.

In addition to lymphocytes are **phagocytes**, cells that engulf and digest nonself materials. The most important phagocytes are the neutrophils and the macrophages. **Macrophages** have the important additional function of presenting partly digested nonself materials to the T cells for inspection and response. Thus macrophages are essential to the functioning of specific defenses.

Immune system cells release proteins that bind pathogens or signal other cells

The various kinds of cells that defend our body work together like cast members in a drama, interacting with one another and with the cells of invading pathogens. These cell–cell interactions are accomplished by a variety of key proteins, including receptors, surface markers, signaling molecules, and toxins. They will be discussed later in the chapter, as they appear in the context of our story. However, let's take a brief look at them here, to help set the scene.

Lymphocytes, the prime agents in our specific defenses, receive protein signals from other cells, have unique receptors on their membranes, and influence the behavior of other cells by secreting special proteins—antibodies or signal molecules. B cells carry antibodies on their plasma membranes as receptors. **Antibodies** are proteins that bind specifically to substances identified by the immune system as nonself or altered self. Descendants of activated B cells secrete antibodies as weapons of defense. T cells also have surface receptors, called T cell receptors. Most cells of the human body display human leukocyte antigen (HLA) proteins on their plasma membranes. The HLA proteins are important "self"-identifying labels and play major parts in coordinating interactions among lymphocytes and macrophages.

T cells and macrophages communicate by secreting a variety of small, soluble signal proteins, called **cytokines**, that alter the behavior of their target cells. Different cytokines activate or inactivate B cells and macrophages, while others stimulate, inhibit, or kill other T cells. Certain cytokines limit tumor growth by killing tumor cells.

There are still other classes of proteins that participate in the body's nonspecific defenses, as we will see in the next section.

Nonspecific Defenses against Pathogens

Animals have defenses that stop pathogens from invading their bodies. Because these initial defenses give general protection against different pathogens, they are called **nonspecific defenses**, or the innate immune response. In humans, these nonspecific defenses include barriers and local conditions (such as the skin, the antibacterial enzyme lysozyme, mucus, and phagocytes) and cellular and chemical defenses (such as natural killer cells, defensive proteins such as interferons and the complement system, and the complex of responses collectively known as inflammation) (Table 18.2). Even bacteria and fungi that normally reside on body surfaces protect against invasion by pathogens.

Barriers and local agents can defend the body

Skin is a primary nonspecific defense against invasion. Bacteria and viruses rarely penetrate healthy, unbroken skin. But damaged skin or other surface tissue is another matter. The sensitivity of surface tissue accounts in part for the greatly increased risk of infection by pathogenic agents in a person who already has a disease that causes breaks in the skin.

The bacteria and fungi that normally live and reproduce in great numbers on our body surfaces without causing disease are referred to as **normal flora** (Figure 18.3). These natural occupants compete with pathogens for locations and nutrients.

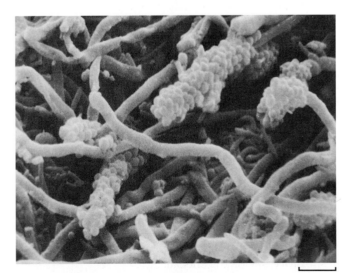

0.8 μm

18.3 Normal Flora Gone Rampant The human mouth harbors a wide variety of microorganisms, most of which cause no damage under normal conditions. When bacteria accumulate on the surfaces of teeth, the result is plaque, which contributes to tooth decay. This electron micrograph shows plaque on a tooth 3 days after the person stopped brushing.

TABLE 18.2   **Nonspecific Defenses**

DEFENSIVE AGENT	FUNCTION
SURFACE BARRIERS	
Skin	Prevents entry of pathogens and foreign substances
Acid secretions	Inhibit bacterial growth on skin
Mucous membranes	Prevent entry of pathogens
Mucus secretions	Trap bacteria and other pathogens in digestive and respiratory tracts
Nasal hairs	Filter bacteria in nasal passages
Cilia	Move mucus and trapped materials away from respiratory passages
Gastric juice	Concentrated HCl and proteases destroy pathogens in stomach
Acid vagina	Limits growth of fungi and bacteria in female reproductive tract
Tears, saliva	Lubricate and cleanse; contain lysozyme, which destroys bacteria
NONSPECIFIC CELLULAR, CHEMICAL, AND COORDINATED DEFENSES	
Normal flora	Compete with pathogens; may produce substances toxic to pathogens
Phagocytes (macrophages and neutrophils)	Engulf and destroy pathogens that enter body
Natural killer cells	Attack and lyse virus-infected or cancerous body cells
Antimicrobial proteins	
Interferons	Released by virus-infected cells to protect healthy tissue from viral infection; mobilize specific defenses
Complement proteins	Lyse microorganisms, enhance phagocytosis, and assist in inflammatory response
Inflammatory response (involves leakage of blood plasma and phagocytes from capillaries and some local heating)	Limits spread of pathogens to neighboring tissues, concentrates defenses, digests pathogens and dead tissue cells; released chemical mediators attract phagocytes and specific defense lymphocytes to site
Fever	Body-wide response inhibits microbial multiplication and speeds body repair processes
Coughing, sneezing	Expel pathogens from upper respiratory passages

Mucus-secreting tissues found in parts of the visual, respiratory, digestive, excretory, and reproductive systems have other defenses against pathogens. Secretions such as tears, nasal drips, and saliva possess an enzyme called **lysozyme** that attacks the cell walls of many bacteria. Mucus in our noses traps microorganisms in the air we breathe, and most of those that get past this filter end up trapped in mucus deeper in the respiratory tract. Mucus and trapped pathogens are removed by the beating of cilia in the respiratory passageway, which moves a sheet of mucus and the debris it contains up toward the nose and mouth. Another effective means of removing microorganisms from the respiratory tract is the sneeze.

Pathogens that reach the digestive tract (stomach, small intestine, and large intestine) are met by other defenses. The stomach is a deadly environment for many bacteria because of the hydrochloric acid and protein-digesting enzymes that are secreted into it. The intact lining of the small intestine cannot be penetrated by bacteria, and some pathogens are killed by bile salts secreted into this part of the tract. The large intestine

harbors many bacteria, which multiply freely; however, these are usually removed quickly with the feces. Most of these bacteria in the large intestine are normal flora that provide benefits to the host. (The digestive system is described fully in Chapter 47.)

All of these barriers and secretions are *nonspecific* because they are the same for all invading pathogens. But there are more complicated nonspecific cellular chemical defenses.

Nonspecific defenses include cellular and chemical processes

Pathogens that manage to penetrate the surface cells of the body encounter additional nonspecific defenses. These defenses include the secretion of various defensive proteins (complement proteins and interferons) and cellular defenses involving phagocytosis.

COMPLEMENT PROTEINS.   Vertebrate blood contains about 20 different antimicrobial proteins that make up the **complement system**. These proteins, in different combinations, provide three types of defenses. In each type,

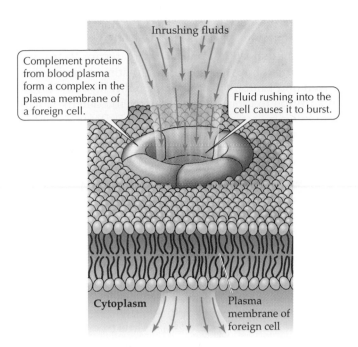

Complement proteins from blood plasma form a complex in the plasma membrane of a foreign cell.

Inrushing fluids

Fluid rushing into the cell causes it to burst.

Cytoplasm

Plasma membrane of foreign cell

18.4 Complement Proteins Destroy a Foreign Cell
Complement proteins attach to foreign cells such as bacteria and can form a porelike structure that allows fluids to pour in, until the foreign cells burst.

the complement proteins act in a characteristic sequence, or cascade, with each protein activating the next.

In one type of nonspecific defense, complement proteins help phagocytes destroy foreign microorganisms. The phagocytes can recognize foreign cells more easily after complement proteins attach to the foreign cells, and the phagocytes engulf complement-coated cells more readily than uncoated ones.

The second defensive activity of the complement system is to attract phagocytes to sites of infection. In concert with this activity, the complement system activates the inflammatory response (which we will describe shortly).

In the third and most impressive defense mounted by the complement system, complement proteins lyse (burst) foreign cells—bacteria, for example. Initial binding of antibodies to the surface of a foreign cell may also bring about the binding of complement proteins to the cell surface. What follows is a cascade of reactions, with different complement proteins acting on one an-

other in succession. The final product of the complement cascade is a doughnut-shaped structure in the membrane of the foreign cell that allows fluids to enter the cell rapidly, causing lysis (bursting) of the foreign cell (Figure 18.4).

INTERFERONS. In the body, virus-infected cells produce small amounts of antimicrobial proteins called **interferons** that increase the resistance of neighboring cells to infection by the same *or other* viruses. Interferons have been found in many vertebrates and are one of the body's lines of nonspecific defense against viral infection.

Interferons differ from species to species, and each vertebrate species produces at least three different interferons. All interferons are glycoproteins (proteins with a carbohydrate component) consisting of about 160 amino acid units. By binding to receptors in the plasma membranes of cells, interferons inhibit the ability of the viruses to replicate. Interferons have been the subject of intensive research because of their possible applications in medicine—for example, the treatment of influenza and the common cold.

PHAGOCYTOSIS AND OTHER CELLULAR ASSAULTS. Phagocytes provide an extremely important nonspecific defense against pathogens that penetrate the surface of the host. Some phagocytes travel freely in the circulatory system; others can move out of blood vessels and adhere to certain tissues. Pathogens become attached to the membrane of a phagocyte (Figure 18.5). Then the phagocyte ingests the pathogens by endocytosis. Once inside an endocytic vesicle in a phagocyte, pathogens are destroyed by enzymes when lysosomes fuse with the vesicle (see Figure 4.20b).

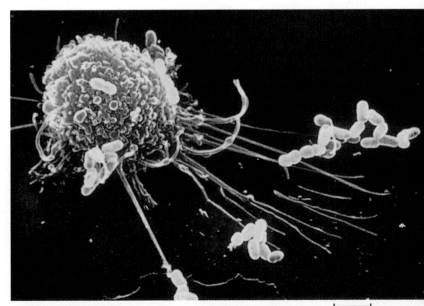

18.5 A Phagocyte and Its Bacterial Prey Some bacteria (appearing bright yellow in this artificially colored scanning electron micrograph) have become attached to the surface of a phagocyte in the human bloodstream. Many of these bacteria will be taken into the phagocyte and destroyed before they can multiply and damage the human host.

2 μm

A single phagocyte can ingest 5 to 25 bacteria before it dies from the accumulation of toxic breakdown products. Even when phagocytes do not destroy all the invaders, they usually reduce the number of pathogens to the point where other defenses can finish the job. So important is the role of the phagocytes that if their functioning is impaired by disease, the animal usually soon dies of infection, not only because of decreased phagocytosis but also because these cells are important in the specific immune response.

There are two major types and one minor type of phagocyte (see Table 18.1). **Neutrophils** are by far the most abundant, but they are relatively short-lived. They recognize and attack pathogens in infected tissue. **Monocytes** mature into *macrophages*, which live longer than neutrophils and can consume large numbers of pathogens. Some macrophages roam through the body; others reside permanently in lymph nodes, the spleen, and certain other lymphoid tissues, "inspecting" the lymph fluid for pathogens. Finally, **eosinophils** are weakly phagocytic. Their primary function is to kill parasites, such as worms, that have been coated by antibodies.

One class of small white blood cells, known as **natural killer cells**, can initiate the lysis of some tumor cells and cells that are infected by a virus. The natural killer cells may seek out cancer cells that appear in the body. The targets of natural killer cells are the body's own cells that have gone awry. In recent years, we have learned how these cells can distinguish virus-infected and tumor cells from normal cells in the body. In addition to their roles in nonspecific defenses, natural killer cells are part of the specific defenses of the immune system.

Inflammation fights infections

Another important nonspecific defense is **inflammation**. The body employs this characteristic response in dealing with infections, and in any other process that causes tissue injury either on the surface of the body or internally. The damaged cells themselves cause the inflammation by releasing various substances, such as the chemical signal **histamine**. Cells adhering to the skin and linings of organs called **mast cells** contain histamine and release it when they are damaged, as do certain white blood cells called **basophils** (see Table 18.1).

You have experienced the symptoms of inflammation: redness and swelling, accompanied by heat and pain. The redness and heat result from histamine-induced dilation of blood vessels in the infected or injured area (Figure 18.6). Histamine also causes the capillaries (the smallest blood vessels) to become leaky, allowing some blood plasma and phagocytes to escape into the tissue, causing characteristic swelling. The pain results from increased pressure (from the

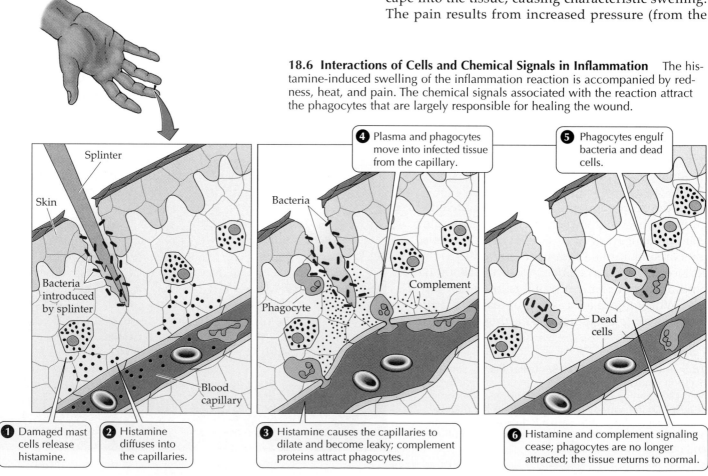

18.6 Interactions of Cells and Chemical Signals in Inflammation   The histamine-induced swelling of the inflammation reaction is accompanied by redness, heat, and pain. The chemical signals associated with the reaction attract the phagocytes that are largely responsible for healing the wound.

④ Plasma and phagocytes move into infected tissue from the capillary.

⑤ Phagocytes engulf bacteria and dead cells.

Splinter

Skin

Bacteria introduced by splinter

Blood capillary

Bacteria

Complement

Phagocyte

Dead cells

❶ Damaged mast cells release histamine.

❷ Histamine diffuses into the capillaries.

❸ Histamine causes the capillaries to dilate and become leaky; complement proteins attract phagocytes.

❻ Histamine and complement signaling cease; phagocytes are no longer attracted; the tissue returns to normal.

leakage) and from the action of some leaked enzymes. Some of the complement proteins and other chemical signals attract other phagocytes—neutrophils first, and then monocytes, which become macrophages. The neutrophils and macrophages are responsible for most of the healing associated with inflammation.

The heat may also play a healing role if it raises the temperature too high for pathogens to multiply effectively. In the aftermath of inflammation, pus may accumulate—it contains dead cells (spent neutrophils and the damaged body cells) and leaked fluid. A normal result of inflammation, pus is gradually cleaned up by macrophages.

Specific Defenses: The Immune Response

Nonspecific defenses are numerous and effective, but some invaders elude them and must be dealt with by defenses targeted against specific threats. The destruction of specific pathogens is an important function of an animal's immune system. In this section, we will identify and discuss the four characteristics and the two types of immune responses.

The immune system uses B and T lymphocytes to recognize and attack specific invaders, such as bacteria and viruses. It also detects normal cells that have been altered by viruses or mutation, and produces signals for their destruction. An animal with a defective immune system can die from infection by even "harmless" bacteria. Some microorganisms normally carried in or on a healthy animal's body are potentially pathogenic and will more readily cause disease if the body's immune system is defective. For example, certain immune deficiencies lead to recurrent, uncontrolled infections with bacteria as *Staphylococcus* or *Streptococcus*.

Four features characterize the immune response

The characteristic features of the immune response are specificity, the ability to respond to an enormous diversity of pathogens, the ability to distinguish self from nonself, and immunological memory.

SPECIFICITY. Pathogens and nonself molecules are diverse, yet their differences are often subtle, as between mutant strains of a single species of bacterium. **Antigens** are organisms or molecules that are recognized and/or interact with the immune system to initiate an immune response. The specific sites on antigens that the immune system recognizes are called **antigenic determinants** (Figure 18.7). Chemically, an antigenic determinant is a group of atoms that may be present on many different large molecules. A large antigen, such as a whole cell, may have many different antigenic determinants on its surface, each capable of being bound by a specific antibody or T cell. Even a single protein has multiple, different antigenic determinants. The host responds to the presence of an anti-

18.7 Each Antibody Matches an Antigenic Determinant
Each antigen has different antigenic determinants that are recognized by specific antibodies. An antibody recognizes and binds to its antigenic determinant to initiate defensive measures against the antigen.

gen by producing highly specific defenses—T cells or antibodies that correspond to the antigenic determinants on that antigen. Each T cell and each antibody is specific for a single antigenic determinant.

THE ENORMOUS SCOPE OF DIVERSITY. Challenges to the immune system are legion: viruses, bacteria, protists, and multicellular parasites. Each of these types of potential pathogens includes many species; each species includes many subtly differing genetic strains; each strain possesses multiple surface features. Estimates vary, but a reasonable guess is that our bodies can respond *specifically* to 10 million different antigenic determinants. On recognition of an antigenic determinant, the immune system responds by activating lymphocytes of the appropriate specificity.

DISTINGUISHING SELF FROM NONSELF. The human body contains tens of thousands of different proteins, each with specific antigenic determinants. Every cell in the body bears a tremendous number of antigenic determinants. A crucial attribute of an individual's immune system is that it recognizes the body's own antigenic determinants and does not attack them. Failure to make this distinction may lead to autoimmune disease—an attack on one's own body.

IMMUNOLOGICAL MEMORY. After responding to a particular type of pathogen once, the immune system "remembers" that pathogen and can usually respond more rapidly and powerfully to the same threat in the future. This memory usually saves us from repeats of childhood diseases such as chicken pox. Vaccination against disease works because the immune system remembers the antigenic determinants that are inoculated into the body.

There are two interactive immune responses

Foreign organisms and substances that invade the animal body and escape the nonspecific defenses come up against the specific defenses of the immune system. The specific immune system has two responses against invaders: the humoral immune response and the cellular immune response. The two responses operate in concert—simultaneously and cooperatively, sharing mechanisms.

In the **humoral immune response** (from the Latin *humor*, "fluid"), the highly specific antibodies react with antigenic determinants on foreign invaders in blood, lymph, and tissue fluids. An animal produces a vast diversity of antibodies that, among them, can react with almost any conceivable antigen in the bloodstream or lymph. Some antibodies travel free in the blood and lymph; others exist as integral membrane proteins on B cells. Antibodies are produced and secreted by specialized B cells that become **plasma cells**. Each antibody recognizes and is capable of binding to a particular antigenic determinant on one or more antigens that invade an animal's body (see Figure 18.7). On stimulation by antigen, a single plasma cell produces multiple soluble copies of antibody with the same specificity as the membrane antibody.

The **cellular immune response** is directed against an antigen that has become established within a cell of the host animal. It destroys virus-infected cells in the animal, as well as some tumor cells. Unlike the humoral response, the cellular immune response does not use antibodies. Instead, it is carried out by T cells within lymph nodes and, to a lesser extent, by T cells that roam through the bloodstream and tissue spaces. The T cells have *T cell receptors*—surface glycoproteins that recognize and bind to antigenic determinants while remaining part of the cell's plasma membrane.

Like antibodies, T cell receptors have specific molecular configurations that bind to specific antigenic determinants. There are two major types of T cells. *Helper T cells* (T_H cells) assist other lymphocytes of both the humoral and cellular immune responses. *Cytotoxic T cells* (T_C cells), or "killer" cells, cause certain other cells to lyse and die. Each of these types will be discussed in some detail later in this chapter. Once bound to a determinant, each type of T cell initiates characteristic activity. Because T cells recognize and mobilize attacks on foreign cells or altered self cells, they are also responsible for the rejection of certain types of organ or tissue transplants.

The specific immune responses act in concert not only with each other but also with the nonspecific de-

fenses. Together, an animal's white blood cells (phagocytes and lymphocytes) defend it against disease.

Clonal selection accounts for the characteristic features of the immune response

Each person possesses an enormous number of different B cells and T cells, apparently capable of dealing with almost any antigen ever likely to be encountered. How does this diversity arise? And why don't our antibodies and T cells attack and destroy our own bodies?

An individual's immune system can mount an immune response against another person's proteins, yet it usually does not mount one against its own. The immune system can distinguish self (one's own antigens) from nonself (those foreign to the body). The versatility of immune responses and immunological memory can be explained satisfactorily by the **clonal selection theory**.

According to the theory, the individual animal contains an enormous variety of different B cells, and each type of B cell is able to produce *only one kind of antibody*. Antigenic determinants do not create an antibody structure and specificity. A surface receptor, already present on a B cell or T cell, binds to the antigen, and this binding signals the particular cell to divide and form a clone. In the case of B cells, the clone consists of plasma cells, all of which secrete a single kind of antibody (Figure 18.8). In the case of T cells, the clone makes signals that lead to various cell-killing mechanisms.

The clonal selection theory accounts nicely for the body's ability to respond rapidly to any of a vast number of different antigens. In the extreme case, even a single B cell might be sufficient for an immunological response by the body, provided that it encounters the antigen and then proliferates into a large clone rapidly enough to combat the invasion. Clonal selection accounts for the proliferation of both B and T cells.

The clonal selection theory also explains two other phenomena: recognition of self and immunological memory (the basis for natural and artificial immunity).

Immunological memory and immunity result from clonal selection

According to the clonal selection theory, an activated lymphocyte (B cell or T cell) produces two types of daughter cells. The ones that carry out the attack on the antigen are **effector cells**—either plasma cells that produce antibodies, or T cells that on binding antigenic determinants release messenger molecules called *interleukins*. The other products of an activated lymphocyte are called **memory cells**—long-lived cells that retain the ability to start dividing on short notice to produce more effector and more memory cells (see Figure 18.8). Effector cells live only a few days, but memory B and T cells may survive for decades.

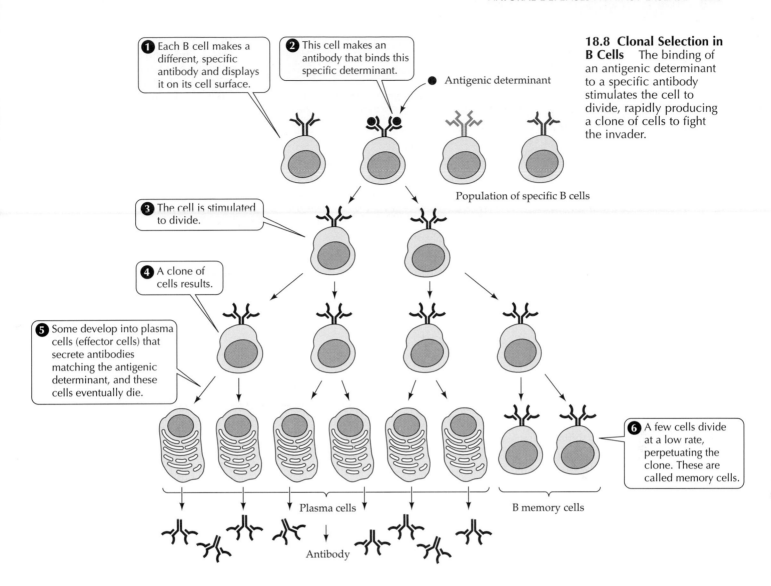

1 Each B cell makes a different, specific antibody and displays it on its cell surface.

2 This cell makes an antibody that binds this specific determinant.

Antigenic determinant

18.8 Clonal Selection in B Cells The binding of an antigenic determinant to a specific antibody stimulates the cell to divide, rapidly producing a clone of cells to fight the invader.

Population of specific B cells

3 The cell is stimulated to divide.

4 A clone of cells results.

5 Some develop into plasma cells (effector cells) that secrete antibodies matching the antigenic determinant, and these cells eventually die.

6 A few cells divide at a low rate, perpetuating the clone. These are called memory cells.

Plasma cells

B memory cells

Antibody

When the body first encounters a particular antigen, a *primary immune response* is activated and one or more types of lymphocytes produce clones of effector and memory cells. The effector cells destroy the invaders at hand and then die, but one or more clones of different memory cells have now been added to the immune system and provide **immunological memory**.

After the body's first immune response to a particular antigen, subsequent encounters with the same antigen will result in a greater and more rapid production of antigen-specific antibody or T cells. This response is called the *secondary immune response*. The first time a vertebrate animal is exposed to a particular antigen, there is a time lag (usually several days) before the number of antibody molecules and T cells slowly increases (Figure 18.9). But for years afterward—sometimes for life—the immune system "remembers" that particular antigen. The secondary immune response has a shorter lag phase, a greater rate of antibody production, and a larger production of total antibody (or T cells).

An early description of immunological memory came from the great historian of ancient Greece, Thucydides. Living in Athens, which was under siege from its rival, Sparta, in 430 B.C., Thucydides described the outbreak of a plague that killed about one-fourth of the Athenians, including their great leader, Pericles. However, some of the ill recovered, and, Thucydides noted, "the same man was never attacked twice, never at least fatally." Those who had been infected once and survived took care of the sick without fear. This pattern of recovery and immunity has been repeated throughout history, for many different diseases.

IMMUNITY: NATURAL AND ARTIFICIAL. Thanks to immunological memory, recovery from many diseases such as chicken pox provides a *natural immunity* to those diseases. However, we don't wait around to gain natural immunity to life-threatening diseases such as typhoid or tetanus. We minimize the risks by seeking *artificial immunity* by way of **immunization**.

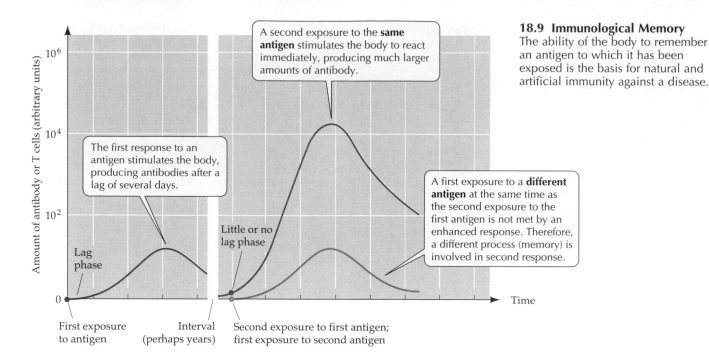

18.9 Immunological Memory
The ability of the body to remember an antigen to which it has been exposed is the basis for natural and artificial immunity against a disease.

The ability of the human body to remember a specific antigen explains why immunization has almost completely wiped out deadly diseases such as smallpox, diphtheria, and polio in medically sophisticated countries. In fact, smallpox has been eliminated worldwide from the spectrum of infectious diseases affecting humans, thanks to a concentrated international effort by the World Health Organization. As far as we know, the only remaining smallpox viruses on Earth are those kept in some laboratories.

Vaccination involves injecting a small amount of virus or bacteria or their proteins (usually treated to make them harmless) into the body. This injection initiates a primary immune response without making the person ill. Later, if the same or very similar disease organisms attack, T and B memory cells already exist. They recognize the antigen and quickly overwhelm the invaders with a massive production of lymphocytes and antibodies.

Animals distinguish self from nonself and tolerate their own antigens

Given the great array of different lymphocytes directed against particular antigens, why doesn't a healthy animal produce self-destructive immune responses? Self-tolerance seems to be based on two mechanisms: clonal deletion and clonal anergy.

Lymphocytes in the bone marrow that have not yet fully differentiated are not capable of attacking antigens. When "antiself" lymphocytes in this undifferentiated state encounter corresponding self antigens, the antiself lymphocytes are eliminated by programmed cell death (apoptosis); thus, no clones of antiself lymphocytes normally appear in the bloodstream. This phe-

nomenon is referred to as **clonal deletion**; it eliminates about 90 percent of all the B cells made in the bone marrow.

If a T cell recognizes an antigenic complex on a body cell and yet does not form cytokines that result in clonal expansion, the cell is said to show **clonal anergy**. Anergy results apparently because the antigenic complex is not the only thing that signals the T cell response; a second, *costimulatory* signal on the surface of the body cell that presents the antigen also signals the T cell. This second signal, a protein called CD28, usually occurs on the cell that presents the antigen to the T cell. But when it is not present or is blocked in some way, anergy results and the clone does not develop.

The phenomenon of **immunological tolerance** was discovered through the observation that some *nonidentical* twin cattle with different blood types contained some of each other's red blood cells. Why did "foreign" blood cells not cause immune responses resulting in their elimination? The hypothesis suggested was that the blood cells had passed between the fetal animals in the womb before the differentiation of immune specificities was complete. Thus each calf regarded the other's red blood cells as self.

This hypothesis was confirmed when it was shown that injecting foreign antigen into an animal early in fetal development caused that animal henceforth to recognize that antigen as self. Two strains of mice were used, each so highly inbred that they were almost a clone. Cells from adult mice of one strain were injected into newborn mice of another strain. Other newborn mice of the second strain served as uninjected controls. Eight to ten weeks later, tolerance was examined in the treated and untreated mice of the second strain by grafting skin from the first strain or from a third strain onto them. The untreated mice rejected grafts from the first strain, but the treated mice accepted them (Figure 18.10). Grafts from the third strain were rejected. Thus it was concluded that immunological tolerance to an antigen can be induced by exposure to the antigen early in development.

Tolerance must be established repeatedly throughout the life of the animal because lymphocytes are produced constantly. Continued exposure to self antigen helps maintain tolerance. For unknown reasons, tolerance to self antigens may be lost. When this happens, the body produces antibodies or T cells against its own proteins, resulting in an *autoimmune disease* (for example, rheumatoid arthritis, psoriasis, or myasthenia gravis).

B Cells: The Humoral Immune Response

Every day, billions of B cells survive the test of clonal deletion and are released from the bone marrow to enter the circulation. The B cells are the basis for the humoral immune response. Since each B cell expresses on its surface an antibody that is specific for a foreign antigen, that antigen can bind to the B cell, causing it to expand to form a clone. In this section on B cells and humoral immunity, we will describe how plasma cells are generated from B cells, how the structure of antibody proteins relates to their functions, and how diverse antibody classes are.

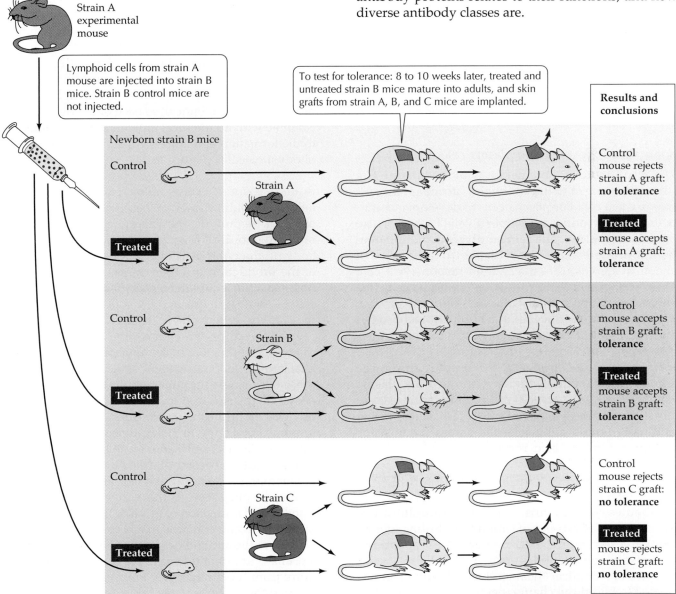

18.10 Making Nonself Seem Like Self   The ability of adult mice to recognize and reject grafts of foreign skin can be overcome by earlier exposure to nonself.

Conclusion: What is recognized as self and nonself depends partly on when it is first encountered.

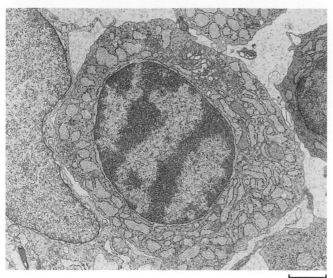

1.2 μm

18.11 A Plasma Cell The prominent nucleus (recognizable by the double membrane), the cytoplasm crowded with rough endoplasmic reticulum, and an extensive Golgi complex are all structural features of a cell actively synthesizing and exporting proteins.

Some B cells develop into plasma cells

The activation of a B cell begins with the binding of a particular antigen to the antibodies carried on the surface of the B cell. For plasma cells to develop and antibodies to be produced, a helper T cell (T_H) must also bind to the antigen. The cellular division and differentiation of the B cells is stimulated by the receipt of chemical signals from the antigen-responsive T cell. These events lead to the formation of **plasma cells** (the effector cells) and memory cells (see Figure 18.8).

As plasma cells develop, the number of ribosomes and the amount of endoplasmic reticulum in their cytoplasm increase greatly (Figure 18.11). These increases prepare the cells for synthesizing large amounts of antibodies for secretion. All the plasma cells arising from a given B cell produce antibodies of specificity identical to that of the surface receptors that originally bound antigen to the parent B cell. Thus specificity is maintained as cells proliferate.

Antibodies share a common structure but may be of different classes

Antibodies are proteins called **immunoglobulins**. There are several types of immunoglobulins, but all contain a tetramer consisting of four polypeptides. Two of these polypeptides are identical "light" chains, and two are identical "heavy" chains. Disulfide bonds (—S—S—) hold the chains together. Each chain consists of a constant region and a variable region (Figure 18.12). The **constant regions** of both light and heavy chains are similar in amino acid sequence from one immunoglobulin to another. The **variable regions** of

the heavy and light chains contribute directly to the binding site and are responsible for the diversity of antibody specificity.

The amino acid sequence of the variable region is unique in each of the millions of immune-specific immunoglobulins. Thus the variable regions of a light and a heavy chain together form a highly specific, three-dimensional structure. This part of a particular immunoglobulin molecule is what binds with a particular, unique antigenic determinant. The enormous range of antibody specificities is accomplished by a combination of genetic rearrangements and mutations, as we will see later in the chapter.

Although the variable regions are responsible for the *specificity* of an immunoglobulin, the constant regions are equally important, for the constant regions are what determine whether the antibody remains part of the cell's plasma membrane or is secreted into the bloodstream. The constant regions also determine the type of action to be taken in eliminating the antigen, as we will see in the next section. The two halves of an antibody, each consisting of one light and one heavy chain, are identical, so each of the two arms can combine with an identical antigen, leading sometimes to the formation of a large complex of antigen and antibody molecules (Figure 18.13).

IMMUNOGLOBULIN CLASSES AND FUNCTIONS. The five immunoglobulin classes are based on differences in the constant region of the heavy chain. One, called immunoglobulin M, or IgM, is always the first antibody product of a plasma cell. Most cells later switch to the production of other classes of immunoglobulins—all with equivalent specificity. The four other classes—IgA, IgD, IgE, and IgG—play different roles in the immune system (Table 18.3).

IgG molecules, which have the γ (gamma) heavy-chain constant region, make up about 85 percent of the total immunoglobulin content of the bloodstream. They consist of a single immunoglobulin unit (two identical heavy chains and two identical light chains) and are produced in greatest quantity during a second immune response (see Figure 18.9). IgG defends the body in several ways. For example, after some IgG molecules bind to antigens, they become attached by their heavy chains to macrophages. This attachment permits the macrophages to phagocytose the antigens (Figure 18.14). Another major function of IgG is to activate the complement system and enhance phagocytosis.

Most of the antibodies produced at the beginning of a primary immune response are IgM molecules. They differ from IgG in that they are composed of five immunoglobulin units (see Table 18.3). Because they have more binding sites, IgM molecules are more active than IgG molecules in activating the complement system and promoting the phagocytosis of antibody-coated cells.

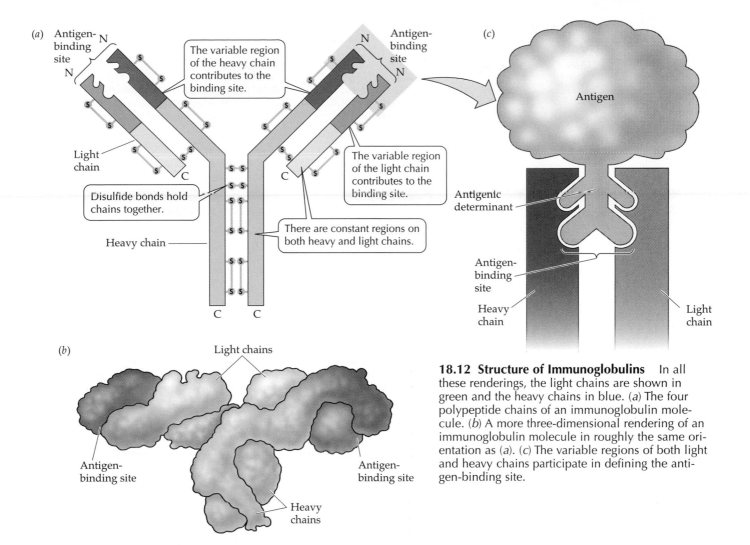

(a) Antigen-binding site

The variable region of the heavy chain contributes to the binding site.

Antigen-binding site

Light chain

Disulfide bonds hold chains together.

The variable region of the light chain contributes to the binding site.

Heavy chain

There are constant regions on both heavy and light chains.

(c)

Antigen

Antigenic determinant

Antigen-binding site

Heavy chain

Light chain

(b)

Light chains

Antigen-binding site

Heavy chains

Antigen-binding site

18.12 Structure of Immunoglobulins In all these renderings, the light chains are shown in green and the heavy chains in blue. (a) The four polypeptide chains of an immunoglobulin molecule. (b) A more three-dimensional rendering of an immunoglobulin molecule in roughly the same orientation as (a). (c) The variable regions of both light and heavy chains participate in defining the antigen-binding site.

Only small amounts of IgD antibody travel free in the bloodstream. The major role of IgD is to serve as membrane receptors on B cells.

IgE antibodies take part in inflammation and allergic reactions. IgE helps kill parasites such as the worms that cause the disease schistosomiasis, which affects some 200 million people in Asia, Africa, and South America (see Chapter 21). At inflammation sites, IgE may participate in bringing white blood cells, components of the complement system, and other factors into the inflamed region. For most people, the effect of IgE is most apparent in allergic reactions.

IgE molecules bind to antigenic determinants on the substances (allergens) that provoke the allergy, and they bind to receptor sites on the surfaces of mast cells. The mast cell–IgE–allergen complex stimulates the release of histamine and other compounds, leading in turn to inflammation, as we have already seen. Hives, hay fever, eczema, and asthma are all common allergic reactions.

Body secretions such as saliva, tears, milk, and gastric fluids all contain IgA. Soluble IgA molecules from

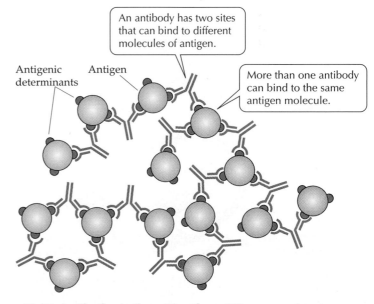

An antibody has two sites that can bind to different molecules of antigen.

Antigenic determinants

Antigen

More than one antibody can bind to the same antigen molecule.

18.13 Antibody–Antigen Complex When more than one antibody binds to the same antigen molecule, large antibody–antigen complexes facilitate neutralization or phagocytosis of the antigen.

TABLE 18.3   Antibody Classes

CLASS	GENERAL STRUCTURE	LOCATION	FUNCTION
IgG	Monomer	Free in plasma; about 80% of circulatory antibodies	Most abundant antibody in primary and secondary responses; crosses placenta and provides passive immunity to fetus
IgM	Pentamer	Surface of B cell; free in plasma	Antigen receptor on B cell membrane; first class of antibodies released by plasma cells during primary response
IgD	Monomer	Surface of B cell	Cell surface receptor of mature B cell; important in B cell activation
IgA	Monomer or polymer	Monomer found in plasma; polymer in saliva, tears, milk, and other body secretions	Protects mucosal surfaces; prevents attachment of pathogens to epithelial cells
IgE	Monomer	Secreted by plasma cells in skin and tissues lining gastrointestinal and respiratory tracts	Found on mast cells and basophils; when bound to antigens, triggers release of histamine from mast cell or basophil that contributes to inflammation and some allergic responses

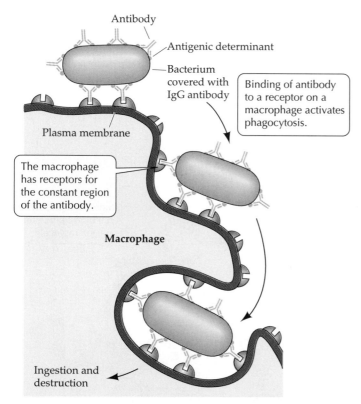

18.14 IgG Antibodies Promote Phagocytosis   IgG antibodies cover a bacterium; receptors on the macrophage recognize the IgG molecules and phagocytose the cell they have coated.

the blood are taken up by epithelial cells, transported across the cells, and then secreted into the mucus that lines digestive, excretory, reproductive, and respiratory organs.

Because it can be a polymer, IgA is very effective at binding to complex antigens, such as bacteria and viruses. This binding prevents pathogens from attaching to the epithelial cells and then entering the tissues. The complexes of IgA and the pathogens are easily eliminated via the mucus. This is an important line of defense against bacteria such as *Salmonella*, which causes food poisoning, and *Neisseria gonorrhoeae*, which causes the sexually transmitted disease gonorrhea, as well as against the viruses that cause influenza and polio.

Hybridomas produce monoclonal antibodies

Because most antigens carry many different antigenic determinants, animals produce a complex mixture of antibodies when injected with a single antigen. It is very difficult to separate the antibodies with different specificities for chemical study. However, a single lymphocyte produces only a single species of antibody. Therefore, in principle, one might expect to be able to cause a single isolated lymphocyte to multiply in pure culture and give rise to a clone of cells, all dedicated to the production of the same antibody.

Unfortunately, cells that produce a single antibody cannot be cultured. On the other hand, cancerous tumors of plasma cells, called **myelomas**, do grow rapidly in culture. Each tumor arises from a single plasma cell. When the antibodies of different tumors are analyzed, each has a unique amino acid sequence. This discovery offered important evidence showing that each B cell clone produces a different antibody molecule.

Some myeloma cells in the laboratory have lost the ability to produce antibodies: These cells live long, but they do not secrete the protein. We now use both cell types—these myeloma cells and normal lymphocytes—to produce hybrid cells called **hybridomas**, which make specific normal antibodies in quantity and which, like the myeloma cells, proliferate rapidly and indefinitely in culture.

To make a clone of hybrid cells, an animal such as a mouse is first inoculated with an antigen to trigger specific lymphocyte proliferation. Later, the spleen is dissected out and B cells are collected from it (Figure 18.15). These B cells are mixed with myeloma cells from a single tumor in the presence of an agent that induces plasma membranes to fuse with one another when cells collide. Thus, some B cells fuse with myeloma cells, combining their contents and giving rise to hybridomas. The cell mixture is then cultured in a manner that destroys all nonhybrid cells.

The hybridomas are grown in a suitable medium so that each activated B cell forms a clone. Individual clones are tested, and the ones that produce the desired antibodies—specific for one antigenic determinant—are selected. These clones produce **monoclonal antibodies** (antibodies for a single antigenic determinant derived from a single clone of cells) in large quantities, either from a mass culture or after being transferred into an animal, where they can grow as a tumor. The hybridomas may be preserved and stored by freezing.

Monoclonal antibodies are ideal for the study of specific antibody chemistry, and they have been used to further our knowledge of cell membranes, as well as for specialized laboratory procedures such as tissue typing for grafts and transplants. Monoclonal antibodies have many practical applications. For example, they have been invaluable in the development of *immunoassays*, which use the great specificity of the antibodies to detect tiny amounts of molecules in tissues and fluids. Most human pregnancy tests use a monoclonal antibody to a hormone made by the developing embryo.

Radioactively tagged monoclonal antibodies are used to target antigens on the surface of cancer cells, enabling precise imaging of the tumor so that the physician can monitor the progress of therapy. The cancer cell–targeted antibody, when attached to a poison, can be used to kill the tumor cells. Monoclonals are also used for *passive immunization*—inoculation with specific antibody rather than with an antigen that causes the patient to develop his or her own antibody

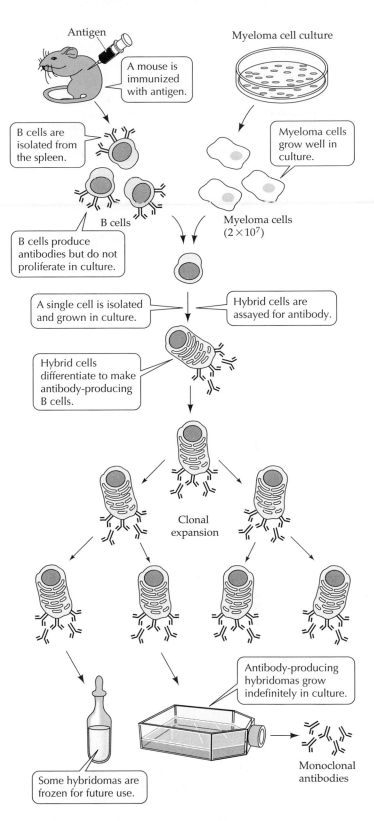

18.15 Creating Hybridomas for the Production of Monoclonal Antibodies Cancerous myeloma cells and normal lymphocytes are hybridized so that the proliferative properties of the myeloma cells can be merged with the antibody-producing lymphocytes. Clones of cells that produce desired monoclonal antibodies are selected.

(as most vaccines are designed to do). Passive immunization is the approach used to treat the early symptoms of rabies infection or a rattlesnake bite, cases in which the toxic nature of the infection is so serious that there is not enough time to allow the person's immune system to mount its own defense.

T Cells: The Cellular Immune Response

Thus far we have been concerned primarily with the humoral immune response, whose effector molecules are the antibodies secreted by plasma cells that develop from activated B cells. T cells are the effectors of the cellular immune response, which is directed against any factor, such as a virus or mutation, that changes a normal cell into an abnormal cell.

In this section, we will describe two types of T cells (helper T cells and cytotoxic T cells) and we will see that the binding of a T cell receptor to an antigenic determinant requires special proteins encoded by the HLA (human leukocyte antigen) genes.* These proteins underlie the immune system's tolerance for cells of its own body and are responsible for the rejection of foreign tissues by the body.

T cells of various types play several roles in immunity

Like B cells, T cells possess specific surface receptors. **T cell receptors** are not immunoglobulins, but glycoproteins with molecular weights about half that of an IgG. They are made up of two polypeptide chains, each encoded by a separate gene (Figure 18.16).

The genes that code for T cell receptors are similar to those for immunoglobulins, suggesting that both are derived from a single, evolutionarily more ancient group of genes. Like the immunoglobulins, T cell receptors include both variable and constant regions. Once formed, the receptors are bound to the plasma membrane of the T cell that produces them. In the sections that follow we discuss how T cell receptors bind antigens.

When T cells are activated by contact with a specific antigenic determinant, they develop and give rise to two types of effector cells. **Cytotoxic T cells**, or **T_C cells**, recognize virus-infected cells and kill them by causing them to lyse (Figure 18.17). **Helper T cells**, or **T_H cells**, assist both the cellular and humoral immune systems.

As mentioned already, a T_H cell of appropriate specificity must bind an antigen before a B cell can be activated by binding the antigen. The helper cell becomes the "conductor" of the "immunological orches-

*In mice, the analogous proteins are called the MHC (major histocompatibility complex).

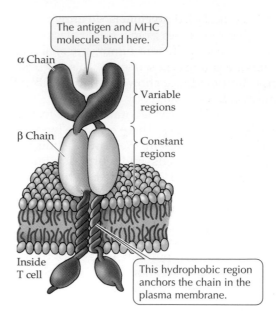

18.16 A T Cell Surface Receptor   T cell receptors are glycoproteins, not immunoglobulins, although their structure is similar. The receptors are bound to the plasma membrane of the T cell that produces them.

tra," as it sends out chemical signals that not only result in its own clonal expansion but set in motion the actions of cytotoxic T cells as well as B cells.

Now that we are familiar with the major types of T cells, we can address the question of how T cells meet the antigenic determinants if the antigens themselves are inside host cells.

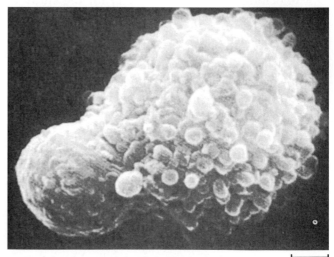

18.17 A Cytotoxic T Cell in Action   A cytotoxic T cell (smaller sphere) has come into contact with a virus-infected cell, causing the infected cell to lyse. The blisters on the infected cell's surface indicate that it is beginning to break up.

The major histocompatibility complex encodes many proteins

We have seen that a body's immune defenses recognize its own cells as self—proteins on our own cell surfaces are tolerated by our immune systems. There are several types of mammalian cell surface proteins, but we will focus on one very important group, the products of a cluster of genes in mice called the **major histocompatibility complex**, or **MHC**. The MHC gene products are plasma membrane glycoproteins, proteins with attached carbohydrate groups. The human counterparts of MHC are the human leukocyte antigens (HLA).

There are three classes of MHC proteins. **Class I MHC molecules** are present on the surface of every nucleated cell in the animal. These proteins function in antiviral T cell immunity. **Class II MHC molecules** are found only on the surfaces of B cells, T cells, and macrophages. This class of MHC proteins is responsible primarily for the interaction of T_H cells, macrophages, and B cells during antibody responses. Classes I and II bind with antigens or antigenic fragments bearing antigenic determinants inside the cell, are exported with antigen to the cell surface, and are then presented along with antigen to T cells. **Class III MHC molecules** include some of the proteins of the complement system that interact with antigen–antibody complexes and result in the lysis of foreign cells (see Figure 18.4).

The HLA complex in humans is highly polymorphic. It consists of more than 50 genes, and most of the genes have many alleles—more than 70 in some cases. Because of the number of HLA genes and the number of their alleles, different individuals are very likely to have different HLA genotypes and phenotypes—and that difference is what leads to the rejection of organ transplant between humans.

Similarities in base sequences between MHC and HLA genes, and the genes coding for antibodies and T cell receptors suggest that all three may have descended from the same ancestral genes and are part of a "superfamily." Major aspects of the immune systems seem to be woven together by a common evolutionary thread.

T cells and MHC molecules contribute to the humoral immune response

When a macrophage phagocytoses an antigen, it breaks the antigen into fragments, each with one or more antigenic determinants. This process is called **antigen processing**. Within the macrophage, class II MHC molecules bind processed antigen in the endoplasmic reticulum. The resulting complex moves by way of the Golgi and its vesicles to the plasma membrane of the cell, where it is displayed and is available for interaction with T_H cells (Figure 18.18). Because the MHC molecules *present* antigen, the macrophages are thus referred to as **antigen-presenting cells**.

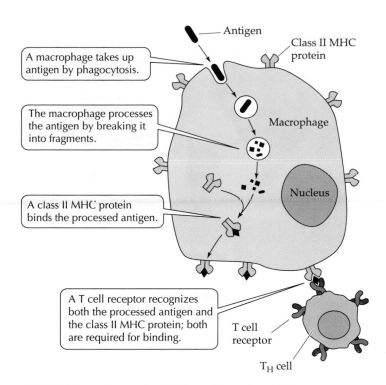

18.18 Macrophages Are Antigen-Presenting Cells
Processed antigen is displayed by MHC protein on the surface of a macrophage. Receptors on the helper T cell can then interact with the processed antigen/MHC protein complex.

A T_H cell can bind to processed antigen only if two criteria are met: (1) The T cell receptors correspond to the displayed antigenic determinant, and (2) the processed antigen is carried by an MHC molecule. The T cell receptor binds both the antigenic determinant and the MHC molecule.

Among the many T cell surface proteins, two—the CD4 and CD8 proteins—participate in the binding of some T cells and processed antigen. CD4, present on all T_H cells, binds to class II MHC molecules; CD8, present on all T_C cells, binds to class I MHC molecules. In each case, the effect is to enhance the binding of T cells to antigen-presenting cells.

When a macrophage phagocytoses a foreign particle, it becomes activated and produces cytokines. Similarly, when a T cell binds to the macrophage, the T cell releases cytokines, such as interleukin-1. These cytokines activate the T_H cell to produce a clone of differentiated cells capable of interacting with B cells. The steps to this point constitute the **activation phase** of the response, and they occur in lymphatic tissue. Next comes the **effector phase**, in which B cells are activated (Figure 18.19a).

B cells are also antigen-presenting cells. B cells take up by endocytosis antigen bound to their receptors, process it, and display it on class II MHC molecules. An activated T_H cell binds only if it recognizes both the dis-

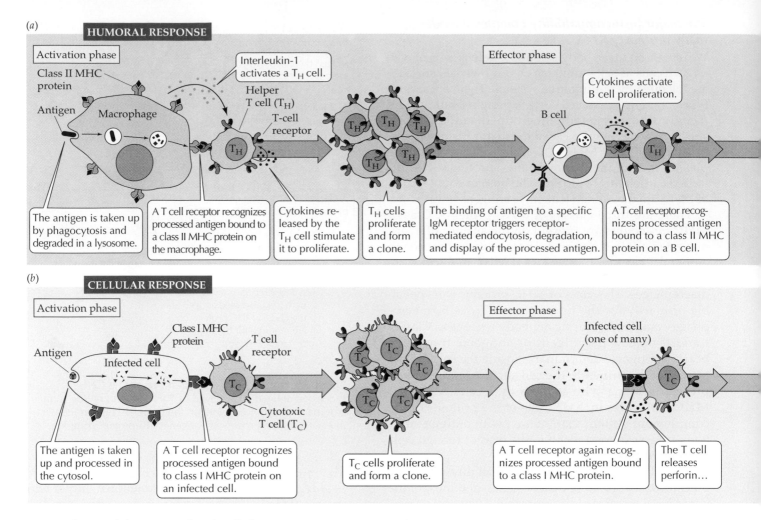

(a)

HUMORAL RESPONSE

Activation phase

Class II MHC protein

Interleukin-1 activates a T_H cell.

Effector phase

Antigen

Macrophage

Helper T cell (T_H)

T-cell receptor

Cytokines activate B cell proliferation.

B cell

The antigen is taken up by phagocytosis and degraded in a lysosome.

A T cell receptor recognizes processed antigen bound to a class II MHC protein on the macrophage.

Cytokines released by the T_H cell stimulate it to proliferate.

T_H cells proliferate and form a clone.

The binding of antigen to a specific IgM receptor triggers receptor-mediated endocytosis, degradation, and display of the processed antigen.

A T cell receptor recognizes processed antigen bound to a class II MHC protein on a B cell.

(b)

CELLULAR RESPONSE

Activation phase

Class I MHC protein

T cell receptor

Effector phase

Infected cell (one of many)

Antigen

Infected cell

Cytotoxic T cell (T_C)

The antigen is taken up and processed in the cytosol.

A T cell receptor recognizes processed antigen bound to class I MHC protein on an infected cell.

T_C cells proliferate and form a clone.

A T cell receptor again recognizes processed antigen bound to a class I MHC protein.

The T cell releases perforin…

18.19 Phases of the Humoral and Cellular Immune Responses Both types of immune response have an activation phase and an effector phase.

played antigenic determinant and the MHC molecule. Once again the bound T_H cell releases interleukins that cause the B cell to produce a clone of plasma cells. Finally, the plasma cells secrete antibody, completing the effector phase of the humoral immune response.

T cells and MHC molecules contribute to the cellular immune response

Class I MHC molecules play a role in the cellular immune response that is similar to the role played by class II MHC molecules in the humoral immune response. In this case the virus-infected or mutated cells themselves are the antigen-presenting cells. Cellular DNA that has been altered by viral infection or mutation leads to the production of "foreign" proteins or peptide fragments, which combine with MHC class I molecules. As in the macrophage, the complex is displayed on the cell surface and presented to T_C cells. When a T_C cell binds the complex of processed antigen and class I MHC molecule, the cell becomes activated to proliferate and differentiate (Figure 18.19*b*).

In the effector phase of the cellular immune response, T cell receptors on the activated T_C cells recognize processed viral antigen displayed by class I MHC molecules on the surface of other virus-infected cells. The T_C cells then produce molecules that insert themselves into plasma membranes and result in cell lysis (see Figure 18.17). Because T cell receptors recognize self MHC molecules complexed with *nonself* antigens, they help rid the body of its own virus-infected cells. Because they also recognize MHC molecules complexed with *altered self* antigens (as a result of mutations), they help eliminate tumor cells, since most tumor cells have mutations (see Chapter 17).

Clearly, T_H cells have a central role in both humoral and cellular immunity.

MHC molecules underlie the tolerance of self

MHC molecules play a key role in establishing tolerance to self, without which an animal would be destroyed by its own immune system. Throughout the animal's life, developing T cells are "tested" in the thy-

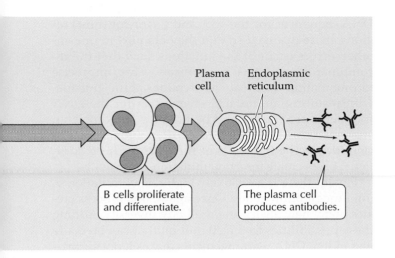

B cells proliferate and differentiate.

The plasma cell produces antibodies.

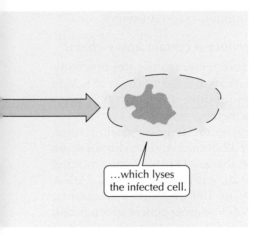

…which lyses the infected cell.

mus. One "question" the thymus "asks" is, Can this cell recognize the body's MHC proteins? A T cell unable to recognize self MHC proteins would be useless to the animal because it could not participate in any immune reactions. Such a cell fails the test and dies within about 3 days.

The second and more crucial question is, Does this cell bind to self MHC protein *and* to one of the body's own antigens? A T cell that satisfied both of these criteria would be harmful or lethal to the animal; it also fails the test and undergoes apoptosis (see Chapter 9).

T cells that survive this pair of tests mature into either T_C cells or T_H cells.

MHC molecules are responsible for transplant rejection

In humans, a major side effect of the HLA molecules (our version of the mouse's MHC) became important with the development of organ transplant surgery, sometimes with devastating results. Because the proteins produced by the HLA are specific to each individual, they act as antigens if transplanted into another individual. An organ or a piece of skin trans-

planted from one person to another is recognized as nonself and soon provokes an immune response; the tissue then is killed, or "rejected," by the cellular immune system. But if the transplant is performed immediately after birth or if it comes from a genetically identical person (an identical twin), the material is recognized as self and is not rejected.

Physicians can temporarily overcome the rejection problem by treating the patient with drugs, such as cyclosporin, that suppress the immune system. However, this approach compromises the ability of patients to defend themselves against bacteria and viruses. Cyclosporin and some other immunosuppressants interfere with communication between cells of the immune system. Specifically, they inhibit the production of interleukins.

The Genetic Basis of Antibody Diversity

A newborn mammal possesses a full set of genetic information for immunoglobulin synthesis. At each of the loci coding for the heavy and light chains, it has an allele from the mother and one from the father. Throughout the animal's life, each of its cells begins with the same full set. However, the genomes of B cells become modified during development in such a way that each cell eventually can produce one—and only one—specific type of antibody. Different B cells develop different antibody specificities. How can a single organism produce millions of different specific immunoglobulins with antibody specificities?

Research in recent years has answered this question. One suggestion was that we simply have millions of antibody genes. However, a simple calculation (the number of base pairs needed per antibody gene multiplied by millions) shows that our entire genome would thus be taken up by antibody genes. More than 30 years ago, an alternative hypothesis was proposed: that a relatively small number of genes recombine at the DNA level to produce many unique combinations, and it is this shuffling of the genetic deck that produces antibody diversity. Research since has proved this hypothesis.

In this section, we will describe the unusual DNA events that generate the enormous antibody diversity that normally characterizes each individual mammal. Then we will see how similar DNA events produce the five classes of antibodies by producing slightly different "constant regions" that have special properties.

Antibody diversity results from unusual genetic processes

In an unusual genetic process, functional immunoglobulin genes assemble from DNA segments that initially are spatially separate. Every cell in the body has hundreds of DNA segments potentially capable of participating in the synthesis of the variable regions, the parts of the antibody molecule that confer immunological specificity. However, during B cell development, these DNA segments are *rearranged*.

Pieces of the DNA are deleted, and DNA segments formerly distant from one another are joined together. In this fashion, a gene is assembled from randomly selected pieces of DNA. Each B cell precursor in the animal assembles its own unique set of immunoglobulin genes. This remarkable process generates many diverse antibodies from the same starting genome. The same type of process also accounts for the diversity of T cell receptors.

In both humans and mice, the DNA segments coding for immunoglobulin heavy chains are on one chromosome and those for light chains are on others. The variable region of the light chain is encoded by two families of DNA segments, and the variable region of the heavy chain is encoded by three families. (We discussed gene families in Chapter 14; see Figure 14.10.)

It is random which of the two variable light-chain families and which of the three variable heavy-chain families are used to make the final antibody. So the diversity afforded by several hundred DNA segments is multiplied by the combination of different families. Furthermore, since light and heavy chains are synthesized independently of one another, the combination of light and heavy chains introduces more diversity.

For example, in the human genome there are 100 different variable light-chain genes in the kappa family and another 100 genes in the lambda family. The heavy-chain family also has 100 members in its variable region. So the total number of possible variable regions is: (100 heavy × 100 light kappas) + (100 heavy × 100 light lambdas) = 20,000 different variable regions in immunoglobulins just by the random association of V (for *variable*) gene sequences.

But there are not just constant and variable regions in antibodies. Light chains have a third DNA region, which codes for amino acids between the variable and constant regions. This is the J (for *joining*) region, and in humans the kappa light-chain family has five members and the lambda family has six members. Thus the number of possible different kappa light chains is 5 × 100 = 500, and the number of lambda possibilities is 6 × 100 = 600. Similarly, there are six J family members in the genes for the variable region for the heavy chain, as well as 30 D (for *diversity*) genes. So the possible heavy chains are 6 × 30 × 100 = 18,000.

Now the possible antibodies that can be made through a random selection of one variable region from the heavy and light families is

18,000 heavy × 500 light kappas = 9 million

and

18,000 heavy × 600 light lambdas = 10.8 million

or a total of almost 20 million.

But there are more diversity mechanisms. When the DNA sequences for the V, J, and C (for *c*onstant) regions are rearranged so that they are next to one another, errors occur at the junctions such that the recombination event is not precise. This imprecise recombination can create new codons at the junction, with resulting amino acid changes. After the DNA fragments are cut out and before they are joined, an enzyme, **terminal transferase**, often adds some nucleotides to the free ends of the DNAs. These additional bases create insertion mutations.

Finally, there is a relatively high rate of mutation in immunoglobulin genes. Once again, this process creates many new alleles and antibody diversity. Adding these possibilities to the 20 million that can be made by random DNA rearrangements makes it not surprising that the immune system can mount a response to almost any natural or human-made substance.

How does a B cell produce a certain heavy chain?

To see how DNA rearrangement generates antibody diversity, let's consider how the heavy chain of IgM is produced. After B cells produce this antibody, it is inserted into the plasma membrane.

The gene families governing all heavy-chain synthesis are on chromosome 12 of mice and on chromosome 14 of humans. In mice, the gene families are arranged as shown in Figure 18.20, with a long stretch of DNA occupied by a family of 100 or more V segments. Humans have about 300 V segments. In a given B cell, only *one* of these segments is used to produce part of the variable region of the heavy chain; the remaining V segments are discarded or rendered inactive.

At some distance from the V segments is a family of ten or more D segments. Again, only one of these is used to produce part of the variable region of the heavy chain of a given B cell, as is only one J segment from the family of four such segments (in mice) lying yet farther along the chromosome. This combination of one each of V, D, and J segments forms a complete variable region for a functional gene.

Still farther along the chromosome, and separated from the suite of J segments, is a family of eight segments, one of which codes for the constant region of the heavy chain. Light chains are produced from similar families of DNA segments, but they lack D segments.

How does order emerge from this seeming chaos of DNA segments? Two important steps impose order: DNA rearrangements and RNA splicing. First, substantial chunks of DNA are deleted from the chromosome during rearrangement of the segments. As a result of these deletions, a particular D segment is rearranged to be directly beside a particular J segment, and then the DJ segment is rearranged to be adjacent to one of the V segments; thus a single "new" sequence, consisting of one V, one D, and one J segment, can now code for the variable region of the heavy chain. All the progeny of

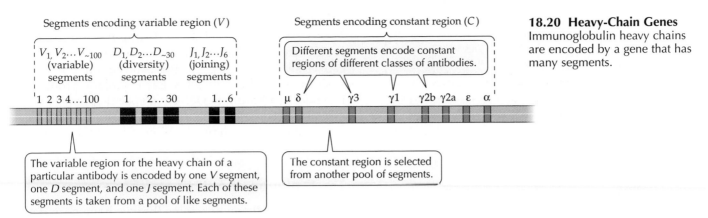

Segments encoding variable region (V)

Segments encoding constant region (C)

$V_1, V_2...V_{~100}$ (variable) segments $D_1, D_2...D_{~30}$ (diversity) segments $J_1, J_2...J_6$ (joining) segments

Different segments encode constant regions of different classes of antibodies.

1 2 3 4...100   1   2...30   1...6   μ δ   γ3   γ1   γ2b γ2a ε α

18.20 Heavy-Chain Genes
Immunoglobulin heavy chains are encoded by a gene that has many segments.

The variable region for the heavy chain of a particular antibody is encoded by one *V* segment, one *D* segment, and one *J* segment. Each of these segments is taken from a pool of like segments.

The constant region is selected from another pool of segments.

this cell constitute a clone having the same sequence for the variable region (Figure 18.21a).

The second step in organizing the synthesis of an immunoglobulin chain follows transcription. Splicing of the RNA transcript (see Chapter 14) removes introns and any *J* segments lying between the selected *J* segment and the first constant-region segment (Figure 18.21b). The result is an mRNA that can be translated, directly yielding the heavy chain of the cell's specific antibody.

In summary, two distinct types of nucleic acid rearrangements contribute to the formation of an antibody. DNA rearrangements, before transcription, join the V, D, and J segments. RNA splicing, after transcription, joins the variable region (*VDJ*) to the constant region.

The constant region is involved in class switching

Early in its life, a B cell produces IgM molecules that are responsible for the specific recognition of a particular antigenic determinant. At this time, the constant region of the antibody's heavy chain is encoded by the

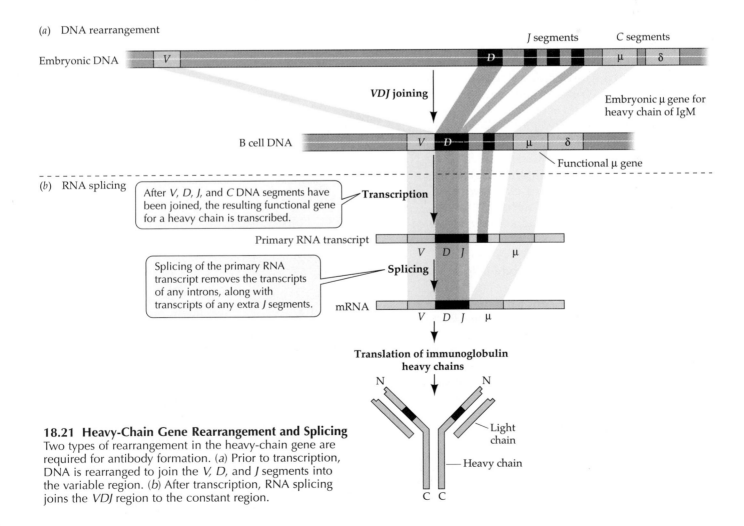

(a) DNA rearrangement

J segments   *C* segments

Embryonic DNA   *V*   *D*   μ   δ

***VDJ* joining**

B cell DNA   V *D*   μ   δ

Embryonic μ gene for heavy chain of IgM

Functional μ gene

(b) RNA splicing

After *V*, *D*, *J*, and *C* DNA segments have been joined, the resulting functional gene for a heavy chain is transcribed.

Transcription

Primary RNA transcript   V D J   μ

Splicing of the primary RNA transcript removes the transcripts of any introns, along with transcripts of any extra *J* segments.

Splicing

mRNA   V D J μ

Translation of immunoglobulin heavy chains

N   N

Light chain

Heavy chain

C C

18.21 Heavy-Chain Gene Rearrangement and Splicing
Two types of rearrangement in the heavy-chain gene are required for antibody formation. (*a*) Prior to transcription, DNA is rearranged to join the *V, D,* and *J* segments into the variable region. (*b*) After transcription, RNA splicing joins the *VDJ* region to the constant region.

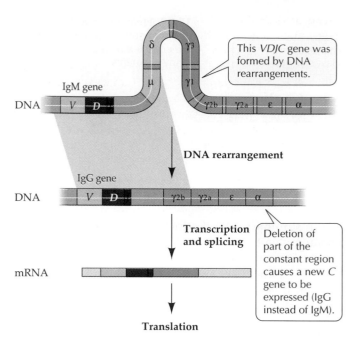

This *VDJC* gene was formed by DNA rearrangements.

IgM gene

DNA | V | D |

DNA rearrangement

IgG gene

DNA | V | *D* | γ2b | γ2a | ε | α |

Transcription and splicing

Deletion of part of the constant region causes a new *C* gene to be expressed (IgG instead of IgM).

mRNA

Translation

18.22 Class Switching The gene produced by joining *V*, *D*, *J*, and *C* segments (see Figure 18.21) may later be modified, causing a different *C* segment to be transcribed. This modification, known as class switching, is accomplished by deletion of part of the constant region. Shown here is class switching from an IgM gene to an IgG gene.

first constant-region segment, the μ segment (see Figure 18.20). During an immunological response—later in the life of the plasma cell—another deletion commonly occurs in the plasma cell's DNA, positioning the heavy-chain variable-region gene (consisting of the same *V*, *D*, and *J* segments) next to a constant-region segment farther down the original DNA, such as the γ, ε, or α constant region (Figure 18.22).

Such a DNA deletion, called **class switching**, results in the production of an antibody with a different *C* region and *function*, but the same *antigen specificity*. The new antibody has the same variable regions of the light and heavy chains but a different constant region of the heavy chain. This new antibody falls into one of the four other immunoglobulin classes (IgA, IgD, IgE, or IgG; see Table 18.3), depending on which of the constant-region segments is placed adjacent to the variable-region gene.

After switching classes, the plasma cell cannot go back to making the previous immunoglobulin class, because that part of the DNA has been deleted and lost. On the other hand, if additional constant-region segments are still present, the cell may switch classes again.

What triggers class switching, and what determines the class to which a given B cell will switch its antibody production? T_H cells influence these tasks, thus directing the course of an antibody response and de-

termining the nature of the attack on the antigen. These T cells induce class switching by sending cytokine signals (Figure 18.23). Interleukin-4 (IL-4) causes switching to IgG or IgE production; other cytokines cause other switches. Each cytokine produces its effect by binding to DNA at a specific point near a constant-region segment and changing the conformation of the chromosome, enabling an enzyme to bind and cause genetic rearrangement at that point.

The Evolution of Animal Defense Systems

The complex defense systems that we have been discussing are found in virtually all vertebrate animals, although there are some significant differences between most vertebrate immune systems and those of sharks, rays, and other evolutionarily ancient vertebrates (see Chapter 30). But did these defense systems first appear in the vertebrates? Not at all.

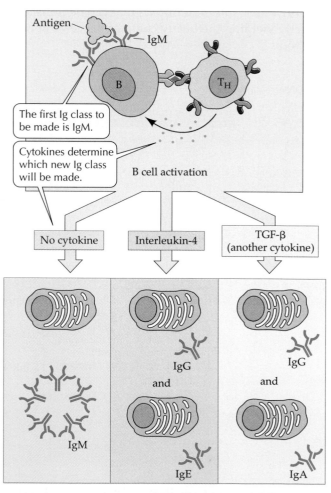

18.23 Cytokines Determine How the Antibody Switches Class A helper T cell initiates class switching in a B cell by secreting a cytokine. Different cytokines produce different switches.

The invertebrates, in all their diversity, have sturdy defense systems, and certain defense system elements are found even in unicellular protists. Many protists carry on phagocytosis, as do our own macrophages, and some protists use phagocytosis as a defense mechanism. Multicellular animals, both invertebrate and vertebrate, employ mobile phagocytic cells to patrol their bodies.

Like vertebrates, invertebrates (and probably some protists) distinguish between self and nonself. Making such distinctions enables invertebrates to reject tissue grafted from other individuals of the same species. Unlike vertebrates, however, invertebrates reject a second graft no more rapidly than a first graft—indicating that invertebrates lack immunological memory. This and other observations show that although immunological functions of invertebrates and vertebrates may be similar, their mechanisms often differ.

Invertebrates do not produce immunoglobulins, lymphocytes, or the complement system. However, they achieve similar protective goals by analogous methods, and the analogs are probably evolutionary precursors of the systems found in vertebrates. Many invertebrates make proteins very similar to interleukin-1, interleukin-2, and other vertebrate cytokines, and the proteins play regulatory roles similar to those in humans.

The evolutionary shift to new defense mechanisms probably occurred in the very earliest vertebrates, which have been extinct for 400 million years. The most ancient vertebrate groups still present today have clearly "vertebrate" immune systems. Studies of the immune systems of invertebrates and evolutionarily ancient vertebrates will help us understand our own immune systems better, and they will likely be helpful in understanding and dealing with human diseases.

Disorders of the Immune System

Immune deficiency diseases such as AIDS show us how much we depend on our immune system to protect us from pathogens. However, sometimes the immune system fails us in one way or another: It may overreact, as in an allergy; it may attack self antigens, as in an autoimmune disease; or it may function weakly or not at all, as in an immune deficiency disease. After a look at allergies and autoimmune conditions, we will examine the acquired immune deficiency that characterizes AIDS.

An inappropriately active immune system can cause problems

A common type of condition arises when the human immune system overreacts to a dose of antigen. Although the antigen itself may present no danger to the host, the immune response may produce inflammation and other symptoms that can cause serious illness or death. **Allergies** are the most familiar examples of such a problem. There are two types of allergic reaction. *Immediate hypersensitivity* occurs when an individual makes IgE that binds to a molecule or structure in a food, pollen, or venom of an insect. Mast cells and basophils bind the IgE on their surfaces, and when the antigen complexes with the IgE, these cells release amines such as histamine. The result is symptoms such as dilation of blood vessels, inflammation, and difficulty breathing. If untreated with antihistamines, a severe reaction can lead to death.

Delayed hypersensitivity does not begin until hours after exposure to the noxious antigen. In this case, the antigen is processed by antigen-presenting cells and a T cell response is initiated. This response can be so massive that the cytokines released cause macrophages to become activated and damage tissues. This is what happens when the bacteria that cause tuberculosis colonize the lung.

Sometimes prevention of the immune recognition of self fails, resulting in the appearance of one or more "forbidden clones" of B and T cells directed against self antigens. This failure does not always result in disease, but in some instances it can be disastrous. Among the **autoimmune diseases** of our species—diseases in which components of the body are attacked by its own immune system—are rheumatoid arthritis, ulcerative colitis, and myasthenia gravis (severe muscle weakness). Multiple sclerosis may result from a failure of tolerance induction in a type of T cell. The abnormal T cells mount an immune attack on the myelin sheath, an insulating material that surrounds many nerve cells (see Chapter 41). When the myelin sheath is damaged, the result can be loss of nerve function, including blindness and loss of motor control.

It now appears that most people have a few lymphocytes that react with self antigens, but this condition does not impair health. However, in autoimmune diseases these lymphocytes divide and become numerous, attacking the self cells and causing disease. In some cases, the underlying reason may be *molecular mimicry*, in which T cells that recognize a nonself antigen also recognize something on the self that has a similar structure.

For example, T cells that recognize an antigen on Coxsackie virus, a common infection, sometimes also react with part of an enzyme in the cells of the pancreas that make insulin. These T cells then attack and lead to the destruction of the pancreatic cells, leading to insulin-dependent diabetes, a disease that affects more than a million Americans.

AIDS is an immune deficiency disorder

There are various immune deficiency disorders, such as those in which T or B cells never form and others in which B cells lose the ability to give rise to plasma cells. In either case, the affected individual is unable to mount an immune response and thus lacks a major line of defense against microbial pathogens. The first human disorder to be treated in part with gene therapy was a hereditary immune deficiency caused by a mutation in the gene coding for the enzyme adenosine deaminase (see Figure 17.18).

Because of its essential roles in both antibody responses and cellular immune responses, the T_H cell is perhaps the most central of all the components of the immune system—a significant cell to lose to an immune deficiency disorder. This cell is the target of HIV (*human immunodeficiency virus*), the virus that eventually results in AIDS (*acquired immune deficiency syndrome*).

HIV is a retrovirus that infects mostly lymphoid tissues, where B cells mature, T cells reside, and the fluid that makes up lymph is filtered. Lymph is filtered on the highly folded cells called *follicular dendritic cells*. Soon after HIV infection, the virus spreads rapidly among T_H cells, since an immune response has not yet been mounted. But within days or weeks, when T cells recognize infected lymphocytes, an immune response is mounted and antibodies appear in the blood (Figure 18.24). These antibodies effectively block any further infections by circulating viruses.

By this time, several months after initial infection, the patient has a lot of circulating HIV complexed with antibodies that is gradually filtered out by the dendritic cells. But before they are filtered out, these antibody-complexed viruses can still infect T_H cells that come in contact with them. This secondary infection process is very slow and inefficient, which explains why it may take years after infection before an HIV carrier shows the clinical signs of severe immune deficiency.

The clinical signs include susceptibility to infections that cause otherwise rare cancers—lymphoma caused by Epstein–Barr virus and Kaposi sarcoma caused by a herpesvirus—as well as infections by *Pneumocystis carinii*, a fungus that causes severe pneumonia.

HIV is a retrovirus that infects immune system cells

As a retrovirus, HIV uses RNA as its genetic material and is capable of inserting a cDNA copy of its genome into the host cell's DNA. The structure of HIV is shown in Figure 13.2. A central core, with a protein coat (p24 core protein), contains two identical molecules of RNA, as well as the enzymes reverse transcriptase, integrase, and a protease. An envelope, derived from the plasma membrane of the cell in which the virus was formed, surrounds the core. The envelope is studded with envelope proteins (gp120 and gp41, where "gp" stands for *glycoprotein*) that enable the virus to infect its target cells.

HIV attaches to host cells at the membrane protein CD4, which is found primarily on T_H cells and macrophages. The initial infection may be in macrophages, and somewhat later, T_H cells are infected. CD4 acts as the receptor for the viral envelope protein gp120. The binding of gp120 to CD4 appears to require several additional cell surface molecules, and successful binding initiates a complex series of events (Figure 18.25; see also Figure 13.7). When HIV infects a cell, the viral core enters the cell and the core "uncoats" itself in the cytoplasm, releasing its contents (RNA and proteins).

Among the enzymes in the core is **reverse transcriptase**, which catalyzes the formation of a double-stranded DNA molecule encoding the same information as the viral RNA (see Chapters 13 and 14). The DNA transcript enters the nucleus of the host cell and is inserted into a chromosome, much as bacteriophage DNA may become incorporated into a bacterial chromosome as a prophage (see Chapter 12). Another HIV enzyme, called **integrase**, catalyzes this insertion. The DNA transcript of the HIV RNA thus becomes a permanent part of the T_H cell's DNA, replicating with it at each cell division.

Soon after the initial HIV infection, the immune system destroys most virus.

The T cell concentration falls and HIV concentration rises, accompanied by some symptoms, such as swollen lymph nodes.

As T cells are further reduced, immune function is impaired and opportunistic infections (such as those caused by yeasts) occur.

Finally, almost all natural immunity is lost.

T cell concentration

HIV concentration

HIV and T cell concentration

Years

18.24 The Course of an HIV Infection HIV infection may be carried, unsuspected, for many years before the onset of symptoms. This long "dormant" period means the infection is often spread by people who are unaware they carry the virus.

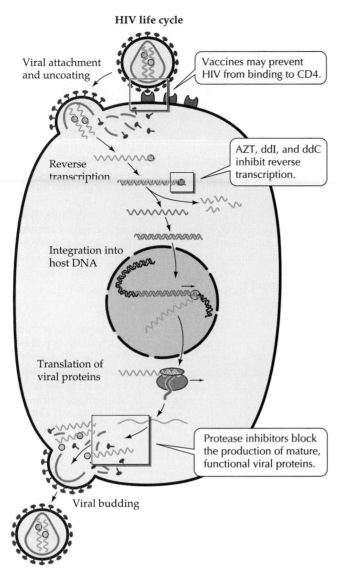

HIV life cycle

Viral attachment and uncoating

Vaccines may prevent HIV from binding to CD4.

Reverse transcription

AZT, ddI, and ddC inhibit reverse transcription.

Integration into host DNA

Translation of viral proteins

Protease inhibitors block the production of mature, functional viral proteins.

Viral budding

18.25 Strategies to Combat HIV Reproduction These widely used drugs block specific steps in the HIV life cycle.

A DNA transcript of the HIV RNA, once incorporated into the genome of a T_H cell, may remain there latent for days, months, or even a decade or more. The latent period ends if the HIV-infected T_H cell becomes activated, for example by infection or immunization. Then the viral DNA is transcribed, yielding many molecules of viral RNA. Some of these RNA transcripts are translated, forming the enzymes and structural proteins of a new generation of viruses. The **protease** encoded by HIV is needed to complete the formation of individual viral proteins from larger products of translation. Other viral RNA molecules are incorporated directly into the new viruses as their genetic material.

Formation of new viruses may be slow. The viruses bud from the infected cell, surrounding themselves with modified plasma membrane from the host.

Several viral genes control the rate of production of individual proteins and of whole viruses. HIV reverse transcriptase is a very fallible enzyme, making a mistake in one out of every 10,000 bases. Since the entire viral genome has about 10,000 bases, this means that nearly every HIV particle is genetically unique. For the host, this is good news—mutated viruses may not reproduce—and bad news—mutated viruses may escape antibodies made to specific structures.

HIV is a relatively fragile virus that needs the moist environment of body fluids in order to remain infective. HIV is transmitted with blood and certain other body fluids, including semen, vaginal fluid, and breast milk of infected persons. HIV can be transmitted by blood transfusions, by sexual activity (either homosexual or heterosexual), and from mother to fetus. HIV is not transmitted by mosquitoes or other insect vectors, or by casual contact. On a worldwide basis, the most common mode of transmission is heterosexual intercourse. However, in the United States the highest incidence of AIDS occurs among drug addicts (from the use of shared, contaminated needles) and male homosexuals.

At this writing, most of the HIV-infected people in the world live in Africa, but the virus is spreading fastest among people living in southern and Southeast Asia. By the year 2000, residents of developing countries will bear 90 percent of new HIV infections as the AIDS epidemic rages on throughout the world. Millions of people will have died before the AIDS epidemic ends. In the meantime, medical and biological research on the subject is proceeding intensely.

Are effective treatments for HIV infection on the horizon?

Prospects for a *cure* for AIDS appear dim at this time. What would an AIDS cure entail? It would include the detection and elimination of every HIV-infected cell in the body. Given the capacity of the viral genetic information to integrate into the host DNA, this task appears overwhelming. So what then?

Many investigators are seeking a vaccine to forestall HIV infection. One of the greatest problems is the exceptional genetic variation of HIV, largely because of unrepaired errors in the activity of reverse transcriptase. Dr. David Ho, a leading physician involved in AIDS treatment, estimates that in an infected person with replicating virus, more than 10 billion new viruses are made every day. It is not surprising that new strains are constantly appearing, and a promising new vaccine against one strain may afford no protection against others.

A second problem with a vaccine is the unusual life cycle of HIV. The rapid integration of the virus into T_H cells and the long life of infected cells make protective vaccines unlikely. One possibility, however, is treatment to diminish the impact of the virus on the body.

There is substantial agreement that we can learn ways to *control the replication* of HIV within the infected body and thus delay the onset of AIDS symptoms. There are many steps in the HIV life cycle. If we can block just *one* of them completely, we break the cycle and hold the infection in check. Potential therapeutic agents are being tested against the major steps shown in Figure 18.26. Of course, it is crucial to block steps that are unique to the virus, so that drug therapies do not harm the patient by blocking a step in the patient's own metabolism.

The first drugs in wide use that prolong the lives of HIV-infected people (AZT, ddI, and ddC) are all directed at the reverse transcription step. Protease inhibitors (saquinovir, indinovir, and ritinovir) can dramatically reduce HIV levels in the blood (recall that some HIV proteins are cut from larger precursor molecules by viral protease). HIV replication can be delayed best by a combination of these two approaches. Multiple-drug therapy using a protease inhibitor along with other drugs provides substantial relief, including reduction in HIV level and restored capability of the immune system—but the monetary cost of such multiple-drug therapy is high.

What can be done until biomedical science provides the tools to bring the worldwide AIDS epidemic to an end? Above all, people must recognize that they are in danger whenever they have sex with a partner whose total sexual history is not known. The danger rises as the number of sex partners rises, and the danger is much greater if partners participating in sexual intercourse are not protected by a latex condom. The danger that heterosexual intercourse will transmit HIV rises ten- to a hundredfold if either partner has another sexually transmitted disease.

Summary of "Natural Defenses against Disease"

• Our bodies defend against pathogens by both nonspecific and specific means.

Defensive Cells and Proteins

• Many of our defenses are implemented by cells and proteins carried in the bloodstream and in the lymphatic system. These cells and proteins also find their way into other parts of the immune system, including the thymus, spleen, and bone marrow. **Review Figure 18.1**
• White blood cells, including lymphocytes (B and T cells) and phagocytes such as neutrophils and macrophages, play many defensive roles. **Review Figure 18.2 and Table 18.1**
• Antibodies bind pathogens, HLA proteins help coordinate the recognition of pathogens and the activation of defensive cells, and cytokines produced by lymphocytes and macrophages alter the behavior of other cells.

Nonspecific Defenses against Pathogens

• An animal's nonspecific defenses include physical barriers, competing resident microorganisms, and hostile local environments that may be acidic or contain an antibacterial enzyme. **Review Table 18.2**
• Some nonspecific defenses are based on proteins. The complement system, composed of about 20 proteins, assembles itself in a cascade of reactions to cooperate with phagocytes or to lyse foreign cells directly. **Review Figure 18.4**
• Interferons produced by virus-infected cells inhibit the ability of viruses to replicate in neighboring cells.
• Macrophages and neutrophils phagocytose invading bacteria. Natural killer cells attack tumor cells and virus-infected body cells.
• Macrophages play an important role in the inflammation response. Activated mast cells release histamine, which causes blood capillaries to swell and leak. Complement proteins attract macrophages to the site, where they leak into the surrounding tissue and phagocytose bacteria and dead cells. **Review Figure 18.6**

Specific Defenses: The Immune Response

• Four features characterize the immune response: specificity, the ability to respond to an enormous diversity of pathogens, the ability to distinguish self from nonself, and memory.
• The immune response is directed against pathogens that evade the nonspecific defenses. The immune response is highly specific, being directed against antigenic determinants on the surfaces of antigens. Each antibody or T cell is directed against a particular antigenic determinant. **Review Figure 18.7**
• The immune response is highly diverse and can respond specifically to perhaps 10 million different antigenic determinants.
• The immune response distinguishes its own cells from foreign cells, attacking only the cells recognized as nonself.
• The immune system remembers; it can respond rapidly and effectively to a second exposure to an antigen.
• There are two interactive immune responses: the humoral immune response and the cellular immune reponse. The humoral immune response employs antibodies secreted by plasma cells to target pathogens in body fluids. The cellular immune response employs T cells to attack body cells that have been altered by viral infection or mutation or to target antigens that have invaded the body's cells.
• Clonal selection accounts for the rapidity, specificity, and diversity of the immune response. It also accounts for immunological memory, which is based on the production of both effector and memory cells as T cell and B cell clones expand. **Review Figure 18.8**
• Immunological memory plays roles in both natural immunity and artificial immunity based on vaccination. **Review Figure 18.9**
• Clonal selection also accounts for the immune system's recognition of self. Tolerance of self results from clonal deletion of antiself lymphocytes. **Review Figure 18.10**

B Cells: The Humoral Immune Response

• Activated B cells differentiate into memory cells and into plasma cells, which synthesize and secrete specific antibodies.
• Antibodies, or immunoglobulins, are of several classes. The basic immunoglobulin unit is a tetramer of four polypeptides: two identical light chains and two identical

heavy chains, each chain consisting of a constant and a variable region. **Review Figure 18.12**
• The variable regions of the light and heavy chains collaborate to form the antigen-binding sites of an antibody. The variable regions determine an antibody's specificity; the constant region determines the destination and function of the antibody. **Review Figure 18.13**
• The immunoglobulin classes are IgM, IgG, IgD, IgE, and IgA. IgM, formed first, is a membrane receptor on B cells, as is IgD. IgG is the most abundant antibody class and performs several defensive functions. IgE takes part in inflammation and allergies. IgA is present in various body secretions. **Review Figure 18.14 and Table 18.3**
• Monoclonal antibodies, produced by hybridomas, are useful in research and in medicine. A sample of a monoclonal antibody consists of identical immunoglobulin molecules directed against a single antigenic determinant. Hybridomas are produced by fusing B cells with myeloma cells (from cancerous tumors of plasma cells). **Review Figure 18.15**

T Cells: The Cellular Immune Response

• The cellular immune response is directed against altered or antigen-infected cells of the body. T_C cells attack virus-infected or tumor cells, causing them to lyse. T_H cells activate B cells (in the humoral immune response) and influence the development of other T cells and macrophages. **Review Figure 18.16**
• T cell receptors in the cellular immune response are analogous to immunoglobulins in the humoral immune response.
• The major histocompatibility complex (MHC) in mice encodes many membrane proteins. The human counterpart of the MHC is called human leukocyte antigen (HLA). In an immune response, MHC molecules in macrophages or lymphocytes bind processed antigen and present it to T cells.
• The humoral immune response requires collaboration of class II MHC molecules, the T cell surface protein CD4, and interleukin-1—all working together to activate T_H cells. The effector phase of the humoral immune response involves T cells, class II MHC molecules, B cells, and interleukins, resulting in the formation of active plasma cells. **Review Figures 18.18, 18.19**
• In the cellular immune response, class I MHC molecules, T_C cells, CD8, and interleukin-2 collaborate to activate T_C cells with the appropriate specificity. **Review Figures 18.17, 18.19**
• Developing T cells undergo two tests: They must be able to recognize self MHC molecules, but they must *not* bind to both self MHC and to any of the body's own antigens. T cells that fail either of these tests die.
• The rejection of organ transplants results from the great diversity of MHC molecules—so great that one individual's cells almost certainly differ antigenically (because of differing MHC molecules) from a given other individual's cells.

The Genetic Basis of Antibody Diversity

• Families of DNA segments underlie the incredible diversity of antibody and T cell receptor specificities.
• Antibody heavy-chain genes are constructed by unusual genetic processes from one each of numerous *V*, *D*, *J*, and *C* segments. The *V*, *D*, and *J* segments combine by DNA rearrangements, and transcription yields an RNA molecule that is spliced to form a translatable mRNA. **Review Figures 18.20, 18.21**
• Other gene families give rise to the light chains. There are millions of possible antibodies as a result of these DNA combinations.

• Imprecise DNA rearrangements, mutations, and random addition of bases to the ends of the DNAs before they are joined contribute even more diversity. **Review Figures 18.20, 18.21**
• A plasma cell produces IgM first, under the influence of cytokines released by T cells, but later it may switch to the production of other classes of antibodies. This class switching, resulting in antibodies with the same antigen specificity but a different function, is accomplished by DNA cutting and rejoining. **Review Figure 18.22, 18.23**

The Evolution of Animal Defense Systems

• Cellular defensive mechanisms such as phagocytosis are older than the animal kingdom, appearing even in some unicellular protists.
• Invertebrate animals reject nonself tissues but lack immunological memory. They possess cells and molecules analogous, but not identical, to lymphocytes, immunoglobulins, and cytokines.
• Even the most evolutionarily ancient groups among today's vertebrates have immune systems more similar to those of humans than to those of invertebrates.

Disorders of the Immune System

• Allergies result from an overreaction of the immune system to an antigen.
• Autoimmune diseases result from a failure in the immune recognition of self, with the appearance of antiself B and T cells that attack the body's own cells. Immune deficiency disorders result from failures of one or another part of the immune system.
• AIDS is an immune deficiency disorder arising from depletion of the body's T_H cells as a result of infection with the HIV retrovirus. Depletion of the T_H cells weakens and eventually destroys the immune system, leaving the host subject defenseless against "opportunistic" infections. **Review Figure 18.24**
• HIV inserts a copy of its genome into a chromosome of a macrophage or T_H cell, where it may lie dormant for years and many cell generations. When the viral genome is transcribed and translated, new viruses form slowly by budding from the infected cell—or rapidly, with lysis of the infected cell. The HIV reproductive cycle is complex. **Review Figures 13.7, 18.25**
• All steps in the reproductive cycle of HIV are under investigation as possible targets for drugs. Currently the most effective drugs are those directed against reverse transcriptase and protease.
• Some treatments may provide a dramatic reduction in HIV levels, but there is as yet no indication that we can prevent infection with HIV, as by vaccination. The only currently available strategy is for people to avoid behaviors that place them at risk.

Self-Quiz

1. Which statement about phagocytes is *not* true?
    a. Some travel in the circulatory system.
    b. They ingest microorganisms by endocytosis.
    c. A single phagocyte can ingest 5 to 25 bacteria before it dies.
    d. Although they are important, an animal can do perfectly well without them.
    e. Lysosomes play an important role in their function.

2. Which statement about immunoglobulins is true?
   a. They help antibodies do their job.
   b. They recognize and bind antigenic determinants.
   c. They encode some of the most important genes in an animal.
   d. They are the chief participants in nonspecific defense mechanisms.
   e. They are a specialized class of white blood cells.

3. Which statement about an antigenic determinant is *not* true?
   a. It is a specific chemical grouping.
   b. It may be part of many different molecules.
   c. It is the part of an antigen to which an antibody binds.
   d. It may be part of a cell.
   e. A single protein has only one on its surface.

4. T cell receptors
   a. are the primary receptors for the humoral immune system.
   b. are carbohydrates.
   c. cannot function unless the animal has previously encountered the antigen.
   d. are produced by plasma cells.
   e. are important in combating viral infections.

5. According to the clonal selection theory
   a. an antibody changes its shape according to the antigen it meets.
   b. an individual animal contains only one type of B cell.
   c. the animal contains many types of B cells, each producing one kind of antibody.
   d. each B cell produces many types of antibodies.
   e. many clones of antiself lymphocytes appear in the bloodstream.

6. Immunological tolerance
   a. depends on exposure to antigen.
   b. develops late in life and is usually life-threatening.
   c. disappears at birth.
   d. results from the activities of the complement system.
   e. results from DNA splicing.

7. The extraordinary diversity of antibodies results in part from
   a. the action of monoclonal antibodies.
   b. the splicing of protein molecules.
   c. the action of cytotoxic T cells.
   d. the rearrangement of gene segments.
   e. their remarkable nonspecificity.

8. Which of the following play(s) no role in the antibody response?
   a. Helper T cells
   b. Interleukins
   c. Macrophages
   d. Reverse transcriptase
   e. Products of class II MHC gene loci

9. The major histocompatibility complex
   a. codes for specific proteins found on the surface of cells.
   b. plays no role in T cell immunity.
   c. plays no role in antibody responses.
   d. plays no role in skin graft rejection.
   e. is encoded by a single locus with multiple alleles.

10. Which of the following plays no role in HIV reproduction?
    a. Integrase
    b. Reverse transcriptase
    c. gp120
    d. Interleukin-1
    e. Protease

Applying Concepts

1. Describe the part of an antibody molecule that interacts with an antigenic determinant. How is it similar to the active site of an enzyme? How does it differ from the active site of an enzyme?

2. Contrast immunoglobulins and T cell receptors with respect to structure and function.

3. Discuss the diversity of antibody specificities in an individual in relation to the diversity of enzymes. Does every cell in an animal contain genetic information for all the organism's enzymes? Does every cell contain genetic information for all the organism's immunoglobulins?

4. Discuss the roles of monoclonal antibodies in medicine and in biological research.

Readings

Ada, G. L. and G. Nossal. 1987. "The Clonal-Selection Theory." *Scientific American*, August. A fascinating historical account of the development of the central concept of immunology.

Beck, G. and G. S. Habicht. 1996. "Immunity and the Invertebrates." *Scientific American*, November. A description of the sophisticated immune systems of invertebrates and how they are similar to and different from the immune systems of vertebrates.

Feldman, M. and L. Eisenbach. 1988. "What Makes a Tumor Cell Metastatic?" *Scientific American*, November. A description of how MHC molecules on tumor cells enable the cells to evade the immune system.

Haynes, B. F. 1993. "Scientific and Social Issues of Human Immunodeficiency Virus Vaccine Development." *Science*, vol. 260, pages 1279–1286. A discussion of the enormous technical—as well as social and ethical—difficulties in conquering AIDS.

Janeway, C. A. and P. Travers. 1994. *Immunobiology: The Immune System in Health and Disease*. Garland, New York. An authoritative, concise text, with hundreds of clear diagrams.

Kuby, J. 1997. *Immunology*, 3rd Edition. W. H. Freeman, New York. A superbly written and illustrated text.

"Life, Death and the Immune System." 1993. *Scientific American*, September. A series of articles, each written by an expert, relating to the topics of this chapter.

Marrack, P. and J. Kappler. 1986. "The T Cell and Its Receptor." *Scientific American*, February. A detailed consideration of the key actors in the cellular immune system.

Tonegawa, S. 1985. "The Molecules of the Immune System." *Scientific American*, October. A beautifully illustrated account of the structures of antibodies and T cell receptors and of how they form.

von Boehmer, H. and P. Kisielow. 1991. "How the Immune System Learns about Self." *Scientific American*, October. A description of the experiments with transgenic mice that established the clonal deletion theory.

Glossary

Abdomen (ab' duh mun) [L.: belly]   In arthropods, the posterior portion of the body; in mammals, the part of the body containing the intestines and most other internal organs, posterior to the thorax.

Abomasum (ab' oh may' sum)   The true stomach of ruminants (animals such as cattle, sheep, and goats).

Abscisic acid (ab sighs' ik) [L. *abscissio*: breaking off]   A plant growth substance having growth-inhibiting action. Causes stomata to close.

Abscission (ab sizh' un) [L. *abscissio*: breaking off]   The process by which leaves, petals, and fruits separate from a plant.

Absolute temperature scale   A temperature scale in which the degree is the same size as in the Celsius (centigrade) scale, and zero is the state of no molecular motion. Absolute zero is –273° on the Celsius scale.

Absorption   (1) Of light: complete retention, without reflection or transmission. (2) Of liquids: soaking up (taking in through pores or cracks).

Absorption spectrum   A graph of light absorption versus wavelength of light; shows how much light is absorbed at each wavelength.

Abyssal zone (uh biss' ul) [Gr. *abyssos*: bottomless]   That portion of the deep ocean where no light penetrates.

Accessory pigments   Pigments that absorb light and transfer energy to chlorophylls for photosynthesis.

Acclimatization   Changes in an organism that improve its ability to tolerate seasonal changes in its environment.

Acellular   Not composed of cells.

Acetylcholine   A neurotransmitter substance that carries information across vertebrate neuromuscular junctions and some other synapses. **Acetylcholinesterase** is an enzyme that breaks down acetylcholine.

Acetyl CoA (acetyl coenzyme A)   Compound that reacts with oxaloacetate to produce citrate at the beginning of the citric acid cycle; a key metabolic intermediate in the formation of many compounds.

Acid [L. *acidus*: sharp, sour]   A substance that can release a proton. (Contrast with base.)

Acid precipitation   Precipitation that has a lower pH than normal as a result of acid-forming precursors introduced into the atmosphere by human activities.

Acidic   Having a pH of less than 7.0 (a hydrogen ion concentration greater than 10^{-7} molar).

Acoelomate   Lacking a coelom.

Acquired Immune Deficiency Syndrome   See AIDS.

Acrosome (a' krow soam) [Gr. *akros*: highest or outermost + *soma*: body]   The structure at the forward tip of an animal sperm which is the first to fuse with the egg membrane and enter the egg cell.

ACTH (adrenocorticotropin)   A pituitary hormone that stimulates the adrenal cortex.

Actin [Gr. *aktis*: a ray]   One of the two major proteins of muscle; it makes up the thin filaments. Forms the microfilaments found in most eukaryotic cells.

Action potential   An impulse in a neuron taking the form of a wave of depolarization or hyperpolarization imposed on a polarized cell surface.

Action spectrum   A graph of biological activity versus wavelength of light. It compares the effectiveness of light of different wavelengths.

Activating enzymes (also called aminoacyl-tRNA synthetases)   These enzymes catalyze the addition of amino acids to their appropriate tRNAs.

Activation energy (E_a)   The energy barrier that blocks the tendency for a set of chemical substances to react. A reaction is speeded up if this energy barrier is surmounted by adding heat, or if the barrier is lowered by providing a different reaction pathway with the aid of a catalyst.

Active site   The region on the surface of an enzyme where the substrate binds, and where catalysis occurs.

Active transport   The transport of a substance across a biological membrane against a concentration gradient—that is, from a region of low concentration (of that substance) to a region of high concentration. Active transport requires the expenditure of energy and is a saturable process. (Contrast with facilitated diffusion, free diffusion; see primary active transport, secondary active transport.)

Adaptation (a dap tay' shun)   In evolutionary biology, a particular structure, physiological process, or behavior that makes an organism better able to survive and reproduce. Also, the evolutionary process that leads to the development or persistence of such a trait.

Adenine (a' den een)   A nitrogen-containing base found in nucleic acids, ATP, NAD, etc.

Adenosine triphosphate   See ATP.

Adenylate cyclase   Enzyme catalyzing the formation of cyclic AMP from ATP.

Adhesion molecules   See cell adhesion molecules.

Adrenal (a dree' nal) [L. *ad-*: toward + *renes*: kidneys]   An endocrine gland located near the kidneys of vertebrates, consisting of two glandular parts, the cortex and medulla.

Adrenaline   See epinephrine.

Adrenocorticotropin   See ACTH.

Adsorption   Binding of a gas or a solute to the surface of a solid.

Aerenchyma (air eng' kyma) [Gr. *aer*: air + *enchyma*: infusion]   Modified parenchyma tissue, with many air spaces, found in shoots of some aquatic plants. (See parenchyma.)

Aerobic (air oh' bic) [Gr. *aer*: air + *bios*: life]   In the presence of oxygen, or requiring oxygen.

Afferent (af' ur unt) [L. *ad*: to + *ferre*: to bear]   To or toward, as in a neuron that carries impulses to the central nervous system, or a blood vessel that carries blood to a structure. (Contrast with efferents.)

Age distribution   The proportion of individuals in a population belonging to each of the age categories into which the population has been divided. The number of divisions is arbitrary.

AIDS (Acquired immune deficiency syndrome)   Condition in which the body's helper T lymphocytes are destroyed, leaving the victim subject to opportunistic diseases. Caused by the HIV-I virus.

Air sacs   Structures in the avian respiratory system that facilitate unidirectional flow of air through the lungs.

Alcohol   An organic compound with one or more hydroxyl (–OH) groups.

Aldehyde (al' duh hide) A compound with a –CHO functional group. Many sugars are aldehydes. (Contrast with ketone.)

Aldosterone (al dahs' ter own) A steroid hormone produced in the adrenal cortex of mammals. Promotes secretion of potassium and reabsorption of sodium in the kidney.

Aleurone layer (al' yur own) [Gr. *aleuron*: wheat flour] In grass seeds, a specialized cell layer just between the seed coat and the endosperm, synthesizing hydrolytic enzymes under the influence of gibberellin, and thus helping mobilize reserves for the developing embryo.

Alga (al' gah) (plural: algae) [L.: seaweed] Any one of a wide diversity of protists belonging to the phyla Pyrrophyta, Chrysophyta, Phaeophyta, Rhodophyta, and Chlorophyta (and, formerly, Cyanophyta—"blue-green algae"). Most live in the water, where they are the dominant autotrophs; most are unicellular, but a minority are multicellular ("seaweeds" and similar protists).

Allele (a leel') [Gr. *allos*: other] The alternate forms of a genetic character found at a given locus on a chromosome.

Allele frequency The relative proportion of a particular allele in a specific population.

Allergy [Ger. *allergie*: altered reaction] An overreaction to an antigen in amounts that do not affect most people; often involves IgE antibodies.

Allometric growth A pattern of growth in which some parts of the body of an organism grow faster than others, resulting in a change in body proportions as the organism grows.

Allopatric (al' lo pat' rick) [Gr. *allos*: other + *patria*: fatherland] Pertaining to populations that occur in different places.

Allopatric speciation See geographical speciation.

Allopolyploid A polyploid in which the chromosome sets are derived from more than one species.

Allostery (al' lo steer' y) [Gr. *allos*: other + *stereos*: structure] Regulation of the activity of an enzyme by binding, at a site other than the catalytic active site, of an effector molecule that does not have the same structure as any of the enzyme's substrates.

Alpha helix Type of protein secondary structure; a right-handed spiral.

Alternation of generations The succession of haploid and diploid phases in a sexually reproducing organism. In most animals (male wasps and honey bees are notable exceptions), the haploid phase consists only of the gametes. In fungi, algae, and plants, however, the haploid phase may be the more prominent phase (as in fungi and mosses) or may be as prominent as the diploid phase (see the life cycle of *Ulva*, for example). In vascular plants, the diploid phase is more prominent.

Altruistic act A behavior whose performance harms the actor but benefits other individuals.

Alveolus (al ve' o lus) (plural: alveoli) [L. *alveus*: cavity] A small, baglike cavity, especially the blind sacs of the lung.

Amensalism (a men' sul ism) Interaction in which one animal is harmed and the other is unaffected. (Contrast with commensalism, mutualism.)

Ames test A test for mutagens (and possible carcinogens) based on the ability of a test compound to cause mutations in the bacterium, *Salmonella*.

Amine An organic compound with an amino group (see Amino acid).

Amino acid An organic compound of the general formula $H_2N–CHR–COOH$, where R can be one of 20 or more different side groups. An amino acid is so named because it has both a basic amine group, $–NH_2$, and an acidic carboxyl group, $–COOH$. Proteins are polymers of amino acids.

Ammonotelic (am moan' o teel' ic) [Gr. *telos*: end] Describes an organism in which the final product of breakdown of nitrogen-containing compounds (primarily proteins) is ammonia. (Contrast with ureotelic, uricotelic.)

Amniocentesis A medical procedure in which cells from the fetus are obtained from the amniotic fluid. The genetic material of the cells is then examined. (Contrast with chorionic villus sampling.)

Amniotic egg The eggs of birds and reptiles, which can be incubated in air because the embryo is enclosed by a fluid-filled sac.

Amoeba (a mee' bah) [Gr. *amoibe*: change] Any one of a large number of different kinds of unicellular protists belonging to the phylum Rhizopoda, characterized among other features by its ability to change shape frequently through the protrusion and retraction of cytoplasmic extensions called pseudopods.

Amoeboid (a mee' boid) Like an amoeba; constantly changing shape by the protrusion and retraction of pseudopodia.

Amphi- [Gr.: both] Prefix used to denote a character or kind of organism that occupies two or more states. For example, amphibian (an animal that lives both on the land and in the water).

Amphipathic (am' fi path' ic) [Gr. *amphi*: both + *pathos*: emotion] Of a molecule, having both hydrophilic and hydrophobic regions.

amu (atomic mass unit, or **dalton)** The basic unit of mass on an atomic scale, defined as one-twelfth the mass of a carbon-12 atom. There are 6.023×10^{23} amu in one gram. This number is known as Avogadro's number.

Amylase (am' ill ase) Any of a group of enzymes that digest starch.

Anabolism (an ab' uh liz' em) [Gr. *ana*: up, throughout + *ballein*: to throw] Synthetic reactions of metabolism, in which complex molecules are formed from simpler ones. (Contrast with catabolism.)

Anaerobic (an ur row' bic) [Gr. *an*: not + *aer*: air + *bios*: life] Occurring without the use of molecular oxygen, O_2.

Anagenesis Evolutionary change in a single lineage over time.

Analogy (a nal' o jee) [Gr. *analogia*: resembling] A resemblance in function, and often appearance as well, between two structures which is due to convergence in evolution rather than to common ancestry. (Contrast with homology.)

Anaphase (an' a phase) [Gr. *ana*: indicating upward progress] The stage in nuclear division at which the first separation of sister chromatids (or, in the first meiotic division, of paired homologues) occurs. Anaphase lasts from the moment of first separation to the time at which the moving chromosomes converge at the poles of the spindle.

Anaphylactic shock A precipitous drop in blood pressure caused by loss of fluid from capillaries because of an increase in their permeability stimulated by an allergic reaction.

Ancestral trait Trait shared by a group of organisms as a result of descent from a common ancestor.

Androgens (an' dro jens) The male sex steroids.

Aneuploidy (an' you ploy dee) A condition in which one or more chromosomes or pieces of chromosomes are either lacking or present in excess.

Angiosperm (an' jee oh spurm) [Gr. *angion*: vessel + *sperma*: seed] One of the flowering plants; literally, one whose seed is carried in a "vessel," which is the fruit. (See fruit.)

Angiotensin (an' jee oh ten' sin) A peptide hormone that raises blood pressure by causing peripheral vessels to constrict; maintains glomerular filtration by constricting efferent glomerular vessels; stimulates thirst; and stimulates the release of aldosterone.

Animal [L. *animus*: breath, soul] A member of the kingdom Animalia. In general, a multicellular eukaryote that obtains its food by ingestion.

Animal hemisphere The metabolically active upper portion of some animal eggs, zygotes, and embryos, which does *not* contain the dense nutrient yolk. The **animal pole** refers to the very top of the egg or embryo. (Contrast with vegetal hemisphere.)

Anion (an' eye one) An ion with one or more negative charges. (Contrast with cation.)

Anisogamy (an' eye sog' a mee) [Gr. *aniso*: unequal + *gamos*: marriage] The existence of two dissimilar gametes (egg and sperm).

Annual Referring to a plant whose life cycle is completed in one growing season. (Contrast with biennial, perennial.)

Anorexia nervosa (an or ex' ee ah) [Gr. *an*: not + *orexis*: appetite] Severe malnutrition and body wasting brought on by a psychological aversion to food.

Antennapedia complex A group of homeotic genes that control the development of the head and anterior thorax of the fruit fly, *Drosophila melanogaster*.

Anterior Toward the front.

Anterior pituitary The portion of the vertebrate pituitary gland that derives from gut epithelium and produces tropic hormones.

Anther (an' thur) [Gr. *anthos*: flower] A pollen-bearing portion of the stamen of a flower.

Antheridium (an' thur id' ee um) (plural: antheridia) [Gr. *antheros*: blooming] The multicellular structure that produces the sperm in bryophytes and ferns.

Antibody One of millions of blood proteins, produced by the immune system, that specifically recognizes a foreign substance and initiates its removal from the body.

Anticodon A "triplet" of three nucleotides in transfer RNA that is able to pair with a complementary triplet (a codon) in messenger RNA, thus aligning the transfer RNA on the proper place on the messenger. The codon (and, reciprocally, the anticodon) codes for a specific amino acid.

Antidiuretic hormone A hormone that controls water reabsorption in the mammalian kidney. Also called vasopressin.

Antigen (an' ti jun) Any substance that stimulates the production of an antibody or antibodies upon introduction into the body of a vertebrate.

Antigen processing The breakdown of antigenic proteins into smaller fragments, which are then presented on the cell surface, along with MHC proteins, to T cells.

Antigenic determinant A specific region of an antigen, which is recognized by and binds to a specific antibody.

Antiparallel Parallel but running in opposite directions. The two strands of DNA are antiparallel.

Antipodals (an tip' o dulls) [Gr. *anti*: against + *podus*: foot] Cells (usually three) of the mature embryo sac of a flowering plant, located at the end opposite the egg (and micropyle).

Antiport A membrane transport process that carries one substance in one direction and another in the opposite direction. (Contrast with symport.)

Antisense nucleic acid A single-stranded RNA or DNA complementary to and thus targeted against the mRNA transcribed from a harmful gene such as an oncogene.

Anus (a' nus) Opening through which digestive wastes are expelled, located at the posterior end of the gut.

Aorta (a or' tuh) [Gr. *aorte*: aorta] The main trunk of the arteries leading to the systemic (as opposed to the pulmonary) circulation.

Apex (a' pecks) The tip or highest point of a structure, as the apex of a growing stem or root.

Apical (a' pi kul) Pertaining to the apex, as the apical meristem, which is the actively growing tissue at the tip of a stem or root.

Apomixis (ap oh mix' is) [Gr. *apo*: away from + *mixis*: sexual intercourse] The asexual production of seeds.

Apoplast (ap' oh plast) In plants, the continuous meshwork of cell walls and extracellular spaces through which material can pass without crossing a plasma membrane. (Contrast with symplast.)

Apoptosis (ay' pu toh sis) A series of genetically programmed events leading to cell death.

Appendix A vestigial portion of the human gut at the junction of the ileum with the colon.

Apterous Lacking wings.

Assisted reproductive technologies (ART) Any of a number of technological approaches to improve human fertility. They include in vitro fertilization of eggs, injection of sperm and eggs into the oviduct, and injection of individual sperm into individual eggs.

Archaebacteria (ark' ee bacteria) [Gr. *archaios*: ancient] One of the two kingdoms of prokaryotes; the archaebacteria possess distinctive lipids and lack peptidoglycan. Most live in extreme environments. (Contrast with eubacteria.)

Archegonium (ar' ke go' nee um) [Gr. *archegonos*: first of a kind] The multicellular structure that produces eggs in bryophytes, ferns, and gymnosperms.

Archenteron (ark en' ter on) [Gr. *archos*: beginning + *enteron*: bowel] The earliest primordial animal digestive tract.

Area cladogram A cladogram in which the geographic ranges of the taxa are substituted for the names of the taxa.

Arteriosclerosis See atherosclerosis.

Artery A muscular blood vessel carrying oxygenated blood away from the heart to other parts of the body. (Contrast with vein.)

Artifact [L. *ars, artis*: art + *facere*: to make] Something made by human effort or intervention. In biology, something that was not present in the living cell or organism, but was unintentionally produced by an experimental procedure.

Ascospore (ass' ko spor) A fungus spore produced within an ascus.

Ascus (ass' cuss) [Gr. *askos*: bladder] In fungi belonging to the class Ascomycetes (sac fungi), the club-shaped sporangium within which spores are produced by meiosis.

Asexual Without sex.

Associative learning "Pavlovian" learning, in which an animal comes to associate a previously neutral stimulus (such as the ringing of a bell) with a particular reward or punishment.

Assortative mating A breeding system under which mates are selected on the basis of a particular trait or group of traits.

Assortment (genetic) The random separation during meiosis of nonhomologous chromosomes and of genes carried on nonhomologous chromosomes. For example, if genes *A* and *B* are borne on nonhomologous

chromosomes, meiosis of diploid cells of genotype *AaBb* will produce haploid cells of the following types in equal numbers: *AB*, *Ab*, *aB*, and *ab*.

Asymmetric The state of lacking any plane of symmetry.

Asymmetric carbon atom In a molecule, a carbon atom to which four different atoms or groups are bound.

Atherosclerosis (ath' er oh sklair oh' sis) A disease of the lining of the arteries characterized by fatty, cholesterol-rich deposits in the walls of the arteries. When fibroblasts infiltrate these deposits and calcium precipitates in them, the disease become arteriosclerosis, or "hardening of the arteries."

Atmosphere The gaseous mass surrounding our planet. Also: a unit of pressure, equal to the normal pressure of air at sea level.

Atom [Gr. *atomos*: indivisible] The smallest unit of a chemical element. Consists of a nucleus and one or more electrons.

Atomic mass (also called atomic weight) The average mass of an atom of an element on the amu scale. (The average depends upon the relative amounts of different isotopes of an element on Earth.)

Atomic mass unit See amu.

Atomic number The number of protons in the nucleus of an atom, also equal to the number of electrons around the neutral atom. Determines the chemical properties of the atom.

ATP (adenosine triphosphate) A compound containing adenine, ribose, and three phosphate groups. When it is formed, useful energy is stored; when it is broken down (to ADP or AMP), energy is released to drive endergonic reactions. ATP is a universal energy storage compound.

ATP synthase An integral membrane protein that couples the transport of protons with the formation of ATP.

Atrium (a' tree um) A body cavity, as in the hearts of vertebrates. The thin-walled chamber(s) entered by blood on its way to the ventricle(s). Also, the outer ear.

Autocatalysis An enzymatic reaction in which the inactive form of an enzyme is converted into its active form by the enzyme itself.

Autoimmune disease A disorder in which the immune system attacks the animal's own body.

Autonomic nervous system The system (which in vertebrates comprises sympathetic and parasympathetic subsystems) that controls such involuntary functions as those of guts and glands.

Autopolyploid A polyploid in which the sets of chromosomes are derived from the same species.

Autoradiography The detection of a radioactive substance in a cell or organism by putting it in contact with a photographic emulsion and allowing it to "take its own picture." The emulsion is developed, and the location of the radioactivity in the cell is seen by the presence of silver grains in the emulsion.

Autoregulatory mechanism   A feedback mechanism that enables a structure to regulate its own function.

Autosome   Any chromosome (in a eukaryote) other than a sex chromosome.

Autotroph (au' tow trow' fik) [Gr. *autos*: self + *trophe*: food]   An organism that is capable of living exclusively on inorganic materials, water, and some energy source such as sunlight or chemically reduced matter. (Contrast with heterotroph.)

Auxin (awk' sin) [Gr. *auxein*: increase]   In plants, a substance (indoleacetic acid) that regulates growth and various aspects of development.

Auxotroph (awks' o trofe) [Gr. *auxanein*: to grow + *trophe*: food]   A mutant form of an organism that requires a nutrient or nutrients not required by the wild type, or reference, form of the organism. (Contrast with prototroph.)

Avogadro's number   The conversion factor between atomic mass units and grams. More usefully, the number of atoms in that quantity of an element which, expressed in grams, is numerically equal to the atomic weight in amu; 6.023×10^{23} atoms. (See mole.)

Axon [Gr.: axle]   Fiber of a neuron which can carry action potentials. Carries impulses away from the cell body of the neuron; releases a neurotransmitter substance.

Axon hillock   The junction between an axon and its cell body; where action potentials are generated.

Axon terminals   The endings of an axon; they form synapses and release neurotransmitter.

Axoneme (ax' oh neem)   The complex of microtubules and their crossbridges that forms the motile apparatus of a cilium.

Bacillus (buh sil' us) [L.: little rod]   Any of various rod-shaped bacteria.

Bacteriophage (bak teer' ee o fayj) [Gr. *bakterion*: little rod + *phagein*: to eat]   One of a group of viruses that infect bacteria and ultimately cause their disintegration.

Bacterium (bak teer' ee um) (plural: bacteria) [Gr. *bakterion*: little rod]   A prokaryote. An organism with chromosomes not contained in nuclear envelopes.

Balanced polymorphism [Gr. *polymorphos*: having many forms]   The maintenance of more than one form, or the maintenance at a given locus of more than one allele, at frequencies of greater than one percent in a population. Often results when heterozygotes are superior to both homozygotes.

Baroreceptor [Gr. *baros*: weight]   A pressure-sensing cell or organ.

Barr body   In mammals, an inactivated X chromosome.

Basal body   Centriole found at the base of a eukaryotic flagellum or cilium.

Basal metabolic rate   The minimum rate of energy turnover in an awake (but resting) bird or mammal that is not expending energy for thermoregulation.

Base   A substance which can accept a proton (H^+). (Contrast with acid.) In nucleic acids, a nitrogen-containing base (purine or pyrimidine) is attached to each sugar in the backbone.

Base pairing   See complementary base pairing.

Basic   Having a pH greater than 7.0 (having a hydrogen ion concentration lower than 10^{-7} molar).

Basidium (bass id' ee yum)   In fungi of the class Basidiomycetes, the characteristic sporangium in which four spores are formed by meiosis and then borne externally before being shed.

Batesian mimicry   Mimicry by a relatively harmless kind of organism of a more dangerous one, by which the mimic enjoys protection from predators that mistake it for the dangerous model. (Contrast with Müllerian mimicry.)

B cell   A type of lymphocyte involved in the humoral immune response of vertebrates. Upon recognizing an antigenic determinant, a B cell develops into a plasma cell, which secretes an antibody. (Contrast with a T cell.)

Benefit   An improvement in survival and reproductive success resulting from a behavior. (Contrast with cost.)

Benign (be nine')   A tumor that grows to a certain size and then stops, usually with a fibrous capsule surrounding the mass of cells. Benign tumors do not spread (metastasize) to other organs.

Benthic zone [Gr. *benthos*: bottom of the sea]   The bottom of the ocean. (Contrast with pelagic zone.)

Beta-pleated sheet   Type of protein secondary structure; results from hydrogen bonding between polypeptide regions running antiparallel to each other.

Biennial   Referring to a plant whose life cycle includes vegetative growth in the first year and flowering and senescence in the second year. (Contrast with annual, perennial.)

Bilateral symmetry   The condition in which only the right and left sides of an organism, divided exactly down the back, are mirror images of each other. (Contrast with radial symmetry.)

Bile   A secretion of the liver delivered to the small intestine via the common bile duct. In the intestine, bile emulsifies fats.

Binocular cells   Neurons in the visual cortex that respond to input from both retinas; involved in depth perception.

Binomial (bye nome' ee al)   Consisting of two names; for example, the binomial nomenclature of biology which gives the name of the genus followed by the name of the species.

Bioaccumulation   The ever-increasing concentration of a chemical compound in the tissues of an organism as it is passed up the food chain.

Biodiversity crisis   The current high rate of loss of species, caused primarily by human activities.

Biogenesis [Gr. *bios*: life + *genesis*: source]   The origin of living things from other living things.

Biogeochemical cycles   Movement of elements through living organisms and the physical environment.

Biogeography   The scientific study of the geographic distribution of organisms. Ecological biogeography is concerned with the habitats in which organisms live, historical biogeography with the complete geographic ranges of organisms and the historical circumstances that determine the ranges.

Biological species concept   The view that a species is most usefully defined as a population or series of populations within which there is a significant amount of gene flow under natural conditions, but which is genetically isolated from other populations.

Bioluminescence   The production of light by biochemical processes in an organism.

Biomass   The total weight of all the living organisms, or some designated group of living organisms, in a given area.

Biome (bye' ome)   A major division of the ecological communities of Earth; characterized by distinctive vegetation.

Biota (bye oh' tah)   All of the organisms, including animals, plants, fungi, and microorganisms, found in a given area.

Biotechnology   The use of cells to make medicines, foods and other products useful to humans.

Biradial symmetry   Radial symmetry modified so that only two planes can divide the animal into similar halves.

Bithorax complex   A group of homeotic genes that control the development of the abdomen and posterior thorax of *Drosophila melanogaster*.

Blastocoel (blass' toe seal) [Br. *blastos*: sprout + *koilos*: hollow]   The central, hollow cavity of a blastula.

Blastodisc (blass' toe disk)   A disk of cells forming on the surface of a large yolk mass, comparable to a blastula, but occurring in animals such as birds and reptiles, in which the massive yolk restricts cleavage to one side of the egg only.

Blastomere   A cell produced by the division of a fertilized egg.

Blastopore   The opening from the archenteron to the exterior of a gastrula.

Blastula (blass' chu luh) [Gr. *blastos*: sprout]   An early stage in animal embryology; in many species, a hollow sphere of cells surrounding a central cavity, the blastocoel. (Contrast with blastodisc.)

Blood–brain barrier   A property of the blood vessels of the brain that prevents most chemicals from diffusing from the blood into the brain.

Bohr effect (boar)   The reduction in affinity of hemoglobin for oxygen caused by acidic conditions, usually as a result of increased CO_2.

Bottleneck   A combination of environmental conditions that causes a serious reduction in the size of the population.

Bowman's capsule   An elaboration of kidney tubule cells that surrounds a knot of capillaries (the glomerulus). Blood is filtered across the walls of these capillaries and the filtrate is collected into Bowman's capsule.

Brain stem   The portion of the vertebrate brain between the spinal cord and the forebrain.

Bronchus (plural: bronchi)   The major airway(s) branching off the trachea into the vertebrate lung.

Brown fat   Fat tissue in mammals that is specialized to produce heat. It has many mitochondria and capillaries, and a protein that uncouples oxidative phosphorylation. It is frequently found in newborns, in mammals acclimated to cold, and in hibernators.

Browser   An animal that feeds on the tissues of woody plants.

Bryophyte (bri' uh fite') [Gr. *bruon*: moss + *phyton*: plant]   Any nonvascular plant, including mosses, liverworts, and hornworts.

Bud primordium [L. *primordium*: the beginning]   In plants, a small mass of potentially meristematic tissue found in the angle between the leaf stalk and the shoot apex. Will give rise to a lateral branch under appropriate conditions.

Budding   Asexual reproduction in which a more or less complete new organism simply grows from the body of the parent organism and eventually detaches itself.

Buffering   A process by which a system resists change—particularly in pH, in which case added acid or base is partially converted to another form.

Bulb   In plants, an underground storage organ composed principally of enlarged and fleshy leaf bases.

Bundle sheath   In C_4 plants, a layer of photosynthetic cells between the mesophyll and a vascular bundle of a leaf.

C_3 photosynthesis   The form of photosynthesis in which 3-phosphoglycerate is the first stable product, and ribulose bisphosphate is the CO_2 receptor.

C_4 photosynthesis   The form of photosynthesis in which oxaloacetate is the first stable product, and phosphoenolpyruvate is the CO_2 acceptor. C_4 plants also perform the reactions of C_3 photosynthesis.

Calcitonin   A hormone produced by the thyroid gland; it lowers blood calcium and promotes bone formation. (Contrast with parathormone.)

Calmodulin (cal mod' joo lin)   A calcium-binding protein found in all animal and plant cells; mediates many calcium-regulated processes.

calorie [L. *calor*: heat]   The amount of heat required to raise the temperature of one gram of water by one degree Celsius (1°C) from 14.5°C to 15.5°C. In nutrition studies, "Calorie" (spelled with a capital C) refers to the kilocalorie (1 kcal = 1,000 cal), the amount of heat required to raise the temperature of one kilogram of water by 1°C.

Calvin–Benson cycle   The stage of photosynthesis in which CO_2 reacts with RuBP to form 3PG, 3PG is reduced to a sugar, and RuBP is regenerated, while other products are released to the rest of the plant.

Calyptra (kuh lip' tra) [Gr. *kalyptra*: covering for the head]   A hood or cap found partially covering the apex of the sporophyte capsule in many moss species, formed from the expanded wall and neck of the archegonium.

Calyx (kay' licks) [Gr. *kalyx*: cup]   All of the sepals of a flower, collectively.

CAM   See crassulacean acid metabolism.

Cambium (kam' bee um) [L. *cambiare*: to exchange]   A meristem that gives rise to radial rows of cells in stem and root, increasing them in girth; commonly applied to the vascular cambium which produces wood and phloem, and the cork cambium, which produces bark.

cAMP (cyclic AMP)   A compound, formed from ATP, that mediates the effects of numerous animal hormones. Also needed for the transcription of catabolite-repressible operons in bacteria. Used for communication by cellular slime molds.

Canopy   The leaf-bearing part of a tree. Collectively the aggregate of the leaves and branches of the larger woody plants of an ecological community.

Capacitance vessels   Refers to veins because of their variable capacity to hold blood.

Capillaries [L. *capillaris*: hair]   Very small tubes, especially the smallest blood-carrying vessels of animals between the termination of the arteries and the beginnings of the veins.

Capping   In eukaryote RNA processing, the addition of a modified G at the 5' end of the molecule.

Capsid   The protein coat of a virus.

Capsule   In bryophytes, the spore case. In some bacteria, a gelatinous layer exterior to the cell wall.

Carbohydrates   Organic compounds with the general formula $C_nH_{2m}O_m$. Common examples are sugars, starch, and cellulose.

Carboxylic acid (kar box sill' ik)   An organic acid containing the carboxyl group, –COOH, which dissociates to the carboxylate ion, –COO⁻.

Carcinogen (car sin' oh jen)   A substance that causes cancer.

Cardiac (kar' dee ak) [Gr. *kardia*: heart]   Pertaining to the heart and its functions.

Carnivore [L. *carn*: flesh + *vovare*: to devour]   An organism that feeds on animal tissue. (Contrast with detritivore, herbivore, omnivore.)

Carotenoid (ka rah' tuh noid) [L. *carota*: carrot]   A yellow, orange, or red lipid pigment commonly found as an accessory pigment in photosynthesis; also found in fungi.

Carpel (kar' pel) [Gr. *karpos*: fruit]   The organ of the flower that contains one or more ovules.

Carrier   In facilitated diffusion, a membrane protein that binds a specific molecule and transports it through the membrane. In genetics, a person heterozygous for a recessive trait. In respiratory and photosynthetic electron transport, a participating substance such as NAD that exists in both oxidized and reduced forms.

Carrying capacity   In ecology, the largest number of organisms of a particular species that can be maintained indefinitely in a given part of the environment.

Cartilage   In vertebrates, a tough connective tissue found in joints, the outer ear, and elsewhere. Forms the entire skeleton in some animal groups.

Casparian strip   A band of cell wall containing suberin and lignin, found in the endodermis. Restricts the movement of water across the endodermis.

Catabolism [Ge. *kata*: down + *ballein*: to throw]   Degradational reactions of metabolism, in which complex molecules are broken down. (Contrast with anabolism.)

Catabolite repression   The decreased synthesis of many enzymes that tend to provide glucose for a cell; caused by the presence of excellent carbon sources, particularly glucose.

Catalyst (cat' a list) [Gr. *kata-*, implying the breaking down of a compound]   A chemical substance that accelerates a reaction without itself being consumed in the overall course of the reaction. Catalysts lower the activation energy of a reaction. Enzymes are biological catalysts.

Cation (cat' eye on)   An ion with one or more positive charges. (Contrast with anion.)

Caudal [L. *cauda*: tail]   Pertaining to the tail, or to the posterior part of the body.

cDNA   See complementary DNA.

Cecum (see' cum) [L. *caecus*: blind]   A blind branch off the large intestine. In many nonruminant mammals, the cecum contains a colony of microorganisms that contribute to the digestion of food.

Cell adhesion molecules   Molecules on animal cell surfaces that affect the selective association of cells during development of the embryo.

Cell cycle   The stages through which a cell passes between one division and the next. Includes all stages of interphase and mitosis.

Cell division   The reproduction of a cell to produce two new cells. In eukaryotes, this process involves nuclear division (mitosis) and cytoplasmic division (cytokinesis).

Cell theory   The theory, well established, that organisms consist of cells, and that all cells come from preexisting cells.

Cell wall   A relatively rigid structure that encloses cells of plants, fungi, many protists, and most bacteria. The cell wall gives these cells their shape and limits their expansion in hypotonic media.

Cellular immune system   That part of the immune system that is based on the activities of T cells. Directed against parasites, fungi, intracellular viruses, and foreign tissues (grafts). (Contrast with humoral immune system.)

Cellular respiration See respiration.

Cellulose (sell' you lowss) A straight-chain polymer of glucose molecules, used by plants as a structural supporting material. **Cellulase** is an enzyme that hydrolyzes cellulose.

Central dogma The statement that information flows from DNA to RNA to polypeptide (in retroviruses, there is also information flow from RNA to cDNA).

Central nervous system That part of the nervous system which is condensed and centrally located, e.g., the brain and spinal cord of vertebrates; the chain of cerebral, thoracic and abdominal ganglia of arthropods.

Centrifuge [L. *fugere*: to flee] A device in which a sample can be spun around a central axis at high speed, creating a centrifugal force that mimics a very strong gravitational force. Used to separate mixtures of suspended materials.

Centriole (sen' tree ole) A paired organelle that helps organize the microtubules in animal and protist cells during nuclear division.

Centromere (sen' tro meer) [Gr. *centron*: center + *meros*: part] The region where sister chromatids join.

Centrosome (sen' tro soam) The major microtubule organizing center of an animal cell.

Cephalization (sef' uh luh zay' shun) [Gr. *kephale*: head] The evolutionary trend toward increasing concentration of brain and sensory organs at the anterior end of the animal.

Cerebellum (sair' uh bell' um) [L.: diminutive of *cerebrum*: brain] The brain region that controls muscular coordination; located at the anterior end of the hindbrain.

Cerebral cortex The thin layer of gray matter (neuronal cell bodies) that overlays the cerebrum.

Cerebrum (su ree' brum) [L.: brain] The dorsal anterior portion of the forebrain, making up the largest part of the brain of mammals. In mammals, the chief coordination center of the nervous system; consists of two **cerebral hemispheres**.

Cervix (sir' vix) [L.: neck] The opening of the uterus into the vagina.

cGMP (cyclic guanosine monophosphate) An intracellular messenger that is part of signal transmission pathways involving G proteins. (See G protein.)

Channel A membrane protein that forms an aqueous passageway though which specific solutes may pass by simple diffusion; some channels are gated: they open and close in response to binding of specific molecules.

Chaperone protein A protein that assists a newly forming protein in adopting its appropriate tertiary structure.

Chemical bond An attractive force stably linking two atoms.

Chemiosmotic mechanism According to this model, ATP formation in mitochondria and chloroplasts results from a pumping of protons across a membrane (against a gradient of electrical charge and of pH), followed by the return of the protons through a protein channel with ATPase activity.

Chemoautotroph An organism that uses carbon dioxide as a carbon source and obtains energy by oxidizing inorganic substances from its environment. (Contrast with chemoheterotroph, photoautotroph, photoheterotroph.)

Chemoheterotroph An organism that must obtain both carbon and energy from organic substances. (Contrast with chemoautotroph, photoautotroph, photoheterotroph.)

Chemosensor A cell or tissue that senses specific substances in its environment.

Chemosynthesis Synthesis of food substances, using the oxidation of reduced materials from the environment as a source of energy.

Chiasma (kie az' muh) (plural: chiasmata) [Gr.: cross] An X-shaped connection between paired homologous chromosomes in prophase I of meiosis. A chiasma is the visible manifestation of crossing over between homologous chromosomes.

Chitin (kye' tin) [Gr. *chiton*: tunic] The characteristic tough but flexible organic component of the exoskeleton of arthropods, consisting of a complex, nitrogen-containing polysaccharide. Also found in cell walls of fungi.

Chlorophyll (klor' o fill) [Gr. *chloros*: green + *phyllon*: leaf] Any of a few green pigments associated with chloroplasts or with certain bacterial membranes; responsible for trapping light energy for photosynthesis.

Chloroplast [Gr. *chloros*: green + *plast*: a particle] An organelle bounded by a double membrane containing the enzymes and pigments that perform photosynthesis. Chloroplasts occur only in eukaryotes.

Choanocyte (cho' an oh cite) The collared, flagellated feeding cells of sponges.

Cholecystokinin (ko' lee sis to kai nin) A hormone produced and released by the lining of the duodenum when it is stimulated by undigested fats and proteins. It stimulates the gallbladder to release bile and slows stomach activity.

Chorion (kor' ee on) [Gr. *khorion*: afterbirth] The outermost of the membranes protecting mammal, bird, and reptile embryos; in mammals it forms part of the placenta.

Chorionic villus sampling A medical procedure that extracts a portion of the chorion from a pregnant woman to enable genetic and biochemical analysis of the embryo. (Contrast with amniocentesis.)

Chromatid (kro' ma tid) Each of a pair of new sister chromosomes from the time at which the molecular duplication occurs until the time at which the centromeres separate at the anaphase of nuclear division.

Chromatin The nucleic acid–protein complex found in eukaryotic chromosomes.

Chromatography Any one of several techniques for the separation of chemical substances, based on differing relative tendencies of the substances to associate with a mobile phase or a stationary phase.

Chromatophore (krow mat' o for) [Gr. *chroma*: color + *phoreus*: carrier] A pigment-bearing cell that expands or contracts to change the color of the organism.

Chromosomal aberration Any large change in the structure of a chromosome, including duplication or loss of chromosomes or parts thereof, usually gross enough to be detected with the light microscope.

Chromosome (krome' o sowm) [Gr. *chroma*: color = *soma*: body] In bacteria and viruses, the DNA molecule that contains most or all of the genetic information of the cell or virus. In eukaryotes, a structure composed of DNA and proteins that bears part of the genetic information of the cell.

Chromosome walking A technique based on recognition of overlapping fragments; used as a step in DNA sequencing.

Chylomicron (ky low my' cron) Particles of lipid coated with protein, produced in the gut from dietary fats and secreted into the extracellular fluids.

Chyme (kime) [Gr. *chymus*, juice] Created in the stomach; a mixture of ingested food with the digestive juices secreted by the salivary glands and the stomach lining.

Ciliate (sil' ee ate) A member of the protist phylum Ciliophora, unicellular organisms that propel themselves by means of cilia.

Cilium (sil' ee um) (plural: cilia) [L. *cilium*: eyelash] Hairlike organelle used for locomotion by many unicellular organisms and for moving water and mucus by many multicellular organisms. Generally shorter than a flagellum.

Circadian rhythm (sir kade' ee an) [L. *circa*: approximately + *dies*: day] A rhythm in behavior, growth, or some other activity that recurs about every 24 hours under constant conditions.

Circannual rhythm (sir can' you al) [L. *circa*: approximately + *annus*: year] A rhythm of behavior, growth, or some other activity that recurs on a yearly basis.

Citric acid cycle A set of chemical reactions in cellular respiration, in which acetyl CoA reacts with oxaloacetate to form citric acid, and oxaloacetate is regenerated. Acetyl CoA is oxidized to carbon dioxide, and hydrogen atoms are stored as NADH and FADH$_2$.

Clade (clayd) [Gr. *klados*: branch] All of the organisms, both living and fossil, descended from a particular common ancestor.

Cladistic classification A classification based entirely on the phylogenetic relationships among organisms.

Cladogenesis (clay doh jen' e sis) [Gr. *klados*: branch + *genesis*: source] The formation of new species by the splitting of an evolutionary lineage.

Cladogram The graphic representation of a clade.

Class In taxonomy, the category below the phylum and above the order; a group of related, similar orders.

Class I MHC molecules These cell surface proteins participate in the cellular immune response directed against virus-infected cells.

Class II MHC molecules These cell surface proteins participate in the cell-cell interactions (of helper T cells, macrophages, and B cells) of the humoral immune response.

Class II MHC molecules These proteins do not present processed antigen, as do classes I and II, but instead include some proteins of the complement system and certain cytokines.

Class switching The process whereby a plasma cell changes the class of immunoglobulin that it synthesizes. This results from the deletion of part of the constant region of DNA, bringing in a new C segment. The variable region is the same as before, so that the new immunoglobulin has the same antigenic specificity.

Clathrin A fibrous protein on the inner surfaces of animal cell membranes that strengthens coated vesicles and thus participates in receptor-mediated endocytosis.

Clay A soil constituent comprising particles smaller than 2 micrometers in diameter.

Cleavages First divisions of the fertilized egg of an animal.

Climax In ecology, a community that terminates a succession and which tends to replace itself unless it is further disturbed or the physical environment changes.

Climograph (clime' o graf) Graph relating temperature and precipitation with time of year.

Cline A gradual change in the traits of a species over a geographical gradient.

Clitoris (klit' er us, klite' er us) A structure in the human female reproductive system that is homologous with the male penis and is involved in sexual stimulation.

Cloaca (klo ay' kuh) [L. *cloaca*: sewer] In some invertebrates, the posterior part of the gut; in many vertebrates, a cavity receiving material from the digestive, reproductive, and excretory systems.

Clonal anergy When a naive T cell encounters a self-antigen, the T cell may bind to the antigen but does not receive signals from an antigen-presenting cell. Instead of being activated, the T cell dies (becomes anergic). In this way, we avoid reacting to our own tissue-specific antigens.

Clonal deletion In immunology, the inactivation or destruction of lymphocyte clones that would produce immune reactions against the animal's own body.

Clonal selection The mechanism by which exposure to antigen results in the activation of selected T- or B-cell clones, resulting in an immune response.

Clone [Gr. *klon*: twig, shoot] Genetically identical cells or organisms produced from a common ancestor by asexual means.

Cnidocytes The feeding cells of cnidarians, within which nematocysts are housed.

Coacervate (ko as' er vate) [L. *coacervare*: to heap up] An aggregate of colloidal particles in suspension.

Coacervate drop Drops formed when a mixture of large proteins and polysaccharides is shaken in water. The interiors of these drops, which are often very stable, contain most of the proteins and polysaccharides.

Coated vesicle Vesicle, sometimes formed from a coated pit, with characteristic "bristly" surface; its membrane contains distinctive proteins, including clathrin.

Coccus (kock' us) [Gr. *kokkos*: berry, pit] Any of various spherical or spheroidal bacteria.

Cochlea (kock' lee uh) [Gr. *kokhlos*: a land snail] A spiral tube in the inner ear of vertebrates; it contains the sensory cells involved in hearing.

Codominance A condition in which two alleles at a locus produce different phenotypic effects and both effects appear in heterozygotes.

Codon A "triplet" of three nucleotides in messenger RNA that directs the placement of a particular amino acid into a polypeptide chain. (Contrast with anticodon.)

Coefficient of relatedness The probability that an allele in one individual is an identical copy, by descent, of an allele in another individual.

Coelom (see' lum) [Gr. *koiloma*: cavity] The body cavity of certain animals, which is lined with cells of mesodermal origin.

Coelomate Having a coelom.

Coenocyte (seen' a sight) [Gr.: common cell] A "cell" bounded by a single plasma membrane, but containing many nuclei.

Coenzyme A nonprotein molecule that plays a role in catalysis by an enzyme. The coenzyme may be part of the enzyme molecule or free in solution. Some coenzymes are oxidizing or reducing agents, others play different roles.

Coevolution Concurrent evolution of two or more species that are mutually affecting each other's evolution.

Cohort (co' hort) [L. *cohors*: company of soldiers] A group of similar-age organisms, considered as it passes through time.

Coleoptile (koe' lee op' til) [Gr. *koleos*: sheath + *ptilon*: feather] A pointed sheath covering the shoot of grass seedlings.

Collagen [Gr. *kolla*: glue] A fibrous protein found extensively in bone and connective tissue.

Collecting duct In vertebrates, a tubule that receives urine produced in the nephrons of the kidney and delivers that fluid to the ureter for excretion.

Collenchyma (cull eng' kyma) [Gr. *kolla*: glue + *enchyma*: infusion] A type of plant cell, living at functional maturity, which lends flexible support by virtue of primary cell walls thickened at the corners. (Contrast with parenchyma, sclerenchyma.)

Colon [Gr. *kolon*: large intestine] The large intestine.

Commensalism The form of symbiosis in which one species benefits from the association, while the other is neither harmed nor benefited.

Common bile duct A single duct that delivers bile from the gallbladder and secretions from the pancreas into the small intestine.

Communication A signal from one organism (or cell) that alters the pattern of behavior in another organism (or cell) in an adaptive fashion.

Community Any ecologically integrated group of species of microorganisms, plants, and animals inhabiting a given area.

Companion cell Specialized cell found adjacent to a sieve tube member in some flowering plants.

Comparative analysis An approach to studying evolution in which hypotheses are tested by measuring the distribution of states among a large number of species.

Compensation point The light intensity at which the rates of photosynthesis and of cellular respiration are equal.

Competitive inhibitor A substance, similar in structure to an enzyme's substrate, that binds the active site and thus inhibits a reaction.

Competition In ecology, use of the same resource by two or more species, when the resource is present in insufficient supply for the combined needs of the species.

Competitive exclusion A result of competition between species for a limiting resource in which one species completely eliminates the other.

Competitive inhibitor A substance, similar in structure to an enzyme's substrate, that binds the active site and inhibits a reaction.

Complement system A group of eleven proteins that play a role in some reactions of the immune system. The complement proteins are not immunoglobulins.

Complementary base pairing The A–T (or A–U), T–A (or U–A), C–G and G–C pairing of bases in double-stranded DNA, in transcription, and between tRNA and mRNA.

Complementary DNA (cDNA) DNA formed by reverse transcriptase acting with an RNA template; essential intermediate in the reproduction of retroviruses; used as a tool in recombinant DNA technology; lacks introns.

Complete metamorphosis A change of state during the life cycle of an organism in which the body is almost completely rebuilt to produce an individual with a very different body form. Characteristic of insects such as butterflies, moths, beetles, ants, wasps, and flies.

Compound (1) A substance made up of atoms of more than one element. (2) Made up of many units, as the compound eyes of arthropods (as opposed to the simple eyes of the same group of organisms).

Condensation reaction A reaction in which two molecules become connected by a covalent bond and a molecule of water is released. ($AH + BOH \rightarrow AB + H_2O$.)

Cones (1) In the vertebrate retina: photoreceptors responsible for color vision. (2) In gymnosperms: reproductive structures consisting of many sporophylls packed relatively tightly.

Conidium (ko nid' ee um) [Gr. *konis*: dust] An asexual fungus spore borne singly or in chains either apically or laterally on a hypha.

Conifer (kahn' e fer) [Gr. *konos*: cone + *phero*: carry] One of the cone-bearing gymnosperms, mostly trees, such as pines and firs.

Conjugation (kahn' jew gay' shun) [L. *conjugare*: yoke together] The close approximation of two cells during which they exchange genetic material, as in *Paramecium* and other ciliates, or during which DNA passes from one to the other through a tube, as in bacteria.

Connective tissue An animal tissue that connects or surrounds other tissues; its cells are embedded in a collagen-containing matrix.

Connexon In a gap junction, a protein channel linking adjacent animal cells.

Consensus sequences Short stretches of DNA that appear, with little variation, in many different genes.

Constant region The constant region in an immunoglobulin is encoded by a single exon and determines the function, but not the specificity, of the molecule. The constant region of the T cell receptor anchors the protein to the plasma membrane.

Constitutive enzyme An enzyme that is present in approximately constant amounts in a system, whether its substrates are present or absent. (Contrast with inducible enzyme.)

Consumer An organism that eats the tissues of some other organism.

Continental climate A pattern, typical of the interiors of large continents at high latitudes, in which bitterly cold winters alternate with hot summers. (Contrast with maritime climate.)

Continental drift The gradual drifting apart of the world's continents that has occurred over a period of billions of years.

Contractile vacuole An organelle, often found in protists, which pumps excess water out of the cell and keeps it from being "flooded" in hypotonic environments.

Convergent evolution The evolution of similar features independently in unrelated taxa from different ancestral structures.

Cooperative act Behavior in which two or more individuals interact to their mutual benefit. No conscious awareness by the actors of the effects of their behavior is implied.

Cooption The act of capturing something for a particular use. In ecology refers to the diversion of ecological production for human use. Such production is said to be coopted.

Copulation Reproductive behavior that results in a male depositing sperm in the reproductive tract of a female.

Corepressor A low molecular weight compound that unites with a protein (the repressor) to prevent transcription in a repressible operon.

Cork A waterproofing tissue in plants, with suberin-containing cell walls. Produced by a cork cambium.

Corm A conical, underground stem that gives rise to a new plant. (Contrast with bulb.)

Corolla (ko role' lah) [L.: diminutive of *corona*: wreath, crown] All of the petals of a flower, collectively.

Coronary (kor' oh nair ee) Referring to the blood vessels of the heart.

Corpus luteum (kor' pus loo' tee um) [L. *corpus*: body + *luteum*: yellow] A structure formed from a follicle after ovulation; it produces hormones important to the maintenance of pregnancy.

Cortex [L.: bark or rind] (1) In plants: the tissue between the epidermis and the vascular tissue of a stem or root. (2) In animals: the outer tissue of certain organs, such as the adrenal cortex and cerebral cortex.

Corticosteroids Steroid hormones produced and released by the cortex of the adrenal gland.

Cost See energetic cost, opportunity cost, risk cost.

Cotyledon (kot' ul lee' dun) [Gr. *kotyledon*: a hollow space] A "seed leaf." An embryonic organ which stores and digests reserve materials; may expand when seed germinates.

Countercurrent exchange An adaptation that promotes maximum exchange of heat or any diffusible substance between two fluids by the fluids flow in opposite directions through parallel tubes in close approximation to each other. An example is countercurrent heat exchange between arterioles and venules in the extremities of some animals.

Covalent bond A chemical bond that arises from the sharing of electrons between two atoms. Usually a strong bond.

Crassulacean acid metabolism (CAM) A metabolic pathway enabling the plants that possess it to store carbon dioxide at night and then perform photosynthesis during the day with stomata closed.

Crista (plural: cristae) A small, shelflike projection of the inner membrane of a mitochondrion; the site of oxidative phosphorylation.

Critical night length In the photoperiodic flowering response of short-day plants, the length of night above which flowering occurs and below which the plant remains vegetative. (The reverse applies in the case of long-day plants.)

Critical period The age during which some particular type of learning must take place or during which it occurs much more easily than at other times. Typical of song learning among birds.

Cross-pollination The pollination of one plant by pollen from another plant. (Contrast with self-pollination.)

Cross section (also called a transverse section) A section taken perpendicular to the longest axis of a structure.

Crossing over The mechanism by which linked markers undergo recombination. In general, the term refers to the reciprocal exchange of corresponding segments between two homologous chromatids. However, the reciprocity of crossing-over is problematical in prokaryotes and viruses; and even in eukaryotes, very closely linked markers often recombine by a nonreciprocal mechanism.

CRP The cAMP receptor protein that interacts with the promoter to enhance transcription; a lowered cAMP concentration results in catabolite repression.

Crustacean (crus tay' see an) A member of the phylum Crustacea, such as a crab, shrimp, or sowbug.

Cryptic appearance The resemblance of an animal to some part of its environment, which helps it to escape detection by predators.

Culture (1) A laboratory association of organisms under controlled conditions. (2) The collection of knowledge, tools, values, and rules that characterize a human society.

Cuticle A waxy layer on the outer surface of a plant or an insect, tending to retard water loss.

Cutin (cue' tin) [L. *cutis*: skin] A mixture of long, straight-chain hydrocarbons and waxes secreted by the plant epidermis, providing a water-impermeable coating on aerial plant parts.

Cyanobacteria (sigh an' o bacteria) [Gr. *kuanos*: the color blue] A division of photosynthetic bacteria, formerly referred to as blue-green algae; they lack sexual reproduction, and they use chlorophyll *a* in their photosynthesis.

Cyclic AMP See cAMP.

Cyclins Proteins that activate cyclin-dependent kinases, bringing about transitions in the cell cycle.

Cyclin-dependent kinase (cdk) A kinase is an enzyme that catalzyes the addition of phosphate groups from ATP to target molecules. Cdk's target proteins involved in transitions in the cell cycle and are active only when complexed to additional protein subunits, cyclins.

Cyst (sist) [Gr. *kystis*: pouch] (1) A resistant, thick-walled cell formed by some protists and other organisms. (2) An abnormal sac, containing a liquid or semisolid substance, produced in response to injury or illness.

Cytochromes (sy' toe chromes) [Gr. *kytos*: container + *chroma*: color] Iron-containing red proteins, components of the electron-transfer chains in photophosphorylation and respiration.

Cytokine A small protein that is made by one type of immune cell and stimulates a target cell which has a specific receptor for that cytokine.

Cytokinesis (sy' toe kine ee' sis) [Gr. *kytos*: container + *kinein*: to move] The division of the cytoplasm of a dividing cell. (Contrast with mitosis.)

Cytokinin (sy' toe kine' in) [Gr. *kytos*: container + *kinein*: to move] A member of a class of plant growth substances playing roles in senescence, cell division, and other phenomena.

Cytoplasm The contents of the cell, excluding the nucleus.

Cytoplasmic determinants In animal development, gene products whose spatial distribution may determine such things as embryonic axes.

Cytosine (site' oh seen) A nitrogen-containing base found in DNA and RNA.

Cytoskeleton The network of microtubules and microfilaments that gives a eukaryotic cell its shape and its capacity to arrange its organelles and to move.

Cytosol The fluid portion of the cytoplasm, excluding organelles and other solids.

Cytotoxic T cells Cells of the cellular immune system that recognize and directly eliminate virus-infected cells. (Contrast with helper T cells, suppressor T cells.)

Dalton See amu.

Deciduous (de sid' you us) [L. *decidere*: fall off] Referring to a plant that sheds its leaves at certain seasons. (Contrast with evergreen.)

Degeneracy The situation in which a single amino acid may be represented by any of two or more different codons in messenger RNA. Most of the amino acids can be represented by more than one codon.

Degradative succession Ecological succession occuring on the dead remains of the bodies of plants and animals, as when leaves or animal bodies rot.

Dehydration See condensation reaction.

Deletion (genetic) A mutation resulting from the loss of a continuous segment of a gene or chromosome. Such mutations never revert to wild type. (Contrast with duplication, point mutation.)

Deme (deem) [Gr. *demos*: common people] Any local population of individuals belonging to the same species that interbreed with one another.

Demographic processes The events—such as births, deaths, immigration, and emigration—that determine the number of individuals in a population.

Demographic stochasticity Random variations in the factors influencing the size, density, and distribution of a population.

Demography The study of dynamical changes in the sizes, densities, and distributions of populations.

Denaturation Loss of activity of an enzyme or nucleic acid molecule as a result of structural changes induced by heat or other means.

Dendrite [Gr. *dendron*: a tree] A fiber of a neuron which often cannot carry action potentials. Usually much branched and relatively short compared with the axon, and commonly carries information to the cell body of the neuron.

Denitrification Metabolic activity by which inorganic nitrogen-containing ions are reduced to form nitrogen gas and other products; carried on by certain soil bacteria.

Density dependence Change in the severity of action of agents affecting birth and death rates within populations that are directly or inversely related to population density.

Density independence The state where the severity of action of agents affecting birth and death rates within a population does not change with the density of the population.

Deoxyribonucleic acid See DNA.

Depolarization A change in the electric potential across a membrane from a condition in which the inside of the cell is more negative than the outside to a condition in which the inside is less negative, or even positive, with reference to the outside of the cell. (Contrast with hyperpolarization.)

Derived trait A trait found among members of a lineage that was not present in the ancestors of that lineage.

Dermal tissue system The outer covering of a plant, consisting of epidermis in the young plant and periderm in a plant with extensive secondary growth. (Contrast with ground tissue system and vascular tissue system.)

Desmosome (dez' mo sowm) [Gr. *desmos*: bond + *soma*: body] An adhering junction between animal cells.

Determination Process whereby an embryonic cell or group of cells becomes fixed into a predictable developmental pathway.

Detritivore (di try' ti vore) [L. *detritus*: worn away + *vorare*: to devour] An organism that eats the dead remains of other organisms.

Deuterium An isotope of hydrogen possessing one neutron in its nucleus. Deuterium oxide is called "heavy water."

Deuterostome One of two major lines of evolution in animals, characterized by radial cleavage, enterocoelous development, and other traits.

Development Progressive change, as in structure or metabolism; in most kinds of organisms, development continues throughout the life of the organism.

Dialysis (dye ahl' uh sis) [Gr. *dialyein*: separation] The removal of ions or small molecules from a solution by their diffusion across a semipermeable membrane to a solvent where their concentration is lower.

Diaphragm (dye' uh fram) [Gr. *diaphrassein*, to barricade] (1) A sheet of muscle that separates the thoracic and abdominal cavities in mammals; responsible for the action of breathing. (2) A method of birth control in which a sheet of rubber is fitted over the woman's cervix, blocking the entry of sperm.

Diastole (dye ahs' toll ee) [Gr.: dilation] The portion of the cardiac cycle when the heart muscle relaxes. (Contrast with systole.)

Dicot (short for dicotyledon) [Gr. *dis*: two + *kotyledon*: a cup-shaped hollow] Any member of the angiosperm class Dicotyledones, flowering plants in which the embryo produces two cotyledons prior to germination. Leaves of most dicots have major veins arranged in a branched or reticulate pattern.

Differentiation Process whereby originally similar cells follow different developmental pathways. The actual expression of determination.

Diffuse coevolution The situation in which the evolution of a lineage is influenced by its interactions with a number of species, most of which exert only a small influence on the evolution of the focal lineage.

Diffusion Random movement of molecules or other particles, resulting in even distribution of the particles when no barriers are present.

Digestion Enzyme-catalyzed process by which large, usually insoluble, molecules (foods) are hydrolyzed to form smaller molecules of soluble substances.

Dihybrid cross A mating in which the parents differ with respect to the alleles of two loci of interest.

Dikaryon (di care' ee ahn) [Gr. *dis*: two + *karyon*: kernel] A cell or organism carrying two genetically distinguishable nuclei. Common in fungi.

Dioecious (die eesh' us) [Gr.: two houses] Organisms in which the two sexes are "housed" in two different individuals, so that eggs and sperm are not produced in the same individuals. Examples: humans, fruit flies, oak trees, date palms. (Contrast with monoecious.)

Diploblastic Having two cell layers. (Contrast with triploblastic.)

Diploid (dip' loid) [Gr. *diploos*: double] Having a chromosome complement consisting of two copies (homologues) of each chromosome. A diploid individual (or cell) usually arises as a result of the fusion of two gametes, each with just one copy of each chromosome. Thus, the two homologues in each chromosome pair in a diploid cell are of separate origin, one derived from the female parent and one from the male parent.

Diplontic life cycle A life cycle in which every cell except the gametes is diploid.

Directional selection Selection in which phenotypes at one extreme of the population distribution are favored. (Contrast with disruptive selection; stabilizing selection.)

Disaccharide A carbohydrate made up of two monosaccharides (simple sugars).

Dispersal stage Stage in its life history at which an organism moves from its birthplace to where it will live as an adult.

Displacement activity Apparently irrelevant behavior performed by an animal under conflict situations, especially when tendencies to attack and escape are closely balanced.

Display   A behavior that has evolved to influence the actions of other individuals.

Disruptive selection   Selection in which phenotypes at both extremes of the population distribution are favored. (Contrast with directional selection; stabilizing selection.)

Distal   Away from the point of attachment or other reference point. (Contrast with proximal.)

Disturbance   A short-term event that disrupts populations, communities, or ecosystems by changing the environment.

Diverticulum (di ver tic' u lum) [L. *divertere*: turn away]   A small cavity or tube that connects to a major cavity or tube.

Division   A term used by some microbiologists and formerly by botanists, corresponding to the term phylum.

DNA (deoxyribonucleic acid)   The fundamental hereditary material of all living organisms. In eukaryotes, stored primarily in the cell nucleus. A nucleic acid using deoxyribose rather than ribose.

DNA hybridization   A process by which DNAs from two species are mixed and heated so that interspecific double helixes are formed.

DNA ligase   Enzyme that unites Okazaki fragments of the lagging strand during DNA replication; also mends breaks in DNA strands. It connects pieces of a DNA strand and is used in recombinant DNA technology.

DNA methylation   Addition of methyl groups to DNA; plays role in regulation of gene expression; protects a bacterium's DNA against its restriction endonucleases.

DNA polymerase   Any of a group of enzymes that catalyze the formation of DNA strands from a DNA template.

Dominance   In genetic terminology, the ability of one allelic form of a gene to determine the phenotype of a heterozygous individual, in which the homologous chromosome carries both it and a different allele. For example, if *A* and *a* are two allelic forms of a gene, *A* is said to be dominant to *a* if *AA* diploids and *Aa* diploids are phenotypically identical and are distinguishable from *aa* diploids. The *a* allele is said to be recessive.

Dominance hierarchy   The set of relationships within a group of animals, usually established and maintained by aggression, in which one individual has precedence over all others in eating, mating, and other activities; a second individual has precedence over all but the highest-ranking individual, and so on down the line.

Dormancy   A condition in which normal activity is suspended, as in some seeds and buds.

Dorsal [L. *dorsum*: back]   Pertaining to the back or upper surface. (Contrast with ventral.)

Double fertilization   Process virtually unique to angiosperms in which one sperm nucleus combines with the egg to produce a zygote, and the other sperm nucleus combines with the two polar nuclei to produce the first cell of the triploid endosperm.

Double helix   Of DNA: molecular structure in which two complementary polynucleotide strands, antiparallel to each other, form a right-handed spiral.

Duodenum (doo' uh dee' num)   The beginning portion of the vertebrate small intestine. (Contrast with ileum, jejunum.)

Duplication (genetic)   A mutation resulting from the introduction of an extra copy of a segment of a gene or chromosome. (Contrast with deletion, point mutation.)

Dynein [Gr. *dunamis*: power]   A protein that undergoes conformational changes and thus plays a part in the movement of eukaryotic flagella and cilia.

Ecdysone (eck die' sone) [Gr. *ek*: out of + *dyo*: to clothe]   In insects, a hormone that induces molting.

Ecological biogeography   The study of the distributions of organisms from an ecological perspective, usually concentrating on migration, dispersal, and species interactions.

Ecological community   The species living together at a particular site.

Ecological niche (nitch) [L. *nidus*: nest]   The functioning of a species in relation to other species and its physical environment.

Ecology [Gr. *oikos*: house + *logos*: discourse, study]   The scientific study of the interaction of organisms with their environment, including both the physical environment and the other organisms that live in it.

Ecosystem (eek' oh sis tum)   The organisms of a particular habitat, such as a pond or forest, together with the physical environment in which they live.

Ecto- (eck' toh) [Gr.: outer, outside]   A prefix used to designate a structure on the outer surface of the body. For example, ectoderm. (Contrast with endo- and meso-.)

Ectoderm [Gr. *ektos*: outside + *derma*: skin]   The outermost of the three embryonic tissue layers first delineated during gastrulation. Gives rise to the skin, sense organs, nervous system, etc.

Ectotherm [Gr. *ektos*: outside + *thermos*: heat]   An animal that depends on environmental rather than metabolic sources of heat to regulate its body temperature. (Contrast with endotherm.)

Edema (i dee' mah) [Gr. *oidema*: swelling]   Tissue swelling caused by the accumulation of fluid.

Edge effect   The changes in ecological processes in a community caused by physical and biological factors originating in an adjacent community.

Effector   Any organ, cell, or organelle that moves the organism through the environment or else alters the environment to the organism's advantage. Examples include muscle, bone, and exocrine glands.

Effector cell   A lymphocyte that performs a role in the immune system without further differentiation.

Effector phase   In this phase of the immune response, effector T cells called cytotoxic T cells attack virus-infected cells, and effector helper T cells assist B cells to differentiate into plasma cells, which release antibodies.

Efferent [L. *ex*: out + *ferre*: to bear]   Away from, as in neurons that conduct action potentials out from the central nervous system, or arterioles that conduct blood away from a structure. (Contrast with afferent.)

Egg   In all sexually reproducing organisms, the female gamete; in birds, reptiles, and some other vertebrates, a structure witin which early embryonic development occurs.

Elasticity   The property of returning quickly to a former state after a disturbance.

Electrocardiogram (EKG)   A graphic recording of electrical potentials from the heart.

Electroencephalogram (EEG)   A graphic recording of electrical potentials from the brain.

Electromyogram (EMG)   A graphic recording of electrical potentials from muscle.

Electron (e lek' tron) [L. *electrum*: amber (associated with static electricity), from Gr. *slektor*: bright sun (color of amber)]   One of the three most important fundamental particles of matter, with mass approximately 0.00055 amu and charge –1.

Electron microscope   An instrument that uses an electron beam to form images of minute structures; the transmission electron microscope is useful for thinly-sliced material, and the scanning electron microscope gives surface views of cells and organisms.

Electrophoresis (e lek' tro fo ree' sis) [L. *electrum*: amber + Gr. *phorein*: to bear]   A separation technique in which substances are separated from one another on the basis of their electric charges and molecular weights.

Electrotonic potential   In neurons, a hyperpolarization or small depolarization of the membrane potential induced by the application of a small electric current. (Contrast with action potential, resting potential.)

Elemental substance   A substance composed of only one type of atom.

Embolus (em' buh lus) [Gr. *embolos*: inserted object; stopper]   A circulating blood clot. Blockage of a blood vessel by an embolus or by a bubble of gas is referred to as an **embolism**. (Contrast with thrombus.)

Embryo [Gr. *en-*: in + *bryein*: to grow]   A young animal, or young plant sporophyte, while it is still contained within a protective structure such as a seed, egg, or uterus.

Embryo sac   In angiosperms, the female gametophyte. Found within the ovule, it consists of eight or fewer cells, membrane bounded, but without cellulose walls between them.

Emergent property   A property of a complex system that is not exhibited by its individual component parts.

Emigration   The deliberate and usually oriented departure of an organism from the habitat in which it has been living.

3' end (3-prime)   The end of a DNA or RNA strand that has a free hydroxyl group at the 3'-carbon of the sugar (deoxyribose or ribose).

5' end (5-prime)   The end of a DNA or RNA strand that has a free phosphate group at the 5'-carbon of the sugar (deoxyribose or ribose).

Endemic (en dem' ik) [Gr. *endemos*: dwelling in a place]   Confined to a particular region, thus often having a comparatively restricted distribution.

Endergonic reaction   One for which energy must be supplied. (Contrast with exergonic reaction.)

Endo- [Gr.: within, inside]   A prefix used to designate an innermost structure. For example, endoderm, endocrine. (Contrast with ecto-, meso-.)

Endocrine gland (en' doh krin) [Gr. *endon*: inside + *krinein*: to separate]   Any gland, such as the adrenal or pituitary gland of vertebrates, that secretes certain substances, especially hormones, into the body through the blood.

Endocrinology   The study of hormones and their actions.

Endocytosis   A process by which liquids or solid particles are taken up by a cell through invagination of the plasma membrane. (Contrast with exocytosis.)

Endoderm [Gr. *endon*: within + *derma*: skin]   The innermost of the three embryonic tissue layers first delineated during gastrulation. Gives rise to the digestive and respiratory tracts and structures associated with them.

Endodermis [Gr. *endon*: within + *derma*: skin]   In plants, a specialized cell layer marking the inside of the cortex in roots and some stems. Frequently a barrier to free diffusion of solutes.

Endomembrane system   Endoplasmic reticulum plus Golgi apparatus plus, when present, lysosomes; thus, a system of membranes that exchange material with one another.

Endometrium (en do mee' tree um) [Gr. *endon*: within + *metrios*: womb]   The epithelial cells lining the uterus of mammals.

Endoplasmic reticulum [Gr. *endon*: within + L. *plasma*: form; L. *reticulum*: little net]   A system of membrane-bounded tubes and flattened sacs, often continuous with the nuclear envelope, found in the cytoplasm of eukaryotes. Exists as rough ER, studded with ribosomes, and smooth ER, lacking ribosomes.

Endorphins   Naturally occurring, opiate-like substances in the mammalian brain.

Endoskeleton   A skeleton covered by other, soft body tissues. (Contrast with exoskeleton.)

Endosperm [Gr. *endon*: within + *sperma*: seed]   A specialized triploid seed tissue found only in angiosperms; contains stored food for the developing embryo.

Endosymbiosis [Gr. *endon*: within + *syn*: together + *bios*: life]   The living together of two species, with one living inside the body (or even the cells) of the other.

Endosymbiotic theory   Theory that the eukaryotic cell evolved from a prokaryote that contained other, endosymbiotic prokaryotes.

Endotherm [Gr. *endon*: within + *thermos*: hot]   An animal that can control its body temperature by the expenditure of its own metabolic energy. (Contrast with ectotherm.)

Energetic cost   The difference between the energy an animal would have expended had it rested, and that expended in performing a behavior.

Energy   The capacity to do work.

Enhancer   In eukaryotes, a DNA sequence, lying on either side of the gene it regulates, that stimulates a specific promoter.

Enterocoelous development   A pattern of development in which the coelum is formed by an outpocketing of the embryonic gut (enteron).

Enterokinase (ent uh row kine' ase)   An enzyme secreted by the mucosa of the duodenum. It activates the zymogen trypsinogen to create the active digestive enzyme trypsin.

Entrainment   With respect to circadian rhythms, the process whereby the period is adjusted to match the 24-hour environmental cycle.

Entropy (en' tro pee) [Gr. *en*: in + *tropein*: to change]   A measure of the degree of disorder in any system. A perfectly ordered system has zero entropy; increasing disorder is measured by positive entropy. Spontaneous reactions in a closed system are always accompanied by an increase in disorder and entropy.

Environment   An organism's surroundings, both living and nonliving; includes temperature, light intensity, and all other species that influence the focal organism.

Environmental toxicology   The study of the distribution and effects of toxic compounds in the environment.

Enzyme (en' zime) [Gr. *en*: in + *zyme*: yeast]   A protein, on the surface of which are chemical groups so arranged as to make the enzyme a catalyst for a chemical reaction.

Epi- [Gr.: upon, over]   A prefix used to designate a structure located on top of another; for example: epidermis, epiphyte.

Epicotyl (epp' i kot' il) [Gr. *epi*: upon + *kotyle*: something hollow]   That part of a plant embryo or seedling that is above the cotyledons.

Epidermis [Gr. *epi*: upon + *derma*: skin]   In plants and animals, the outermost cell layers. (Only one cell layer thick in plants.)

Epididymis (epuh did' uh mus) [Gr. *epi*: upon + *didymos*: testicle]   Coiled tubules in the testes that store sperm and conduct sperm from the seiminiferous tubules to the vas deferens.

Epinephrine (ep i nef' rin) [Gr. *epi*: upon + *nephros*: a kidney]   The "fight or flight" hormone. Produced by the medulla of the adrenal gland, it also functions as a neurotransmitter. Also known as adrenaline.

Epiphyte (ep' e fyte) [Gr. *epi*: upon + *phyton*: plant]   A specialized plant that grows on the surface of other plants but does not parasitize them.

Episome   A plasmid that may exist either free or integrated into a chromosome. (See plasmid.)

Epistasis   An interaction between genes, in which the presence of a particular allele of one gene determines whether another gene will be expressed.

Epithelium   In animals, a layer of cells covering or lining an external surface or a cavity.

Equilibrium   (1) In biochemistry, a state in which forward and reverse reactions are proceeding at counterbalancing rates, so there is no observable change in the concentrations of reactants and products. (2) In evolutionary genetics, a condition in which allele and genotype frequencies in a population are constant from generation to generation.

Error signal   In physiology, the difference between a set point and a feedback signal that results in a corrective response.

Erythrocyte (ur rith' row sight) [Gr. *erythros*: red + *kytos*: hollow vessel]   A red blood cell.

Esophagus (i soff' i gus) [Gr. *oisophagos*: gullet]   That part of the gut between the pharynx and the stomach.

Essential amino acid   An amino acid an animal cannot synthesize for itself and must obtain from its diet.

Essential element   An irreplaceable mineral element without which normal growth and reproduction cannot proceed.

Estivation (ess tuh vay' shun) [L. *aestivalis*: summer]   A state of dormancy and hypometabolism that occurs during the summer; usually a means of surviving drought and/or intense heat. Contrast with hibernation.

Estrogen   Any of several steroid sex hormones, produced chiefly by the ovaries in mammals.

Estrous cycle   The cyclical changes in reproductive physiology and behavior in female mammals (other than some primates), culminating in estrus.

Estrus (es' truss) [L. *oestrus*: frenzy]   The period of heat, or maximum sexual receptivity, in some female mammals. Ordinarily, the estrus is also the time of release of eggs in the female.

Ethology (ee thol' o jee) [Gr. *ethos*: habit, custom + *logos*: discourse]   The study of whole patterns of animal behavior in natural environments, stressing the analysis of adaptation and evolution of the patterns.

Ethylene   One of the plant hormones, the gas $H_2C = 2CH_2$.

Etiolation   Plant growth in the absence of light.

Eubacteria (yew bacteria)   Kingdom including the great majority of bacteria, such as the gram negative bacteria, gram positive bacteria, mycoplasmas, etc. (Contrast with Archaebacteria.)

Euchromatin Chromatin that is diffuse and non-staining during interphase; may be transcribed. (Contrast with heterochromatin.)

Eukaryotes (yew car' ry otes) [Gr. *eu*: true + *karyon*: kernel or nucleus] Organisms whose cells contain their genetic material inside a nucleus. Includes all life other than the viruses, Archaebacteria, and Eubacteria.

Eusocial Term applied to insects, such as termites, ants, and many bees and wasps, in which individuals cooperate in the care of offspring, there are sterile castes, and generations overlap.

Eutrophication (yoo trofe' ik ay' shun) [Gr. *eu-*: well + *trephein*: to flourish] The addition of nutrient materials to a body of water, resulting in changes to species composition therein.

Evergreen A plant that retains its leaves through all seasons. (Contrast with deciduous.)

Evolution Any gradual change. Organic evolution, often referred to as evolution, is any genetic and resulting phenotypic change in organisms from generation to generation.

Evolutionary agent Any factor that influences the direction and rate of evolutionary changes.

Evolutionary biology The collective branches of biology that study evolutionary process and their products—the diversity and history of living things.

Evolutionarily conservative Traits of organisms that evolve very slowly.

Evolutionary innovations Major changes in body plans of organisms; these have been very rare during evolutionary history.

Evolutionary radiation The proliferation of species within a single evolutionary lineage.

Excision repair The removal and damaged DNA and its replacement by the appropriate nucleotides. Often, several bases on either side of the damaged base are removed by the action of an endonuclease. Then a DNA polymerase adds the correct bases according to the template still present on the other strand of DNA. DNA ligase catalyzes the sealing up of the repaired strand.

Excitatory postsynaptic potential (EPSP) A change in the resting potential of a postsynaptic membrane in a positive (depolarizing) direction. (Contrast with inhibitory postsynaptic potential.)

Excretion Release of metabolic wastes by an organism.

Exergonic reaction A reaction in which free energy is released. (Contrast with endergonic reaction.)

Exo- (eks' oh) Same as ecto-.

Exocrine gland (eks' oh krin) [Gr. *exo*: outside + *krinein*: to separate] Any gland, such as a salivary gland, that secretes to the outside of the body or into the gut.

Exocytosis A process by which a vesicle within a cell fuses with the plasma membrane and releases its contents to the outside. (Contrast with endocytosis.)

Exon A portion of a DNA molecule, in eukaryotes, that codes for part of a polypeptide. (Contrast with intron.)

Exoskeleton (eks' oh skel' e ton) A hard covering on the outside of the body to which muscles are attached. (Contrast with endoskeleton.)

Experiment A scientific method in which particular factors are manipulated while other factors are held constant so that the potential influences of the manipulated factors can be determined.

Exploitation competition Competition that occurs because resources are depleted. (Contrast with interference competition.)

Exponential growth Growth, especially in the number of organisms in a population, which is a simple function of the size of the growing entity: the larger the entity, the faster it grows. (Contrast with logistic growth.)

Expression vector A DNA vector, such as a plasmid, that carries a DNA sequence that includes the adjacent sequences for its expression into mRNA and protein in a host cell.

Expressivity The degree to which a genotype is expressed in the phenotype— may be affected by the environment.

Extensor A muscle that extends an appendage.

Extinction The termination of a lineage of organisms.

Extrinsic protein A membrane protein found only on the surface of the membrane. (Contrast with intrinsic protein.)

F-duction Transfer of genes from one bacterium to another, using the F-factor as a vehicle.

F-factor In some bacteria, the fertility factor; a plasmid conferring "maleness" on the cell that contains it.

F_1 generation The immediate progeny of a parental (P) mating; the first filial generation.

F_2 generation The immediate progeny of a mating between members of the F_1 generation.

Facilitated diffusion Passive movement through a membrane involving a specific carrier protein; does not proceed against a concentration gradient. (Contrast with active transport, free diffusion.)

Facultative Capable of occurring or not occurring, as in facultative aerobes. (Contrast with obligate.)

Family In taxonomy, the category below the order and above the genus; a group of related, similar genera.

Fat A triglyceride that is solid at room temperature. (Contrast with oil.)

Fatty acid A molecule with a long hydrocarbon tail and a carboxyl group at the other end. Found in many lipids.

Fauna (faw' nah) All of the animals found in a given area. (Contrast with flora.)

Feces [L. *faeces*: dregs] Waste excreted from the digestive system.

Feedback control Control of a particular step of a multistep process, induced by the presence or absence of a product of one of the later steps. A thermostat regulating the flow of heating oil to a furnace in a home is a negative feedback control device.

Fermentation (fur men tay' shun) [L. *fermentum*: yeast] The degradation of a substance such as glucose to smaller molecules with the extraction of energy, without the use of oxygen (i.e., anaerobically). Involves the glycolytic pathway.

Fertilization Union of gametes. Also known as syngamy.

Fertilization membrane A membrane surrounding an animal egg which becomes rapidly raised above the egg surface within seconds after fertilization, serving to prevent entry of a second sperm.

Fetus The latter stages of an embryo that is still contained in an egg or uterus; in humans, the unborn young from the eighth week of pregnancy to the moment of birth.

Fiber An elongated and tapering cell of vascular plants, usually with a thick cell wall. Serves a support function.

Fibrin A protein that polymerizes to form long threads that provide structure to a blood clot.

Filter feeder An organism that feeds upon much smaller organisms, that are suspended in water or air, by means of a straining device.

Filtration In the excretory physiology of some animals, the process by which the initial urine is formed; water and most solutes are transferred into the excretory tract, while proteins are retained in the blood or hemolymph.

First law of thermodynamics Energy can be neither created nor destroyed.

Fission Reproduction of a prokaryote by division of a cell into two comparable progeny cells.

Fitness The contribution of a genotype or phenotype to the composition of subsequent generations, relative to the contribution of other genotypes or phenotypes. (See inclusive fitness.)

Fixed action pattern A behavior that is genetically programmed.

Flagellin (fla jell' in) The protein from which prokaryotic (but not eukaryotic) flagella are constructed.

Flagellum (fla jell' um) (plural: flagella) [L. *flagellum*: whip] Long, whiplike appendage that propels cells. Prokaryotic flagella differ sharply from those found in eukaryotes.

Flexor A muscle that flexes an appendage.

Flora (flore' ah) All of the plants found in a given area. (Contrast with fauna.)

Florigen A plant hormone (not yet isolated) involved in the conversion of a vegetative shoot apex to a flower.

Flower   The total reproductive structure of an angiosperm; its basic parts include the calyx, corolla, stamens, and carpels.

Fluorescence   The emission of a photon of visible light by an excited atom or molecule.

Follicle [L. *folliculus*: little bag]   In female mammals, an immature egg surrounded by nutritive cells.

Follicle-stimulating hormone   A gonadotropic hormone produced by the anterior pituitary.

Food chain   A portion of a food web, most commonly a simple sequence of prey species and the predators that consume them.

Food web   The complete set of food links between species in a community; a diagram indicating which ones are the eaters and which are consumed.

Forb   Any broad-leaved (dicotyledonous), herbaceous plant. Especially applied to such plants growing in grasslands.

Fossil   Any recognizable structure originating from an organism, or any impression from such a structure, that has been preserved over geological time.

Founder effect   Random changes in allele frequencies resulting from establishment of a population by a very small number of individuals.

Fovea [L. *fovea*; a small pit]   The area, in the vertebrate retina, of most distinct vision.

Frame-shift mutation   A mutation resulting from the addition or deletion of a single base pair in the DNA sequence of a gene. As a result of this, mRNA transcribed from such a gene is translated normally until the ribosome reaches the point at which the mutation has occurred. From that point on, codons are read out of proper register and the amino acid sequence bears no resemblance to the normal sequence. (Contrast with missense mutation, nonsense mutation.)

Free diffusion   Diffusion directly across a membrane without the involvement of carrier molecules. Free diffusion is not saturable and cannot cause the net transport from a region of low concentration to a region of higher concentration. (Contrast with facilitated diffusion and active transport.)

Free energy   That energy which is available for doing useful work, after allowance has been made for the increase or decrease of disorder. Designated by the symbol G (for Gibbs free energy), and defined by: $G = H - TS$, where H = heat, S = entropy, and T = absolute (Kelvin) temperature.

Frequency-dependent selection   Selection that changes in intensity with the proportion of individuals having the trait.

Fruit   In angiosperms, a ripened and mature ovary (or group of ovaries) containing the seeds. Sometimes applied to reproductive structures of other groups of plants, and includes any adjacent parts which may be fused with the reproductive structures.

Fruiting body   A structure that bears spores.

Fundamental niche   The range of condition under which an organism could survive if it were the only one in the environment. (Contrast with realized niche.)

Fungus (fung' gus)   A member of the kingdom Fungi, a (usually) multicellular eukaryote with absorptive nutrition.

G_1 phase   In the cell cycle, the gap between the end of mitosis and the onset of the S phase.

G_2 phase   In the cell cycle, the gap between the S (synthesis) phase and the onset of mitosis.

G protein   A membrane protein involved in signal transduction; characterized by binding guanyl nucleotides. The activation of certain receptors activates the G protein, which in turn activates adenylate cyclase. G protein activation involves binding a GTP molecule in place of a GDP molecule.

Gametangium (gam i tan' gee um) [Gr. *gamos*: marriage + *angeion*: vessel or reservoir]   Any plant or fungal structure within which a gamete is formed.

Gamete (gam' eet) [Gr. *gamete*: wife, *gametes*: husband]   The mature sexual reproductive cell: the egg or the sperm.

Gametocyte (ga meet' oh site) [Gr. *gamete*: wife, *gametes*: husband + *kytos*: cell]   The cell that gives rise to sex cells, either the eggs or the sperm. (See oocyte and spermatocyte.)

Gametogenesis (ga meet' oh jen' e sis) [Gr. *gamete*: wife, *gametes*: husband + *genesis*: source]   The specialized series of cellular divisions that leads to the production of sex cells (gametes). (Contrast with oogenesis and spermatogenesis.)

Gametophyte (ga meet' oh fyte)   In plants with alternation of generations, the haploid phase that produces the gametes. (Contrast with sporophyte.)

Ganglion (gang' glee un) [Gr.: tumor]   A group or concentration of neuron cell bodies.

Gap junction   A 2.7-nanometer gap between plasma membranes of two animal cells, spanned by protein channels. Gap junctions allow chemical substances or electrical signals to pass from cell to cell.

Gas exchange   In animals, the process of taking up oxygen from the environment and releasing carbon dioxide to the environment.

Gastrovascular cavity   Serving for both digestion (gastro) and circulation (vascular); in particular, the central cavity of the body of jellyfish and other cnidarians.

Gastrula (gas' true luh) [Gr. *gaster*: stomach]   An embryo forming the characteristic three cell layers (ectoderm, endoderm, and mesoderm) which will give rise to all of the major tissue systems of the adult animal.

Gastrulation   Development of a blastula into a gastrula.

Gated channel   A channel (membrane protein) that opens and closes in response to binding of specific molecules or to changes in membrane potential.

Gel electrophoresis (jel ul lec tro for' eesis)   A semisolid matrix suspended in a salty buffer in which molecules can be separated on the basis of their size and change when current is passed through the gel.

Gene [Gr. *gen*: to produce]   A unit of heredity. Used here as the unit of genetic function which carries the information for a single polypeptide.

Gene amplification   Creation of multiple copies of a particular gene, allowing the production of large amounts of the RNA transcript (as in rRNA synthesis in oocytes).

Gene cloning   Formation of a clone of bacteria or yeast cells containing a particular foreign gene.

Gene family   A set of identical, or once-identical, genes, derived from a single parent gene; need not be on the same chromosomes; classic example is the globin family in vertebrates.

Gene flow   The exchange of genes between different species (an extreme case referred to as hybridization) or between different populations of the same species caused by migration following breeding.

Gene pool   All of the genes in a population.

Gene therapy   Treatment of a genetic disease by providing patients with cells containing wild type alleles for the genes that are nonfunctional in their bodies.

Generalized transduction   The transfer of any bacterial host gene in a virus particle to another bacterium.

Generative nucleus   In a pollen tube, a haploid nucleus that undergoes mitosis to produce the two sperm nuclei that participate in double fertilization. (Contrast with tube nucleus.)

Generator potential   A stimulus-induced change in membrane resting potential in the direction of threshold for generating action potentials.

Genet   The genetic individual of a plant that is composed of a number of nearly identical but repeated units.

Genetic drift   Changes in gene frequencies from generation to generation in a small population as a result of random processes.

Genetic stochasticity   Variation in the frequencies of alleles and genotypes in a population over time.

Genetics   The study of heredity.

Genetic structure   The frequencies of alleles and genotypes in a population.

Genome (jee' nome)   The genes in a complete haploid set of chromosomes.

Genotype (jean' oh type) [Gr. *gen*: to produce + *typos*: impression]   An exact description of the genetic constitution of an individual, either with respect to a single trait or with respect to a larger set of traits. (Contrast with phenotype.)

Genus (jean' us) (plural: genera) [Gr. *genos*: stock, kind]   A group of related, similar species.

Geographical (allopatric) speciation Formation of two species from one by the interposition of (or crossing of) a physical barrier. (Contrast with parapatric, sympatric speciation.)

Geotropism See gravitropism.

Germ cell A reproductive cell or gamete of a multicellular organism.

Germination The sprouting of a seed or spore.

Gestation (jes tay' shun) [L. *gestare*: to bear] The period during which the embryo of a mammal develops within the uterus. Also known as **pregnancy**.

Gibberellin (jib er el' lin) [L. *gibberella*: hunchback (refers to shape of a reproductive structure of a fungus that produces gibberellins)] One of a class of plant growth substances playing roles in stem elongation, seed germination, flowering of certain plants, etc. Named for the fungus *Gibberella*.

Gill An organ for gas exchange in aquatic organisms.

Gill arch A skeletal structure that supports gill filaments and the blood vessels that supply them.

Gizzard (giz' erd) [L. *gigeria*: cooked chicken parts] A very muscular port of the stomach of birds that grinds up food, sometimes with the aid of fragments of stone.

Gland An organ or group of cells that produces and secretes one or more substances.

Glans penis Sexually sensitive tissue at the tip of the penis.

Glia (glee' uh) [Gr.: glue] Cells, found only in the nervous system, which do not conduct action potentials.

Glomerulus (glo mare' yew lus) [L. *glomus*: ball] Sites in the kidney where blood filtration takes place. Each glomerulus consists of a knot of capillaries served by afferent and efferent arterioles.

Glucocorticoids Steroid hormones produced by the adrenal cortex. Secreted in response to ACTH, they inhibit glucose uptake by many tissues in addition to mediating other stress responses.

Glucagon A hormone produced and released by cells in the islets of Langerhans of the pancreas. It stimulates the breakdown of glycogen in liver cells.

Gluconeogenesis The biochemical synthesis of glucose from other substances, such as amino acids, lactate, and glycerol.

Glucose (glue' kose) [Gr. *gleukos*: sweet wine mash for fermentation] The most common sugar, one of several monosaccharides with the formula $C_6H_{12}O_6$.

Glycerol (gliss' er ole) A three-carbon alcohol with three hydroxyl groups, the linking component of phospholipids and triglycerides.

Glycogen (gly' ko jen) A branched-chain polymer of glucose, similar to starch (which is less branched and may be of lower molecular weight). Exists mostly in liver and muscle; the principal storage carbohydrate of most animals and fungi.

Glycolysis (gly kol' li sis) [from glucose + Gr. *lysis*: loosening] The enzymatic breakdown of glucose to pyruvic acid. One of the oldest energy-yielding machanisms in living organisms.

Glycosidic linkage The connection in an oligosaccharide or polysaccharide chain, formed by removal of water during the linking of monosaccharides.by root pressure.

Glyoxysome (gly ox' ee soam) An organelle found in plants, in which stored lipids are converted to carbohydrates.

Golgi apparatus (goal' jee) A system of concentrically folded membranes found in the cytoplasm of eukaryotic cells. Plays a role in the production and release of secretory materials such as the digestive enzymes manufactured in the pancreas. First described by Camillo Golgi (1844–1926).

Gonad (go' nad) [Gr. *gone*: seed, that which produces seed] An organ that produces sex cells in animals: either an ovary (female gonad) or testis (male gonad).

Gonadotropin A hormone that stimulates the gonads.

Grade The level of complexity found in an animal's body plan.

Gram stain A differential stain useful in characterizing bacteria.

Granum Within a chloroplast, a stack of thylakoids.

Gravitropism A directed plant growth response to gravity.

Grazer An animal that eats the vegetative tissues of herbaceous plants.

Green gland An excretory organ of crustaceans.

Gross morphology The sizes and shapes of the major body parts of a plant or animal.

Gross primary production The total energy captured by plants growing in a particular area.

Ground meristem That part of an apical meristem that gives rise to the ground tissue system of the primary plant body.

Ground tissue system Those parts of the plant body not included in the dermal or vascular tissue systems. Ground tissues function in storage, photosynthesis, and support.

Groundwater Water present deep in soils and rocks; may be stationary or flow slowly eventually to discharge into lakes, rivers, or oceans.

Group transfer The exchange of atoms between molecules.

Growth Irreversible increase in volume (probably the most accurate definition, but at best a dangerous oversimplification).

Growth factors A group of proteins that circulate in the blood and trigger the normal growth of cells. Each growth factor acts only on certain target cells.

Growth stage That stage in the life history of an organism in which it grows to its adult size.

Guanine (gwan'een) A nitrogen-containing base found in DNA, RNA, and GTP.

Guard cells In plants, paired epidermal cells which surround and control the opening of a stoma (pore).

Gut An animal's digestive tract.

Guttation The extrusion of liquid water through openings in leaves, caused by root pressure.

Gymnosperm (jim' no sperm) [Gr. *gymnos*: naked + *sperma*: seed] A plant, such as a pine or other conifer, whose seeds do not develop within an ovary (hence, the seeds are "naked").

Gyrus (plural: gyri) The raised or ridged portion of the convoluted surface of the brain. (Contrast to sulcus.)

Habit The form or pattern of growth characteristic of an organism.

Habitat The environment in which an organism lives.

Habituation (ha bich' oo ay shun) The simplest form of learning, in which an animal presented with a stimulus without reward or punishment eventually ceases to respond.

Hair cell A type of mechanosensor in animals.

Half-life The time required for half of a sample of a radioactive isotope to decay to its stable, nonradioactive form.

Halophyte (hal' oh fyte) [Gr. *halos*: salt + *phyton*: plant] A plant that grows in a saline (salty) environment.

Haploid (hap' loid) [Gr. *haploeides*: single] Having a chromosome complement consisting of just one copy of each chromosome. This is the normal "ploidy" of gametes or of asexual spores produced by meiosis or of organisms (such as the gametophyte generation of plants) that grow from such spores without fertilization.

Haplontic life cycle A life cycle in which the zygote is the only diploid cell.

Hardy–Weinberg rule The rule that the basic processes of Mendelian heredity (meiosis and recombination) do not alter either the frequencies of genes or their diploid combinations. The Law also states how the percentages of diploid combinations can be predicted from a knowledge of the proportions of alleles in the population.

Haustorium (haw stor' ee um) [L. *haustus*: draw up] A specialized hypha or other structure by which fungi and some parasitic plants draw food from a host plant.

Haversian systems Units of organization in compact bone that reflect the action of intercommunicating osteoblasts.

Helicase (heel' uh case) An enzyme that unwinds the DNA double helix for DNA relplication.

Helper T cells T cells that participate in the activation of B cells and of other T cells; targets of the HIV-I virus, the agent of AIDS. (Contrast with cytotoxic T cells, suppressor T cells.)

Hematocrit (heme at o krit) [Gr. *haima*: blood + *krites*: judge]   The proportion of 100 cc of blood that consists of red blood cells.

Hemizygous (hem' ee zie' gus) [Gr. *hemi*: half + *zygotos*: joined]   In a diploid organism, having only one allele for a given trait, typically the case for X-linked genes in male mammals and Z-linked genes in female birds. (Contrast with homozygous, heterozygous.)

Hemoglobin (hee' mo glow' bin) [Gr. *haima*: blood + L. *globus*: globe]   The colored protein of vertebrate blood (and blood of some invertebrates) which transports oxygen.

Hepatic (heh pat' ik) [Gr. *hepar*: liver] Pertaining to the liver.

Hepatic duct   The duct that conveys bile from the liver to the gallbladder.

Herbicide (ur' bis ide)   A chemical substance that kills plants.

Herbivore [L. *herba*: plant + *vorare*: to devour]   An animal which eats the tissues of plants. (Contrast with carnivore, detritivore, omnivore.)

Heritable   Able to be inherited; in biology usually refers to genetically determined traits.

Hermaphroditism (her maf' row dite' ism) [Gr. *hermaphroditos*: a person with both male and female traits]   The coexistence of both female and male sex organs in the same organism.

Hertz (abbreviated as Hz)   Cycles per second.

Hetero- [Gr.: other, different]   A prefix used in biology to mean that two or more different conditions are involved; for example, heterotroph, heterozygous.

Heterochromatin   Chromatin that retains its coiling during interphase; generally not transcribed. (Contrast with euchromatin.)

Heterocyst   A large, thick-walled cell in the filaments of certain cyanobacteria; performs nitrogen fixation.

Heterogeneous nuclear RNA (hnRNA) The product of transcription of a eukaryotic gene, including transcripts of introns.

Heterokaryon (het' er oh care' ee ahn) [Gr. *heteros*: different + *karyon*: kernel]   A cell or organism carrying a mixture of genetically distinguishable nuclei. A heterokaryon is usually the result of the fusion of two cells without fusion of their nuclei.

Heteromorphic (het' er oh more' fik) [Gr. *heteros*: different + *morphe*: form]   having a different form or appearance, as two heteromorphic life stages of a plant. (Contrast with isomorphic.)

Heterosporous (het' er os' por us) Producing two types of spores, one of which gives rise to a female megaspore and the other to a male microspore. Heterosporous plants produce distinct female and male gametophytes. (Contrast with homosporous.)

Heterotherm   An animal that regulates its body temperature at a constant level at some times but not others, such as a hibernator.

Heterotroph (het' er oh trof) [Gr. *heteros*: different + *trophe*: food]   An organism that requires preformed organic molecules as food. (Contrast with autotroph.)

Heterozygous (het' er oh zie' gus) [Gr. *heteros*: different + *zygotos*: joined]   Of a diploid organism having different alleles of a given gene on the pair of homologues carrying that gene. (Contrast with homozygous.)

Hfr (for "high frequency of recombination")   Donor bacterium in which the F-factor has been integrated into the chromosome. This produces a bacterium that transfers its chromosomal markers at a very high frequency to recipient (F⁻) cells.

Hibernation [L. *hibernus*: winter]   The state of inactivity of some animals during winter; marked by a drop in body temperature and metabolic rate.

Highly repetitive DNA   Short DNA sequences present in millions of copies in the genome, next to each other (in tandem). In a In a reassociation experiment, denatured highly repetitive DNA reanneals very quickly.

Hippocampus   A part of the forebrain that takes part in long-term memory formation.

Histamine (hiss' tah meen)   A substance released within a damaged tissue by a type of white blood cell. Histamines are responsible for aspects of allergice reactions, including the increased vascular permeability that leads to edema (swelling).

Histology   The study of tissues.

Histone   Any one of a group of basic proteins forming the core of a nucleosome, the structural unit of a eukaryotic chromosome. (See nucleosome.)

Historical biogeography   The study of the distributions of organisms from a long-term, historical perspective.

hnRNA   See heterogeneous nuclear RNA.

Holdfast   In many large attached algae, specialized tissue attaching the plant to its substratum.

Homeobox   A segment of DNA, found in a few genes, perhaps regulating the expression of other genes and thus controlling large-scale developmental processes.

Homeostasis (home' ee o sta' sis) [Gr. *homos*: same + *stasis*: position]   The maintenance of a steady state, such as a constant temperature or a stable social structure, by means of physiological or behavioral feedback responses.

Homeotherm (home' ee o therm) [Gr. *homos*: same + *therme*: heat]   An animal which maintains a constant body temperature by virtue of its own heating and cooling mechanisms. (Contrast with heterotherm, poikilotherm.)

Homeotic genes (home' ee ott' ic)   Genes that determine what entire segments of an animal become.

Homeotic mutation   A drastic mutation causing the transformation of body parts in *Drosophila* metamorphosis. Examples include the *Antennapedia* and *ophthalmoptera* mutants.

Homolog (home' o log') [Gr. *homos*: same + *logos*: word]   One of a pair, or larger set, of chromosomes having the same overall genetic composition and sequence. In diploid organisms, each chromosome inherited from one parent is matched by an identical (except for mutational changes) chromosome—its homolog—from the other parent.

Homology (ho mol' o jee) [Gr. *homologi(a)*: agreement]   A similarity between two structures that is due to inheritance from a common ancestor. The structures are said to be homologous. (Contrast with analogy.)

Homoplasy (home' uh play zee) [Gr. *homos*: same + *plastikos*: to mold]   The presence in several species of a trait not present in their most common ancestor. Can result from convergent evolution, reverse evolution, or parallel evolution.

Homosporous   Producing a single type of spore that gives rise to a single type of gametophyte, bearing both female and male reproductive organs. (Contrast with heterosporous.)

Homozygous (home' o zie' gus) [Gr. *homos*: same + *zygotos*: joined]   Of a diploid organism having identical alleles of a given gene on both homologous chromosomes. An organism may be a "homozygote" with respect to one gene and, at the same time, a "heterozygote" with respect to another. (Contrast with heterozygous.)

Hormone (hore' mone) [Gr. *hormon*: excite, stimulate]   A substance produced in one part of a multicellular organism and transported to another part where it exerts its specific effect on the physiology or biochemistry of the target cells.

Host   An organism that harbors a parasite and provides it with nourishment.

Host–parasite interaction   The dynamic interaction between populations of a host and the parasites that attack it.

Humoral immune system   The part of the immune system mediated by B cells; it is mediated by circulating antibodies and is active against extracellular bacterial and viral infections.

Humus (hew' muss)   The partly decomposed remains of plants and animals on the surface of a soil. Its characteristics depend primarily upon climate and the species of plants growing on the site.

Hyaluronidase (hill yew ron' uh dase)   An enzyme that digests proteoglycans. Found in sperm cells, it helps digest the coatings surrounding an egg so the sperm can penetrate the egg cell membrane.

Hybrid (high' brid) [L. *hybrida*: mongrel] The offspring of genetically dissimilar parents. In molecular biology, a double helix formed of nucleic acids from different sources.

Hybridoma   A cell produced by the fusion of an antibody-producing cell with a myeloma cell; it produces monoclonal antibodies.

Hydrocarbon   A compound containing only carbon and hydrogen atoms.

Hydrogen bond A chemical bond which arises from the attraction between the slight positive charge on a hydrogen atom and a slight negative charge on a nearby fluorine, oxygen, or nitrogen atom. Weak bonds, but found in great quantities in proteins, nucleic acids, and other biological macromolecules.

Hydrological cycle The sum total of movement of water from the oceans to the atmosphere, to the soil, and back to the oceans. Some water is cycled many times within compartments of the system before completing one full circuit.

Hydrolyze (hi' dro lize) [Gr. *hydro*: water + *lysis*: cleavage] To break a chemical bond, as in a peptide linkage, with the insertion of the components of water, –H and –OH, at the cleaved ends of a chain. The digestion of proteins is a hydrolysis.

Hydrophilic [Gr. *hydro*: water + *philia*: love] Having an affinity for water. (Contrast with hydrophobic.)

Hydrophobic [Gr. *hydro*: water + *phobia*: fear] Molecules and amino acid side chains, which are mainly hydrocarbons (compounds of C and H with no charged groups or polar groups), have a lower energy when they are clustered together than when they are distributed through an aqueous solution. Because of their attraction for one another and their reluctance to mix with water they are called "hydrophobic." Oil is a hydrophobic substance; phenylalanine is a hydrophobic amino acid. (Contrast with hydrophilic.)

Hydrophobic interaction A weak attraction between highly nonpolar molecules or parts of molecules suspended in water.

Hydrostatic skeleton The incompressible internal liquids of some animals that transfer forces from one part of the body to another when acted upon by the surrounding muscles.

Hydroxyl group The —OH group, characteristic of alcohols.

Hyperosmotic Having a more negative osmotic potential, as a result of having a higher concentration of osmotically active particles. Said of one solution as compared with another. (Contrast with hypoosmotic, isosmotic.)

Hyperpolarization A change in the resting potential of a membrane so the inside of a cell becomes more electronegative. (Contrast with depolarization.)

Hypersensitive response A defensive response of plants to microbial infection; it results in a "dead spot."

Hypertension High blood pressure.

Hypha (high' fuh) (plural: hyphae) [Gr. *hyphe*: web] In the fungi, any single filament. May be multinucleate (zygomycetes, ascomycetes) or multicellular (basidiomycetes).

Hypocotyl That part of the embryonic or seedling plant shoot that is below the cotyledons.

Hypoosmotic Having a less negative osmotic potential, as a result of having a lower concentration of osmotically active particles. Said of one solution as compared with another. (Contrast with hyperosmotic, isosmotic.)

Hypothalamus The part of the brain lying below the thalamus; it coordinates water balance, reproduction, temperature regulation, and metabolism.

Hypothetico-deductive method A method of science in which hypotheses are erected, predictions are made from them, and experiments and observations are performed to test the predictions. The process may be repeated many times in the course of answering a question.

Icosahedron (eye kos a heed' ron) A 20-sided crystal. Some viruses have coat proteins which form a icosahedron.

Imaginal disc In insect larvae, groups of cells that develop into specific adult organs.

Imbibition [L. *imbibo*: to drink] The binding of a solvent to another molecule. Dry starch and protein will imbibe water.

Immune system A system in mammals that recognizes and eliminates or neutralizes either foreign substances or self substances that have been altered to appear foreign.

Immunization The deliberate introduction of antigen to bring about an immune response.

Immunoglobulins A class of proteins, with a characteristic structure, active as receptors and effectors in the immune system.

Immunological memory Certain clones of immune system cells made to respond to an antigen persist. This leads to a more rapid and massive response of the immune system to any subsequenct exposure to that antigen.

Immunological tolerance A mechanism by which an animal does not mount an immune response to the antigenic determinants of its own macromolecules.

Imprinting A rapid form of learning, in which an animal comes to make a particular response, which is maintained for life, to some object or other organism.

Inclusive fitness The sum of an individual's own fitness (the effect of producing its own offspring: the individual selection component) plus its influence on fitness in relatives other than direct descendants (the kin selection component).

Incomplete dominance Condition in which the heterozygous phenotype is intermediate between the two homozygous phenotypes.

Incomplete metamorphosis Insect development in which changes between instars are gradual.

Incus (in' kus) [L. *incus*: anvil] The middle of the three bones that conduct movements of the eardrum to the oval window of the inner ear. (See malleus, stapes.)

Individual fitness That component of inclusive fitness that results from an organism producing its own offspring. (Contrast with kin selection component.)

Indoleacetic acid See auxin.

Induced fit A change in the tertiary structures of some enzymes, caused by binding of substrate to the active site.

Inducer (1) In enzyme systems, a small molecule which, when added to a growth medium, causes a large increase in the level of some enzyme. (2) In embryology, a substance that causes a group of target cells to differentiate in a particular way.

Inducible enzyme An enzyme that is present in much larger amounts when a particular compound (the inducer) has been added to the system. (Contrast with constitutive enzyme.)

Inflammation A nonspecific defense against pathogens; characterized by redness, swelling, pain, and increased temperature.

Inflorescence A structure composed of several flowers.

Inhibitor A substance which binds to the surface of an enzyme and interferes with its action on its substrates.

Inhibitory postsynaptic potential A change in the resting potential of a postsynaptic membrane in the hyperpolarizing (negative) direction.

Initiation complex Combination of a ribosomal light subunit, an mRNA molecule, and the tRNA charged with the first amino acid coded for by the mRNA; formed at the onset of translation.

Initiation factors Proteins that assist in forming the translation initiation complex at the ribosome.

Inositol triphosphate (IP3) An intracellular second messenger derived from membrane phospholipids.

Insertion sequence A large piece of DNA that can give rise to copies at other loci; a type of transposable genetic element.

Instar (in' star) [L.: image, form] An immature stage of an insect between molts.

Instinct Behavior that is relatively highly sterotyped and self-differentiating, that develops in individuals unable to observe other individuals performing the behavior or to practice the behavior in the presence of the objects toward which it is usually directed.

Insulin (in' su lin) [L. *insula*: island] A hormone, synthesized in islet cells of the pancreas, that promotes the conversion of glucose to the storage material, glycogen.

Integrase An enzyme that integrates retroviral cDNA into the genome of the host cell.

Integrated pest management A method of control of pests in which natural predators and parasites are used in conjunction with sparing use of chemical methods to achieve control of a pest without causing serious adverse environmental side effects.

Integument [L. *integumentum*: covering] A protective surface structure. In gymnosperms and angiosperms, a layer of tissue around the ovule which will become the seed coat. Gymnosperm ovules have one integument, angiosperm ovules two.

Intention movement The preparatory motions that animals go through prior to a complete behavior response; for example, the crouch before flying, the snarl before biting, etc.

Intercalary meristem A meristematic region in plants which occurs not apically, but between two regions of mature tissue. Intercalary meristems occur in the nodes of grass stems, for example.

Intercostal muscles Muscles between the ribs that can augment breathing movements by elevating and suppressing the rib cage.

Interference competition Competition resulting from direct behavioral interactions between organisms. (Contrast with exploitation competition.)

Interferon A glycoprotein produced by virus-infected animal cells; increases the resistance of neighboring cells to the virus.

Interkinesis The phase between the first and second meiotic divisions.

Interleukins Regulatory proteins, produced by macrophages and lymphocytes, that act upon other lymphocytes and direct their development.

Intermediate filaments Fibrous proteins that stabilize cell structure and resist tension.

Internode Section between two nodes of a plant stem.

Interphase The period between successive nuclear divisions during which the chromosomes are diffuse and the nuclear envelope is intact. It is during this period that the cell is most active in transcribing and translating genetic information.

Interspecific competition Competition between members of two or more species.

Intertropical convergence zone The tropical region where the air rises most strongly; moves north and south with the passage of the sun overhead.

Intraspecific competition Competition among members of a single species.

Intrinsic protein A membrane protein that is embedded in the phospholipid bilayer of the membrane. (Contrast with extrinsic protein.)

Intrinsic rate of increase The rate at which a population can grow when its density is low and environmental conditions are highly favorable.

Intron A portion of a DNA molecule that, because of RNA splicing, is not involved in coding for part of a polypeptide molecule. (Contrast with exon.)

Invagination An infolding.

Inversion (genetic) A rare mutational event that leads to the reversal of the order of genes within a segment of a chromosome, as if that segment had been removed from the chromosome, turned 180°, and then reattached.

Invertebrate Any animal that is not a vertebrate, that is, whose nerve cord is not enclosed in a backbone of bony segments.

In vitro [L.: in glass] In a test tube, rather than in a living organism. (Contrast with in vivo.)

In vivo [L.: in the living state] In a living organism. Many processes that occur in vivo can be reproduced in vitro with the right selection of cellular components. (Contrast with in vitro.)

Ion (eye' on) [Gr.: wanderer] An atom or group of atoms with electrons added or removed, giving it a negative or positive electrical charge.

Ionic channel A membrane protein that can let ions pass across the membrane. The channel can be ion-selective, and it can be voltage-gated or ligand-gated.

Ionic bond A chemical bond which arises from the electrostatic attraction between positively and negatively charged ions. Usually a strong bond.

Iris (eye' ris) [Gr. iris: rainbow] The round, pigmented membrane that surrounds the pupil of the eye and adjusts its aperture to regulate the amount of light entering the eye.

Irruption A rapid increase in the density of a population. Often followed by massive emigration.

Islets of Langerhans Clusters of hormone-producing cells in the pancreas.

Isogamy (eye sog' ah mee) [Gr. isos: equal + gamos: marriage] A kind of sexual reproduction in which the gametes (or gametangia) are not distinguishable on the basis of size or morphology.

Isolating mechanism Geographical, physiological, ecological, or behavioral mechanisms that lead to a reduction in the frequency of hybrid matings.

Isomers Molecules consisting of the same numbers and kinds of atoms, but differing in the way in which the atoms are combined.

Isomorphic (eye' so more' fik) [Gr. isos: equal + morphe: form] having the same form or appearance, as two isomorphic life stages. (Contrast with heteromorphic.)

Isosmotic Having the same osmotic potential. Said of two solutions. (Contrast with hyperosmotic, hypoosmotic.)

Isotope (eye' so tope) [Gr. isos: equal + topos: place] Two isotopes of the same chemical element have the same number of protons in their nuclei, but differ in the number of neutrons.

Isozymes Chemically different enzymes that catalyze the same reaction.

Jejunum (jih jew' num) The middle division of the small intestine, where most absorption of nutrients occurs. (See duodenum, ileum.)

Joule (jool, or jowl) A unit of energy, equal to 0.24 calories.

Juvenile hormone In insects, a hormone maintaining larval growth and preventing maturation or pupation.

Karyotype The number, forms, and types of chromosomes in a cell.

Kelvin temperature scale See absolute temperature scale.

Keratin (ker' a tin) [Gr. keras: horn] A protein which contains sulfur and is part of such hard tissues as horn, nail, and the outermost cells of the skin.

Ketone (key' tone) A compound with a C=O group attached to two other groups, neither of which is an H atom. Many sugars are ketones. (Contrast with aldehyde.)

Keystone species A species that exerts a major influence on the composition and dynamics of the community in which it lives.

Kidneys A pair of excretory organs in vertebrates.

Kin selection component The component of inclusive fitness resulting from helping the survival of relatives containing the same alleles by descent from a common ancestor.

Kinase (kye' nase) An enzyme that transfers a phosphate group from ATP to another molecule. Protein kinases transfer phosphate from ATP to specific proteins, playing important roles in cell regulation.

Kinesis (ki nee' sis) [Gr.: movement] Orientation behavior in which the organism does not move in a particular direction with reference to a stimulus but instead simply moves at an increasing or decreasing rate until it ends up farther from the object or closer to it. (Contrast with taxis.)

Kinetochore (kin net' oh core) [Gr. kinetos: moving + khorein: to move] Specialized structure on a centromere to which microtubules attach.

Kingdom The highest taxonomic category in the Linnaean system.

Knockout mouse A genetically engineered mouse in which one or more functioning alleles have been replaced by defective alleles.

Lac operon A region of DNA in E. coli that contains a single promoter and operator controlling the expression of three adjacent genes involved in the utilization of the sugar, lactose.

Lactic acid The end product of fermentation in vertebrate muscle and some microorganisms.

Lagging strand In DNA replication, the daughter strand that is synthesized discontinuously.

Lamella Layer.

Larynx (lar' inks) A structure between the pharynx and the trachea that includes the vocal cords.

Larva (plural: larvae) [L.: ghost, early stage] An immature stage of any invertebrate animal that differs dramatically in appearance from the adult.

Lateral Pertaining to the side.

Lateral inhibition In visual information processing in the arthropod eye, the mutual inhibition of optic nerve cells; results in enhanced detection of edges.

Laterization (lat' ur iz ay shun) The formation of a nutrient-poor soil that is rich in nsoluble iron and aluminum compounds.

Law of independent assortment Alleles of different, unlinked genes assort independently of one another during gamete formation, Mendel's second law.

Law of segregation Alleles segregate from one another during gamete formation, Mendel's first law.

Leader sequence A sequence of amino acids at the N-terminal end of a newly synthesized protein, determining where the protein will be placed in the cell.

Leading strand In DNA replication, the daughter strand that is synthesized continuously.

Leaf axil The upper angle between a leaf and the stem, site of lateral buds which under appropriate circumstances become activated to form lateral branches.

Leaf primordium [L.: the beginning] A small mound of cells on the flank of a shoot apical meristem that will give rise to a leaf.

Lek A traditional courtship display ground, where males display to females.

Lenticel Spongy region in a plant's periderm, allowing gas exchange.

Leukocyte (loo' ko sight) [Gr. *leukos*: clear + *kutos*: hollow vessel] A white blood cell.

Leuteinizing hormone A peptide hormone produced by pituitary cells that stimulates follicle maturation in females.

Lichen (lie' kun) [Gr. *leikhen*: licker] An organism resulting from the symbiotic association of a true fungus and either a cyanobacterium or a unicellular alga.

Life cycle The entire span of the life of an organism from the moment of fertilization (or asexual generation) to the time it reproduces in turn.

Life history The stages an individual goes through during its life.

Life table A table showing, for a group of equal-aged individuals, the proportion still alive at different times in the future and the number of offspring they produce during each time interval.

Ligament A band of connective tissue linking two bones in a joint.

Ligand (lig' and) A molecule that binds to a receptor site of another molecule.

Lignin The principal noncarbohydrate component of wood, a polymer that binds together cellulose fibrils in some plant cell walls.

Limbic system A group of primitive vertebrate forebrain nuclei that form a network and are involved in emotions, drives, instinctive behaviors, learning, and memory.

Limiting resource The required resource whose supply most strongly influences the size of a population.

Linkage Association between genetic markers on the same chromosome such that they do not show random assortment and seldom recombine; the closer the markers, the lower the frequency of recombination.

Lipase (lip' ase; lye' pase) An enzyme that digests fats.

Lipids (lip' ids) [Gr. *lipos*: fat] Substances in a cell which are easily extracted by organic solvents; fats, oils, waxes, steroids, and other large organic molecules, including those which, with proteins, make up the cell membranes. (See phospholipids.)

Litter The partly decomposed remains of plants on the surface and in the upper layers of the soil.

Littoral zone The coastal zone from the upper limits of tidal action down to the depths where the water is thoroughly stirred by wave action.

Liver A large digestive gland. In vertebrates, it secretes bile and is involved in the formation of blood.

Lobes Regions of the human cerebral hemispheres; includes the temporal, frontal, parietal, and occipital lobes.

Locus In genetics, a specific location on a chromosome. May be considered to be synonymous with "gene."

Logistic growth Growth, especially in the size of an organism or in the number of organisms that constitute a population, which slows steadily as the entity approaches its maximum size. (Contrast with exponential growth.)

Loop of Henle (hen' lee) Long, hairpin loop of the mammalian renal tubule that runs from the cortex down into the medulla, and back to the cortex. Creates a concentration gradient in the interstitial fluids in the medulla.

Lophophore A U-shaped fold of the body wall with hollow, ciliated tentacles that encircles the mouth of animals in several different phyla. Used for filtering prey from the surrounding water.

Lordosis (lor doe' sis) [Gk. *lordosis*: curving forward] A posture assumed by females of some mammalian species (especially rodents) to signal sexual receptivity.

Lumen (loo' men) [L.: light] The cavity inside any tubular part of an organ, such as a piece of gut or a kidney tubule.

Lungs A pair of saclike chambers within the bodies of some animals, functioning in gas exchange.

Luteinizing hormone A gonadotropin produced by the anterior pituitary. It stimulates the gonads to produce sex hormones.

Lymph [L. *lympha*: water] A clear, watery fluid that is formed as a filtrate of blood; it contains white blood cells; it collects in a series of special vessels and is returned to the bloodstream.

Lymph nodes Specialized tissue regions that act as filters for cells, bacteria and foreign matter.

Lymphocyte A major class of white blood cells. Includes T cells, B cells, and other cell types important in the immune response.

Lysis (lie' sis) [Gr.: a loosening] Bursting of a cell.

Lysogenic The condition of a bacterium that carries the genome of a virus in a relatively stable form. (Contrast with lytic.)

Lysosome (lie' so soam) [Gr. *lysis*: a loosening + *soma*: body] A membrane-bounded inclusion found in eukaryotic cells (other than plants). Lysosomes contain a mixture of enzymes that can digest most of the macromolecules found in the rest of the cell.

Lysozyme (lie' so zyme) An enzyme in saliva, tears, and nasal secretions that attacks bacterial cell walls, as one of the body's nonspecific defense mechanisms.

Lytic Condition in which a bacterium lyses shortly after infection by a virus; the viral genome does not become stabilized within the bacterial cell. (Contrast with lysogenic.)

Macro- (mack' roh) [Gr. *makros*: large, long] A prefix commonly used to denote something large. (Contrast with micro-.)

Macroevolution Evolutionary changes occurring over long time spans and usually involving changes in many traits. (Contrast with microevolution.)

Macroevolutionary time The time required for macroveolutionary changes in a lineage.

Macromolecule A giant polymeric molecule. The macromolecules are proteins, polysaccharides, and nucleic acids.

Macronutrient A mineral element required by plant tissues in concentrations of at least 1 milligram per gram of their dry matter.

Macrophage (mac' roh faj) A type of white blood cell that endocytoses bacteria and other cells.

Major histocompatibility complex (MHC) A complex of linked genes, with multiple alleles, that control a number of immunological phenomena; it is important in graft rejection.

Malignant tumor A tumor whose cells can invade surrounding tissues and spread to other organs.

Malleus (mal' ee us) [L. *malleus*: hammer] The first of the three bones that conduct movements of the eardrum to the oval window of the inner ear. (See incus, stapes.)

Malpighian tubule (mal pee' gy un) A type of protonephridium found in insects.

Mammal [L. *mamma*: breast, teat] Any animal of the class Mammalia, characterized by the production of milk by the female mammary glands and the possession of hair for body covering.

Mantle A sheet of specialized tissues that covers most of the viscera of mollusks; provides protection to internal organs and secretes the shell.

Map unit In eukaryotic genetics, one map unit corresponds to a recombinant frequency of 0.01.

Mapping In genetics, determining the order of genes on a chromosome and the distances between them.

Marine [L. *mare*: sea, ocean] Pertaining to or living in the ocean. (Contrast with aquatic, terrestrial.)

Maritime climate Weather pattern typical of coasts of continents, particularly those on the western sides at mid latitudes, in which the difference between summer and winter is relatively small. (Contrast with continental climate.)

Marsupial (mar soo' pee al) A mammal belonging to the subclass Metatheria, such as opossums and kangaroos. Most have a pouch (marsupium) that contains the milk glands and serves as a receptacle for the young.

Mass extinctions Geological periods during which rates of extinction were much higher than during intervening times.

Mass number The sum of the number of protons and neutrons in an atom's nucleus.

Mast cells Typically found in connective tissue, mast cells can be provoked by antigens or inflammation to release histamine.

Maternal effect genes These genes code for morphogens that determine the polarity of the egg and larva in the fruit fly, *Drosophila melanogaster*.

Maternal inheritance (cytoplasmic inheritance) Inheritance in which the phenotype of the offspring depends on factors, such as mitochondria or chloroplasts, that are inherited from the female parent through the cytoplasm of the female gamete.

Mating type In some bacteria, fungi, and protists, sexual reproduction can occur only between partners of a different mating type. "Mating type" is not the same as "sex," since some species have as many as 8 mating types; mating may also be between hermaphroditic partners of opposite mating type, with both partners acting as both "male" and "female" in terms of donating and receiving genetic information.

Maturation The automatic development of a pattern of behavior, which becomes increasingly complex or precise as the animal matures. Unlike learning, the development does not require experience to occur.

Mechanosensor A cell that is sensitive to physical movement and generates action potentials in response.

Medulla (meh dull' luh) [L.: narrow] (1) The inner, core region of an organ, as in the adrenal medulla (adrenal gland) or the renal medulla (kidneys). (2) The portion of the brain stem that connects to the spinal cord.

Medusa (meh doo' suh) The tentacle-bearing, jellyfish-like, free-swimming sexual stage in the life cycle of a cnidarian.

Mega- [Gr. *megas*: large, great] A prefix often used to denote something large. (Contrast with micro-.)

Megareserve A large park or reserve; usually has associated buffer areas in which human use of the environment is restricted to activities that do not destroy the functioning of the ecosystem.

Megasporangium The special structure (sporangium) that produces the megaspores.

Megaspore [Gr. *megas*: large + *spora*:seed] In plants, a haploid spore that produces a female gametophyte. In many cases the

megaspore is larger than the male-producing microspore.

Meiosis (my oh' sis) [Gr.: diminution] Division of a diploid nucleus to produce four haploid daughter cells. The process consists of two successive nuclear divisions with only one cycle of chromosome replication.

Membrane potential The difference in electrical charge between the inside and the outside of a cell, caused by a difference in the distribution of ions.

Memory cells Long-lived lymphocytes produced by exposure to antigen. They persist in the body and are able to mount a rapid response to subsequent exposures to the antigen.

Mendelian population A local population of individuals belonging to the same species and exchanging genes with one another.

Menopause The time in a human female's life when the ovarian and menstrual cycles cease.

Menstrual cycle The monthly sloughing off of the uterine lining if fertilization does not occur in the female. Occurs between puberty and menopause.

Meristem [Gr. *meristos*: divided] Plant tissue made up of actively dividing cells.

Mesenchyme (mez' en kyme) [Gr. *mesos*: middle + *enchyma*: infusion] Embryonic or unspecialized cells derived from the mesoderm.

Meso- (mez' oh) [Gr.: middle] A prefix often used to designate a structure located in the middle, or a stage that appears at some intermediate time. For example, mesoderm, Mesozoic.

Mesoderm [Gr. *mesos*: middle + *derma*: skin] The middle of the three embryonic tissue layers first delineated during gastrulation. Gives rise to skeleton, circulatory system, muscles, excretory system, and most of the reproductive system.

Mesoglea The jelly-like middle layer that constitutes the bulk of the bodies of the medusae of many cnidarians; not a true cell layer.

Mesophyll (mez' a fill) [Gr. *mesos*: middle + *phyllon*: leaf] Chloroplast-containing, photosynthetic cells in the interior of leaves.

Mesosome (mez' o soam') [Gr. *mesos*: middle + *soma*: body] A localized infolding of the plasma membrane of a bacterium.

Messenger RNA (mRNA) A transcript of one of the strands of DNA, it carries information (as a sequence of codons) for the synthesis of one or more proteins.

Meta- [Gr.: between, along with, beyond] A prefix used in biology to denote a change or a shift to a new form or level; for example, as used in metamorphosis.

Metabolic compensation Changes in biochemical properties of an organism that render it less sensitive to temperature changes.

Metabolic pathway A series of enzyme-catalyzed reactions so arranged that the product of one reaction is the substrate of the next.

Metabolism (meh tab' a lizm) [Gr. *metabole*: to change] The sum total of the chemical reactions that occur in an organism, or some subset of that total (as in "respiratory metabolism").

Metamorphosis (met' a mor' fo sis) [Gr. *meta*: between + *morphe*: form, shape] A radical change occurring between one developmental stage and another, as for example from a tadpole to a frog or an insect larva to the adult.

Metaphase (met' a phase) [Gr. *meta*: between] The stage in nuclear division at which the centromeres of the highly supercoiled chromosomes are all lying on a plane (the metaphase plane or plate) perpendicular to a line connecting the division poles.

Metapopulation A population divided into subpopulations, among which there are occasional exchanges of individuals.

Metastasis (meh tass' tuh sis) The spread of cancer cells from their original site to other parts of the body.

Methanogen Any member of a group of Archaebacteria that release methane as a metabolic product. This group is considered to be an extremely ancient one.

MHC See major histocompatibility complex.

Micelles (my sells') [L. *mica*: grain, crumb] The small particles of fat in the small intestine, resulting from the emulsification of dietary fat by bile.

Micro- (mike' roh) [Gr. *mikros*: small] A prefix often used to denote something small. (Contrast with macro-, mega-.)

Microbiology [Gr. *mikros*: small + *bios*: life + *logos*: discourse] The scientific study of microscopic organisms, particularly bacteria, unicellular algae, protists, and viruses.

Microevolution The small evolutionary changes typically occurring over short time spans; generally involving a small number of traits and minor genetic changes. (Contrast with macroevolution.)

Microevolutionary time The time required for microevolutionary changes within a lineage of organisms.

Microfilament Minute fibrous structure generally composed of actin found in the cytoplasm of eukaryotic cells. They play a role in the motion of cells.

Micromorphology The structure of the macromolecules of an organism.

Micronutrient A mineral element required by plant tissues in concentrations of less than 100 micrograms per gram of their dry matter.

Microorganism Any microscopic organism, such as a bacterium or one-celled alga.

Micropyle (mike' roh pile) [Gr. *mikros*: small + *pyle*: gate] Opening in the integument(s) of a seed plant ovule through which pollen grows to reach the female gametophyte within.

Microsporangium The special structure (sporangium) that produces the microspores.

Microspores [Gr. *mikros*: small + *spora*: seed] In plants, a haploid spore that produces a male gametophyte. In many cases the microspore is smaller than the female-producing megaspore.

Microtubules Minute tubular structures found in centrioles, spindle apparatus, cilia, flagella, and other places in the cytoplasm of eukaryotic cells. These tubules play roles in the motion and maintenance of shape of eukaryotic cells.

Microvilli (singular: microvillus) The projections of epithelial cells, such as the cells lining the small intestine, that increase their surface area.

Middle lamella A layer of derivative polysaccharides that separates plant cells; a common middle lamella lies outside the primary walls of the two cells.

Migration The regular, seasonal movements of animals between breeding and nonbreeding ranges.

Mimicry (mim' ik ree) The resemblance of one kind of organism to another, or to some inanimate object; serves the function of making the organism difficult to find, of discouraging potential enemies or of attracting potential prey. (See Batesian mimicry and Müllerian mimicry.)

Mineral An inorganic substance other than water.

Mineralocorticoid A hormone produced by the adrenal cortex that influences mineral ion balance; aldosterone.

Minimal medium A medium for the growth of bacteria, fungi, or tissue cultures, containing only those nutrients absolutely required for the growth of wild type cells.

Minimum viable population. The smallest number of individuals required for a population to persist in a region.

Mismatch repair When a single base in DNA is changed into a different base, or the wrong base inserted during DNA replication, there is a mismatch in base pairing with the base on the opposite strand. A repair system removes the incorrect base and inserts the proper one for pairing with the opposite strand.

Missense mutation A mutation that changes a codon for one amino acid to a codon for a different amino acid. (Contrast with frame-shift mutation, nonsense mutation.)

Mitochondrial matrix The fluid interior of the mitochondrion, enclosed by the inner mitochondrial membrane.

Mitochondrion (my' toe kon' dree un) (plural: mitochondria) [Gr. *mitos*: thread + *chondros*: cartilage, or grain] An organelle that occurs in eukaryotic cells and contains the enzymes of the ctric acid cycle, the respiratory chain, and oxidative phosphorylation. A mitochondrion is bounded by a double membrane.

Mitosis (my toe' sis) [Gr. *mitos*: thread] Nuclear division in eukaryotes leading to the formation of two daughter nuclei each with a chromosome complement identical to that of the original nucleus.

Mitotic center Cellular region that organizes the microtubules for mitosis. In animals a centrosome serves as the mitotic center.

Mobbing Gathering of calling animals around a predator; their calls and the confusion they create reduce the probability that the predator can hunt successfully in the area.

Moderately repetitive DNA DNA sequences that appear hundreds to thousands of times in the genome. They include the DNA sequences coding for rRNAs and tRNAs, as well as the DNA at telomeres.

Modular organism An organism which grows by producing additional units of body construction that are very similar to the units of which it is already composed.

Mole A quantity of a compound whose weight in grams is numerically equal to its molecular weight expressed in atomic mass units. Avogadro's number of molecules: 6.023×10^{23} molecules.

Molecular formula A representation that shows how many atoms of each element are present in a molecule.

Molecular weight The sum of the atomic weights of the atoms in a molecule.

Molecule A particle made up of two or more atoms joined by covalent bonds or ionic attractions.

Molting The process of shedding part or all of an outer covering, as the shedding of feathers by birds or of the entire exoskeleton by arthropods.

Monoecious (mo nee' shus) [Gr.: one house] Organisms in which both sexes are "housed" in a single individual, which produces both eggs and sperm. (In some plants, these are found in different flowers within the same plant.) Examples: corn, peas, earthworms, hydras. (Contrast with dioecious, perfect flower.)

Moneran (moh neer' un) A bacterium. This term is coined when both archaebacteria and eubacteria were considered to be members of a single kingdom, Monera.

Mono- [Gr. *monos*: one] Prefix denoting a single entity. (Contrast with poly.)

Monoclonal antibody Antibody produced in the laboratory from a clone of hybridoma cells, each of which produces the same specific antibody.

Monocot (short for monocotyledon) [Gr. *monos*: one + *kotyledon*: a cup-shaped hollow] Any member of the angiosperm class Monocotyledones, plants in which the embryo produces but a single cotyledon (seed leaf). Leaves of most monocots have their major veins arranged parallel to each other.

Monocytes White blood cells that produce macrophages.

Monohybrid cross A mating in which the parents differ with respect to the alleles of only one locus of interest.

Monomer A small molecule, two or more of which can be combined to form oligomers (consisting of a few monomers) or polymers (consisting of many monomers).

Monophyletic (mon' oh fih leht' ik) [Gk. *monos*: single + *phylon*: tribe] Being descended from a single ancestral stock.

Monosaccharide A simple sugar. Oligosaccharides and polysaccharides are made up of monosaccharides.

Monosynaptic reflex A neural reflex that begins in a sensory neuron and makes a single synapse before activating a motor neuron.

Morphogens Diffusible substances whose concentration gradients determine patterns of development in animals and plants.

Morphogenesis (more' fo jen' e sis) [Gr. *morphe*: form + *genesis*: origin] The development of form. Morphogenesis is the overall consequence of determination, differentiation, and growth.

Morphology (more fol' o jee) [Gr. *morphe*: form + *logos*: discourse] The scientific study of organic form, including both its development and function.

Mosaic development Pattern of animal embryonic development in which each blastomere contributes a specific part of the adult body. (Contrast with regulative development.)

Motor end plate The modified area on a muscle cell membrane where a synapse is formed with a motor neuron.

Motor neuron A neuron carrying information from the central nervous system to an effector such as a muscle fiber.

Motor unit A motor neuron and the set of muscle fibers it controls.

mRNA (See messenger RNA.)

Mucosa (mew koh' sah) An epithelial membrane containing cells that secrete mucus. The inner cell layers of the digestive and respiratory tracts.

Müllerian mimicry The resemblance of two or more unpleasant or dangerous kinds of organisms to each other.

Multicellular [L. *multus*: much + *cella*: chamber] Consisting of more than one cell, as for example a multicellular organism. (Contrast with unicellular.)

Muscle Contractile tissue containing actin and myosin organized into polymeric chains called microfilaments. In vertebrates, the tissues are either cardiac muscle, smooth muscle, or striated (skeletal) muscle.

Muscle fiber A single muscle cell. In the case of striated muscle, a syncitial, multinucleate cell.

Muscle spindle Modified muscle fibers encased in a connective sheat and functioning as stretch sensors.

Mutagen (mute' ah jen) [L. *mutare*: change + Gr. *genesis*: source] An agent, especially a chemical, that increases the mutation rate.

Mutation In the broad sense, any discontinuous change in the genetic constitution of an organism. In the narrow sense, the word usually refers to a "point mutation," a change along a very narrow portion of the nucleic acid sequence.

Mutation pressure   Evolution (change in gene proportions) by different mutation rates alone.

Mutualism   The type of symbiosis, such as that exhibited by fungi and algae or cyanobacteria in forming lichens, in which both species profit from the association.

Mycelium (my seel′ ee yum) [Gr. *mykes*: fungus]   In the fungi, a mass of hyphae.

Mycorrhiza (my′ ka rye′ za) [Gr. *mykes*: fungus + *rhiza*: root]   An association of the root of a plant with the mycelium of a fungus.

Myelin (my′ a lin)   A material forming a sheath around some axons. It is formed by Schwann cells that wrap themselves about the axon. It serves to insulate the axon electrically and to increase the rate of transmission of a nervous impulse.

Myofibril (my′ oh fy′ bril) [Gr. *mys*: muscle + L. *fibrilla*: small fiber]   A polymeric unit of actin or myosin in a muscle.

Myogenic (my oh jen′ ik) [Gr. *mys*: muscle + *genesis*: source]   Originating in muscle.

Myoglobin (my′ oh globe′ in) [Gr. *mys*: muscle + L. *globus*: sphere]   An oxygen-binding molecule found in muscle. Consists of a heme unit and a single globin chain, and carrys less oxygen than hemoglobin.

Myosin [Gr. *mys*: muscle]   One of the two major proteins of muscle, it makes up the thick filaments. (See actin.)

NAD (nicotinamide adenine dinucleotide)   A compound found in all living cells, existing in two interconvertible forms: the oxidizing agent NAD$^+$ and the reducing agent NADH.

NADP (nicotinamide adenine dinucleotide phosphate)   Like NAD, but possessing another phosphate group; plays similar roles but is used by different enzymes.

Natal group   The group into which an individual was born.

Natural selection   The differential contribution of offspring to the next generation by various genetic types belonging to the same population. The mechanism of evolution proposed by Charles Darwin.

Nauplius (no′ plee us) [Gk. *nauplios*: shellfish]   The typical larva of crustaceans. Has three pairs of appendages and a median compound eye.

Necrosis (nec roh′ sis)   Tissue damage resulting from cell death.

Negative control   The situation in which a regulatory macromolecule (generally a repressor) functions to turn off transcription. In the absence of a regulatory macromolecule, the structural genes are turned on.

Negative feedback   A pattern of regulation in which a change in a sensed variable results in a correction that opposes the change.

Nekton [Gr. *nekhein*: to swim]   Animals, such as fish, that can swim against currents of water. (Contrast with plankton.)

Nematocyst (ne mat′ o sist) [Gr. *nema*: thread + *kystis*: cell]   An elaborate, thread-like structure produced by cells of jellyfish and other cnidarians, used chiefly to paralyze and capture prey.

Nephridium (nef rid′ ee um) [Gr. *nephros*: kidney]   An organ which is involved in excretion, and often in water balance, involving a tube that opens to the exterior at one end.

Nephron (nef′ ron) [Gr. *nephros*: kidney]   The basic component of the kidney, which is made up of numerous nephrons. Its form varies in detail, but it always has at one end a device for receiving a filtrate of blood, and then a tubule that absorbs selected parts of the filtrate back into the bloodstream.

Nephrostome (nef′ ro stome) [Gr. *nephros*: kidney + *stoma*: opening]   An opening in a nephridium through which body fluids can enter.

Nerve   A structure consisting of many neuronal axons and connective tissue.

Net primary production   Total photosynthesis minus respiration by plants.

Neural plate   A thickened strip of ectoderm along the dorsal side of the early vertebrate embryo; gives rise to the central nervous system.

Neural tube   An early stage in the development of the vertebrate nervous system consisting of a hollow tube created by two opposing folds of the dorsal ectoderm along the anterior–posterior body axis.

Neuromuscular junction   The region where a motor neuron contacts a muscle fiber, creating a synapse.

Neuron (noor′ on) [Gr. *neuron*: nerve, sinew]   A cell derived from embryonic ectoderm and characterized by a membrane potential that can change in response to stimuli, generating action potentials. Action potentials are generated along an extension of the cell (the axon), which makes junctions (synapses) with other neurons, muscle cells, or gland cells.

Neurotransmitter   A substance, produced in and released by one neuron, that diffuses across a synapse and excites or inhibits the postsynaptic neuron.

Neurula (nure′ you la) [Gr. *neuron*: nerve]   Embryonic stage during formation of the dorsal nerve cord by two ectodermal ridges.

Neutral alleles   Alleles that differ so slightly that the proteins for which they code function identically.

Neutron (new′ tron) [E.: neutral]   One of the three most fundamental particles of matter, with mass approximately 1 amu and no electrical charge.

Nicotinamide adenine dinucleotide   (See NAD.)

Nicotinamide adenine dinucleotide phosphate   (See NADP.)

Nitrification   The oxidation of ammonia to nitrite and nitrate ions, performed by certain soil bacteria.

Nitrogenase   In nitrogen-fixing organisms, an enzyme complex that mediates the stepwise reduction of atmospheric N$_2$ to ammonia.

Nitrogen fixation   Conversion of nitrogen gas to ammonia, which makes nitrogen available to living things. Carried out by certain prokaryotes, some of them free-living and others living within plant roots.

Node [L. *nodus*: knob, knot]   In plants, a (sometimes enlarged) point on a stem where a leaf or bud is or was attached.

Node of Ranvier   A gap in the myelin sheath covering an axons, where the axonal membrane can fire action potentials.

Noncompetitive inhibitor   An inhibitor that binds the enzyme at a site other than the active site. (Contrast with competitive inhibitor.)

Nondisjunction   Failure of sister chromatids to separate in meiosis II or mitosis, or failure of homologous chromosomes to separate in meiosis I. Results in aneuploidy.

Nonpolar molecule   A molecule whose electric charge is evenly balanced from one end of the molecule to the other.

Nonsense (chain-terminating) mutation   Mutations that change a codon for an amino acid to one of the codons (UAG, UAA, or UGA) that signal termination of translation. The resulting gene product is a shortened polypeptide that begins normally at the amino-terminal end and ends at the position of the altered codon. (Contrast with frame-shift mutation, missense mutation.)

Nonspecific defenses   Immunologic responses directed against most or all pathogens, generally without reference to the pathogens′ antigens. These defenses include the skin, normal flora, lysozyme, the acidic stomach, interferon, and the inflammatory response.

Nontracheophytes   Those plants lacking well-developed vascular tissue; the liverworts, hornworts, and mosses. (Contrast with tracheophytes.)

Normal flora   The bacteria and fungi that live on animal body surfaces without causing disease.

Norepinephrine   A neurotransmitter found in the central nervous system and also at the postganglionic nerve endings of the sympathetic nervous system. Also called noradrenaline.

Notochord (no′ tow kord) [Gr. *notos*: back + *chorde*: string]   A flexible rod of gelatinous material serving as a support in the embryos of all chordates and in the adults of tunicates and lancelets.

Nuclear envelope   The surface, consisting of two layers of membrane, that encloses the nucleus of eukaryotic cells.

Nucleic acid (new klay′ ik) [E.: nucleus of a cell]   A long-chain alternating polymer of deoxyribose or ribose and phosphate groups, with nitrogenous bases—adenine, thymine, uracil, guanine, or cytosine (A, T, U, G, or C)—as side chains. DNA and RNA are nucleic acids.

Nucleoid (new' klee oid)   The region that harbors the chromosomes of a prokaryotic cell. Unlike the eukaryotic nucleus, it is not bounded by a membrane.

Nucleolar organizer (new klee' o lar)   A region on a chromosome that is associated with the formation of a new nucleolus following nuclear division. The site of the genes that code for ribosomal RNA.

Nucleolus (new klee' oh lus) [from L. diminutive of *nux*: little kernel or little nut] A small, generally spherical body found within the nucleus of eukaryotic cells. The site of synthesis of ribosomal RNA.

Nucleoplasm (new' klee o plazm)   The fluid material within the nuclear envelope of a cell, as opposed to the chromosomes, nucleoli, and other particulate constituents.

Nucleosome   A portion of a eukaryotic chromosome, consisting of part of the DNA molecule wrapped around a group of histone molecules, and held together by another type of histone molecule. The chromosome is made up of many nucleosomes.

Nucleotide   The basic chemical unit (monomer) in a nucleic acid. A nucleotide in RNA consists of one of four nitrogenous bases linked to ribose, which in turn is linked to phosphate. In DNA, deoxyribose is present instead of ribose.

Nucleus (new' klee us) [from L. diminutive of *nux*: kernel or nut]   (1) In chemistry, the dense central portion of an atom, made up of protons and neutrons, with a positive charge. Surrounded by a cloud of negatively charged electrons. (2) In cells, the centrally located chamber of eukaryotic cells that is bounded by a double membrane and contains the chromosomes. The information center of the cell.

Nutrient   A food substance; or, in the case of mineral nutrients, an inorganic element required for completion of the life cycle of an organism.

Obligate (ob' li gut)   Necessary, as in obligate anaerobe. (Contrast with facultative.)

Obligate anaerobe   An animal that can live only in oxygenated environments.

Oil   A triglyceride that is liquid at room temperature. (Contrast with fat.)

Okazaki fragments   Newly formed DNA strands making up the lagging strand in DNA replication. DNA ligase links the Okazaki fragments to give a continuous strand.

Olfactory   Having to do with the sense of smell.

Oligomer   A compound molecule of intermediate size, made up of two to a few monomers. (Contrast with monomer, polymer.)

Ommatidium [Gr. *omma*: an eye]   One of the units which, collected into groups of up to 20,000, make up the compound eye of arthropods.

Omnivore [L. *omnis*: all, everything + *vorare*: to devour]   An organism that eats both animal and plant material. (Contrast with carnivore, detritivore, herbivore.)

Oncogenic (ong' co jen' ik) [Gr. *onkos*: mass, tumor + *genes*: born]   Causing cancer.

Oocyte (oh' eh site) [Gr. *oon*: egg + *kytos*: cell]   The cell that gives rise to eggs in animals.

Oogenesis (oh' eh jen e sis) [Gr. *oon*: egg + *genesis*: source]   Female gametogenesis, leading to production of the egg.

Oogonium (oh' eh go' nee um)   In some algae and fungi, a cell in which an egg is produced.

Operator   The region of an operon that acts as the binding site for the repressor.

Operon   A genetic unit of transcription, typically consisting of several structural genes that are transcribed together; the operon contains at least two control regions: the promoter and the operator.

Opportunity cost   The sum of the benefits an animal forfeits by not being able to perform some other behavior during the time when it is performing a given behavior.

Opsin (op' sin) [Gr. *opsis*: sight]   The protein protion of the visual pigment rhodopsin. (See rhodopsin.)

Optic chiasm   Stucture on the lower surface of the vertebrate brain where the two optic nerves come together.

Optical isomers   Isomers that differ in the configuration of the four different groups attached to a single carbon atom; so named because solutions of the two isomers rotate the plane of polarized light in opposite directions. The two isomers are mirror images of one another.

Optimality models   Models developed to determine the structures or behaviors that best solve particular problems faced by organisms.

Order   In taxonomy, the category below the class and above the family; a group of related, similar families.

Organ   A body part, such as the heart, liver, brain, root, or leaf, composed of different tissues integrated to perform a distinct function for the body as a whole.

Organ identity genes   These plant genes specify the various parts of the flower.

Organ of Corti   Structure in the inner ear that transforms mechanical forces produced from pressure waves ("sound waves") into action potentials that are sensed as sound.

Organelles (or' gan els') [L.: little organ] Organized structures that are found in or on cells. Examples: ribosomes, nuclei, mitochrondria, chloroplasts, cilia, and contractile vacuoles.

Organic   Pertaining to any aspect of living matter, e.g., to its evolution, structure, or chemistry. The term is also applied to any chemical compound that contains carbon.

Organism   Any living creature.

Organizer, embryonic   A region of an embryo which directs the development of nearby regions. In amphibian early gastrulas, the dorsal lip of the blastopore.

Origin of replication   A DNA sequence at which helicase unwinds the DNA double helix and DNA polymerase binds to initiate DNA replication.

Osmoregulation   Regulation of the chemical composition of the body fluids of an organism.

Osmosensor   A neuron that converts changes in the osmotic potential of interstial fluids into action potentials.

Osmosis (oz mo' sis) [Gr. *osmos*: to push] The movement of water through a differentially permeable membrane from one region to another where the water potential is more negative. This is often a region in which the concentration of dissolved molecules or ions is higher, although the effect of dissolved substances may be offset by hydrostatic pressure in cells with semi-rigid walls.

Osmotic potential   A property of any solution, resulting from its solute content; it may be zero or have a negative value. A negative osmotic potential tends to cause water to move into the solution; it may be offset by a positive pressure potential in the solution or by a more negative water potential in a neighboring solution. (Contrast with turgor pressure.)

Ossicle (ah' sick ul) [L. *os*: bone]   The calcified construction unit of echinoderm skeletons.

Osteoblasts   Cells that lay down the protein matrix of bone.

Osteoclasts   Cells that dissolve bone.

Otolith (oh' tuh lith) [Gk.*otikos*: ear + *lithos*: stone]   Structures in the vertebrate vestibular apparatus that mechanically stimulate hair cells when the head moves or changes position.

Outgroup   A taxon that separated from another taxon, whose lineage is to be inferred, before the latter underwent evolutionary radiation.

Oval window   The flexible membrane which, when moved by the bones of the middle ear, produces pressure waves in the inner ear

Ovary (oh' var ee)   Any female organ, in plants or animals, that produces an egg.

Oviduct [L. *ovum*: egg + *ducere*: to lead]   In mammals, the tube serving to transport eggs to the uterus or to outside of the body.

Oviparous (oh vip' uh rus)   Reproduction in which eggs are released by the female and development is external to the mother's body. (Contrast with viviparous.)

Ovulation   The release of an egg from an ovary.

Ovule (oh' vule) [L. *ovulum*: little egg]   In plants, an organ that contains a gametophyte and, within the gametophyte, an egg; when it matures, an ovule becomes a seed.

Ovum (oh' vum) [L.: egg]   The egg, the female sex cell.

Oxidation (ox i day' shun)   Relative loss of electrons in a chemical reaction; either outright removal to form an ion, or the sharing of electrons with substances having a greater affinity for them, such as oxygen. Most oxidation, including biological ones, are associated with the liberation of energy. (Contrast with reduction.)

Oxidative phosphorylation ATP formation in the mitochondrion, associated with flow of electrons through the respiratory chain.

Oxidizing agent A substance that can accept electrons from another. The oxidizing agent becomes reduced; its partner becomes oxidized.

P generation Also called the parental generation. The individuals that mate in a genetic cross. Their immediate offspring are the F_1 generation.

Pacemaker That part of the heart which undergoes most rapid spontaneous contraction, thus setting the pace for the beat of the entire heart. In mammals, the sinoatrial (SA) node. Also, an artificial device, implanted in the heart, that initiates rhythmic contraction of the organ.

Pacinian corpuscle A sensory neuron surrounded by sheaths of connective tissue. Found in the deep layers of the skin, where it senses touch and vibration.

Pair rule genes Segmentation genes that divide the *Drosophila* larva into two segments each.

Paleontology (pale' ee on tol' oh jee) [Gr. *palaios*: ancient, old + *logos*: discourse] The scientific study of fossils and all aspects of extinct life.

Palisade parenchyma In leaves, one or several layers of tightly packed, columnar photosynthetic cells, frequently found just below the upper epidermis.

Pancreas (pan' cree us) A gland, located near the stomach of vertebrates, that secretes digestive enzymes into the small intestine and releases insulin into the bloodstream.

Pangaea (pan jee' uh) [Gk. *pan*: all, every] The single land mass formed when all the continents came together in the Permian period.

Parabronchi Passages in the lungs of birds through which air flows.

Paradigm A general framework within which some scientific discipline (or even the whole Earth) is viewed and within which questions are asked and hypotheses are developed. Scientific revolutions usually involve major paradigm changes.

Parapatric speciation Development of reproductive isolation among members of a continuous population in the absence of a geographical barrier. (Contrast with geographic, sympatric speciation.)

Paraphyletic taxon A taxon that includes some, but not all, of the descendants of a single ancestor.

Parasite An organism that attacks and consumes parts of an organism much larger than itself. Parasites sometimes, but not always, kill the host.

Parasitoid A parasite that is so large relative to its host that only one individual or at most a few individuals can live within a single host.

Parasympathetic nervous system A portion of the autonomic (involuntary) nervous system. Activity in the parasympathetic nervous system produces effects such as decreased blood pressure and decelerated heart beat. (Contrast with sympathetic nervous system.)

Parathormone Hormone secreted by the parathyroid glands. Stimulates osteoclast activity and raises blood calcium levels.

Parathyroids Four glands on the posterior surface of the thyroid that produce and release parathormone.

Parenchyma (pair eng' kyma) [Gr. *para*: beside + *enchyma*: infusion] A plant tissue composed of relatively unspecialized cells without secondary walls.

Parental investment Investment in one offspring or group of offspring that reduces the ability of the parent to assist other offspring.

Parsimony The principle of preferring the simplest among a set of plausible explanations of a phenomenon. Commonly employed in evolutionary and biogeographic studies.

Parthenocarpy Formation of fruit from a flower without fertilization.

Parthenogenesis (par' then oh jen' e sis) [Gr. *parthenos*: virgin + *genesis*: source] The production of an organism from an unfertilized egg.

Partial pressure The portion of the barometric pressure of a mixture of gases that is due to one component of that mixture. For example, the partial pressure of oxygen at sea level is 20.9% of barometric pressure.

Pasteur effect The sharp decrease in rate of glucose utilization when conditions become aerobic.

Pastoralism A nomadic form of human culture based on the tending of herds of domestic animals.

Patch clamping A technique for isolating a tiny patch of membrane to allow the study of ion movement through a particular channel.

Pathogen (path' o jen) [Gr. *pathos*: suffering + *gignomai*: causing] An organism that causes disease.

Pattern formation In animal embryonic development, the organization of differentiated tissues into specific structures such as wings.

Pedigree The pattern of transmission of a genetic trait in a family.

Pelagic zone (puh ladj' ik) [Gr. *pelagos*: the sea] The open waters of the ocean.

Penetrance Of a genotype, the proportion of individuals with that genotype who show the expected phenotype.

Penis (pee' nis) [L.: tail] The male organ inserted into the female during sexual intercourse.

PEP carboxylase The enzyme that combines carbon dioxide with PEP to form a 4-carbon dicarboxylic acid at the start of C_4

photosynthesis or of Crassulacean acid metabolism (CAM).

Pepsin [Gr. *pepsis*: digestion] An enzyme, in gastric juice, that digests protein.

Peptide linkage The connecting group in a protein chain, –CO–NH–, formed by removal of water during the linking of amino acids, –COOH to –NH$_2$. Also called an amide linkage.

Peptidoglycan The cell wall material of many prokaryotes, consisting of a single enormous molecule that surrounds the entire cell.

Perennial (per ren' ee al) [L. *per*: through + *annus*: a year] Referring to a plant that lives from year to year. (Contrast with annual, biennial.)

Perfect flower A flower with both stamens and carpels, therefore hermaphroditic.

Pericycle [Gr. *peri*: around + *kyklos*: ring or circle] In plant roots, tissue just within the endodermis, but outside of the root vascular tissue. Meristematic activity of pericycle cells produces lateral root primordia.

Periderm The outer tissue of the secondary plant body, consisting primarily of cork.

Period (1) A minor category in the geological time scale. (2) The duration of a cyclical event, such as a circadian rhythm.

Peripheral nervous system Neurons that transmit information to and from the central nervous system and whose cell bodies reside outside the brain or spinal cord.

Peristalsis (pair' i stall' sis) [Gr. *peri*: around + *stellein*: place] Wavelike muscular contractions proceeding along a tubular organ, propelling the contents along the tube.

Peritoneum The mesodermal lining of the coelom among coelomate animals.

Permease A protein in membranes that specifically transports a compound or family of compounds across the membrane.

Peroxisome An organelle that houses reactions in which toxic peroxides are formed. The peroxisome isolates these peroxides from the rest of the cell.

Petal In an angiosperm flower, a sterile modified leaf, nonphotosynthetic, frequently brightly colored, and often serving to attract pollinating insects.

Petiole (pet' ee ole) [L. *petiolus*: small foot] The stalk of a leaf.

pH The negative logarithm of the hydrogen ion concentration; a measure of the acidity of a solution. A solution with pH = 7 is said to be neutral; pH values higher than 7 characterize basic solutions, while acidic solutions have pH values less than 7.

Phage (fayj) Short for bacteriophage.

Phagocyte A white blood cell that ingests microorganisms by endocytosis.

Phagocytosis [Gr.: *phagein* to eat; cell-eating] A form of endocytosis, the uptake of a solid particle by forming a pocket of plasma membrane around the particle and pinching off the pocket to form an intracellular particle bounded by membrane. (Contrast with pinocytosis.)

Pharynx [Gr.: throat]   The part of the gut between the mouth and the esophagus.

Phenotype (fee' no type) [Gr. *phanein*: to show]   The observable properties of an individual as they have developed under the combined influences of the genetic constitution of the individual and the effects of environmental factors. (Contrast with genotype.)

Pheromone (feer' o mone) [Gr. *phero*: carry + *hormon*: excite, arouse]   A chemical substance used in communication between organisms of the same species.

Phloem (flo' um) [Gr. *phloos*: bark]   In vascular plants, the food-conducting tissue. It consists of sieve cells or sieve tubes, fibers, and other specialized cells.

Phosphate group   The functional group $-OPO_3H_2$; the transfer of energy from one compound to another is often accomplished by the transfer of a phosphate group.

Phosphodiester linkage   The connection in a nucleic acid strand, formed by linking two nucleotides.

3-Phosphoglycerate   The first product of photosynthesis, produced by the reaction of ribulose bisphosphate with carbon dioxide.

Phospholipids   Cellular materials that contain phosphorus and are soluble in organic solvents. An example is lecithin (phosphatidyl choline). Phospholipids are important constituents of cellular membranes. (See lipids.)

Phosphorylation   The addition of a phosphate group.

Photoautotroph   An organism that obtains energy from light and carbon from carbon dioxide. (Contrast with chemoautotroph, chemoheterotroph, photoheterotroph.)

Photoheterotroph   An organism that obtains energy from light but must obtain its carbon from organic compounds. (Contrast with chemoautotroph, chemoheterotroph, photoautotroph.)

Photon (foe' tohn) [Gr. *photos*: light]   A quantum of visible radiation; a "packet" of light energy.

Photoperiod (foe' tow peer' ee ud)   The duration of a period of light, such as the length of time in a 24-hour cycle in which daylight is present. The regulation of processes such as flowering by the changing length of day (or of night) is known as photoperiodism.

Photophosphorylation   Photosynthetic reactions in which light energy trapped by chlorophyll is used to produce ATP and, in noncyclic photophosphorylation, is used to reduce $NADP^+$ to NADPH.

Photorespiration   Light-driven uptake of oxygen and release of carbon dioxide, the carbon being derived from the early reactions of photosynthesis.

Photosensor   A cell that senses and responds to light energy. Also called a **photoreceptor**.

Photosynthesis (foe tow sin' the sis) [literally, "synthesis out of light"]   Metabolic processes, carried out by green plants, by which visible light is trapped and the energy used to synthesize compounds such as ATP and glucose.

Phototropism [Gr. *photos*: light + *trope*: a turning]   A directed plant growth response to light.

Phylogenetic tree   Graphic representation of lines of descent among organisms.

Phylogeny (fy loj' e nee) [Gr. *phylon*: tribe, race + *genesis*: source]   The evolutionary history of a particular group of organisms; also, the diagram of the "family tree" that shows genetic linkages between ancestors and descendants.

Phylum (plural: phyla) [Gr. *phylon*: tribe, stock]   In taxonomy, a high-level category just beneath kingdom and above the class; a group of related, similar classes.

Physiology (fiz' ee ol' o jee) [Gr. *physis*: natural form + *logos*: discourse, study]   The scientific study of the functions of living organisms and the individual organs, tissues, and cells of which they are composed.

Phytoalexins   Substances toxic to fungi, produced by plants in response to fungal infection.

Phytochrome (fy' tow krome) [Gr. *phyton*: plant + *chroma*: color]   A plant pigment regulating a large number of developmental and other phenomena in plants; can exist in two different forms, one of which is active and the other is not. Different wavelengths of light can drive it from one form to the other.

Phytoplankton (fy' tow plangk' ton) [Gr. *phyton*: plant + *planktos*: wandering]   The autotrophic portion of the plankton, consisting mostly of algae.

Pigment   A substance that absorbs visible light.

Pilus (pill' us) [Lat. *pilus*: hair]   A surface appendage by which some bacteria adhere to one another during conjugation.

Pinocytosis [Gr.: drinking cell]   A form of endocytosis; the uptake of liquids by engulfing a sample of the external medium into a pocket of the plasma membrane followed by pinching off the pocket to form an intracellular vesicle. (Contrast with phagocytosis and endocytosis.)

Pistil [L. *pistillum*: pestle]   The female structure of an angiosperm flower, within which the ovules are borne. May consist of a single carpel, or of several carpels fused into a single structure. Usually differentiated into ovary, style, and stigma.

Pith   In plants, relatively unspecialized tissue found within a cylinder of vascular tissue.

Pituitary   A small gland attached to the base of the brain in vertebrates. Its hormones control the activities of other glands. Also known as the hypophysis.

Placenta (pla sen' ta) [Gr. *plax*: flat surface]   The organ found in most mammals that provides for the nourishment of the fetus and elimination of the fetal waste products.

Placental (pla sen' tal)   Pertaining to mammals of the subclass Eutheria, a group characterized by the presence of a placenta; contains the majority of living species of mammals.

Plankton [Gr. *planktos*: wandering]   The free-floating organisms of the sea and fresh water that for the most part move passively with the water currents. Consisting mostly of microorganisms and small plants and animals. (Contrast with nekton.)

Plant   A member of the kingdom Plantae. Multicellular, gaining its nutrition by photosynthesis.

Planula (plan' yew la) [L. *planum*: something flat]   The free-swimming, ciliated larva of the cnidarians.

Plaque (plack) [Fr.: a metal plate or coin]   (1) A circular clearing in a turbid layer (lawn) of bacteria growing on the surface of a nutrient agar gel. Produced by successive rounds of infection initiated by a single bacteriophage. (2) An accumulation of prokaryotic organisms on tooth enamel. Acids produced by the metabolism of these microorganisms can cause tooth decay.

Plasma (plaz' muh) [Gr. *plassein*: to mold]   The liquid portion of blood, in which blood cells and other particulates are suspended.

Plasma cell   An antibody-secreting cell that developed from a B cell. The effector cell of the humoral immune system.

Plasma membrane   The membrane that surrounds the cell, regulating the entry and exit of molecules and ions. Every cell has a plasma membrane.

Plasmid   A DNA molecule distinct from the chromosome(s); that is, an extrachromosomal element. May replicate independently of the chromosome.

Plasmodesma (plural: plasmodesmata) [Gr. *plasma*: formed or molded + *desmos*: band]   A cytoplasmic strand connecting two adjacent plant cells.

Plasmodium   In the noncellular slime molds, a multinucleate mass of protoplasm surrounded by a membrane; characteristic of the vegetative feeding stage.

Plasmolysis (plaz mol' i sis)   Shrinking of the cytoplasm and plasma membrane away from the cell wall, resulting from the osmotic outflow of water. Occurs only in cells with rigid cell walls.

Plastid   Organelle in plants that serves for food manufacture (by photosynthesis) or food storage; bounded by a double membrane.

Platelet   A membrane-bounded body without a nucleus, arising as a fragment of a cell in the bone marrow of mammals. Important to blood-clotting action.

Pleiotropy (plee' a tro pee) [Gr. *pleion*: more]   The determination of more than one character by a single gene.

Pleural membrane [Gk. *pleuras*: rib, side]   The membrane lining the outside of the lungs and the walls of the thoracic cavity. Inflammation of these membranes is a condition known as *pleurisy*.

Podocytes   Cells of Bowman's capsule of the nephron that cover the capillaries of the glomerulus, forming filtration slits.

Poikilotherm (poy' kill o therm) [Gr. *poikilos*: varied + *therme*: heat]   An animal whose body temperature tends to vary with the surrounding environment. (Contrast with homeotherm, heterotherm.)

Point mutation A mutation that results from a small, localized alteration in the chemical structure of a gene. Such mutations can give rise to wild-type revertants as a result of reverse mutation. In genetic crosses, a point mutation behaves as if it resided at a single point on the genetic map. (Contrast with deletion.)

Polar body A nonfunctional nucleus produced by meiosis, accompanied by very little cytoplasm. The meiosis which produces the mammalian egg produces in addition three polar bodies.

Polar molecule A molecule in which the electric charge is not distributed evenly in the covalent bonds.

Polar nucleus One of two nuclei derived from each end of the angiosperm embryo sac, both of which become centrally located. They fuse with a male nucleus to form the primary triploid nucleus that will prduce the endosperm tissue of the angiosperm seed.

Polarity In development, the difference between one end and the other. In chemistry, the property that makes a polar molecule.

Pollen [L.: fine powder, dust] The fertilizing element of seed plants, containing the male gametophyte and the gamete, at the stage in which it is shed.

Pollination Process of transferring pollen from the anther to the receptive surface (stigma) of the ovary in plants.

Poly- [Gr. poly: many] A prefix denoting multiple entities.

Polygamy [Gr. poly: many + gamos: marriage] A breeding system in which an individual acquires more than one mate. In polyandry, a female mates with more than one male, in polygyny, a male mates with more than one female.

Polygenes Multiple loci whose alleles increase or decrease a continuously variable phenotypic trait.

Polymer A large molecule made up of similar or identical subunits called monomers. (Contrast with monomer, oligomer.)

Polymerase chain reaction (PCR) A technique for the rapid production of millions of copies of a particular stretch of DNA.

Polymerization reactions Chemical reactions that generate polymers by means of condensation reactions.

Polymorphism (pol' lee mor' fiz um) [Gr. poly: many + morphe: form, shape] (1) In genetics, the coexistence in the same population of two distinct hereditary types based on different alleles. (2) In social organisms such as colonial cnidarians and social insects, the coexistence of two or more functionally different castes within the same colony.

Polyp The sessile, asexual stage in the life cycle of most cnidarians.

Polypeptide A large molecule made up of many amino acids joined by peptide linkages. Large polypeptides are called proteins.

Polyphyletic group A group containing taxa, not all of which share the most recent common ancestor.

Polyploid (pol' lee ploid) A cell or an organism in which the number of complete sets of chromosomes is greater than two.

Polysaccharide A macromolecule composed of many monosaccharides (simple sugars). Common examples are cellulose and starch.

Polysome A complex consisting of a threadlike molecule of messenger RNA and several (or many) ribosomes. The ribosomes move along the mRNA, synthesizing polypeptide chains as they proceed.

Polytene (pol' lee teen) [Gr. poly: many + taenia: ribbon] An adjective describing giant interphase chromosomes, such as those found in the salivary glands of fly larvae. The characteristic, reproducible pattern of bands and bulges seen on these chromosomes has provided a method for preparing detailed chromosome maps of several organisms.

Pons [L. pons: bridge] Region of the brain stem anterior to the medulla.

Population Any group of organisms coexisting at the same time and in the same place and capable of interbreeding with one another.

Population density The number of individuals (or modules) of a population in a unit of area or volume.

Population dynamics Changes in the distribution and abundance of individuals in a population.

Population structure The proportions of individuals in a population belonging to different age classes (age structure). Also, the distribution of the population in space.

Population vulnerability analysis (PVA) A determination of the risk of extinction of a population given its current size and distribution.

Portal vein A vein connecting two capillary beds, as in the hepatic portal system.

Positive control The situation in which a regulatory macromolecule is needed to turn transcription of structural genes on. In its absence, transcription will not occur.

Positive cooperativity Occurs when a molecule can bind several ligands and each one that binds alters the conformation of the molecule so that it can bind the next ligand more easily. The binding of four molecules of O_2 by hemoglobin is an example of positive cooperativity.

Positive feedback A regulatory system in which an error signal stimulates responses that increase the error.

Postabsorptive period When there is no food in the gut and no nutrients are being absorbed.

Posterior Toward or pertaining to the rear.

Postsynaptic cell The cell whose membranes receive the neurotransmitter released at a synapse.

Postzygotic isolating mechanism Any factor that reduces the viability of zygotes resulting from matings between individuals of different species.

Predator An organism that kills and eats other organisms. Predation is usually thought of as involving the consumption of animals by animals, but it can also mean the eating of plants.

Presynaptic excitation/inhibition Occurs when a neuron modifies activity at a synapse by releasing a neurotransmitter onto the presynaptic nerve terminal.

Prey [L. praeda: booty] An organism consumed as an energy source.

Prezygotic isolating mechanism A mechanism that reduces the probability that individuals of different species will mate.

Primary active transport Form of active transport in which ATP is hydrolyzed, yielding the energy required to transport ions against their concentration gradients. (Contrast with secondary active transport.)

Primary growth In plants, growth produced by the apical meristems. (Contrast with secondary growth.)

Primary producer A photosynthetic or chemosynthetic organism that synthesizes complex organic molecules from simple inorganic ones.

Primary succession Succession that begins in an areas initially devoid of life, such as on recently exposed glacial till or lava flows.

Primary structure The specific sequence of amino acids in a protein.

Primary wall Cellulose-rich cell wall layers laid down by a growing plant cell.

Primate (pry' mate) A member of the order Primates, such as a lemur, monkey, ape, or human.

Primer A short, single-stranded segment of DNA serving as the necessary starting material for the synthesis of a new DNA strand, which is synthesized from the 3' end of the primer.

Primitive streak A line running axially along the blastodisc, the site of inward cell migration during formation of the three-layered embryo. Formed in the embryos of birds and fish.

Primordium [L. primordium: origin] The most rudimentary stage of an organ or other part.

Pro- [L.: first, before, favoring] A prefix often used in biology to denote a developmental stage that comes first or an evolutionary form that appeared earlier than another. For example, prokaryote, prophase.

Probe A segment of single stranded nucleic acid used to identify DNA molecules containing the complementary sequence.

Procambium Primary meristem that produces the vascular tissue.

Progesterone [L. pro: favoring + gestare: to bear] A vertebrate female sex hormone that maintains pregnancy.

Prokaryotes (pro kar' ry otes) [L. pro: before + Gk. karyon: kernel, nucleus] Organisms whose genetic material is not contained within a nucleus. The bacteria. Considered an earlier stage in the evolution of life than the eukaryotes.

Prometaphase The phase of nuclear division that begins with the disintegration of the nuclear envelope.

Promoter The region of an operon that acts as the initial binding site for RNA polymerase.

Proofreading The correction of an error in DNA replication just after an incorrectly paired base is added to the growing polynucleotide chain.

Prophage (pro' fayj) The noninfectious units that are linked with the chromosomes of the host bacteria and multiply with them but do not cause dissolution of the cell. Prophage can later enter into the lytic phase to complete the virus life cycle.

Prophase (pro' phase) The first stage of nuclear division, during which chromosomes condense from diffuse, threadlike material to discrete, compact bodies.

Proplastid [Gr. *pro*: before + *plastos*: molded] A plant cell organelle which under appropriate conditions will develop into a plastid, usually the photosynthetic chloroplast. If plants are kept in the dark, proplastids may become quite large and complex.

Prostaglandin Any one of a group of specialized lipids with hormone-like functions. It is not clear that they act at any considerable distance from the site of their production.

Prosthetic group Any nonprotein portion of an enzyme.

Protease (pro' tee ase) See proteolytic enzyme.

Protein (pro' teen) [Gr. *protos*: first] One of the most fundamental building substances of living organisms. A long-chain polymer of amino acids with twenty different common side chains. Occurs with its polymer chain extended in fibrous proteins, or coiled into a compact macromolecule in enzymes and other globular proteins.

Proteolytic enzyme An enzyme whose main catalytic function is the digestion of a protein or polypeptide chain. The digestive enzymes trypsin, pepsin, and carboxypeptidase are all proteolytic enzymes (proteases).

Protist A member of the kingdom Protista, which consists of those eukaryotes not included in the kingdoms Animalia, Fungi, or Plantae. Many protists are unicellular. The kingdom Protista includes protozoa, algae, and fungus-like protists.

Protoderm Primary meristem that gives rise to epidermis.

Proton (pro' ton) [Gr. *protos*: first] One of the three most fundamental particles of matter, with mass approximately 1 amu and an electrical charge of +1.

Proton motive force The proton gradient and electric charge difference produced by chemiosmotic proton pumping. It drives protons back across the membrane, with the concomitant formation of ATP.

Protonema (pro' tow nee' mah) [Gr. *protos*: first + *nema*: thread] The hairlike growth form that constitutes an early stage in the development of a moss gametophyte.

Proto-oncogenes The normal alleles of genes possessing oncogenes (cancer-causing genes) as mutant alleles. Proto-oncogenes encode growth factors and receptor proteins.

Protoplast A cell that would normally have a cell wall, from which the wall has been removed by enzymatic digestion or by special growth conditions.

Protostome One of two major lines of animal evolution, characterized by spiral, determinate cleavage of the egg, and by schizocoelous development. (Contrast with deuterostome.)

Prototroph (pro' tow trofe') [Gr. *protos*: first + *trophein*: to nourish] The nutritional wild type, or reference form, of an organism. Any deviant form that requires growth nutrients not required by the prototrophic form is said to be a nutritional mutant, or auxotroph.

Protozoa A group of single-celled organisms classified by some biologists as a single phylum; includes the flagellates, amoebas, and ciliates. This textbook follows most modern classifications in elevating the protozoans to a distinct kingdom (Protista) and each of their major subgroups to the rank of phylum.

Provincialized A biogeographic term referring to the separation, by environmental barriers, of the biota into units with distinct species compositions.

Provirus See prophage.

Proximal Near the point of attachment or other reference point. (Contrast with distal.)

Pseudocoelom A body cavity not surrounded by a peritoneum. Characteristic of nematodes and rotifers.

Pseudogene A DNA segment that is homologous to a functional gene but contains a nucleotide change that prevents its expression.

Pseudoplasmodium [Gr. *pseudes*: false + *plasma*: mold or form] In the cellular slime molds such as *Dictyostelium*, an aggregation of single amoeboid cells. Occurs prior to formation of a fruiting structure.

Pseudopod (soo' do pod) [Gr. *pseudes*: false + *podos*: foot] A temporary, soft extension of the cell body that is used in location, attachment to surfaces, or engulfing particles.

Pulmonary Pertaining to the lungs.

Punctuated equilibrium An evolutionary pattern in which periods of rapid change are separated by longer periods of little or no change.

Pupa (pew' pa) [L.: doll, puppet] In certain insects (the Holometabola), the encased developmental stage that intervenes between the larva and the adult.

Pupil The opening in the vertebrate eye through which light passes.

Purine (pure' een) A type of nitrogenous base. The purines adenine and guanine are found in nucleic acids.

Purkinje fibers Specialized heart muscle cells that conduct excitation throughout the ventricular muscle.

Pyramid of biomass Graphical representation of the total masses at different trophic levels in an ecosystem.

Pyramid of energy Graphical representation of the total energy contents at different trophic levels in an ecosystem.

Pyrimidine (peer im' a deen) A type of nitrogenous base. The pyrimidines cytosine, thymine, and uracil are found in nucleic acids.

Pyrogen A substance that causes fever.

Pyruvate A three-carbon acid; the end product of glycolysis and the raw material for the citric acid cycle.

Q_{10} A value that compares the rate of a biochemical process or reaction over a 10°C range of temperature. A process that is not temperature-sensitive has a Q_{10} of 1. Values of 2 or 3 mean the reaction speeds up as temperature increases.

Quantum (kwon' tum) [L. *quantus*: how great] An indivisible unit of energy.

Quaternary structure Of aggregating proteins, the arrangement of polypeptide subunits.

R factor (resistance factor) A plasmid that contains one or more genes that encode resistance to antibiotics.

Radial symmetry The condition in which two halves of a body are mirror images of each other regardless of the angle of the cut, providing the cut is made along the center line. Thus, a cylinder cut lengthwise down its center displays this form of symmetry. (Contrast with bilateral symmetry.)

Radioisotope A radioactive isotope of an element. Examples are carbon-14 (^{14}C) and hydrogen-3, or tritium (^{3}H).

Radiotherapy Treatment, as of cancer, with X or gamma rays.

Rain shadow A region of low precipitation on the leeward side of a mountain range.

Ramet The repeated morphological units of sessile, modular organisms. (Contrast with genet.)

Random drift Evolution (change in gene proportions) by chance processes alone.

Rate constant Of a particular chemical reaction, a constant which, when multiplied by the concentration(s) of reactant(s), gives the rate of the reaction.

Reactant A chemical substance that enters into a chemical reaction with another substance.

Reaction, chemical A process in which atoms combine or change bonding partners.

Reaction wood Modified wood produced in branches in response to gravitational stimulation. Gymnosperms produce compression wood that tends to push the branch up; angiosperms produce tension wood that tends to pull the branch up.

Realized niche The actual niche occupied by an organism; it differs from the fundamental niche because of the presence of other species.

Receptacle [L. *receptaculum*: reservoir] In an angiosperm flower, the end of the stem to which all of the various flower parts are attached.

Receptive field Of a neuron, the area on the retina from which the activity of that neuron can be influenced.

Receptor-mediated endocytosis A form of endocytosis in which macromolecules in the environment bind specific receptor proteins in the plasma membrane and are brought into the cell interior in coated vesicles.

Receptor potential The change in the resting potential of a sensory cell when it is stimulated.

Recessive See dominance.

Reciprocal altruism The exchange of altruistic acts between two or more individuals. The acts may be separated considerably in time.

Reciprocal crosses A pair of crosses, in one of which a female of genotype A mates with a male of genotype B and in the other of which a female of genotype B mates with a male of genotype A.

Recognition site (also called a restriction site) A sequence of nucleotides in DNA to which a restriction enzyme binds and then cuts the DNA.

Recombinant An individual, meiotic product, or single chromosome in which genetic materials originally present in two individuals end up in the same haploid complement of genes. The reshuffling of genes can be either by independent segragation, or by crossing over between homologous chromosomes. For example, a human may pass on genes from both parents in a single haploid gamete.

Recombinant DNA technology The application of genetic tools (restriction endonucleases, plasmids, and transformation) to the production of specific proteins by biological "factories" such as bacteria.

Rectum The terminal portion of the gut, ending at the anus.

Redox reaction A chemical reaction in which one reactant becomes oxidized and the other becomes reduced.

Reducing agent A substance that can donate electrons to another substance. The reducing agent becomes oxidized, and its partner becomes reduced.

Reduction (re duk' shun) Gain of electrons; the reverse of oxidation. Most reductions lead to the storage of chemical energy, which can be released later by an oxidation reaction. Energy storage compounds such as sugars and fats are highly reduced compounds. (Contrast with oxidation.)

Reflex An automatic action, involving only a few neurons (in vertebrates, often in the spinal cord), in which a motor response swiftly follows a sensory stimulus.

Refractory period Of a neuron, the time interval after an action potential, during which another action potential cannot be elicited.

Regulative development A pattern of animal embryonic development in which the fates of the first blastomeres are not absolutely fixed. (Contrast with mosaic development.)

Regulatory gene A gene that contains the information for making a regulatory macromolecule, often a repressor protein.

Releaser A sensory stimulus that triggers a fixed action pattern.

Releasing hormone One of several hypothalamic hormones that stimulates the secretion of anterior pituitary hormone.

REM sleep A sleep state characterized by dreaming, skeletal muscle relaxation, and rapid eye movements.

Renal [L. *renes*: kidneys] Relating to the kidneys.

Replica plating A technique used in the selection of colonies of cells with a desired genotype.

Replication fork A point at which a DNA molecule is replicating. The fork forms by the unwinding of the parent molecule.

Repressible enzyme An enzyme whose synthesis can be decreased or prevented by the presence of a particular compound. A repressible opren often controls the sytthesis of such an enzyme.

Repressor A protein coded by the regulatory gene. The repressor can bind to a specific operator and prevent transcription of the operon.

Reproductive isolating mechanism Any trait that prevents individuals from two different populations from producing fertile hybrids.

Reproductive isolation The condition in which a population is not exchanging genes with other populations of the same species.

Reproductive value The expected contribution of an individual of a particular age to the future growth of the population to which it belongs.

Rescue effect The avoidance of extinction by immigration of individuals from other populations.

Resolving power Of an optical device such as a microscope, the smallest distance between two lines that allows the lines to be seen as separate from one another.

Resource Something in the environment required by an organism for its maintenance and growth that is consumed in the process of being used.

Resource defense polygamy A breeding system in which individuals of one sex (usually males) defend resources that are attractive to individuals of the other sex (usually females); individuals holding better resources attract more mates.

Respiration (res pi ra' shun) [L. *spirare*: to breathe] (1) Cellular respiration; the oxidation of the end products of glycolysis with the storage of much energy in ATP. The oxidant in the respiration of eukaryotes is oxygen gas. Some bacteria can use nitrate or sulfate instead of O_2. (2) Breathing.

Respiratory chain The terminal reactions of cellular respiration, in which electrons are passed from NAD or FAD, through a series of intermediate carriers, to molecular oxygen, with the concomitant production of ATP.

Respiratory uncoupler A substance that allows protons to cross the inner mitochondrial membrane without the concomitant formation of ATP, thus uncoupling respiration from phosphorylation.

Resting potential The membrane potential of a living cell at rest. In cells at rest, the interior is negative to the exterior. (Contrast with action potential, electrotonic potential.)

Restoration ecology The science and practice of restoring damaged or degraded ecosystems.

Restriction endonuclease Any one of several enzymes, produced by bacteria, that break foreign DNA molecules at very specific sites. Some produce "sticky ends." Extensively used in recombinant DNA technology.

Restriction map A partial genetic map of a DNA molecule, showing the points at which particular restriction endonuclease recognition sites reside.

Reticular system A central region of the vertebrate brain stem that includes complex fiber tracts conveying neural signals between the forebrain and the spinal cord, with collateral fibers to a variety of nuclei that are involved in autonomic functions, including arousal from sleep.

Retina (rett' in uh) [L. *rete*: net] The light-sensitive layer of cells in the vertebrate or cephalopod eye.

Retinal The light-absorbing portion of visual pigment molecules. Derived from β-carotene.

Retrovirus An RNA virus that contains reverse transcriptase. Its RNA serves as a template for cDNA production, and the cDNA is integrated into a chromosome of the mammalian host cell.

Reverse transcriptase An enzyme that catalyzes the production of DNA (cDNA), using RNA as a template; essential to the reproduction of retroviruses.

Reversion (genetic) A mutational event that restores wild type phenotype to a mutant.

RFLP (Restriction fragment length polymorphism) Coexistence of two or more patterns of restriction fragments (patterns produced by restriction enzymes), as revealed by a probe. The polymorphism reflects a difference in DNA sequence on homologous chromosomes.

Rhizoids (rye' zoids) [Gr. *rhiza*: root] Hairlike extensions of cells in mosses, liverworts, and a few vascular plants that serve the same function as roots and root hairs in vascular plants. The term is also applied to branched, rootlike extensions of some fungi and algae.

Rhizome (rye' zome) [Gr. *rhizoma*: mass of roots]   A special underground stem (as opposed to root) that runs horizontally beneath the ground.

Rhodopsin   A photopigment used in the visual process of transducing photons of light into changes in the membrane potential of photosensory cells.

Ribonucleic acid   See RNA.

Ribose (rye' bose)   A sugar of chemical formula $C_5H_{10}O_5$, one of the building blocks of ribonucleic acids.

Ribosomal RNA (rRNA)   Several species of RNA that are incorporated into the ribosome.

Ribosome   A small organelle that is the site of protein synthesis.

Ribozyme   An RNA molecule with catalytic activity.

Ribulose 1,5-bisphosphate (RuBP)   The compound in chloroplasts which reacts with carbon dioxide in the first reaction of the Calvin-Benson cycle.

Risk cost   The increased chance of being injured or killed as a result of performing a behavior, compared to resting.

RNA (ribonucleic acid)   A nucleic acid using ribose. Various classes of RNA are involved in the transcription and translation of genetic information. RNA serves as the genetic storage material in some viruses.

RNA polymerase   An enzyme that catalyzes the formation of RNA from a DNA template.

RNA splicing   The last stage of RNA processing in eukaryotes, in which the transcripts of introns are excised through the action of small nuclear ribonucleoprotein particles (snRNP).

Rods   Light-sensitive cells (photosensors) in the retina. (Contrast with cones.)

Root cap   A thimble-shaped mass of cells, produced by the root apical meristem, that protects the meristem and that is the organ that perceives the gravitational stimulus in root gravitropism.

Root hair   A specialized epidermal cell with a long, thin process that absorbs water and minerals from the soil solution.

Round dance   The dance performed on the vertical surface of a honeycomb by a returning honeybee forager when she has discovered a food source less than 100 meters from the hive.

Round window   A flexible membrane between the middle and inner ear that distributes pressure waves in the fluid of the inner ear.

rRNA   See ribosomal RNA.

Rubisco (RuBP carboxylase)   Enzyme that combines carbon dioxide with ribulose bisphosphate to produce 3-phosphoglycerate, the first product of C_3 photosynthesis. The most abundant protein on Earth.

Rumen (rew' mun)   The first division of the ruminant stomach. It stores and initiates bacterial fermentation of food. Food is regurgitated from the rumen for further chewing.

Ruminant   An herbivorous, cud-chewing mammal such as a cow, sheep, or deer, having a stomach consisting of four compartments.

S phase   In the cell cycle, the stage of interphase during which DNA is replicated. (Contrast with G_1 phase, G_2 phase.)

Sap   An aqueous solution of nutrients, minerals, and other substances that passes through the xylem of plants.

Saprobe [Gr. *sapros*:rotten + *bios*: life]   An organism (usually a bacterium or fungus) that obtains its carbon and energy directly from dead organic matter.

Sarcomere (sark' o meer) [Gr. *sark*: flesh + *meros*: a part]   The contractile unit of a skeletal muscle.

Saturated hydrocarbon   A compound consisting only of carbon and hydrogen, with the hydrogen atoms connected by single bonds.

Schizocoelous development   Formation of a coelom during embryological development by a splitting of mesodermal masses.

Schwann cell   A glial cell that wraps around part of the axon of a peripheral neuron, creating a myelin sheath.

Sclereid   A type of sclerenchyma cell, commonly found in nutshells, that is not elongated.

Sclerenchyma (skler eng' kyma)   A plant tissue composed of cells with heavily thickened cell walls, dead at functional maturity. The principal types of sclerenchyma cells are fibers and sclereids.

Second messenger   A signaling molecule that is created or actived inside the cell in response to activation of a receptor on the cell surface. The second messenger molecule then triggers the cell's response. An example is cyclic AMP.

Secondary active transport   Form of active transport in which ions or molecules are transported against their concentration gradient using energy obtained by relaxation of a gradient of sodium ion concentration rather than directly from ATP. (Contrast with primary active transport.)

Secondary compound   A compound synthesized by a plant that is not needed for basic cellular metabolism. Typically has an antiherbivore or antiparasite function.

Secondary growth   In plants, growth produced by vascular and cork cambia, contributing to an increase in girth. (Contrast with primary growth.)

Secondary structure   Of a protein, localized regularities of structure, such as the α helix and the β pleated sheet.

Secondary wall   Wall layers laid down by a plant cell that has ceased growing; often impregnated with lignin or suberin.

Second law of thermodynamics   States that in any real (irreversible) process, there is a decrease in free energy and an increase in entropy.

Second messenger   A compound, such as cyclic AMP, that is released within a target cell after a hormone or other "first messenger" has bound to a surface receptor on a cell; the second messenger triggers further reactions within the cell.

Secretin (si kreet' in)   A peptide hormone secreted by the upper region of the small intestine when acidic chyme is present. Stimulates the pancreatic duct to secrete bicarbonate ions.

Section   A thin slice, usually for microscopy, as a tangential section or a transverse section.

Seed   A fertilized, ripened ovule of a gymnosperm or angiosperm. Consists of the embryo, nutritive tissue, and a seed coat.

Seed crop   The number of seeds produced by a plant during a particular bout of reproduction.

Seedling   A young plant that has grown from a seed (rather than by grafting or by other means).

Segmentation genes   In insect larvae, genes that determine the number and polarity of larval segments.

Segment polarity genes   Genes that determine the boundaries and front-to-back organization of the segments in the *Drosophila* larva.

Segregation (genetic)   The separation of alleles, or of homologous chromosomes, from one another during meiosis so that each of the haploid daughter nuclei produced by meiosis contains one or the other member of the pair found in the diploid mother cell, but never both.

Selective permeability   A characteristic of a membrane, allowing certain substances to pass through while other substances are excluded.

Self-differentiating   Behavior that develops without experience with the normal objects toward which it is usually directed and without any practice. (See also instinct.)

Selfish act   A behavioral act that benefits its performer but harms the recipients.

Self-pollination   The fertilization of a plant by its own pollen. (Contrast with cross-pollination.)

Semelparous organism   An organism that reproduces only once in its lifetime. (Contrast with iteroparous.)

Semen (see' men) [L.: seed]   The thick, whitish liquid produced by the male reproductive organ in mammals, containing the sperm.

Semicircular canals   Part of the vestibular system of mammals.

Semiconservative replication   The common way in which DNA is synthesized. Each of the two partner strands in a double helix acts as a template for a new partner strand. Hence, after replication, each double helix consists of one old and one new strand.

Seminiferous tubules The tubules within the testes within which sperm production occurs.

Senescence [L. *senescere*: to grow old] Aging; deteriorative changes with aging.

Sensor A sensory cell; a cell transduces a physical or chemical stimulus into a membrane potential change.

Sensory neuron A neuron leading from a sensory cell to the central nervous system. (Contrast with motor neuron.)

Sepal (see' pul) One of the outermost structures of the flower, usually protective in function and enclosing the rest of the flower in the bud stage.

Septum [L.: partition] A membrane or wall between two cavities.

Sertoli cells Cells in the seminiferous tubules that nuture the developing sperm.

Serum That part of the blood plasma that remains after clots have formed and been removed.

Sessile (sess' ul) [L. *sedere*: to sit] Permanently attached; not moving.

Sertoli cells Cells in the seminiferous tubules that nuture the developing sperm.

Set point In a regulatory system, the threshold sensitivity to the feedback stimulus.

Sex chromosome In organisms with a chromosomal mechanism of sex determination, one of the chromosomes involved in sex determination. One sex chromosome, the X chromosome, is present in two copies in one sex and only one copy in the other sex. The autosomes, as opposed to the sex chromosomes, are present in two copies in both sexes. In many organisms, there is a second sex chromosome, the Y chromosome, that is found in only one sex—the sex having only one copy of the X.

Sexduction See F-duction.

Sex linkage The pattern of inheritance characteristic of genes located on the sex chromosomes of organisms having a chromosomal mechanism for sex determination. The sex that is diploid with respect to sex chromosomes can assume three genotypes: homozygous wild type, homozygous mutant, or heterozygous carrier. The other sex, haploid for sex chromosomes, is either hemizygous wild type or hemizygous mutant.

Sexuality The ability, by any of a multitude of mechanisms, to bring together in one individual genes that were originally carried by two different individuals. The capacity for genetic recombination.

Sexual selection Selection by one sex of characteristics in individuals of the opposite sex. Also, the favoring of characteristics in one sex as a result of competition among individuals of that sex for mates.

Shoot The aerial part of a vascular plant, consisting of the leaves, stem(s), and flowers.

Sieve plate In sieve tubes, the highly specialized end walls in which are concentrated the clusters of pores through which the protoplasts of adjacent sieve tube members are interconnected.

Sieve tube A column of specialized cells found in the phloem, specialized to conduct organic matter from sources (such as photosynthesizing leaves) to sinks (such as roots). Found principally in flowering plants.

Sieve tube member A single cell of a sieve tube, containing cytoplasm but relatively few organelles, with highly specialized perforated end walls leading to elements above and below.

Sign stimulus The single stimulus, or one out of a very few stimuli, by which an animal distinguishes key objects, such as an enemy, or a mate, or a place to nest, etc.

Signal sequence The sequence of a protein that directs the protein through a particular cellular membrane.

Signal transduction pathway The series of biochemical steps whereby a stimulus to a cell (such as a hormone or neurotransmitter binding to a receptor) is translated into a response of the cell.

Silencer A sequence of eukaryotic DNA that binds proteins that inhibit the transcription of an associated gene.

Silent mutations Genetic changes that do not lead to a phenotypic change. At the molecular level, these are DNA sequence changes that, because of the redundancy of the genetic code, result in the same amino acids in the resulting protein.

Similarity matrix A matrix to compare the structures of two molecules constructed by adding the number of their amino acids that are identical or different

Sinoatrial node (sigh' no ay' tree al) The pacemaker of the mammalian heart.

Sinus (sigh' nus) [L. *sinus*: a bend, hollow] A cavity in a bone, a tissue space, or an enlargement in a blood vessel.

Skeletal muscle See striated muscle.

Sliding filament theory A proposed mechanism of muscle contraction based on formation and breaking of crossbridges between actin and myosin filaments, causing them to slide together.

Small intestine The portion of the gut between the stomach and the colon, consisting of the duodenum, the jejunum, and the ileum.

Small nuclear ribonucleoprotein particle (snRNP) A complex of an enzyme and a small nuclear RNA molecule, functioning in RNA splicing.

Smooth muscle One of three types of muscle tissue. Usually consists of sheets of mononucleated cells innervated by the autonomic nervous system.

Society A group of individuals belonging to the same species and organized in a cooperative manner; in the broadest sense, includes parents and their offspring.

Sodium cotransport Carrier-mediated transport of molecules across membranes driven by sodium ions binding to the same carrier and moving down their concentration gradient.

Sodium–potassium pump The complex protein in plasma membranes that is responsible for primary active transport; it pumps sodium ions out of the cell and potassium ions into the cell, both against their concentration gradients.

Solute A substance that is dissolved in a liquid (solvent).

Solution A liquid (solvent) and its dissolved solutes.

Solvent A liquid that has dissolved or can dissolve one or more solutes.

Somatic [Gr. *soma*: body] Pertaining to the body, or body cells (rather than to germ cells).

Somite (so' might) One of the segments into which an embryo becomes divided longitudinally, leading to the eventual segmentation of the animal as illustrated by the spinal column, ribs, and associated muscles.

Southern blotting Transfer of DNA fragments from an electrophoretic gel to a sheet of paper or other absorbent material for analysis with a probe.

Spatial summation In the production or inhibition of action potentials in a postsynaptic neuron, the interaction of depolarizations and hyperpolarizations produced by several terminal boutons.

Spawning The direct release of sex cells into the water.

Specialized transduction In some types of bacteriophage (e.g., lambda), a prophage inserts at a specific location in the genome. When the prophage is induced to become lytic, it leaves the host chromosome and may take only the adjacent bacterial genes along with its phage DNA.

Speciation (spee' shee ay' shun) The process of splitting one population into two populations that are reproductively isolated from one another.

Species (spee' shees) [L.: kind] The basic lower unit of classification, consisting of a population or series of populations of closely related and similar organisms. The more narrowly defined "biological species" consists of individuals capable of interbreeding freely with each other but not with members of other species.

Species diversity A weighted representation of the species of organisms living in a region; large and common species are given greater weight than are small and rare ones. (Contrast with species richness.)

Species pool All the species potentially available to colonize a particular habitat.

Species richness The number of species of organisms living in a region. (Contrast with species diversity.)

Specific heat The amount of energy that must be absorbed by a gram of a substance to raise its temperature by one degree centigrade. By convention, water is assigned a specific heat of one.

Sperm [Gr. *sperma*: seed] A male reproductive cell.

Spermatocyte (spur mat' oh site) [Gr. *sperma*: seed + *kytos*: cell] The cell that gives rise to the sperm in animals.

Spermatogenesis (spur mat' oh jen' e sis) [Gr. *sperma*: seed + *genesis*: source] Male gametogenesis, leading to the production of sperm.

Spermatogonia Undifferentiated germ cells that give rise to primary spermatocytes and hence to sperm.

Sphincter (sfingk' ter) [Gr. *sphinkter*: that which binds tight] A ring of muscle that can close an orifice, for example at the anus.

Spindle apparatus An array of microtubules stretching from pole to pole of a dividing nucleus and playing a role in the movement of chromosomes at nuclear division. Named for its shape.

Spiracle (spy' rih kel) [L. *spirare*: to breathe] An opening of the treacheal respiratory system of terrestrial arthorpods.

Spiteful act A behavioral act that harms both the actor and the recipient of the act.

Spliceosome An RNA–protein complex that splices out introns from eukaryotic pre-mRNAs.

Splicing The removal of introns and connecting of exons in eukaryotic pre-mRNAs.

Spongy parenchyma In leaves, a layer of loosely packed photosynthetic cells with extensive intercellular spaces for gas diffusion. Frequently found between the palisade parenchyma and the lower epidermis.

Spontaneous generation The idea that life is generated continually from nonliving matter. Usually distinguished from the current idea that life evolved from nonliving matter under primordial conditions at an early stage in the history of earth.

Spontaneous reaction A chemical reaction which will proceed on its own, without any outside influence. A spontaneous reaction need not be rapid.

Sporangiophore [Gr. *phore*: to bear] Any branch bearing one or more sporangia.

Sporangium (spor an' gee um) [Gr. *spora*: seed + *angeion*: vessel or reservoir] In plants and fungi, any specialized stucture within which one or more spores are formed.

Spore [Gr. *spora*: seed] Any asexual reproductive cell capable of developing into an adult plant without gametic fusion. Haploid spores develop into gametophytes, diploid spores into sporophytes. In prokaryotes, a resistant cell capable of surviving unfavorable periods.

Sporophyll (spor' o fill) [Gr. *spora*: seed + *phyllon*: leaf] Any leaf or leaflike structure that bears sporangia; refers to carpels and stamens of angiosperms and to sporangium-bearing leaves on ferns, for example.

Sporophyte (spor' o fyte) [Gr. *spora*: seed + *phyton*: plant] In plants with alternation of generations, the diploid phase that produces the spores. (Contrast with gametophyte.)

Stabilizing selection Selection against the extreme phenotypes in a population, so that the intermediate types are favored. (Contrast with disruptive selection.)

Stamen (stay' men) [L.: thread] A male (pollen-producing) unit of a flower, usually composed of an anther, which bears the pollen, and a filament, which is a stalk supporting the anther.

Starch [O.E. *stearc*: stiff] An α-linked polymer of glucose; used by plants as a means of storing energy and carbon atoms.

Start codon The mRNA triplet (AUG) that acts as signals for the beginning of translation at the ribosome. (Compare with stop codons. There are a few mnior exceptions to these codons.)

Stasis Period during which little or no evolutionary change takes place within a lineage or groups of lineages.

Statocyst (stat' oh sist) [Gk. *statos*: stationary + *kystos*: pouch] An organ of equilibrium in some invertebrates.

Statolith (stat' oh lith) [Gk. *statos*: stationary + *lithos*: stone] A solid object that responds to gravity or movement and stimulates the mechanosensors of a statocyst.

Stele (steel) [Gr. *stele*: pillar] The central cylinder of vascular tissue in a plant stem.

Stem cell A cell capable of extensive proliferation, generating more stem cells and a large clone of differentiated progeny cells, as in the formation of red blood cells.

Step cline A sudden change in one or more traits of a species along a geographical gradient.

Steroid Any of numerous lipids based on a 17-carbon atom ring system.

Sticky ends On a piece of two-stranded DNA, short, complementary, one-stranded regions produced by the action of a restriction endonuclease. Sticky ends allow the joining of segments of DNA from different sources.

Stigma [L.: mark, brand] The part of the pistil at the apex of the style, which is receptive to pollen, and on which pollen germinates.

Stimulus Something causing a response; something in the environment detected by a receptor.

Stolon A horizontal stem that forms roots at intervals.

Stoma (plural: stomata) [Gr. *stoma*: mouth, opening] Small opening in the plant epidermis that permits gas exchange; bounded by a pair of guard cells whose osmotic status regulates the size of the opening.

Stop codons Triplets (UAG, UGA, UAA) in mRNA that act as signals for the end of translation at the ribosome. (See also start codon. There are a few mnior exceptions to these codons.)

Stratosphere The part of the atmosphere above the troposphere; extends upward to approximately 50 kilometers above the surface of the earth; contains very little water.

Stratum (plural strata) A layer or sedimentary rock laid down at a particular time in a past.

Striated muscle Contractile tissue characterized by multinucleated cells containing highly ordered arrangements of actin and myosin microfilaments. Also known as **skeletal muscle**.

Strobilus (strobe' a lus) [Gr. *strobilos*: a cone] The cone, or characteristic fruit, of the pine and other gymnosperms. Also, a cone-shaped mass of sprophylls found in club mosses.

Stroma The fluid contents of an organelle, such as a chloroplast.

Stromatolite A composite, flat-to-domed structure composed of successive mineral layers. Some are known to be produced by the action of bacteria in salt or fresh water, and some ancient ones are considered to be evidence for early life on the earth.

Structural formula A representation of the positions of atoms and bonds in a molecule.

Structural gene A gene that encodes the primary structure of a protein.

Style [Gr. *stylos*: pillar or column] In flowering plants, a column of tissue extending from the tip of the ovary, and bearing the stigma or receptive surface for pollen at its apex.

Sub- [L.: under] A prefix often used to designate a structure that lies beneath another or is less than another. For example, subcutaneous, subspecies.

Suberin A waxy material serving as a waterproofing agent in cork and in the Casparian strips of the endodermis in plants.

Submucosa (sub mew koe' sah) The tissue layer just under the epithelial lining of the lumen of the digestive tract. (Contrast with mucosa.)

Substrate (sub' strayte) The molecule or molecules on which an enzyme exerts catalytic action.

Substrate level phosphorylation ATP formation resulting from direct transfer of a phosphate group to ADP from an intermediate in glycolysis. (Contrast with oxidative phosphorylation.)

Succession In ecology, the gradual, sequential series of changes in species composition of a community following a disturbance.

Sulcus (plural: sulci) The valleys or creases between the raised portions of the convoluted surface of the brain. (Contrast to gyrus.)

Sulfhydryl group The —SH group.

Summation The ability of a neuron to fire action potentials in response to numerous subthreshold postsynaptic potentials arriving simultaneously at differentiated places on the cell, or arriving at the same site in rapid succession.

Suppressor T cells T cells that inhibit the responses of B cells and other T cells to antigens. (Contrast with cytotoxic T cells, helper T cells.)

Surface-to-volume ratio For any cell, organism, or geometrical solid, the ratio of surface area to volume; this is an important factor in setting an upper limit on the size a cell or organism can attain.

Surfactant A substance that decreases the surface tension of a liquid. Lung surfactant, secreted by cells of the alveoli, is mostly phospholipid and decreases the amount of work necessary to inflate the lungs.

Survivorship curve A plot of the logarithm of the fraction of individuals still alive, as a function of time.

Suspensor In plants, a cell or group of cells derived from the zygote, but not actually part of the embryo proper, which in some seed plants pushes the young embryo deeper into nutritive gametophyte tissue or endosperm by its growth.

Swim bladder An internal gas-filled organ that helps fishes maintain their position in the water column; later evolved into an organ for gas exchange in some lineages.

Symbiosis (sim' bee oh' sis) [Gr.: to live together] The living together of two or more species in a prolonged and intimate ecological relationship. (See parasitism, commensalism, mutualism.)

Symmetry In biology, the property that two halves of an object are mirror images of each other. (See bilateral symmetry and radial symmetry.)

Sympathetic nervous system A division of the autonomic (involuntary) nervous system. Its activities include increasing blood pressure and acceleration of the heartbeat. The neurotransmitter at the sympathetic terminals is epinephrine or norepinephrine. (Contrast with parasympathetic nervous system.)

Sympatric (sim pat' rik) [Gr. syn: together + patria: homeland] Referring to populations whose geographic regions overlap at least in part.

Sympatric speciation Formation of new species even though members of the daughter species overlap in their distribution during the speciation process. (Contrast with geographic, parapatric speciation.)

Symplast The continuous meshwork of the interiors of living cells in the plant body, resulting from the presence of plasmodesmata. (Contrast with apoplast.)

Symport A membrane transport process that carries two substances in the same direction across the membrane. (Contrast with antiport.)

Synapse (sin' aps) [Gr. syn: together + haptein: to fasten] The narrow gap between the terminal bouton of one neutron and the dendrite or cell body of another.

Synapsis (sin ap' sis) The highly specific parallel alignment (pairing) of homologous chromosomes during the first division of meiosis.

Synaptic vesicle A membrane-bounded vesicle, containing neurotransmitter, which is produced in and discharged by the presynaptic neuron.

Synergids (sin nur' jids) Two cells found close to the egg cell in the angiosperm embryo sac; they disappear shortly after fertilization.

Syngamy (sing' guh mee) [Gr. sun-: together + gamos: marriage] Union of gametes. Also known as fertilization.

Syrinx (sear' inks) [Gr.: pipe, cavity] A specialized structure at the junction of the trachea and the primary bronchi leading to the lungs. The vocal organ of birds.

Systematics The scientific study of the diversity of organisms.

Systemic circulation The part of the circulatory system serving those parts of the body other than the lungs or gills.

Systole (sis' tuh lee) [Gr.: contraction] Contraction of a chamber of the heart, driving blood forward in the circulatory system.

T cell A type of lymphocyte, involved in the cellular immune response. The final stages of its development occur in the thymus gland. (Contrast with B cell; see also cytotoxic T cell, helper T cell, suppressor T cell.)

T cell receptor A protein on the surface of a T cell that recognizes the antigenic determinant for which the cell is specific.

T tubules A system of tubules that runs throughout the cytoplasm of muscle fibers, through which action potentials spread.

Target cell A cell with the appropriate receptors to bind and respond to a particular hormone or other chemical mediator.

Taste bud A structure in the epithelium of the tongue that includes a cluster of chemosensors innervated by sensory neurons.

TATA box An eight-base-pair sequence, found about 25 base pairs before the starting point for transcription in many eukaryotic promoters, that binds a transcription factor and thus helps initiate transcription.

Taxis (tak' sis) [Gr. taxis: arrange, put in order] The movement of an organism in a particular direction with reference to a stimulus. A taxis usually involves the employment of one sense and a movement directly toward or away from the stimulus, or else the maintenance of a constant angle to it. Thus a positive phototaxis is movement toward a light source, negative geotaxis is movement upward (away from gravity), and so on.

Taxon A unit in a taxonomic system.

Taxonomy (taks on' oh me) [Gr. taxis: arrange, classify] The science of classification of organisms.

Telomeres (tee' lo merz) [Gr. telos: end] Repeated DNA sequences at the ends of eukaryotic chromosomes.

Telophase (tee' lo phase) [Gr. telos: end] The final phase of mitosis or meiosis during which chromosomes became diffuse, nuclear envelopes reform, and nucleoli begin to reappear in the daughter nuclei.

Template In biochemistry, a molecule or surface upon which another molecule is synthesized in complementary fashion, as in the replication of DNA. In the brain, a pattern that responds to a normal input but not to incorrect inputs.

Template strand In a stretch of double-stranded DNA, the strand that is transcribed.

Temporal summation In the production or inhibition of action potentials in a postsynaptic neuron, the interaction of depolarizations or hyperpolarizations produced by rapidly repeated stimulation of a single point.

Tendon A collagen-containing band of tissue that connects a muscle with a bone.

Tepal In an angiosperm flower, a sterile modified leaf. This term is used to refer to such flower parts when one is unable to distinguish between petals and sepals.

Terminal transferase An enzyme that adds nucleotides to free ends of DNA, without reference to a template strand.

Terrestrial (ter res' tree al) [L. terra: earth] Pertaining to the land.

Territory A fixed area from which an animal or group of animals excludes other members of the same species by aggressive behavior or display.

Tertiary structure In reference to a protein, the relative locations in three-dimensional space of all the atoms in the molecule. The overall shape of a protein. (Contrast with primary, secondary, and quaternary structures.)

Test cross A cross of a dominant-phenotype individual (which may be either heterozygous or homozygous) with a homozygous-recessive individual.

Testis (tes' tis) (plural: testes) [L.: witness] The male gonad; that is, the organ that produces the male sex cells.

Testosterone (tes toss' tuhr own) A male sex steroid hormone.

Tetanus [Gr. tetanos: stretched] (1) In physiology, a state of sustained, maximal muscular contraction caused by rapidly repeated stimulation. (2) In medicine, an often-fatal disease ("lockjaw") caused by the bacterium Clostridium tetani.

Thalamus A region of the vertebrate forebrain; involved in integration of sensory input.

Thallus (thal' us) [Gr.: sprout] Any algal body which is not differentiated into root, stem, and leaf.

Thermocline In a body of water, the zone where the temperatures change abruptly to about 4°C.

Thermoneutral zone The range of temperatures over which an endotherm does not have to expend extra energy to thermoregulate.

Thermosensor A cell or structure that responds to changes in temperature.

Thoracic cavity The portion of the mammalian body cavity bounded by the ribs, shoulders, and diaphragm. Contains the heart and the lungs.

Thorax In an insect, the middle region of the body, between the head and abdomen. In mammals, the part of the body between the neck and the diaphragm.

Thrombin An enzyme that converts fibrinogen to fibrin, thus triggering the formation of blood clots.

Thrombus (throm' bus) [Gk. thrombos: clot] A blood clot that forms within a blood vessel and remains attached to the wall of the vessel. (Contrast with embolus.)

Thylakoid A flattened sac within a chloroplast. The membranes of the numerous thylakoids contain all of the chlorophyll in a plant, in addition to the electron carriers of photophosphorylation. Thylakoids stack to form grana.

Thymine A nitrogen-containing base found in DNA.

Thymus A ductless, glandular portion of the lymphoid system, involved in development of the immune system of vertebrates.

Thyroid [Gr. *thyreos*: door-shaped] A two-lobed gland in vertebrates. Produces the hormone thyroxin.

Thyrotropic hormone A hormone that is produced in the pituitary gland of amphibia such as frogs and transported in the blood-stream to the thyroid gland, inducing the thyroid gland to produce the thyroid hormone that regulates metamorphosis from tadpole to adult frog.

Tight junction A junction between epithelial cells, in which there is no gap whatever between the adjacent cells. Materials may get through a tight junction only by entering the epithelial cells themselves.

Tissue A group of similar cells organized into a functional unit and usually integrated with other tissues to form part of an organ such as a heart or leaf.

Tonus A low level of muscular tension that is maintained even when the bodyis at rest.

Tornaria (tor nare' e ah) [L. *tornus*: lathe] The free-swimming ciliated larva of certain echinoderms and hemichordates; its existence indicates the evolutionary relationship of these two groups.

Totipotency In a cell, the condition of possessing all the genetic information and other capacities necessary to form an entire individual.

Toxigenicity The ability of a bacterium to produce chemical substances injurious to the tissues of the host organism.

Trachea (tray' kee ah) [Gr. *trakhoia*: a small tube] A tube that carries air to the bronchi of the lungs of vertebrates, or to the cells of arthropods.

Tracheid (tray' kee id) A distinctive conducting and supporting cell found in the xylem of nearly all vascular plants, characterized by tapering ends and walls that are pitted but not perforated.

Tracheophytes [Gr. *trakhoia*: a small tube + *phyton*: plant] Those plants with xylem and phloem, including psilophytes, club mosses, horsetails, ferns, gymnosperms, and angiosperms. (Contrast with nontracheophytes.)

Trade winds The winds that blow toward the intertropical convergence zone from the northeast and southeast.

Trait One form of a character: Eye color is a character; brown eyes and blue eyes are traits.

Transcription The synthesis of RNA, using one strand of DNA as the template.

Transcription factors Proteins that assemble on a eukaryotic chromosome, allowing RNA polymerase II to perform transcription.

Transdetermination Alteration of the developmental fate of an imaginal disc in *Drosophila*.

Transduction (1) Transfer of genes from one bacterium to another, with a bacterial virus acting as the carrier of the genes. (2) In sensory cells, the transformation of a stimulus (e.g., light energy, sound pressure waves, chemical or electrical stimulants) into action potentials.

Transfection Uptake, incorporation, and expression of recombinant DNA.

Transfer cells A modified parenchyma cell that transports mineral ions from its cytoplasm into its cell wall, thus moving the ions from the symplast into the apoplast.

Transfer RNA (tRNA) A category of relatively small RNA molecules (about 75 nucleotides). Each kind of transfer RNA is able to accept a particular activated amino acid from its specific activating enzyme, after which the amino acid is added to a growing polypeptide chain.

Transformation Mechanism for transfer of genetic information in bacteria in which pure DNA extracted from bacteria of one genotype is taken in through the cell surface of bacteria of a different genotype and incorporated into the chromosome of the recipient cell. By extension, the term has come to be applied to phenomena in other organisms in which specific genetic alterations have been produced by treatment with purified DNA from genetically marked donors.

Transgenic Containing recombinant DNA incorporated into its genetic material.

Translation The synthesis of a protein (polypeptide). This occurs on ribosomes, using the information encoded in messenger RNA.

Translocation (1) In genetics, a rare mutational event that moves a portion of a chromosome to a new location, generally on a nonhomologous chromosome. (2) In vascular plants, movement of solutes in the phloem.

Transpiration [L. *spirare*: to breathe] The evaporation of water from plant leaves and stem, driven by heat from the sun, and providing the motive force to raise water (plus ions) from the roots.

Transposable element A segment of DNA that can move to, or give rise to copies at, another locus on the same or a different chromosome. May be a single insertion sequence or a more complex structure (transposon) consisting of two insertion sequences and one or more intervening genes.

Trichocyst (trick' o sist) [Gr. *trichos*: hair + *kystis*: cell] A threadlike organelle ejected from the surface of ciliates, used both as a weapon and as an anchoring device.

Triglyceride A simple lipid in which three fatty acids are combined with one molecule of glycerol.

Triplet See codon.

Triplet repeat Occurrence of repeated triplet of bases in a gene, often leading to genetic disease, as does excessive repetition of CGG in the gene responsible for fragile-X syndrome.

Triploblastic Having three cell layers. (Contrast with diploblastic.)

Trisomic Containing three, rather than two members of a chromosome pair.

tRNA See transfer RNA.

Trochophore (troke' o fore) [Gr. *trochos*: wheel + *phoreus*: bearer] The free-swimming larva of some annelids and mollusks, distinguished by a wheel-like band of cilia around the middle, and indicating an evolutionary relationship between these two groups.

Trophic level A group of organisms united by obtaining their energy from the same part of the food web of a biological community.

Tropic hormones Hormones of the anterior pituitary that control the secretion of hormones by other endocrine glands.

Tropism [Gr. *tropos*: to turn] In plants, growth toward or away from a stimulus such as light (phototropism) or gravity (gravitropism).

Tropomyosin (troe poe my' oh sin) A protein that, along with actin, constitutes the thin filaments of myofibrils. It controls the interactions of actin and myosin necessary for muscle contraction.

Troposphere The atmospheric zone reaching upward approximately 17 km in the tropics and subtropics but only to about 10 km at higher latitudes. The zone in which virtually all the water vapor in the atmosphere is located.

Trypsin A protein-digesting enzyme. Secreted by the pancreas in its inactive form (trypsinogen), it becomes active in the duodenum of the small intestine.

T-tubules A set of transverse tubes that penetrates skeletal muscle fibers and terminates in the sarcoplasmic reticulum. The T-system transmits impulses to the sacs, which then release Ca^{2+} to initiate muscle contraction.

Tube foot In echinoderms, a part of the water vascular system. It grasps the substratum, prey, or other solid objects.

Tube nucleus In a pollen tube, the haploid nucleus that does not participate in double fertilization. (Contrast with generative nucleus.)

Tuber [L.: swelling] A short, fleshy underground stem, usually much enlarged, and serving a storage function, as in the case of the potato.

Tubulin A protein that polymerizes to form microtubules.

Tumor A disorganized mass of cells, often growing out of control. Malignant tumors spread to other parts of the body.

Tumor suppressor genes Genes which, when homozygous mutant, result in cancer. Such genes code for protein products that inhibit cell proliferation.

Turgor pressure   The actual physical (hydrostatic) pressure within a cell. (Contrast with osmotic potential, water potential.)

Twitch   A single unit of muscle contraction.

Tympanic membrane [Gr. *tympanum*: drum]   The eardrum.

Umbilical cord   Tissue made up of embryonic membranes and blood vessels that connects the embryo to the placenta in eutherian mammals.

Uncoupler   See respiratory uncoupler.

Understory   The aggregate of smaller plants growing beneath the canopy of dominant plants in a forest.

Unicellular (yoon' e sell' yer ler) [L. *unus*: one + *cella*: chamber]   Consisting of a single cell; as for example a unicellular organism. (Contrast with multicellular.)

Uniport   A membrane transport process that carries a single substance. (Contrast with antiport, symport.)

Unitary organism   An organism that consists of only one module.

Unsaturated hydrocarbon   A compound containing only carbon and hydrogen atoms. One or more pairs of carbon atoms are connected by double bonds.

Upwelling   The upward movement of nutrient-rich, cooler water from deeper layers of the ocean.

Urea   A compound serving as the main excreted form of nitrogen by many animals, including mammals.

Ureotelic   Describes an organism in which the final product of the breakdown of nitrogen-containing compounds (primarily proteins) is urea. (Contrast with ammonotelic, uricotelic.)

Ureter (your' uh tur) [Gr. *ouron*: urine]   A long duct leading from the vertebrate kidney to the urinary bladder or the cloaca.

Urethra (you ree' thra) [Gr. *ouron*: urine]   In most mammals, the canal through which urine is discharged from the bladder and which serves as the genital duct in males.

Uric acid   A compound that serves as the main excreted form of nitrogen in some animals, particularly those which must conserve water, such as birds, insects, and reptiles.

Uricotelic   Describes an organism in which the final product of the breakdown of nitrogen-containing compounds (primarily proteins) is uric acid. (Contrast with ammonotelic, ureotelic.)

Urinary bladder   A structure structure that receives urine from the kidneys via the ureter, stores it, and expels it periodically through the urethra.

Urine (you' rin) [Gk. *ouron*: urine]   In vertebrates, the fluid waste product containing the toxic nitrogenous by-products of protein and amino acid metabolism.

Uterus (yoo' ter us) [L.: womb]   The uterus or womb is a specialized portion of the female reproductive tract in certain mammals. It receives the fertilized egg and nurtures the embryo in its early development.

Vaccination   Injection of virus or bacteria or their proteins into the body, to induce immunization. The injected material is usually attenuated (weakened) before injection.

Vacuole (vac' yew ole) [Fr.: small vacuum]   A liquid-filled cavity in a cell, enclosed within a single membrane. Vacuoles play a wide variety of roles in cellular metabolism, some being digestive chambers, some storage chambers, some waste bins, and so forth.

Vagina (vuh jine' uh) [L.: sheath]   In female mammals, the passage leading from the external genital orifice to the uterus; receives the copulatory organ of the male in mating.

Van der Waals interaction   A weak attraction between atoms resulting from the interaction of the electrons of one atom with the nucleus of the other atom. This attraction is about one-fourth as strong as a hydrogen bond.

Variable regions   The part of an immunoglobulin molecule or T-cell receptor that includes the antigen-binding site.

Vascular (vas' kew lar)   Pertaining to organs and tissues that conduct fluid, such as blood vessels in animals and phloem and xylem in plants.

Vascular bundle   In vascular plants, a strand of vascular tissue, including conducting cells of xylem and phloem as well as thick-walled fibers.

Vascular ray   In vascular plants, radially oriented sheets of cells produced by the vascular cambium, carrying materials laterally between the wood and the phloem.

Vascular tissue system   The conductive system of the plant, consisting primarily of xylem and phloem. (Contrast with dermal tissue system, ground tissue system.)

Vasopressin   See antidiuretic hormone.

Vector   (1) An agent, such as an insect, that carries a pathogen affecting another species. (2) A plasmid or virus that carries an inserted piece of DNA into a bacterium for cloning purposes in recombinant DNA technology.

Vegetal hemisphere   The lower portion of some animal eggs, zygotes, and embryos, in which the dense nutrient yolk settles. The **vegetal pole** refers to the very bottom of the egg or embryo. (Contrast with animal hemisphere.)

Vegetative   Nonreproductive, or nonflowering, or asexual.

Vein [L. *vena*: channel]   A blood vessel that returns blood to the heart. (Contrast with artery.)

Vena cava [L.: hollow vein]   One of a pair of large veins that carry blood from the systemic circulatory system into the heart.

Ventral [L. *venter*: belly, womb]   Toward or pertaining to the belly or lower side. (Contrast with dorsal.)

Ventricle   A muscular heart chamber that pumps blood through the body.

Vernalization [L. *vernalis*: belonging to spring]   Events occurring during a required chilling period, leading eventually to flowering. Vernalization may require many weeks of below-freezing temperatures.

Vertebral column   The jointed, dorsal column that is the primary support structure of vertebrates.

Vertebrate   An animal whose nerve cord is enclosed in a backbone of bony segments, called vertebrae. The principal groups of vertebrate animals are the fishes, amphibians, reptiles, birds, and mammals.

Vessel [L. *vasculum*: a small vessel]   In botany, a tube-shaped portion of the xylem consisting of hollow cells (vessel elements) placed end to end and connected by perforations. Together with tracheids, vessel elements conduct water and minerals in the plant.

Vestibular apparatus (ves tib' yew lar) [L. *vestibulum*: an enclosed passage]   Structures associated with the vertebrate ear; these structures sense changes in position or momentum of the head, affecing balance and motor skills.

Vestigial (ves tij' ee al) [L. *vestigium*: footprint, track]   The remains of body structures that are no longer of adaptive value to the organism and therefore are not maintained by selection.

Vicariance (vye care' ee unce) [L. *vicus*: change]   The splitting of the range of a taxon by the imposition of some barrier to dispersal of its members. May lead to cladogenesis.

Vicariant distribution   A distribution resulting from the disruption of a formerly continuous range by a vicariant event.

Villus (vil' lus) (plural: villi) [L.: shaggy hair]   A hairlike projection from a membrane; for example, from many gut walls.

Virion (veer' e on)   The virus particle, the minimum unit capable of infecting a cell.

Viroid (vye' roid)   An infectious agent consisting of a single-stranded RNA molecule with no protein coat; produces diseases in plants.

Virus [L.: poison, slimy liquid]   Any of a group of ultramicroscopic infectious particles constructed of nucleic acid and protein (and, sometimes, lipid) that can reproduce only in living cells.

Visceral mass   The major internal organs of a mollusk.

Vitamin [L. *vita*: life]   Any one of several structurally unrelated organic compounds that an organism cannot synthesize itself, but nevertheless requires in small quantity for normal growth and metabolism.

Viviparous (vye vip' uh rus) [L. *vivus*: alive]   Reproduction in which fertilization of the egg and development of the embryo occur inside the mother's body. (Contrast with oviparous.)

Waggle dance   The running movement of a working honey bee on the hive, during which the worker traces out a repeated figure eight. The dance contains elements that transmit to other bees the location of the food.

Water potential   In osmosis, the tendency for a system (a cell or solution) to take up water from pure water, through a differentially permeable membrane. Water flows toward the system with a more negative water potential. (Contrast with osmotic potential, turgor pressure.)

Water vascular system   The array of canals and tubelike appendages that serves as the circulatory system, locomotory system, and food-capturing system of many echinoderms; is in direct connection with the surrounding sea water.

Wavelength   The distance between successive peaks of a wave train, such as electromagnetic radiation.

Wild type   Geneticists' term for standard or reference type. Deviants from this standard, even if the deviants are found in the wild, are said to be mutant.

Xanthophyll (zan' tho fill) [Gr. *xanthos*: yellowish-brown + *phyllon*: leaf]   A yellow or orange pigment commonly found as an accessory pigment in photosynthesis, but found elsewhere as well. An oxygen-containing carotenoid.

X chromosome   See sex chromosome.

X-linked (also called sex-linked)   A character that is coded for by a gene on the X chromosome.

Xerophyte (zee' row fyte) [Gr. *xerox*: dry + *phyton*: plant]   A plant adapted to an environment with a limited water supply.

Xylem (zy' lum) [Gr. *xylon*: wood]   In vascular plants, the woody tissue that conducts water and minerals; xylem consists, in various plants, of tracheids, vessel elements, fibers, and other highly specialized cells.

Y chromosome   See sex chromosome.

Yeast artificial chromosome   A laboratory-made DNA molecule containing sequences of yeast chromosomes (origin or replication, telomeres, centromere, and selectable markers) so that it can be used as a vector in yeast.

Yolk   The stored food material in animal eggs, usually rich in protein and lipid.

Z-DNA   A form of DNA in which the molecule spirals to the left rather than to the right.

Zooplankton (zoe' o plang ton) [Gr. *zoon*: animal + *planktos*: wandering]   The animal portion of the plankton.

Zoospore (zoe' o spore) [Gr. *zoon*: animal + *spora*: seed]   In algae and fungi, any swimming spore. May be diploid or haploid.

Zygospore   A highly resistant type of fungal spore produced by the zygomycetes (conjugating fungi).

Zygote (zye' gote) [Gr. *zygotos*: yoked]   The cell created by the union of two gametes, in which the gamete nuclei are also fused. The earliest stage of the diploid generation.

Zymogen   An inactive precursor of a digestive enzyme secreted into the lumen of the gut, where a protease cleaves it to form the active enzyme.

Answers to Self-Quizzes

Chapter 2

1. b	6. a
2. e	7. c
3. c	8. b
4. c	9. e
5. d	10. d

Chapter 3

1. e	6. a
2. d	7. c
3. c	8. e
4. d	9. a
5. b	10. d

Chapter 4

1. a	6. e
2. e	7. a
3. c	8. d
4. e	9. b
5. c	10. d

Chapter 5

1. e	6. b
2. d	7. c
3. a	8. b
4. d	9. e
5. c	10. c

Chapter 6

1. c	6. a
2. e	7. e
3. b	8. b
4. c	9. d
5. c	10. e

Chapter 7

1. a	6. d
2. d	7. a
3. c	8. e
4. e	9. c
5. c	10. e

Chapter 8

1. c	6. d
2. b	7. c
3. d	8. d
4. b	9. b
5. e	10. b

Chapter 9

1. e	6. a
2. c	7. e
3. b	8. d
4. d	9. b
5. c	10. a

Chapter 10*

1. d	6. d
2. a	7. b
3. e	8. a
4. d	9. b
5. d	10. c

Chapter 11

1. c	6. b
2. a	7. d
3. c	8. d
4. b	9. a
5. e	10. c

Chapter 12

1. c	6. d
2. d	7. b
3. c	8. d
4. d	9. d
5. b	10. a

Chapter 13

1. c	6. d
2. d	7. c
3. b	8. a
4. b	9. b
5. e	10. d

Chapter 14

1. c	6. c
2. d	7. c
3. c	8. b
4. a	9. e
5. c	10. d

Chapter 15

1. c	6. c
2. a	7. d
3. d	8. b
4. a	9. a
5. b	10. b

Chapter 16

1. b	6. b
2. d	7. c
3. a	8. a
4. c	9. c
5. b	10. e

Chapter 17

1. a	6. b
2. c	7. e
3. b	8. d
4. b	9. c
5. e	10. b

Chapter 18

1. d	6. a
2. b	7. d
3. e	8. d
4. e	9. a
5. c	10. d

Chapter 19

1. d	6. b
2. e	7. c
3. a	8. e
4. a	9. e
5. c	10. b

Chapter 20

1. d	6. e
2. c	7. b
3. d	8. e
4. b	9. d
5. d	10. e

Chapter 21

1. c	6. a
2. a	7. b
3. e	8. a
4. d	9. c
5. c	10. a

Chapter 22

1. e	6. d
2. c	7. a
3. a	8. d
4. c	9. b
5. a	10. d

Chapter 23

1. b	6. a
2. e	7. a
3. c	8. c
4. a	9. a
5. a	10. c

Chapter 24

1. e	6. c
2. d	7. a
3. e	8. a
4. c	9. d
5. e	10. b

Chapter 25

1. e	6. b
2. e	7. d
3. b	8. a
4. c	9. c
5. e	10. b

Chapter 26

1. a	6. d
2. e	7. c
3. c	8. b
4. d	9. b
5. a	10. d

Chapter 27

1. d	6. b
2. c	7. b
3. e	8. c
4. b	9. b
5. c	10. c

Chapter 28

1. b	6. a
2. d	7. e
3. e	8. a
4. c	9. c
5. d	10. c

Chapter 29

1. c	6. d
2. d	7. d
3. b	8. e
4. e	9. d
5. b	10. a

Chapter 30

1. b	7. d
2. c	8. e
3. a	9. a
4. d	10. a
5. c	11. c
6. b	12. c

Chapter 31

1. d	6. b
2. b	7. b
3. e	8. c
4. e	9. a
5. a	10. d

*Answers to the "Applying Concepts" questions in Chapter 10 appear at the end of this section.

Chapter 32
1. c	6. d
2. d	7. e
3. b	8. a
4. e	9. d
5. b	10. d

Chapter 33
1. e	6. a
2. b	7. b
3. c	8. c
4. c	9. d
5. d	10. a

Chapter 34
1. d	6. e
2. c	7. a
3. a	8. b
4. e	9. d
5. c	10. e

Chapter 35
1. a	6. c
2. e	7. e
3. c	8. c
4. d	9. a
5. b	10. b

Chapter 36
1. d	6. e
2. b	7. a
3. e	8. b
4. b	9. c
5. d	10. d

Chapter 37
1. c	6. e
2. a	7. a
3. d	8. e
4. b	9. d
5. b	10. c

Chapter 38
1. c	6. b
2. b	7. a
3. d	8. e
4. b	9. c
5. a	10. e

Chapter 39
1. (i) b	4. a
(ii) a	5. d
(iii) c	6. d
(iv) a,b,c	7. d
(v) a,b,c	8. d
2. c, e	9. c
3. e	10. d

Chapter 40
1. a	6. c
2. c	7. b
3. e	8. d
4. c	9. b
5. d	10. c

Chapter 41
1. d	6. e
2. a	7. e
3. d	8. c
4. c	9. d
5. c	10. a

Chapter 42
1. d	6. e
2. d	7. e
3. a	8. c
4. b	9. c
5. e	10. c

Chapter 43
1. c	6. c
2. a	7. a
3. e	8. b
4. d	9. a
5. d	10. b

Chapter 44
1. d	6. d
2. e	7. b
3. a	8. a
4. b	9. a
5. b	10. e

Chapter 45
1. e	6. b
2. d	7. c
3. a	8. c
4. b	9. a
5. c	10. d

Chapter 46
1. d	6. d
2. a	7. b
3. c	8. d
4. d	9. c
5. c	10. e

Chapter 47
1. b	6. d
2. e	7. a
3. c	8. b
4. a	9. d
5. b	10. d

Chapter 48
1. b	6. b
2. a	7. e
3. d	8. a
4. c	9. c
5. d	10. e

Chapter 49
1. a	6. c
2. e	7. d
3. b	8. c
4. b	9. d
5. a	10. a

Chapter 50
1. e	6. e
2. c	7. d
3. d	8. e
4. c	9. a
5. c	10. c

Chapter 51
1. b	6. e
2. e	7. a
3. d	8. c
4. a	9. d
5. d	10. a

Chapter 52
1. b	6. c
2. e	7. e
3. c	8. c
4. b	9. b
5. e	10. e

Chapter 53
1. b	6. a
2. d	7. e
3. d	8. a
4. b	9. d
5. c	10. e

Chapter 54
1. a	6. a
2. d	7. d
3. b	8. e
4. c	9. a
5. d	10. c

Chapter 55
1. b	6. a
2. c	7. d
3. c	8. e
4. e	9. e
5. a	10. a

Answers to "Applying Concepts" for Chapter 10, "Transmission Genetics: Mendel and Beyond"

1. Each of the eight boxes in the Punnet squares should contain the genotype Tt, regardless of which parent was tall and which dwarf.

2. Yellow parent = $s^Y s^b$; offspring 3 yellow (s^Y–): 1 black ($s^b s^b$). Black parent = $s^b s^b$; offspring all black ($s^b s^b$). Orange parent = $s^O s^b$; offspring 3 orange (s^O–): 1 black ($s^b s^b$). Both s^O and s^Y are dominant to s^b.

3. See Figure 10.5.

4. The trait is autosomal. Mother $dp\, dp$, father $Dp\, dp$. If the trait were sex-linked, all daughters would be wild-type and sons would be *dumpy*.

5. All females wild-type; all males spotted.

6. F_1 all wild-type, $PpSwsw$; F_2 9:3:3:1 in phenotypes. See Figure 10.8 for analogous genotypes.

7a. Ratio of phenotypes in F_2 is 3:1 (double dominant to double recessive).

7b. The F_1 are $Pby\, pB^Y$; they produce just two kinds of gametes (Pby and pBy). Combine them carefully and see the 1:2:1 phenotypic ratio fall out in the F_2.

7c. Pink-blistery.

7d. See Figures 9.15 and 9.17. Crossing over took place in the F_1 generation.

8. The genotypes are $PpSwsw$, $Ppswsw$, and $ppswsw$ in a ratio of 1:1:1:1.

9a. 1 black:2 blue:1 splashed white

9b. Always cross black with splashed white.

10a. $w^+ > w^e > w$

10b. Parents $w^e w$ and $w^+ Y$. Progeny $w+w^e$, $w+w$, $w^e Y$, and wY.

11. All will have normal vision because they inherit Dad's wild-type X chromosome, but half of them will be carriers.

12. Agouti parent $AaBb$. Albino offspring $aaBb$ and $aabb$; black offspring $Aabb$; agouti offspring $AaBb$.

13. Because the gene is carried on mitochondrial DNA, it is passed through the mother only. Thus if the woman does not have the disease but her husband does, their child will not be affected. On the other hand, if the woman has the disease but her husband does not, their child *will* have the disease.

Illustration Credits

Authors' Photographs
William K. Purves by Mark Cameron
Gordon H. Orians by Elizabeth N. Orians
H. Craig Heller by Meera Heller
David Sadava by Mark Cameron

Table of Contents
Cell division (mitosis) in a bean: Spike Walker/Tony Stone Images
Galapagos iguana: Frans Lanting/Minden Pictures
Star coral: Linda Pitkin
Poppies: Jan Tove Hohansson
Frost-covered plants: Darrell Gulin/Tony Stone Images
Rainbow lorikeet: Hans Christian Heap/TCL
Iceberg: B. and C. Alexander/Photo Researchers, Inc.

Part-Opener Photographs
Part One: Dr. Kari Lounatmaa/Photo Researchers, Inc.
Part Two: C. L. Rieder
Part Three: Staffan Widstrand
Part Four: Ted Mead/Woodfall Wild Images
Part Five: Adam Jones
Part Six: James Beveridge/Visuals Unlimited
Part Seven: Carr Clifton/Minden Pictures

Chapter 1 *Opener*: William M. Smithey Jr./Planet Earth Pictures. 1.1: T. Stevens and P. McKinley/Photo Researchers, Inc. 1.2: B. Dowsett/Photo Researchers, Inc. 1.3: Fred Marsik/Visuals Unlimited. 1.4: Michael Abbey/Visuals Unlimited. 1.5 *larva*: Valorie Hodgson/Visuals Unlimited. 1.5 *pupa*: Dick Poe/Visuals Unlimited. 1.5 *butterfly*: Bill Beatty/Visuals Unlimited. 1.6: K. and K. Amman/Planet Earth Pictures. 1.7a: Staffan Widstrand. 1.7b: David Kjaer/Masterfile. 1.7c: Art Wolfe. 1.7d: Jack Fields/Photo Researchers, Inc. C. *Darwin*: American Philosophical Society. 1.10: Marty Snyderman/Planet Earth Pictures. 1.11: Bruce S. Cushing/Visuals Unlimited. 1.12: Levi, W. 1965. *Encyclopedia of Pigeon Breeds*. T. F. H. Publications, Jersey City, NJ. (a,b: photos by R. L. Kienlen, courtesy of Ralston Purina Company; c,d: photos by Stauber.). 1.13: Doug Perrine/Planet Earth Pictures. 1.14a: D. Cavanaugh/Visuals Unlimited. 1.14b: J. H. Robinson/Photo Researchers, Inc. 1.15a: Gregory G. Dimijian/Photo Researchers, Inc. 1.15b: Sinclair Stammers/Photo Researchers, Inc.

Chapter 2 *Opener*: Alex Williams/Masterfile. 2.3: Courtesy of Walter Gehring. 2.17: Art Wolfe. 2.20: P. Armstrong/Visuals Unlimited.

Chapter 3 *Opener*: Dan Richardson. 3.1: Christopher Small. 3.5: Gavriel Jecan/Art Wolfe Inc. 3.13a: Biophoto Associates/Photo Researchers, Inc. 3.13b: W. F. Schadel, BPS*. 3.13c: CNRI, Science World Enterprises/BPS. 3.18a,b,c; 3.19; 3.23: Dan Richardson.

Chapter 4 *Opener*: Jeremy Burgess/Photo Researchers, Inc. 4.3: J. J. Cardamone Jr. & B. K. Pugashetti/BPS. 4.4a,b: Stanley C. Holt/BPS. 4.5a: J. J. Cardamone Jr./BPS. 4.5c: S. Abraham & E. H. Beachey, VA Medical Center, Memphis, TN. Table 4.1 *upper row*: David M. Phillips/Visuals Unlimited. Table 4.1 *lower left*: Conly L. Rieder/BPS. Table 4.1 *lower center*: David Albertini, Tufts Univ. School of Medicine. Table 4.1 *lower right*: M. Abbey/Photo Researchers, Inc. 4.7 *centrioles*: Barry F. King/BPS. 4.7 *mitochondrion*: K. Porter, D. Fawcett/Visuals Unlimited. 4.7 *rough ER*: Fred E. Hossler/Visuals Unlimited. 4.7 *plasma membrane*: J. David Robertson, Duke Univ. Medical Center. 4.7 *nucleolus*: Richard Rodewald/BPS. 4.7 *golgi apparatus*: Gary T. Cole/BPS. 4.7 *smooth ER*: David M. Phillips/Visuals Unlimited. 4.7 *cell wall*: David M. Phillips/Visuals Unlimited. 4.7 *chloroplast*: W. P. Wergin, courtesy of E. H. Newcomb/BPS. 4.8 *upper*: Richard Rodewald/BPS. 4.8 *lower*, 4.9: Larry Gerace, Scripps Research Institute. 4.10: Jim Solliday/BPS. 4.11: K. Porter, D. Fawcett/Visuals Unlimited. 4.12: Alfred Owezarzak/BPS. 4.13: W. P. Wergin, courtesy of E. H. Newcomb/BPS. 4.14: Chuck Davis/Tony Stone Images. 4.16: Don Fawcett/Visuals Unlimited. 4.17a: Gary T. Cole/BPS. 4.18a: K. G. Murti/Visuals Unlimited. 4.19: E. H. Newcomb & S. E. Frederick/BPS. 4.20: M. C. Ledbetter, Brookhaven National Laboratory. 4.22a: H. W. Beams and R. G. Kessel. 1976. *Am. Sci.* 64: 279–290. 4.22b: Stanley Flegler/Visuals Unlimited. 4.23: Fred E. Hossler/Visuals Unlimited. 4.25a,b: W. L. Dentler/BPS. 4.27c: Barry F. King/BPS. 4.28: David M. Phillips/Visuals Unlimited.

*BPS = Biological Photo Service

Chapter 5 *Opener* and 5.3: L. Andrew Staehelin, Univ. of Colorado. 5.6 *top*: D. S. Friend, Univ. of California, SF. 5.6 *center*: Darcy E. Kelly, Univ. of Washington. 5.6 *bottom*: Courtesy of C. Peracchia. 5.15: M. M. Perry. 1979. *J. Cell Sci.* 39, p. 26.

Chapter 6 *Opener*: Greg Epperson/Adventure Photo & Film. 6.2: Jonathan Scott/Masterfile. 6.8: E. R. Degginger/Photo Researchers, Inc. 6.17: Clive Freeman, The Royal Institution/Photo Researchers, Inc. 6.19: Dan Richardson.

Chapter 7 *Opener*: Antonia Deutsch/Tony Stone Images. 7.9b: Francis Leroy/Photo Researchers, Inc. 7.14: Ephraim Racker/BPS.

Chapter 8 *Opener*: Susan McCartney/Photo Researchers, Inc. 8.1: C. G. Van Dyke/Visuals Unlimited. 8.16a: Lawrence Radiation Lab., Univ. of California. 8.16b: J. A. Bassham, Lawrence Berkeley Lab., Univ. of California. 8.22; 8.23b: E. H. Newcomb & S. E. Frederick/BPS. 8.25 *left*: Arthur R. Hill/Visuals Unlimited. 8.26 *right*: David Matherly/Visuals Unlimited.

Chapter 9 *Opener*: Steve Rogers & Vladimir Gelfand. 9.1a: John D. Cunningham/Visuals Unlimited. 9.1b: David M. Phillips/Visuals Unlimited. 9.1c: John D. Cunningham/Visuals Unlimited. 9.2: Ruth Kavenoff, Designergenes Ltd., P.O. Box 100, Del Mar, CA 90214. 9.3: John J. Cardamone Jr./BPS. 9.6: G. F. Bahr/BPS. 9.8 *upper inset*: A. L. Olins/BPS. 9.8 *lower inset*: David Ward, Yale Univ. School of Medicine. 9.9: Andrew S. Bajer, Univ. of Oregon. 9.10b,c: C. L. Rieder/BPS. 9.11a: T. E. Schroeder/BPS. 9.11b: B. A. Palevitz & E. H. Newcomb/BPS. 9.12: Gary T. Cole/BPS. 9.14: Courtesy of Applied Spectral Imaging. 9.15: C. A. Hasenkampf/BPS. 9.16: B. John Cabisco. 9.20b: Gopal Murti/Photo Researchers, Inc.

Chapter 10 *Opener*: Russell & Sons, ca. 1894. Gernsheim Collection, courtesy of the Harry Ransom Humanities Research Center, Univ. of Texas, Austin. G. *Mendel*: Leslie Holzer/Photo Researchers, Inc. 10.3: R. W. Van Norman/Visuals Unlimited. 10.14: NCI/Photo Researchers, Inc. 10.16: MERO/JACANA/Photo Researchers, Inc. T. H. *Morgan*: Corbis Bettmann. 10.26: Science

VU/Visuals Unlimited. *Bay scallops*: Barbara J. Miller/BPS.

Chapter 11 *Opener*: Dan Richardson. 11.2: Lee D. Simon/Photo Researchers, Inc. 11.4: Courtesy of Prof. M. H. F. Wilkins, Dept. of Biophysics, King's College, Univ. of London. 11.6b *left*: Dan Richardson. 11.6b *right*: A. Barrington Brown/Photo Researchers, Inc.

Chapter 12 *Opener* and 12.6: Dan Richardson. 12.12b: Courtesy of J. E. Edstrom and *EMBO J.* 12.14: Michael Abbey/Photo Researchers, Inc.

Chapter 13 *Opener*: Rosenfeld Images LTD/Photo Researchers, Inc. 13.1a: D. L. D. Caspar, Brandeis Univ. 13.1b: Omikron/Photo Researchers, Inc. 13.1c: A. B Dowsett/Photo Researchers, Inc. 13.7 *upper*: Hans Gelderblom/Visuals Unlimited. 13.7 *lower*: Science VU/Visuals Unlimited. 13.9: Rich Hambert/BPS. 13.11: Courtesy of L. Caro and R. Curtiss.

Chapter 14 *Opener*: Andrew Syred/Tony Stone Images. 14.9: Tiemeier et al., *Cell* 14:237–246, 1978. 14.19: Karen Dyer, Vivigen. 14.20: O. L. Miller, Jr.

Chapter 15 *Opener*: Courtesy of E. B. Lewis. 15.1: AP Photo/PA Files/Wide World Photo. 15.7: J. E. Sulston and H. R. Horvitz. 1977. *Dev. Bio.* 56, p. 100. 15.8: Courtesy of S. Strome. 15.17: Courtesy of W. Driever and C. Nüsslein-Volhard. 15.18: Courtesy of C. Rushlow and M. Levine.

Chapter 16 *Opener*: James Holmes/Fulmer Research/Photo Researchers, Inc. 16.3: Philippe Plailly/Photo Researchers, Inc. 16.6b *upper*: N. Y. State Agricultural Experiment Station, Cornell Univ. 16.6b *lower*: J. S. Yun & T. E. Wagner, Ohio Univ. 16.8: Courtesy of In Vitrogen Corporation. 16.16: Courtesy of Novartis Seeds.

Chapter 17 *Opener*: UPI/Corbis-Bettmann. 17.4: C. Harrison et al. 1983. *J. Med. Genet.* 20, p. 280. 17.9: Courtesy of Harvey Levy and Cecelia Walraven, New England Newborn Screening Program. 17.12: P. P. H. De-Bruyn, Univ. of Chicago. 17.19: E. McCabe, UCLA Medical School

Chapter 18 *Opener*: Andrew Syred/Photo Researchers, Inc. 18.2: G. W. Willis/Tony Stone Images. 18.3: Ziedonis Skobe/BPS. 18.5: Courtesy of Lennart Nilsson/Boehringer Ingelheim GmbH. 18.11: R. Rodewald/BPS. 18.17: A. Liepins, Sloan-Kettering Research Inst.

Chapter 19 *Opener*: National Institutes of Health/Photo Researchers, Inc. 19.5: Peter Ward, Univ. of Washington. 19.6: W. B. Saunders/BPS. 19.9 *left*: Ken Lucas/BPS. 19.9 *right*: Stanley M. Awramik/BPS. 19.10b: S. Conway Morris. 19.12b: Tom McHugh/Field Museum, Chicago/Photo Researchers, Inc. 19.13: B. Miller/BPS. 19.15: Ludek Pesek/Photo Researchers, Inc.

Chapter 20 *Opener*: Jon Riley/Tony Stone Images. 20.7: Frank S. Balthis. 20.12a,b,c: Tom Vezo. 20.13: From R. Dawkins. 1969. *The Blind Watchmaker*. W. W. Norton, New York.

Chapter 21 *Opener*: Ray Coleman/Photo Researchers, Inc. 21.1a,b: Tom Vezo. 21.7: Anthony D. Bradshaw, Univ. of Liverpool. 21.8a: Virginia P. Weinland/Photo Researchers, Inc. 21.8b: Heather Angel/BIOFOTOS. 21.10: M. Patterson/Photo Researchers, Inc. 21.11: Tom Vezo. 21.12 *left*, *right*: Peter J. Bryant/BPS. 21.12 *center*: Kenneth Y. Kaneshiro, Univ. of Hawaii. 21.15 *left*: Elizabeth Orians. 21.15 *center*, *right*: Jim Denny.

Chapter 22 *Opener*: Gary Brettnacher/Adventure Photo & Film. 22.1a: Michael Giannechini/Photo Researchers, Inc. 22.1b: Helen Carr/BPS. 22.1c: Nigel Downer/Planet Earth Pictures. 22.3 *left*: Adam Jones. 22.3 *right*: Staffan Widstand. 22.7 *left*, *right*: Peter J. Bryant/BPS. 22.9 *upper*, *lower*: Art Wolfe.

Chapter 23 *Opener*; 23.1: Richard Alexander, Univ. of Pennsylvania. 23.5: Courtesy of E. B. Lewis.

Chapter 24 *Opener*: NASA/Photo Researchers, Inc. 24.8a: Sinclair Stammers/Photo Researchers, Inc. 24.8b: Elizabeth Orians.

Chapter 25 *Opener*: Kari Lounatmaa/Photo Researchers, Inc. 25.2: ASM/Visuals Unlimited. 25.3a: David Phillips/Photo Researchers, Inc. 25.3b: R. Kessel-G. Shih/Visuals Unlimited. 25.3c: Stanley Flegler/Visuals Unlimited. 25.4: T. J. Beveridge/BPS. 25.5a: J. A. Breznak and H. S. Pankratz/BPS. 25.5b: J. Robert Waaland/BPS. 25.6: George Musil/Visuals Unlimited. 25.7a *left*: S. C. Holt/BPS. 25.7a *center*: David M. Phillips/Visuals Unlimited. 25.7b *left*: Leon J. LeBeau/BPS. 25.7b *center*: A. J. J. Cardamone, Jr./BPS. 25.8: Alfred Pasieka/Photo Researchers, Inc. 25.9: H. W. Jannasch/BPS. 25.12: S. C. Holt/BPS. 25.13a: Paul W. Johnson/BPS. 25.13b: H. S. Pankratz/BPS. 25.13c: Bill Kamin/Visuals Unlimited. 25.14a: Paul W. Johnson/BPS. 25.14b: K. Stephens/BPS. 25.15: Science VU/Visuals Unlimited. 25.16: G. W. Willis/BPS. 25.17: D. A. Glawe/BPS. 25.18: Randall C. Cutlip/BPS. 25.19: T. J. Beveridge/BPS. 25.20: G. W. Willis/BPS. 25.21: Science VU/Visuals Unlimited. 25.22: Michael Gabridge/Visuals Unlimited. 25.24: Krafft-Explorer. 25.25: Martin G. Miller/Visuals Unlimited.

Chapter 26 *Opener*: M. Abbey/Visuals Unlimited. 26.1a: David Phillips/Visuals Unlimited. 26.1b: J. Paulin/Visuals Unlimited. 26.1c: Hal Beral/Visuals Unlimited. 26.7a: E. R. Degginger/Photo Researchers, Inc. 26.7b: Cabisco/Visuals Unlimited. 26.7c: M. Abbey/Photo Researchers, Inc. 26.8: David M. Phillips/Visuals Unlimited. 26.9: G. W. Willis/BPS. 26.11a: Robert Brons/BPS. 26.11b: A. M. Siegelman/Visuals Unlimited. 26.13a: Mike Abbey/Visuals Unlimited. 26.13b: M. Abbey/Photo Researchers, Inc. 26.13c,d: Paul W. Johnson/BPS. 26.14b: M. A. Jakus, NIH. 26.16: John D. Cunningham/Visuals Unlimited. 26.18a: Barbara J. Miller/BPS. 26.18b: Cabisco/Visuals Unlimited. 26.19a: D. W. Francis, Univ. of Delaware. 26.19b: David Scharf. 26.20: James W. Richardson/Visuals Unlimited. 26.21: Sanford Berry/Visuals Unlimited. 26.22a: Jan Hinsch/Photo Researchers, Inc. 26.22b: Biophoto Assoc./Photo Researchers, Inc. 26.24a: J. N. A. Lott/BPS. 26.24b: J. Robert Waaland/BPS. 26.25a: Paul Gier/Visuals Unlimited. 26.25b: J. N. A. Lott/BPS. 26.27a: Maria Schefter/BPS. 26.27b: J. N. A. Lott/BPS. 26.28a: Cabisco/Visuals Unlimited. 26.28b: Andrew J. Martinez/Photo Researchers, Inc. 26.28c: M. Abbey/Visuals Unlimited.

Chapter 27 *Opener upper*: Staffan Widstrand. *Opener lower*: Michael Busselle/Masterfile. 27.1: Peter F. Zika/Visuals Unlimited. 27.2: Ron Dengler/Visuals Unlimited. 27.4: Brian Enting/Photo Researchers, Inc. 27.5a,b: J. Robert Waaland/BPS. 27.6a: John D. Cunningham/Visuals Unlimited. 27.6b: William Harlow/Photo Researchers, Inc. 27.6c: Science VU/Visuals Unlimited. 27.8b: J. H. Troughton. 27.9: Farrell Grehan/Photo Researchers, Inc. 27.11: Figure information provided by Hermann Pfefferkorn, Dept. of Geology, Univ. of Pennsylvania. Original oil painting by John Woolsey. 27.16a: Bill Beatty/Visuals Unlimited. 27.16b: Cabisco/Visuals Unlimited. 27.17a: J. N. A. Lott/BPS. 27.17b: David Sieren/Visuals Unlimited. 27.18: W. Ormerod/Visuals Unlimited. 27.19a: Adam Jones. 27.19b: Nuridsany et Perennou/Photo Researchers, Inc. 27.19c: Dick Keen/Visuals Unlimited. 27.20: L. West/Photo Researchers, Inc. 27.23: Phil Gates/BPS. 27.24a: Richard Shiell. 27.24b: Bernd Wittich/Visuals Unlimited. 27.24c: M. Graybill/J. Hodder/BPS. 27.24d: Louisa Preston/Photo Researchers, Inc. 27.27a: Dick Poe/Visuals Unlimited. 27.27b,c; 27.28a: Richard Shiell. 27.28b: Noboru Komine/Photo Researchers, Inc. 27.30a: Ross Frid/Visuals Unlimited. 27.30b: Holt Studios/Photo Researchers, Inc. 27.30c: Catherine M. Pringle/BPS. 27.30d: Richard Shiell. 27.32a: Dora Lambrecht/Visuals Unlimited. 27.32b: Helen Carr/BPS. 27.32c; 27.33a: Richard Shiell. 27.33b: Adam Jones. 27.33c: Richard Shiell. 27.34: Rod Planck/Photo Researchers, Inc.

Chapter 28 *Opener*: Adam Jones. 28.1a: Jim W. Grace/Photo Researchers, Inc. 28.1b: L. E. Gilbert/BPS. 28.1c: G. L. Barron/BPS. 28.2: G. T. Cole/BPS. 28.3: N. Allin and G. L. Barron/BPS. 28.4: Milton H. Tierney, Jr./Visuals Unlimited. 28.5 *upper*: R. Calentine/Visuals Unlimited. 28.5 *lower*: John D. Cunningham/Visuals Unlimited. 28.7: J. Robert Waaland/BPS. 28.8: Gary R. Robinson/Visuals Unlimited. 28.9: James W. Richardson/Visuals Unlimited. 28.10: John D. Cunningham/Visuals Unlimited. 28.11*a*: Jim Solliday/BPS. 28.11*b*: L. West, National Audubon Society/Photo Researchers, Inc. 28.11*c*: Ray Coleman/Photo Researchers, Inc. 28.12: Centers for Disease Control, Atlanta. 28.14*a*: Angelina Lax/Photo Researchers, Inc. 28.14*b*: Manfred Danegger/Photo Researchers, Inc. 28.14*c*: Stan Flegler/Visuals Unlimited. 28.15: Biophoto Associates. 28.16: R. L. Peterson/BPS. 28.17: Paul A. Zahl/Photo Researchers, Inc. 28.18*a*: Gregory G. Dimijian/Photo Researchers, Inc. 28.18*b*: George Herben Photo/Visuals Unlimited. 28.18*c*: David Sieren/Visuals Unlimited. 28.19*a*: J. N. A. Lott/BPS.

Chapter 29 *Opener*: Rod Salm/Planet Earth Pictures. 29.1: Courtesy of R. L. Trelsted. 29.6 *left*: Gillian Lythgoe/Planet Earth Pictures. 29.6 *right*: Christian Petron/Planet Earth Pictures. 29.13*a*: Georgette Douwma/Planet Earth Pictures. 29.13*b*: Rod Salm/Planet Earth Pictures. 29.14*a*: Robert Brons/BPS. 29.14*b,c*: Chuck Davis Photo. 29.15: David J. Wrobel/BPS. 29.17*b*, 29.19*b*: James Solliday/BPS. 29.19*c*: Jim Solliday/BPS. 29.21*a*: Fred McConnaughey/Photo Researchers, Inc. 29.21*c*: Robert Brons/BPS. 29.25*c*: R. R. Hessler, Scripps Institute of Oceanography. 29.27*a*: Georgette Douwma/Planet Earth Pictures. 29.27*c*: Roger K. Burnard/BPS. 29.27*d*: Roger & Linda Mitchell. 29.29*a*: Ken Lucas/Planet Earth Pictures. 29.29*b*: Pete Atkinson/Masterfile. 29.29*c*: Chuck Davis Photo. 29.29*d*: Richard Humbert/BPS. 29.29*e*: Alex Kerstitch/Planet Earth Pictures. 29.29*f*: David J. Wrobel/BPS. 29.32: J. N. A. Lott/BPS. 29.33*a*: Joel Simon. 29.33*b*: Barbara Miller/BPS. 29.34*a*: Peter J. Bryant/BPS. 29.34*b*: David Maitland/Masterfile. 29.34*c*: L. E. Gilbert/BPS. 29.34*d*: Robert Brons/BPS. 29.36*a*: Gary Bell/Masterfile. 29.36*b*: Peter J. Byrant/BPS. 29.36*c*: David Wrobel/BPS. 29.36*d*: Geoff du Feu/Planet Earth Pictures. 29.37*a*: Charles R. Wyttenbach/BPS. 29.37*b*: Roger K. Burnard/BPS. 29.38*a*: R. F. Ashley/Visuals Unlimited. 29.38*b*: From Kristensen and Hallas, 1980. 29.39*a*: Richard Humbert/BPS. 29.39*b,c*: Peter J. Bryant/BPS. 29.39*d,g*: David Maitland/Masterfile. 29.39*e*: Steve Nicholls/Planet Earth Pictures. 29.39*f*: Brian Kenney/Planet Earth Pictures. 29.39*h*: Steve Hopkin/Planet Earth Pictures.

Chapter 30 *Opener*: Tim Davis/Tony Stone Images. 30.3; 30.4: David Wrobel/BPS. 30.5*a*: Ken Lucas/Planet Earth Pictures. 30.9*a*: Doug Perrine/DRK PHOTO. 30.9*b–e*: David Wrobel/BPS. 30.10: C. R. Wyttenbach/BPS. 30.11: Gary Bell/Masterfile. 30.12: M. Laverack/Planet Earth Pictures. 30.15: H. W. Pratt/BPS. 30.17*a*: David Wrobel/BPS. 30.17*b*: Marty Snyderman/Masterfile. 30.18*a,d*: David Wrobel/BPS. 30.18*b*: Doug Perrine/Masterfile. 30.18*c*: Ken Lucas/Planet Earth Pictures. 30.19: Peter Scoones/Planet Earth Pictures. 30.21*a*: Ken Lucas/BPS. 30.21*b*: Nick Garbutt/Indri Images. 30.21*c*: Art Wolfe. 30.24*a*: David J.

Wrobel/BPS. 30.24*b*: Carl Gans/BPS. 30.24*c,d*: Brian Kenney/Masterfile. 30.25*a*: Gavriel Jecan/Art Wolfe Inc. 30.25*b*: Art Wolfe. 30.26: Courtesy of Carnegie Museum of Natural History, Pittsburgh. 30.28*a*: Peter Scoones/Masterfile. 30.28*b*: Larry Tackett/Masterfile. 30.28*c,d*: Adam Jones. 30.30: Fritz Prenzel/Animals Animals. 30.31*a*: Art Wolfe. 30.31*b*: Jany Sauvanet/Photo Researchers, Inc. 30.32*a*: Staffan Widstrand. 30.32*b*: Merlin D. Tuttle, Bat Conservation International. 30.32*c*: Carol Farneti/Masterfile. 30.32*d*: Theo Allofs. 30.34*a,b,c*: Art Wolfe. 30.35*a*: Steve Kaufman/DRK PHOTO. 30.35*b*: John Bracegirdle/Masterfile. 30.36*a*: Art Wolfe. 30.36*b*: Kennan Ward/DRK Photo. 30.36*c*: K. & K. Ammann/Masterfile. 30.36*d*: Brian Kenney/Planet Earth Pictures. 30.38*a*: The American Museum of Natural History. 30.38*b,c*: Terraphotographics/BPS.

Chapter 31 *Opener*: Rod Planck/Photo Researchers, Inc. 31.1*a*: Hubertus Kanus/Photo Researchers, Inc. 31.2*a*: Michael P. Gadomski/Bruce Coleman Inc. 31.3*a*: D. Waugh/Tony Stone Images. 31.4: Nigel Cattlin/Holt Studios International/Photo Researchers, Inc. 31.7*a*: John Kaprielian/Photo Researchers, Inc. 31.7*b*, 31.8: John D. Cunningham/Visuals Unlimited. 31.9*a*: J. Robert Waaland/BPS. 31.9*b*: Joyce Photographics/Photo Researchers, Inc. 31.9*c*: Renee Lynn/Photo Researchers, Inc. 31.13*a,d*: Phil Gates/BPS. 31.13*b*: Biophoto Associates/Photo Researchers, Inc. 31.13*c*: Jack M. Bostrack/Visuals Unlimited. 31.13*e*: John D. Cunningham/Visuals Unlimited. 31.13*f*; 31.15*b*; 31.16 *upper*, *lower*: J. Robert Waaland/BPS. 31.18*a*: Jim Solliday/BPS. 31.18*b*: Microfield Scientific LTD/Photo Researchers, Inc. 31.18*c*: Alfred Owczarzak/BPS. 31.18*d*: John D. Cunningham/Visuals Unlimited. 31.20*a upper*: J. Robert Waaland/BPS. 31.20*a lower*: Cabisco/Visuals Unlimited. 31.20*b upper*: J. Robert Waaland/BPS. 31.20*b lower*: Cabisco/Visuals Unlimited. 31.22: J. N. A. Lott/BPS. 31.23: Jim Solliday/BPS. 31.24*a,b*: Phil Gates/BPS. 31.25*b*: Jeff Lepore/Photo Researchers, Inc. 31.25*c*: C. G. Van Dyke/Visuals Unlimited.

Chapter 32 *Opener*: Peter K. Ziminski/Visuals Unlimited. 32.2: Biophoto Associates/Photo Researchers, Inc. 32.7: John D. Cunningham/Visuals Unlimited. 32.10*a*: David M. Phillips/Visuals Unlimited. 32.12 *right*: Derrick Ditchburn/Visuals Unlimited. 32.13: Thomas Eisner, Cornell Univ.

Chapter 33 *Opener* and 33.2: Nigel Cattlin/Holt Studios International/Photo Researchers, Inc. 33.4: Jess R. Lee/Photo Researchers, Inc. 33.6: Stephen Parker/Photo Researchers, Inc. 33.8: Thomas Eisner, Cornell Univ. 33.9: Jon Mark Stewart/BPS. 33.10: J. N. A. Lott/BPS. 33.11, 33.12: Richard Shiell. 33.13: Janine Pestel/Visuals Unlimited. 33.14: Barbara J. Miller/BPS. 33.15: J. N. A. Lott/BPS. 33.16: Robert & Linda Mitchell. 33.17: Budd Titlow/Visuals Unlimited.

Chapter 34 *Opener*: Cabisco/Visuals Unlimited. 34.1: The Photo Works/Photo Researchers, Inc. 34.4: Kathleen Blanchard/Visuals Unlimited. 34.7: Jeremy Burgess/Photo Researchers, Inc. 34.8: Barbara J. O'Donnell/BPS. 34.10: E. H. Newcomb and S. R. Tandon/BPS. 34.12: Richard C. Johnson/Visuals Unlimited. 34.13: Nuridsany et Pérennou/Photo Researchers, Inc.

Chapter 35 *Opener*: Mary Clay/Masterfile. 35.5: Tom J. Ulrich/Visuals Unlimited. 35.6: Barbara J. Miller/BPS. 35.7: J. N. A. Lott/BPS. 35.11: J. A. D. Zeevaart, Michigan State Univ. 35.19*a*: Biophoto Associates/Photo Researchers, Inc. 35.21: Cabisco/Visuals Unlimited. 35.22: Adam Jones. 35.23; 35.25: J. N. A. Lott/BPS. 35.29: R. Last, Cornell Univ. Courtesy of the Society for Plant Physiology.

Chapter 36 *Opener*: David Woodfall/Woodfall Wild Images. 36:2: Stan Flegler/Visuals Unlimited. 36.3: Nigel Cattlin, Holt Studios International/Photo Researchers, Inc. 36.4: Bowman, J. (ed.). 1994. *Arapidopsis: An Atlas of Morphology and Development*. Springer-Verlag, New York. Photos by S. Craig & A. Chaudhury, Plate 6.2. 36.8*a*: John Kaprielian/Photo Researchers, Inc. 36.8*b*: Renee Lynn/Photo Researchers, Inc. 36.17*a*: Nigel Cattlin, Holt Studios International/Photo Researchers, Inc. 36.17*b*: James W. Richardson/Visuals Unlimited.

Chapter 37 *Opener*: Daniel J. Cox/Natural Exposures. 37.15*a*: B. & C. Alexander/Photo Researchers, Inc. 37.15*b*: Timothy Ransom/BPS. 37.20 *left*, *center*: G. W. Willis/BPS. 37.20 *right*: Fran Thomas, Stanford Univ. 37.21*a*: Barbara Gerlach/Visuals Unlimited. 37.21*b*: Art Wolfe.

Chapter 38 *Opener*: R. D. Fernald, Stanford Univ. 38.6*a*: Associated Press Photo. 38.6*b*: The Bettmann Archive, Inc.

Chapter 39 *Opener*: Ron Austing/Photo Researchers, Inc. 39.1*a*: Biophoto Associates/Photo Researchers, Inc. 39.1*b*: Andrew J. Martinez/Photo Researchers, Inc. 39.1*c*: Peter J. Bryant/BPS. 39.2: David M. Phillips/Photo Researchers, Inc. 39.4: Paul W. Johnson/BPS. 39.5: Fletcher & Baylis/Photo Researchers, Inc. 39.6: James Beveridge/Visuals Unlimited. 39.7*a*: Jim Merli/Visuals Unlimited. 39.7*b*: Robert W. Hernandez. 39.8: Renee Lynn/Photo Researchers, Inc. 39.13 *inset*: P. Bagavandoss/Photo Researchers, Inc. 39.17: Lara Hartley/BPS. 39.19: CC Studio/Photo Researchers, Inc. 39.21: C. Eldeman/Photo Researchers, Inc. 39.22: Nestle/Photo Researchers, Inc. 39.24: S. I. U. School of Med./Photo Researchers, Inc.

Chapter 40 *Opener*: Courtesy of John Morrill, New College. 40.5 *inset*: Courtesy of Richard Elinson, Univ. of Toronto.

Chapter 41 *Opener*: Stephen Krasemann, Nature Conservancy/Photo Researchers, Inc. 41.4: C. Raines/Visuals Unlimited.

Index